HISTORY AND HERITAGE OF COASTAL ENGINEERING

A COLLECTION OF PAPERS ON THE HISTORY OF COASTAL ENGINEERING IN COUNTRIES HOSTING THE INTERNATIONAL COASTAL ENGINEERING CONFERENCE 1950 - 1996

PREPARED UNDER THE AUSPICES OF THE COASTAL ENGINEERING RESEARCH COUNCIL OF THE AMERICAN SOCIETY OF CIVIL ENGINEERS

EDITED BY NICHOLAS C. KRAUS

Published by

ASCE American Society of Civil Engineers

345 East 47th Street
New York, NY 10017-2398

ABSTRACT:

Coastal engineering is a relatively new discipline within the field of civil engineering. The first International Coastal Engineering Conference was convened in Long Beach, California, in 1950. The theme of the twenty-fifth or "silver" ICCE held in Orlando, Florida, in September, 1996, was History and Heritage of Coastal Engineering to honor those individuals and institutions contributing to the foundations of the discipline. As part of the celebration at this premier technical conference, this volume was conceived to document the history of coastal engineering in the 15 countries which have hosted the ICCE. Coastal engineering works have been conducted for hundreds and even thousands of years for port development, coastal hazard protection, and reclamation of land from the sea. The needs of the different countries and the approaches taken are unique and document the evolution of society and its relation with the coast from 15 perspectives. The reader will be fascinated by the ingenuity and resolution of our ancestors, as well as the accomplishments made in modern times as documented in this volume rich with information, colorful anecdotes, and citations to many almost-forgotten original references worthy of re-examination.

Library of Congress Cataloging-in-Publication Data

History and heritage of coastal engineering : a collection of papers on the history of coastal engineering in countries hosting the international Coastal Engineering Conference 1950-1996 ; prepared under the auspices of the Coastal Engineering Research Council of the American Society of Civil Engineers / edited by Nicholas C. Kraus.
p. cm.
Includes indexes.
ISBN 0-7844-0196-9
1. Coastal engineering—History—Congresses. I. Kraus, Nicholas C. II. Coastal Engineering Research Council. III. Coastal Engineering Conference.
TC215.H57 1996 96-32238
627'.58'09—dc20 CIP

Library of Congress Catalog Card No: 96-32238
ISBN 0-7844-0196-9
Manufactured in the United States of America.

Cover photograph provided by Professor Robert L. Wiegel, Professor Emeritus, University of California, Berkeley, showing "field party with Dukw, Winkler, Bascom, Goodwin, and Grabber, May 7, 1947."

PREFACE

Coastal engineering emerged some 50 years ago as a distinct technical area of civil engineering. In 1950, the International Conference on Coastal Engineering series began in Long Beach, California, and, after a period of convening annually, assumed the present biennial format in 1968. The Twenty-Fifth International Coastal Engineering Conference—the *Silver Conference*—celebrated our "Coastal Engineering History and Heritage" in recognition of the important milestone that has been achieved.

As part of the celebratory theme of ICCE96, this volume was conceived to document the history of coastal engineering in the 15 countries which have hosted the conference. Coastal engineering works have been conducted for hundreds and even thousands of years for port development, coastal hazard protection, and reclamation of land from the sea. The needs and solution approaches of the various countries are unique, and the individual chapters document the activities of society and its relation with the coast from 15 perspectives.

Despite cultural differences, common themes are found. One such theme is evolution of coastal engineering practice through the three cycles of (1) exploitation and utilization of the coast, (2) development of protection from coastal hazards such as flooding and erosion, and (3) preserving and creating harmony between nature and coastal uses. Some countries have entered the final cycle of this process, whereas others are in either the first or second cycles. A fascinating theme is that of pivotal workers in the field, those who have influenced generations that followed in their own countries and, sometimes, in the world. Of course, we are also humbled to learn that wise observers and thinkers had discovered many basic processes and solutions hundreds of years before their rediscovery in modern times. A valuable aspect of this collection of papers is the plentiful citations to the original literature, the true source of our knowledge.

I would like to express my deep appreciation, gratitude, and admiration to the authors of the chapters comprising this volume. They labored as volunteers for the love of their field and the desire to hold it in the highest esteem. Their efforts document a pivotal era in coastal engineering and related sciences, one in which the pioneers are now turning over the mantle of responsibility to those who were not in direct contact with the founders. Through this volume it is my hope that this contact will be forever maintained.

Nicholas C. Kraus, editor

June 26, 1996

CONTENTS

HISTORY AND HERITAGE OF COASTAL ENGINEERING

HISTORY OF COASTAL ENGINEERING IN AUSTRALIA

Michael R. Gourlay[1]

ABSTRACT: After brief reviews of Australia's coastal environment and its political and administrative divisions, the historical development of maritime and coastal engineering in Australia is reviewed from early times to the post second world war period. The implementation of coastal protection works and the development of coastal management in each of the five mainland states since the second world war is then considered with particular attention to the role of coastal engineers. Several significant coastal engineering projects of various kinds from different parts of the country are described. These are followed by an account of some of the coastal engineering research, as well as a brief mention of relevant marine science, undertaken in Australia. Finally the activities of The National Committee on Coastal and Ocean Engineering in providing a technical focus for coastal engineering and as a lobbying group for significant matters affecting the coast are recorded. The paper includes a wide ranging bibliography on coastal engineering in Australia.

INTRODUCTION

"The coastal zone has a special place in the lives of Australians. Most Australians want to live or take their holidays there. It is a priceless national asset." (DEST 1995)

With two thirds of the Australian continent lying in an arid zone it is not surprising that 86% of Australia's population live within its relatively well watered coastal zone. During the last 20 years there have been significant increases in population, development and tourism in this zone. About half of the total population growth during this period has occurred in regions away from the older population centres based on the state capital cities. The most rapidly developing coastal areas are the southeast and far north of Queensland, the southwest of Western Australia and the central and north

---

1) Senior Lecturer in Civil Engineering, The University of Queensland, Queensland, 4072, Australia.

coasts of New South Wales (DEST 1995). In such a situation coastal engineering has had and should be continuing to have a significant role in meeting the needs of the community.

The history of coastal engineering in Australia is very much a reflection of the development of this island continent nation within the constraints of its varied coastal environments and its complex political and administrative arrangements. The topic is an enormous one and it has been necessary to be selective in the presentation.

Short reviews of the Australian Coastal Environment and Political and Administrative Divisions are followed by an overview of the Historical Development of Maritime and Coastal Engineering in Australia during the nineteenth and twentieth centuries. The diverse approaches to Coastal Protection and Coastal Management in Australia are then discussed for each mainland state with the emphasis on the contribution of (coastal) engineers. Some Significant Projects or aspects of coastal engineering in Australia are then presented. These cover various kinds of works constructed in different parts of Australia during the period from the beginning of the twentieth century up to the present time. A review of Coastal Engineering Research in several defined fields follows together with a brief review of related Australian Marine Science. The paper concludes with a discussion of the role of the National Committee on Coastal and Ocean Engineering and its activities, including the Australian Coastal and Port Engineering Conferences.

As Australia is a large country and many readers may be unfamiliar with its geography, Appendix A includes a series of maps showing the location of places referred to in the text.

AUSTRALIAN COASTAL ENVIRONMENT

"Australia is an island continent with some 35 000 km of coastline, which is large by any national standard. The margins stretch from latitudes of 11 degrees to 44 degrees south and experience the complete range of sea conditions possible, except ice. They vary from almost zero tide to some of the largest in the world... The strongest swell possible impinges on our southern margins originating from distant weather systems in the southern ocean. Our coasts can experience the severest storm wave action possible as, for example, generated by fetches formed by south west winds between high pressure systems tracking across the Australian continent and low pressure systems tracking across south of the continent. The northern coastlines in latitudes less than 25 degrees experience the ravages of tropical cyclones with their associated devastating storm surges. Most of our population lives in coastal cities and towns and we depend on shipping for export of our primary mineral and secondary products and import of petroleum, specialised materials and manufactured goods. The coastal waters provide fishing grounds, recreation areas, sought after residential sites, and are exploited for undersea gas oil and minerals. This enormous range of activity is carried on by government agencies at national, state, regional and local levels, by

manufacturers, shipping companies, engineering firms and consultants, universities, learned societies, groups of citizens and individuals. Conflicts of interest are numerous. The need to assist in the resolution of these conflicts, and to solve the numerous problems associated with coastal development gave birth to a new species of engineer in the mid to late 1960's, the coastal engineer." (Nelson 1982).

Australian coastal engineers not only have to contend with the complete range of sea conditions possible, except ice, and with tides of varied magnitudes but also with very varied geomorphological and geological conditions as well as diverse ecosystems. Australia's climate is the driest and most variable of any of the inhabited continents. The surface of Australia is flatter than any other continent; tectonic activity is low and isostatic uplift small. Its rivers contribute relatively small quantities of sediment but relatively high loads of dissolved salts to its coasts. In few places is the sediment supply sufficient to be causing significant accretion of the coastline and with the anticipated rise in sea level in coming years most coastlines will be under threat of erosion.

The broad geographical characteristics of Australia's coasts compared with the coasts of other continents are presented by Davies (1972). More specifically the coast of Australia can be broadly classified from a geological and geomorphological viewpoint into the following eight different coastal regions (Gill, 1982)(Fig. 1).

1. The Great Barrier Reef coast (Queensland).

2. The heavy mineral sand coast (Queensland and New South Wales).

3. The rocky shore/inlet coast (Southeastern Australia).

4. The aeolianite coast (Victoria and South Australia).

5. The arid riverless coast (Great Australian Bight).

6. The coral reef/aeolianite coast (Western Australia).

7. The tropical Northwestern coast.

8. The Gulf of Carpentaria coast.

Numerous coral reefs with their diverse ecology protect a large part of the northeastern Australian coast (coast 1). Coral reefs are also very significant along portions of the coast of Western Australia (coast 6). Mangrove forests dominate estuaries in northern Australia (coasts 7 and 8) and extend with decreasing diversity around the eastern coastline from Queensland to Victoria (coasts 1, 2 and 3). Large sand dunes of Pleistocene age form barrier islands enclosing large bays in southern Queensland (coast 2) and northward alongshore transport dominates the northern New South Wales and southern Queensland coasts (coast 2). In central and southern New South Wales and

Tasmania (coast 3) beaches and coastal lakes are flanked by rocky headlands and littoral drift systems tend to be compartmentalised. Sediments on coasts 2 and 3 are generally predominantly siliceous. In western Victoria and southeastern South Australia (coast 4) there are extensive deposits of carbonate sands formed on the continental shelf from the broken shells of marine biota. The combined actions of strong winds and ocean swells during the various sea level cycles of the Pleistocene have formed large sand dunes and much of this material has become calcarenite (aeolianite) rock.

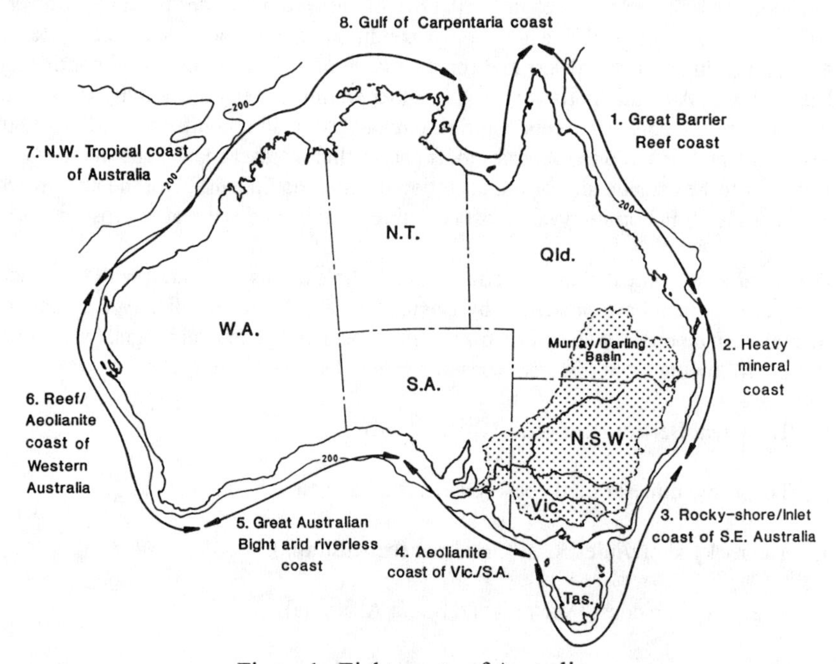

Figure 1. Eight coasts of Australia

These different coastal regions are subjected to varied climatic conditions as well as varied tidal conditions (Radok 1976, Mitchell and Silvester 1981). In the case of the tides not only does the tidal range vary along the coast but tides vary from semidiurnal to mixed to diurnal. For example, tides along the eastern coast of Queensland vary from less than 2m up to 9m in range and from semidiurnal to mixed while in the southern Gulf of Carpentaria tides are diurnal. In Western Australia micro tides (1m) tending to diurnal predominate in the south while in the northwest semidiurnal macro tides (10m) occur.

An overview of the Australian coast has been presented through aerial photographs, descriptions and feature articles (Pullan 1983). A comprehensive overview of coastal evolution and coastal erosion of the New South Wales portions of coasts 2 and 3 has been given by Chapman et al. (1982) and descriptions of specific portions of coastlines are available for other places. For example, three quite different sections of the Great Barrier Reef coast (coast 1) are described in a series of reports by Queensland's Beach Protection Authority (BPA 1979, 1984, 1989).

POLITICAL AND ADMINISTRATIVE DIVISIONS

As well as the highly varied climatic, marine, geological and ecological conditions around Australia's coasts, there are also different political and administrative arrangements. Australia is a federation of six states, New South Wales, Tasmania, Western Australia, South Australia, Victoria and Queensland together with two mainland territories, the Australian Capital Territory and the Northern Territory, a few small island territories such as Norfolk Island, and the Australian Antarctic Territory. Each state retains responsibility for all areas of government not specifically ceded to the Commonwealth[2] government by the provisions of the Australian Constitution adopted at Federation in 1901. In practice the Commonwealth government exercises overall control through its fiscal policies as it collects all customs and excise dues and income tax revenues but returns a portion of these to the states.

Ports and harbours, coastal management and protection and related matters are essentially subject to state jurisdiction. However, management of landbased defence establishments in the coastal zone as well as management of the waters and the utilisation of resources on the continental shelf beyond 5.6km (3 nautical miles) from the coast are the concern of the Commonwealth. Environmental matters concerning the coast are becoming increasingly subject to the constraints of international agreements and treaties entered into by the Australian government, for example the London Convention.

Within states, ports and harbours may be administered by either a single state government department or by semiautonomous local port authorities. Most Australian states had a Public Works Department (PWD) from mid to late nineteenth century up to relatively recent times. There was no administrative connection between the various PWDs, although they generally performed similar functions under various Acts of their State Parliaments. In New South Wales, Victoria and Western Australia these functions included maritime and coastal engineering activities for some or all of the period of existence of the particular PWD. In Queensland, the Department of Harbours and Marine performed these functions while in South Australia the relevant agency was the Department of Marine and Harbours. In recent years the trend has been from single state authorities to corporatised local port authorities and also for the contracting of engineering services to consultants but the situation varies from state to state as it continues to evolve.

Coastal protection and management are essentially the responsibility of various local (municipal) government authorities within each state. Throughout Australia there are

---

2) Sometimes referred to as the Federal government or the Australian government.

250 to 300 local authorities with coastal management responsibilities. These local authorities are subject to overall direction and support from state instrumentalities responsible for planning and coastal management. The structure and nomenclature of those instrumentalities vary from state to state.

When European settlement began in Australia all lands were deemed to belong to the Crown and the early colonial governors had powers to make grants of land based upon the principles of English property law.[3] In 1825 Lord Bathurst, Secretary of State for colonies, directed Governor Brisbane to make reservations of land for public purposes. These public purposes included reservations *"...for the purpose of health and recreation; and lands in the neighbourhood of navigable streams or the sea coast ..."* Later Governor Darling was directed *"...not to allow private persons to occupy any such land for private purpose."* This principle of protecting Crown land on the coast for public use is still one of the main coastal planning principles applied today in Australia. Moreover, where freehold title does extend to the coast, its seaward limit is high water mark. The beach remains in public ownership.

In recent years political pressures have led to the Commonwealth government taking increasing interest in the coast and its management. A series of enquiries by both the Commonwealth parliament and its agencies has reflected the growing environmental concern of ordinary Australians for the protection of the coastal environment. These culminated in 1995 in the formulation of a Commonwealth coastal policy and action plan to support state and local authorities in their efforts to manage Australia's coast effectively for both present and future generations (DEST 1995).

HISTORICAL DEVELOPMENT OF MARITIME AND COASTAL ENGINEERING

Early Times to the Second World War

Forty to fifty thousand years ago when the first humans arrived in Australia from the north, it was a much larger continent including New Guinea and, from time to time as sea level rose and fell, Tasmania, all in one continuous land surface. The seas dividing Australia and New Guinea from the Indonesian Archipelago were much narrower then and the ancestors of the Australian Aborigines were able to cross them in their small boats. At the peak of the last ice age, 18 000 to 20 000 years ago, this would have been easier. On the other hand, following the melting of the ice, the sea had risen and flooded the continental shelf by the time the first European navigators reached Australia about 400 years ago. Their much more sophisticated sailing vessels, capable of sailing around the world, were still relatively small and were able to use the many natural harbours in bays, estuaries and river entrances around the island continent.

---

3) The prior ownership and rights of indigenous peoples have been recognised in recent judicial decisions and legislation by Commonwealth and State parliaments.

The early European port settlements, Sydney 1788 and Hobart 1804, were located on such natural harbours selected for shelter for ships and a good water supply. Settlement generally expanded inland in the search for productive agricultural land which could supply the settler's food and other material needs. Subsequently export industries developed, initially wool and coal. In time new settlements were developed along the coast to service the hinterland areas. For example, in New South Wales the discovery of coal in 1797 at Newcastle about 100 km north of Sydney led to the establishment of a new settlement using the mouth of the Hunter River as its port.

Subsequent exploration and particularly the search for the coveted red cedar timber along the northern coast of New South Wales led to the establishment of towns in the various coastal river valleys. These were almost always located on the estuaries of the rivers, often at the upstream limit of navigation at that time. Transport to and from these centres was by sea with coastal shipping entering the rivers and using their tidal waterways to reach the towns. As ships became larger and more numerous and shipwrecks more frequent the need to provide for the safe navigation of the often unstable, shallow and dangerous river entrances became urgent. During the late nineteenth and early twentieth centuries most of these coastal rivers, such as the Hastings, Clarence, Richmond and Tweed Rivers, required the construction of entrance breakwaters and estuarine training walls.

Increased activities arising from the gold rushes in New South Wales and Victoria during the 1850s stimulated commerce and required the expansion of infrastructure. The first railways were constructed, linking the major ports with their nearby urban centres. Roads further inland were still primitive and competition to supply the Victorian goldfields developed between bullock teams negotiating the 100 to 150 km of winter mud from Melbourne and river paddle steamers navigating the 1700km of shallow meandering, snag infested waterway of the Murray River, part of Australia's largest river system. As the South Australians battled for this trade they were confronted with the problem that the mouth of the Murray was completely unsuitable for navigation. To solve this problem, in 1854 the river port of Goolwa on Lake Alexandrina inside the Murray mouth was linked with Port Elliot, 11km away on the coast, by Australia's first public railway, albeit initially only horse powered. To create Port Elliot, a breakwater was constructed to provide shelter for the relatively small oceangoing ships of that time. This breakwater was one of the earliest coastal engineering structures built in Australia. Its construction was described by the engineer who supervised the contractor building it at a discussion of a paper on breakwater design and construction at the Institution of Civil Engineers in London during 1858 (Hays 1858-59). This is apparently the first record of Australian civil engineering in an overseas publication (Corbett 1973). As will be discussed later, the method of construction was a typically Australian one.

Expansion of mining and pastoral activities in the late nineteenth century required new ports, particularly in northern Queensland where facilities were provided at Rockhampton, Mackay, Townsville and at other locations. These developments were located in the tropical cyclone[4] affected Great Barrier Reef region. The potential for destruction of coastal towns and port facilities with accompanying loss of life gradually became evident. Significant events were the Mahina cyclone in 1899 which destroyed a pearling fleet at Bathurst Bay with a loss of over 300 lives. This was accompanied by a storm surge believed to be of the order of 12m. Fortunately this occurred in a sparsely populated area. Two decades later in January 1918 a severe tropical cyclone passed close to Mackay and a 3.8m storm surge, followed two days later by a record river flood, inundated the city which was devastated by the combined effects of winds and flooding from both the sea and the river. Thirty lives were lost. Fortunately the storm surge did not coincide with a 6m spring high tide.

During this period the particular difficulties associated with building structures on coral reefs became apparent as manned lightships and lighthouses began to be replaced by unattended navigation lights marking the inner shipping channel sheltered by the northern portion of the Great Barrier Reef (Mehaffey, 1919-20).

On the western side of the continent the decade following the achievement of responsible government in Western Australia in 1890 coincided with major gold discoveries and the new government embarked upon a large scale programme of public works. Chief among these works was the construction of a new harbour at Fremantle, the entry port for southwestern Australia and the nearby capital city, Perth. This major project, involving construction of two new breakwaters, removal of a rock bar, dredging and provision of wharves, was undertaken by C.Y. O'Connor, the newly appointed Irish born Chief Engineer, recently arrived from New Zealand. In the ensuing decade prior to his tragic death he was responsible for not only Fremantle port but the upgrading of the state railway system and the conception and implementation of the goldfields water supply system involving a 565 km pipeline from the coastal ranges near Perth to inland mining centres (LePage 1986).[5]

Nineteenth century coastal engineering in Australia was largely port and harbour engineering. The engineers were almost always British - English, Welsh, Scots and Irish. They were civil engineers in the broad sense of the word, generally with experience in more than one field of public works. While Australian trained engineers were becoming more numerous and influential towards the end of the century, the engineering profession and governments still looked to Britain for specialist expertise. For example, port developments in several states were strongly influenced by the recommendations of Sir John Coode, engineer in chief for the British Admiralty, who

---

4) Tropical cyclone is the term used by Australian meteorologists for rotating weather systems known in other parts of the world as hurricanes or typhoons. In the southern hemisphere these systems rotate in a clockwise direction, that is the opposite direction to the northern hemisphere systems.
5) The original Fremantle port development was described by Palmer (1910-1911). A full biography of O'Connor has been published by Tauman (1978).

made two visits to Australia in 1877 and 1885. Port developments in Queensland, New South Wales, Victoria and South Australia were strongly influenced by his recommendations. Only in Western Australia was his advice largely ignored in favour of O'Connor's more suitable concept.

These British engineers were for the most part competent, well experienced with tides and their influence and conscious of the destructive force of waves but not always appreciative of the significance of wave-induced sediment transport.

Development of railway systems radiating from state capital cities and a few other coastal ports, e.g. Newcastle, Rockhampton, Townsville and Cairns, during the second half of the nineteenth century changed the economic and social situation. In 1865 the railway from Melbourne to Echuca on the Murray River captured most of the inland navigation trade on the Murray-Darling River system for Melbourne and Victoria. Later railway extensions in New South Wales diverted some of this to Sydney. Suburban railways and tramways in Sydney, Melbourne and Adelaide made seaside living possible as well as holiday excursions and picnics. Railways from Brisbane, Sydney, Melbourne, Adelaide and Rockhampton to nearby coastal and bayside areas led to the development of holiday towns, often with homes and public facilities located right on the shoreline, creating potential erosion and other coastal engineering problems on Port Phillip Bay, the Gold Coast and other areas.

Construction of coastal railway lines linking the coastal cities in New South Wales and Queensland in the 1920s mortally wounded the coastal shipping services which finally died during the second world war (passenger services) or during the subsequent two decades (cargo services).

Difficulties maintaining water depths in estuarine ports for the increasingly larger ships now using them resulted in the port of Brisbane building new wharves progressively further downstream in the Brisbane River estuary. At Mackay in northern Queensland the port was moved in the 1930s from the Pioneer River, where navigation was only possible at high tide, to a new harbour on the coast formed from two breakwaters. This also occurred at Port Adelaide with berths moving downstream to Outer Harbour.

Post Second World War - Ports and Harbours

The cessation of the coastal shipping services at least in eastern Australia during the second world war was accompanied by increased usage of the major ports. Expansion of mineral and agricultural exports in the post war era, - coal, iron ore, bauxite, wheat, sugar, etc. created the demand for new specialised ports for handling these bulk cargos.

In New South Wales the supply of raw materials to the steelworks at Newcastle, an estuarine river port, and Port Kembla, an artificial harbour formed by two breakwaters, and the subsequent expansion of coal exports required extensive modifications to existing facilities. At Port Kembla ship ranging at the jetties[6] in what is now the outer harbour was a significant problem causing ship operations to stop on an average of 11 days a year. Design of a new inner harbour to service the expanded steelworks and increased coal exports stimulated extensive investigations to minimise ranging problems and the extensive physical model experiments for Port Kembla harbour were among the first made during the early 1950s at the Manly Vale hydraulics laboratory operated by the New South Wales Public Works Department.

The problems at Newcastle were more complex. Twin parallel breakwaters defined the entrance channel of the Hunter River and provided protection for vessels entering the port. The southern breakwater was exposed to the full force of waves generated by southeasterly storms in the Tasman Sea. A rock bar restricted the depth in the main channel and a sand shoal formed from time to time near the seaward end of the northern breakwater. Inside the entrance the port areas required increasing dredging as they were enlarged to accommodate more and larger ships, the estuarine part of the river being subject to siltation initiated by the effects of salinity upon the fine silts brought down by the river during floods from a catchment largely alienated by agricultural activities. This siltation became of increasing economic importance during the 1950s which was a period of above average rainfall, increased runoff and more frequent flood events.

In Victoria a new artificial harbour was constructed at Portland in the 1950s to expedite trade and exports from the western districts of Victoria. In the subsequent decades expansion of mineral exports such as iron ore from northwestern Australia, bauxite from Weipa on the Gulf of Carpentaria and coal from central Queensland required the provision of new heavy haulage railway systems and appropriate port facilities for the large bulk carriers used to ship these raw materials overseas or to other Australian ports for processing. In Queensland bulk shipping of sugar for export also required new or expanded port facilities. In both Brisbane and Sydney the major port facilities were relocated. The twin problems of increased draft of vessels and increased costs of maintenance dredging led to the removal of Brisbane's port facilities downstream to a new reclaimed site at the mouth of the Brisbane River where the latter discharged into Moreton Bay. In Sydney the rapidly expanding general cargo trade in container ships led to the development of new port facilities in Botany Bay, the site originally proposed for the first European settlement in 1788 but rejected because of lack of fresh water and exposure to ocean waves.

Offshore oil production commencing off the Victorian coast in Bass Strait during the 1970s and subsequently on the North West Shelf in Western Australia during the 1980s provided opportunities for both coastal and ocean engineering. The activities of

---

6) In Australian usage a jetty is a wharf structure projecting from the shoreline. The term is never used in the American sense to describe breakwaterlike structures used for stabilising river entrances or tidal inlets. The latter structures are referred to in Australia as either breakwaters or training walls.

Woodside Offshore Petroleum Pty Ltd on the North West Shelf involved not only the construction of offshore exploration and collection structures but also mainland port facilities for export of liquified natural gas. Located in the tropical cyclone belt, design, construction and operation required considerable new research and the development of methods for predicting tropical cyclone generated waves both for the design of structures directly exposed to storm waves (tropical cyclone Orson (905hPa) the most intense cyclone recorded in the Australian region passed directly over the North Rankine platform in 1989) and for the operation of coastal port facilities affected by swells generated by cyclones offshore. In the latter case real time wave predictions are required.

Post Second World War - Coastal Protection and Management

Increased use of private cars and the gradual improvement of roads greatly accelerated the use of beaches, coastal lakes and estuaries as holiday and recreation areas. Increasingly coastal regions north and south of Sydney and Brisbane, as well as those near the other mainland capital cities, were overtaken by holiday developments - many subdivisions extending to highwater mark and houses being constructed on frontal dunes.

Subsequently movement of people, particularly retirees from Melbourne and Sydney, to northern New South Wales and southern Queensland accelerated this development, putting increased pressure on the coastal zone. Conflicts also arose from expansion of the export oriented mineral sands industry which was extracting rutile to provide titanium for the United States aerospace industry from the black sands found on the beaches and in the dunes of northern New South Wales and southern Queensland. In the early decades of this activity, mining of heavy mineral sands was undertaken with little regard for its environmental consequences and there were no requirements for the sand mining companies to rehabilitate the vegetation or restore the landforms of mined dunal areas.

By the 1960s it was becoming evident that coastal erosion and its effects were a serious problem in many places. Private homes, commercial buildings, public facilities - roads, parks, bathing sheds and toilet blocks were all endangered. Public opinion wanted the government to do something about this problem which was interfering with individuals' right to use the beach and dunes however and whenever they chose to do so.

The problems had been evident to engineers, particularly local authority and highway engineers for some time. The first paper published by the Institution of Engineers Australia on coastal erosion concerned problems along the eastern shore of Port Phillip Bay where Melbourne's bayside suburbs are located (Mackenzie 1939). Coastal erosion problems also were occurring in other states. In Queensland the Main Roads Department had been battling since the early 1920s to protect the Pacific Highway (the main highway between Sydney and Brisbane) at Narrowneck on the Gold Coast from the twin threats of erosion from the sea to the east and erosion from the Nerang River to the west. The highway had been constructed on the very recent Holocene frontal dune separating the river from the sea. In Western Australia erosion of Perth's metropolitan beaches was an increasing problem following the second world war and action was needed to restore many of them (Gillespie 1965).

In New South Wales Foster and Stone (1965) published the results of a historical survey of beaches at Cronulla, 32km south of Sydney. They identified three periods of long term change since 1905 and observed marked short term fluctuations of the beaches with cycles of storm and swell waves. The adverse impacts upon these beaches of seawall and swimming pool construction as well as shell grit mining were noted also.

Awareness of coastal engineering problems and their solutions was increased during 1964 when Dr Per Bruun visited Australia, particularly Queensland, for the first time. At this time the Queensland government invited the Delft Hydraulics Laboratory to advise on what action should be taken to solve the increasingly serious beach erosion problems on the Gold Coast.

In South Australia various sea defence works had been utilised to combat coastal erosion at several locations (Beverley 1966) and the problems of Adelaide's metropolitan beaches were becoming serious. Investigations by the University of Adelaide (Culver 1970) provided the basis for future coastal management in that state. Similar action was taken in due course in different ways in all the mainland states. The role of Australian coastal engineers in coastal management is recorded in the following section.

COASTAL PROTECTION AND COASTAL MANAGEMENT

The Queensland Way

The need for coastal protection only becomes evident as human activities become established in the coastal zone and fixed structures and buildings are constructed on potentially unsuitable coastal landforms. On Queensland's South Coast (now known as the Gold Coast) storm events in 1936 and 1950 endangered some buildings and roads and the problems were sufficiently serious for a government report to be prepared on the subject (Kindler and O'Connor 1951). Its recommendations were summarised as follows.

> *"This report recommends the adoption of development planned to counter the effects of coast erosion, suggests some modification to present practices in beach mining, and requests the future systematic recording of certain fundamental information related to South Coast beach movements".*

However the problems were not yet sufficiently serious for significant action to follow.

In May 1960 beach erosion problems in Queensland became the responsibility of the Co-ordinator Generals Department (COG). B.L. McGrath, then a junior engineer in that department, was given responsibility for this aspect of COG's functions and he continued in this role for the following eleven years. His first task was to study and catalogue the literature which had been acquired from the United States Beach Erosion Board (BEB). In 1963, COG's chief engineer, J. Kindler, made an overseas tour of various institutions, following which the decision was made to invite the Delft Hydraulics Laboratory (DHL) to advise the Queensland government on coastal protection. During the same year the COG acquired new hydrographic surveying resources including the latest position fixing and echo sounding equipment for use in the development of the Port of Brisbane and these were subsequently used to great advantage in offshore surveys for coastal investigations. In 1965, following a visit to Queensland by J.G.H.R. Diephius, DHL presented their first report recommending a comprehensive investigation to provide basic data to enable the design of *"technically sound and economically justified coastal engineering works"* to be undertaken, together with the preferred methods and equipment to be used. Also recommended were the creation of a central body for applied hydraulic research and planning and a comprehensive system of laws on coastal protection that would allow co-ordinated planning, finance and administration and define the rights, obligations and responsibilities of all parties concerned.

The Gold Coast investigation was initiated in 1966 and continued to 1969 with some further work thereafter. In April 1968 the Queensland government commissioned DHL to assist in data evaluation and subsequently McGrath spent fifteen months at DHL's de Voorst Laboratory working on this project. In December 1970 the Delft Report (DHL 1970), as it was locally referred to, was received by the Queensland government. This report provided the basis for subsequent coastal protection work and investigations on the Gold Coast and in other regions of Queensland as the expertise of that report was assimilated, reviewed and applied by Queensland coastal engineers.

On the administrative side, concerned citizens had lobbied the then Queensland premier, G.F.R. Nicklin, in 1966 for action on coastal erosion. Political will was significantly strengthened by a series of storms during the first six months of 1967 culminating in severe erosion all along the Gold Coast in late 1967 (McGrath 1968). After consultation with all interested parties and examination of overseas practices, particularly in the Netherlands, the Queensland Beach Protection Act came into force on 1 July 1968, a few days after the death of its principal architect, J. Kindler. This act set up the Beach Protection Authority (BPA) and a Beach Protection Advisory Board with representation from a wide range of interested state government authorities, coastal local authorities, sand mining companies and, indicative of the role of the engineering profession at that time, a representative of the Department of Civil Engineering of the University of Queensland, a position held first by Professor G.R. McKay and then by Professor C.J. Apelt.

In January 1971 the Beach Protection Act was amended to allow the Department of Harbours and Marine (DHM) to supply administrative and technical services to the BPA instead of COG. Significant amendments were made to the Act in 1984 in relation to coastal subdivisions. Following a change of state government in 1989 and the reorganisation of government departments the administration of the Beach Protection Act became the responsibility of the newly established Department of Environment and Heritage (DEH), with the coastal engineers and other staff from DHM being transferred to the new department. After several years preparation a new Coastal Protection and Management Act was passed by the Queensland Parliament in November 1995 and became law on 1 February 1996. When the transition to this act is completed the BPA is expected to be no more.

The Beach Protection Act provides for the declaration of Coastal Management Control Districts (CMCD) and for land to be identified as an Erosion Prone Area (EPA). Local authorities are required to obtain the views of the BPA with respect to town planning matters affecting land in CMCDs and EPAs. Property holders require a permit to erect or alter buildings in a CMCD. Once EPAs have been identified, buffer zones of appropriate width are determined to define the areas where controls are required. While the BPA has been somewhat constrained in its ability to limit coastal development already on subdivided land - it is required to pay compensation to landholders prevented from developing seafront properties and no funds have ever been provided for it to pay such compensation - it has performed a very significant investigative and monitoring role for Queensland coasts. This technical role has been supplemented by various activities aimed at educating the general community and particularly local authority decision makers as to the most appropriate ways for managing the coastal zone.

Following the completion of the initial Gold Coast coastal investigations, several other sections of the Queensland coast where there were actual coastal erosion or potential coastal management problems were investigated. These included the Capricorn coast in central Queensland, Mulgrave Shire coast north of Cairns and the Hervey Bay coast 250 km north of Brisbane (BPA 1979, 1984, 1989, Patterson and Ford 1980). These were comprehensive investigations drawing on a range of professional expertise and materials, covering historical, geological and geomorphological, botanical and meteorological aspects of the coast as well as normal coastal engineering aspects. The coastal engineering work was based upon an extensive data collection programme which includes a statewide wave recording system using wave rider buoys with several long term permanent stations at representative locations along the Queensland coast, supplemented by short term project stations operating as required for a few years (Fig. 2). This system has been maintained on a continuing basis since 1968 with equipment and analysis procedures being upgraded to conform with best contemporary practice. The largest wave recorded to date was a 13.09m maximum height wave which occurred off the southeastern Queensland coast during tropical cyclone Roger in March 1993. The wave recording network is supplemented by a network of voluntary observers making daily beach and surf zone observations at specific locations.

This programme is know as the COPE programme - Coastal Observation Programme Engineering - and is based upon the CERC LEO programme (Robinson and Jones 1977). It contributes, not only to the technical aspects of BPA's operations, but also to its community education objectives as individuals, school and other community groups become involved in it. Data from both the wave recording and COPE programmes are published in special report series.

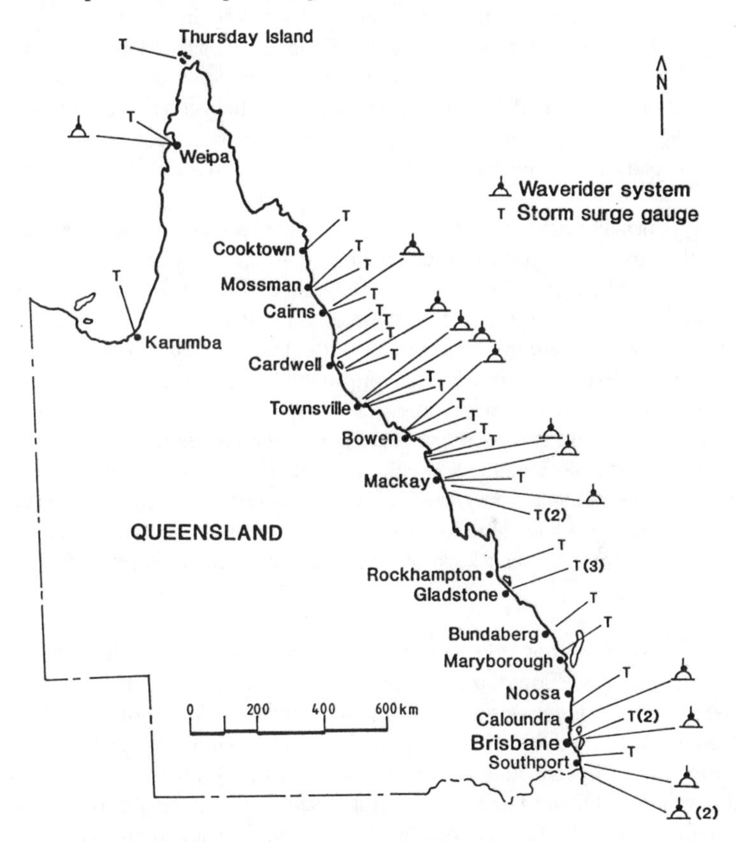

Figure 2. Queensland's wave recording and storm surge prediction network

The BPA also installed a network of storm tide recorders at all major coastal population centres as the basis of a warning and response, and evacuation scheme for impending disasters from tropical cyclone-induced storm surges.

An early realisation of the BPA coastal engineers was the need to prevent erosion rather than protect against it. The importance of maintaining natural coastal sand dunes and rehabilitating them after sand mining led to the establishment of a sand dune research station on South Stradbroke Island just north of the Gold Coast. There various dune planting systems are trialed, the effectiveness of different plant species and their fertilizer needs studied and plants propagated and grown for supply to coastal local authorities for use in local dune conservation schemes. A comprehensive manual was prepared describing recommended practice for conservation of sand dune vegetation and demonstration projects undertaken at a number of locations.

The educative functions of the BPA were promoted by its engineers and scientists visiting various local community groups along the coast. These activities were supported by BPA's newsletter, Beach Conservation, which provided news and advice on all matters relating to beach conservation including concise articles written by coastal engineers and scientists to explain various coastal processes to the general reader. A total of 69 issues was published between 1970 and 1990.

The achievements of BPA in coastal engineering have been impressive. A steady stream of technical papers has been presented to both Australian and International coastal engineering conferences. For example, McGrath and Robinson (1972) reviewed the major findings of the Delft report. Robinson and Patterson (1975) recorded the case history of Kirra Point groyne clearly documenting the erosion and accretion effects of the groyne including the increased erosion from the updrift Tweed River training walls. Political pressure at that time favoured the use of coastal structures to solve coastal erosion problems and it took some years before the adverse effects of groynes and seawalls were generally understood by the general public and local politicians. In time the accumulated evidence from over twenty years of comprehensive beach surveys for various coastal works on the Gold Coast clearly showed the detrimental effect that inappropriate structures can have on a natural sand beach system (Macdonald and Patterson 1984).

The Delft report did not have access to recorded wave data and had relied on ship-based observed data. Alongshore sediment transport calculations using this data led to the conclusion that there was a significant differential in alongshore sand transport on the Gold Coast with the transport rate at the southern Gold Coast being significantly lower than both updrift and downdrift. The erosion problems of the Gold Coast beaches were attributed to this transport rate differential. After wave rider buoys had been in operation for ten years off the Gold Coast, Pattearson and Patterson (1983) made new alongshore sediment calculations showing that there was no significant alongshore sediment transport differential at the Gold Coast and therefore most of the southern Gold Coast erosion problems were directly attributable to the construction of training walls at the Tweed River mouth in northern New South Wales.

Implementation of the major recommendations of the Delft report with respect to beach replenishment and sand bypassing involved major coastal engineering projects which will be described separately.

The South Australian Way

The need for coastal protection works was evident to engineers concerned with ports and harbours or public facilities located on the shoreline. South Australian experiences during the immediate post war period were reviewed by Beverley (1966) who recorded the effects of groynes and seawalls constructed at a number of locations in that state. During the mid 1950s laboratory experiments on wave mass transport effects and offshore bar formation were made at the University of Adelaide by R. Culver who subsequently visited many overseas hydraulic laboratories to gain further understanding of coastal processes.

During the early 1960s beach erosion and damage to coastal facilities increased significantly on the Adelaide coastline and public concern led the Seaside Councils' Committee, representing the local authorities affected, and the South Australian government to commission a five year Beach Erosion Assessment Study by the Civil Engineering Department of the University of Adelaide. This study was carried out from 1964 to 1970 and was directed by Culver. Its report (Culver 1970), generally known as the Culver report, was based upon a detailed study of the Adelaide coastline and came to the conclusion that the beaches were running out of sand. There was irregular northward littoral drift with an average alongshore transport rate of about 30 000 m^3/a but there was no major continuing source of sand for the beach system. The problem was made more difficult by an apparent slow increase in sea level. Substantial wind blown sand loss also was occurring as was depletion of offshore sand reserves at the southern (updrift) end of the coast. The Culver report also concluded that, while the solutions to the erosion problem were relatively simple technically, the administrative system was complex. The need for a properly constituted Beach Protection and Planning Authority was real and URGENT. Such an Authority would need to be given responsibility, adequate funding and total support if the continuing beach problems were to be solved. A copy of the Queensland Beach Protection Act was appended to the Culver report as an example of the type of legislation required. The state government responded with the formation of an interim Foreshore and Beaches Committee and during the period 1970-1972 the South Australian Act setting up the Coast Protection Board (CPB) was drafted, Culver assisting in compiling the briefing draft. The South Australian Act was proclaimed and the CPB established in late 1972. Its executive functions were supplied by the Department of Environment and Planning, the first executive engineer being the late L. Buenfeld. Culver was a member of the Coast Protection Board from its inception to 1987 and was its chairman from 1987 to 1989.

Coast Protection Districts were proclaimed over the 4000 km of the South Australian coastline, consultative committees established, and study reports and management plans prepared for most of the seven Districts. Like the Queensland reports these were comprehensive investigations covering all aspects of the coastal zone then known to be relevant. In the mid 1980s it became apparent that the State's planning system provided a more effective framework for most coastal policies, including control of development in erosion or flood prone areas. This led to legislative amendment in 1993 when development control powers of the CPB were extended and relocated in regulations under the new Development Act. Recognising that coast protection works can adversely affect adjacent land and beaches, the CPB now has a power of direction over decisions on all coast protection works as well as on the excavation or dumping of more than 9 m^3 of sand or other material, including dredging.

South Australia was the first Australian State to give serious consideration to the question of greenhouse climate change and sea level rise, this being first addressed in a 1984 review of protection strategy for the Adelaide coast, and subsequently by an Advisory Committee on Mean Sea Level established in 1989. In 1991 the State Government officially endorsed a new set of coastal hazard policies which established risk criteria for coastal erosion and flooding and included allowances for sea level rise. These subsequently formed part of a coastal planning document proclaimed in 1994 and having effect in all the State's coastal regions. South Australia has placed emphasis on avoiding future problems through planning and development control processes.

The initial coast protection works undertaken by the CPB on the Adelaide coast were mainly the construction of rock seawalls which were the only immediate way to stop catastrophic damage to roads and buildings. In accord with Culver's 1970 recommendations, beach replenishment progressively became the main protection strategy with sand mainly being trucked southward from the downdrift northern beaches. Despite concerted investigations, no suitable offshore sand source was found until the late 1980s. Since 1989 the primary beach replenishment has been carried out biennially by trailer suction dredge, mainly using a sand reserve off Port Stanvac a few kilometres to the south. Beach and nearshore changes at Adelaide and other problem areas in the State have been monitored by a survey program which dates back to 1974.

The protection strategy has been reviewed several times. In 1983 an extensive reworking of the earlier Culver study was made with assessment and comparative costing of groynes, offshore breakwaters and different methods of beach replenishment. The study found that beach replenishment by trucking was still the preferred and most economic option (Wynne 1984). A further review in 1992 confirmed this conclusion, notwithstanding the increased costs associated with a change from sand trucking to dredging. In late 1995 the South Australian government appointed a reference group to provide an independent report on strategy for protection and management of the Adelaide beaches. New coastal legislation is being deferred pending the reference group's report.

The New South Wales Way

In New South Wales coastal management has developed in an ongoing process implemented through the New South Wales Public Works Department (PWD).[7]

---

7) Since 1995 the Department of Land and Water Conservation has been responsible for coastal management in New South Wales.

During the post second world war period the department's Harbours and Rivers Branch was responsible for planning and construction of all ports in the state except Sydney. Developments at Newcastle and Port Kembla harbour and flooding problems in estuarine rivers such as the Hunter and Tweed provided the subjects for the early hydraulic model studies at PWD's Manly Vale hydraulic laboratory. Some river entrance training works, e.g. Tweed River in 1962-1964, were undertaken for improving navigation conditions for commercial fishing boats or for flood mitigation purposes. In the early 1970s responsibility for Newcastle and Port Kembla ports was transferred to the Maritime Services Board and PWD's activities began to address the growing public concern over the relationship between coastal processes and development.

Stormy weather threatening coastal development during the mid 1970s led to the establishment within the PWD of a group to study coastal processes. Destruction of houses had occurred at Wamberal in the Gosford area between Sydney and Newcastle and on the far north coast, the village of Sheltering Palms, located a few kilometres north of Brunswick Heads and its entrance training walls, was abandoned because of erosion. The first major investigation was a comprehensive study of this section of the north coast extending from Byron Bay to Hastings Point. The study was completed in 1979 and revealed average rates of coastal recession of 0.5 to 1 m/a with greater rates approaching 2 m/a in some places (Gordon et al. 1978). Many technical innovations were introduced during this investigation (see section on Coastal Engineering Research). Similar studies were subsequently made for other parts of the New South Wales coast and more recent work has indicated that the period of time (1950s to 1970s) investigated in the earlier studies may have been unusually stormy and hence the erosion rates then were greater than the long term average. Studies were also made of estuarine processes with a view to restoring environmental degradation caused by intensive development in their catchments[8]. The first estuarine investigation (Druery and Curedale 1979) made in the late 1970s concerned the Tweed River near the New South Wales-Queensland border, where marine sand had been dredged from the estuary in 1975 for replenishment of the southern beaches of the Gold Coast.

In 1975 the New South Wales government introduced the Beach Improvement Programme to assist local authorities in restoring and improving the recreational amenity of the state's beaches. This program was based upon technical work carried out by PWD and was used also as a means for increasing public awareness of coastal problems.

To ensure that technical advice on coastal processes was actually used in planning and management of the coast, the Coastal Protection Act was passed in 1979. The Minister

---

8) The term catchment is commonly used in Australia when referring to drainage basins of rivers. This reflects Australian engineers' ongoing concerns with the lack of water on the driest of continents and the consequent importance of hydrology and water supply.

for Public Works now had the power, if he desired, to intervene in planning decisions on coastal process grounds. While the formal provisions of the Act were used infrequently, its existence provided an incentive, along with an increasing awareness of "duty of care", for councils to seek advice. In 1988 a Coastline Hazard Policy was adopted which provided for state government financial (50%) and technical assistance for protecting assets at risk, for developing management plans to ensure that future planning took account of coastal processes as well as other planning issues, and to continue to improve and restore the coastal amenity. The Coastline Management Manual (NSWG 1990), together with the Estuary Management Manual (NSWG 1994), sets out a system for managing the coast and provides technical background for local authorities. The management system is designed to assist local authorities toward merit based decisions taken in partnership with the local community and relevant state agencies.

An interesting example of the application of the Coastal Hazard Policy was undertaken at Lennox Head, few kilometres north of the Richmond River entrance in northern New South Wales. Coastal recession had been a problem since the first subdivision in the 1920s. The local authority had made a commitment to protect development and a coastal management scheme has been implemented using a balance of hard protection structures, dune works and planning controls. Much of the work was on private property and the individual approval of 27 property owners was required together with their 50% contribution to works on their property (Lord et al. 1993).

The Victorian Way

In accordance with the powers it had inherited from the early colonial governors the Victorian government proclaimed in 1879 a general reservation along the coastline which set aside in public ownership a coastal reserve of at least 30m. Some 95% of the state's coastline was included in this general reservation, the remaining 5% was already in private ownership to high water mark. There was no overall planning and management of the coastline until almost one hundred years later. Local committees of management, unfunded by the state government, encouraged foreshore activities which generated revenue. Camping areas, private bathing boxes and boat sheds, amusement parks and later car parks, toilet blocks and sporting facilities were allowed in the public foreshore reservation. Haphazard siting of facilities in dynamic areas led to considerable coastal erosion problems, destruction of coastal vegetation and dune systems and, in due course, to the loss of both many of these facilities and the attractive natural coastal environment.

The first public enquiry into erosion of Melbourne's recreational beaches bordering Port Phillip Bay was held in 1935. During the next twenty years a large number of coastal protection structures, seawalls, groynes, etc. were constructed. In 1966 the Port Phillip Authority, one of the earliest coastal planning authorities in the world, was established. Its members included representatives of government agencies and community groups and it had the responsibility for co-ordinating development on public land 600m to seaward and 200m inland. In 1978 it released a major coastal strategy for Port Phillip Bay but this lacked adequate enforcement provisions.

During this period coastal protection works were the responsibility of the Ports and Harbours Division of the Department of Public Works. In Port Phillip Bay these works were initially seawalls augmented by groynes but, as the extent of suitable submarine sand deposits became known, beach replenishment has been increasingly implemented on the beaches along the eastern side of Port Phillip Bay since the mid 1970s (Jennings and Barraclough 1978). Coastal protection within Port Phillip Bay was transferred to the Port of Melbourne Authority in 1990.

A review of coastal management in Victoria in 1982 identified a large number of government agencies at Commonwealth, state, and local authority level and a substantial number of Acts of Parliament which covered various aspects of coastal planning and management in that state. This confused state of affairs continued until August 1995 when a new coastal management body, the Coastal and Bay Management Council (CBMC) was appointed by the state government as part of the cooperative Commonwealth-State Coast Action programme. The CBMC will be a peak body responsible for defining a new set of priorities for Victoria's coastline and will work closely with three Regional Coastal Boards to be established in 1996 (Wescott 1995). Associated with this was restoration of funding for beach protection works which had been cancelled as an economy measure by a previous government.

The Western Australian Way

Over 40% of Australia's coastline is in Western Australia and that portion of the coast includes a vast array of physical and biological environments. At the present time 80% of the state's population live within 10km of the shore and 80% of recreation and tourism activities occur on the coast. At the beginning of this century use of the Perth metropolitan beaches followed the English and continental European pattern with the emphasis on the promenade and enjoying the sea air rather than the beach and water activities favoured today. Consequently structures were located too close to the water and beach restoration and coastal protection works became necessary (Gillespie 1965).

During this period coastal protection works in Western Australia were the responsibility of the Harbours and Rivers Branch of the Public Works Department (PWD). In 1967 J. Butcher was appointed investigations engineer and the development of PWD's capacity to collect and analyse coastal data for design purposes commenced. Physical models were used as design tools. A coastal investigations section was set up in 1976 under M. Paul and during the next 10 years this section's multidisciplinary team of specialists was engaged in the planning and development of a wide range of maritime facilities including a number of boat harbours (eg. Paul 1981). At the same time, the PWD investigations engineer, W.Andrew, was working closely with local authorities and the state planning and environmental authorities to overcome various coastal erosion problems and to prepare guidelines for artificial waterways. He was also responsible for the detailed conceptual planning for the facilities at Fremantle for the 1987 America's Cup challenge.

The PWD was disbanded in mid 1985 and all its maritime activities and resources transferred to the Department of Marine and Harbours. In 1994 further organisational restructuring led to these functions being transferred to the Maritime Division of the Department of Transport. During the period 1983 to 1994 the major coastal engineering activity was the Dawesville Channel, a major coastal and estuarine management project with the objective of improving the water quality in the Peel-Harvey estuarine system, south of Perth.

Western Australia has no state government organisation with statutory authority for coastal zone management. There is a number of government agencies, acting under a variety of Acts of Parliament, which have the task of planning and implementing different aspects of coastal policy. In the early 1970s erosion was affecting houses built close to the shoreline at a number of locations. There was considerable expenditure on coastal protection works. In 1972 an Interdepartmental Committee on Sand Drift and Sea Erosion was formed and it reported in 1974 that *"There is evidently much to be said for avoiding erosion rather than trying to cure it after it has happened."* In 1975 this committee was renamed the Coastal Development Committee (CDC) and it became the responsibility of the Town Planning Department.

At the same time the state Environmental Protection Authority conducted an internal review of coastal management and produced draft guidelines for a policy on coastal zone management in Western Australia. Following considerable debate coastal management was reviewed by another committee, the report of which highlighted the government's fragmented and uncoordinated approach to coastal management. It also concluded that the state government was likely to face mounting costs for repair of degraded coastal areas if coastal planning and management were not improved.

In 1982 a Coastal Management Coordinating Committee (CMCC) comprising senior officials from relevant departments was formed, together with a coastal management section within the Department of Conservation and Environment. The CMCC operated in parallel with the older CDC. The operations of the CMCC were reviewed in 1985. A proposal at that time for a Coastal Management Council was not proceeded with and in 1986 the CMCC became the responsibility of the State Planning Commission. The CDC was disbanded in the following year.

Coastal management in Western Australia was reviewed during 1994-95 by a government appointed committee. The report of this committee (WACMRC 1995) made recommendations on the following matters: Defining the coastal zone; State government goals, principles and objectives for coastal management, including environmental, social, economic and administrative objectives; Administration and coordination of coastal management, including the formation of a high level coordinating body, the Coastal Zone Management Council; and Funding sources, funding requirements and funding distribution.

SOME SIGNIFICANT PROJECTS

River Ports in New South Wales

The development of river entrances for coastal shipping during the nineteenth century has been referred to previously. Many of these entrances were unstable, shallow and dangerous for the ships using them. Efforts were soon made to correct these problems by the dredging of estuarine shoals and the construction of training walls. This early work was carried out under the direction of E.O. Moriarty, the first chief engineer for Harbours and Rivers. Initial attempts at improvement did not always work either because the resources available were too small to have much effect or because the natural processes were not clearly understood. In particular, the influence of waves, as opposed to tides and ocean currents, as the main cause of sand movement along the coast was not recognised by all engineers of that time. The earliest British engineers were well aware of the influence of tides and carried that experience with them to Australia but it took some time before the significant influence of wave action was recognised. As late as 1910-11 Halligan in a description of these bar harbours of New South Wales presents the idea that the dominant alongshore sand movement was southward because of the southward flowing ocean current on the continental shelf and that northward sand movement occurred only where there were current eddies adjacent to projecting headlands. Others were of a different opinion and the reasons for the formation of river entrance sand bars and how to eliminate them were debated at length by engineers from the 1880s.

One of the first to accept that *"the battle is to be fought with the waves"* was W. Shellshear, a railway engineer, who studied the improvement of *"these bar-bound rivers"* in his spare time, corresponding with Sir John Coode. He came to the conclusions that *"the formation of bars at the entrances to our rivers is mainly due to the action of the waves in lifting large quantities of sand as they pass into shallow water, the sand being carried up the estuary by the incoming tide, and is deposited as soon as it is beyond the influence of the waves; the ebb tide, being unassisted by the waves, is unable to cope with the incoming sand, and thus we see, when the tide and waves are left to themselves, the tendency is to close the entrance altogether ... certain entrances are only open after a heavy fresh*[9]." (Shellshear 1884). To cope with the quantity of sand stirred up by the waves he deduced that *"it is necessary to extend artificially the entrance to a point where the depth of water is such that the waves are unable to heap up the sand, to make the entrance of such form that the force of the waves will be expended before they advance into shallow water, and at the same time directing and concentrating the action of the ebb tide and upland waters, so that their force may be used to the best advantage in combating the mischievous action of the*

---

9) "fresh" - freshwater flow, that is runoff from the upland catchment.

waves." Quoting Coode, he indicated that, where practical, breakwaters should extend out to at least 6 m depth. Recognising that each river needed to be considered individually he concluded that the key to success in resolving this problem involves *"The close observation of physical features and effects, and the adoption of means to assist the operations of Nature instead of opposing them."*

Shellshear applied his principles to propose a solution for the problems of the Richmond River entrance at Ballina. This was a typical unstable downdrift river mouth which shifted its location from time to time. In its northernmost position it ran along the southern side of a rocky headland. Four times in 35 years flood flows had broken through the sand spit as much as 2200m further south to form a new entrance which then moved northwards under the influence of northward alongshore sand transport. It was the most dangerous entrance on the coast but it also provided the longest length of navigable waterway - 100 km from Ballina to Lismore was available to ships of 3 m draft. Shellshear's ideas clearly influenced Coode as the latter's recommendations to the New South Wales government for the Richmond River entrance were very similar to Shellshear's proposals (Fig. 3). With some modifications these were constructed over a period of 20 years commencing in the early 1890s. Work, using day labour rather than contractors, proceeded at a more or less constant rate controlled by the amount of funds made available each year. By 1904 the essential parts of the scheme had been implemented. The water level over the bar at low tide had increased from 2.7 m to 4.4 m and the entrance had been transformed from being dangerous and frequently impossible to navigate into one of the most easily accessible on the coast. Changes to the tides in the river were also noted with the range at Lismore increasing by 0.46m (Burrows 1904-05).

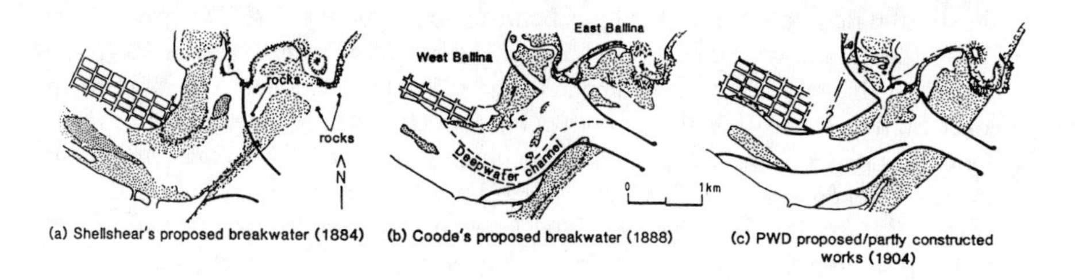

Figure 3. Richmond River entrance

Not all the river entrances were located in the downdrift position. Some were located at the updrift (southern) end of a section of sandy coast in the lee of a headland. In this case there is a local reversal of the inshore alongshore transport with sand tending to come from the north and push the entrance against the headland. This process was most noticeable at the Hunter River entrance to the port of Newcastle where the coastline to the north trends more easterly than anywhere else along the New South

Wales coast. Here southward sand transport into the entrance channel was a problem, particularly after a southern breakwater was constructed linking an island to the mainland and then extended further seaward from the island. When a northern breakwater was constructed to prevent sand entering the navigation channel, the sand movement was deflected seaward and formed a shoal at the end of the breakwater, again spilling into the entrance channel. The alternatives of continuing maintenance dredging or extending the breakwater were both costly. The latter option requiring a parallel extension of the southern breakwater to prevent reflection of wave energy from southeasterly storms into the harbour was ultimately adopted (King 1910-11, Allan 1920-21).

A feature of Newcastle harbour and some other New South Wales river entrance works is the use of wave traps formed by increasing the width between the breakwaters at their landward end and extending them so they overlapped the internal estuarine training walls (Fig. 4). A natural sand beach usually forms in the gaps between the two walls on either side of the entrance allowing diffracted/refracted waves to be dissipated (Ford 1963).

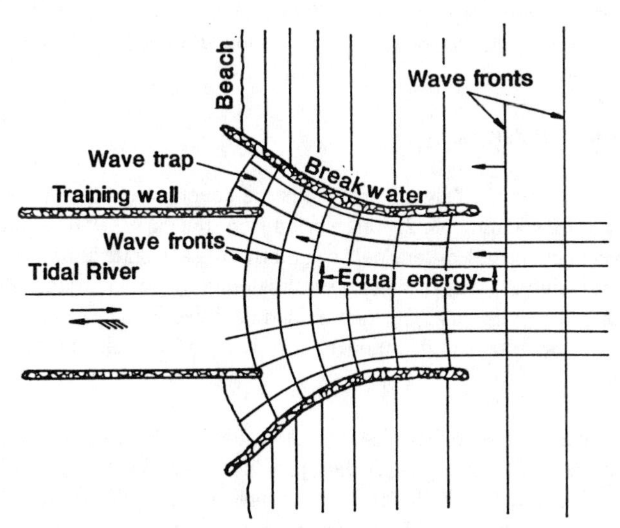

Figure 4. Wave traps at river entrances

Most of these river entrance works were completed by 1910 with some minor changes being carried out up to 1930. A major training scheme was initiated for the Clarence River entrance during the 1950s while in the 1960s there was further development of some small river entrances as fishing harbours. The latter included new extended training walls for the Tweed River entrance just south of the New South Wales-Queensland border. Experience with the river entrances was reviewed by Ford in 1963. Subsequently an analysis of the various river training schemes, sixteen in total, formed the basis of the first paper on an Australian coastal engineering topic presented by an Australian author, C.D. Floyd, at one of what are now the International Conferences on Coastal Engineering (Floyd 1968). It was concluded that while the training works had improved conditions for navigation they had not resulted in any appreciable increase in bar depths, since the bars had eventually reformed seaward of the training walls. Furthermore, despite the complex mechanisms involved in bar formation, there was a consistent simple correlation between channel and bar depths - the latter were approximately half the former and this relationship was supported by data from river entrances in other parts of the world.

The history of the ports and coastal waterways of New South Wales is the subject of a forthcoming book (Coltheart, in prep.). All through the period since 1810, when the first training works were constructed in the Hunter River at Newcastle, until the 1970s the focus was on the coastal engineering works required to maintain stable channels for navigation and little if any thought was given to the disruption to alongshore sand transport caused by these works and their consequent impact upon the adjoining coastline. Floyd's (1968) final comment on these river entrances was both understated and prophetic - *"Detailed examination of littoral drift behaviour in the areas adjacent to the entrances where bars have been removed would probably reveal some interesting information."*

The Hydraulic Models Era

Hydraulic models played a significant role in the development of coastal engineering expertise in Australia during the 25 year period from the early 1950s to the late 1970s. Several universities and at least one public authority in each state established some kind of laboratory facility. Many of these were small one model laboratories dedicated to a particular harbour or estuary. Others provided facilities for a large number of models and in some cases these included models of rivers and hydraulic structures as well as coastal models.

The most significant of these facilities is located at Manly Vale in the northern suburbs of Sydney where an old water supply dam provides water for the hydraulic laboratory operated by the New South Wales Public Works Department (now Department of Public Works and Services) and the Water Research Laboratory of the University of New South Wales (UNSW). Two other government departments concerned with water supply and irrigation also had laboratories on this site. The PWD facility was established in the early 1940s. Work on the first coastal model, a wave penetration study of Port Kembla Harbour, commenced in 1949. A.B. Sinclair, who later established a well known Australian consulting engineering firm, was the engineer responsible for the laboratory at that time. Following an overseas tour in 1951, during which he visited a number of hydraulic laboratories in U.S.A., U.K. and Europe, Sinclair in his report to PWD recommended that the best balance between intellectual stimulation and practicality would be achieved by a marriage of the PWD laboratory with proposed University of New South Wales facilities planned for the same site. This proposal was rejected firmly by PWD at that time and it was not until twenty five years later that this union was achieved. Sinclair left the PWD laboratory in the mid 1950s. His place was taken by D.N. Foster, who in 1957 joined the UNSW to establish the Water Research Laboratory (WRL) on the opposite side of the creek. WRL commenced operations with a flood mitigation model for the city of Launceston in Tasmania. While this was not a coastal model in itself it was in fact a development from a previous model study begun almost 20 years earlier but never completed.

Launceston, one of the oldest European settlements in Australia (it was founded in 1806), is located 64km from the sea where the North and South Esk Rivers join to form the estuarine Tamar River. In the 1930s larger ships using the port and the difficult tidal currents at constricted portions of the estuary were creating the need for action to improve navigation conditions and maintain the port's viability. In 1939 the Marine Board of Launceston (now Port of Launceston Authority) appointed T.A. Lang as hydraulic research engineer with instructions *"to design and construct, a hydraulic model of the River Tamar and by use of the model, to investigate the following:*

a) Weir and Lock - The most suitable position and the effects on the Port of a weir or lock in the River Tamar, so as to maintain a water level in the Port of Launceston at approx HWST.

b) Upper Reaches - The most suitable method other than a weir or lock whereby a suitable navigable waterway may be maintained in the Upper Reaches, from say Rosevears to Launceston.

c) The general effects of the above measures on the River as a whole, and on the floods that periodically occur in the North and South Esk Rivers." (Lang 1947).

Model construction commenced in April 1939 and work was suspended in September 1940 because of the second world war. During this period design and construction were completed and sufficient experiments were made to demonstrate that the model would fulfill its purpose. In the event none of the above problems were investigated and the model was eventually dismantled. Indeed the problems were "solved" in due course by the removal of the port facilities from Launceston to Bell Bay at the mouth of the Tamar River.

The importance of this model lies, not in the results as none of significance were obtained, but in the very advanced electrical and electronic technology which was developed to generate, control and measure the tides.[10] Tidal flows into and out of the model ocean were controlled by automatic valves with inflow supplied from an elevated tank to which water was returned by pumping from a temporary storage tank at model level (Fig. 5). The control of the tide used an ingenious photoelectric system in which

10) Sir John Madsen, Professor of Electrical Engineering at the University of Sydney, was specialist consultant for this part of the project.

the tide curve, reproduced on 35mm film, moved at the speed corresponding to the model tidal period. Light from a common source passed through two identical horizontal slits, the exposed opening of one of which indicated the tide height on the 35mm film and that of the other the position of a mechanical shutter which was connected by fine piano wire to the horizontal plate tide controller. The latter was maintained at a constant distance above the actual model water surface.

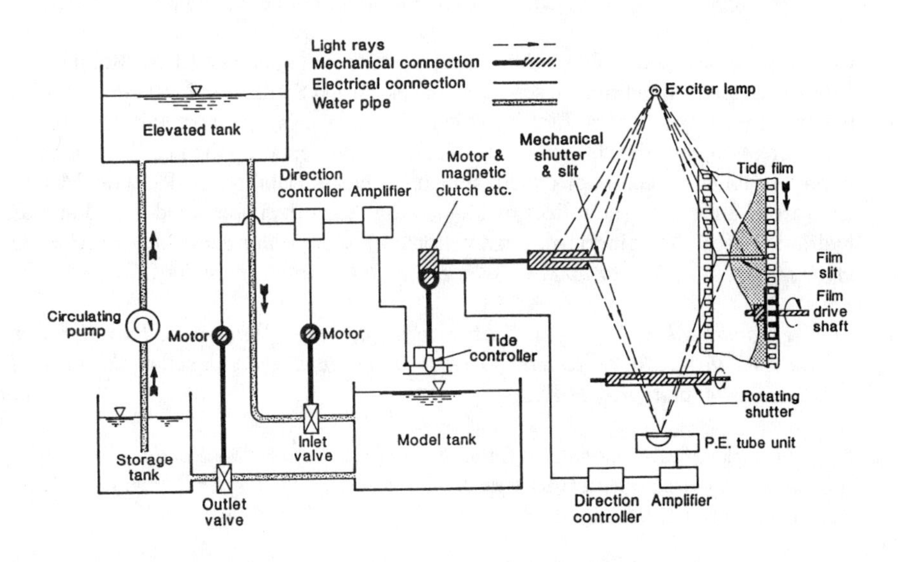

Figure 5. Tamar River model - tide generator control system

A rotating shutter allowed light from each slit to alternatively fall on the photoelectric cell which detected any difference in the signals. The difference signal was then used to raise or lower the tide controller to reduce the difference to zero. As the tide controller moved, the varying distance between it and the water surface changed the capacitance of one arm of a potential divider circuit. The resulting out of balance voltage then operated the valves controlling inflow and outflow to the model so as to maintain the water level at a constant distance from the tide controller plate which was moving up and down with the required vertical rise and fall determined by the tide curve on the 35mm film.

Horizontal plate capacitance devices similar to the tide controller were used to record tidal water levels at various locations along the model estuary. This measurement method produced no disturbance of the flow.

This application of local expertise, informed by the latest developments in other countries and utilising local resources and facilities, was the forerunner of the various facilities for coastal hydraulic models developed in Australia during the 1950s. In many cases the funds and expertise available were not as generous as those provided for the Tamar River model and considerable ingenuity had to be employed to develop efficient wave and tide generating equipment and their associated measuring systems.

The increasing navigation problems of the estuarine port of Rockhampton in central Queensland provided the initial impetus for the development of hydraulic laboratory facilities in Queensland during the 1950s. Port facilities had been provided at at least three locations within the Fitzroy River estuary during the previous 90 years in attempts to overcome the impacts of tropical cyclones and devastating floods which made it impossible to maintain the unstable shallow navigation channels. In 1949 a Committee of Inquiry recommended a hydraulic model study to investigate another proposed port development (Davenport 1986).

This model investigation was undertaken by G.R. McKay assisted by R.D. Watkins both of the University of Queensland. The model was constructed at Cairncross Dock alongside the Brisbane River. There were several similarities with the Tamar River model. Firstly, a novel tide generating system was developed; secondly, existing charts were not related to a common level datum and surveys of local chart and gauge datums were required before the model could be constructed; and thirdly, no significant results were produced.

The tide generating system had three basic components - water control valve and pump, tide synthesiser and electronic controller. Water flow in and out of the model was controlled by a specially manufactured four way valve and pump system (Fig. 6). Movement of the rotating vertical plate within the valve through 90° caused a change from inflow to the model to outflow to the reservoir. Control of the valve by a servo controlled electric motor was achieved using a specially constructed mechanical tide synthesiser similar to mechanical tide prediction machines used before the advent of digital computers. Six tidal components could be reproduced at an appropriate time scale for the model using a variable speed drive. The synthesiser was constructed with two series of excentrics, one reproducing the integrated vertical motion of the six tidal components. The other series of excentrics, set at 90° to the first set, reproduced a motion proportional to the rate of rise of the tide and hence approximated the flow rate of the incoming/outgoing tide. The vertical motion signal actuated a lever on which was mounted a horizontal plate capacitance water level sensor similar in principle to those used in the Tamar River model. The rate of rise movement actuated a potentiometer which fed a signal to the electronic control system to actuate the vane of the four way valve. Fine adjustment of the valve position was obtained using the out of balance signal from the water level sensor.

Subsequently G.R. McKay developed hydraulic laboratory facilities at the University of Queensland's St Lucia site which provided for the needs of both state government and consultants for over 20 years to the mid 1970s. An early model study, in which the tide generating system designed for the Fitzroy River model was used, was one for

Mourilyan Harbour, a small natural sheltered tidal basin in northern Queensland which was to be developed for bulk sugar exports. The shallow entrance was constrained by a rock bar while deep water occurred in the harbour. Experiments with various tidal conditions revealed the existence of a previously unknown vertical current circulation which was responsible for scouring the bed within the harbour. Considerable ingenuity was required to devise special neutrally buoyant floats for measuring the change in vertical current strength with variations in tidal range and as the entrance bar was deepened. The complete project, including model investigations, harbour development and bulk sugar terminal, is described in McKay et al. (1960).

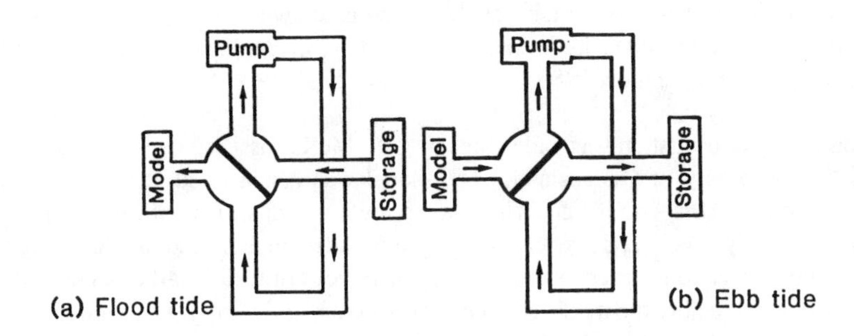

Figure 6. University of Queensland tide generator control system

In the following years a number of tidal and wave models for proposed harbour works were constructed at St Lucia. One long running project concerned the various problems associated with the Pioneer River estuary in northern Queensland. Three separate hydraulic models were built over a period of 25 years and extensive sedimentological and historical investigations were also made (Gourlay and Hacker 1986).

The early wave models were associated with breakwater protected harbours and their development to meet the expanding needs of the post second world war era. As previously stated, the first models built at the NSWPWD laboratory were of Port Kembla Harbour, the port for the city of Wollongong. This is an artificial harbour formed by two rubble mound breakwaters, each about a kilometre in length. Commenced at the beginning of the twentieth century, the original harbour and its various cargo jetties were completed before the second world war. Coal exports, raw materials for and products from the steelworks and general cargo expanded during the second world war and the following years. During the late 1940s port operations were being increasingly interrupted during storms and sometimes during relatively calm weather as waves and surge action within the harbour caused moored ships to move at their berths with consequent interference to cargo handling, breaking of mooring springs and at times damage to both ships and jetties. During the more severe conditions ships had to leave their berths and either anchor in the harbour or put to sea to ride out the storm. It was the well known ship ranging phenomenon which was causing problems in a number of artificial harbours around the world at that time.

The first Port Kembla model was constructed in 1951 and was used to study short period storm and swell waves within the harbour. A second distorted scale model was subsequently constructed to study long period waves both in the existing (now outer) harbour and in the future inner harbour proposed to accommodate increasing trade arising from the enlarged steelworks and expanding coal exports. A wave flume was also provided for model studies of breakwater stability and wave filter design. At the same time the meteorological conditions in the western Tasman Sea were studied; historical storms identified and the newly developed techniques of wave hindcasting and wave refraction analysis used to determine ocean wave heights and the effects of two offshore islands which could not be included in the models. While offshore wave recording was not possible at that time, waves and ship movements within the harbour were measured using simple devices involving floats and mechanical chart recorders. Statistics of interruptions to harbour operations were also compiled. As the model experiments proceeded the importance of providing nonreflective spending beaches along the outer harbour foreshores became evident while the complexity and very sharp peaked response of the inner harbour to long period waves of various periods showed that ship ranging problems were unlikely to be solved by harbour geometry modifications.

The problems investigated at Port Kembla were concerned with reduction of wave action and ship movement within the harbour. Sediment movement was not a significant problem. On the other side of the continent at Bunbury in Western Australia, a breakwater had been constructed to protect an open jetty. Over the years the jetty was extended to accommodate the needs of more and larger ships using the port. The breakwater protecting the jetty from westerly and northwesterly waves was also extended. However in this case there was also continuing sedimentation of the harbour from wave-induced alongshore sediment transport. A physical model investigation of this process was undertaken at the University of Western Australia by R. Silvester (1956). Wave hindcasting analysis for the year 1951 showed that both storm waves and swell approached the breakwater from various directions between WSW and NNW and the model was designed accordingly to reproduce waves from these directions. Alongshore sediment transport was reproduced by feeding fine sand at the updrift end of the breakwater. Comparison between model sediment movements and deposits with historical hydrographic surveys and other field observations indicated good agreement. The effectiveness of various modifications to the breakwater in preventing sand movement into the harbour was assessed. Furthermore the different actions of storm waves and swell in moving sand at this location were clarified.

The coast between Melbourne and Adelaide has been a dangerous one for shipping and does not provide many places where safe harbour facilities can be constructed. Portland is the oldest European settlement in Victoria and the natural outlet for the dominantly rural industries of western Victoria. An artificial harbour similar to Port Kembla was proposed before the first world war but only small facilities were available until after the second world war. An artificial harbour was constructed by the 1950s; its investigation and design is described by Hughes (1957). The new coastal engineering techniques involving hydrographic surveys, meteorological analysis and wave hindcasting, and wave refraction analysis were applied. Three dimensional model studies of the proposed breakwater alignments together with two dimensional breakwater stability tests were undertaken. The question of littoral sediment transport was considered during the investigation as there were know sedimentation problems at Warrnambool Harbour about 75km to the east of Portland. Mineralogical studies of sediment movement using heavy mineral tracers were made and it was concluded that there was no net alongshore transport at the harbour site. Design and construction of the harbour proceeded on the basis that no significant updrift accretion or downdrift erosion would occur. Unfortunately, subsequent experience showed that the conclusions of the tracer study were wrong. Sand accumulated against the southern breakwater and beaches to the north of the harbour eroded. Recently sand bypassing facilities have been provided to overcome these problems.

During the 1960s more complex problems needed investigation and more advanced hydraulic model technology was required. This led to Australian laboratories seeking advice from overseas laboratories. In New South Wales two large models were constructed for port developments both using technology based upon that developed by the Hydraulics Research Station at Wallingford in the United Kingdom. The first was the Newcastle Harbour model constructed by PWD at Manly Vale. The second was the Botany Bay model constructed by the Maritime Services Board (MSB) of New South Wales.

The problems of the Hunter River entrance at Newcastle have already been described and the development of the sand shoal at the end of the northern breakwater was the subject of another very early movable bed wave model operating in Newcastle in the late 1940s and early 1950s. The 1960s model was a tidal model representing 19km of the lower estuary including the port area. The problem was the large amount of maintenance dredging required to keep the port clear of silt brought down by river floods. The silt flocculates in the presence of salt water and is concentrated and deposited in the port area by salt water wedge action. The model reproduced tides using a pneumatic tide generator and also the density currents arising as the upland fresh water flow encountered the incoming saline tidal flow. A range of possible solutions was investigated including mixing of fresh and salt water by aeration, a new river entrance bypassing silt laden floods and making the port a simple tidal inlet, isolation of a shallow tidal lake connected to the estuary which was acting as a silt storage, etc.

During this period, i.e. early 1950s to late 1970s, a number of hydraulic laboratory facilities were established in the various states. Some of these were central laboratories for a state government authority, for example the Queensland Government Hydraulics Laboratory and the Victorian Marine Models Laboratory. Others were associated with universities, while in some cases, e.g. Portland Harbour, models were constructed at the location of the project. With the increasing use of mathematical modelling and reducing funding to support facilities many of these laboratory facilities no longer exist.

The Queensland Department of Harbours and Marine (DHM) commenced its own hydraulic model investigations in the early 1970s. The first model was a movable bed tidal model of the entrance channel to the sugar port at Lucinda Point in northern Queensland. Sediment transport was studied using crushed macadamia nut shells as the model bed material. The model and its facilities were inundated during the 1974 Brisbane floods and DHM subsequently established the Queensland Government Hydraulics Laboratory at Deagon near Brisbane.

In Victoria the Ports and Harbours Division of the Public Works Department established a hydraulic models laboratory at Port Melbourne in 1966 with a single wave basin and a staff of three. By 1971 staff had trebled and were increasingly involved in field investigation measuring winds, waves and currents. In 1972 a programme of sending engineers to study in the Netherlands was commenced. Between 1974 and 1980 the laboratory facilities were expanded and both field and laboratory equipment upgraded, including the introduction of computers. Random wave generators were installed and a flume for coastal structure tests constructed. Both fixed bed and movable bed model studies were undertaken and field work, including baseline data collection along the Victorian coastline, became a large proportion of the overall work effort. Contracts were undertaken for projects in Victoria and other states as well as overseas for both government and private clients. At the peak of the laboratory's activities, 18 staff were employed. From 1980 there was a steady decline in the laboratory's capabilities as funding then staffing were reduced following changes in the state government administrative structure. In 1993 the Marine Laboratory was closed and all equipment and facilities disposed of. Base level data collection ceased and specialist staff dispersed to other positions.

In Western Australia the coastal engineering laboratory facilities originally set up by the Public Works Department were acquired in 1984 by the Centre for Water Research (CWR), a multidisciplinary research centre originally established under the direction of J. Imberger at the University of Western Australia with the support of the state government. These facilities were redeveloped by CWR and a new Coastal and Hydraulic Engineering Laboratory established at Floreat Park, a Perth suburb, in 1987. The activities of this laboratory currently include a wide range of field studies in coastal, estuarine and continental shelf waters as well as physical and numerical modelling studies.

At the present time the major coastal engineering laboratories in Australia are the Queensland Government Hydraulics Laboratory in Brisbane, the University of New South Wales Water Research Laboratory and the New South Wales Department of Public Works and Services (formerly PWD) Manly Hydraulics Laboratory in Sydney and the CWR Coastal and Hydraulic Engineering Laboratory in Perth. The two

Sydney-based laboratories combine their resources in a joint consulting role as Australian Water and Coastal Studies Pty Ltd (AWACS).

Airports and Seaports in Botany Bay

Botany Bay was the site of the first landing by James Cook on the eastern coast of Australia in 1770. It was the destination in 1788 of the first fleet bringing the first, mostly unwilling, European inhabitants to this country. It was shallow, exposed and lacked good freshwater and the first settlement of Sydney and later its port was established on Port Jackson (Sydney Harbour) 11km to the north. Botany Bay was left largely undisturbed until after the second world war.

Sydney's airport was established adjacent to the northwestern side of the bay in the 1930s. A new runway to accommodate jet aircraft was constructed during the 1960s by reclamation projecting into the bay (Fig. 7). Material for the reclamation was dredged from the shallow western portion of the bay. This was the first major coastal engineering project undertaken by the then Commonwealth Department of Housing and Construction and led to the establishment of a Maritime Works Branch within that department with A. Hicks as Chief Engineer. With no prior coastal engineering expertise, technical advice was sought from the Hydraulics Research Station (HRS) at Wallingford in England. The runway extension was protected from wave action by a seawall which was constructed in the dry in a trench cut along the axis of an initial sand bund enclosing the area to be reclaimed. Tribar armour units were used on the seawall which was designed using model tests (Foster 1984). An unanticipated and undesirable side effect of this reclamation project was that the location and configuration of the hole, from which the sand for reclamation was obtained, were such that wave refraction patterns were altered and a recreational beach on the western side of the bay severely eroded. This accident was not without subsequent benefits.

At the same time as the airport runway was being extended the New South Wales Maritime Services Board was investigating the establishment on the northern side of the bay of a new modern deepwater port with facilities for container vessels. However, there were some problems to be solved. Botany Bay is large, about 8km at its widest point, and its entrance is wide with headlands about 1.2km apart. Considerable wave energy penetrates into the bay during predominant eastsoutheasterly to southerly storms and maximum wave action occurred along the alignment of the proposed entrance channel. Development of the port depended upon the control and reduction of waves in the bay to provide safe conditions along the entrance channel. The solution adopted was special configuration dredging of the entrance channel to control wave refraction and redirect wave energy away from both the entrance channel and the port area. The dredge spoil was used for reclaiming areas for port facilities and storage. Learning from the earlier problem the dredging for this project was planned to give beneficial effects.

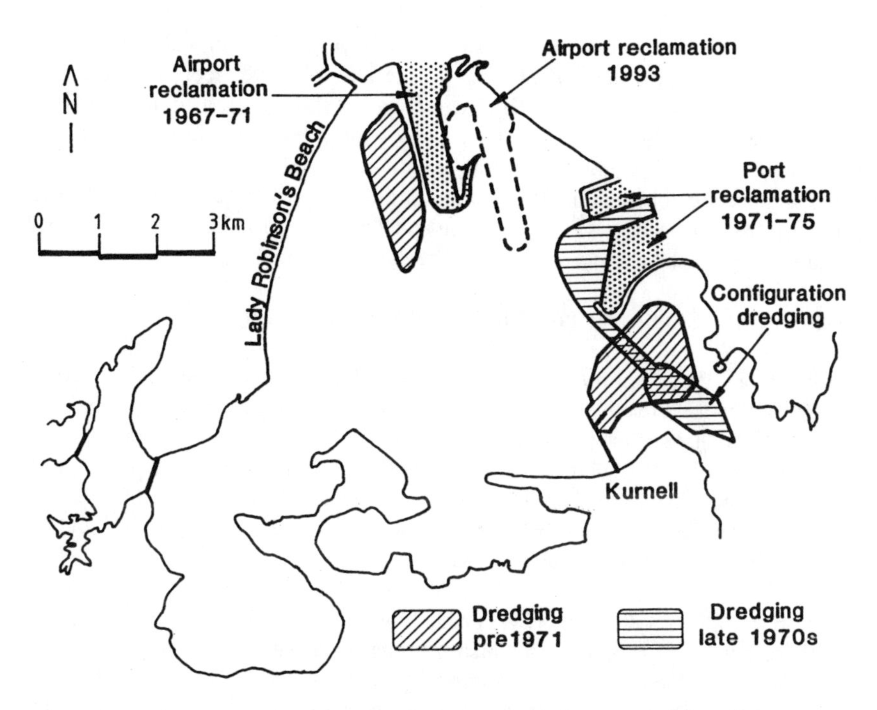

Figure 7. Botany Bay showing Sydney Airport runway and Port Botany

Again HRS was retained as specialist consultants but the major part of the investigation and design was done by Australian coastal engineers. The extensive investigations are described by Wallace (1977) using material supplied by N.V. Lawson. These included wave recording, hydraulic model tests, mathematical modelling of wave refraction and bed movement studies with both radioactive and fluorescent tracers. It was important that the dredged configuration be stable and require minimum maintenance dredging.

A fixed bed model of the whole bay and adjoining ocean to 82m depth was constructed adjacent to the site to study wave action within the bay. It covered an area of about 1ha and reproduced waves from the east to the south as well as alongshore currents outside the bay, tides and river flows into it. There were 100 wave recording stations in the model. Computer programs were used to process model wave data, correcting it for bed friction effects and determining long term wave statistics at each location in the bay throughout a tidal cycle for all wave period and direction combinations considered important. These statistics were then combined with wave height, period and direction exceedance data from offshore wave measurements to give wave height exceedance curves for each location in the bay. Comparison of experiments made with the special configuration dredging with those made with the natural channel conditions allowed the effectiveness of the dredging in reducing wave heights to be assessed.

Offshore wave conditions have been recorded using wave rider buoys off the entrance of Botany Bay since 8 April 1971 - one of the longest wave recording stations in continuous operation in the world. The techniques and skills developed in the analysis and application of this wave data have been widely used and further developed throughout Australia by both governmental authorities and private consultants. Two of the engineers involved in the Port Botany project, N.V. Lawson and P.D. Treloar subsequently formed the consultancy Lawson and Treloar Pty Ltd, an expanding group specialising in coastal engineering and marine data acquisition.

Environmental Studies for Port Phillip and Westernport Bays

Port Phillip and Westernport Bays are two large enclosed bays on the Victorian coast. Port Phillip Bay has a relatively narrow entrance to Bass Strait and relatively open bay waters. Melbourne is on the northern side of the bay, 55km northnortheast of its entrance. The Port of Melbourne is located both in docks adjoining the Yarra River downstream of Melbourne and on the shore of Hobsons Bay, the northernmost part of Port Phillip Bay. Residential suburbs extend along the eastern side of the bay and the beaches there have heavy recreational usage.

Westernport Bay lies to the east of Port Phillip Bay. It has two entrances to Bass Strait and there are two major islands which give its water surface a "figure of eight" configuration in which tidal flows dominate. There is a system of sandy channels and banks together with extensive, densely vegetated intertidal areas. Westernport Bay remained substantially unchanged throughout the nineteenth century but in recent times, since the 1960s, has been increasingly developed for industrial purposes and as an alternative port for Melbourne.

The first major Australian marine environmental study was the Port Phillip Bay study undertaken in 1968-71. It was a multidisciplinary study jointly sponsored by the Melbourne and Metropolitan Board of Works (now Melbourne Water) and the Fisheries and Wildlife Department of Victoria (now Conservation and Natural Resources). The study included a detailed review of the characteristics of Port Phillip Bay including catchment geology, climate, land use and commercial activities, runoff and waste discharges into the Bay. Substantial water quality and biological sampling of bay waters were made and the physical studies included the first major use of mathematical modelling of water movement in Australia. Coastal processes and sediment movement were not investigated, although coastal engineering data collection within Port Phillip was subsequently conducted over several years by the Ports and Harbours Division of the Department of Public Works through its Marine Models Laboratory.

A second major multidisciplinary study was conducted between 1973 and 1975 in Westernport Bay. This was a very ambitious study involving specialists with expertise in various areas and implemented through a large number of consultancies with consulting firms, universities and government agencies. The marine studies involved physical processes, including water and sediment movements; chemistry of water and sediments; estuarine botany, particularly mangroves and seagrasses and their ecology; estuarine fauna, their population and community structure; toxicology of heavy metals to marine flora and fauna and modelling, particularly water quality modelling. Physical studies of the land areas surrounding the estuary were also made, together with social and economic studies of the region. All were reported in a comprehensive report (Shapiro 1975), as well as in numerous individual reports and papers. Of particular interest to coastal engineers are the water and sediment studies described in a special issue of the journal Marine Geology (Marsden 1979). Nine papers describe the geological and sedimentary environment, hydrodynamic data, bottom current measurements, seabed drifter studies, circulation as deduced from marine chemistry and numerical hydrodynamic modelling studies. The general movement of water and sediment was deduced from a synthesis of all those studies and was confirmed using a range of data (Harris et al. 1979). This study has provided the basis for subsequent investigations and for planning decisions. These have included activities such as oil spill modelling, extension of port areas and channel dredging.

A third study was the Port of Melbourne environmental study carried out in 1977-78 by the Centre for Environmental Studies, University of Melbourne, under the direction of J.B. Hinwood. This study focussed on the northern part of Port Phillip Bay, Hobsons Bay and the port of Melbourne, including the lower Yarra River estuary. Water movement and water quality, beaches and sediments and marine ecology were studied using a wide range of techniques to characterise the typical, seasonal, diurnal and transient states of this marine environment. Some of the detailed components of the study were:

- determination of the heavy metals present and their likely resolution following dredging;
- fluorescent tracer studies of sand movement on Sandridge Beach, combined with beach profiling over the two years;
- sedimentation from the Yarra River plume and resuspension of the subtidal delta by wind waves;
- extensive long term current metering combined with intensive exercises utilising direct reading current meters and radar tracked drogues;
- studies of the flushing of the existing Webb Dock;
- investigations of the stratification and flow regimes in the lower Yarra estuary and of the river plume discharged into Hobsons Bay;
- studies of the bottom stability including sidescan sonar surveys;
- very extensive seasonal sampling of the benthic ecology, including the infaunal benthos;
- characterisation of the intertidal invertebrates on both hard and soft substrates and their relationship to the physical environment;
- production of algae within Hobsons Bay; and
- a survey of amateur angling in Hobsons Bay including types of fish caught, eating preferences of respondents.

This study was completed by the publication of a major report (Hinwood et al. 1975). The data gathered provided important background information for the approval of port extensions and has recently been used as a source of information for planning the redevelopment of the Webb Dock area following the impending closure of the docks with the removal of port facilities from the Yarra River.

Sydney Ocean Outfalls

Ocean disposal of Sydney's sewage was initiated in 1888 with the completion of the Bondi ocean outfall. Subsequently, the Long Bay (now known as Malabar) outfall was completed in 1916 and the North Head outfall ten years later. Peake (1920-21) gives a description of the Long Bay outfall. Raw sewage effluent was discharged at the cliff face through two 1.5m diameter tunnels with an outlet invert level approximately 8m below low water level. There was considerable engineering controversy before construction of this outfall commenced as to whether the flat grade of the sewer conveying the sewage to the outlet would be self cleansing and also whether wave action would hinder the flow of sewage and, particularly during storms, prevent its discharge. The Long Bay outlet system was duplicated during the late 1930s and primary treatment facilities subsequently were provided at each outfall. Over the years these outfalls became overloaded and there was increasing concern with pollution of nearby beaches and general environmental degradation within the nearshore region in the vicinity of the outfall.

Investigations of the ocean disposal of sewage were made in Queensland for the Gold Coast in the 1960s and in New South Wales for various locations during the 1970s (Higham and Robson 1981, Harper and Greentree 1981). In 1972 the Metropolitan Water Sewage and Drainage Board (now known as Sydney Water) initiated investigations for the disposal of Sydney's sewage and a feasibility study in 1976 recommended the adoption of deepwater ocean outfalls. A 1979 environmental impact statement concluded that the beneficial effects of such outfalls would outweigh their adverse effects. While there was controversy as to whether this was the most desirable solution, economic assessment indicated that the deepwater ocean outfall solution would be cheaper than other options (Higham and Robson 1981).

The conceptual design of the system (Carroll 1985) involved consideration as to whether tunnels or pipelines should be adopted. The unfavourable prevailing sea conditions, the rough and rocky seafloor, as well as 50m high coastal cliffs, all indicated that tunnels would be preferable to an underwater pipeline. The tunnels vary in length from 2.1km to 3.5km and the effluent is released through a number of vertical risers capped with diffusers and located along the outer 400 to 800m of the tunnels (Fig. 8). The latter are 2.2m (at Bondi) or 3.5m in diameter and are located 45m below the upper surface of the rocky seabed. The effluent from the outfalls is initially diluted in the ocean, then dispersed by currents and waves, during which process bacterial die off occurs with the expectation that contamination of beaches and nearshore areas will be reduced to acceptable limits.

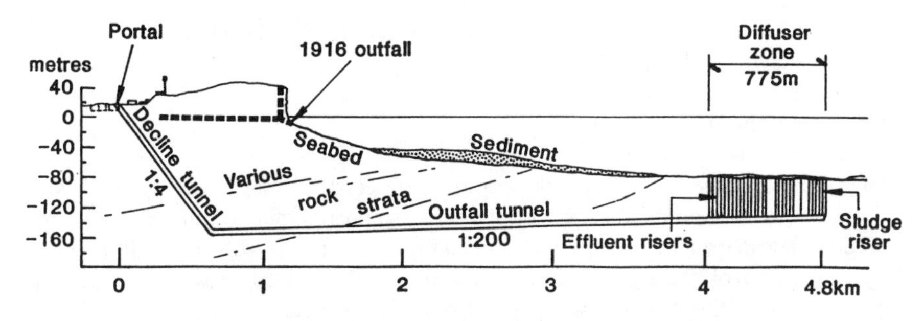

Figure 8. Sydney ocean outfalls - Malabar outfall

The design of the submarine outfalls posed significant engineering problems as virtually nothing was known of the seabed and waters off the Sydney coast (Ebner and French 1985). The effectiveness of the dispersion process required a knowledge of ocean currents and many measurements using current meters and drogues at various depths and locations were made prior to commencement of construction (Wallis 1983). Temperature and salinity measurements over a three year period showed that during the summer bathing season, September to April, the effluent field will generally be at levels 7m below the ocean surface. Geophysical surveys, offshore drilling and seabed sediment investigations were made to determine the condition of the rock through which the tunnels were to be excavated and risers drilled, as well as the possible sediment movements in the vicinity of diffusers. Particular design aspects requiring consideration were the hydraulics of the tunnel/riser system, including its ability to remove saltwater from the system at start up and minimise seawater circulation during operation (Wilkinson 1984, 1985), detailed design of diffusers and their protection against anchor damage and seabed scour. The minimum design life of the system was set at 100 years. This led to the selection of fibreglass as the material for the risers and diffusers. Construction of the Malabar outfall commenced in October 1984 and all three outfalls were completed by June 1991.

The disposal of sewage in the ocean has been a matter of considerable controversy in recent years and the approval for the three deepwater ocean outfalls by the State Pollution Control Commission in 1984 required an environmental monitoring programme to be carried out over a five year period. This monitoring programme was required to address three perceived public concerns about the outfalls:

- what effect will the outfalls have on the marine environment?
- will the bathing waters be suitable to swim in?
- will the fish and invertebrates be safe to eat?

This environmental monitoring programme involves studies on fish and macroinvertebrate community structures, accumulation of contaminants in marine organisms and sediments, water quality, assessment of the suitability of beaches for recreation, oceanographic studies, and development of mathematical models of sewage plume behaviour (Philip 1991).[11]

Offshore Loading Facilities

Unprotected timber jetties for coal loading were constructed on the exposed New South Wales coast north and south of Sydney from the mid nineteenth century. Kerle (1886-87) describes the construction of a 265m long structure at North Bulli between Sydney and Wollongong. Timber piles were dowelled to the rock bottom and the seaward end of the structure was supported by four concrete filled cast iron columns set in holes blasted into the rock bottom.

Much longer unprotected timber jetties were used in the more sheltered waters of South Australia and at a number of locations in Western Australia. Many of these structures were at least 0.5km long with the longest being 1.7km. The long length was required because of very shallow flat sea beds sometimes combined with high tidal ranges. Structures erected on the northwestern Australian coast were also prone to tropical cyclone damage. For example, at Onslow the original jetty was severely damaged by a storm in 1897 when its construction was almost completed. Its later relocated successor was damaged by cyclones on several occasions before being abandoned in the 1960s (Le Page, 1986).

In Queensland, Sir John Coode's proposals in the late 19th century for the port of Mackay involved an iron viaduct about 2km long linking wharves to be constructed in the lee of an offshore island to the mainland. It was never built. However, during the last 25 years several offshore loading facilities of comparable length have been built for the export of bulk cargos, mainly coal but also sugar. These facilities essentially comprise a long jetty like structure[12] between the land and the offshore berth located in sufficient water depth, approximately 15m at low tide. Tide ranges are up to 6m. A belt conveyer and a service roadway run out along the jetty and the actual loading facility is located at the ship berth at the seaward end. The coal loading facilities have capacities of the order of 20 Mt/a and the vessels using them are up to approximately 200 000 dwt capacity.

The first such structure at the coal port of Hay Point 22km south of Mackay is 2.1km long and is of conventional concrete pile construction. It was completed in 1971. A second berth was added in 1975. This latter structure is of considerable coastal engineering interest, being supported on large concrete caisson units with a plan size 46.7m by 41.4m and four 12.2m square vertical columns rising above the water surface

---

11) Progress on monitoring studies was recorded in a series of six papers published in 1993 - Preprints: 11th Australasian Conf. Coastal and Ocean Eng., Townsville, Qld., Inst. Eng. Aust., Nat. Conf. Publ. No. 93/4, Vol. 1, pp.37-58, 79-98.

12) Jetty in the Australian sense as defined previously.

at high tide. The caissons were constructed in the dry at Mackay harbour and towed to Hay Point. Physical model tests were undertaken to verify numerical computations of wave forces and to test proposed scour protection of the sea bed around the caissons (Apelt and Macknight 1976).

The longest of these structures is at Lucinda, 100km north of Townsville, where the only offshore bulk sugar loading facility is located at the end of a 5.76km long jetty. This structure, which was completed in 1978, is the first of three constructed using open ended thin walled tubular steel piles, with diameters between 850 and 1100mm. At Lucinda these are driven through clay layers into the underlying sand bed 40m below low tide level. All structural members are steel with appropriate corrosion protection. The decking is precast prestressed concrete planking and aluminium is used for nonstructural cladding. In this case the offshore facility option was chosen after considerations of other options, including a long dredged channel through a region of shifting sand banks and complex current patterns.

Subsequently two coal loading facilities of similar steel construction were completed in 1984. One at Dalrymple Bay adjacent to Hay Point is 3.8km long and the other at Abbot Point near Bowen is 2.9km long. In the latter case the 100m long causeway at the landward end is protected using Seabees, an Australian designed interlocking unit for constructing revetments (Brown 1978, Foster 1984). An integrated approach to design and construction techniques was required to ensure economic construction consistent with the requirements of structural sufficiency and in-service performance (Miller and Tranberg 1986). Geotechnical conditions were significantly different at each site and influenced the structural design of each jetty. Sea state was a significant factor limiting the choice of construction methods and led to cantilevering forward from previous construction rather than the use of floating plant. Dynamic response of the structures under wave loading was an important design consideration (Hooper et al. 1985). The superstructure above water level was subject to high wind loads.

The design of structures of this type, together with their associated foreshore reclamations, requires careful selection of design wave and water level conditions. Design wave heights at the berths may be at least 7m and as much as 12m while inshore causeways and revetments may be subjected to depth-limited wave heights of the order of 4m. Tidal ranges are up to 6m and storm surges of several metres may occur. The possible worst conditions are very severe. Fortunately, their probability of occurrence is also very small. To date none of these structures has been subjected to conditions exceeding those used for its design.

Beach Replenishment and Sand Bypassing for the Gold Coast

In 1984 Macdonald and Patterson reviewed the response of Gold Coast beaches to coastal engineering works. These works included the training walls at the entrance to the Tweed River in northern New South Wales updrift of the Gold Coast, several groynes on the southern Gold Coast, seawalls at various locations and beach replenishment both at the southern and northern ends of the coast. They concluded that: -

> *"The Gold Coast erosion problems, namely the threat of property damage and the loss of adjacent beaches, are the result of inappropriately located development (leading to the construction of seawalls) exacerbated by works (groynes) intended to overcome these problems together with the construction of the Tweed River training walls aimed at improving navigation to the river."*

Furthermore, the variety of works already undertaken and their observed effects provided decision makers with clear evidence of their impacts on the coast. The fundamental recommendation of the Delft Hydraulics Laboratory (DHL 1970), namely restoration of the beaches by extensive sand replenishment, was still the most appropriate measure to resolve the coastal management problems of this developed coast.

By this time continuing problems of the Nerang River entrance at the northern end of the Gold Coast had reached the stage where action was required. The river discharges into a large tidal waterway, the Broadwater, which is connected by a series of tidal channels to Moreton Bay to the north. The Broadwater is separated from the sea by a relatively narrow, low, Holocene sand dune barrier. Monitoring since 1840 indicated that the Nerang River entrance, which cut through this barrier, had moved several kilometres northward and was continuing to do so at an average rate of between 20 and 40 m/a. While the Broadwater provided safe shelter for small craft, the entrance was not a safe one. However, any action taken to stabilise it and improve its navigability was certain to produce significant impact upon alongshore sand transport if the necessary works were not carefully designed. The average net northward alongshore sand transport is 500 000 m^3/a and highly variable. South Stradbroke Island to the north of the entrance, while undeveloped, was vulnerable to erosion by waves and a possible breakthrough into the Broadwater. With the continuing consequences of the Tweed River training walls at the southern end of the Gold Coast still unresolved, it was clear that the new Nerang River entrance would require the installation of a sand bypassing system.

An initial investigation was undertaken for the Queensland Government in the 1970s by the Delft Hydraulics Laboratory, using a movable bed laboratory model, and it was recommended that a new entrance be constructed through the sand spit to the south of the existing entrance which would then be closed by sand pumping. Experience in undertaking such work was gained during the late 1970s on a smaller scale project at Noosa north of Brisbane (Lloyd 1980). In 1979 the Queensland Government established the Gold Coast Waterways Authority which, in 1981, released a strategy plan for the Broadwater. This plan required the stabilisation of the Nerang River entrance.

The Nerang project or Gold Coast Seaway, as it was named on its completion, is described by Witt and Hill (1987) and Coughlan and Robinson (1990). Briefly it consists of two entrance training walls, a fixed sand bypassing system and a wave absorbing island within the Broadwater (Fig. 9). Associated with it is an outfall for secondary treated sewage effluent which is discharged on the ebb tide.

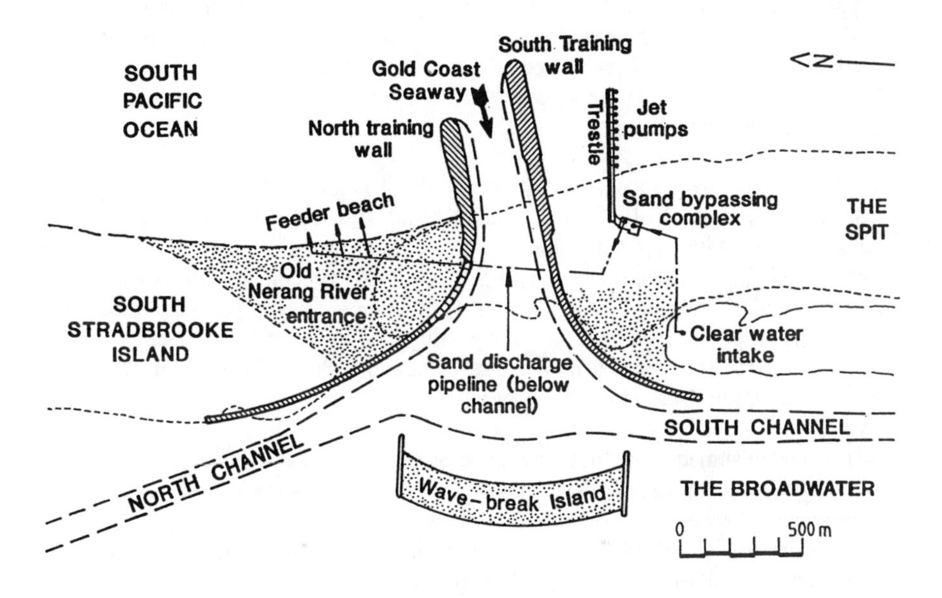

Figure 9. Gold Coast Seaway

Physical models were tested at the Queensland Government Hydraulics Laboratory for the stability design of the training walls and for wave penetration and the design of Wave Break Island. The training walls are constructed with a rockfill core, a roadway formed from two layers of 20t concrete cubes, and they are armoured with randomly placed 25t concrete cubes. Project construction began in April 1984 and the works were substantially completed in May 1986. The old entrance was closed by continuous sand pumping by three dredges over a period of about one month.

The sand bypassing system, constructed on a separate contract, consists of 10 jet pumps mounted on a jetty extending 490m into the sea and located about 300m south (updrift) of the southern training wall. Water to power the jet pumps is pumped from the Broadwater and the sand/water mix from the jet pumps is discharged into a slurry pit from where two high pressure pumps pump it through a pipeline under the entrance channel to the northern (downdrift) side of the seaway. Operation of the sand bypassing system has required careful consideration of the beach alignment south of the seaway to ensure that sand does not travel around the southern training wall and cause regrowth of the entrance bar. Monitoring of the project over ten years has shown that it has fulfilled its purpose in stabilising the river entrance with minimal impact upon the adjoining coastline. The entrance channel is navigable and depths have been maintained below the minimum design bed level of -5.5m below mean sea level compared with natural depths generally considerably less than -3m. Indeed some problems have occurred in the seaway channel with scouring and exposure of the sand bypassing pipeline. Moreover, the increased hydraulic efficiency of the entrance has increased tidal flows and, as was experienced at the Richmond River entrance in New South Wales 80 years earlier, tidal ranges have increased with lower low tide levels within the estuarine Broadwater.

This project is a particularly significant one for Australian coastal engineering. Its design involved not only normal hydraulic and structural design but also a full consideration of the project's impact upon coastal processes and the consequent installation of a permanent sand bypassing system. Experimental design tools included three hydraulic models and design was based upon more than 10 years recorded wave data. The jet pump/fixed trestle sand bypassing system was the first of its type in the world. From the commencement of operation in mid 1986 until the end of 1995 more than 4 300 000 m^3 of sand was bypassed.

Beach replenishment was first attempted on the Gold Coast during 1974-75 when 765 000 m^3 of sand was pumped from the lower reaches of the Tweed River estuary to beaches at Kirra on the southern Gold Coast and 1 150 000 m^3 was pumped from the Broadwater southwards to Surfers Paradise. The sand was placed directly on the upper beach. The work was undertaken by the Gold Coast City Council with a 20% cost subsidy from the Queensland Government.

The pumping from the Tweed River was intended to counter erosion caused by the trapping of littoral sand by the river's entrance training walls. It was not particularly successful because subsequent investigations[13] showed that the sand had been taken from estuarine shoals which were part of the alongshore sand transporting system. Moreover the quantity of sand was insufficient to counteract the continuing erosion. On the other hand, the replenishment at Surfers Paradise, which was undertaken to restore the beach in front of the boulder wall protecting the esplanade, was successful since the sand source was not part of the presently active alongshore system. In this case the sand replenishment was accompanied by reestablishment of a vegetated frontal sand dune to stabilise sand above the reach of normal wave action from wind blowing. It took some time before this was successfully completed (Smith 1977).

The need to restore beaches on the southern Gold Coast led to the Gold Coast City Council investigating more economical methods of beach replenishment. With a depleted nearshore profile from continuing erosion, placement of sand on the upper beach might not be a long term solution. Study of the behaviour of natural offshore

---

13) See section on Coastal Engineering Research - Coastal Processes.

storm bars and numerical modelling indicated that nearshore replenishment could be an efficient method for restoring beaches if offshore sand resources were used (Jackson and Tomlinson 1990). A small scale trial had been made in 1985 when 315 000 m^3 of sand had been dredged from offshore deposits in 20 to 30m depth using a trailing suction hopper dredge and placed both on the beach and in the nearshore zone at North Kirra and Bilinga. Monitoring indicated that onshore transport of the nearshore deposit occurred. A larger scale project was undertaken in the same area in 1988 using nearshore replenishment only to restore the eroded seabed and hence protect the upper beach. 1 500 000 m^3 of sand were placed at a cost much reduced compared with that for pumping the sand ashore (Jackson 1989).

In mid 1989 there was still little visible sand at Kirra and southern Bilinga beaches and erosion was progressing northwards. Full natural sand bypassing has not yet been restored around the Tweed training walls. Over 7 000 000 m^3 of sand had been trapped on the updrift side of the walls and in the bar and action was required to make up for this deficiency. During the period from October 1989 to May 1990 a total of 3 600 000 m^3 of sand was dredged from offshore depths between 18 and 28m. 395 000 m^3 was deposited in the nearshore regions by a trailing suction hopper dredge and 3 230 600 m^3 was pumped onto the upper beach using a similar but larger dredge discharging at an inshore single buoy mooring through a submerged pipeline. An extensive monitoring programme was undertaken and it was found that two years after completion of replenishment 87% of the material remained within the replenishment zone. The nearshore replenishment had gradually moved shoreward and upcoast over three years. However, there was still the need for additional replenishment sand to restore the beaches to their 1962 "natural" condition (Murray et al. 1993, 1994).

The solution to the Gold Coast sand replenishment problem became possible after sand began to pass around the Tweed training walls and the entrance bar began to be reestablished. This resulted in a deterioration in navigability, reduced tidal circulation in the estuary and potential increase in river flood levels. Pressure to take action was exerted on the New South Wales Government by affected members of the community. A study by Tomlinson and Foster (1987) of the mechanisms of sand bypassing at the Tweed River entrance clearly indicated that dredging of sand from the bar (ebbtide delta) or estuary shoals would result in immediate infilling of these areas and a consequent reduction in the amount of sand bypassing the river entrance.

Following earlier discussions, in July 1990 the New South Wales and Queensland Governments initiated a joint study of the Tweed River entrance, its problems and solutions for them. Consultants - Australian Water and Coastal Studies Pty Ltd and WBM Pty Ltd - were commissioned to prepare a discussion paper on known littoral processes, costing of works, benefits to each state and administration of any cooperative management scheme. The study was to focus on an artificial sand bypassing system for the Tweed River entrance. An essential aspect was the identification of matters on which there was agreement and of those where there was disagreement. Professor C.J. Apelt of the University of Queensland had the role of overviewing methodology, reviewing findings and assisting in the attainment of consensus and resolution of outstanding issues in the finalisation of the draft discussion papers. A joint report was submitted in May 1991 and this provided the basis for top level New South Wales/Queensland Government discussions which led to eventual agreement on the work to be done and the basis for cost sharing.

The objectives of the joint Tweed River Entrance Sand Bypassing Project are to establish and maintain a navigable entrance to the Tweed River and to enhance and maintain the southern Gold Coast beaches, with the objectives to be achieved in perpetuity. The principles and progress of this work have been reported by Murray et al. in May 1995. At that time dredging had just commenced to remove the bar and deepen the entrance channel of the Tweed River with 1 700 000 m^3 of sand to be placed on the upper beach and in downdrift nearshore areas off the southernmost Gold Coast beaches. Environmental impact assessment and design studies for the sand bypassing system are continuing.

Dawesville Channel

The Dawesville Channel in Western Australia was constructed as part of a management strategy for the Peel-Harvey Inlet system. The latter is a shallow coastal lake system with a water surface area of 133km^2 and is located about 70km south of Perth. Three rivers discharge into the lakes which were linked from Peel Inlet to the ocean at Mandurah by a single narrow shallow channel which restricted tidal inflow into the system. Catchment degradation as a consequence of agricultural activities, including clearing, drainage and use of fertilisers had been creating problems within the inlet system since the mid 1960s. Too much nutrient, particularly phosphorus, led to eutrophic conditions, initially with growth of various water weeds and then, beginning in 1974, infestation by the blue green alga *Nodularia*. By 1984 the condition of the system was a major political and environmental issue. An environmental review was undertaken for the state government by Kinhill Engineers Pty Ltd and various management strategies for reducing the influx of phosphorus and improving flushing of the inlet system recommended. Ultimately a three-pronged management strategy was adopted. This involved management of the catchment area with controls on fertiliser use and land clearing; continuing mechanical weed clearing; and the cutting of the Dawesville Channel.

The function of the Dawesville Channel is to improve the flushing of the system. This involved constructing a second channel to connect Harvey Inlet to the ocean. This will increase the amount of phosphorus leaving the system and make the estuary generally more marine in character with an increased tidal range and saltier water, the latter inhibiting growth of blue green algae. Detailed engineering studies were conducted for the new channel (DMH 1987). These included channel stability, geotechnical investigations, oceanography, modelling, littoral drift, channel construction and land redevelopment.

Study of various approaches to the stability of tidal inlet channels led to the selection of basic channel dimensions which were modified to accommodate local constraints such as a limestone outcrop and the need to allow some sedimentation of the entrance before maintenance dredging is required. In the end the 2.5km long, 200m wide sinuous channel was designed with an increasing width and a depth decreasing from 6.5m at the seaward end to 4.5m at the landward end (Hutton 1987). Assessment of the littoral drift by several methods indicated an average rate of 80 000 m^3/a. As this was in excess of the estimated 30 000 m^3/a for channel stability a sand bypassing plant, using a fixed jet pump on a timber jetty located updrift of the entrance, was proposed. Modelling for the project included a physical wave model to determine the alignment of the entrance breakwaters to control wave penetration and ensure navigability. Mathematical modelling was used to determine the changes to the tides and salinity regime within the estuary.

Construction of the channel, its breakwaters, training walls and sand bypassing system, as well as a highway bridge across it, commenced in January 1992. The channel, apart from its entrance, was excavated in the dry. This operation required a massive dewatering system. The total volume excavated, sand, calcarenite and limestone, was approximately 4 500 000 m^3. The breakwaters have a total length of 725m and are armoured with 3 tonne granite rocks. The project was opened in April 1994, nine months ahead of schedule (Fig. 10).

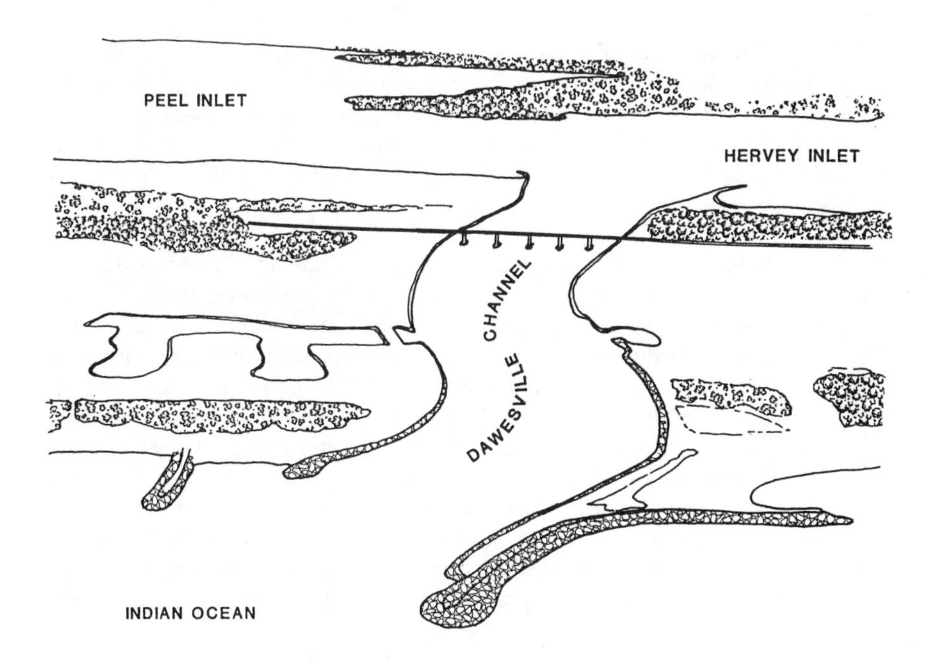

Figure 10. Dawesville Channel

COASTAL ENGINEERING RESEARCH

Coastal Processes

Prior to the second world war coastal engineering in Australia was primarily concerned with ports and their requirements, e.g. breakwaters, safe navigable entrances, control of siltation, etc. and occasionally with protecting buildings, promenades, or other facilities located too close to the shoreline. There was little general research as such since all works were undertaken in connection with specific projects and were investigated, designed and constructed by engineers mainly employed by governmental authorities.

Following the second world war and the establishment of hydraulic laboratories by both governmental organisations and universities, the need for basic research became both obvious and possible. As practical problems were investigated using models, the gaps in understanding of fundamental processes of sediment transport by waves and tides became evident. Quantitative scaling of movable bed models was virtually impossible without large amounts of data concerning prototype conditions and these usually did not exist nor could they be obtained with the resources and time available. Initially the hydraulic laboratories became not only a focus for experimental research but also, in many cases, for the field investigations required to understand coastal and estuarine processes.

In 1965 R. Silvester in a wide ranging paper on coastal sediment movement and research support advocated *"that Australia should commence to bear a greater share of the responsibility in the field of Coastal Engineering, both in solving its current and future problems and in training its personnel for this activity"* (Silvester 1965). In this and following sections some of Australia's coastal engineering research work is described.

In Western Australia both alongshore sand transport at Bunbury Harbour and beach erosion at Cottesloe near Fremantle were investigated at the University of Western Australia using hydraulic models (Silvester 1956, 1961). At this time Silvester (1959) reviewed current knowledge of the engineering aspects of coastal sediment movement and carried out some simple laboratory experiments with an idealised model of two sandy beaches located between headlands in which oblique waves were allowed to reshape the shoreline into a stable equilibrium form, the crenulate-shaped bay (Silvester 1960). He subsequently showed that the crenulate-shaped bay could be used as a geomorphological indicator of the direction of the net alongshore sediment transport along coasts throughout the world (Silvester 1962) and particularised it for the Japanese coast (Silvester 1966). The latter paper was the first paper presented by an Australian author at one of what are now the International Conferences on Coastal Engineering.

Silvester (1970) presented the characteristics of these equilibrium shaped bays and in combination with Japanese colleagues developed the concept of headland control for stabilising coastlines prone to erosion (Silvester et al. 1980). Artificial headlands were to be constructed offshore of the coast and the beach between them allowed to develop as a stable equilibrium crenulate-shaped bay. The size of the bays would be determined by the spacing between headlands, the wave direction and the alongshore sediment supply. This principle was applied with some success, using three offshore breakwaters, to protect a cooling water intake from sand inflow in a situation where the wave action causing sand supply to one side of a cuspate foreland had been diminished by construction of a causeway to an offshore island (Silvester and Searle 1981). However, apart from a few similar small scale applications, headland control has not been adopted widely by coastal engineers.

Silvester's colleague, J.R.C. Hsu, expanded the work on equilibrium bay forms and evolved a universal equation for the shape of these bays which he applied to many natural bays and to the shape of the salients (tombolos) formed behind offshore breakwaters (Hsu and Silvester 1990). Hsu also investigated the oblique reflection from seawalls and breakwaters of wave trains arriving as swell and, after reflection, forming complex three dimensional short-crested waves (Hsu 1990). The high velocities and vortices generated within these waves provided an explanation for the scour problems observed in front of sea walls (Hsu and Silvester 1989).

The concept of headland control and the potential for scour by short-crested waves were new developments in the understanding of coastal processes. These and other research at the University of Western Australia have been summarised in Silvester (1974) and Silvester and Hsu (1993).

Hydraulic model studies at the University of Queensland of small harbours located in the lee of headlands or in the vicinity of reefs revealed the occurrence of significant wave-generated current systems which often caused local reversals of the alongshore current in the lee of a structure or headland (Gourlay 1965). Generalised physical modelling showed that currents of this type could be generated by an alongshore gradient of wave set-up in the surf zone, associated with alongshore gradients in breaking wave height caused by wave diffraction behind an offshore structure (Gourlay 1974, 1976). Such current systems are of particular significance in the local sediment transport and development of nearshore morphology close to headlands and offshore reefs or engineering structures such as offshore breakwaters (Gourlay 1981).

In May and June 1974 a series of storms caused widespread damage to coastal structures along the central New South Wales coast (Foster et al. 1975). As already indicated these events triggered significant changes in administrative arrangements for the coast in New South Wales. They also provided the stimulus for research into coastal processes. Moreover, the prevailing problems at Newcastle Harbour, involving a submarine rock bar across the harbour entrance and continuous maintenance dredging to control siltation had encouraged the search for alternative port facilities. One of these alternatives was a proposed artificial harbour behind the beach and frontal dunes north of Newcastle with a channel linking it to the sea. An understanding of coastal processes and particularly sand movement in the vicinity of the proposed entrance was essential. The investigation, reported by Gordon and Roy (1977), is important in that it involved active collaboration between coastal engineers and marine geologists. The geological setting of the coast was studied and the entire 25 km long beach between headlands was considered as a coastal compartment for sediment budget purposes. Wave data from the Botany Bay offshore wave rider buoy made long term alongshore sand transport computations possible. It was concluded that there was negligible net alongshore sand transport along this beach.

The input of marine geologists to this study was important, as it was in a parallel geological study into the nature of sediments in the estuaries of the major coastal rivers of northern New South Wales (Roy and Crawford 1977). The latter work showed that the sand in these estuaries is not fluvial sand recently brought down by river floods but older reworked coastal sand deposited in the estuary by wave and tidal action. The northern New South Wales coast is not being replenished with sand from its rivers and hence the consequence of the dominant northward alongshore transport and landward wind transport is continuing long term coastal retreat.

This broad concept of a retreating coastline put forward by the marine geologists was investigated by coastal engineers in a study of the section of the northern New South Wales coast from Byron Bay to Hastings Point (Gordon et al. 1978). Cape Byron at the southern end of this coastal compartment is the most easterly part of the Australian mainland. A methodology was developed for the study which involved a three staged approach: -

"1. *An assessment of the existing social, geomorphological and economic situation;*
2. *The development of an understanding of coastal process mechanisms, past and present;*
3. *The formulation of rational management strategies based on a knowledge of the existing situation and the mechanisms of the natural processes.*"

Historical, geological and sedimentological, bathymetric data, etc. obtained during the first stage of the investigation provided input for understanding coastal processes and for mathematical modelling so that the sediment budget could be assessed and a predictive model developed. Once the existing situation and the nature of its processes were understood then rational coastal management strategies could be developed for the region being studied.

Technical innovations in this project included the use of photogrammetry to provide a historical record of the position of coastal features. This historical analysis showed that the coastline in this region had been receding at an average rate of 0.5 to 1 m/a over the thirty year period 1947 to 1977. Sedimentological investigations clearly defined active and inactive zones of sand movement in the nearshore region, including significant offshore sand loss (50 000 m^3/a) at Cape Byron. Calculated alongshore sand transport varied from 15 000 m^3/a entering the compartment from the south around Cape Byron to 200 000 m^3/a leaving at the northern end of the compartment. Clearly the marine geologists' general model was verified for this particular coastal compartment.

This approach adopted in New South Wales for understanding coastal processes has been reviewed by Gordon and Lord (1980) using the above two examples as specific case histories. The technique involves the formation of a regional coastal model and its adaptation to the specific site under investigation. This site specific conceptual model is then tested and modified on the basis of theoretical calculations and field data collection. The resulting numero-descriptive model is used as the basis for assessing various coastal management options.

The accumulation of sediments within the New South Wales estuaries was investigated in a study of the Tweed River estuary (Druery and Curedale 1979, Druery 1980). Sand had been dredged from estuarine shoals during 1974 and 1975 for beach replenishment on the southern Gold Coast in Queensland and a comprehensive field data programme was subsequently established to monitor the consequent changes in the hydraulic processes in the Tweed River. Again innovative methods were used. Sidescan sonar surveys were used to locate bed forms. Sediment bed load transport rates were estimated from detailed underwater measurements of bed form movement and these rates were used to calibrate a sediment transport formula. This formula was used in conjunction with a one dimensional model of tidal hydraulics to simulate estuarine shoal dynamics using simple sediment routing techniques. The results of the field and mathematical investigations showed that the dredged hole was being infilled with sediments, predominantly (85%) brought in through the river entrance. Tidal ranges in the lower estuary increased after dredging and in general it was found that tidal conditions in a shallow estuary are very sensitive to changes in the average depth, particularly relatively close to its entrance. This conclusion is consistent with the behaviour of the Richmond River entrance recorded by Burrows (1904-05).

Proposals for offshore sand extraction, sewerage outfalls, spoil disposal, etc. on the inner continental shelf in the vicinity of Sydney led to a comprehensive investigation and research concerning sediment transport on the shelf (Gordon and Hoffman 1984). The work involved definition of the geological, sedimentological and morphological characteristics of the continental shelf out to 80m depth. The magnitudes of the various currents, both oscillatory and steady, were determined, together with their potential to cause sediment movement. It was found that transport of fine sand mainly occurs during storm events when wave activity produces sufficient bed shear to cause movement. The dominant northward current is seldom sufficiently large to move sand on its own but when wave-induced transport occurs it is to the north. In a subsequent investigation Nielsen (1994) was able to quantify the gross annual alongshore and shore-normal transport rates at different depths on the continental shelf and concluded that they were negligible beyond a depth of 35m.

Flood studies of coastal rivers require an estimate of the effect of wave set-up on river levels while sand dune water quality and beach stability are affected by coastal groundwater levels. Early research on wave set-up tended to concentrate upon the surf zone and very little attention was paid to the swash zone. Similarly studies of the coastal water table were restricted to the area landward of the runup limit. Field research to understand the relationship between these processes was carried out by Nielsen (1988, 1991). Simple experimental techniques using manometer tubes and stilling wells were used to obtain the shape of the set-up profile extending from the surf zone through the swash zone and into the beach. The actual coastal water table was found to be significantly affected by the tide through the asymmetry of the infiltration/draining process during a tidal cycle, as well as by waves and rainfall. On the New South Wales coast the average coastal groundwater level is of the order of 1.5m above mean sea level. However, this level does not necessarily represent the magnitude of the wave set-up affecting flood levels within a river channel entrance. Laboratory studies, involving a Hele-Shaw cell to model water table response to tides and a wave flume to investigate the effects of wave runup on the water table, were used to clarify the separate effects of tides and waves upon the coastal groundwater level (Aseervatham et al. 1993).

Numerical Modelling of Tides in Estuaries and Bays

The behaviour of tides in estuaries and bays has been of concern to Australian coastal engineers for a long time. In 1958 H.A. Scholer presented a general mathematical treatment of the tides in rivers and coastal inlets together with formulae and methods of computation for predicting tidal behaviour. His one dimensional impedance method for tidal computations assumed sinusoidal tidal fluctuations and its equations were analogous to the "telegraph equations" used by electrical engineers (Scholer 1958).

In recent years the increasing importance of environmental concerns in coastal and particularly estuarine regions has required the development of techniques for numerical modelling of tidal flows in estuaries and shallow bays. This began in the early 1970s with studies such as the Port Phillip Bay study previously described. Because this early work concerned shallow bays much of the early numerical modelling of tides focussed on vertically averaged two-dimensional in plan models.

The early Australian two dimensional models were based upon models originally developed in the United Kingdom or the United States. Apelt et al. (1974) used a model based on an explicit finite difference scheme from a U.K. source to study the effects of a proposed causeway on tidal phenomena in a shallow inlet, Hays Inlet, near Brisbane. This model had the capability of modelling the wetting and drying of intertidal flats and is one of the earliest published developments including this feature.

The long estuaries connecting to the inlet were modelled as one dimensional segments which were linked dynamically to the two dimensional system.

Tronson and Noye (1974) studied the characteristics of tides in Spencer Gulf and St Vincent's Gulf in South Australia using a model developed from the same source as Apelt et al.'s model. Another two dimensional model developed by Williams and Hinwood (1974) was based upon a semi-implicit finite difference scheme from a U.S. source. This model was used to study water movement and water quality in Westernport Bay, Victoria.

Much of the development of two dimensional numerical tidal models in Australia was done by research workers who had little interest in or incentive for developing user-friendly software of the kind needed by professional engineers. Consequently these early models, which were relatively well advanced for their time, did not result in the development of significant commercial software which was competitive with that developed overseas, particularly by DHI and SOGREAH. Nevertheless, some research groups (eg., Bode et al. 1985) have continued to develop two dimensional models for their own use.

Many research workers and some engineering consultancies developed a variety of one dimensional models. Such a model has become a virtually indispensable tool in investigations and design associated with estuaries. Applications include design of tidal barrages (Fox and Wilke 1981), study of the hydraulics of a mangrove swamp system (Bunt and Wolanski 1980), effects of sand extraction on salt intrusion in an estuary (Dyson and Druery 1985) and simulation of long term sedimentary processes in estuaries (Apelt and Ryall 1992).

Satisfactory modelling of processes such as contaminant transport requires three dimensional models in many cases. The great increase in computing power in recent years has made three dimensional modelling in estuaries and shallow seas economically feasible and a variety of models is in use. Only a few examples can be cited here. Numerical modelling of the impact on water quality from deepwater sewage outfalls offshore from Sydney was done with a three dimensional finite element model (Peirson et al. 1993). Three dimensional finite difference models have been used to model effluent dispersion in the Derwent estuary, Tasmania (Walker and Hunter 1995) and oil spill movement in Exmouth Gulf, Western Australia (Hubbert 1993).

The contemporary situation with regard to three dimensional models has some similarities to that with regard to two dimensional models in the 1970s. A question to ponder is whether any of the three dimensional models developed by individuals will be still in use a decade hence and to what extent commercially developed and supported software will have replaced them.

Rubblemound Breakwaters

As mentioned earlier Australia's first recorded coastal engineering structure was a breakwater at Port Elliott in South Australia. The then apparently novel construction method was simply to quarry granite rocks (2 to 7 tonnes) from a nearby headland and tip them into the sea. *"As the work progressed, lines of rail were laid down on the top of the embankment, and the whole was carried forward, precisely like a railway embankment, being finished at once to its full height. The blocks took their own natural slope which was found to be about two horizontal to one vertical."* (Hays 1858-59).[14] Similar construction methods were employed subsequently for river entrance training walls and for artificial harbour breakwaters in New South Wales and elsewhere. Variations occurred with the selection of larger stones for the armour layers and the use of large (40 to 60 tonne) concrete rectangular prism units where sufficiently large rock was not economically available. At Mackay in northern Queensland initial breakwater construction in the late 1930s was followed by about 50 years of continuous maintenance as the breakwaters, originally formed of quarry stone placed at its angle of repose, were repaired after successive cyclonic storms. Model testing was rarely undertaken for any of these breakwaters.

Since 1970 the trend has been to design breakwaters more carefully so as to minimise or eliminate maintenance and to optimise costs. A series of laboratory experiments by Foster and Gordon (1973) compared the behaviour of tribars and dolosse with randomly placed rock. In Botany Bay a successful application of tribars to armour the seawall protecting the reclamation for a new runway for Sydney airport led to their choice for the seawall to protect the new Port Botany facilities. The airport runway seawall had been constructed in the dry but the construction of the Port Botany one was undertaken in the wet. Placement difficulties were experienced and storms caused major damage to unprotected core and secondary armour. Model tests were used to redesign the breakwater using dolosse below water and surplus tribars above. A satisfactory design was achieved but composite construction was not recommended as the junction between unit types is a weak point (Foster 1984).

A refinement of the dumped rock breakwater was achieved using available resources and careful design at Grassy on King Island at the western end of Bass Strait (Burren 1975). Large quantities of waste rock were available from a nearby mine - 5% was between 2 and 10 tonnes and 95% was less than 2 tonnes. A core of run of quarry material was pushed out to an offshore island in water depths up to 18m below low water. Waves were allowed to form this material into a rocky beach and 6 to 10 tonne armour rock was placed on the reformed surface to stabilise the structure. During construction the seaward face was reshaped rapidly to a slope of 1 in 3.5. Sufficient material had been placed for adjustment of the slope to 1 in 10 as model tests had

---

14) Hays omitted to report that the breakwater was severely damaged in 1856 shortly after its completion and was not repaired (Tolley 1978).

indicated that the equilibrium slope of the rock face would be between 1 in 7 and 1 in 10. Field experience indicates that actual slopes are steeper than these.

Significant problems occur in the design of breakwaters for tropical cyclone conditions in northern Australia. The possibility of storm surges of several metres added to astronomical tides of similar magnitude makes it impossible to design economically a nonovertopping breakwater for extreme conditions. At Rosslyn Bay in central Queensland an eight year old conventional rubble mound breakwater suffered severe damage during tropical cyclone David in 1976 (Foster et al. 1978). The catastrophic failure of the breakwater after heavy overtopping at high tide occurred within a few hours, the crest of the breakwater being lowered by about 4m over most of its length. Nevertheless, the breakwater still gave substantial protection after failure and the harbour was still usable during normal conditions after the storm. Model testing after the event was able to closely simulate both the failure of the breakwater and its post failure profile. These tests also showed that damage to the crest was a function, not only of wave height, but also of wave period and storm tide level and that there was a critical combination of these variables which caused most damage. The redesigned breakwater, constructed with commonly available rock sizes intermixed with modified concrete cubes with the highest possible permeability, successfully withstood a subsequent tropical cyclone.

The experience at Rosslyn Bay was subsequently applied to the design of a new breakwater to protect a reclamation adjoining Townsville harbour. This led to the dual breakwater concept (Bremner et al. 1980, 1981). Briefly, a shore parallel submerged offshore breakwater was found to be the most economical structure to protect the reclamation. However, such structures are difficult to construct in Australia because suitable floating plant is generally not available and all construction is done with land based plant. So an offshore breakwater was constructed with crest level above high tide level with the expectation that under severe storm conditions this would be overtopped and flattened to become a submerged breakwater. Detailed model tests were made with various crest heights and berm widths to determine the reshaped profile and the wave transmission characteristics of the failed structure. The inshore revetment protecting the reclamation could then be designed to withstand the smaller waves in the stilling basin behind the overtopped submerged breakwater.

Experience from the previous structures was utilised in the design of another northern Queensland breakwater for the Hay Point tug boat harbour (Bremner and Foster 1987, Bremner et al. 1987). Initial designs using dolosse in both conventional and overtopped structures were too expensive. However, a geotechnical investigation located sources of large quantities of 3 to 7 tonne rock and a design similar to that used at Grassy was adopted. That is a sufficiently large mound of rock was placed for the waves to reshape into a stable S shaped profile providing the required protection. The important difference from a conventional breakwater is that , instead of two layers of armour stone overlying an impermeable core of smaller material, the whole breakwater cross section of the mass armoured breakwater, as it has been called, is armour stone with a grading of the highest possible permeability. This type of breakwater has considerable advantages. It uses natural rock in its available sizes. It attains its own stable profile as it is reshaped by natural forces and has very high tolerance to conditions exceeding the design conditions. It is easily maintained and modified while there are large savings in initial capital cost.

Further research on permeable breakwaters and their performance, including modification of the design formulae used for conventional rubble mound structures, was initiated but lack of funds has prevented its completion (Harper 1987).

Tropical Cyclone Phenomena

Northern Australia lies within the regions affected by tropical cyclones. The impacts of such cyclones were well known by coastal engineers from examples such as those previously discussed. Designing coastal structures to resist these impacts was difficult using information produced in other countries so methods for estimating design waves based upon Australian meteorological conditions were developed (Nelson 1972). In December 1971 cyclone Althea crossed the Queensland coast just north of Townsville and subjected that city to wind gusts of the order of 55m/s. It caused substantial damage and disruption of services to the community which at that time was of the order of 100 000 persons. It was accompanied by a storm surge with a maximum height of 3.7m which fortunately coincided closely with a predicted low tide with the result that the total storm tide water level was only about 0.5 m above highest astronomical tide. No significant damage was caused by the surge.

The occurrence of cyclone Althea provided an opportunity for the Department of Civil and Systems Engineering at Townsville's James Cook University (JCU) to gain support for a wide range of engineering research projects relating to the impacts of tropical cyclones. One of the first of those projects involved the prediction of storm surge and the related implications for civil defence planning. This work was initiated by K.P. Stark, who already had a background in numerical flow modelling, and his colleague R.J. Sobey. The coastal engineering applications subsequently involved not only storm surge modelling but later spectral wave modelling, shallow water mixing and tidal modelling. The devastation of Darwin in the Northern Territory by tropical cyclone Tracy in December 1974 provided further stimulation for tropical cyclone research at JCU.

The primary support for the storm surge research came from the Beach Protection Authority (BPA) which supported the project from 1975 to 1978. A generalised 2-D vertically integrated numerical hydrodynamic model, SURGE, was produced. This was an explicit "leap-frog" model but with many additional features such as dynamic open boundaries, reef controls and one of the first fully parametric tropical cyclone windfield models (Sobey et al. 1977, 1982). The research student at the centre of the development was B.A. Harper, then a recent graduate of JCU. A series of twelve research reports addressing the likely envelope of storm surge response at all major population centres along the Queensland coast was produced for the BPA. Later in 1984 when Harper was working with consulting engineers Blain, Bremner and Williams (BBW), the initial JCU storm surge studies were extended to full scale statistical studies of storm tides including astronomical tide and storm surge with an allowance for wave set-up (BBW 1985). Again this work was funded by BPA as part of its coastal planning and protection activities.

The development of spectral wave modelling was initiated by a request from BBW for assistance in estimating long term wind and wave frequencies in Cleveland Bay for the design of a breakwater to protect a new reclamation located next to Townsville Harbour. Sobey realised that the best available wave prediction techniques of the time were seriously deficient in representing the dynamic fetch, duration and direction effects of waves in the Great Barrier Reef region under the influence of tropical cyclones. I.R.Young, working as a research student under Sobey, developed the second generation 2-D spectral wave model, SPECT, a particular feature of which was its "fractional step" algorithm which avoided the numerical dispersion which commonly afflicted similar models (Sobey and Young 1986, Young 1988). The first major application of the SPECT model was undertaken by BBW and JCU in 1981 for Woodside Offshore Petroleum Pty Ltd and involved a detailed tropical cyclone hindcast study for offshore oil installations on the North West Shelf adjoining the northern Western Australian coast. Young subsequently studied the mechanisms of spectral wave decay and after leaving JCU developed a new spectral wave model ADFA1, based upon SPECT.

The capabilities of these spectral wave models were given a severe testing when they were used to hindcast waves generated by tropical cyclone Orson, the most severe storm of this type yet recorded in the Australian region (Harper et al. 1993). Winds of 60 m/s were estimated to have produced maximum wave heights of 21 to 24m during the cyclone's passage directly over the North Rankine A platform where platform damage indicated the occurrence of 20m waves.

Waves on Coral Reefs

The problems of constructing navigation aids on coral reefs have been referred to earlier. In the 1970s coastal engineering problems involving structures located on reefs and resort developments on reef-top islands (cays) within the Great Barrier Reef region off northeastern Australia became apparent. The need for improved meteorological data for cyclone prediction required automatic weather stations to be constructed on reefs or on small potentially unstable cays. Tourist resort facilities on Heron and Green Islands were endangered as these cay's shorelines moved in response to changing weather patterns and climate trends. Hydrographer's Passage, a new navigation channel through the central Great Barrier Reef for bulk carriers, required new navigation aids to be constructed on reefs. Dredging of a navigation channel and boat harbour at Heron Island changed reef-top ecology and beach stability while silt dispersed over the reef during dredging operations created further environmental damage (Gourlay 1983, Gourlay and Jell 1993).

The design of navigation aids for Hydrographers Passage required information concerning wave heights and water levels on the reef-top. Nelson and Lesleighter (1985) carried out model experiments on wave transformation and set-up on a horizontal reef-top. Subsequently Young (1989) attempted to measure wave transformation across one of the outer barrier reefs in the northern Great Barrier Reef but was only partly successful, losing the wave rider buoy measuring the deepwater offreef wave conditions. Successful experiments were made subsequently at John Brewer Reef near Townsville (Hardy et al. 1990, 1991) and these provided the basis for a better understanding of wave transformation over reefs. Hardy and Young (1991) developed a method for modelling spectral wave transformation on a coral reef flat and Nelson (1993) derived reef-top currents from wave observations.

A particularly significant result from this field work was the confirmation of earlier laboratory work by Nelson (1985, 1987) showing that the maximum wave height obtainable on a horizontal bottom never exceeds 0.55 times the water depth (Nelson 1994). This result contradicts the commonly used value of 0.8 based upon solitary wave theory. Laboratory experiments on model reefs (Gourlay 1993, 1994) provided further confirmation of this 0.55 maximum wave height to water depth ratio for reef-top waves as well as showing how the magnitudes of wave set-up on a reef and wave-induced flow across it vary with wave conditions and relative submergence of the reef. Theoretical studies are continuing to provide an explanation as to why maximum depth limited waves on horizontal bottoms are lower than previously predicted (Massel 1996).

MARINE AND COASTAL SCIENCE

Prior to the second world war Australian engineers were largely dependent upon their own resources and upon general scientific knowledge obtained overseas for their understanding of the coastal and marine environments. In some cases engineers were active participants in the various Royal Societies and other scientific bodies in each state, some of which such as the Royal Society of New South Wales had special engineering sections. Some engineers, for example W. Shellshear referred to earlier, studied coastal or marine science in their spare time and published their findings in local scientific journals. An outstanding example was Professor R.W. Chapman of the University of Adelaide who made significant contributions to the study of tides in South Australian waters. On the negative side the absence of locally based scientific research into coastal phenomena meant that engineers and others often had little choice except to use scientific understandings and theories based upon conditions occurring elsewhere, principally in the United Kingdom or the United States of America.

Since the second world war marine science in Australia has developed in a number of areas and coastal engineers, while still at the forefront of certain fields of research, for example wind generation of waves in shallow water (Young and Verhagen 1995) and higher order wave theory (Fenton 1990), have the benefit of several active high quality marine science research organisations. These include CSIRO[15] Division of Oceanography originally located at Cronulla near Sydney but now at Hobart, Australian Institute of Marine Science located at Cape Ferguson south of Townsville, the National Tidal Facility established within Flinders University of South Australia in Adelaide, the Victorian Institute of Marine Science centred in Melbourne and the Coastal Studies Unit of the University of Sydney.

Research by these organisations covers various fields relevant to coastal engineers including the ocean's role in affecting climate, current circulations on the Australian continental shelf, application of remote sensing for the measurement of waves and currents, tidal analysis and prediction, sea level monitoring and climate change, estuarine hydrodynamics and mangrove ecology, marine ecology, ocean currents within coral reef systems and waves on reefs, coastal processes including the behaviour of beach and dune systems.

Coastal engineers and marine scientists are working together in various projects of common interest and efforts are being made to decrease areas of overlap and unnecessary competition and to increase cooperation between them.

NATIONAL COMMITTEE ON COASTAL AND OCEAN ENGINEERING[16]

Formation of Committee

During the 1960s there was increasing coastal engineering activity in the various states. This activity was spread over a large number of government instrumentalities and universities and there was little coordination between them and often no continuity from one project to the next. Within the engineering profession there was no committee or group which covered the full scope of coastal and ocean engineering activities. For some years R. Silvester of the University of Western Australia had advocated the formation of such a committee and in April 1971 a paper written by him discussing Australia's interest and needs in Coastal and Ocean Engineering was published by the Institution of Engineers Australia (IEA). A little earlier in September 1970 J. Ewers and R. Culver of the University of Adelaide had responded positively to an enquiry from the IEA Council concerning likely interest in a Technical Committee on Coastal Engineering. The interest was sufficient for IEA Council on 1 September 1971 to appoint Culver, who was one of its members, as chairman of the new committee. The

---

15) CSIRO = Commonwealth Scientific and Industrial Research Organisation.

16) This account of the activities of the National Committee on Coastal and Ocean Engineering is based upon committee minutes, annual reports, news items in Journal of Institution of Engineers Australia and its successor Engineers Australia and various publications of the committee. The latter are not referred to explicitly in the text but are listed in chronological order in Appendix B.

first meeting of the twelve man committee was held in Sydney on 19 October 1971. Its terms of reference as approved by IEA Council were:

1. To promote and advance the science and practice of Coastal Engineering.

2. To arrange conferences, symposia and meetings at suitable intervals.

3. To encourage engineers in the field to write, present and publish papers.

4. To recommend Council action to meet new or increasing requirements in the field.

5. To act as a focus of communication for engineers in the field.

The committee, initially called the Technical Committee on Coastal Engineering, was renamed National Committee on Coastal and Ocean Engineering (NCCOE) at its third meeting on 14 August 1972. Initially the committee reported directly to the IEA Council but following the establishment in 1975 of the college structure for learned society activities within IEA, NCCOE has reported to the Civil College Board.

Issues considered at the committee's first meeting included the initiation of a review of the then existing situation of Coastal Engineering in Australia; the holding of an Australian Coastal Engineering Conference in May 1973; the possibility of holding an international conference in Australia in 1978; and potential coincidence of interest with IEA's Environmental Engineering Committee on matters such as pollution of coastal waters. Significant activities of NCCOE are reviewed in the next section while Coastal Engineering conferences in Australia are discussed separately in a following section.

Activities of the Committee

The review of Coastal and Ocean Engineering in Australia was published in 1973. It covered Administrative Aspects of Practice of Coastal Engineering, Projects being Undertaken, Technical Aspects of the Practice of Coastal Engineering and Education and Research. Of the needs identified the most urgent one was for *"a broad national policy for management of the coastal zone."* To achieve this and to be able to identify areas of special value would require *"a full enquiry involving all bodies interested in the coastal zone."* The primary recommendation was therefore that a national policy for the coastal zone be formulated and that this could only be done after a full enquiry. This was an urgent necessity.

The review also identified action which could be undertaken by NCCOE, including making informed opinions available to government relating to national policy for utilisation of the coastal zone; formulating standards for physical data collection and storage - a matter of great urgency; and improving the standard of practice of Coastal

Engineering by organising meetings and through the planned conference series. Improved courses were required in Coastal Engineering because existing university courses were not ideal. Moreover, as many bodies and disciplines were involved in the utilisation of the coastal zone, liaison between them and the engineering profession was essential to ensure that all the appropriate information was obtained for a particular project and that the best use was made of the coastal zone.

In late 1973 NCCOE prepared a submission to a Committee of Inquiry established by the Australian Government to enquire into the nature and state of the National Estate and related matters. This was followed by a letter from the President of IEA to the Prime Minister requesting that "*the Australian Government give earnest consideration to undertaking a national inquiry into the research, specialised education, and development needs of the nation in the rapidly expanding field of coastal and ocean engineering.*" The accompanying statement suggested that an Australian Institute of Coastal and Ocean Engineering was required to meet the various needs already identified and that this could have a similar form to the recently established Australian Institute of Marine Science. The Prime Minister's reply was received one year later. It said that the general problem of managing Australia's coastline would require a much wider based enquiry encompassing not only engineering but other scientific and technical disciplines which would be involved in a management programme. Moreover, the Australian Government had recently taken a number of initiatives in the general field of marine science and was unlikely to make further commitments in this broad field in the near future. It was almost twenty years before a wider based enquiry into the management of Australia's coastline was undertaken and when it produced its draft report there was no mention of engineering.

Following a letter from IEA to the Minister for the Northern Territory warning about the dangers of tropical cyclones (sent 16 days before Cyclone Tracy devastated Darwin, the territory's capital city, on 24 December 1974) the NCCOE issued a statement in 1976 on the danger of cyclonic storm surges in the tropical cyclone areas of Australia. The need for reliable early warning systems was identified as well as the need for engineering studies to identify surge prone areas and to study possible protective works.

During 1976 NCCOE first considered the need for standardised practice for dune management and control. This eventually resulted in the publication in 1983 of a series of papers on sand dune management in Australia.

In 1978 NCCOE summarised the availability of tidal data in Australia and also reported on alternative marine sources of energy including tidal and wave power. In the following year representatives of NCCOE were involved with members of other IEA national committees in a working group on Guidelines for Offshore Structures. This working group in due course recommended in 1982 that the Standards Association of Australia prepare a code for fixed offshore structures. However, this recommendation was not accepted.

In 1979 submissions were made to both the House of Representatives Inquiry into Management of the Coastal Zone and to a Senate Inquiry into Marine Science in Australia. In the progress report of the latter enquiry containing recommendations for government expenditure on marine science and technology only one recommendation of twenty was concerned with coastal and ocean engineering.

A draft report prepared by the Australian Science and Technology Council (ASTEC) received a mixed reception from NCCOE in late 1976 largely because engineering and technological aspects had received little consideration. A position paper was submitted in October 1977 reemphasising the need for a national coastal and ocean engineering centre.

The establishment by the Commonwealth Government in the mid 1970s of the Australian Marine Science and Technologies Advisory Committee (AMSTAC) provided an opportunity for NCCOE to have input into the development of government policy for marine research and the development of marine based industry. NCCOE was represented initially on AMSTAC's Working Group on Coastal and Ocean Engineering by its then chairman A. Hicks. A year later he became a member of AMSTAC itself. As he was also secretary of the Australian National Section of PIANC there were many opportunities for cooperation. In 1982 and 1983 an AMSTAC member, R. Smith, attended NCCOE meetings as an observer and subsequently NCCOE member R. Male became a member of AMSTAC continuing opportunities for cooperation.

In 1980 J.B. Hinwood, then chairman of NCCOE, prepared a report on Coastal and Ocean Engineering Laboratory Facilities in Australia which was later published by IEA. This report was used subsequently as a basis for a 1983 report by AMSTAC's Working Group on Coastal and Ocean Engineering to the Commonwealth Government on National Needs and Priorities for Coastal and Ocean Engineering Hydraulic Laboratory Facilities. The latter report recommended the upgrading of Australia's Coastal and Ocean Engineering capability by building on existing regional facilities and expertise to provide a minimum network of Coastal and Ocean Engineering hydraulic laboratory facilities to meet basic national needs. Specific recommendations were made for the improvement of several university and government laboratories and for the establishment of an independent umbrella organisation to provide a permanent secretariat for the Coastal and Ocean Engineering laboratories. Despite the relatively modest cost of these proposals, they were not accepted by the government.

Following an approach by NCCOE, IEA's Civil College Board established in late 1980 a Working Party to consider the implications of the 200 mile (320km) Exclusive Economic Zone (EEZ). This Working Party was chaired by J.B. Hinwood who had just completed his term as NCCOE chairman. Other NCCOE members were involved in the Working Party which in 1984 produced a very comprehensive report on the opportunities which the 200 mile EEZ presented for Australian marine industries.

During the mid 1980s NCCOE produced or encouraged others to produce a number of reports covering various matters including a review of Australian tertiary education courses in Coastal and Ocean Engineering, Coastal protection, Planning and design of marinas, Professional engineer's role in coastal zone management and Investigation and research projects in coastal engineering. Then, following considerable public and scientific concern about the greenhouse effect and related issues such as sea level rise, NCCOE in May 1988 published a position paper concerning Engineers and the Greenhouse Effect. This stimulated response from coastal engineers, including D.B. Lord and A.F. Nielsen, who pointed out some of the practical implications of the climatic change scenarios for designers and planners concerned with works in the coastal zone. NCCOE responded by holding a workshop in Sydney on 31 May to 1 June 1990 to initiate the formulation of guidelines for the inclusion of the effects of climatic change on engineering in the coastal zone. The report stating these guidelines was published in the following year and was widely distributed throughout Australia to coastal engineers and other engineers concerned with the coast.

During 1991-92 NCCOE made submissions to the House of Representatives inquiry on protection of the coastal environment, the Commonwealth's Ecologically Sustainable Development process and to the Commonwealth Government's Resource Assessment Commission's Coastal Zone Inquiry. The draft report of the latter was released at the 11th Australasian Conference on Coastal and Ocean Engineering in Townsville in August 1993 but it received a mixed reception because it did not recognise the role of coastal engineering in various aspects of coastal management. Following submissions from the committee, the Inquiry's final report presented a more balanced viewpoint. The report proposed a national approach to coastal management in cooperation with the states. This is now being implemented by the Commonwealth Government.

In 1993 NCCOE, with funding assistance from the Department of Industry, Trade and Regional Development, held a workshop to discuss issues relating to the collection and management of coastal engineering data. Changes in governmental administrative structures, including the contracting out of many engineering services, were creating a situation where the continuation of many marine and coastal data programmes was being threatened by lack of continuing resources to support them. Following the workshop the committee, assisted by former member A.D. Gordon, prepared and published At What Price Data?, a provocative report focussing attention on the question of accountability of nonengineering decision makers in their need for and use of data. This report was widely distributed and has stimulated discussion of this question by not only coastal engineers but also other groups of engineers.

In 1994 NCCOE was commissioned by the Department of Environment, Sport and Territories (DEST) to assess research and development priorities for coastal zone management. This was achieved through a workshop convened with marine science and local authority representatives, followed by submission of a report which has been utilised by DEST in coastal action planning.

Conferences

One of the actions initiated at the committee's first meeting was to seek approval to hold a Conference on Coastal Engineering in Sydney during May 1973. This conference proved to be very successful with 145 persons attending. 35 papers selected from over twice that number of proposals were included in the programme for the two and a half day conference held at the Sydney seaside suburb of Manly, the location of a much publicised incident of open ocean bathing early this century. The chairman of the organising committee was NCCOE member J.G. Betty and the conference was opened by IEA President A.H. Corbett. The success of this first conference encouraged NCCOE to plan a biennial series to be held in odd numbered years, alternating with the series sponsored by ASCE, the International Conferences on Coastal Engineering.

The second conference at Broadbeach on the Gold Coast in Queensland attracted 200 registrants and in general each succeeding conference has continued to attract at least 150 to 200 registrants. The number of papers presented has increased substantially and conferences are now run in several parallel sessions over five days with keynote speakers, workshops and half day technical tours. Conferences have now been held in all Australian states as well as New Zealand (Table 1).

The Coastal and Ocean Engineering Conference series was extended across the Tasman Sea to include New Zealand in 1985 when the seventh conference in the series was held in Christchurch in the South Island. After two more conferences in Australia, the tenth conference was held in 1991 in Auckland in the North Island and the series currently is planned to continue with the 1997 conference in Christchurch. The Institution of Professional Engineers New Zealand (IPENZ) is joint sponsor with IEA for these conferences.

In the mid 1980s a second conference series on Port and Harbour Engineering (PHE)[17] was established in cooperation with the Australian section of the Permanent International Association of Navigation Congresses (PIANC). The first PHE conference was held in Sydney during 1986 with subsequent conferences in Brisbane and Melbourne, and returning to Sydney in 1992. These conferences were more specifically focussed than the broad based COE conferences. Engineers and other interested persons were involved in discussion of various issues concerning ports and harbours and associated maritime industries. Conferences were well supported by industry and attendances of up to 200 registrants were attained.

---

17) The first two conferences in this series were described as Port Harbour and Offshore Engineering Conferences.

Table 1. Conferences Supported by NCCOE

A.	Coastal and Ocean Engineering Conferences			
	Date	Location	Chairman Organising Committee	No. of Papers/ Posters (Preprints)
1st	1973 14 - 17 May	Sydney, NSW (Manly)	J.G. Betty	35
2nd	1975 27 Apr - 1 May	Gold Coast, Qld (Broadbeach)	B.L. McGrath	31
3rd	1977 18 - 21 Apr	Melbourne, Vic	A.B. Hicks	38
4th	1978 8 - 10 Nov	Adelaide, SA	L.S. Buenfeld	51
5th[1]	1981 25 - 27 Nov	Perth, WA	R. Silvester	81 abstracts in depth 55 proceedings papers
6th	1983 13 - 15 July	Gold Coast, Qld[2] (Surfers Paradise)	H.V. Macdonald	61
7th	1985 2 - 6 Dec	Christchurch, NZ	R.W. Morris	109
8th	1987 30 Nov - 4 Dec	Launceston, Tas	D.F.E. Bowen	90
9th	1989 4 - 8 Dec	Adelaide, SA	R.K. Tucker	80
10th	1991 2 - 6 Dec	Auckland, NZ	J. Duder	95
11th	1993 23 - 27 Aug	Townsville, Qld	D.C. Patterson	115
12th	1995 28 May - 2 June	Melbourne, Vic	G. Byrne	55 coasts 26 ports 5 keynote
B.	Port Harbour (and Offshore) Engineering Conferences			
	Date	Location	Chairman Organising Committee	No. of Papers/ Posters (Preprints)
1st	1986 29 Sept - 2 Oct	Sydney, NSW	B. Robertson	63
2nd	1988 23 - 28 Oct	Brisbane, Qld	J. Beath	43
3rd	1990 28 - 30 Aug	Melbourne, Vic	R. Jones	34
4th	1992 25 - 27 Aug	Sydney, NSW	A. Patterson	49
5th[3]	1995	Melbourne, Vic		

1 17th International Conference on Coastal Engineering was held in Sydney on 23-28 March 1980.

2 Conference Preprints volume incorrectly shows conference location as Brisbane.

3 5th Port Harbour Engineering Conference was combined with 12th Coastal and Ocean Engineering Conference.

Government restructuring with downsizing of engineering based public authorities as well as tightening economic conditions led to a review of NCCOE conference activities. The 1995 conference in Melbourne combined the twelfth COE conference with the fifth PHE conference. While the conference itself was very successful both organisationally and financially and the programme was interesting with several diverse keynote speakers and the usual selection of good papers on a wide range of topics, the total attendance was only 220, significantly less than the combined total of a COE conference and a PHE conference in previous years. Moreover, commercial and other pressures meant that increasing numbers of registrants attended part time, while some nonengineers who previously attended COE conferences were absent. An increasing number of regional Coastal Management conferences and the initiation of a national Coastal Management conference series in 1994 may have been factors affecting attendance.

A comparison of topics of papers and background of authors for the first two and last two conferences is interesting. In 1973 and 1975 coastal processes and wave data/wave climate were the most popular topics accounting for about 30% of the papers. However in 1993, 30% of the papers were on Ocean outfalls, Coastal management and Environmental impact of dredging, while in 1995 a third of the COE papers were on either specific Coastal Engineering projects or wave generation, propagation, etc., and half the PHE papers related to specific projects.

Not only has the number of papers presented increased over the years but also the number of authors for each paper. In 1973 and 1975 more than half of all papers were written by a single author and about a third had two authors. In 1995 only a quarter of all papers had a single author, just over two fifths had two authors and just over a quarter had three authors. The backgrounds of authors also show some interesting trends (Table 2). 75 to 80% of authors have been from Australia while nonengineers from Australia have varied between 20 and 30%. There has been a significant decrease in the number of authors who are engineers from Government instrumentalities and an even greater proportional increase in the number from consultants. The same change is reflected in the employment status of registrants. Comparison of the 1973 and 1993 conferences shows that the proportion of engineers from Government and universities has decreased from 49% to 32% whereas the proportion from consultants and industry has increased from 26 to 42%.

Table 2. Employment Status of Authors of Conference Papers

Conference	Number of Authors	Percentage (%) of Total Number of Authors		
1973 and 1975	50 and 46	80% Australian 20% Overseas →	55% engineers 25% others →	33% government 17% university 5% consultants
1993	204	75% Australian 25% Overseas →	45% engineers 30% others →	13% government 17% university 15% consultants
1995 Coasts	93	81% Australian 19% Overseas →	61% engineers 20% others →	19% government 23% university 19% consultants
1995 Ports	49	100% Australian →	92% engineers 8% others →	22% government 55% consultants 12% industry 2% university

An island nation such as Australia can be somewhat isolated from developments in other parts of the world. Over the years Australian coastal engineers have benefitted from visitors from overseas countries. Virtually all Australian COE conferences have had at least one prominent overseas visitor. On a number of occasions the visitor also made a tour of all or most states as an IEA Civil College eminent speaker thus allowing engineers who were unable to attend the conference to benefit from the speaker's visit to Australia. Notable overseas visitors are listed in Table 3. In one case, R.G. Bea, NCCOE was able to publish the visitor's various lectures in a single publication of particular interest to engineers designing coastal and offshore structures.

Table 3. Overseas Coastal and Ocean Engineering Distinguished Visitors

1879 1885	J. Coode (U.K.)	1983	Y. Goda (Japan)
1964	P. Bruun (U.S.A.)	1985	F. Gerritsen (U.S.A.) - N.Z. Conf.
1973	J.W. Kamphuis (Canada)	1987	R.G. Dean (U.S.A.)
1975	C.L. Bretschneider (U.S.A.)	1989	R.G. Bea (U.S.A.)
1977	E.W. Bijker (Netherlands) R.L. Wiegel (U.S.A.)	1991	T. Sørensen (Denmark) - N.Z. Conf.
1980	M.S. Longuet-Higgins (U.K.)	1993	J.R. Houston (U.S.A.)
1981	W.A. Price (U.K.)	1995	K. Pilarcyzk (Netherlands)

In March 1980 the 17th International Conference on Coastal Engineering was held at the Hilton Hotel in Sydney, J.G. Betty being chairman of the organising committee. About 220 papers and posters were presented to 550 registrants, approximately half of whom were overseas visitors. Subsequently in 1994 after Sydney had been awarded the 2000 Olympic Games, New South Wales members of NCCOE applied to the ASCE Coastal Engineering Research Council for the right to hold the 27th International Conference on Coastal Engineering in Sydney during 2000. Their proposal has been accepted and an organising committee has been formed to handle this event.

Committee Membership and Awards

Over the years membership of NCCOE has varied between 12 and 16 persons. Over 50 Australian coastal engineers have been members - some for a few years, others for periods of ten years of more. Membership has always been distributed between the various states with a weighting towards the three most populous eastern states. Likewise a balance is maintained between engineers employed by universities, government agencies and consultants. Committee members are appointed by the chairman who is selected by committee consensus. Table 4 lists members who have served as chairman. All members serve in a honorary capacity adding their NCCOE activities to their already demanding full time professional responsibilities with their employers or in their own businesses.

Table 4. Chairman of National Committee on Coastal and Ocean Engineering

1971 (Sept)[18]	R. Culver	1985	F.L. Wilkinson
1975	R. Silvester	1987	H.V. Macdonald
1977	A.B. Hicks	1989	R. Male
1979	J.B. Hinwood	1989 (July)[19]	G. Byrne
1981	D.N. Foster	1991 (July)	R.K. Tucker
1983	W.M. Lewis	1993 (Nov)	M.R. Gourlay
1996	B.A. Harper		

18) Period of appointment commenced at beginning of year unless otherwise indicated.

19) An overseas transfer prevented Male serving his full two year term.

The committee's first chairman, Dr Robert Culver, was, until his retirement in 1991, an engineering academic who consulted widely with Governments and Industry on hydraulic and coastal matters for some 40 years. These contributions were acknowledged in several awards - election to the Fellowship of the Academy of Technological Sciences and Engineering in 1987, Member of the Order of Australia 1991, Doctor of the University (for his wide University service) 1992 and in 1994 he received the prestigious IEA award of the Peter Nicol Russell Memorial medal. The summary commentary submitted upon his nomination for this last award stated *"Robert Culver is a fine example of the classical gentleman and scholar with an acute appreciation of life, its pleasures and its pains, but above all, of his obligation to use his gifts to serve his fellow man whenever the opportunity arises."* NCCOE was very fortunate to have such a man as its first chairman.

In 1992 the committee decided to institute the Kevin Stark Memorial Award. While he was never a member of NCCOE, Professor Stark was a pioneer in important areas of coastal engineering research and in engineering systems and his untimely death on 28 April 1989 was a great loss not only to the engineering community but to the wider marine science community of Australia. As Foundation Professor of Systems Engineering at James Cook University of North Queensland in Townsville he sought to break down the barriers between disciplines, particularly in the marine science and technology area. He served on many marine science committees at international, national, state and local levels. As a researcher one of his primary interests was in numerical modelling of coastal and ocean hydrodynamic phenomena. Following cyclone Althea in 1971, he initiated studies on storm surges and became acknowledged as Australia's authority in this area. This led to the establishment of a Marine Modelling Unit to study aspects of ocean circulation in the Great Barrier Reef region. Later he became increasingly concerned with the potential dangers due to the Greenhouse Effect. In 1987 he received IEA's Sir John Monash award.

The Kevin Stark Memorial Award recognises excellence in coastal and ocean engineering and is awarded to the best paper presented at each COE conference, the selection criteria including the following considerations:

- potential contribution to state of knowledge and practice;
- extent of multi-disciplinary expertise applied ;
- quality and clarity of the written presentation;
- potential for fostering further enquiry;
- technical quality.

The award is in the form of a bronze medal engraved with the recipient's name and date of award.

This award was inaugurated in 1993 at the 11th COE Conference in Townsville where Professor Stark's former colleague R.J. Sobey presented the special Kevin Stark Memorial Keynote Address (Sobey 1993). The inaugural award was made to M.R.

Fitzmaurice and A.J. Johnston for their paper "Can oil slicks be contained?". The 1995 award went to E. Couriel, E. Griffiths and N. Tomkins for a paper on "Effective management of coastal lagoons through real time monitoring and predictive modelling - Case study for Narrabeen Lagoon."

Achievements of Committee

NCCOE will celebrate the 25th anniversary of its formation in September 1996. During its 25 years it has sought to promote and advance the science and practice of Coastal and Ocean Engineering in Australia in many different ways. It has established and maintained a very successful series of biennial conferences. It has endeavoured to focus on specific issues relevant to coastal engineers and has produced a number of significant reports recommending specific action to decision makers in government and elsewhere. In particular, it has continually drawn attention to the need to maintain and expand basic marine and coastal data collection networks so that both coastal engineers and coastal managers will have adequate information to make decisions affecting the coastal environment and hence fulfill their "duty of care" to the community.

NCCOE was not successful in getting the Commonwealth Government to establish an Australian Institute for Coastal and Ocean Engineering nor even to establish a more limited coordinating body. Its 1973 call for the formulation of a national policy for the coastal zone was long ignored but now has been achieved through the efforts of many community groups and individuals. Many of those involved in implementing the Commonwealth Coastal Policy and its Coastal Action Plan would be unaware of the existence of NCCOE as would most of the Australian people. Yet timely lobbying of the Commonwealth Government early last year by IEA, instigated by NCCOE members, is believed to have been a crucial factor in gaining political acceptance for the Commonwealth Coastal Policy.

In April 1982 when NCCOE was 10 years old Robin Henderson, IEA College Secretary responsible for servicing national committees, commented that NCCOE was *"one of our most successful and busy National Committees"* and also was the *"only national body taking an interest in the coastline as a whole - and this is a country with the longest coastline in the world!"* That sole responsibility is now being shared with, indeed has been seized by, others. The National Committee on Coastal and Ocean Engineering now has the responsibility of ensuring that coastal engineers with their expertise in devising practical, relevant and affordable solutions to the problems of the coast continue to exercise a key role in meeting the Australian peoples' expectations for their coasts.

CONCLUSION

Coastal engineering in Australia developed in the nineteenth century from the need to provide port and harbour facilities at numerous locations around the nation's long coastline. Initially engineers were concerned with maintaining and expanding these facilities to provide for safe navigation and shelter for shipping.

In the twentieth century, increasing use of coastal regions for recreational and residential purposes resulted in many short sighted developments within the active shoreline zone. As these developments were threatened from time to time protective works became necessary. Moreover, in many places the impacts of previously constructed harbour works upon the adjoining coast became apparent and additional protective works were required.

To solve these problems Australian engineers drew on experience in overseas countries, principally the United Kingdom, The Netherlands and the United States of America. Different climatic, geographical and administrative conditions, as well as isolation and lack of money, required the adaptation of overseas practices to the various local situations. Construction methods were developed to use available local resources such as dumped rock, hardwood timber and shore connected construction procedures.

Coastal management in several Australian states was largely pioneered by engineers who in most cases initially had great difficulty in convincing politicians that coastal structures such as groynes would not by themselves solve the problems of eroding beaches. It took some time for the community to accept that the best management of the coast involved prevention rather than cure of erosion.

Successful Australian coastal engineering projects have made significant contributions to the nation's economy in providing ports and harbours for exporting bulk commodities, in restoring beaches in major coastal tourist regions and in improving the quality of the natural environment in estuarine and coastal waters.

Both in research and in practice Australian coastal engineering has been innovative and has produced significant new developments. Some of these initiatives have been widely published; others, particularly those originating from some government and consulting offices are not widely known. In many cases Australian coastal engineers also have been the leaders in establishing multidisciplinary teams to address the impacts of infrastructure and development upon coastal and estuarine environments.

The National Committee on Coastal and Ocean Engineering has provided a focus for coastal engineering in Australia by voluntary effort with minimal resources. Through its conference series it has provided opportunities for engineers and others concerned with the coast to meet and discuss technical and related matters. Through its reports and other activities it has endeavoured to ensure that engineers and others concerned with activities affecting the coast are fully aware of their responsibilities to the community.

ACKNOWLEDGEMENTS

The constraints of time and space prevent a fully comprehensive treatment of the topic so it has been necessary to be selective. Initially a number of colleagues in various parts of the country were circularised and invited to assist with this project. Many responded with comments, suggestions, reports, minutes of meetings, draft texts, etc. This material was supplemented by independent research by the author, including specific inquiries to colleagues about various matters. The following persons supplied material or otherwise provided assistance.

Colin Apelt, Phil Barlow, Bill Bremner, Gerry Byrne, Michael Clarke, Neil Collins, Bob Culver, Lee Donnelly, Doug Foster, Angus Gordon, Bruce Harper, Arthur Hicks, Jon Hinwood, Angus Jackson, Neil Lawson, Brian McGrath, Russell Murray, Ray Nelson, Mike Paul, Don Ross, Dick Silvester, Bruce Sinclair, Will Strachan, John Tainsh, Rod Thomas, Rob Tucker, John Van der Peyl, Tony Wynne, Ian Young.

Following completion of the draft text, copies were sent to a number of persons for comment.

The text has been prepared by Janna Seto and the figures by Reg Stonard.

The author wishes to express his appreciation for the support received from his colleagues and acknowledges that without that support this paper could not have been written.

REFERENCES

Allan, P., 1920-21. "Port Improvements at Newcastle, New South Wales," *Min. Proc. Inst. Civ. Eng.*, Vol. 212, pp. 1-34.

Apelt, C.J., and Baddiley, P., 1981. "Breaking Wave Forces on Vertical Cylinders," *Proc. 5th Australian Conf. Coastal and Ocean Eng.*, Perth, W.A., Inst. Eng. Aust., Nat. Conf. Publ., No.81/16, pp. 85-89.

Apelt, C.J., Gout, J.J., and Szewczyk, A.A., 1974. "Numerical Modelling of Pollutant Transport and Dispersion in Bays and Estuaries," In Brebbia, C.A., and O'Connor, J.J. (eds.), *Numerical Methods in Fluid Dynamics*, Pentech Press, London, pp. 307-324.

Apelt, C.J., and Macknight, A., 1976. "Wave Action on Large Offshore Structures," *Proc. 15th Int. Conf. on Coastal Eng.*, Honolulu, Hawaii, Am. Soc. Civ. Eng., New York, Vol. 3, pp. 2228-2247.

Apelt, C.J., and Piorewicz, J., 1987. "Laboratory Studies of Breaking Wave Forces Acting on Vertical Cylinders in Shallow Water," *Coastal Eng.*, Vol. 11, pp. 263-282.

Apelt, C.J., and Piorewicz, J., 1991. "Impact Force as a Part of the Total Breaking Wave Force on a Vertical Cylinder, *Preprints: 10th Australasian Conf. Coastal and Ocean Eng.*, Auckland, N.Z., DSIR Marine and Freshwater, Water Quality Centre, Publ. No 21, pp. 429-434.

Apelt, C.J., and Ryall, G.L., 1992. "Modelling the Sedimentary Processes in Real Estuaries," *Aust. Civ. Eng. Trans.*, Vol. CE34, pp. 1-7.

Aseervatham, A.M., Kang, H.Y., and Nielsen, P., 1993. "Groundwater Movement in Beach Watertables," *Preprints: 11th Australasian Conf. Coastal and Ocean Eng.*, Townsville, Qld., Inst. Eng. Aust., Nat. Conf., Publ. No. 93/4, Vol. 2, pp. 589-594.

BBW, 1985. "Storm Tide Statistics Methodology," A report prepared for the Beach Protection Authority Queensland by Blain Bremner & Williams Pty. Ltd., ISSN 0814-8295.

Beverley, R.F., 1966. "Sea Defence Works in South Australia," *Jour. Inst. Eng. Aust.*, Vol. 38, pp. 19-26.

Bode, L., Volker, R.E., and Sobey, R.J., 1985. "Two-dimensional Convection-dispersion with Moving Coordinates, " *Preprints: 1985 Australasian Conf. Coastal and Ocean Eng.*, Christchurch, N.Z., Conf. Org. Comm., Christchurch, Vol. 2, pp. 433-443.

BPA, 1979. "Capricorn Coast Beaches," Beach Protection Authority Queensland.

BPA, 1984. "Mulgrave Shire Northern Beaches", Beach Protection Authority Queensland. ISBN 0 7307 0007 0.

BPA, 1989. "Hervey Bay Beaches", Beach Protection Authority Queensland. ISBN 0 7242 3459 4.

Bremner, W., and Foster, D.N., 1987. "Use of Physical Models as an Aid in the Design and Construction of Breakwaters," *Preprints: 8th Australasian Conf. Coastal and Ocean Eng.*, Launceston, Tas., Inst. Eng. Aust., Nat. Conf. Publ. No. 87/17, pp. 244-248.

Bremner, W., Harper, B.A., and Foster, D.N., 1987. "The Design and Construction of a Mass Armoured Breakwater at Hay Point, Australia," Nat. Res. Council, Canada, Seminar on Unconventional Rubble-Mound Breakwater, Ottawa, Canada.

Bremner, W., Foster, D.N., Miller, C.A., and Wallace, B.C., 1981. "The Design Concept of Dual Breakwaters and its Application to Townsville Harbour, Queensland, Australia," *Inst. Eng. Aust. Civ. Eng. Trans.*, Vol. CE23, pp. 283-288. Also published in Proc. 17th Int. Coastal Eng. Conf., Sydney, Australia, Am. Soc. Civ. Eng., New York, Vol. 2, pp. 1898-1908.

Brown, C.T., 1978. "Blanket Theory and Low Cost Revetments," *Proc. 16th Int. Conf. on Coastal Eng.*, Hamburg, Germany, Am. Soc. Civ. Eng., New York, Vol. 3, pp. 2510-2527.

Bunt, J.S., and Wolanski, E., 1980. "Hydraulics and Sediment Transport in a Creek-Mangrove Swamp System," *Proc. 7th Australasian Hydraulics and Fluid Mechanics Conf.*, Brisbane, Qld., Inst. Eng. Aust., Nat. Conf. Publ. No. 80/4, pp. 492-495.

Burren, K.R., 1975. "Investigation, Design and Construction of a Harbour on King Island, Tasmania," *Inst. Eng. Aust. Civ. Eng. Trans.*, Vol. 17, pp. 64-68.

Burrows, T.E., 1904-05. "Improvements at the Entrance to the Richmond River, New South Wales," *Min. Proc. Inst. Civ. Eng.*, Vol. 160, pp. 326-339.

Carroll, D.J., 1985. "Conceptual Design of the Sydney Tunnelled Ocean Outfalls," *Preprints: 1985 Australasian Conf. Coastal and Ocean Eng.*, Christchurch, N.Z., Conf. Org. Comm., Christchurch, Vol. 1, pp. 157-167.

Chapman, D.M., Geary, M., Roy, P.S., and Thom, B.G., 1982. "Coastal Evolution and Coastal Erosion in New South Wales," Coastal Council of New South Wales, Sydney, 341p., ISBN 0 7240 6582 2.

Coltheart, L., in prep. "Between Wind and Water - A History of the Ports and Coastal Waterways of New South Wales," Unpublished manuscript.

Corbett, A.H., 1973. "The Institution of Engineers Australia - History of First 50 Years," *Inst. Eng. Aust.* and *Angus and Robertson*, Sydney.

Coughlan, P.M., and Robinson, D.A., 1990. "The Gold Coast Seaway, Queensland, Australia," *Shore and Beach*, Vol. 58, No. 1, pp. 9-16.

Couriel, E.D., and Cox, R.J. 1991. "Breaking Wave Forces on Piles," *Preprints: 10th Australasian Conf. Coastal and Ocean Eng.*, Auckland, N.Z., DSIR Marine and Freshwater, Water Quality Centre, Publ. No. 21, pp. 423-427.

Couriel, E.D., Griffiths, E., and Tomkins, N., 1995. "Effective Management of Coastal Lagoons Through Real Time - Monitoring and Predictive Modelling - Case Study for Narabeen Lagoon", *Preprints: 12th Australasian Conf. Coastal and Ocean Eng.*, Melbourne, Vic, Inst. Eng. Aust., pp. 47-52.

Culver, R., 1970. "Beach Erosion Assessment Study - Summary Report," University of Adelaide, Department of Civil Engineering.

Davenport, W., 1986. "Harbours and Marine: Port and Harbour Development in Queensland from 1824 to 1985," Department of Harbours and Marine, Brisbane, 883p. ISBN 0 7242 1638 3.

Davies, J.L., 1972. "Geographical Variation in Coastal Development," Oliver & Boyd, Edinburgh, 204p. ISBN 0 05 002597 X.

DEST, 1995. "Living on the Coast - The Commonwealth Coastal Policy," Department of Environment, Sport and Territories, Australia, 70p. ISBN 0 642 22772 1.

DHL, 1965. "Queensland Coastal Erosion - Recommendations for a Comprehensive Coastal Investigation," Delft Hydraulics Laboratory, The Netherlands, Report R257.

DHL, 1970. "Gold Coast, Queensland, Australia - Coastal Erosion and Related Problems," Three volumes, Delft Hydraulics Laboratory, The Netherlands, Report R257.

DMH, 1987. "Peel Inlet and Harvey Estuary Management Strategy, Dawesville Channel Engineering Investigations," Department of Marine and Harbours, Western Australia, Rep. No. DMH 5/88.

Druery, B.M., 1980. "Estuarine Response to Dredging in the Tweed River, Australia," *Proc. 17th Conf. Coastal Eng.*, Sydney, Australia, Am. Soc. Civ. Eng., New York, Vol. 2, pp. 1599-1618.

Druery, B.M., and Curedale, J.W., 1979. "Tweed River Dynamics Study," Department of Public Works, NSW, Rep. No. 78009. ISBN 7240 2715 7.

Dyson, A.R., and Druery, B.M., 1985. "The Impact of Sand Extraction on Salt Intrusion in the Hawkesbury River," *Preprints: 1985 Australasian Conf. Coastal and Ocean Eng.*, Christchurch, N.Z., Conf. Org. Comm, Christchurch, Vol. 1, pp. 557-567.

Ebner, J.G., and French, S., 1985. "Investigation and Design of Seabed Works for Sydney Ocean Outfalls,"*Preprints: 1985 Australasian Conf. Coastal and Ocean Eng.*, Christchurch, N.Z., Conf. Org. Comm, Christchurch, Vol. 1, pp. 123-133.

Fenton, J., 1990. "Nonlinear Wave Theories," In "*The Sea - Volume 9, Ocean Engineering Science, Part A*," Le Méhauté, B., and Hanes, D.M. (eds.), Wiley, New York.

Fitzmaurice, M.R., and Johnston, A.J., 1993. "Can Oil Slicks be Contained?", *Preprints: 11th Australasian Conf. Coastal and Ocean Eng.,* Townsville, Qld., Inst. Eng. Aust., Nat. Conf. Publ. No. 93/4, Vol. 2, pp. 683-688.

Floyd, C.D., 1968. "River Mouth Training in New South Wales, Australia," *Proc. 11th Conf. on Coastal Eng.*, London, England, Am. Soc. Civ. Eng., New York, Vol. 2, pp. 1267-1282.

Ford, A.R., 1963. "River Entrances of N.S.W.", *Jour. Inst. Eng. Aust.*, Vol. 35, pp. 313-320.

Foster, D.N., 1984. "A Review of Breakwater Development in Australia," *Proc. 19th Int. Coastal Eng. Conf.*, Houston, Texas, Am. Soc. Civ. Eng., New York, Vol. 3, pp. 2751-2759.

Foster, D.N., and Gordon, A.D., 1973. "Stability of Armour Units Against Breaking Waves," *Preprints: 1st Australian Conf. Coastal Eng.*, Inst. Eng. Aust., Nat. Conf. Publ. No. 73/1, pp. 98-107.

Foster, D.N., Gordon, A.D., and Lawson, N.V., 1975. "The Storms of May-June 1974, Sydney, N.S.W.," *Preprints: 2nd Aust. Conf. Coastal and Ocean Eng.*, Gold Coast, Qld., Inst. Eng. Aust., Nat. Conf. Publ. No.75/2, pp. 1-11.

Foster, D.N., McGrath, B.L., and Bremner, W., 1978. "Rosslyn Bay Breakwater, Queensland, Australia," *Proc. 16th Int. Coastal Eng. Conf.*, Hamburg, Germany, Am. Soc. Civ. Eng., New York, Vol. 3, pp. 2086-2103.

Foster, D.N. and Stone, D.M., 1965. "Historical Evidence of Erosion at Cronulla", *Jour. Inst. Eng. Aust.,*Vol. 37, pp. 275-286.

Fox, I.B., and Wilke, M.A., 1981. "Hydraulic Considerations in the Design of Mary River Tidal Barrage," *Proc. Conf. on Hydraulics in Civil Eng.*, Sydney, N.S.W., Inst. Eng. Aust., Nat. Conf. Publ. No. 81/12, pp. 70-74.

Gill, E.D., 1982. "Eights Coasts of Australia," CSIRO, Inst. Energy and Earth Resources, Div. Appl. Geomechs., Tech. Rep. No. 119, 66p. ISBN 0 643 02689 4.

Gillespie, J.D., 1965. "Beach Erosion and Restorative Works, Perth, W.A.," *Journ. Inst. Eng. Aust.*, Vol. 37, pp. 137-142.

Gordon, A.D., and Hoffman, J.G., 1984. "Sediment Transport on the South-East Australian Continental Shelf," *Proc. 19th Int. Conf. Coastal Eng.*, Houston, Texas, Am. Soc. Civ. Eng., New York, Vol. 2, pp. 1952-1967.

Gordon, A.D., and Lord, D.B., 1980. "An Approach to Understanding Coastal Processes," *Proc. 17th Conf. Coastal Eng.*, Sydney, Australia, Am. Soc. Civ. Eng., New York, Vol. 2, pp. 1235-1254. Also published in *Inst. Eng. Aust. Civ. Eng. Trans.*, Vol. CE23, pp. 178-185.

Gordon, A.D., and Roy, P.S., 1977. "Sand Movements in Newcastle Bight," *Preprints: 3rd Aust. Conf. Coastal and Ocean Eng.*, Melbourne, Vic., Inst. Eng. Aust., Nat. Conf. Publ. No. 77/2, pp. 64-69.

Gordon, A.D., Lord, D.B., and Nolan,M.W., 1978. "Byron Bay - Hastings Point Erosion Study," Department of Public Works, N.S.W., Coastal Engineering Branch, Rep. No. PWD78026. ISBN 7240 2691 6.

Gourlay, M.R., 1965. "Wave Generated Currents - Some Observations Made in Fixed Bed Hydraulic Models," University of Queensland, Department of Civil Engineering, Bul. No. 7., 21p.

Gourlay, M.R., 1974. "Wave Set-up and Wave-Generated Currents in the Lee of a Breakwater or Headland," *Proc. 14th Conf. Coastal Eng.*, Copenhagen, Denmark, Am. Soc. Civ. Eng., New York, Vol. 3, pp. 1976-1987.

Gourlay, M.R., 1976. "Nonuniform Alongshore Currents," *Proc. 15th Conf. Coastal Eng.*, Honolulu, Hawaii, Am. Soc. Civ. Eng., New York, Vol. 1, pp. 701-720.

Gourlay, M.R., 1981. "Beach Processes in the Vicinity of Offshore Breakwaters," *Proc. 5th Aust. Conf. Coastal and Ocean Eng.*, Perth, W.A., Inst. Eng. Aust., Nat. Conf. Publ. No. 81/16, pp. 129-134.

Gourlay, M.R., 1983. "Interaction Between Natural Processes and Engineering Works on the Leeward Side of a Coral Cay: A Case Study of Heron Island on the Great Barrier Reef," *Proc. Int. Conf. Coastal and Port Eng. in Developing Countries*, Colombo, Sri Lanka, Vol. 2, p. 1468-1482.

Gourlay, M.R., 1993. "Wave Set-up and Wave-Generated Currents on Coral Reefs," *Preprints: 11th Australasian Conf. Coastal and Ocean Eng.*, Townsville, Qld., Inst. Eng. Aust., Nat. Conf. Publ. No. 93/4, Vol. 2, pp. 479-484.

Gourlay, M.R., 1994. "Wave Transformation on a Coral Reef," *Coastal Eng.*, Vol. 23, pp. 17-42.

Gourlay, M.R., and Hacker, J.L.F., 1986. "Pioneer River Estuary: Sedimentation Studies," Univ. Qld., Dept. Civ. Eng., St Lucia, Qld., 207p., 98 figs. ISBN 0 86776 1431.

Gourlay, M.R., and Jell, J.S., 1993. "Heron Island Spoil Dump, " Great Barrier Reef Marine Park Authority, Townsville, Qld., Res. Publ. No. 28, 62p., 60 figs. ISBN 0 642 17434 2.

Halligan, G.H., 1910-11. "The Bar Harbours of New South Wales," *Min. Proc. Inst. Civ. Eng.*, Vol. 184, pp. 128-148.

Hardy, T.A., and Young, I.R., 1991. "Modelling Spectral Wave Transformation on a Coral Reef Flat," *Proc. 10th Australasian Conf. Coastal and Ocean Eng.*, Auckland, N.Z., DSIR Marine and Freshwater, Water Quality Centre, Publ. No. 21, pp. 345-350.

Hardy, T.A., Young, I.R., Nelson, R.C., and Gourlay, M.R., 1990. "Wave Attenuation on an Offshore Coral Reef," *Proc. 22nd Int. Coastal Eng. Conf.*, Delft, The Netherlands, Am. Soc. Civ. Eng., New York, Vol. 1., pp. 330-340.

Hardy, T.A., Young, I.R., Nelson, R.C., and Gourlay, M.R., 1991. "Wave Attenuation on a Coral Reef," *Aust. Civ. Eng. Trans.*, Vol. CE33, pp. 17-22.

Harper, B.A., 1987. "Stability of Highly Permeable Breakwaters," Report for Aust. Mar. Sc. and Tech. Grants Scheme, Blain Bremner & Williams Pty. Ltd., Milton, Qld.

Harper, B.A., and Greentree, G.S.,1981. "Dispersion Characteristics of Nearshore Coastal Waters," *Preprints: Conf. Environmental Eng. 1981*, Townsville, Qld., Inst. Eng. Aust., Nat. Conf. Publ. No. 81/6, pp. 125-129.

Harper, B.A., Mason, L.B., and Bode, L., 1993. "Tropical Cyclone Orson - A Severe Test for Modelling," *Preprints: 11th Australasian Conf. Coastal and Ocean Eng.*, Townsville, Qld., Inst. Eng. Aust., Nat. Conf. Publ. No. 93/4, Vol. 1, pp. 59-64.

Harris, J.E., Hinwood, J.B., Marsden, M.A.H., and Sternberg, R.W., 1979. "Water Movements, Sediment Transport and Deposition, Western Port, Victoria," *Mar. Geol.*, Vol. 30, pp. 131-161.

Hays, W.B., 1858-59. "Discussion on Construction of Breakwaters," *Min. Proc. Inst. Civ. Eng.*, Vol. 18, pp .143-145.

Higham, L., and Robson, F., 1981. "Sewage Treatment and Effluent Discharge to the Ocean from N.S.W.," *Preprints: Conf. Environmental Eng. 1981*, Townsville, Qld., Inst. Eng. Aust., Nat. Conf. Publ. No. 81/6, pp. 118-124.

Hinwood, J.B., Watson, J.E., Mills, D.A., Dandy, G.C. and Burridge, D.M., 1979. "Port of Melbourne Environmental Study: Marine Study - Webb Dock". University of Melbourne.

Hooper, G.R., Miller, C.W., and Tranberg, C.H., 1985. "Dynamic Response of Wave Loaded Structures," *Preprints: 1985 Australasian Conf. Coastal and Ocean Eng.*, Christchurch, N.Z., Conf. Org. Comm., Christchurch, Vol. 2, pp. 171-181.

Hsu, J.R.C., 1990. "Short-Crested Waves," Chapter 3 In *Handbook of Coastal and Ocean Engineering*, (ed. J.B. Herbich), Gulf Publishing, Texas, Vol. 1, pp. 95-174.

Hsu, J.R.C., and Silvester, R., 1989. " Model Test Results of Scour Along Breakwaters," *Jour. Waterway, Port, Coastal and Ocean Eng.*, Am. Soc. Civ. Eng.,Vol. 115, No. 1, pp. 66-85.

Hsu, J.R.C., and Silvester, R., 1990. "Accretion Behind Single Offshore Breakwater," *Journ. Waterway, Port, Coastal and Ocean Eng.*, Am. Soc. Civ. Eng., Vol. 116, pp. 362-380.

Hubbert, G.D., 1993. "Oil Spill Trajectory Modelling with a Fully Three-Dimensional Ocean Model," *Preprints: 11th Australasian Conf. Coastal and Ocean Eng.*, Townsville, Qld., Inst. Eng. Aust., Nat. Conf. Publ. No. 93/4, Vol. 2, pp. 677-682.

Hughes, E.P.C., 1957. "The Investigation and Design for Portland Harbour, Victoria," *Jour. Inst. Eng. Aust.*, Vol. 29, pp. 55-68.

Hutton, I.M., 1987. "Dawesville Channel - Ocean Entrance". *Preprints: 8th Australasian Conf. Coastal and Ocean Eng.,* Launceston, Tas., Inst. Eng. Aust., Nat. Conf. Publ. No. 87/17, pp. 330-334.

Jackson, L.A., 1989. "Implementation and Monitoring of 1.5 x $10^6 m^3$ Nearshore Nourishment at Kirra/Bilinga Gold Coast," *Preprints: 9th Australasian Conf. Coastal and Ocean Eng.,* Adelaide, S.A., Inst. Eng. Aust., Nat. Conf. Publ. No. 89/20, pp. 188-192.

Jackson, L.A., and Tomlinson, R.B., 1990. "Nearshore Nourishment Implementation, Monitoring and Model Studies of 1.5 Mm^3 at Kirra Beach," *Proc. 22nd Int. Conf. Coastal Eng.*, Delft, The Netherlands, Am. Soc. Civ. Eng., New York, Vol. 3, pp. 2241-2254.

Jennings, D.S. and Barraclough, J.R., 1978. "Port Phillip - The Practicalities of Coast Protection," *Preprints: 4th Australian Conf. Coastal and Ocean Eng.*, Adelaide, S.A., Inst. Eng. Aust., Nat. Conf. Publ. No 78/11, pp. 13-18.

Kerle, H.W., 1886-87. "Ocean Jetty for the North Illawarra Coal Mining Company, North Bulli," *Min. Proc. Eng. Assoc. NSW.,* Vol. 2, pp. 108-118, plus Plates V, VI and VII.

Kindler, J.E., and O'Connor, C., 1951. "Erosion of Beaches in the Town of South Coast," Unpublished report prepared for the Chief Engineer, Co-ordinator General's Department, Queensland, 57p.

King, C.W., 1910-11. "Sand Movements at Newcastle Entrance, New South Wales," *Min. Proc. Inst. Civ. Eng.*, Vol. 184, pp. 149-156.

Lang, T.A., 1947. "Report on Tidal Model of River Tamar Tasmania," Marine Board of Launceston, Tas., Unpublished report.

Le Page, J.S.H., 1986. "Building a State - The Story of the Public Works Department of Western Australia 1829-1985," Water Authority of W.A., Leederville, W.A., 670pp. ISBN 0 7244 6862 5.

Lloyd, R.J., 1980. "Noosa Beach Restoration Scheme," *Proc. 17th Conf. Coastal Eng.*, Sydney, N.S.W., Am. Soc. Civ. Eng., New York, Vol. 2, pp. 1619-1635.

Lord, D., Nielsen, A.F., and Moratti, M., 1993. "Coastal Zone Management: Lennox Head, New South Wales," *Preprints: 11th Australasian Conf. Coastal and Ocean Eng.*, Townsville, Qld., Inst. Eng. Aust., Nat. Conf. Publ. No. 93/4, Vol. 2, pp. 547-552.

Macdonald, H.V., and Patterson, D.C., 1984. "Beach Response to Coastal Works, Gold Coast, Australia," *Proc. 19th Int. Coastal Eng. Conf.*, Houston, Texas, Am. Soc. Civ. Eng., New York, Vol. 2, pp. 1522-1538.

Mackenzie, A.D., 1939. "Coastal Erosion in Victoria," *Jour. Inst. Eng. Aust.*, Vol. 11, pp. 229-236.

McGrath, B.L., 1968. "Erosion of Gold Coast Beaches, 1967 (with discussion)," *Jour. Inst. Eng. Aust.*, Vol. 40, pp. 155-166. (Discussion Vol. 41, 1969, pp. 8-10, 72, 182-3).

McGrath, B.L., and Robinson, D.A., 1972. "Erosion and Accretion of Gold Coast, Queensland, Beaches," *Inst. Eng. Aust. Civ. Eng. Trans.*, Vol., CE14, pp. 158-163.

McKay, G.R., Fison, E.C., and Tranberg, C.R., 1960. "Mourilyan Harbour Development," *Jour. Inst. Eng. Aust.,* Vol. 32, pp. 313-333.

Marsden, M.A.H., ed., 1979. "Sedimentary Processes and Water Circulation in Western Port, Victoria, Australia", *Mar. Geol.*, Vol. 30, pp. 1-173.

Massel, S.R., 1996. "On the Largest Wave Height in Water of Constant Depth," *Ocean Eng.*, In press.

Mehaffey, M.W., 1919-20. "New Lighthouses in Queensland," *Min. Proc. Inst. Civ. Eng.*, Vol. 210, pp. 383-393.

Miller, C.W., and Tranberg, C.H., 1986. "Economic Construction of Offshore Berths," *Macdonald Wagner Tech. Quarterly*, No. 13.

Mitchell, H.L., and Silvester, R., 1981. "Tides Around the Australian Coastlines," *Inst. Eng. Aust. Civ. Eng. Trans.*, Vol. CE23, pp. 289-296.

Murray, R.J., Brodie, R.P, Jackson, L.A., Porter, M., Robinson, D.A., Lawson, S., and Perry, M.R., 1995. "Tweed River Entrance Sand Bypassing Project: Principles and Progress," *Preprints: 12th Australasian Conf. Coastal and Ocean Eng.*, Melbourne, Vic., Inst. Eng. Aust., pp. 7-12.

Murray, R.J., Robinson, D.A., and Soward, C.L., 1993. "Monitoring of the Southern Gold Coast Beach Nourishment Project,"*Preprints: 11th Australasian Conf. Coastal and Ocean Eng.*, Townsville, Qld., Inst. Eng. Aust., Nat. Conf. Publ. No. 93/4, Vol. 1, pp. 119-125.

Murray, R.J., Robinson, D.A., and Soward, C.L., 1994. "Southern Gold Coast Beach Nourishment Project: Implementation, Results, Effectiveness," *Terra et Aqua*, No. 56, pp. 12-23.

Nelson, R.C., 1972. "Design Wave Characteristics for Tropical Cyclones in the Australian Region," *Inst. Eng. Aust. Civ. Eng. Trans.*, Vol. CE14, pp. 37-41.

Nelson, R.C., 1982. "Coastal Engineering - An Overview," Paper presented to the Civil/Structural Branch, Inst. Eng. Aust., Canberra Division, 14 Sept., 1982.

Nelson, R.C., 1985. "Wave Heights in Depth-Limited Conditions," *Aust. Civ. Eng. Trans.*, Vol. CE27, pp. 210-215.

Nelson, R.C., 1987. "Design Wave Heights on Very Mild Slopes - An Experimental Study," *Aust. Civ. Eng. Trans.*, Vol. CE29, pp. 157-161.

Nelson, R.C., 1993. "Derivation of Reef Top Currents from Wave Observations," *Preprints: 11th Australasian Conf. Coastal and Ocean Eng.*, Townsville, Qld., Inst. Eng. Aust., Nat. Conf. Publ. No 93/4, Vol. 2, pp. 473-478.

Nelson, R.C., 1994. "Depth Limited Design Wave Heights in Very Flat Regions," *Coastal Eng.*, Vol. 23, pp. 43-59.

Nelson, R.C., and Lesleighter, E.J., 1985. "Breaker Height Attenuation Over Platform Coral Reefs," *Preprints: 1985 Australasian Conf. Coastal and Ocean Eng.*, Christchurch, N.Z., Conf. Org. Comm., Christchurch, Vol. 2, pp. 9-16.

Nielsen, A.F., 1994. "Sediment Transport on the Inner Continental Shelf, Sydney, Australia," *Aust. Civ. Eng. Trans.*, Vol. CE36, pp. 309-317.

Nielsen, P., 1988. "Wave Set-up: A Field Study," *Jour. Geophys. Res.*, Vol. 93, No. C12, pp. 15643-15651.

Nielsen, P., 1991. "Measurements of Wave Set-up and the Coastal Water Table," *Aust. Civ. Eng. Trans.*, Vol. CE33, pp. 23-27.

NSWG, 1990. "Coastline Management Manual," New South Wales Government. ISBN 0730575063.

NSWG, 1994. "Estuary Management Manual," New South Wales Government. ISBN 073 1009339.

Palmer, C.S.R., 1910-11. Fremantle Harbour-works, Western Australia, *Min. Proc. Inst. Civ. Eng.* Vol. 184, pp. 157-182. Discussion pp.183-283 plus Plate 7.

Pattearson, C.C., and Patterson, D.C., 1983. "Gold Coast Longshore Transport," *Sixth Australian Conf. Coastal and Ocean Eng.,* Surfers Paradise, Qld., Inst. Eng. Aust., Nat. Conf. Publ. No. 83/6, pp. 251-256. ISBN 0 85825 197 3.

Patterson, D.C., and Ford, L.R., 1980. "Capricorn Coast Beaches," *Proc. 17th Coastal Eng. Conf.*, Sydney, Australia, Am. Soc. Civ. Eng., New York, Vol. 2, pp. 1649-1668.

Paul, M.J., 1981. "Denison Fishing Boat Harbour," *Proc. 5th Australian Conf. Coastal and Ocean Eng.,* Perth, W.A., *Inst. Eng. Aust.,* Nat. Conf. Publ. No. 81/16, pp. 39-43.

Peake, A., 1920-21. "The Southern and Western Suburbs Ocean Outfall Sewer, Sydney, New South Wales", *Min. Proc. Inst. Civ. Eng.,* Vol. 212, pp.109-130. Discussion pp. 143-166 plus Plate 4.

Peirson, W.L., Cathers, B., and King, I.P., 1993. "Numerical Modelling of Deepwater Outfall Plumes at Sydney," *Preprints: 11th Australasian Conf. Coastal and Ocean Eng.*, Townsville, Qld., Inst. Eng. Aust., Nat. Conf. Publ. No. 93/4 pp. 91-98.

Philip, N.A., 1991. "Sydney Deepwater Outfalls Environmental Monitoring Programme - An Overview," *Preprints: 10th Australasian Conf. Coastal and Ocean Eng.*, Auckland, N.Z., DSIR Marine and Freshwater, Water Quality Centre, Publ. No. 21, pp. 317-321.

Pullan, R., ed., 1983. "Guide to the Australian Coast," Readers Digest, Surrey Hills, N.S.W., 479p.

Robinson, D.A., and Jones C.M., 1977. "Queensland Volunteer Coastal Observation Programme - Engineering (COPE)", *Preprints: 3rd Australian Conf. Coastal and Ocean Eng.,* Melbourne, Vic., Inst. Eng. Aust., Nat. Conf. Publ. No. 77/2, pp. 88-95.

Robinson, D.A., and Patterson, D.C., 1975. "The Kirra Point Groyne - A Case History," *2nd Australian Conf. Coastal and Ocean Eng.,* Gold Coast, Qld., Inst. Eng. Aust., Nat. Conf. Publ. No. 75/2, pp. 46-52. ISBN 0 85825 048 9.

Roy, P.S, and Crawford, E.A., 1977. "Significance of Sediment Distribution in Major Coastal Rivers, Northern N.S.W.," *Preprints: 3rd Aust. Conf. Coastal and Ocean Eng.*, Melbourne, Vic, Inst. Eng. Aust., Nat. Conf. Publ. No. 77/2, pp. 177-184.

Scholer, H.A., 1958. "Tides in Rivers and Coastal Inlets," *Jour. Inst. Eng. Aust.,* Vol. 30, pp.125-136.

Shapiro, M.A., ed., 1975. "Westernport Bay Environmental Study: Report for the Period 1973-1974," Ministry for Conservation., Victoria, 654p.

Shellshear, W., 1884. "On the Removal of Bars from the Entrances to our Rivers," *Jour. and Proc. Roy. Soc. NSW,* Vol. 18, pp. 25-36. Reprinted in Coltheart, L., and Fraser, D.J., 1987. Landmarks in Public Works: Engineers and their Works in New South Wales 1884-1914, Hale & Iremonger, Sydney.

Silvester, R., 1956. "A Model Study of Littoral Drift at Bunbury Harbour, W.A.," *Jour. Inst. Eng. Aust.*, Vol. 28, pp. 219-230.

Silvester, R., 1959. "Engineering Aspects of Coastal Sediment Movement," *Proc. Am. Soc. Civ. Eng., Jour. Waterways and Harbours Div.*, Vol. 85, No. WW3, pp. 11-39.

Silvester, R., 1960. "Stabilization of Sedimentary Coastlines," *Nature*, Vol. 188, No. 4749, pp. 467-469.

Silvester, R., 1961. "Beach Erosion at Cottesloe, W.A.," *Inst. Eng. Aust. Civ. Eng. Trans.*, Vol. CE3, pp. 27-33.

Silvester, R., 1962. "Sediment Movement Around the Coastlines of the World," *Conf. on Civ. Eng. Problems Overseas*, London, June 1962, Inst. Civ. Eng., London, pp. 289-304.

Silvester, R., 1965. "Coastal Sediment Movement - Some Fundamental Problems with Discussion of Research Support," *Jour. Inst. Eng. Aust.*, Vol. 37, pp. 311-323.

Silvester, R., 1966. "Sediment Transport and Accretion Around the Coastlines of Japan," *Proc. 10th Conf. Coastal Eng.*, Tokyo, Japan, Chap 29., pp. 469-487.

Silvester, R., 1970. "Development of Crenulate Shaped Bays to Equilibrium," *Proc. Am. Soc. Civ. Eng. Jour. Waterways and Harbors Div.*, Vol. 96, No. WW2, pp. 275-287.

Silvester, R., 1974. "Coastal Engineering," Elsevier, Amsterdam, 2 volumes, 457pp. and 338 pp.

Silvester, R., and Hsu, J.R.C., 1993. "Coastal Stabilization: Innovative Concepts," Prentice-Hall Inc. Englewood Cliffs, New Jersey, 578 pp.

Silvester, R., and Searle, M., 1981. "Headland Control to Prevent Cooling Water Sand Incursion," *Proc. 5th Aust. Conf. Coastal and Ocean Eng.*, Perth, W.A., Inst. Eng. Aust., Nat. Conf. Publ. No. 81/16, pp. 135-138.

Silvester, R., Tsuchiya, Y., and Shibano, Y., 1980. "Zeta Bays, Pocket Beaches and Headland Control," *Proc. 17th Int. Conf. Coastal Eng.*, Sydney, N.S.W., Am. Soc. Civ. Eng., New York, Vol. 2, pp. 1306-1319.

Smith, A.W., 1977. "The Surfers Paradise Beach Replenishment Project - A Case Study," *Preprints: 3rd Australian Conf. Coastal and Ocean Eng.*, Melbourne, Vic., Inst. Eng. Aust., Nat. Conf. Publ. No. 77/2, pp .190-195.

Sobey, R.J., 1993. "Quantifying Coastal and Ocean Processes," *Preprints: 11th Australasian Conf. Coastal and Ocean Eng.*, Townsville, Qld., Inst. Eng. Aust., Nat. Conf. Publ. No. 93/4, Vol. 1, pp. 1-10.

Sobey, R.J., and Young, I.R., 1986. "Hurricane Wind Waves - A Discrete Spectral Model," *Jour. Waterway, Port, Coastal and Ocean Eng.*, Am. Soc. Civ. Eng., Vol. 112, pp. 370-389.

Sobey, R.J., Harper, B.A., and Mitchell, G.M., 1982. "Numerical Modelling of Tropical Cyclone Storm Surge," *Inst. Eng. Aust. Civ. Eng. Trans.*, Vol. 24, pp. 151-161.

Sobey, R.J., Harper, B.A., and Stark, K.P., 1977. "Numerical Simulation of Tropical Cyclone Storm Surge," James Cook University, N. Qld., Department Civil & Systems Eng., Res. Bul. No. CS14.

Tauman, M., 1978. "The Chief - C.Y. O'Connor," University Western Australia Press, Nedlands, W.A., 290 p. ISBN 0 85564 123 1.

Tolley, J.C., 1978. "South Coast Story - The History of Goolwa, Pt Elliot, Middleton and the Murray Mouth," Roy. Geograph. Soc. S.A., Reprint.

Tomlinson, R.B., and Foster, D.N., 1987. Sand Bypassing at the Tweed River Entrance," *Preprints: 8th Australasian Conf. Coastal and Ocean Eng.*, Launceston, Tas., Inst. Eng. Aust., Nat. Conf. Publ. No. 87/17, pp. 85-89.

Tronson, K.C.S., and Noye, B.J., 1974. "Propagation of Tides into the South Australian Gulf System," In Lindley, D., and Sutherland, A.J. (eds.), *Proc. Fifth Australasian Conf. on Hydraulics and Fluid Mechanics*, Christchurch, N.Z., University of Canterbury, Christchurch, N.Z., Vol. 2, pp. 258-266.

WACMRC, 1995. "Final Report of the Review of Coastal Management in Western Australia", Western Australian Coastal Management Review Committee, Report prepared for Minister of Planning.

Walker, S.J., and Hunter, J.R., 1995. "Modelling Effluent Dispersion in the Derwent Estuary," *Preprints: 12th Australasian Conf. Coastal and Ocean Eng.*, Melbourne, Vic., Inst. Eng. Aust., pp. 145-149. ISBN 0 85825 639 8.

Wallace, J.M., 1977. "The Control of Wave Action by Configuration Dredging at the Entrance to Botany Bay, Sydney, Australia," *Proc. 24th Int. Nav. Congress,* Leningrad, USSR, Sec. 2, Subj. 2, pp. 5-14.

Wallis, I.G., 1983. "Ocean Currents Offshore from Sydney," *Preprints: 6th Australian Conf. Coastal and Ocean Eng.*, Gold Coast, Qld., Inst. Eng. Aust., Nat. Conf. Publ. No. 83/6, pp. 206-210.

Wescott, G., 1995. "Victoria's Major Review of Coastal Policy: The Establishment of a Co-ordinating Coastal Council," *Envtal. and Plann. Law Journal*, Vol. 12, pp. 288-295.

Wilkinson, D.L., 1984. "Purging of Saline Wedges From Ocean Outfalls," *Jour. Hydr. Eng.*, Am. Soc. Civ. Eng., Vol. 110, pp. 1815-1829.

Wilkinson, D.L., 1985. "Seawater Circulation in Sewage Outfall Tunnels," *Jour. Hydr. Eng.*, Am. Soc. Civ. Eng., Vol. 111, pp. 846-858.

Williams, B.J., and Hinwood, J.B., 1974. "On the Development and Calibration of a Large Numerical Model," In Lindley, D., and Sutherland, A.J. (eds.), *Proc. Fifth Australasian Conf. on Hydraulics and Fluid Mechanics*, Christchurch, N.Z., University of Canterbury, Christchurch, N.Z., Vol. 2, pp. 244-249.

Witt, C.L., and Hill, P.C.,1987. "Gold Coast Seaway: An Overview of the Project's Design and Construction and its Subsequent Performance," *Preprints: 8th Australasian Conf. Coastal and Ocean Eng.*, Launceston, Tas., Inst. Eng. Aust., Nat. Conf. Publ. No. 87/17, pp. 259-266.

Wynne, A.A., 1984. "Adelaide Coast Protection Strategy Review," A report prepared for the Coast Protection Board of South Australia, Coastal Management Branch, Department of Environment and Planning SA, March 1984.

Young, I.R., 1988. "A Shallow Water Spectral Wave Model," *Jour. Geophys. Res.*, Vol. 93, No. C5, pp. 5113-5129.

Young, I.R., 1989. "Wave Transformation Over Coral Reefs," *Jour. Geophys. Res.*, Vol. 94, No. C7, pp. 9779-9789.

Young, I.R., and Verhagen, L.A., 1995. "The Evolution of Wind-generated Waves in Water of Finite Depth," *Preprints: 12th Australasian Conf. Coastal and Ocean Eng.*, Melbourne, Vic, Inst. Eng Aust., pp. 241-245.

APPENDIX A

Location Maps

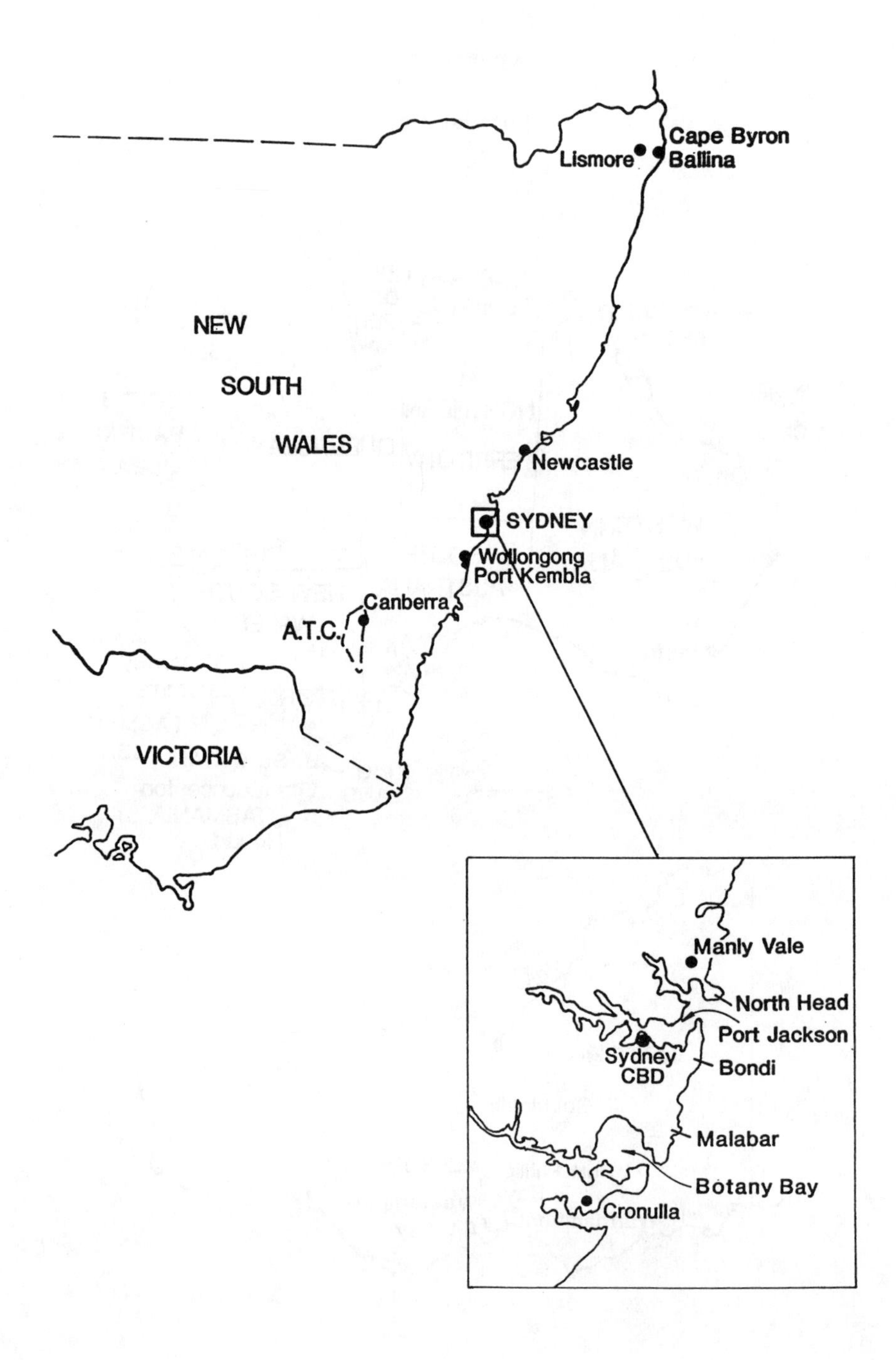
Cape Byron
Lismore
Ballina
NEW
SOUTH
WALES
Newcastle
SYDNEY
Wollongong
Port Kembla
Canberra
A.T.C.
VICTORIA
Manly Vale
North Head
Port Jackson
Sydney
CBD
Bondi
Malabar
Botany Bay
Cronulla

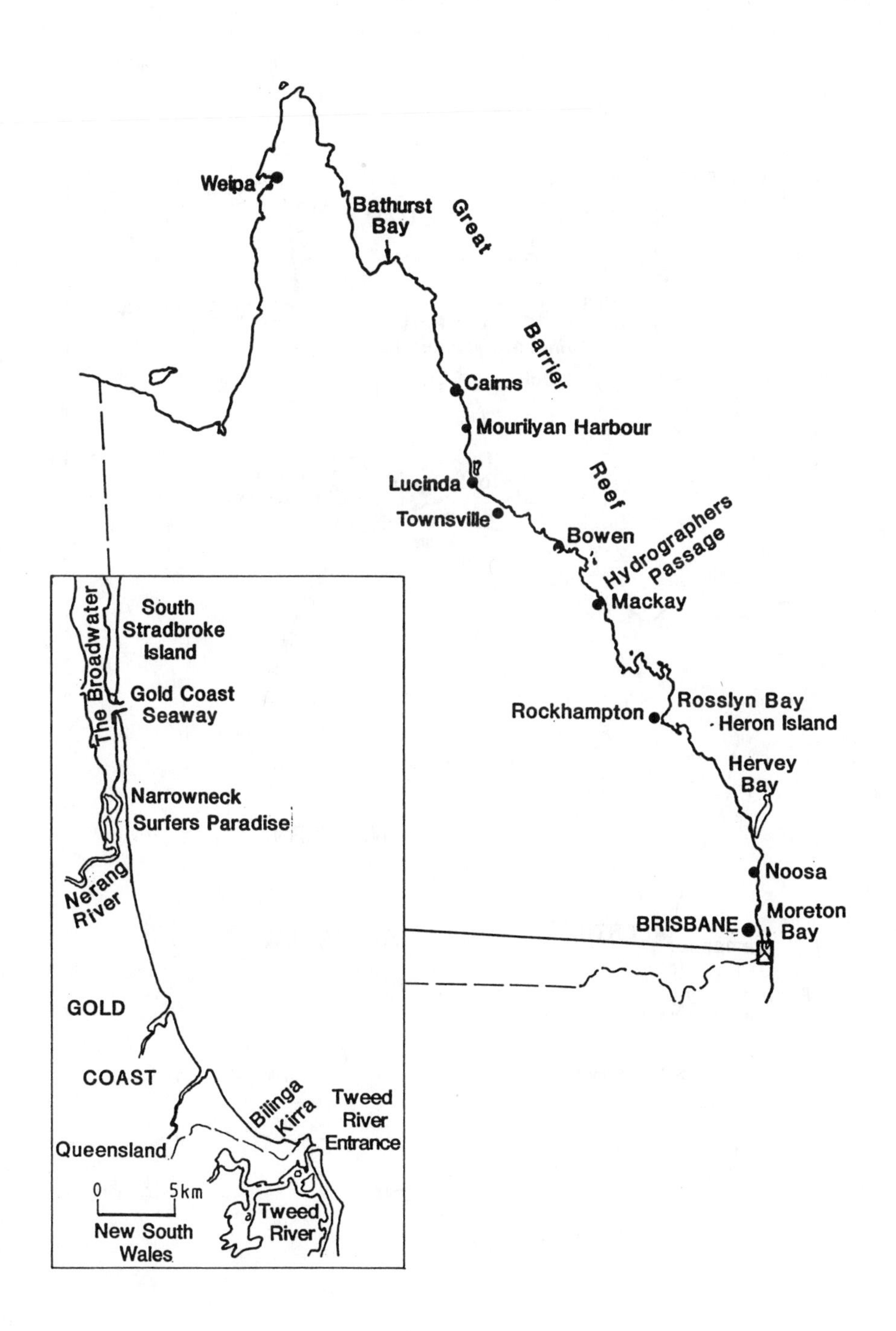

Weipa
Bathurst Bay
Great Barrier Reef
Cairns
Mourilyan Harbour
Lucinda
Townsville
Bowen
Hydrographers Passage
Mackay
Rockhampton
Rosslyn Bay
Heron Island
Hervey Bay
Noosa
BRISBANE
Moreton Bay
The Broadwater
South Stradbroke Island
Gold Coast Seaway
Narrowneck
Surfers Paradise
Nerang River
GOLD COAST
Bilinga
Kirra
Tweed River Entrance
Queensland
0 5km
New South Wales
Tweed River

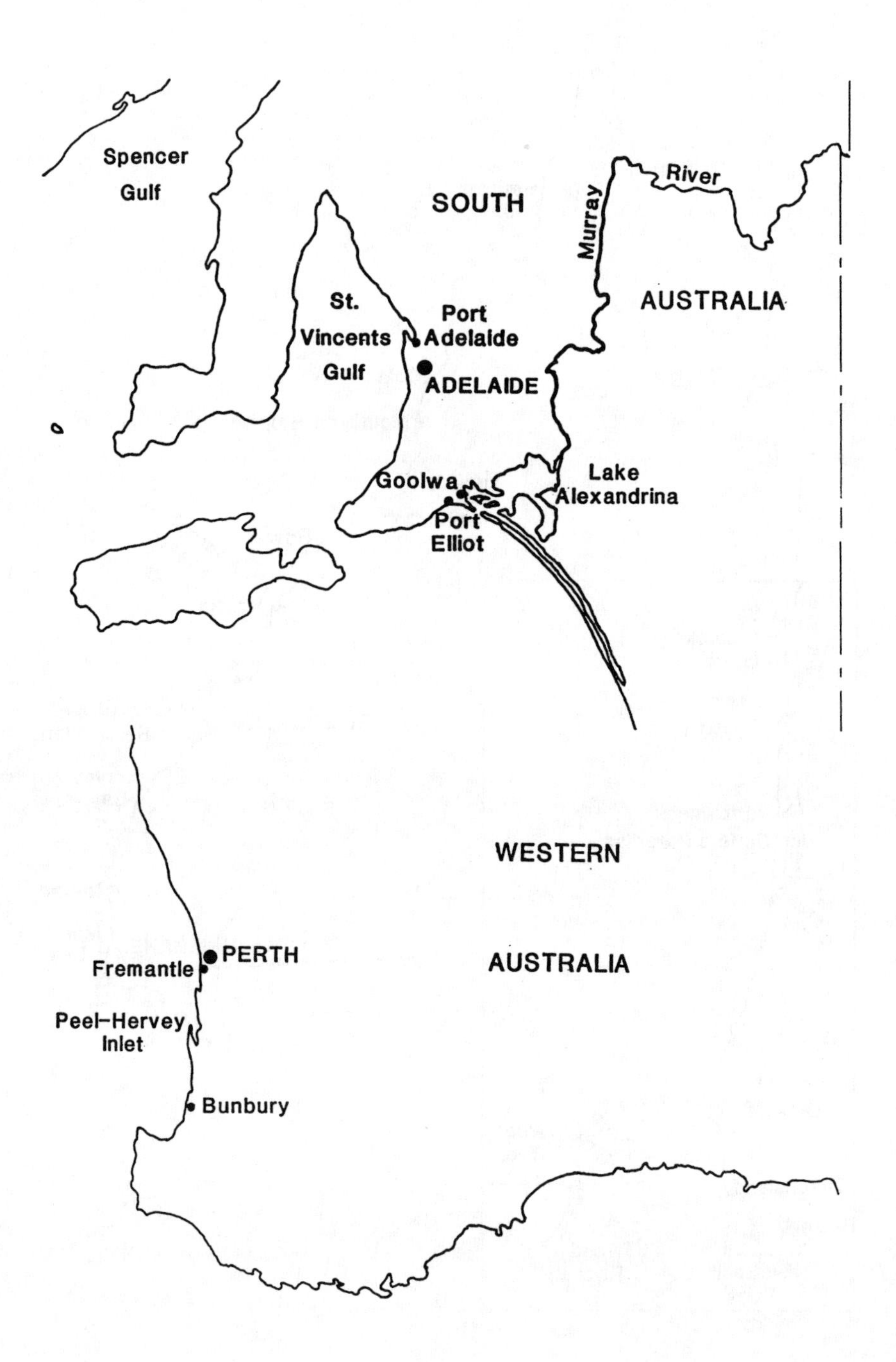
Spencer
Gulf
SOUTH
River
Murray
AUSTRALIA
St.
Vincents
Gulf
Port
Adelaide
ADELAIDE
Goolwa
Lake
Alexandrina
Port
Elliot
WESTERN
AUSTRALIA
PERTH
Fremantle
Peel-Hervey
Inlet
Bunbury

APPENDIX B

Significant Publications by National Committee on Coastal and Ocean Engineering and Related Groups

Silvester, R, 1971. "Australia's Interest in Ocean Engineering," IEA, *Civ. Eng. Trans.*, Vol. CE 13, pp. 71-75.

NCCOE, 1973. "Coastal and Ocean Engineering in Australia - A Survey by the National Committee on Coastal and Ocean Engineering of the Institution of Engineers, Australia," *IEA*, Sydney, NSW, 52p. ISBN 085825 046 2.

NCCOE, 1974. "The National Estate," *Jour. IEA*, Vol. 46, p. 19.

NCCOE, 1974-75. "Proposal for National Inquiry into Coastal and Ocean Engineering," *Jour. IEA*, Vol. 46, pp. 17-18 and Vol. 47, p. 32.

NCCOE, 1976. "The Danger of Cyclone Storm Surges in the Tropical Cyclone Areas of Australia," *Jour. IEA,* Vol. 48, p. 33.

Hinwood, J.B., 1980. "Coastal and Ocean Engineering Laboratory Facilities in Australia," Monash University, Department of Mechanical Engineering, MMER Rep. No. 18, Prepared for NCCOE, IEA, Barton, ACT, 58p.

Hinwood, J.B., 1982. "Coastal and Ocean Engineering Laboratory Facilities. Classification, Australia's Needs and Present Facilities," IEA, *Civ. Eng. Trans.*, Vol. CE 24, pp. 97-106.

NCCOE, 1982. "Lecture Courses in Coastal and Ocean Engineering at Australian Tertiary Institutions," IEA, NCCOE, 1st ed., Sept. 1982, 19p.

AMSTAC, 1983. "National Needs and Priorities for Coastal and Ocean Engineering Hydraulic Laboratory Facilities," Report of AMSTAC Working Group on Coastal and Ocean Engineering, 193p.

Foster, D., and Geary, M., 1983. "Coastal Protection," IEA, *Civil College Board Tech. Rep.*

Sand Dune Management in Australia 1983," Papers presented at 5th Australian Conference on Coastal and Ocean Engineering, Perth, W.A., 25-27 Nov. 1981, IEA, Barton, ACT.

IEA, 1984. "Engineering Offshore - Engineering Implications of an Australian 200 Mile Exclusive Economic Zone," IEA, Barton, ACT, 157p. ISBN 085825 235 X.

Hinwood, J., 1985. "Planning and Design of Marinas," IEA, *Civil College Board Tech. Rep.*

DOS, 1985. "1985 Directory of Australian Coastal and Ocean Engineering Hydraulics Laboratory Capability," Department of Science, Canberra.

Griffin, A., and Tooker, M., 1986. "The Professional Engineer's Role in Coastal Zone Management," IEA, *Civil College Board Tech. Rep.*

NCCOE, 1987. "Investigation and Research Projects in Coastal Engineering Australia 1987," NCCOE, IEA, Barton, ACT, 24p.

NCCOE, 1988. "Engineers and the Greenhouse Effect," Engrs. Aust., Vol. 60, No. 10, pp. 54-55.

Bea, R.G., 1990. "Reliability Based Design Criteria for Coastal and Ocean Structures," NCCOE, IEA, Barton, ACT, 219p. ISBN 085825 518 9.

NCCOE, 1991. "Guidelines for Responding to the Effects of Climatic Change in Coastal Engineering Design," IEA, Barton, ACT. ISBN 85825 564 2.

NCCOE, 1992. "Australian State's Coastal Flooding and Erosion Policies," IEA, Barton, ACT.

NCCOE, 1993. "At What Price Data?," IEA, Barton, ACT. ISBN 85825 596 0.

NCCOE, 1994. "Coastal Zone Management Research Needs," Report on the Outcomes from a Priority Setting Workshop for the Department of Environment, Sport, and Territories convened by IEA, NCCOE, 11-12 Oct. 1994, Sydney, NSW.

HISTORY OF COASTAL ENGINEERING IN CANADA

J. William Kamphuis[1], M. ASCE.

ABSTRACT: The development of *Coastal Engineering* in Canada is closely linked to Canadian government projects related to trade, transportation and fisheries. The National Research Council of Canada, has played a leading role in the development of coastal research through its own work, through the establishment of study committees and through the provision of funding. Canada's coastal engineering history is also shown to be closely related to physical facilities, initially built to study specific projects for the government of Canada. Once the studies were completed, these facilities became centres of coastal research. Along the Great Lakes, periods of increased coastal engineering activity can be related to periodic high water levels. Finally, a discipline of *Coastal Science and Engineering* has developed in Canada, growing from hard Coastal Engineering (structural designs related to transportation and power generation) in the 1950s to include physical sciences such as oceanography and geomorphology in the 1960s and '70s, coastal zone management in the 1980s and biological sciences in the 1990s.

INTRODUCTION

Canada has the longest coastline of any nation in the world. It spans four very different coastal environments, the Atlantic, Pacific and Arctic Oceans, and the Great Lakes. In addition, there are the shores of the Gulf of St. Lawrence, the St. Lawrence River and the Mackenzie River, two of the world's longest rivers, and large Lakes, such as Great Bear, Great Slave, Athabasca and Lake Winnipeg (Figure 1).

Canada is also dominated meteorologically by the Polar Front, where the cold arctic air and the warmer air from the mid-latitudes meet to produce a relentless parade of depression storms, which ensure almost constant and sometimes violent wave action along the Pacific, Atlantic and Great Lakes shores. The West Coast (Pacific) and theEast Coast (Atlantic) are both subject to large swell. At the East Coast the warm Gulf Stream meets the cold Arctic waters to produce impenetrable fog and periodically a hurricane sweeps through. That coast

[1] Department of Civil Engineering, Queen's University, Kingston, Ontario, Canada, K7L 3N6

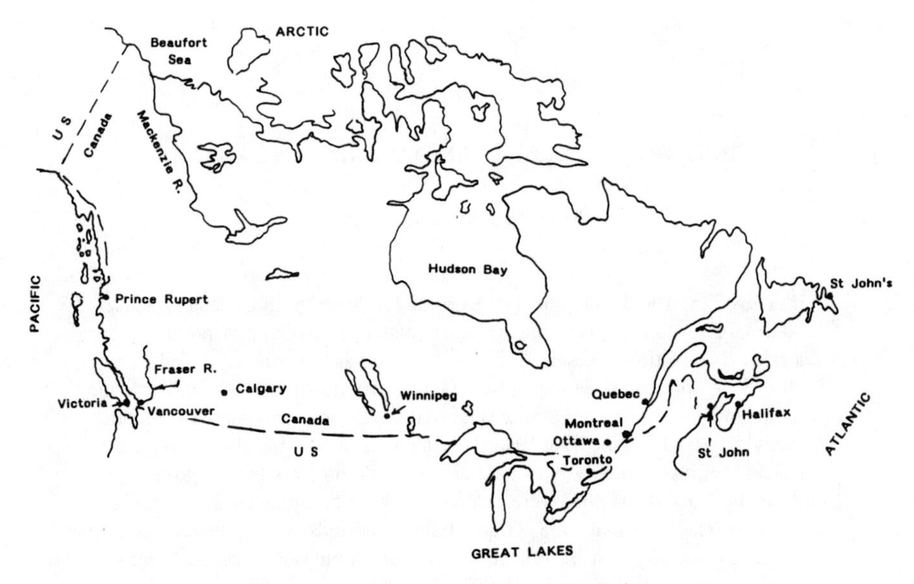

Figure 1 Location Map - Canada

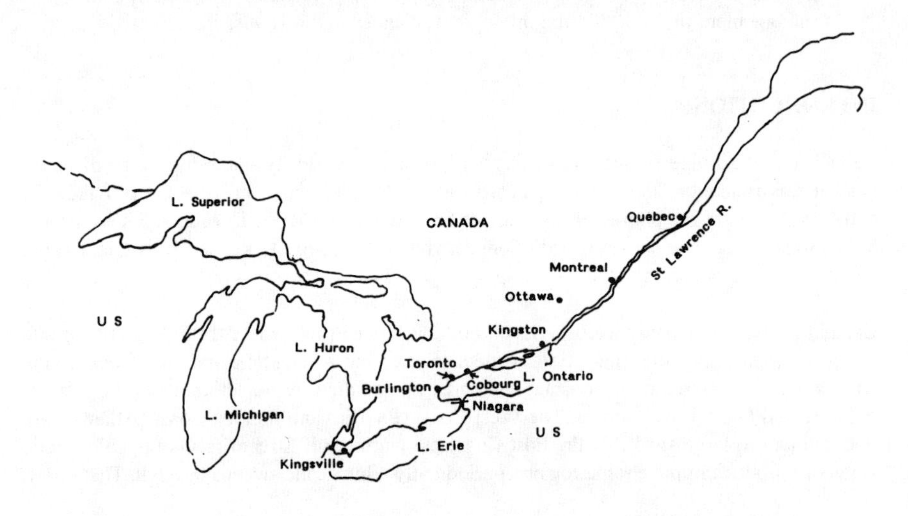

Figure 2 Location Map - Great Lakes - St Lawrence

also experiences some of the highest tides in the world and all Canadian shores are subjected to ice action for parts of the year. These are indeed some of the world's most inhospitable shores.

The East and West Coasts of Canada host(ed) some of the world's most prolific fish habitat. Thus it is little wonder that early *Coastal Engineering* with respect to these coasts primarily consisted of providing harbours and refuge for fishing vessels. Some large trading ports also sprang up at the natural harbour sites. Notable are Halifax, St John, St John's, Quebec, Montreal, Vancouver and Prince Rupert.

The Great Lakes-St Lawrence River system (Figure 2) provides ready access to the centre of the continent, the industrial heartland of both Canada and the United States. Shipping along this waterway has been brisk since the original "coureurs de bois" moved beaver pelts from the hinterland to the Port of Hochelaga (Montreal). Thus the early coastal engineering focus along this convenient 3800 km long trade route was the provision of port facilities to be used for trade and travel. Ports were established every 30-50 kms along the lower Great Lakes and St Lawrence River. Canada's sparse population is concentrated along this Great Lakes - St Lawrence corridor and therefore it later was the first area to become concerned with provision of recreational beaches and marinas, coastal flooding, shore erosion, coastal management and eventually environmental pollution and ecosystem design.

The large Arctic shoreline was never considered to be important, except that the Beaufort Sea and the Mackenzie Delta enjoyed a brief period of attention. This was the result of the world oil crisis of 1973, which made oil exploration in the far north of Canada economically viable.

The early years were mainly concerned with hard Coastal Engineering (design and construction of port and docking facilities). Later developments required a more thorough and in-depth understanding of scientific principles related to the coast. Thus, research into the physical processes, involving, oceanography, geology and geomorphology became more important to the profession. It was only natural that these physical sciences became integrally linked with Coastal Engineering in the 1960s and '70s. In the 1970s and '80s it became obvious that proper management of the coastal resource was needed and thus planning, geography and law became major linkages. Environmental and biological considerations became part of the discipline in the 1980s and '90s. In the remainder of this paper, the term *Coastal Science and Engineering* will include these related fields.

Any treatment of history is necessarily influenced by the purview of the writer. He has attempted to network with as many people as possible, but the final description of events is no doubt biased toward what he knows best. In this paper, the historical aspects, rather than research and project development have been stressed. Therefore only those names specifically needed to trace the historical development of the discipline in Canada are mentioned. Apologies are offered to all those diligent and innovative engineers and scientists who have contributed so much individually to specific research programmes and engineering projects, but who are not mentioned here.

THE EARLY YEARS

In the early years, Public Works Canada was the primary agency responsible for harbours and related structures. These consisted mainly of timber crib breakwaters, which made use of two plentiful resources, timber and rock. Many of these were built with a concrete superstructure, which was often sloped to provide less resistance to ice forces.

Along the Great Lakes - St Lawrence trade route, the maintenance of ports for trade and travel was of major concern. One such project, maintaining access to Toronto harbour, involved many of the day's leading experts, including Sir Sandford Fleming (he who invented the time zones and was a long-time chancellor of Queen's University), who studied the geomorphology of the islands offshore of Toronto and developed strategies to keep an entrance open into the harbour (Langevin, 1881).

The National Research Council of Canada (NRCC) was established in Ottawa in 1916. According to Middleton (1984), NRCC operated, among others, a large ship towing tank and a glass-sided flume in the 1930s. Much of the initial research concerned design of floats for aircraft and investigation of navigation locks, stoplogs, etc. for the canals along the St Lawrence waterway. In 1935, NRCC established a Committee on Hydraulic Research, consisting of members of several government departments to consider:

> "the requirements of the several departments of the Dominion Government and of industry in respect to the provision of laboratory facilities for research and testing in connection with stream flow, power development, harbour works, river control, canals and channels, hydraulic machinery and equipment"

The above quotation is indicative of the pattern of development of coastal science and engineering over the next half century. The impetus was usually work sponsored by the departments of the Government of Canada. NRCC was the major research arm of this government. NRCC was also the distributor of research funding, until the 1970s, when a new organization - the Natural Sciences and Engineering Research Council (NSERC) took over funding of science and engineering. Thus, by the 1940s, the Canadian government departments sponsored much of the coastal design and NRCC had assumed a leading role in coastal research and hydraulic modeling.

By 1945, the NRCC hydraulics laboratory and ship laboratory had been formed. Facilities included a ship towing tank and flumes, used for both ship models and hydraulic models. One of the first coastal studies in the new facilities was commissioned by Public Works Canada. The sanding up of the entrance of the small harbour at Kingsville on Lake Erie was investigated in a model with scales of 1:60 vertical and 1:240 horizontal. This model sported NRCC's first wave generator. The main conclusion was that none of the 40 different pier layouts tested gave satisfactory results! A very ambitious project was the study of the Fraser River and Estuary. Begun in 1946, this study lasted until 1962. The objectives were to reduce the dredging quantities and to design control dikes to regulate flow. This mobile bed tidal model was built to horizontal and vertical scales of 1:400 and 1:100. The initial study was completed in 1949.

After the Second World War considerable capital funding was applied to marine facilities in support of marine transportation and fishing. Numerous docks and wharves were built as part of the infrastructure required to deliver goods to remote inland communities and to provide protected harbours for large inshore fishing fleets. Most of the design and construction of these works was simply undertaken by the engineering and support staff of Public Works Canada. Harbours developed on the West Coast, for example, during the late 1940s and early 1950s were at Gibsons, Powell River, Campbell River and Queen Charlotte City.

1950s

In the late 1940s, it was decided that a much larger model of the Fraser Estuary was needed. An outdoor model (1:600 horizontal and 1:70 vertical) was constructed by NRCC on the campus of the University of British Columbia (UBC) in Vancouver in 1950. The model simulated tidal flow over 80 km and included a part of the Straits of Georgia and the Fraser River up to the head of tide. It was built and initially operated by George Ashe. Following him, Doug Baines, who later became professor at University of Toronto, was in charge of the model. Professor Ted Pretious of the Department of Civil Engineering at UBC was involved and Tom Blench, who later became professor at the University of Alberta, worked on the sediment transport aspects of the model.

In the 1950s, a much improved St. Lawrence Seaway was designed and built in co-operation with the United States. This multi-purpose hydro-electric and navigation project was probably the largest coastal-related undertaking in Canadian history. It involved the building of large control dams, power generation plants and deeper and wider canals and locks above Montreal. Construction took place between 1954 and 1959. Most of the hydraulic modeling for these projects and other routine hydraulic modeling required by the Government of Canada's coastal designs was performed at the hydraulics laboratory of NRCC in Ottawa. Thus Canadian government institutions confirmed their leadership role in the development of coastal engineering.

The early 1950s were a period of the record high water levels along the lower Great Lakes. This event instigated the first major investigation of Lake levels, shore erosion and inundation by the Province of Ontario (Ontario, 1953). This report is the first evidence that high water levels along the Great Lakes result in heightened concern. Ontario (1953) discusses shore protection, but also jurisdictional aspects and land use restrictions. At about the same time and for the same reason, Public Works Canada became very interested in reducing the dredging costs for their harbours, particularly in the lower Great Lakes. To investigate solutions to the erosion and sedimentation along Lakes Ontario and Erie, NRCC established the Associate Committee on Waves and Littoral Drift in 1958. The members of this committee were representatives from Public Works Canada, NRCC, the University of Toronto and Queen's University at Kingston. Both universities are located on Lake Ontario.

The Department of Civil Engineering at Queen's University had traditionally emphasized hydraulics during the late 1940s and the 1950s. Its department head (Professor Doug Ellis), later the Dean of Engineering was a well-known hydraulic engineer and the author of a much-used textbook on the subject. This strong hydraulics background was still evident in 1957 when Professor Arthur Brebner joined Queen's from the University of Aberdeen, where he had been involved in many, varied hydraulic model studies. In 1959, NRCC, through its Associate Committee on Waves and Littoral Drift decided to fund some field investigations and hydraulic model studies at Queen's University to solve some of the sedimentation and erosion problems along Lakes Ontario and Erie.

By the end of the 1950s, *Coastal Engineering* had become a distinct discipline in Canada.

1960s

These beginnings of the discipline of Coastal Engineering were further developed in the 1960s and the start of coastal engineering research is seen. The West Coast has large sections of deep, semi-protected inland waters and ideal protection for vessels can be provided by floating breakwaters. Early designs consisted of bundles of logs, but later concepts were much more sophisticated, such as the A-frame structure placed at Lund in 1964. On the UBC campus, the work on the Fraser Estuary model continued but other coastal research was carried out as well. Professor Pretious built a wave machine and began testing various structures, including floating breakwaters and Professor Michael Quick was hired in 1962. Coastal engineering consulting started on the West Coast with Swan Wooster (Frank Leighton and Jack Wood). The design of a long causeway over tidal flats to accommodate the ferry to Vancouver Island was one major project.

At Queen's University a state-of-the-art wave research flume was built in Ellis Hall and Professor Brebner's first Ph.D. student (Ian Collins) defined the transition from laminar to turbulent regime under waves there. Queen's also proceeded with a field investigation to relate wind, waves and sediment transport along the north shore of Lake Ontario. In 1960, a field survey and hydraulic model study were started at Queen's to reduce sedimentation of Cobourg harbour on Lake Ontario. This model study was conducted by Brebner, Bernard Le Méhauté, who had recently arrived from France, and Ian Collins and Milne Dick, who were graduate students. There was insufficient space in Ellis Hall to build such a large model and thus, in 1960, NRCC rented a property in Kingston from the Department of National Defence and Queen's University's first coastal laboratory was born in an empty navy drill hall. After the Cobourg harbour study was completed, this laboratory space and equipment became an invaluable resource for coastal research at Queen's. The first graduate student to use this facility for research was Bill Kamphuis. The research topics studied in this laboratory during its existence from 1960 to 1968 included wave breaking, longshore currents, sediment transport under waves and the use of log booms as floating breakwaters. Public Works Canada, in the mid 1960s, also sponsored Queen's to do a series of field investigations and hydraulic model studies of erosion of shorelines along the St Lawrence River and the St Lawrence Seaway by ship-generated waves.

In the early 1960s, NRCC worked on several harbour projects and Gerry Jarlan developed the perforated breakwater. It was found necessary to investigate navigation on the St Lawrence River, downstream of Montreal. The first study of extensive dredging proposals resulted in the construction of a large hydraulic model and the establishment of the LaSalle Hydraulics Laboratory in Montreal. In 1965 an expanded study was set up at NRCC. The hydraulics laboratory at NRCC had just completed the tidal study of the Fraser Estuary and this placed it in a unique position to embark on a large hydraulic model study of the St Lawrence. Major funding was obtained to expand the existing NRCC hydraulics laboratory in 1966 and a 250 m long tidal hydraulic model covering the St Lawrence from Montreal to the Gulf of St Lawrence was built. This large hydraulic model study turned out to be another crucial development in the history of Coastal Engineering in Canada. It resulted in major improvements in hydraulic modeling techniques and equipment, but equally important, it left, as a legacy, a very large indoor modeling facility, which has since been used for hydraulic modeling of all kinds of coastal research and engineering problems. This facility has permitted NRCC to become a world-class player in coastal research. It is interesting to note that coastal research at NRCC, UBC and Queen's is so closely related to facilities initially acquired or built to accommodate models to solve engineering problems for the government of Canada.

NRCC carried out a numerical study of the St Lawrence in parallel with the hydraulic model study. This one-dimensional numerical model of the St Lawrence was the beginning of what would become world-class numerical modeling expertise at NRCC. This combined effort also led to the development of hybrid modeling, a technique in which hydraulic and numerical models operate interactively. The hydraulic model study of the St Lawrence was carried out by Joe Ploeg and later by Bruce Pratte. Bill Kamphuis conducted the numerical model study. In 1968, Kamphuis left NRCC to become professor of Civil Engineering at Queen's University and in the following year Milne Dick left NRCC to become director of the new hydraulics facility at the Canada Centre for Inland Waters (CCIW), a government research laboratory at Burlington, Ontario. Ploeg stayed at NRCC and conducted a large, state-of-the-art wave climate study, involving comprehensive measurements of wind profiles and waves on the Great Lakes and the Gulf of St Lawrence. This study resulted in substantial development of new techniques, equipment and computer software for the acquisition and analysis of wave data. It was the first of many studies that involved Ed Funke. Ploeg later would become director of the NRCC hydraulics laboratory and eventually Vice-President (Engineering) of NRCC, a position from which he would retire in 1994.

In 1965, Professor Brebner became head of the Civil Engineering department at Queen's. It was a time when universities were expanding rapidly and he was able to create several new teaching and research positions. When the lease on the Queen's coastal research facility ran out, he was also able to obtain funding (mostly from NRCC) for a completely new research facility, the Queen's University Coastal Engineering Research Laboratory which was built in 1968. He attracted Kamphuis from NRCC into one of the new positions, specifically to concentrate on coastal engineering. An undergraduate 4th year course and two graduate courses in Coastal Engineering were started in 1968 and these continue until today.

Coastal Engineering began at Université Laval in Quebec in 1962 with Professor Hans-Werner Partenscky. He left Laval in 1965 to join Ecole Polytechnique in Montreal, where he stayed until 1972. One of his research efforts was to develop a tidal harmonic model of the St Lawrence. Coastal work was carried on at Laval by Professor Yvon Ouellet and at Ecole Polytechnique by Professor Claude Marche.

On the East Coast, the Canadian government established another research facility, the Bedford Institute of Oceanography (BIO), in Halifax in 1962. It performs applied research on management of freshwater and marine environments, hydrographic surveys, map production and response to marine environmental emergencies. BIO's area of geographic concern extends from the Western Arctic to the temperate Atlantic.

1970s

The discipline of Coastal Engineering and the research associated with coastal design were now firmly established. The 1970s saw a continued development as well as an integration of physical science disciplines, such as oceanography and geomorphology, so that *Coastal Engineering* became *Coastal Science and Engineering.* Two government research facilities were started in the 1970s. The new hydraulics facility at the Canada Centre for Inland Waters (CCIW) in Burlington at the west end of Lake Ontario was officially opened in 1972. Its early coastal work was on air-sea interaction with Mark Donelan, wave spectra with Michael Skafel and nearshore processes along the lower Great Lakes with John Coakley, Norm Rukavina and Alex Zeman. The Pacific Geoscience Centre was opened near Victoria in 1978. Its mandate is to investigate the geological evolution and sedimentary processes on the continental shelf and in the coastal and nearshore areas along the Pacific coast. Its work directly related to coastal engineering has been mainly concerned with computation of tides and storm surges. Several consulting engineering firms were established at that time, for example, by Keith Philpott, who came from NRCC, Duncan Hay and Bill Baird who left Public Works Canada and by Don Hodgins who studied at UBC.

On the West Coast, the coastal engineers and scientists of the region met in a seminar at UBC in 1970. Following this, Canada sponsored the 13th International Conference on Coastal Engineering in Vancouver in 1972. At UBC, work with the Fraser Estuary continued and projects were carried out on salt water intrusion, effluent discharge and mixing by Peter Ward. With the arrival of Professor Michael Isaacson in 1976, work started on wave loading of structures.

At Queen's, the new research facility quickly became a hive of activity. It attracted other faculty who performed their research there (Professors Yalin, Wilson, Mitchell and Turcke). Some of the projects at Queen's consisted of contract work, but the bulk of the work was research, funded by NRCC and later by NSERC. The research topics were invariably practical, often the direct result of experience from consulting work or contract projects. Typical projects included the generation of waves by landslides, wave dynamics in the bottom boundary, sediment transport (under waves, along coasts, in inlets, in rivers, near coastal structures, in reservoirs and in pipelines), formation of bedforms under waves, artificial beach nourishment, stability of coastal structures (particularly floating breakwaters), wave agitation in small craft harbours, etc. Part of the research effort was consciously directed to study and improve hydraulic modeling procedures and theory. In 1975, the Queen's laboratory set up its own mini-computing facility. It used software developed at NRCC and hardware that was compatible with the nearby NRCC laboratory at Ottawa.

At Laval, studies were carried out on breakwater stability, floating breakwaters and on hydrodynamics and transport modelling for small rivers and for the St Lawrence. Much of the work was done in close co-operation with Public Works Canada. Several other universities also became active in Coastal Science and Engineering. University of Toronto began field and laboratory research on sediment transport and morphodynamics in 1970 with Professor Brian Greenwood. Specific projects are sediment transport by both long and short waves, formation of bedforms and shore morphodynamics under storm surge. Université du Québec at Rimouski began work on dispersion of oil in 1972 with Professor Georges Drapeau. Dalhousie University in Halifax organized its department of Physical Oceanography in 1974 with Professors Tony Bowen and David Huntley. Their work is field related and focuses on development of methodology and instrumentation used in research on fluid-sediment interaction in the coastal environment. They have been important participants in several large field studies. University of Guelph, near Toronto, began its work on foreshore erosion and on beach and dune dynamics with Professor Robin Davidson-Arnott in 1976. Other, particularly ocean structures related activity started at Memorial University of Newfoundland at St John's with Professors Allen, Bruno, Muggeridge and Chari and at the Technical University of Nova Scotia in Halifax. Extensive research facilities were built at Memorial University.

At NRCC, the hydraulic and numerical modeling of the St. Lawrence River were completed. Once the laboratory was cleared of the large hydraulic model, the space was occupied by numerous coastal model studies executed for the Canadian government, for engineering firms and for basic research. New equipment and methodologies were developed and irregular (long-crested) wave generators were first used.

The 1970s were an exciting time with rapid expansion of knowledge. In 1977, NRCC established the Associate Committee for Research on Shoreline Erosion and Sedimentation (ACROSES), later to become the Associate Committee on Shorelines (ACOS). This committee consisted of members from engineering firms, government and the universities and included coastal engineers, oceanographers, geologists and geographers. It was multi-disciplinary and helped the coastal profession move along in the direction of integration of the various disciplines. It was also representative of Canada's geographic regions. Its first chair was Milne Dick of CCIW. Its original members were from BIO, Dalhousie University, Université Laval, LaSalle Hydraulics Laboratory, Public Works, Environment and Transport Canada, NRCC, Queen's University, Crysler and Lathem Consultants Ltd of Toronto, the Canadian Freshwater Institute in Winnipeg, and Western Canada Hydraulic Laboratory and Swan Wooster Engineering, both from Vancouver. The primary function of ACROSES (ACOS) was to strengthen coastal knowledge and research in Canada. It did this through the organization of seminars, workshops and the Canadian Coastal Conferences of which the first was held in 1980 at CCIW in Burlington. The information was disseminated through the publication of the proceedings of all these meetings and listing research projects conducted in Canada (Willis and Kamphuis, 1990). Dave Willis of NRCC was the committee's secretary and bulletin editor for many years and he has been primarily responsible for the committee's success in achieving its goals.

It was quite clear in the 1970s that the discipline of *Coastal Science and Engineering* was severely hampered by the fact that many departments at three levels of government (federal, provincial and municipal) all had legal jurisdiction over different aspects of the shoreline.

There was not one organization that even came close to carrying prime responsibility (such as the US Army Corps of Engineers or the Netherlands Rijkswaterstaat). To introduce some sort of reason into the morass of government agencies, each protecting its own turf, the Canadian Council of Resource and Energy Ministers called together a conference of the resource and energy ministers and their advisors, representing the Canadian federal government and all 10 Provincial governments. This meeting, held in 1978 in Victoria, British Columbia, produced many good ideas (Canadian Council of Resource and Energy Ministers, 1978), in particular some strong calls for rationalization. Unfortunately, not much has been accomplished in that direction to date, leaving the management of the shorelines of Canada literally in the hands of hundreds of agencies.

Along the Great Lakes, the water levels were very high once again in the mid 1970s. This resulted in renewed concern over shore erosion and the (Ontario) Ministry of Natural Resources began to play a leading role in the management of these shores. Guidelines (Ontario, 1981), public information (Ontario, 1981a) and a shore damage atlas (Environment Canada, 1976) were some of the results of this concern. Local Conservation Authorities were also beginning to take a substantial interest in what was occurring at their lake shores and so Coastal Zone Management was born in Ontario.

AFTER 1980

At NRCC since 1980, major engineering and research projects on estuary dynamics, sediment transport, breakwater stability and wave forces on coastal and ocean structures were carried out. Of particular note is the complex investigation of the overturning of the semi-submersible drilling rig "Ocean Ranger" carried out by Geoff Mogridge. This study in 1983-'84 involved wind and wave forces and was the first model to use the new multi-directional wave basin at NRCC. In fact, research on wave generation became a major topic of research at NRCC for many years with Ed Funke, Etienne Mansard and Joe Ploeg. This research benefitted greatly from the field work done 15 years earlier. Many of the laboratory techniques for wave generation and active wave absorption were developed by Mike Miles and Etienne Mansard in collaboration with Professor Isaacson and his students at UBC. In 1987, NRCC enhanced its ship research by developing a completely new research facility (the Institute for Marine Dynamics) in St John's, Newfoundland. This facility concentrates on ships and ocean structures to be used for oil exploration. It contains a long towing tank with a wave generator, a very large directional wave basin and a large refrigerated tank to model ice forces on ships and structures. Together with the research facilities at Memorial University and its Centre for Cold Ocean Resources Engineering, they form a substantial research centre on the East Coast.

Since 1980, ACROSES has set up many inter-disciplinary seminars and workshops as well as five Canadian Coastal Conferences (1980, 1983, 1985 and 1987 and 1990). Its membership was changed to include the coastal management discipline. It has also organized the Canadian Coastal Sediment Study (C^2S^2), an extensive three-year field study (Willis, 1987). This study was sponsored by government departments concerned with the coast and it was managed by NRCC. The participants hailed from various government departments and universities across Canada. The results of this inter-disciplinary exercise accomplished for field research what the earlier decade had accomplished for laboratory research; it produced Canadian field research of world class. In 1992, after many years of vital involvement in the development of scientific and engineering expertise, NRCC divested itself of all its associate committees including ACROSES. It was thought, however, by the coastal profession that the opportunities and the progress made through the technical forum provided by ACROSES should not disappear and thus, the Canadian Coastal Science and Engineering Association (CC-SEA) was organized, once again with substantial help from NRCC, who provided the initial secretariat (Mike Davies). This coastal "learned society" encompasses all the earlier disciplines and the biological sciences appeared for the first time as a result of the greening of coastal science and engineering. CC-SEA's founding president is Bill Kamphuis and it has sponsored workshops, seminars and the 6th and 7th Canadian Coastal Conferences (in 1993 and 1995).

The early 1980s witnessed an explosion of interesting coastal engineering projects. One such major project was the design of artificial drilling islands for oil exploration in the Canadian Beaufort Sea, adjacent to the Canadian Arctic Ocean. The design of these islands by the various oil companies and their consultants was mainly based on very simple concepts. Ice acts as protection for most of the year and the major problem was to protect the islands from erosion by waves during the short open water season. However, there was a decided lack of physical data on which to base any designs and there is a complete absence of protective stone in that part of the Arctic. Thus, innovative solutions and much laboratory work were needed. Both design and basic research models were carried out at Queen's. Similar experimental work at larger scales was carried out by ESSO Resources in Calgary, where the structures were also tested for ice forces. NRCC tested several structures and developed a new method for modelling sea ice. Baird and Associates tested island models at NRCC and CCIW and Seaconsult was instumental in developing the wave and surge climate.

Because of the abundant occurrence of glacial tills throughout Canada, Queen's investigated the erosion mechanisms of such consolidated formations under currents and waves. The early work was done to define erosion by currents and waves in connection with the design of oil production islands in the Mackenzie River. A short time later, work was carried out in connection with litigation involving bluff erosion on Lake Erie. After that, work was done in connection with bridge pier scour in channels with a glacial till bottom. The litigation, referred to above, involved accelerated erosion of consolidated cohesive material (till bluffs) by wave action in the vicinity of a long breakwater at Port Burwell on Lake Erie. At the time, it was generally assumed that a cohesive shore eroded in a similar way to a sandy shore. The in-depth studies connected with this litigation drew together many individuals from federal and provincial government institutions, universities and consulting engineering firms. Eventually a concept was developed that describes shore processes in this entirely different environment, which is so prevalent along the lower Great Lakes.

As the 1990s approached, the major topic of research at Queen's became the overall modeling of all the processes involved in alongshore sediment transport, first without any structures and later with seawalls and groins. This topic was pursued with both physical and numerical models. In 1989, Professor Kevin Hall joined the Queen's coastal group, reviving a major interest in the area of breakwater research. Professor Brebner retired in 1988. The latest research directions at Queen's include the study of aquatic habitat and of structures to improve water quality. Environmentally sensitive design of coastal structures is also studied.

At Laval, research was carried out on shore erosion in connection with the reservoirs of the gigantic James Bay hydro-power project. Numerical modeling of refraction, diffraction and agitation in harbours, sediment transport, shore protection and the statistics of winds, waves and water levels were also studied. At Rimouski, work was carried out on coastal and fluvial erosion (Professor Drapeau), formation of longshore bars (Professor Barbara Karakiewicz) and the use of radioactive tracers (Professor Bernard Long). University of Toronto, Dalhousie and Memorial Universities and NRCC co-operated on field and laboratory investigations of wave-sediment interaction. At UBC, forces on structures, wave generation, wave agitation in harbours and dispersion of pollutants continued to be studied and research began on coastal processes.

BIO, operates 6 oceanographic research vessels with 6 additional vessels available in the east coast region. Research relevant to coastal engineering by BIO and the Atlantic Geoscience Centre is the work on gravel beach morphodynamics, sea level rise, sediment transport and coastal permafrost processes. Innovative coastal applications of multibeam bathymetric surveying have resulted from recent collaborative efforts between BIO and the University of New Brunswick. The Coastal Oceanography Group of BIO has also contributed significantly to the dynamics of unconsolidated cohesive sediments in the coastal zone. Coastal Engineering research continued at CCIW with further work on air-sea interaction and directional waves. Mark Donelan developed a 2D wave forecasting model for the Great Lakes. The nearshore sediments in the lower Great Lakes were documented and current and past processes for understanding the evolution of the Great Lakes were studied. Craig Bishop and Michael Skafel conducted studies in support of specific engineering projects along the Great Lakes and contributed to the understanding of floating breakwaters and of the erosion of cohesive sediments by waves.

The high water in the lower Great Lakes around 1973 was followed by a period of lower water levels (and consequently relatively little concern about the shorelines). The latest wake-up call came from the high water levels of 1986-'87. Each high water period has intensified the concerns about erosion (Ontario, 1987), most likely because additional development has taken place along the shorelines. This time the Ontario Ministry of Natural Resources really became involved in shore management. This group has initiated, supported and carried out many studies in coastal zone management and now has in place an Integrated Coastal Zone Management (ICZM) policy, based on adequate scientific and engineering evidence. In British Columbia, the province where the large symposium on shore management was held in 1978, there still appears to be a lack of federal, provincial and inter-jurisdictional co-ordination, resulting in continued damage to habitat and environment (Day and Gamble, 1990). The Fraser River Environmental Management Plan (FREMP) is a positive step, but in British Columbia and elsewhere in Canada, without the periodic reminders from large water level fluctuations such as occur on the Great Lakes, the move toward integrated shore zone management continues to be slow. Coastal Zone management research and teaching now takes place at the University of Waterloo, near Toronto and at Simon Fraser University in greater Vancouver (Professor Chad Day). Active participation from some federal government departments has recently resulted in the development of national Integrated Coastal Zone Management concepts and the first Canadian Coastal Zone Conference was held in Halifax in 1994.

The 1990s have also seen the development, particularly along the Great Lakes, of an ecosystem approach to coastal design (Waterfront Regeneration Trust, 1995). The ecological system is considered, rather than simply the physical system (concern is about salamanders, as well as sand). Innovative shoreline structures that improve fish and bird habitat and human interaction with the waterfront are now being designed. Finally, new coastal experimental facilities were opened in the 1990s at Université de Sherbrooke (Professors Guy Lefebvre and Karol Rohan) and at the University of Manitoba, with Professor Jay Doering.

SUMMARY

The discipline of *Coastal Science and Engineering* in Canada has developed from hard Coastal Engineering (structural designs related to transportation and power generation) in the 1950s to include physical sciences such as oceanography and geomorphology in the 1960s and '70s, coastal zone management in the 1980s and biological sciences in the 1990s.

Coastal Engineering was born as a discipline in the 1950s. A rapid and exciting expansion in knowledge, facilities and funding took place during the 1960s and '70s. Some very challenging problems such as the capsize of the Ocean Ranger, the development of irregular and multi-directional wave facilities, the design and construction of the Arctic drilling platforms and the research on erosion of cohesive shorelines presented themselves in the 1980s. The 1990s have, however, seen a severe cutback on activity, primarily because the discipline has been (and is) very dependent on spending by the Government of Canada. As financial restraints at all levels of government persist, NRCC, the universities and engineering consultants will continue to face some difficult times. But a re-birth is also taking place. Many firms and institutions are using the knowledge acquired over the past 50 years and applying it globally and are thus coping with the great unsettling changes of the 1990s. The profession looks forward with a renewed sense of purpose to the 21st century, which will be a different world. It will be much more interdisciplinary, involve much less design and construction and more maintenance and infrastructure replacement, and it will require much more reflection on the environment, habitat and quality of life.

ACKNOWLEDGMENTS

This historical perspective probably is biased to reflect the author's familiarity and it is also probably is not complete. I acknowledge gratefully the help of W. Baird, V. Barrie, T. Bowen, A. Brebner, R. Davidson-Arnott, C. Day, G. Drapeau, B. Greenwood, D. Hay, M. Isaacson, P. McKeen, R. Nairn, Y. Ouellet, M. Paul, J. Ploeg, B. Pratte, O. Sayao, S. Solomon, M. Skafel and others on whom I called for help.

As a profession, we acknowledge that the Canadian Government departments have been the driving force for most of the engineering research and innovation on Canada's coasts, and that NRCC has been the leader in the development of Coastal Science and Engineering in

Canada. With the support of other Government departments, NRCC was instrumental in the formation of this profession through its Associate Committees. Its own research has added substantially to the body of knowledge available in Canada and throughout the world. Also several research groups and sometimes their facilities are direct spin-offs of NRCC encouragement and funding.

References

Canadian Council of Resource and Energy Ministers (1978), *Shore Management Symposium Proceedings*, Victoria.

Day, J.C. and D.B. Gamble (1990), "Coastal Zone Management in British Columbia: An Institutional Comparison with Washington, Oregon and California.", *Coastal Management*, Vol. 18, pp. 115-141.

Environment Canada (1976), *"Canada/Great Lakes Shore Damage Survey, Technical Report and Coastal Zone Atlas"*, Canada Department of Supply and Services.

Langevin, H.L. (1881), "*Memorandum with Accompanying Plans and Documents, Relative to the Past and Present State of the Harbour of Toronto, Province of Ontario"*, Department of Public Works, Ottawa, pp. 53-76.

Middleton, W.E.K (1984), *"Mechanical Engineering at the National Research Council of Canada 1929-1951"*, Wilfrid University Press, Waterloo.

Ontario (1953), *"Lake Levels Report 1953"*, Report of the Select Committee of the Ontario Legislature on Lake Levels of the Great Lakes, Queen's Printer.

Ontario (1981), *"Great Lakes Shore Management Guide"*, Ontario Ministry of Natural Resources.

Ontario (1981a), *"Great Lakes Shore Processes and Shore Protection"*, Ontario Ministry of Natural Resources.

Ontario (1988), *"First Annual Report"*, Ontario Shoreline Management Advisory Council, Ontario Ministry of Natural Resources, Queen's Printer.

Willis, D.H. (1987), "The Canadian Coastal Sediment Study - an Overview", *Proc. Canadian Coastal Conf.*, (NRCC), Quebec, pp. 395-408.

Willis, D.H. and Kamphuis, J.W. (1990), "The NRCC Committee on Shorelines", *Proc. Canadian Coastal Conf. '90,* (NRCC), Kingston, pp 1-12.

Waterfront Regeneration Trust (1995), *"Lake Ontario Greenway Strategy"*, Waterfront Regeneration Trust, Toronto.

HISTORY OF COASTAL ENGINEERING IN DENMARK

Torben Sørensen[1], Jørgen Fredsøe[2], Per Roed Jakobsen[3]

ABSTRACT: The geography of Denmark has conditioned the Danes to become a nation of seafarers already in the early Middle-Ages, where the Danes, together with the Swedes and the Norwegians conquered and settled near and far - from North America over the British Isles, Russia, and Normandy to Spain and Italy. Likewise, fishing has always played an important role in the life of the Danes.

These activities and the fact that the land consists almost exclusively of sedimentary soils and is located in a region with a rough climate in terms of winds and waves created the need to observe and understand the interaction between sea and land - i.e. the need for coastal science and engineering.

The concept of coastal engineering developed in Denmark during the 19th century led in the early part of the 20th century to the clear understanding that coastal sediment transport - littoral drift - was heavily dominated by wave action. During this period authorities had the foresight to take initiatives and established a continuous monitoring of key coastal reaches, based upon which long-term analyses of coastal problems were performed. This tradition is continued today by the Danish Coastal Authority. Danish engineers also experimented with innovative concepts of coastal harbours, such as island harbours, from which further lessons in coastal engineering were learned.

Truly scientific progress in coastal hydraulics and engineering was initiated in the 1950s and 60s driven by people like Professors Helge Lundgren and Frank Engelund of the Technical University of Denmark, while the Danish Hydraulic Institute became the platform for the introduction by Professor Michael B. Abbott of modern computational hydraulics in the field of coastal engineering, including the calculation of non-steady sediment transport.

In recent years the realisation of the need for protecting and restoring the environment has further stimulated analytical research as well as the development and application of analytical and predictive methods in the field of coastal hydraulics and engineering in Denmark and, indeed, world-wide. Denmark has participated actively in international exchange and collaboration in this field both through the International Conferences on Coastal Engineering and through the joint European research programme on Marine

---

1) Director, M.Sc., Danish Hydraulic Institute, Agern Allé 5, DK-2970 Hørsholm, Denmark.
2) Professor, Dr.techn. J. Fredsøe, Department of Hydrodynamics and Water Resources (ISVA), Technical University of Denmark, DK-2800 Lyngby, Denmark.
3) Director, M.Sc., Danish Coastal Authority, Højbovej 1, DK-7620 Lemvig, Denmark

Science and Technology. If anything, the experience from these activities demonstrates the need for and benefit to be obtained from even more intensive international exchange and collaboration.

INTRODUCTION

Denmark has maintained a strong position in coastal engineering throughout 150 years. Its geography and settings offer great challenges. Though the country is small by area and population the coastline is long and varied in nature and exposure.

Only the main issues will be mentioned in this position paper.

Presumably, there are a number of readers on the international level, who are not too familiar with our national situation. Others will know many of the cases which over the years have been presented at international conferences and likewise, over the years many colleagues have visited our country, notably at the International Conference on Coastal Engineering held in Copenhagen in 1974. In view of the various interest of the potential readers, the following structure of the presentation has been chosen:

- Firstly, a description of the setting, geology, demography as a basis for understanding the coastal engineering issues we are facing.

- Secondly, an account of the history of coastal engineering in the country, emphasizing the latest 150 years.

- Finally, we shall concentrate on the latest 25 years which coincide with the development of Danish Coastal Authority and Danish Hydraulic Institute, as well as important development on the academic level, subsequently identifying trends and activities, which are believed to set agendas in the years to come.

Projects in Coastal Engineering never end. The dynamic coastal processes always call for better understanding and modification of approaches.

THE SETTING AND LAND SEA INTERACTION

Denmark on the European Continent

Denmark is a member of the European Union since 1973. This reflects not only political and economic realities, but has more than anything else its outspring in the geographical facts. Denmark is a fairly small country - about 43,000 km² - the land, however, is distributed over one larger peninsula, Jutland, and about 500 islands of which 200 are inhabited. On the largest island - Sealand - you find the capital, Copenhagen. This geography implies that the country resembles a large archipelago, which all together covers a fairly large area, cf. Fig. 2. For comparison Fig. 1 gives figures for a few European countries.

Thus, the geography implies that no Danish citizen lives more than 50 km away from the sea. Since the total length of coast is about 7,300 km, it also means that there is 1.4 m of coastline per inhabitant.

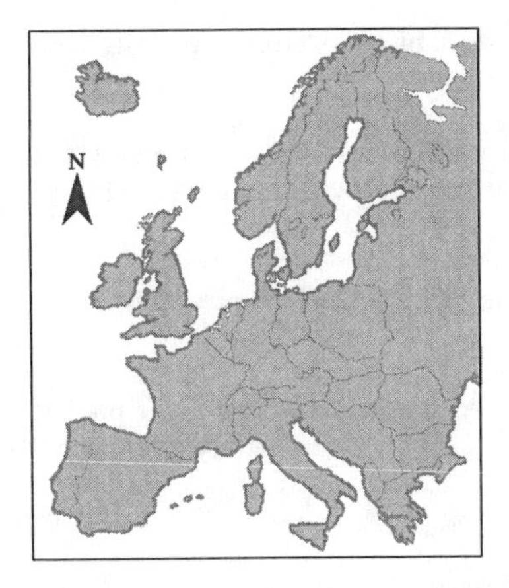

Table 1. Inhabitants, area and length of coastline in Denmark and some other european countries.

	Inhabitants mio	Area 1000 km^2	Coastal length km
Denmark	5.2	43	7,300
Germany	80.6	357	3,400
UK	58.0	242	17,500
Spain	39.1	505	5,800
France	57.4	544	4,700

Fig 1. Map of Western Europe.

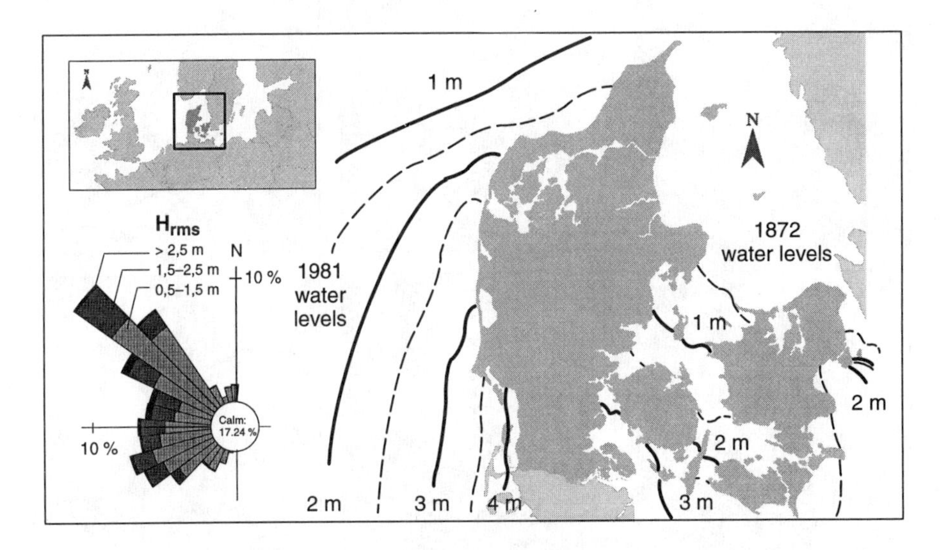

Fig 2. Denmark. Water levels during particularly severe storms (1872 at the islands and 1981 at the West Coast). Frequency and directions of the waves at the West Coast.

Geological Background

The genesis of the contemporary Danish coast has by and large its origin in the late Quaternary which means, that the coast is dominated by soft marine deposits, except at certain points, where chalk or moraine clay headlands are found, between which mostly long, uninterrupted sandy beaches are suspended in gently curved forms.

The changing battle between land and sea may be understood by considering three postglacial extremes (Danmarks Natur, 1979):

- The Ice-Sea, which left the northern Jutland as an archipelago, the evidence of which is found today in old coastlines at elevations of 10 - 60 m above present sea level (Fig. 3).

- The Continental period, during which land extended far beyond the present coastline (Fig. 4).

- The Littorina or Stone Age sea, which transgressed considerable parts of the present coastal landscape (Fig. 5). The coasts of this period were contemporary to the stone age civilization.

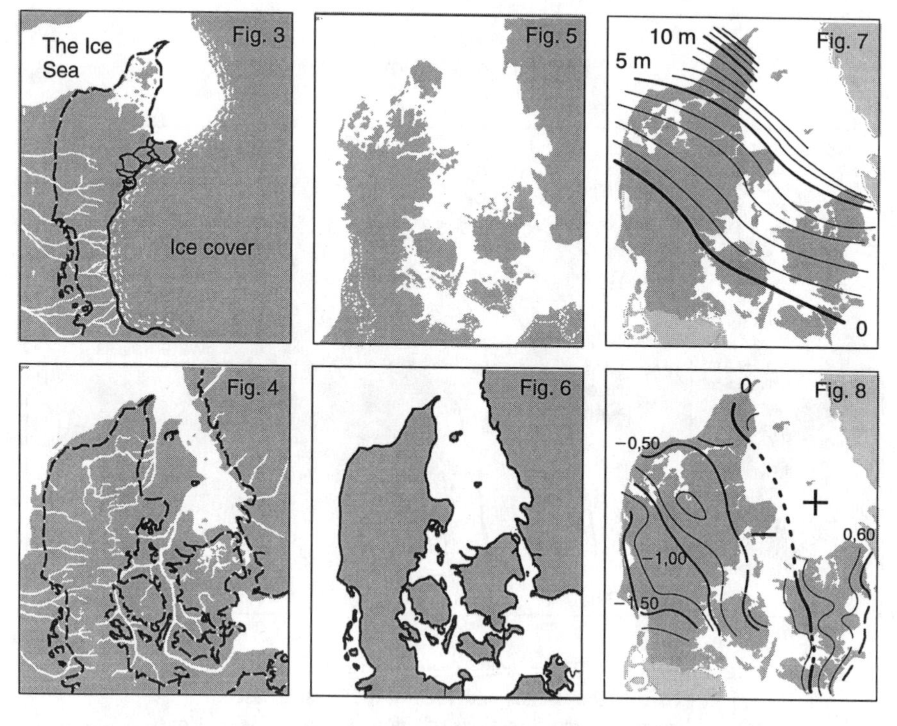

Fig. 3 Postglacial Period approx. 10,000 B.C.

Fig. 5 Littorina Sea Period approx. 5-2,000 B.C.

Fig. 7 Relative landrise for the Littorina Sea

Fig. 4 Continental Period approx. 8-6,000 B.C.

Fig. 6 Present situation

Fig. 8 The present vertical land movements in mm/year

These dramatic changes in land levels are mostly the result of tectonic movements caused by the disappearance of the ice cover of the last glacial period. To a large extent the present coastline thus represents a fragile balance between possible extremes.

Land and Sea

The post Littorina relative land rise has been the predominant process over the last 6,000 years in the northern part of the country. During the same period the south-western part, which includes the Danish part of the Wadden Sea, has been subject to a relative land depression in the order of 0.10 - 0.15 m per century, a trend which is also known from Northern Germany.

The former tilting line theory, which is illustrated in Fig. 7 (Mertz, 1924), does no longer apply to the present day processes. The balance between the global trend of relative sea level rise and the slowing down of the glacial rebound, has changed in such a way, that the zero line runs almost through the northern part of Jutland (Fig. 8) (Remmer, 1991) and (Miljøministeriet, 1992). The consequence of this is, that all along the already exposed Danish North Sea coast there is a relative sea level rise of 0 - 2 mm/year.

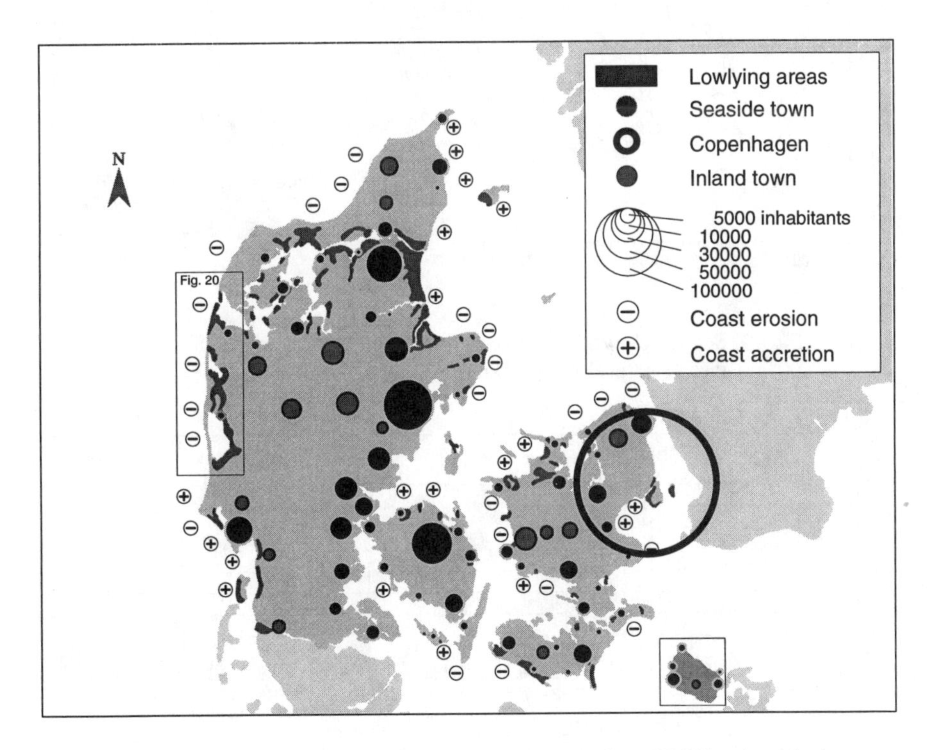

Fig 9. Lowland protected by 500 km of dikes comprises 65,000 private lots. Demographic development and coastline development scenario. (Danmarks Natur, 1979).

Climatic Changes

Any coastal engineer is very much aware of the influence of varying climate on the coastal conditions. For instance, it is known that a Little Ice Age occurred during the Middle Ages, apparently followed by a certain stabilization of coastal conditions and even barrier growth.

This definitely changed in the 19th century and was felt most severely on the Jutland North Sea coast, where a breach finally established a permanent gap in the Lime Fiord barrier (Fig. 20). Fig. 10 shows an analysis of sea level changes over the latest one hundred years (Binderup et al, 1993).

It appears (Fig. 10) that conditions have grown worse. This can also be illustrated in another way by showing - for selected water level stations - the number of maximum storm flood levels above certain threshold values (fig. 11). The three stations shown are Esbjerg on the Jutland west coast, Hornbæk in the Kattegat, and Gedser in the Baltic. These graphs suggest an increasing dominance of stronger westerly winds and may be related to the overall condition in the North Atlantic.

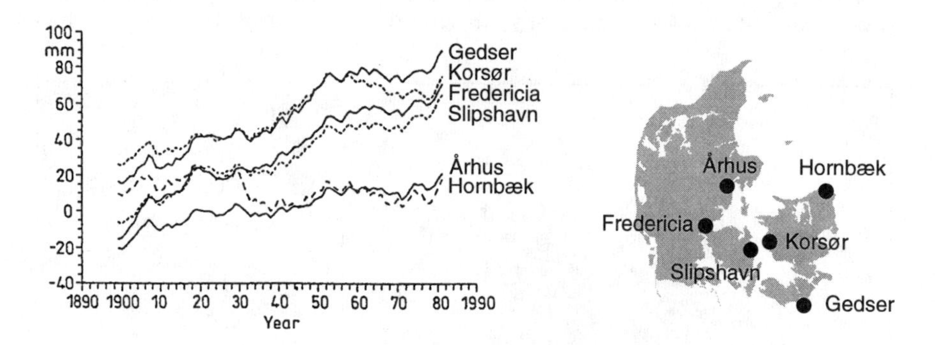

Fig 10. Sea Level trends. Nineteen-year running mean values of annual mean sea level for six Danish tide gauge stations.

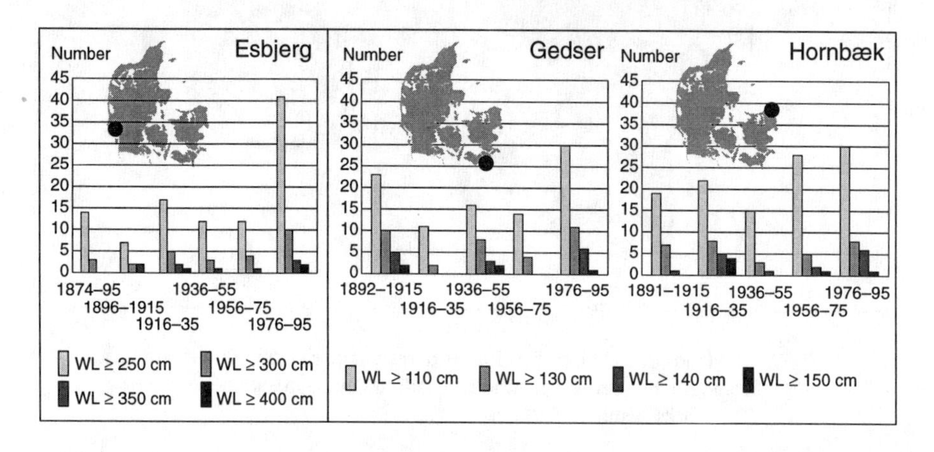

Fig. 11. Frequencies of severe water levels.

Littoral Environment
Combining the geological background, the geographical conditions and the climatic impacts, the Danish coasts can be classified in two major categories and, further, for each of these in three distinct subgroups (Jakobsen et al., 1993)

North Sea Coasts

- The Northern Jutland Headland Coast along which erosion is predominantly in the order of 2 - 4 m/year over the last 20 years. Littoral drift is northward due to the sheltering effect of Norway. It grows from zero in the south to about 1 mio m^3/ year at the Skaw Spit at the extreme North.

- The Central West Coast is dominated by barrier beaches, which are separating major lagoons: Lime Fiord, Nissum Fiord and Ringkøbing Fiord (Fig. 20), from the sea. Erosion rates used to be high but are now balanced by large beach nourishments.

- The Wadden Sea coasts: Despite the relative rise in sea level, which has been the dominant process over several thousand years, it appears that this area is in an overall sediment balance. The situation may be even better, because the two major barrier islands of Fanø and Rømø have grown considerably over the last several hundred years.

The total length of the North Sea coasts is 450 km, of which the Headland Coast is 200 km, the Central West Coast 140 km, and the Wadden Sea outer coast 110 km long.

Inner Coasts
This term is defined as all other coasts than the North Sea coasts. They account altogether for about 7000 km and show, naturally, much variation (Jakobsen et al., 1987).

They fall in three major categories:

- Northern medium exposed coasts along the Kattegat with littoral drift rates of an order of magnitude less than along the North Sea, i.e. 10 - 75,000 m^3/year.

- Eastern and southern medium exposed coasts in the Baltic.

- All remaining coasts of low exposure along straits, belts and fiords or on sheltered islands. Littoral drift less than 10,000 m^3/year.

Coastal Balance and Greenhouse Effects
The relative sea level rise of 0 - 2 mm/year along the West Coast results in an increase in shoreline recession rate of about 0.2 m/year. This shall be viewed in relation to the fact that, till the middle of the seventies, the erosion rate of unprotected beaches along the West Coast was about 0.5 m/year. Hence the secular relative movements of land and sea may explain about half of the coastal erosion.

However, the situation during the latest 20 years has been different since the coast has been subject to an increasing number of erosion events, notably high water levels combined with high waves. During these 20 years "Bad Weather Period" the shore recession rates on unprotected North Sea coasts have grown, typically from 0.5 to 2 m/year.

On top of this might come a greenhouse effect, or is it already there, of the same order of magnitude. An analysis indicates that beach nourishment should be increased by a factor two to compensate for the predicted sea level rise.

DEMOGRAPHIC DEVELOPMENT

Following deglaciation hunters started moving into the country.

The first civilization really to be identified was the Stone Age culture about 10,000 years ago. The Stone Age man's coasts were those of the Littorina Sea, during a period of high water levels, so we can imagine how he was sitting on the now fossil moraine cliffs watching the carving out of the landscape.

However, it was during the Viking Age A.D. 900 - 1100, that the Danes and the other Scandinavians founded the tradition of being great sea farers and tradesmen. They operated out of natural harbours easily found along the coasts, sometimes backed by rather large ringforts of which three have been excavated. To our knowledge they did not build harbours as such, but they protected bays and fiords by building submerged sills often from sunken ships, to protect their natural harbours against intrusion.

They entered the history of coastal engineering by telling us that they sailed out west from the Lime Fiord on their raids to England, Ireland and France. In other words, there was still passage through the barrier system, which had started emerging around 2000 years ago.

The end of the Viking era coincided with the final barrier consolidation and the closing of the Lime Fiord, and the barrier system then remained dynamically stable for about 700 years. This situation also left the major part of the Danish North Sea coast without any natural harbours.

During the beginning of the present millennium most of the Danish cities that exist to day were founded. They are with rather few exceptions coastal towns. Copenhagen was founded in 1166. The background for its successful development was good natural harbour conditions and a strategic position as a major trading centre for the Baltic Sea. In recent times, especially after the second world war the Copenhagen region has attracted a great number of people, Fig. 9.

The overall coastal environment in the Copenhagen region and the coastal situations in this region are thus characterised by a high demographic pressure for recreation and vacation and a medium wave exposure - Kattegat/Baltic coasts - alternating with less exposed urbanised coasts with high demand for recreational facilities. This combination has led to a set of coastal situations, the handling of which is important on a national level, only surpassed by the problems encountered along the North Sea coast, which is characterised by the high erosion level, storm flood hazards and intensive fishing and tourist activities. With this short introduction we shall leave the discussion of the demography and its impact on the coastal situation.

Nevertheless, it is important since - without human development and expansion - there is no real need for coastal engineering!

Table 2. Area, population, vacation housing and lengths of coasts in Denmark

	Area		Population		Vacation Housing (1982)		Coast
	1000 km²	%	Mio	%	1000 ha	%	km
Denmark	43.1	100	5.1	100	42	100	7.314
West Coast: Ringkjøbing & Ribe Counties	8	18	0.5	9	5	12	804
Copenhagen: Metropolitan Region	2.8	7	1.7	34	7	17	539

COASTAL ENGINEERING THROUGH 200 YEARS (1783 - 1973)

The oldest well established dike scheme in Denmark dates back from about 1550, when it was executed just north of the present Danish-German border to protect low-lying marshlands along the Wadden Sea coast.

Claus Hinrich Christensen was an engineer and a colonel major in the Royal Danish Corps of Engineers and lived from 1768 - 1841. He did a large amount of work on the dike protection schemes in the southwest, the North-Baltic Sea Canal and other harbour and coastal engineering projects at the time (Meiners, 1995).

In the 19th century the modern industrialization gradually gained momentum. which led to the establishment of the Technical University of Denmark in 1829.

It was during this period, the young Carl Carlsen grew up, who was to become in 1868 the first coastal and harbour director in Denmark in the newly established authority. In 1839 he had graduated as the 35th engineer from the Technical University, in mechanical engineering. Teaching in civil engineering only started in 1857.

After successful consulting in the beginning Carlsen went abroad for practical studies of river and coastal problems in western France and along the Channel coasts. He attended lectures at "Ecole des Ponts et Chaussees", Paris 1842 - 43. This background qualified him for nomination as the first Danish public inspector for coastal works in 1845. He had, however, practically no staff, but started nevertheless the first engineering studies on the North Sea coast with the purpose of improving the harbour conditions at Ringkøbing Fiord, which had deteriorated due to barrier growth. Finally, in 1868 he was appointed director for the first Harbour and Coastal Authority in Denmark, which became the leading agency for the development within the field for the next 105 years (Geil, 1995).

The first problem to deal with was the breach of the Thyborøn Barrier in 1862, which was stabilized by the end of the last century by construction of groynes. The inlet thus formed constitutes - and still is - the largest coastal engineering challenge in Denmark. It will be described separately in the chapter on Hanstholm and Thyborøn.

In the 19th century also a tremendous effort was made in the building of new harbours at more hostile locations than previously thought of as feasible. The fields of harbour and coastal engineering boomed. Much debate concerned the building of island harbours which are detached from the coast connected to it by a trestle-bridge. From 1860 to 1890 three such harbours were built. They are still existing and are even today good cases for studying tombolo formation. In 1903 a commission sorted out the pros and cons of the concept.

Somewhat later, between 1910 and 1930 several fairly large detached coastal breakwaters were built to improve landing conditions for fishing boats. Today also these structures exhibit perfect examples of tombolos.

In this context it is of interest to mention that from 1939 exposed landing places on the open North Sea coast were provided with beach landing systems i.e. motorised winches and that such systems still survive at a few locations despite the competition from the modern fishing ports (Nielsen, 1983).

Professor J. Munch-Petersen wrote in 1927 a review of the development in Danish harbour and coastal engineering through 50 years on the 50th anniversary of the Danish Society of Civil Engineering (Munch-Petersen, 1927).

Figs. 12-14 illustrate three present situations at the North Sea coast, their characteristic layouts and coastal engineering implications.

Finally, in 1967 the port of Hanstholm - probably the last port to be built on the Danish North Sea coast - was inaugurated. This port, which was completed only after a century of tense discussions and thorough studies is of particular interest and will later be described separately.

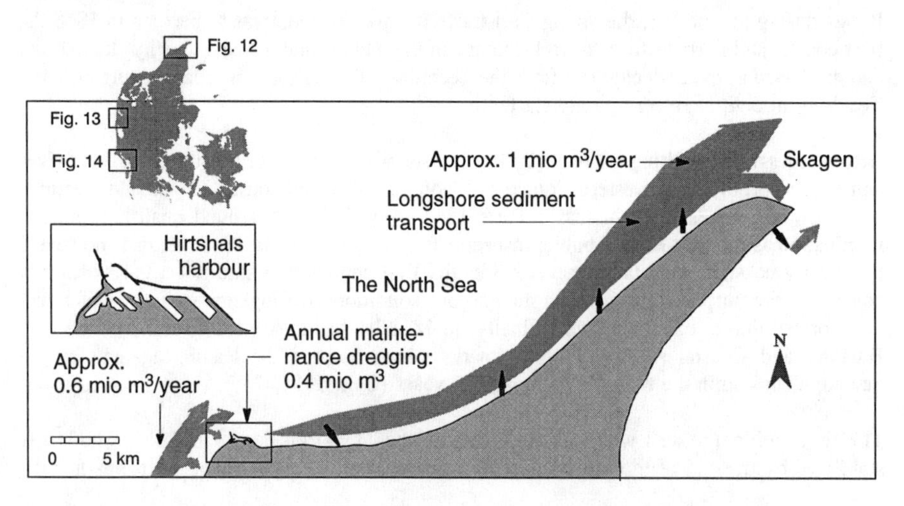

Fig. 12 Maintenance dredging at Hirtshals and longshore sediment transport.

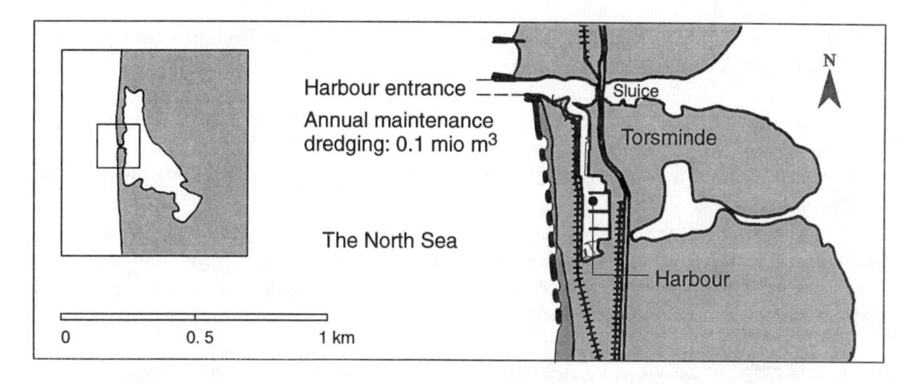

Fig. 13 Torsminde Harbour. Harbour protected by breakwaters and yearly beach nourishment. Harbour on the down drift side of the entrance.

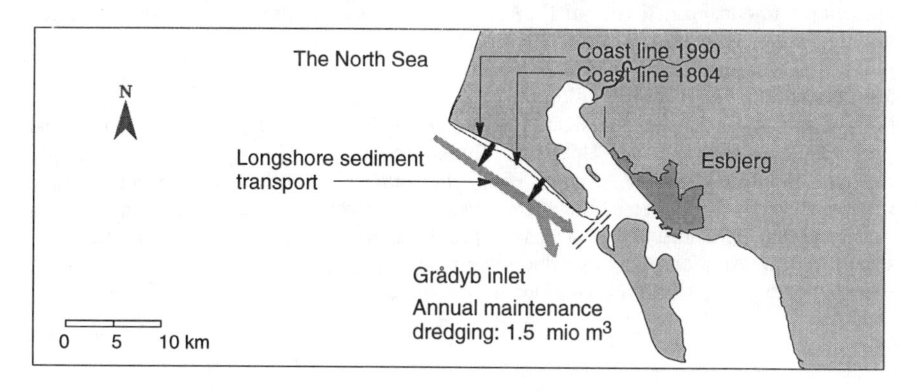

Fig. 14 Grådyb inlet to Esbjerg.

DANISH COASTAL AUTHORITY - DCA

During the 1960s, a period of considerable economic growth, a new structure for the major state owned ports along the North Sea was established. The running and development of the ports required so much attention that it was difficult at the same time to serve the coasts. Since 1973 the administration of the Government owned harbours has been placed under two State Harbour Authorities (in Esbjerg and Frederikshavn), while the then established Danish Coastal Authority (DCA) in Lemvig takes care of all coastal projects, also serving directly under the Ministry of Transport. Coast-harbour interactions are coordinated in an informal council in which also the Danish Hydraulic Institute (DHI) and the Technical University of Denmark (DTU) participate.

The main tasks of DCA are:

- Consultancy to the Ministry of Transport concerning coastal protection issues, with emphasis on problems along the North Sea coast and Wadden Sea.

- Administration of Government owned coastal works, notably the Thyborøn Barriers, Bovbjerg, Blåvand, and the Rømø embankment.

- Administration of the Coast Protection Act, applicable to all coasts in the country.

- Administration of the sovereignty over the Sea Territory, including structures in the territorial sea.

The daily activities cover the full range of coastal engineering, such as monitoring of the coast, coastal studies, planning and design, execution of major coastal works and a considerable amount of regulatory work.

Due to this range of activities DCA has privatized a number of disciplines, especially in the construction field where private contractors account for 80% of the work. Furthermore, DCA has to maintain a comprehensive network, including other authorities, institutes and universities, municipalities, regional councils and central environmental authorities. In the following a few major activities of DCA will be mentioned in order to give an impression of the nature of its work (Jakobsen, 1994).

Monitoring and Surveying Tradition

More than 100 years ago foresighted coastal engineers established a system of annual surveying of coastal profiles along the Thyborøn Barriers. The system was gradually expanded to cover a total of 30 km of coast. Re-examination of the barrier problem in the fifties and sixties demonstrated the value of a good monitoring system as prerequisite for a satisfactory modern analysis of a coastal problem. As a consequence of this practical, scientific demonstration of its usefulness the monitoring system was expanded in 1969 to cover 300 km of the North Sea coast on an annual basis.

Fig. 15. Old and new sounding techniques: The DCA survey boats 1899 and today.

The backbone of the system is the continuous work of two full-fledged survey boats that can operate in the North Sea. They are equipped with multibeam echosounding and navigation by Syledis. The monitoring system includes on-line data from 15 water level recorders along the 450 km North Sea coastline, three wave recording stations and the necessary meteorological back-up. All data are stored in an Oracle data base from which they are ready available for studies and design and also for modelling.

Modern Analysis and Design

DCA has as primary design philosophy that modern computational hydraulics in the practical applicable form should be part of daily routines. A close collaboration is therefore maintained with DHI to ensure that computer tools such as LITPACK (DHI, 1991) are transferred whenever user ready. As permanent user of such tools DCA is in a position currently to provide feed back to DHI from a major user.

An illustrative example of the modern approach to coastal analysis and the subsequent inclusion of findings into the design is the NOURTEC project - Innovative Nourishment Techniques (Madsen et al., 1995) and (Niemeyer et al., 1995).

Beach nourishment is a tool for combating coastal erosion which is applied increasingly in recent years. Also the scale of these measures has gradually increased. Therefore it has become more and more important to require economic improvement and technical optimization. With a view to this a joint project was initiated by the Dutch National Institute for Coastal and Marine Management (Rijksinstituut voor Kust en Zee), DCA (Kystinspektoratet) and the German Coastal Research Station (Forschungsstelle Küste), being sponsored from national sources as well as from the Commission of the European Union in the framework of the MAST-program. The object of the NOURTEC project is to study the beach and shore face nourishments in Denmark, German and in the Netherlands that have been carried out in 1992 and 1993 and have been monitored since then. The major scope of the project is to develop basic criteria for the design of nourishments and, particularly, for the necessicity of repetition of nourishment which is especially important in cases with structural erosion. The formulation of these criteria shall be in terms of common design parameters which, on the one hand, shall be representative for nourishments and easily derivable from monitoring and, on the other hand, shall consider effectiveness and design optimization.

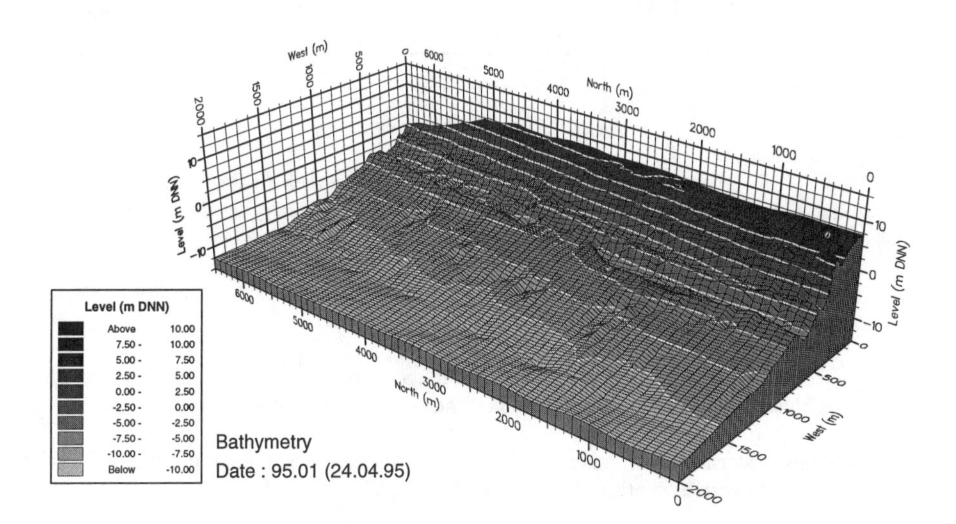

Fig. 16. Block diagram of nourishment site. NOURTEC - Torsminde Barrier.

The project will be finished by the end of 1996. Apart from the concrete results, it has also demonstrated the value of international cooperation in this field.

Within the last few years the study has been extended with an environmental baseline study RIACON, which focuses on the ecological effects of nourishments, covering the entire work processes from the borrow site to the beaches.

Dike Building and Rehabilitation

Shortly after the establishment of DCA in 1973 a severe storm flood struck the Danish Wadden Sea in 1976. It caused much destruction to the system of dike protection, though without catastrophic consequences. Later the event became to be seen as the start of a "Bad Weather Period", which since then has set the agenda for many activities of DCA. The flood also urged the execution of already prepared reinforcements plans developed by a National Storm Flood Committee.

New dikes were built in front of Tønder (1981) and Ribe (1980). The previous rather steep dike profiles were rebuilt by more bulky ones with very flat front slopes (1:10), Fig. 17. Apart from providing a much higher safety, the concept of dike protection with classical sand and clay materials could be maintained, resulting in a very attractive appearance from a landscape point of view.

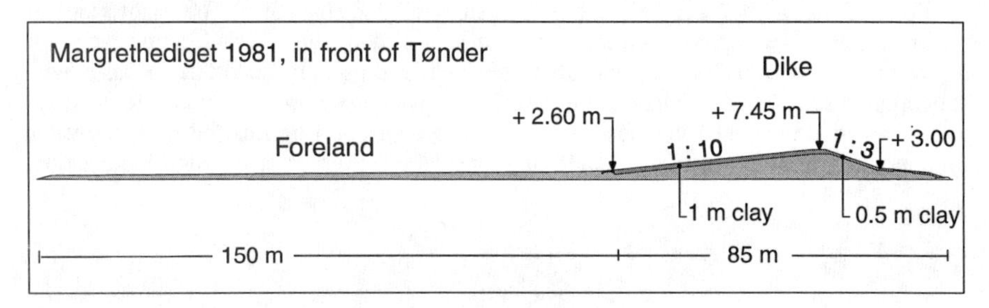

Fig. 17. New dike design.

Already in 1981 - before the just-mentioned dikes had reached their final settlements - a new severe flood struck the entire North Sea coast. This storm has actually set the agenda for the subsequent activities for more than 10 years:

- It necessitated the reinforcement of more storm flood protections than the Storm Flood Committee originally had found feasible.

- It gave rise to the foundation of a more thorough coast protection program along the central North Sea coast, jointly financed by the Government and local authorities.

- The Danish storm flood warning system, which had been active along the Wadden Sea coast since the early seventies, was extended to cover the central North Sea coast as well.

Dike construction and repair were thus major activities through the 1980s, gradually, however, the maintenance and protection of the central West Coast came into focus.

The Modern Coast Protection Approach

The increase in severity of the coastal climate since 1976 required new and more intensive efforts on the Central West Coast. Previously, governmental schemes had dealt only with reaches where, typically, government owned groyne groups from earlier periods existed. This was the case for about 25% of this exposed coast with a total length of about 125 km. As a consequence the coast had been in a squeeze between steadily increasing attacks from the sea side and a growing demographic pressure from the interior. It was therefore decided to establish cooperation between state, regional, and municipal authorities along the coast to finance and execute a program, the objectives of which were:

- To restore and enhance the storm flood protection of the area.

- To control coastal erosion reducing it to acceptable and agreed-upon limits.

- To enhance technical benefits and environment values by introducing beach nourishment on a larger scale.

This new program took off in 1982 and, with two major revisions, it is still in force. The present investment level is about 85 million Danish kroner/year. In the beginning of the 1980s nourishment played a much smaller role than today. This was due to a number of reasons, some of the important ones are mentioned below.

Especially the local politicians believed in and, hence, preferred solid structures such as groynes and breakwaters. The main reason for this was the successful design and constrution of the large groynes around the turn of the century. Later on it was realised that the building of long groynes in a lee side erosion area would accelerate erosion. It was therefore decided to build coast-parallel breakwaters with the main function of protecting the beach.

The principle of beach nourishment was new to politicians and the argument behind ("the erosion of the nourishment during storm is part of the plan") was hard to bring across. (Laustrup, 1993). The actual development is illustrated in Fig. 19, in which the phasing out of the use of solid structures and the increasing use of beach nourishment are illustrated.

Fair enough, it can now be stated that, since 1990 where the continuous nourishment efforts showed their value in the field, there is now a good back-up from the political field and that, at its revision in the early 1990s, the entire program was augmented by 20%. A recent analysis also demonstrates - for the first time in several hundred years - a positive development of the coastal profile from -6 m to +4 m, corresponding to a net advance of the profile of 0.3 m/year over the last 10 years (Fig. 20).

It will be seen that beach nourishment to day accounts for about 80% of all coast protection activities along the Central West Coast. The annual quantity of sand moved amounts to approx 3 million m^3/year. On top of this a considerable effort is also spent on reinforcement and maintenance of the somewhat deteriorated dune systems along the coast.

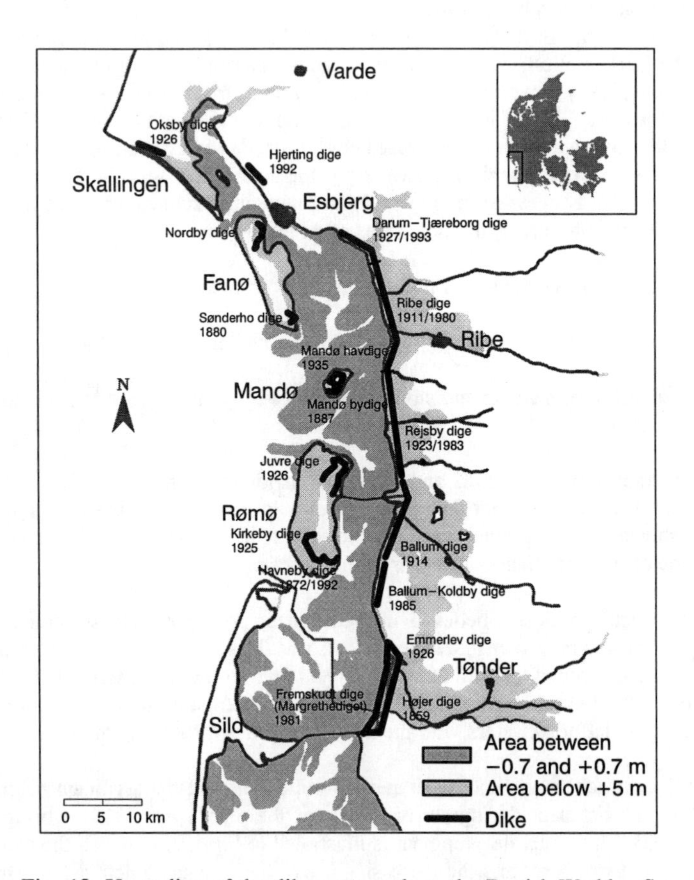

Fig. 18. Upgrading of the dike system along the Danish Wadden Sea.

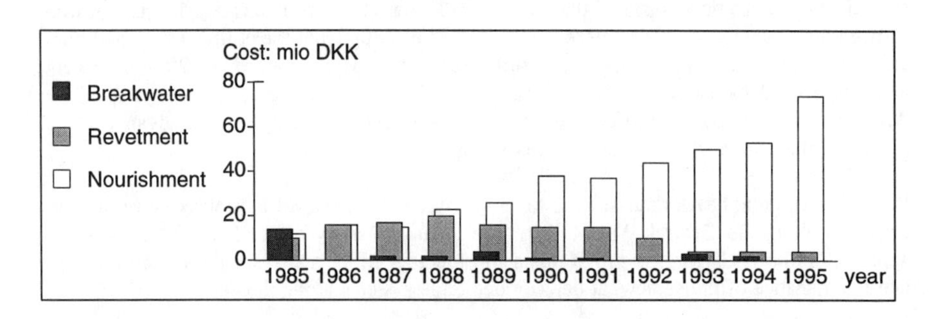

Fig. 19. Coast protection works.

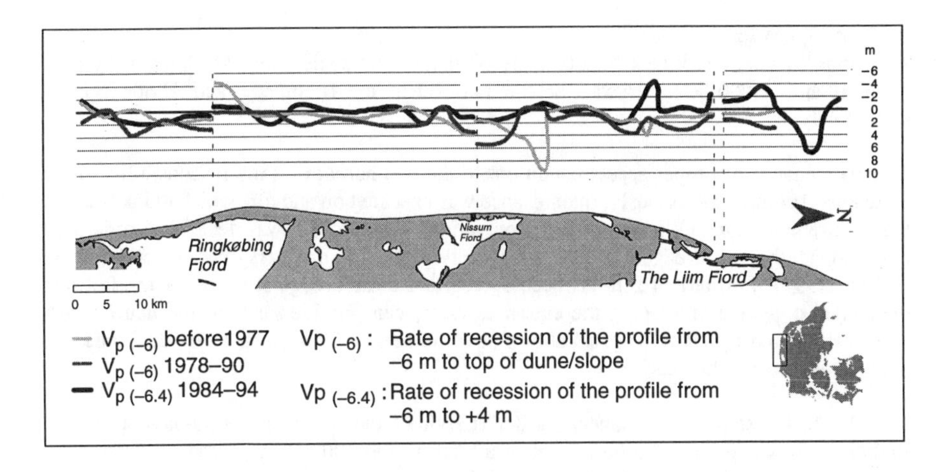

Fig. 20. Recent coastal development.

DCA applies a suite of methods in its effort of a comprehensive beach nourishment program. The various amounts of sand supplies are:

- Offshore method 500,000 m^3/year
- Onshore by submerged pipeline 2,000,000 m^3/year
- By-pass method 500,000 m^3/year
- Miscellaneous 200,000 m^3/year
- Total 3,200,000 m^3/year

In conclusion, the nourishment has now reached a scale that makes the favourable results evident to decision makers and the public. The methods will therefore be applied to increasing measure in the future.

Fig. 21. Sand dredging onto the beach through pipeline

Fig. 22. Sand dredging by the rainbow method at the coast

By-Pass Operation

According to a new sediment budget analysis, nature and society together have to by-pass 700,000 and 1,000,000 m³/year, at the two inlets of Torsminde and Hvide Sande, respectively.

The maintenance of proper access conditions to the two harbours is the responsibility of the State Port Authority. Previously, maintenance was executed on and off, which did not satisfy the requirements for safe and continuous operations. Therefore it was decided to order two custom-made hopper dredgers of 200 m³ capacity each. The dredgers operate permanently. (Frisch, 1991). Furthermore, it has been agreed that these dredgers, when having fulfilled their main objective of keeping the entrances clear, can provide sand for the nourishment work carried out by DCA. Fig. 23 shows how the sand is brought in at Hvide Sande and pumped ashore for the beach.

In 1991 DCA installed a permanent booster at Hvide Sande to assist the by-pass operation. In 1995 it thus boosted 200,000 m³ out of a total of 300,000 m³ by-passed artificially. At Torsminde north of Hvide Sande the man-made by-pass quantity is about 200,000 m³/year. (Jakobsen et al., 1993).

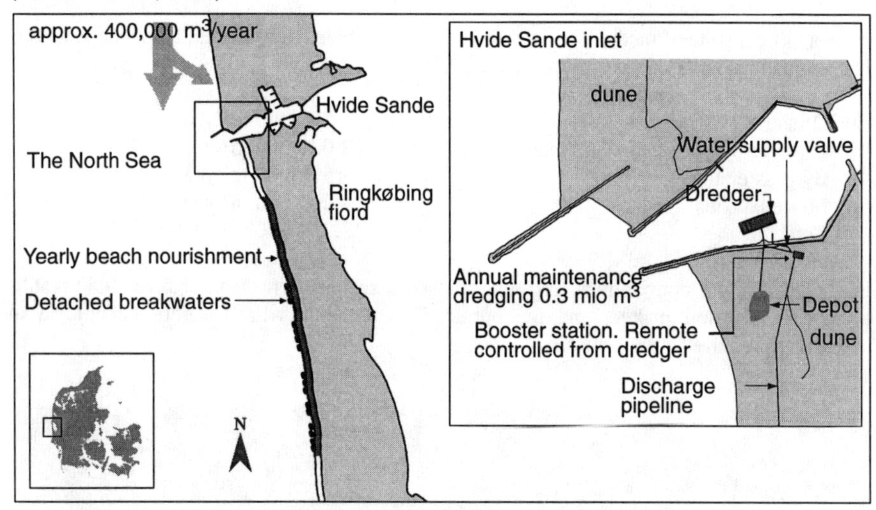

Fig. 23. By-passing at Hvide Sande.

New Initiatives Along the Inner Coasts

Orginally the role of DCA regarding the protection of the Inner Coasts - i.e. coasts inside the Skaw (Skagen) - was primarily of a regulatory nature. However, with the introduction of the Coast Protection Act of 1988, a mode of cooperation was prescribed, whereby DCA should support regional councils in their efforts to coordinate larger coastal erosion management schemes (see also the section "Coastal Zone Management").

This approach may have its restrictions with respect to encouraging a modern and environmentally acceptable attitude among the public, notably front property owners, especially because of lack of funds. Today the situation is (Fig. 24):

- About 10 major schemes coordinated by regional councils have been prepared with DCA assistance

- Three major reaches are handled by private associations, supported by DCA, viz. on the islands of Funen (north coast), Lolland (south coast) and Falster (east coast).

- One large beach protection and beach park has been established along an 8 km long stretch south of Copenhagen (CPH).

- Numerous individual, seldomly coordinated schemes of minor size are still being permitted and executed. There are probably also a number of works that have never been authorized. With 7,000 km of coast one can not supervise everything!

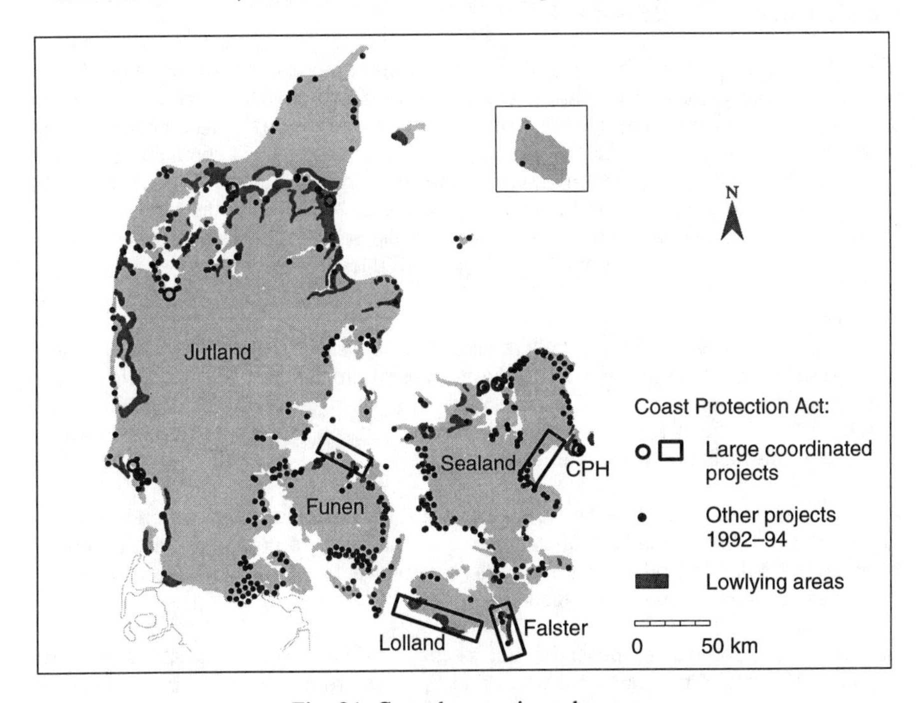

Fig. 24. Coastal protection schemes

DCA has a good majority of all projects in their files, especially after the implementation of the Act of 1988. Before then, revetments and similar structures on the dry beach were under different jurisdictions. Recently, in order to be able to set goals for a new formulated and environmentally acceptable effort within coastal protection of the Inner Coasts, DCA has worked out at new program - "The Inner Coasts" - to be launched in 1996. This shall run for the next four years.

The program will re-examine all 7,000 km of the Inner Coasts and update their status. Subsequently, the coasts will be classified according to their coastal qualities and protection status. The work will be based on modern photogrammetry, supplemented by a considerable number of field inspections. A data base will be built up. At the same time the Ministry of Environment and Energy will set up an exploratory regional program, from which a set of environmental key values and parameters will be developed.

It is belived that, within a few years and even before the study has been completed, it will be possible to present the issue in such a way to the central administration and the politicians, that a reformulated approach to the handling of coast erosion problems along the Inner Coasts can be presented:

For the benefit of those living along the coasts or using the beaches and for the benefit of society as a whole!

HANSTHOLM AND THYBORØN

In the 1950s and 1960s two major projects in the field of coastal and harbour engineering presented a major challenge to Danish coastal engineers and, through the work with these two projects, coastal engineering science and technology in Denmark made unprecedented progress. One project concerned the plan to build a major fishing and commercial port on the very exposed site of Hanstholm located on the NW corner of Jutland (Fig. 2); the other one the plan to close the Thyborøn channel connecting the North Sea to the Limfjord 50 km south of Hanstholm in order to stabilize the coasts of the narrow sand barriers separating the Limfjord from the North Sea. Both these projects had a long history.

Hanstholm

In the case of Hanstholm plans to build a harbour at this site were conceived early in this century, and in the 1930s the western breakwater was constructed, see Fig. 25. Construction was, however, interrupted by the second world war when the German occupation army evacuated the village of Hanstholm and transformed it into a fortress provided with heavy guns that could shoot half the way to Norway!

After the war it was found that the breakwater had suffered serious deterioration and, more importantly, by 1948 some 2 million m^3 of sand had been deposited in the lee of the breakwater (on the East side). This caused the government to set up a committee to analyse the project and recommend future action.

In 1951 this committee recommended not to complete the project because it found that sedimentation around the completed port could be so severe that it could be impossible to maintain adequate navigation depths in an entrance channel to the harbour.

In 1955 Helge Lundgren, who had been appointed Professor of Harbour and Foundation Engineering of the Technical University of Denmark in 1950, found the conclusion of the committee to be highly questionable. Upon his recommendation the minister of public works reopened the matter and commissioned Lundgren's laboratory to carry out new studies, including model tests with a movable bed.

A few basic facts about site conditions at Hanstholm will be given. The headland at Hanstholm owes its stability to the fact that the sea bed consists of rock (limestone) on the otherwise completely sedimentary west coast of Jutland. The rock is partially covered with sand and shingle.

The astronomical tide is insignificant, less than 0.3 m. Extreme water levels and currents are therefore completely atmospherically generated.

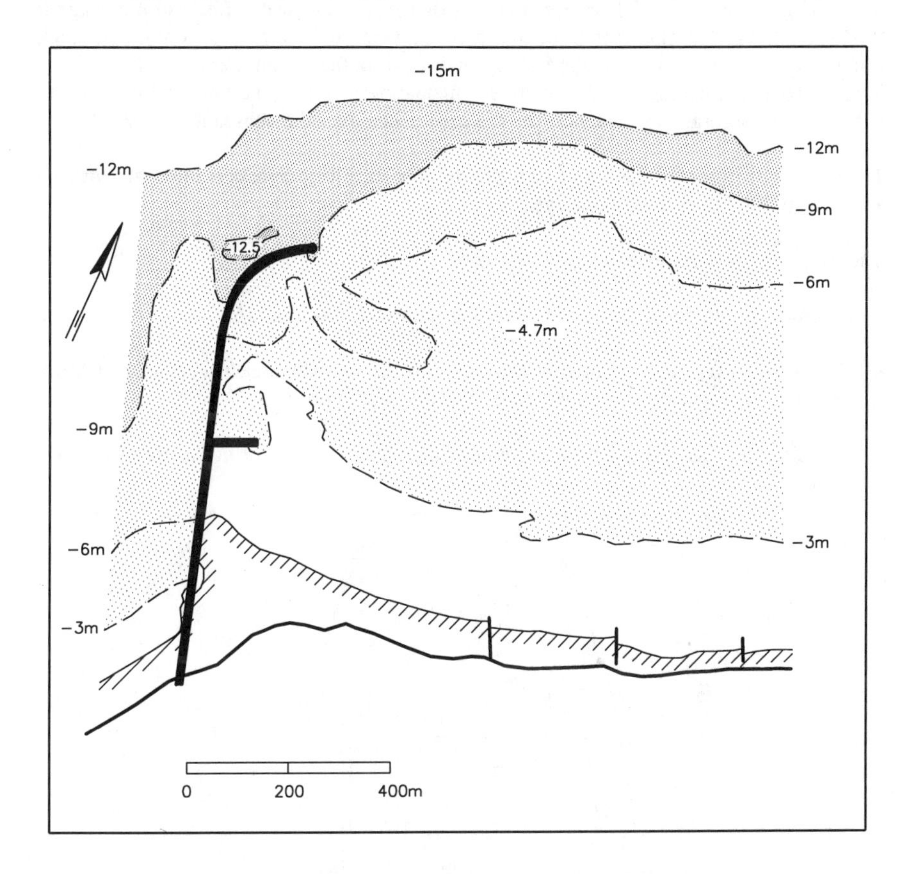

Fig. 25. Hanstholm breakwater and depth contours, 1951.

The coastal current at Hanstholm generated by wind set-up effects runs very close to the shore and even dominates the wave generated currents. Westerly storms dominate in the very rough wind climate and generate nearshore extreme waves of significant wave heights of up to 8 m.

The studies generated a few interesting innovations:

The first oscillating water tunnel in the world was developed and used to study the behaviour of a sand bed under large waves at the scale of 1:1 (Lundgren and Sørensen, 1958).

Wind-generated currents at the site were measured in storm conditions using rockets at night provided with a battery powered light and surveyed from land by means of two theodolites.

The movable bed model tests were designed using a new concept whereby only the outermost parts of the breakwaters were reproduced in the model. This made it possible to use the scales of 1:25 vertical and 1:35 horizontal. This reduced the scale effects to a manageable level such that Helge Lundgren through theoretical work and the results from the oscillating water tunnel could predict both the future water depth in the entrance area involving little or no maintenance dredging and the future sedimentation inside the harbour (50,000 m^3/year) with amazing accuracy as evidenced by the experience over the subsequent 25 years.

The key issues related to the design of the harbour were thus to bypass the littoral drift which amounts to 700,000-1,000,000 m^3/year.

The bypassing was achieved by:

- Carefully streamlined design of the harbour entrance, see Fig. 26.

- Taking advantage of the wave reflection from the vertical breakwaters to increase the sediment in suspension at the entrance, and thus the bypass.

- Avoiding the classical upstream accumulation; again an effect of the vertical breakwater design.

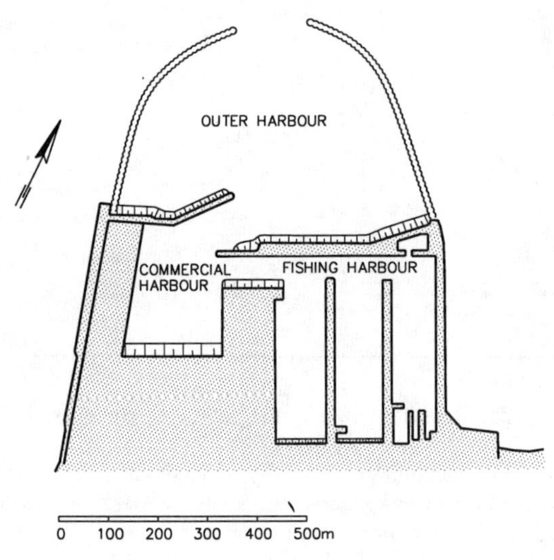

Fig. 26. General plan of Hanstholm Harbour.

After the completion of these studies the Danish government decided to build the port of Hanstholm. This led to a number of additional studies (at the university laboratory), which produced further important innovations, the most important ones being:

a) Development, in the close cooperation between Helge Lundgren and Torben Sørensen, of a new type of vertical face breakwater consisting of reinforced concrete cylindrical shells of wall thickness 250 mm, with a backward sloping upper part of the seaward face, see Fig. 27 (Lundgren, 1962).

b) Development of the irregular waves model testing technology for the study of wave disturbance in harbours including the use of moored model ships in the harbour, see Fig. 28 (Sørensen, 1973).

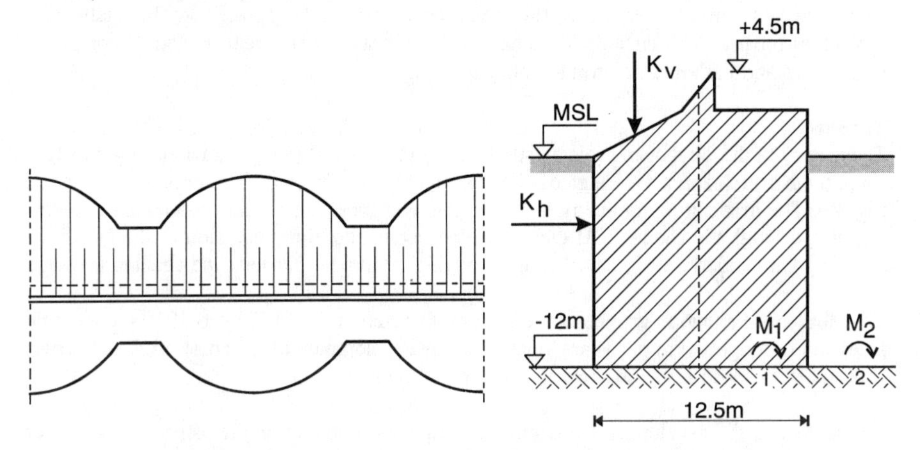

Fig. 27. Plan and cross-section of Hanstholm breakwater structure.

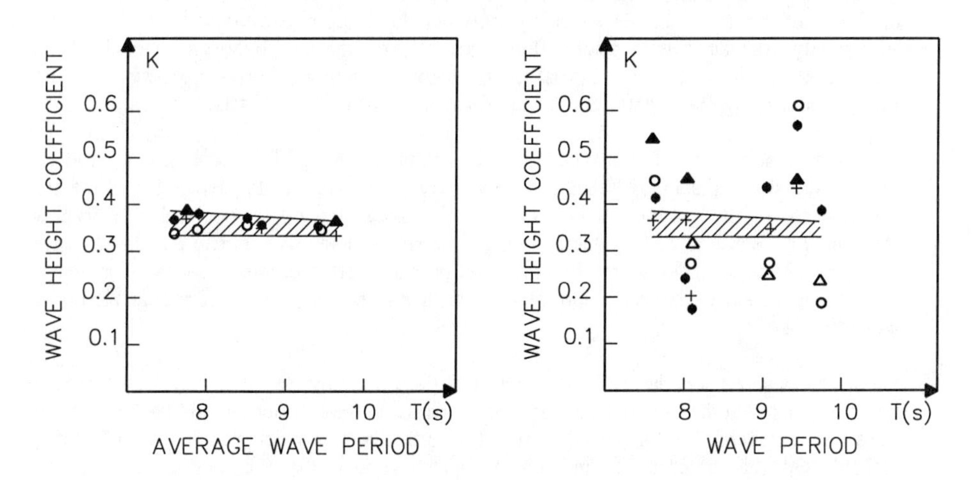

Fig. 28. Hanstholm harbour. Wave height coefficients obtained at four locations along a 50 m long section of the most exposed wharf. Left: Irregular waves. Right: Regular waves.

The new type of vertical face breakwater, which substantially reduced the wave forces and overturning moments on the structure as compared with traditional vertical face breakwaters, was later used for the Brighton Marina in the UK and for the extension of the Marsa el Brega breakwater in Libya. These concrete breakwaters have required little repair.

The new technology of model tests with wave disturbance in harbours has since gained general acceptance all over the world. Thus the innovations generated by the challenging project for the port of Hanstholm have been of benefit not only to Denmark but to engineers, researchers and harbours on an international scale.

Thyborøn

The history of the Thyborøn project dates back more than 100 years to a winter night in 1862 when the North Sea in a severe gale broke through the narrow sand barrier at Thyborøn, see Fig. 29. The breakthrough quickly increased into a channel wide and deep enough to permit shipping through the channel. At the same time severe shoreline recession of the North Sea coasts of the barriers set in, amounting to roughly 20 m per year near the channel entrance.

To stabilize the situation groynes were built on these coasts from 1884 to 1910 reducing the recession to roughly 2 m per year. Since 1897 the development of the coast profiles has been kept under close observation through regular soundings.

To understand the development it is important to keep in mind that the astronomical tide on this coast is very small, max. approx. 0.3 m, whereas extreme high water levels caused by wind set-up with westerly winds may reach 2 m or more. Due to the large water surface area of the Limfjord (see Fig. 2) strong onshore winds result in severe wave action being consistently correlated with strong flood current through the Thyborøn channel. As a consequence of this the heavy littoral drifts on the barriers are invariably swept into the Limfjord during westerly gales and thus permanently lost to these coasts.

Since the construction of the groyne system the initial flattening of the coast profiles caused by the excessive recession of the shoreline was replaced by a gradual steepening of the sea bed seaward of the groynes, which went on for more than 60 years. In the 1930s this steepening gave rise to deep concern among the responsible engineers who felt that disaster was looming ahead. They feared that the coast profiles were becoming so steep that even a short series of severe storms could generate such effects that control of the development would be lost.

On this basis an act was passed by Parliament in 1946 whereby the Thyborøn channel would be closed such that the loss of sand from the sea coasts to the Limfjord would be stopped and the sea coasts thereby stabilized. The project included the construction of 16 km of "safety" embankments placed approx. 2 km behind the coastline, as well as the building of two major breakwaters and two sluices, for vessels and salt water inflow, respectively.

Before the works to close the channel had been commenced Per Bruun in his Doctor's thesis (Bruun, 1954) raised serious doubts as to the "disaster theory" which was the basis for the project and proposed new investigations of the problems. Along similar lines Helge Lundgren in 1954 and 1956 made specific proposals for new investigations, including scale model tests with movable bed to be carried out in Holland, particularly with a view to keeping the channel open and to saving a major part of the project costs.

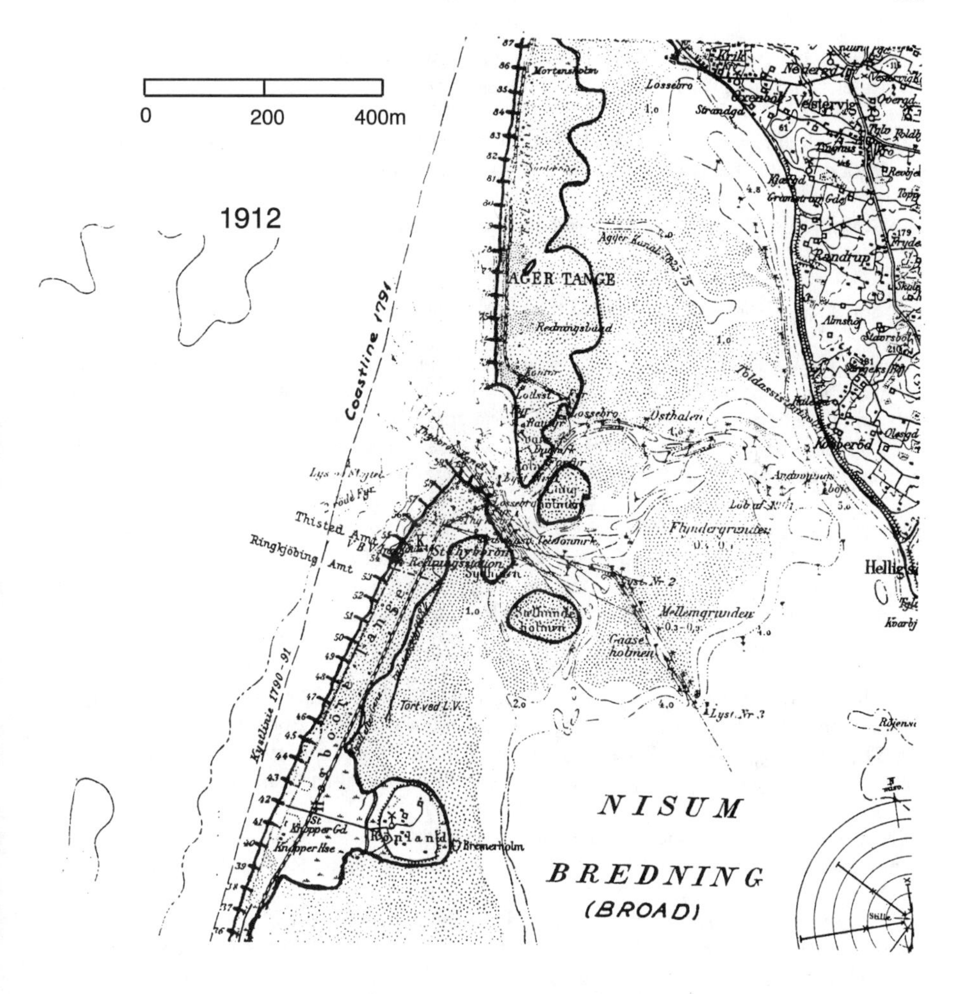

Fig. 29. Thyborøn Barriers and Channels 1912.

The government followed these recommendations and a range of studies of various aspects of the problem were initiated, the most important ones of which were carried out at the Technical University of Denmark headed by Helge Lundgren and Torben Sørensen.

The crucial question of whether the steepening of the coast profiles indicated a risk of disastrous development was resolved by an analysis demonstrating that the steepening was simply a question of the coast profiles adjusting to the reduced shoreline recession achieved by the construction of the groyne system (Sørensen, 1960). This realization in fact eliminated the basis for this very expensive and in many ways controversial project. The soundings of coast profiles over the subsequent 40 years have confirmed the correctness of this analysis.

However, the project had a number of other aspects which also required analysis and consideration. The principal ones of these were the following:

a) The hydraulic resistance of the Thyborøn channel: Was it decreasing, stable, or increasing with possible effects on storm flood levels in the Limfjord?

b) Storm flood levels in the Limfjord, especially with regard to extreme water levels and flooding.

c) Statistics of storm flood levels in the North Sea at Thyborøn.

During the years 1958-63 all of these aspects were analysed by means of numerical modelling (often - somewhat misleadingly - referred to as mathematical modelling) using the first electronic radio valve based computer in Denmark, DASK, which has just been installed. A one-dimensional explicit numerical model was developed at the Technical University of Denmark and calibrated and verified with very satisfactory accuracy.

The application to develop storm flood statistics for the North Sea at Thyborøn was motivated by the fact that the water level recorder placed on one of the groynes was malfunctioning, especially during severe storms. Instead, the storm flood levels were determined by the numerical model using the reliable records from a recorder placed in the port of Thyborøn inside the Thyborøn Channel.

Finally, a two-dimensional explicit numerical model of the entire North Sea developed by Professor Walter Hansen in Hamburg around 1960 was used to calculate storm flood levels at Thyborøn during scaled-up extreme storm conditions so as to arrive at deterministically calculated extreme storm surges to supplement the statistical analysis.

These studies marked the introduction of numerical modelling of hydraulics in coastal engineering in Denmark at a very early time - probably among the first in the world.

As a result of the new scientific and engineering studies the act of 1946 was repealed in 1970. It was now concluded that the channel could remain open for at least 30-50 years provided that DCA followed the development of the barriers and inlet very carefully and continued the normal procedures of protection.

This has thus been done regularly and the 25th anniversary has been marked by a more thorough examination of all aspects of the development in collaboration between DCA and DHI. Some of the results of this examination are:

a) The number of higher storm flood levels has increased significantly from 1940 to 1990. The annual recession of unprotected profiles is multiplied by 3-4 in the latest 20 years.

b) In general, the protected profiles have become steeper with a gradual reduction of the sediment transport.

c) The cross-sectional areas of the Thyborøn Channel have been increasing steadily, by 40-60% from 1935 to 1995.

DANISH HYDRAULIC INSTITUTE (DHI)

One of the most important, more recent events in the field of coastal engineering in Denmark - and that of hydraulic and hydrological engineering in general - has turned out to be the establishment of the Danish Hydraulic Institute in 1964 by the initiative of Helge Lundgren. The basis for this on the professional level was not least the experience and technology developed at the Technical University of Denmark through the work with the projects of Hanstholm and Thyborøn described above. However, it was clear from the beginning that a relevant and sustainable operation in this field could not be based on Danish projects alone, so international activities had to be actively pursued.

At that time the important and highly professional Danish engineering and construction company of Christiani & Nielsen, who had subsidiaries in numerous countries all over the world, for many of its projects needed advanced hydraulic technology and expertise and chose to rely upon Danish engineers for support of this kind.

This provided the newly established DHI with the opportunities and challenges on the international scene that it needed to start a sustainable development leading up to the present situation where DHI is recognised as one of the leading centers in the world in hydraulic engineering, including coastal engineering.

Many people and companies in Denmark and abroad have contributed to this development. Most important, however, has been the continued strong collaboration with the Technical University of Denmark as well as with Dr. Michael B. Abbott, whom one might call the father of modern computational hydraulics.

DHI owes much of its successful development to the fact that it was from the beginning established in the domain of private enterprise as a non-profit operation. The founding organization was the Danish Academy of Technical Sciences, which is a private foundation with the objective to promote application of new scientific knowledge and technology in private businesses and government activities. To achieve this objective the Academy has established a range of institutes dealing with widely different areas of technology, all of them with the purpose of serving private business and government agencies with advanced technology and expertise that they normally would not or could not develop themselves.

It is a direct consequence of this concept that the institutes, including DHI, shall not compete with private business, but serve and complement it. The institutes therefore constitute a very convenient vehicle for the government to channel financial support for technological development to private business without infringing on national or international rules for fair competition. Thus the institutes, while supplying their services and technology on a strictly commercial basis, receive on the average approx. 20% of their turnover as government funding of their R&D activities under strict control by the government to ensure that these funds are not used to subsidize their commercial activities.

This structure has proven to be an excellent framework for the development of DHI. Contrary to other leading centers in the field of hydraulics and coastal engineering DHI has never had to go through the agonizing process of privatization. The attitudes of efficiency and purposefulness characteristic of successful commercial enterprises have always been at the core of DHI's company culture.

At the same time the government's financial support for its R&D has enabled DHI to take the lead on a number of crucial points over the last 30 years, such as the introduction of irregular waves in wave disturbance physical modelling (1972), the development of generalized numerical modelling systems - computational hydraulics -for rivers, estuaries and shallow seas (1970), and the introduction of hydroinformatics in coastal and hydraulic engineering (1985), just to name a few examples.

Without any question DHI has contributed greatly to strengthen the position of Danish engineers on the international market and to the solution of environmental and engineering problems in Denmark and abroad in collaboration with authorities and engineers in Denmark and countries throughout the world.

BASIC RESEARCH IN COASTAL ENGINEERING

The basic research in Coastal Engineering in Denmark has mainly been carried out at the Technical University of Denmark in Copenhagen. Formerly the environment was divided between the Harbour and Coastal Engineering Laboratory and the Hydraulic Laboratory. In 1971, where these two laboratories were headed by Professors H. Lundgren and F. Engelund, respectively, it was agreed to merge the two laboratories into one institute named the Institute of Hydrodynamics and Hydraulic Engineering, or ISVA which is an abbrevation of its Danish name.

In the sixties a new university was formed in Aalborg (AAU) and here extended research on breakwaters and other subjects has been made under the guidance of Professor Hans F. Burcharth.

It must also be mentioned that at the Geographical Institute of the University of Copenhagen, significant contributions have been made to the understanding of the morphology of coastal plains and coastal development. Here the Professors Axel Schou (1945) and Kingo Jacobsen were pioneers.

Outside the university environment, much basic research has been carried out at DHI, which in recent times has spent 20-30% of its total budget on research and development. In the field of coastal engineering the DHI-research has especially concentrated on the development of integrated computational models to describe waves and currents and their impact on structures and sediment transport.

In the following the development in basic research is described a little more in depth in the fields where the Danish contributions are most significant, namely sediment transport, wave hydrodynamics, breakwaters and computational modelling.

Sediment Transport and Wave Hydrodynamics

Energetic Models. Today it is very popular (especially in the US) to relate the local sediment transport to the dissipation of wave energy. One well-known example is the Bailard-Inmann formula (Bailard and Inman (1981)). They adapted their approach from Bagnold (1963). The famous CERC-formula for longshore sediment transport is an integrated version of the above mentioned approaches: Here the longshore sediment transport rate is related to the so-called 'longshore energy flux factor'.

In Denmark, a forerunner of the CERC-formula was proposed by Professor J. Munch-Petersen already in 1914 (see Svendsen (1938) or Rang (1944)). Munch-Petersen's work was inspired by his predecessor Professor Palle Bruun, who outlined some basic ideas in a lecture already in 1909. Munch-Petersen based his analysis on the energy of waves and without knowing concepts such as wave energy flux, let alone radiation stresses, Munch-Petersen concluded that the wave energy is lost in the surf zone. This loss can be taken as a measure of the wave attack on the sea floor and beach. Actually, he also concluded that the loss in energy would create the wave-driven longshore current, and that this current was the main factor responsible for the longshore transportation of sediment, a statement with which everybody agrees today. This happened in 1914.

Recent Development. The theoretical development in the description of sediment transport in the coastal environment has later been characterized by a close interaction between improved understanding of sediment transport mechanisms and wave hydrodynamics. It was Professor Engelund who in the sixties started to consider the sediment transport from a hydro-mechanic point of view and introduced a very detailed description of sediment transport, which for instance included the influence from bed waves on bed roughness and sediment transport (Engelund and Hansen, 1972). Since 1950 Professor Lundgren was responsible for the very high level of wave research in Denmark (Lundgren, 1984).

Wave Boundary Layers. It was very early realized by Professor Lundgren that an improved understanding of sediment transport in the coastal environment required a detailed knowledge of the near-bed flow processes under waves. Here the turbulent wave boundary layer is located, which has a thickness of 2-10 cm. This layer is of major importance to keep sediment in suspension. To study this on an experimental basis to a scale of 1:1, a large pulsating water tunnel, today named an oscillatory flume, was constructed in the late fifties (Lundgren and Sørensen, 1958). This facility allowed experiments to be performed in the fully turbulent range. Among many famous experiments in this flume the large number of wave boundary layer experiments must be mentioned, which include those by Jonsson (1963), Jonsson and Carlsen (1976). These formed the basis for Jonsson's expressions for the bed-friction factor of a rough bed, which were the first of its kind when they were presented. Although many smaller corrections have been introduced by other researchers later, the expressions by Jonsson are still widely accepted. The oscillatory flume is still very much in use, and now supplied with modern instruments like LDA- and hot-film equipment, so more detailed information on wave boundary layer properties can be achieved (Jensen et al. (1989), Fredsøe et al. (1993)).

Waves Plus Current. When waves co-exist with a current, then waves will interact with the current, modifying the vertical distribution of the flow velocity and in addition, increasing the flow resistance. One of the best known papers on this topic is that by Madsen and Grant (1976), who applied an eddy-viscosity concept to model the combined flow. Lundgren (1972), however, several years earlier described the physics and put forward a procedure to calculate the increased flow resistance due to the presence of waves. His idea was that the increased eddy viscosity near the bed would, for the same bed shear stress, decrease the vertical gradient in the flow velocity. This leads to a smaller water discharge. The eddy viscosity was based on Jonsson's measurements obtained in the oscillatory flume. Lundgren's work was inspired by a large study in Karachi, Pakistan (Kirkegaard and Sørensen, 1972), where DHI was commissioned to predict the siltation of a new access channel. Later,

Lundgren's model has been improved and extended by Danish researchers such as Christoffersen and Jonsson (1985) and Fredsøe (1984).

Radiation Stress (or Wave Thrust). The radiation stress concept is today very important to coastal engineering. It forms the basis for modelling wave-driven currents and wave-induced variations in the mean water level - including bound long waves. This force, which turns out as a time-averaged quantity (mean value over one wave period), consists of two contributions: one from the pressure and another one from the flux of momentum. The concept was first published by Longuet-Higgins and Stewart (1962). In the Scandinavian countries, Lundgren already in 1951 in his lectures was teaching his students (including one of the authors of this article, Torben Sørensen) about this same concept (which he at the time called the "wave thrust"). Unfortunately, he did not publish his ideas, including the mean energy level, until the IAHR Congress in London, 1963 (Lundgren 1963).

Surf Zone Dynamics. The surf zone has always fascinated coastal engineers (and human beings in general). To achieve a real understanding of the physics of this phenomenon, however, is a major challenge. To properly describe the violent turbulence formed by the breaking waves will always require some idealization (schematization) of the physical processes involved. Much effort has been spent on the description of the behaviour of waves in the surf zone by Danish researchers such as Svendsen, Madsen, and Deigaard. All these researchers have taken their basis in the analogy between a broken wave and a bore, described as a hydraulic jump. Svendsen and Madsen (1984) used turbulence modelling to describe the properties of broken waves, while Deigaard (1989) later utilized a small note by Engelund (1981) to get an amazingly good description of the waves in the surf zone. Engelund's note also described the hydraulic jump, and from the momentum equation he was able to directly calculate the thickness of the 'dead' roller. Deigaard (1989) applied this principle to explain under which conditions a wave will start to break and when it will stop to be broken and, further, to calculate the volume of the surface roller in broken waves. This principle has been implemented in the Boussinesq-modelling to get reliable models for waves in the surf zone. Realistic modelling of plunging breakers is, however, still a challenge for the future.

Sediment Transport Modelling. Detailed descriptions of the sediment transport in the marine environment of combinations of waves and currents were made by Fredsøe et al. (1984) and Deigaard et al. (1986). The former work uses an intra-wave model to describe the behaviour of suspended sediment in the wave-boundary layer for the general wave-current case. In the latter work, the description is extended to include the influence of surface-generated turbulence from broken waves. These works form a natural extension of the basic research on the sediment transport in steady currents carried out within the "Engelund School". Thus, the Engelund-Fredsøe sediment transport model (1976) is the basis for both the above mentioned works.

Morphological Modelling. With a detailed sediment transport description in hand, much attention has been given to morphological modelling in a fruitful cooperation between DHI and ISVA during the latest decade. For some years profile modelling was in focus, and here an accurate description of on/offshore sediment transport is needed. Svendsen (1984) modelled the resulting near-bed flow velocities in the cross-shore direction inside and outside the surf zone. This work was later extended by Deigaard and Fredsøe (1989). Subsequently, the hydrodynamic description was combined with the so-called STP-module (sediment transport module) to describe bar development and related cross-shore phenomena (Brøker et al. 1991).

Rubble Mound Breakwater

The major part of the breakwaters constructed in Denmark is of the rubble mound type. In the inner Danish waters, they are mostly protected by large boulders as armour, whereas suitable quarry stones may be obtained either from a Danish island in the Baltic Sea or imported from Norway or Sweden. Research has focused on wave overtopping, armour layer stability, reliability analysis, and berm breakwaters.

Wave Overtopping. The amount of water overtopping breakwaters is of interest due to the risk of damage to the rear side of the breakwater or to persons, buildings, structures, and equipment located on and behind the breakwater as well as moored ships; and due to waves generated by overtopping. These aspects have always had a high priority in the research at the DHI (Jensen and Juhl, 1987) and, recently, expressions for the mean overtopping discharge and for the individual overtopping volumes of traditional rubble mound breakwaters were established.

Armour Stability. The research by Bruun and Günbak (1977) has triggered subsequent studies on the armour layer stability due to the effect of wave sequences and individual waves.

Following the catastrophic failures of some large breakwaters armoured with slender types of armour units, research in this field was initiated worldwide. At Aalborg University extensive research on both hydraulic stability and structural integrity of Dolos armour units has been carried out, and design diagrams have been presented by Burcharth (1993).

Reliability Analysis. All parameters governing the stability of breakwaters are stochastic of nature and thus associated with standard deviations. Burcharth has played a major role in the application of reliability analysis to coastal structures and the establishment of partial coefficients of safety for rubble mound breakwaters (PIANC TCII, Report of Working Group No. 12). At present he is working with the same aspects for vertical breakwaters.

Berm Breakwaters. Without being named so, berm breakwaters have actually been applied since the middle of the nineteenth century. In 1978 the DHI developed a berm breakwater alternative for Skopen Harbour, Faroe Islands (Jensen and Sørensen, 1987). Since then, an increasing interest has been shown in this type of breakwater.

At present, a comprehensive joint European research project is carried out by seven institutions with the DHI as the coordinator and Aalborg University as a partner. The main objective is to arrive at a better design basis for berm breakwater structures, especially with respect to the reshaping of the trunk and the breakwater round-head, which is an essential process in arriving at a structure with satisfactory long term stability.

Vertical Breakwaters

Wave Forces. The research on vertical breakwaters was initiated by the intensive hydraulic studies carried out for the port of Hanstholm, which was completed in 1970 (Lundgren, 1962). In order to eliminate the impact forces, the caissons were cylindrical with a 30° slope at the top, starting from still water level (Lundgren and Gravesen, 1974). The costs were further reduced by applying 250 mm thick cylindrical shells of reinforced concrete for the caissons. This type of structure, for which structural considerations were of prime importance, was developed in a close cooperation between H. Lundgren and T. Sørensen. It is widely known as the 'Hanstholm' type of breakwater.

Research has continued, particularly, on various measures (in cross-section and plan) that can be taken with a view to reducing both the wave forces and the amount of overtopping, with due regard to constructional, structural, foundation and architectural considerations (Lundgren and Juhl, 1995).

Interaction between Breakwater and Foundation. Geotechnical problems, including parameters for dynamic analysis, instantaneous pore pressures, pore pressure build-up and soil degradation, play an important role in the design of vertical breakwaters. This applies not only to the safety against sliding and foundation failure, but also to the interaction of rocking, shifting and sliding. The breakwater-foundation interaction is part of a joint European research project in which Aalborg University participates.

Computational Hydraulics

One of the decisive steps forward in hydraulics practice in Denmark was introduced by Professor Michael B. Abbott. In the sixties he approached the director of the Danish Hydraulic Institute, Torben Sørensen, and convinced him that the future in coastal engineering would rely heavily on robust, sophisticated numerical models. This "computational hydraulics" started with the so-called 'box models' where the net and gross transports of fluid volumes or other properties were estimated on a quite crude basis.

Real progress was first introduced by Abbott in the so-called System 11 which provided a numerical solution of the partial differential equations of conservation of mass and momentum for rivers, i.e. for one-dimensional flow of a one-layer (vertically homogeneous) fluid. This was quickly followed by a two-dimensional model, System 21, and later by a three-dimensional version, System 3.

These numerical modelling systems were gradually expanded to cover more hydraulics-related processes such as advection-dispersion, waves, sediment transport and water quality. Of particular interest to coastal engineering is the rapid development during the latest 10-15 years of the coastal sediment transport modules. A variety of wave models allow the computation of the near-shore wave field and thus the radiation stresses, which can be used as one of the driving forces in the 2D hydrodynamic model. This model can then compute the total near-shore current pattern in response to waves, tide and wind. Hence, all the necessary data are available for computing the 2D sediment transport pattern and the erosion/deposition around, e.g., groynes, offshore breakwaters, and harbours.

A coastline and coastal profile model, LITPACK, was developed at DHI in parallel with the above work on 2D models. LITPACK allows studies of the long term (5-50 years) development of coasts in response to all possible coastal protection works such as groynes, revetments, and offshore breakwaters.

The latest scientific development, supported by the European MAST programme and the Danish Technical Research Council, is the extension of the 2D modelling complex to enable true, short term morphological modelling where the sea bed level changes interactively under the influence of the wave field and current patterns (Brøker et al., 1996).

The recent investigation of the morphological evolution around the Thyborøn Channel on the West Coast of Jutland, Denmark, is an illustrative example of the application of the numerical modelling tools. Fig. 30 shows the initial bathymetry and the bathymetry and sediment transport field 2 days and 15 hours after the start of a historical storm. (In the figure, only a small part of the model area is shown).

As a logical consequence of the fact that today powerful computers are available to and can be used by all engineers, as well as the fact that present day education provides engineers with a high level of knowledge and understanding of hydraulics and related processes, the numerical models developed by centers of excellence, such as DHI, are now available in user-friendly forms to practicing engineers in Denmark and on a world-wide basis.

COASTAL ZONE MANAGEMENT (CZM).

Management of the coastal zone has been in existence in Denmark, as in several other countries for many decades, indeed through centuries. Physical planning as such has been applied in this country for about 100 years. CZM has never been defined as a separate issue but has organically matured through gradual harmonisation and coordination of administrative and legislative frameworks.

An example of an early CZM initiative provoked by necessity (and, at present, actually an integral part of CZM), was the Dune Protection Law, compelled by the migration of dunes in coastal zones, which aggravated in the 16th century and continued into modern times. The first decree was issued in 1539, and the first law was enacted in 1792.

The Nature Preservation Act 1917 reinstated the right of public access to all Danish Beaches. The act was revised in 1937, when a definitive stop for placing buildings on the beach proper and the adjacent 100 m of the hinterland was enforced. A directive issued 1978 by the Ministry of Environment and Energy has provided a general stop for the planning of new vacation housing and hotel in a 3 km wide zone along open beaches. Finally, in 1994 the protection of nearshore areas in the whole country was extended to ban all construction - except coastal protection works - in a 300 m wide zone along all Danish coasts (Vaaben 93).

Coast Protection Act 1988. Ministry of Transport.
Since 1874 and 1927, respectively, Denmark has had a Dike Protection Law and a Coastal Defence Law based on the landowners individual responsibility for the coastal activity. The integrated act of 1988 substitutes the previous laws and takes into account the new dimensions in the coastal situation, including changed economic relations.

In principle the protection of land and property affects only the front property owners, while coast erosion management relates to areas and interests in considerable depth of the hinterland. Furthermore, especially beach nourishment schemes require conceptual thinking, planning and financing on a larger scale than previously imagined.

The regional authority has a leading role in the approval process with a solid background in its role in the general planning process, and is thus well suited to combine a good understanding of the local values at stake, with properly balanced problem assessment.

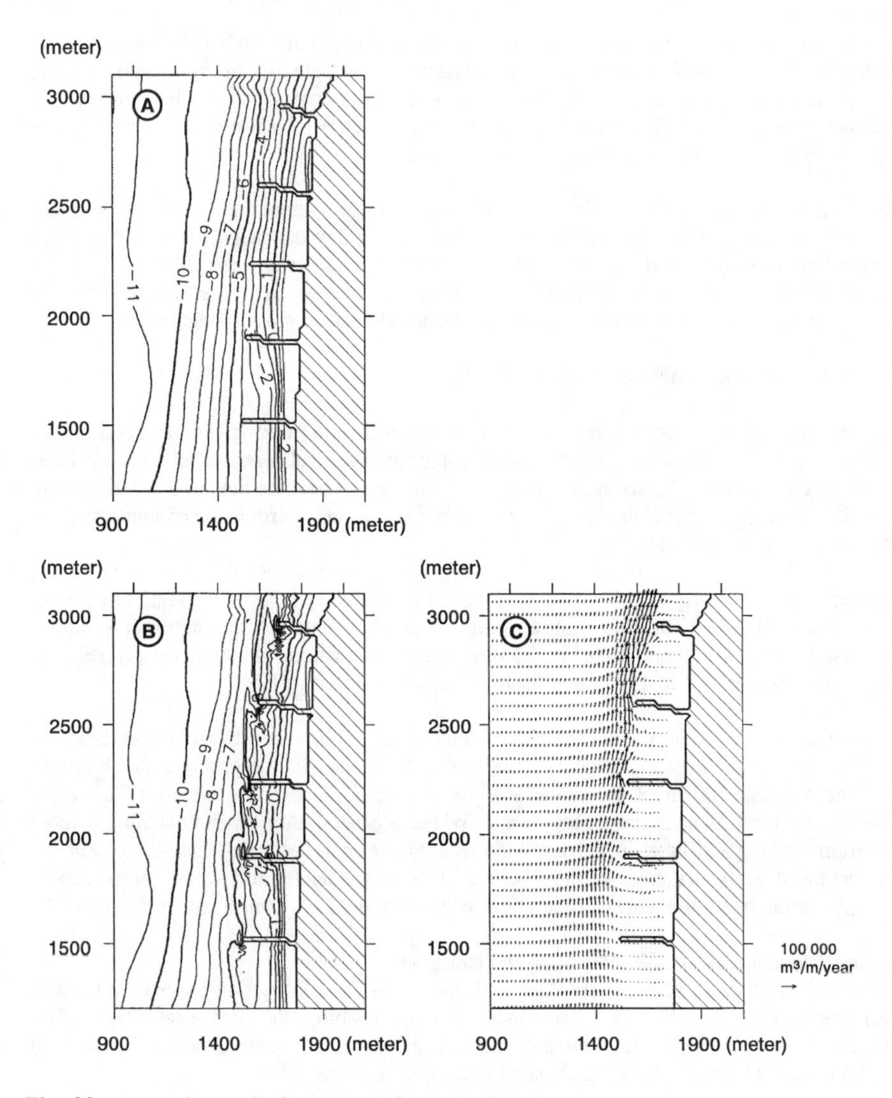

Fig. 30
a) Bathymetry at the start of a storm.
b) Bathymetry after 2 days and 15 hours, simulated by a morphological model.
c) Calculated instantaneous sediment transport field, 2 days and 15 hours after the start of the storm.

Often the regional authority lacks, however, the expertise required for a proper technical evaluation of proposed actions. Therefore it is required that DCA shall be involved directly in examination of the projects and subsequent approval.

In addition, the regulatory procedures still require final permission from DCA for all coast protection works and other man made changes in a zone 100 m landward of the coastline as well as seaward in the territorial waters.

The financing of coast protection schemes remains the responsibility of individuals. municipalities and regional authorities, except along the exposed North Sea coast, where considerable governmental resources are allocated to schemes of general and regional importance. (Jakobsen, 1994)

LOOKING AHEAD

In the future, the development of coastal engineering will be fast. New technologies in coastal protection will be introduced, and to a large extent hard structures will be replaced by soft ones (beach nourishment) being less harmful to nature. To optimize such structures a more profound understanding of the sediment transport pattern on/off shore as well as long-shore will be required. With respect to the modelling of these processes a large step was taken when 2D (depth-averaged) flow modelling of the coastal zone was introduced recently (in Europe mainly by very close collaboration between the large hydraulic institutes and the universities in the EU-funded research programme MAST). On a longer view real 3D modelling seems unavoidable. In river morphology 3D modelling has progressed very much during the last decade, and many morphological features can only be explained by the 3D flow structure (secondary currents). There is no reason not to expect a similar (delayed) development in the numerical/mathematical modelling of coastal morphology. Although the introduction of 3D modelling computationally will be very demanding, the computer capacity will certainly increase fast also in the future, and so 3D modelling will become feasible.

Sophistication of sediment transport models in the coastal environment is also needed. The knowledge of the behaviour of suspended sediment is sparse with respect to irregular waves, and the effect of long bound waves on the cross-shore sediment transport is today a white spot on the map.

Mechanisms such as grain sorting have also reached a much more sophisticated level in river engineering than in coastal engineering. In rivers, the transverse sorting often has significant impact on the morphology and sediment transport rates. In coastal engineering this problem has got almost no attention even though sorting is obvious, at least on beaches.

Intensive research in coastal hydrodynamics has been devoted to the description of regular and irregular waves up to breaking, mainly by application of potential flow theory. Much less attention has been given to the turbulent flow processes that dominate the surf zone and are extremely important for the sediment transport pattern.

The future will also see many attempts to make shortcuts. An example: If you are interested in the long-term changes of a coast caused by some external impact like a possible sea level rise, then you might ask whether very detailed modelling is really necessary? This has introduced the concept "long-term modelling". Actually, the long-term behaviour is the result of many short-term events, which one way or the other must be introduced in the long-term modelling. However, it seems possible to model the long-term behaviour by using a simplified description. Here the short-term behaviour is included in the coefficients that appear in the equations governing the long-term behaviour. In this way it is likely that the detailed deterministic modelling and the more overall modelling must go hand in hand in the future, and that sophistication of both types of modelling still will be needed.

In the field of physical modelling the technology has now reached a stage where the hydrodynamic processes in nature can be reproduced in all their significant aspects in scale modelling using 3-D multidirectional waves. The main obstacle in this context is that of cost related to the large size of model basins and machinery required for such modelling.

When we look at physical modelling of coastal sediment transport, however, the problem of scale effects is of course still with us. The only answer to this is that of large model scales, indeed so large that they are unattainable for practical applications. Therefore numerical modelling remains the only tool that can be developed for practical applications in the future.

However, further development of the numerical modelling tools are in urgent need of reliable and consistent data from physical experiments, data that could be obtained from experiments in a very large physical modelling facility where boundary conditions can be controlled at will and kept constant for the duration of the experiment. Such a facility would be so costly that it could probably be built only through a collaboration between a number of countries and be available to researchers through an international structure of collaboration.

Is this a fantasy - or is it a realisable dream?

REFERENCES

Bagnold, R.A. (1963): Mechanics of marine sedimentation. In: M.N. Hill (Editor), The Sea: Ideas and Observations. Interscience, New York, 3:507-553.

Bailard, J.A. and Inman, D.L. (1981): An energetics bed load transport model for a plane sloping beach; local transport. J. Geophys. Res., 86C:2035-2043.

Binderup, M. and Frich, P. (1993): Sea-level variations, trends and cycles, Denmark 1890-1990; Proposal for a reinterpretation. Annales Geoghysicae 11, 753-760.

Bruun, P. (1954): Coast stability, Copenhagen, Danish Technical Press.

Bruun, P. and Günbak, A.R. (1978): Stability of sloping structures in relation to $\xi = tan\alpha/\sqrt{H/L_o}$ risk criteria in design. Coastal Engineering, 1(4):287-322.

Brøker, I., Deigaard, R. and Fredsøe, J. (1991): Onshore/offshore sediment transport and morphological modelling of coastal profiles. *Proc. Coastal Sediments '91*, ASCE, Seattle, pp. 643-657.

Brøker, I., Zyserman, J. and Jakobsen, P. Roed (1996): Thyborøn coastal investigations 1995: New lessons from an old coastal problem. Accepted for the *ICCE '96 Conference.*

Burcharth, H.F. (1993): Structural integrity and hydraulic stability of Dolos armour layers. Series paper 9, Dept. of Civil Engineering, Aalborg University, Denmark.

Christoffersen, J.B. and Jonsson, I.G. (1985): Bed friction and dissipation in a combined current and wave motion. Ocean Eng., 12(5):387-423.

Danmarks Meteorologiske Institut (DMI) (1988): Danmarks klima 1988. DMI.

Danmarks Natur (1979): Vol. IV.

Deigaard, R., Fredsøe, J. and Brøker, I. (1986): Suspended sediment in the surf zone. J. Waterw. Port Coastal Ocean Eng., ASCE, 112(1):115-128.

Deigaard, R. (1989): Mathematical modelling of waves in the surf zone. Progress Report No. 69, Inst. of Hydrodynamics and Hydraulic Engineering, ISVA, Techn. Univ. Denmark, pp. 47-60.

Deigaard, R. and Fredsøe, J. (1989): Shear stress distribution in dissipative water waves. Coastal Engineering, 13:357-378.

DHI 1991. LITPACK. Danish Hydraulic Institute.

Engelund, F. and Fredsøe, J. (1976): A sediment transport model for straight alluvial channels. Nord. Hydrol., 7:293-306.

Engelund, F. (1981): A simple theory of weak hydraulic jumps. Progress Report No. 54, Inst. of Hydrodynamics and Hydraulic Engineering, ISVA, Techn. Univ. Denmark, pp. 29-32.

Engelund, F. and Hansen, E. (1972): A monograph on sediment transport in alluvial streams. Teknisk Forlag, Copenhagen.

European Commission Services (1996): Demonstration programme on integrated management of coastal zones.

Fredsøe, J. (1984): Turbulent boundary layers in wave-current motion. J. Hydraul. Eng., ASCE, 110(8):1103-1120.

Fredsøe, J., Andersen, O.H. and Silberg, S. (1985): Distribution of suspended sediment in large waves. J. Waterw. Port Coastal Ocean Eng., ASCE, 111(6):1041-1059.

Fredsøe, J., Sumer, B.M., Laursen, T.S. and Pedersen, C. (1993): Experimental investigation of wave boundary layers with a sudden change in roughness. J. Fluid Mech., 252:117-145.

Frisch, P.H. (1991): Coastal inlets on the Danish coast, curses or blessings. *PIANC* Bulletin no. 72, pp. 52-57.

Geil, B. (1995): Carlsen. Unpublished.

Jakobsen, P.R., Tovgaard, N. and Larsen, K. (1987): Copenhagen metropolitan region, coast erosion management. *Coastal Zone '87*, Seattle, Washington.

Jakobsen, P.R. and Madsen, H.T. (1993): Morphology of the Danish North Sea Coast. Coastlines of the Southern North Sea, pp. 41-51 (Coastlines of the World), *Coastal Zone '93*, New Orleans.

Jakobsen, P.R. and Sandgrav, B. (1993): Modern Beach Nourishment in Denmark - how do you do it! Kystinspektoratet. Unpublished.

Jakobsen, P.R. (1994): Aspect of coastal zone management in Denmark. Unpublished.

Jensen, B.L., Sumer, B.M. and Fredsøe, J. (1989): Turbulent oscillatory boundary layers at high Reynolds numbers. J. Fluid Mech., 206:265-297.

Jensen, O. Juul and Juhl, J. (1987): Wave overtopping on breakwaters and sea dikes. *Second Int. Conf. on Coastal and Port Engineering in Developing Countries*, Beijing, China.

Jensen, O. Juul and Sørensen, T. (1987): Hydraulic performance of berm breakwaters. *Seminar on Berm Breakwaters*: Unconventional Rubble Mound Breakwaters, Ottawa, Canada.

Jonsson, I.G. (1963): Measurements in the turbulent wave boundary layer. *Int. Ass. Hydr. Res., 10th Congr.*, London 1963, 1:85-92.

Jonsson, I.G. and Carlsen, N.A. (1976): Experimental and theoretical investigations in an oscillatory turbulent boundary layer. J. Hydraul. Res., 14(1):45-60.

Kirkegaard Jensen, J. and Sørensen, T. (1972): Measurement of sediment suspension in combinations of waves and currents. Proc. 13th. *International Coastal Engineering Conference*, Vancouver 1972, 2:1097-1104.

Kystinspektoratet (1988): Stormflodsrisikoanalyse, de indre danske farvande - hovedrapport. Kystinspektoratet.

Laustrup, C. (1993): Coastal erosion management of sandy beaches in Denmark. *Coastal Zone 93*. New Orleans.

Longuet-Higgins, M.S. and Stewart, R.W. (1962): Radiation stress and mass transport in gravity waves, with application to 'surf beats'. J. Fluid Mech., 13:481-504.

Lundgren, H. and Sorensen, T., 1957. "A Pulsating Water Tunnel", *Proceedings 6th International Conference on Coastal Engineering*, Florida, pp 356-358.

Lundgren, H. (1962): A new type of breakwater for exposed locations. Dock and Harbour Authority, Vol. 43, No. 505, pp 228-231.

Lundgren, H. (1963): Wave thrust and wave energy level. *Proc. 10th IAHR Congress*, London 1963, 1:147-151.

Lundgren, H. (1973): Turbulent currents in the presence of waves. *Proc. 13th Coastal Engineering Conference*, Vancouver 1972, 1:623-634.

Lundgren, H. and Gravesen, H. (1974): Vertical face breakwaters. *6th International Harbour Congress*, Antwerp, Section 2.11.

Lundgren, H. (1984): Scientific Engineering: Selected papers 1942 to 1984. Published by DHI.

Lundgren, H. and Juhl, J. (1995): Optimisation of caisson breakwater design. In 'Wave Forces on Inclined and Vertical Wall Structures', ASCE, pp 181-204.

Madsen, H.T., Laustrup, C. and Sørensen, P. (1995): A full scale comparison of beach and shoreface nourishment. *COPEDEC IV*. Rio de Janeiro.

Madsen, O.S. and Grant, W.D. (1976): Sediment Transport in the Coastal Environment. M.I.T. Ralph M. Parsons Lab. Report 209.

Meiners, J. (1995): Claus Hinrich Christensen - Festungen, Deiche, Schleusen in Schleswig-Holstein und Dänemark. Canal-Vereins mit der Schleswig-Holsteinischen Landesbibliotek.

Mertz, E.L. (1924): Oversigt over de sen- og postglaciale niveauforandringer i Danmark. Danmarks Geologiske Undersøgelse. II række nr. 41.

Miljøministeriet (1992): The greenhouse effect and climate change - implications for Denmark. In Danish with English summary. Miljøministeriet.

Munch-Petersen, J. (1927): Danmarks Vandbygning i de sidste 50 aar. Den Tekniske fore nings Tidsskrift, pp 109-162.

Nielsen, A.H., Molkte, E. and Jakobsen, P.R. (1983): The Danish beach landing system. *COPEDEC 83*. Colombo.

Niemeyer, H.D., Biegel, E., Kaiser, R., Knaack, H., Laustrup, C., Mulder, J.P.M., Spanhoff, R. and Toxvig, H. (1995): General aims of the NOURTEC-project - effectiveness and execution of beach and shoreface nourishments. *COPEDEC IV*. Rio de Janeiro.

PIANC TCII, Report of Working Group No. 12. (1992): Analysis of rubble mound breakwaters.

Rang, V. (1944): Munch-Petersens formel för sanddrift - en ny tillämpning. Den Tekniske Forenings Tidsskrift, September 1944, pp. 1101-1102.

Remmer, O. (1991): Kort- og Matrikelstyrelsen. Unpublished.

Schou, A. (1945): Det Marine Forland. Hagerups Forlag. (With an English summary).

Svendsen, I.A. (1984): Mass flux and undertow in a surf zone, Coastal Engineering, 8(4):347-366.

Svendsen, I.A. and Madsen, P.A. (1984): A turbulent bore on a beach. J. Fluid Mech., 148: 73-96.

Svendsen, Sv. (1938): Munch-Petersens formel for materialvandring. Stads- og Havneingeniøren, Nr. 12, pp. 1-18.

Sorensen, T., 1973. Model Testing with Irregular Waves. Dock and Harbour Authority, May 1973.

Sorensen, T., 1960. The Development of Coast Profiles on a Receding Coast Protected by Groynes, *Proceedings 7th International Conference on Coastal Engineering*, The Hague, Vol. 2, pp. 836-846.

Vaaben, I. (1993): Coastal planning for recreation in Denmark along the North Sea Coast. Coastlines of the Southern North Sea, pp. 227-232 (Coastlines of the World), *Coastal Zone 93*, New Orleans.

HISTORY OF COASTAL ENGINEERING IN FRANCE

Luc Hamm[1]

ABSTRACT: This article begins with physical and demographic data relating to the French coast. A description is then given of the various institutes and figures who were instrumental in establishing coastal engineering in France. This focuses on two complementary centres, a State-run one in the Paris area and an industrial one in Grenoble. These centres underwent considerable development in the years following the Second World War, leading to the organisation of the 5th ICCE in Grenoble in 1954. The most outstanding scientific and technical contributions from the 1940s, 50s and 60s are then summarised and references given for the available literature (mainly in French) concerning these topics.

INTRODUCTION

Some data concerning the French coast

In 1995, France was the world's foremost tourist destination. Every year, some 30 million tourists flock to its shores, including 20 million French people and 10 million foreigners. Most of these (20 - 25 million) come during the summer months (Gerlier 1992). Figure 1 shows the coastal areas concerned, along with the names commonly associated with them, the most famous being undoubtedly the Côte d'Azur. The total length of coastline in mainland France is estimated at 5533 km (from 1:100 000 maps), including 766 km along estuaries (Ministère de la Qualité de la Vie 1977). The western coast of France runs for 3830 km along the shores of the North Sea, the Channel, and the Atlantic Ocean. It consists of 40% beaches, 30% rocky shore, and 30% mud flats and marshes. There is a significant tide, which reaches exceptional ranges in Mont-Saint-Michel bay (14 m during spring tides). The southern coast runs for 1705 km along the Mediterranean, where tides are negligible. The southern coast consists of 25% beaches, 65% rocky shore, and 10% mud flats and marshes. In addition, there are the coasts of France's overseas territories and departments (1703 km). Historically, France also managed the shores of its colonies in Africa, Asia, and Oceania during the 19th and early part of the 20th centuries.

---

[1] Senior civil engineer, Sogreah Ingénierie, 6 rue de Lorraine, 38130 Echirolles, France

Almost everywhere in France, the coastline is now retreating, though at very different rates. Erosion is taking place at the rate of more than 1 m a year along about 850 km of coast and 0.5 m per year along another 1000 km.

The damage caused to German military installations built during the Second World War (along the Atlantic Wall) is the most visible evidence of coastal retreat along the western shores of France.

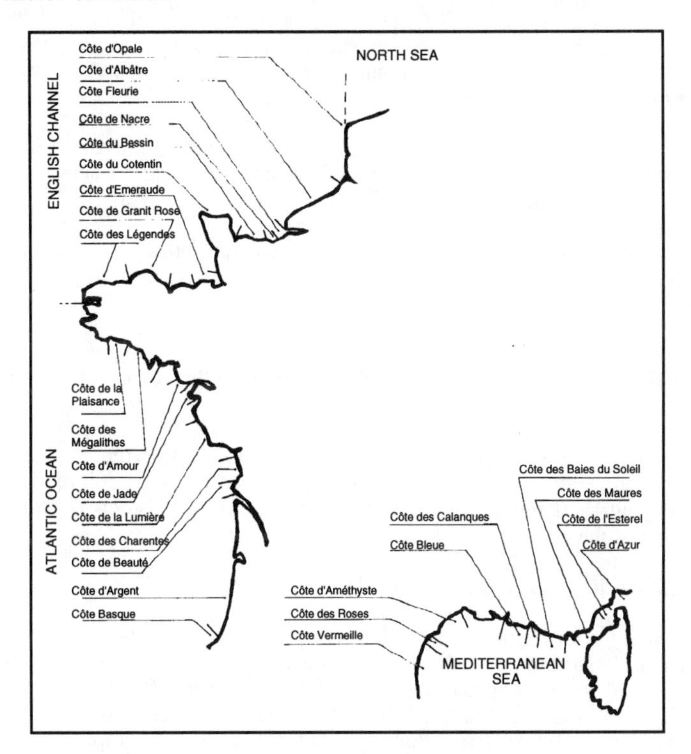

Figure 1 - Coastal areas and common tourist names
(Ministère de la Qualité de la Vie)

Legal aspects concerning the coast

In 1681, King Louis XIV's chief minister Colbert issued a decree to the effect that all coastal public land was inalienable and indefeasible, stating that "all that the sea covers and exposes during new and full moons and as far as the highest tide of March can reach on the strand shall be deemed shoreline." Traditionally, this public land ("*Domaine Public Maritime*" in French) comprised both the sea shore and inland stretches of salt water. A law of November 28th, 1963, extended this definition to include the ground and subsoil below the sea up to 12 sea miles offshore. A further law of December 31st, 1976, provided a 3 m wide right of way for pedestrians running along all coastal public land, and referred to as the "Customs officers' path." However, there is still no precise definition of what constitutes the shore, even though a law was passed on this subject on January 3rd, 1986.

The law of September 16th, 1807, specifies that all costs incurred for coastal defence works shall be borne by the landowners who are thereby protected and in proportion to their interests, except in cases where the Government decides that subsidies from public funds would be advisable or merited. In practice, such subsidies have usually been extremely small, owing to the limited financial resources generally devoted to coastal defence works. This law also sets out guidelines for the so-called "compulsory" associations that are responsible for having these works carried out and maintained. It has always been difficult to put these laws into effect, which has given rise to the saying that "France has no coastal defence system, only expenses." At the 5th ICCE in 1954, de Rouville pinpointed the three causes of poor coastal defence work in France:

a) The urban development brought about by tourism has resulted in a large number of individual points (hotels, casinos, villas) that no one wishes to leave to their fate, and this prevents the establishment of rational coastal defence plans.

b) The administrative set-up described above has not led to the creation of strong associations managed by specialised and experienced technicians, determined to implement effective coastal defence works.

c) The two World Wars that have affected France made it impossible to get to or work on the coast, and this has considerably aggravated the consequences of the other factors.

It was not until 1972, however, that the conclusions of a report handed to the Government and entitled "Long-term prospects for the French Coast" really alerted the public and politicians regarding the need for intervention, in order to prevent further deterioration of the coast.

Aims and limits of the present article

An initial analysis of the general conditions described above shows that France is a geographically maritime country that culturally faces inland. For many years, the development of coastal engineering was therefore a matter for specialists, and their scope of action was often limited by the absence of any active policy and the concomitant lack of funds. The following chapter is devoted to a rapid overview of the institutions and people who were at the origin of coastal engineering development in France between 1930 and 1970, approximately. A few comments will then be made on the 5th ICCE held in Grenoble in 1954. The final part of the article will contain a short assessment of the scientific and technical heritage of that period.

INSTITUTIONS AND PIONEERS

The development of coastal engineering in France may be described by discussing the complementary work carried out at two major geographical centres:

- Firstly, the Paris area, including at the outset engineers of the *"Ponts et Chaussées"* (the State Corps of Engineers overseeing the Ministry of Public Works) and hydrographic engineers (Ministry of the Navy). These State departments were created in the 18th century and two hundred years later were at the origin of the two maritime hydraulics laboratories set up in the Paris area, namely the Laboratoire National d'Hydraulique (LNH) at Chatou and the Laboratoire Central d'Hydraulique de France (LCHF) situated at Maisons-Alfort.

• Secondly, the Grenoble area, where institutions grew up around the development of hydroelectricity in the second half of the 19th century. Local industries were to help towards the creation of a technical university of hydraulics, stimulate the development of university research laboratories in the field of fluid mechanics, and create the Laboratoire Dauphinoise d'Hydraulique (LDH) in 1918. This was to become the Société Grenobloise d'Etudes et d'Aménagements Hydrauliques (Sogreah) in 1955.

Mention should also be made of the Toulouse area, which has a technical university of hydraulics dating from 1910.

The Paris Area

Navy Hydrographic Department

From a historical point of view, it is appropriate to recognize the French Hydrographic Department, which has systematically mapped the French coast and sea bed, thus providing a data base of vital importance for coastal engineers. The figure below is just one example among many others. It shows the changes that occurred in the mouth of the river Rhône, where it joins the Mediterranean, between 1905, when a new artificial mouth was dug, and 1980.

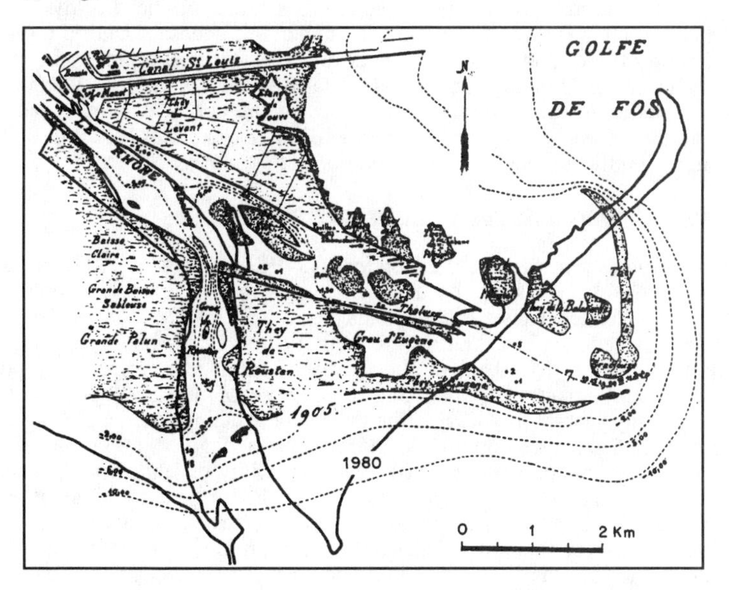

Figure 2 - Changes in the mouth of the Rhône between 1905 and 1980

According to historians, marine hydrography in France began in Dieppe in the 15th century (Massoud and Piboubès 1994). In 1720 the *Dépôt des Cartes et Plans de la Marine* (Naval Maps and Drawings Department) was created. The *"Pilote Français,"* a collection of 150 maps published between 1822 and 1844, is a systematic survey of the French coast carried out under the supervision of Beautemps-Beaupré (1766-1854). Through the use of precise, reliable surveying techniques, this marks the dawn of modern hydrography in France. The maps are often used as a starting point for studying changes in the sea bed in coastal areas. An example of the use of such data for validating a mobile-bed scale model is described by Parthiot (1981). The *"Cahiers de Recherches Hydrographiques sur le Régime des Côtes"* first appeared in 1838. These contain essays by hydrographers from the *Dépôt* concerning conditions in river mouths, banks, and coastal erosion and tides during particular periods. The *"Annales Hydrographiques,"* which are of a more scientific nature, were published at the same time, beginning in 1848.

In 1886, the *Dépôt* became the *Service Hydrographique de la Marine,* and in 1971 it assumed its present form, under the title of *Service Hydrographique et Océanographique de la Marine* (SHOM). This change was in recognition of the important oceanographic work it had carried out, particularly with regard to tides and associated currents.

Lastly, mention should be made of the hydrographic work done in the former French colonies in Africa, South-East Asia, and Oceania during the first half of this century, which made a significant contribution to the development of this Department.

The most significant contemporary figure in the Department was without doubt André Gougenheim (1902-1975), an outstanding engineer, scientist, and teacher. Entering the Department in 1922, he became its Director between 1957 and 1964. Grousson (1975) gives a summary of this remarkable man's career. As head of the "Tides and Geophysics" section from 1947 to 1958, he undertook intense scientific research into the prediction of tides and stimulated the development of oceanography in France. In 1947 he created the Oceanography and Coastal Studies Committee, with the aim of drawing attention to the increasing importance of the ocean in human activities. A periodical, the *"Cahiers Océanographiques et d'Etude des Côtes,"* followed in 1949, containing scientific articles by French authors on this subject.

The Corps of Engineers (*Corps des Ponts et Chaussées*)

The *Ecole Nationale des Ponts et Chaussées* (ENPC), created in 1747, is the oldest technical university in France. Its aim has always been to train civil engineers for service to the State. These engineers all form part of the *Corps des Ponts et Chaussées*, created in 1716. During this century, the Corps has also admitted civil engineering students. Members of the Corps of Engineers working in provincial Maritime Affairs Departments thus play a key role in the development of marine works. For many years, this participation meant constructing and developing harbours and improving their access conditions. Indeed, coastal engineering in the strict sense, involving coastal erosion control, was the responsibility of landowners, according to the law of September 16th, 1807, as described in the Introduction. Nevertheless, the State undertook coastal defence works in certain areas in collaboration with local authorities, whenever this was of general public interest. A famous example is the Pointe de la Grave at the south of the Gironde estuary, where defence works began more than 150 years ago. The State also carries out any work required to protect public property (lighthouses, roads, etc.).

On the international scene, the Corps of Engineers' work in maritime development schemes was publicised mainly through the Permanent International Association of Navigation Congresses (PIANC). The 1930 Congress held in Venice led to the publication of "conclusions" concerning coastal defence projects, that were to serve as basic teaching material on coastal engineering subjects in France (Blosset 1951 p. 412).

The name of André Gervais de Rouville (1882-1979) should be mentioned here. For more than half a century, he was an active member of the PIANC, contributing considerably to its development (PIANC 1979) as well as to that of coastal engineering in France. A member of the Corps of Engineers, he began his career in the Port of Cherbourg in 1907. He was manager of the Lighthouses and Beacons Department from 1930 to 1952 and at the same time Professor of Maritime Works at the ENPC from 1937 to 1952. He acted as President of the PIANC from 1948 to 1955 and for 40 years chaired the PIANC's First International Commission on Waves. The author of numerous technical papers, he was in particular responsible for the publication of an outstanding general report on vertical breakwaters presented to the PIANC Congress in 1935, and another work on conditions around the French coasts (de Rouville 1950). He delivered a general overview of coastal defence works in France at the 5th ICCE held in Grenoble in 1954. Even after his retirement from active service the following year, he continued to show interest in maritime works and to travel around the world, fired by a passion that only death would extinguish.

The Laboratoire National d'Hydraulique at Chatou

Built between 1936 and 1946, the Laboratoire National d'Hydraulique (LNH) was officially inaugurated in 1947 as a facility shared jointly by the Ministry of Public Works and the French Electricity Board (EDF). It was intended as an experimental centre for studying any hydraulics problem affecting ports (and particularly those in the African colonies), navigation, and energy production using tides or river (Blosset 1951). Maritime hydraulics models occupied a significant place in the LNH's activity right from the outset, and this led in 1954 to the creation of a Maritime Hydraulics Division. The first person to head this division was Jean Valembois, who participated in the 5th ICCE and was also senior editor of the periodical *"La Houille Blanche"* from 1963 to 1993. He was succeeded by René Bonnefille, who was also Professor of Maritime Works at the ENPC. In addition to physical scale models, the LNH developed instrumentation, particularly for measuring waves in situ (Valembois 1954). Numerical modelling in all areas of fluid mechanics began to take on considerable importance, particularly in relation to thermodynamics studies for conventional and nuclear power stations. These highly sophisticated models were also of great benefit for maritime hydraulics, with respect to long gravity waves (Benqué et al. 1982).

The Laboratoire Central D'Hydraulique de France at Maisons-Alfort

After being approached by the public authorities in 1939, Jean Laurent, Chief Engineer at the *Société Hydrotechnique de France*, began to build a laboratory in his garden. In 1943 this was to become the Laboratoire Central d'Hydraulique de France (LCHF). The first scale models were used to study ports in the French colonies in Africa, especially in Morocco. Particular importance was attached to the sedimentological and morphodynamic aspects of port and coastal studies, by developing semi-empirical rules using artificial low-density materials such as bakelite to represent natural sand. In 1952, the LCHF set up the Chilean National Hydraulics Laboratory in Santiago, followed by another in Colombia in 1954.

Initially the laboratory had the status of a public limited company, with the State holding 60% of the shares. In 1970 it was bought out by private investors and expanded its activities to include river and subsurface hydraulics. In 1986, the laboratory amalgamated with Sogreah, and the two companies' testing installations were transferred to a new site in Grenoble.

After its founder, Jean Laurent, the most outstanding LCHF figure on the international scene was undoubtedly Claude Migniot, who specialised in the modelling of sandy coasts (Migniot et al 1975). His detailed experimental study of the mechanical properties of muds (Migniot 1968) laid the foundations of this type of highly complex modelling, from both the theoretical and practical standpoints.

The Grenoble area

The Laboratoire Dauphinois d'Hydraulique

In 1854 Casimir Brenier founded a lock-making and general machinery workshop near Grenoble. By 1867 the company employed 200 people and it began manufacturing equipment for generating electricity by harnessing the many mountain streams in the area. The first full-scale turbine testing station was built in 1906, and in 1917 the company moved to Beauvert on the outskirts of Grenoble. At that period it was known as the *"Ateliers Neyret-Beylier-Piccard et Pictet,"* a mouthful that was condensed in 1948 to Neyrpic. The firm's hydraulic testing laboratory was transformed at that time into the Laboratoire Dauphinois d'Hydraulique (LDH). Specialising in experimental studies on scale models, its premises grew from 65 m^2 in 1918 to 60 000 m^2 in 1955 (Plénet, 1996). In 1929, the *Ecole des Ingénieurs Hydrauliciens de Grenoble* was founded as a section of the *Institut Polytechnique de Grenoble* (INPG). This technical university was set up to nurture the hydraulics skills required for developing hydroelectricity, and possessed its own hydraulics laboratory.

In 1928, Pierre Danel (1902-1966) joined Neyrpic, becoming manager of the LDH. From then on until his death in 1966, this exceptional man's life was to be fused with the development of the laboratory. A detailed description of P. Danel's life and work spanning almost 40 years can be found in *"La Houille Blanche"* (Dagallier 1970). The main events are as follows.

It was under Pierre Danel's management that the LDH progressively expanded its physical modelling activities to river and maritime hydraulics and irrigation networks. In 1934, the problems of river scour were studied in relation to Jons dam. The first maritime hydraulics model (for the port of Boulogne-sur-Mer) was built in 1940, using a wave generator that had been under development since 1937.

In November 1944, the American 6th Army drew up at the Rhine. It was feared that, if they tried to cross the river, the Germans might destroy the dams upstream and create a flood wave that would carry away the bridges and boats and flood the plain. The military authorities consulted the LDH and a 450 m long model was built within 10 days using a pre-existing model. The studies were completed by March 20th, 1945, in the record time of four months. Thus it was that the army safely crossed the Rhine downstream of Strasbourg on March 27th, 1945, earning the Laboratory the highest praise from the Allied military command (Neyrpic 1951). Although this episode is not directly related to coastal engineering, it is important to mention in relation to the 5th ICCE, which was held in Grenoble in 1954. It illustrates the close relations built up between Pierre Danel and American hydraulics specialists, dating back to 1936 at least. His frequent trips to the United States led to exchanges between scientists, and these developed particularly after the Second World War thanks to the help of Hunter Rouse, who visited the LDH in 1946 (Rouse 1970). These exchanges in turn led to several theses being presented in Grenoble by American engineers, some in the field of maritime hydraulics (namely studies of seiches). Similarly, a number of engineers and researchers from Grenoble went to lecture in the United States, or even settled there, such as Bernard Le Méhauté.

The LDH's international reputation was to be considerably enhanced after the War under the impetus of Pierre Danel. The first issue of *"La Houille Blanche"* appeared in January 1946; this was followed by six and then eight issues a year from 1948 onwards. Issue No. 3 contained the first of many international contributions. These became more frequent with the years, as the publication made increasing use of both French and English in its articles. The *Société Hydrotechnique de France* was one of the pillars of this periodical, publishing reports on the learned society's sessions in two extra issues each year (Dagallier 1970, p. 514). Another important vector of the LDH's international reputation at the time was the International Association on Hydraulic Research (IAHR), which had been founded in 1935 at the instigation of Professors Fellenius and Rehbock. Pierre Danel became a member in 1937, the year of its first Congress, and was elected to the Council in 1948. He organised the 3rd Congress in Grenoble in 1949, was elected Vice-President in 1951, and then President from 1955 to 1959. He was responsible for setting up committees within the IAHR, including the Committee for Coastal Studies (Lisbon, 1957), which enabled researchers to work in teams. He always defended the idea of the IAHR being an association of people, and emphasised his preference for individual presentations of papers at the various Congresses (Dagallier 1970, p. 517). This decidedly Anglo-Saxon approach was in contrast with the usual French practice, which gave greater importance to institutions (cf. the PIANC congresses, for example).

Maritime hydraulics studies and coastal engineering activities underwent considerable development after the end of the War. The invention of the Tetrapod block for breakwater armourings was a major innovation that led the LDH to carry out far more coastal engineering work outside France from 1950 onwards. Physical model studies called for the development of wave generators and were accompanied by theoretical work on gravity waves. From 1951 to 1956, studies were also performed to improve navigation conditions along the river Seine, using a very large mobile-bed model. Lastly, the 1950s also witnessed major research into the propagation of tides in coastal areas, with a view to designing tidal power stations. One of these was actually built on the river Rance and commissioned in 1966. The decade following the Second World War marked the apogee of hydraulic scale modelling at the LDH.

Due to the considerable development of river and coastal engineering studies, along with those for irrigation and drainage schemes, the LDH was increasingly required to provide complete consultancy services. As a consequence, in 1955 the LDH split from Neyrpic, which henceforth concentrated on turbine manufacturing, and became an independent company under the name of Sogreah.

Sogreah

The transition from isolated special studies (for turbines or irrigation devices, etc.) to regional development studies covering huge areas inevitably raised the problem of the inherent limitations of scale models resulting from their size. The transformation of the LDH into Sogreah therefore brought about a spectacular explosion in numerical modelling, and the name most closely associated with this is that of Francis Biesel (1920-1993). Thanks to the biographical research work by Plénet (1966), it is possible to retrace his career precisely. The main stages are summarised below.

Francis Biesel joined the LDH in 1942 and discovered that he had a passion for hydraulics. Unfortunately, he was deported to Germany a few months later and it was only after the end of the War that he was gradually able to resume his activities. At that time, he worked mainly on the first snake wave generator (Biesel and Suquet 1951, Biesel 1954a). At the same period (1948), he had the idea of taking the IBM punched card machine that had been used in the accounts department since 1945 and adapting it to perform scientific calculations. The tests carried out by the department and contacts with IBM convinced Francis Biesel that he had stumbled upon something with a fantastic potential. He was soon to obtain the first "calculator" (an IBM 604) for the LDH. In 1952, the LDH received the second punched-card calculator to be installed in France, and this was used to perform the first flood forecasting calculations. In 1955, on completion of this work, IBM organised a conference on numerical calculations in Grenoble, the "*Journées Alpines de Calcul Numérique,*" with the participation of the LDH and the University's Calculations Laboratory.

The year 1955 was also the one in which Sogreah was created, and Pierre Danel soon set up a Scientific Division with Francis Biesel in charge. In 1957, Sogreah became pilot user for the first IBM 650 to be delivered in Europe. In view of the prospects offered by this powerful machine, a computational hydraulics department was set up in 1958, headed by the mathematician Alexandre Preissmann, along with a management and operations research department to ensure that the Scientific Division would be profitable and to pay for its hardware. This machine also spurred the invention of what was to become numerical analysis (Biesel 1959). The Division's growth was spectacular, and its staff increased from 10 to 80 between 1958 and 1962, reaching as many as 120 by 1968. Abbott and Cunge (1981) give a detailed description of this period and of the decisive scientific role played in the development of numerical methods from 1958 onwards by Alexandre Preissmann. On the international scene, this work was revealed at the 9th IAHR Congress held in Dubrovnik in 1961. For the first time, one of the topics discussed at this Congress was indeed "hydraulics problems for calculating machines," and Sogreah presented 12 papers in all areas of hydraulics. In 1968, the activities of the Scientific Division had expanded to such an extent that a computer services company, *L'Institut International Informatique,* was set up in Grenoble. This was headed by Francis Biesel, who remained Scientific Manager at Sogreah. A few years later, this company was bought out by an industrial group as part of a strategy to develop computers in France, and in 1972 Francis Biesel left Grenoble to head a scientific calculations department at the LCHF in Maisons-Alfort.

During this crucial period from 1950 to 1970, Francis Biesel was an outstanding pioneer, playing a decisive role as both an engineer and manager in introducing the new data processing technology into industry.

Grenoble University

The development of industrial hydraulics applications in the Grenoble area at the beginning of the century led to an increased interest in hydraulics and fluid mechanics at Grenoble University (Bouvard 1994). In addition to the technical university that has already been mentioned, the University opened a testing and research laboratory named the *Institut de Mécanique de Grenoble* (IMG). In 1955 this Institute opened a unique installation for oceanographic experiments, the "Coriolis platform." This device could be rotated so as to reproduce the effects of the Coriolis force on currents and sea levels in large physical models. The installation was built originally to reproduce the complex propagation of tides in the English Channel and in particular in the Normandy-Brittany bay area, where the tidal range can reach as much as 14 m at

Mont-Saint-Michel. At the time, this site was being intensively studied as a possible location for tidal power plants.

Experiments performed on this installation are in fact only one aspect of the work carried out at the IMG, which deals with all fields of mechanics. The IMG had already acquired an international reputation by the 1960s and 70s. It was Professor Julien Kravtchenko (1911-1994) who was at the origin of this institute. He was a professor of mechanics at the University and at the INPG by 1948, and in 1952 was appointed head of the INPG testing laboratory. At that period, about 10 people were assigned to work in this new institute. By the time Professor Kravtchenko left in 1970, there were 180. This spectacular growth was the result of the way in which he viewed mechanics. At the time, this field received little attention in French universities. Kravtchenko viewed it as an interdisciplinary science, with its own deeply inherent unity but having a wide range of applications. He was in close contact with engineers working in Grenoble, and in particular Pierre Danel and Sogreah. This was why he started to research the subject of wakes around obstacles (1935-1952) and then turned his attention to gravity waves. He supervised several theses on the question of seiches in harbours, including those of the American professor John McNown of Iowa, as well as work carried out at the LDH (Le Méhauté and Biesel 1956). He was also interested in waves from both the theoretical and experimental points of view. Notable features of his work included investigations into the generation of regular waves in wave flumes, and an in-depth study of interference phenomena that could only be eliminated by second-order generation (Fontanet 1961). He initiated research on tide propagation using the Coriolis platform and then the development of numerical methods, which have led to the recent implementation of a tide propagation model covering the entire world ocean (Le Provost et al. 1994). At the same time, his research led him to consider other fields of hydraulics such as flows in porous media and plasticity in soil mechanics. He was also extremely active on the international scene, in North America, Europe, and more particularly, in the Eastern bloc countries, where he established close scientific relations, notably in Poland. With Pierre Danel and Francis Biesel, he was one of the three "pillars" of hydraulics in Grenoble. He published relatively little, and evidence of his tremendous influence, which is still evident today, must be sought in the publications of his former students.

THE FIFTH CONFERENCE

The fifth International Conference on Coastal Engineering (ICCE) was held in Grenoble in September 1954 after a series of four conferences in the United States between 1950 and 1953. At that time, 43 papers were presented. French engineers were most prominent, presenting 28 papers, with most of them (25) being written in French. The topics dealt with at the time were classical ones which are still relevant today in terms of their theoretical, experimental and practical aspects. The only major absence was numerical simulation, which was to appear in the 1960s. An examination of the origin of the papers presented by the French participants (namely institutes) and of the composition of the Organising Committee gives a clear idea of the situation of coastal engineering in France at the time, and was a starting point for this article. However, one question remained unanswered after this analysis: why had the Americans chosen Grenoble as venue for the conference? The Proceedings give no indication at all.

The answer came from Professor Wiegel as this article was being written. During the 4th ICCE, held in Chicago, Pierre Danel's and several Dutch papers were judged to be excellent by the U.S. participants. Pierre Danel suggested that the next conference be convened in Europe to introduce the European state of coastal engineering practice. His offer to host the conference was accepted and Prof. Wiegel took the responsibility for coordinating it, assisted locally by Professor Kravtchenko.

At a more general level, the Proceedings of this Conference represent the largest French contribution ever to the ICCE. Since then, France has contributed just a few papers or has been completely absent. But this should not be construed as a sign of ostracism with regard to the ASCE. Indeed, after having met for a major international congress on sea energy in Paris in 1956 (SHF, 4èmes Journées de l'Hydraulique), the French coastal engineering community did not organise another meeting until 1984, in Marseilles. It was not until the 1990s that national coastal engineering conferences were organised on a regular basis. In contrast, the French have been constantly represented within the PIANC, which has in the past dealt with coastal engineering subjects.

They have also been well represented within the IAHR, particularly with regard to numerical modelling, which includes coastal engineering questions. The fact that publications by these two international organisations appear in both English and French bears witness to French influence.

In conclusion, then, the fact that the fifth Conference was held in Grenoble was clearly a milestone in the major contributions made by French hydraulics specialists between 1950 and 1970 to the development of modern coastal engineering methods, marked by the intensive use of physical and numerical modelling techniques.

SCIENTIFIC AND TECHNICAL HERITAGE

This section contains a summary of the most significant contributions made by French specialists to coastal engineering. These were published in the years 1950-1970, which was indeed the most productive period in this field.

These contributions concerned a limited number of subjects, which makes it easy to discuss them under various headings. In the field of hydrodynamics, they concern waves from both the theoretical and physical/numerical modelling aspects, plus the various aspects of long and tidal waves. In the field of morphodynamics, the major contribution was that of physical modelling, plus numerical modelling of the coastline. Then the subject of artificial armouring blocks and other man-made coastal protection structures will be discussed and, finally, environmental aspects are considered.

Waves

Theoretical aspects

French mathematicians played an active role in the development of wave theories as early as the 19th century (Boussinesq, Gertsner). In the period of interest here, the most significant works to be used by coastal engineers prior to the development of scientific computing were those of Miche (1944) and Biesel (1951-52).

The series of articles published by Miche (1944) tackles the question of wave transformation along a plane beach profile. For that purpose, an amplitude second-order harmonic solution is derived both in Eulerian and Lagrangian variables. The partial reflection of incident waves is included in the derivation, giving rise to an expression of the reflection coefficient for a plane beach. The use of Lagrangian variables enabled Miche to enter the shallow water domain up to the theoretical breaking point, defined as a singularity of the free surface which appears from the solution derived. He also demonstrated the energy flux conservation equation commonly used in shoaling computations. Another well-known part of this work is the derivation of a breaking criterion for progressive waves in finite water depth. A theoretical solution for the limiting wave profile was derived, including a rotational component which influences the 120° angle already found by Stokes for an irrotational wave in deep water. This theoretical solution was difficult to apply and Miche suggested an approximate expression which became widespread.

The world-wide dissemination of that work could be measured as far as Australia, where Miche wavelets are very popular (Smith and Jackson, 1995). This is in stark contrast with the material environment in which Miche carried out his solitary work during the Second World War. At that time, he was under house arrest because the Germans needed his advice on running the port he directed.

In a difficult paper following on Miche's work, Biesel (1951) proposes a first-order and then second-order solution for shoaling up to breaking point. Assuming irrotationality to investigate a velocity potential in Eulerian variables, he then studies the solution he has obtained using Lagrangian variables. The results that he presents for a wave of 0.013 steepness on a 10% gradient (Figure 3) are forerunners of present work being carried out in this area.

An interesting point to note with regard to these authors is their use of Lagrangian variables. These have since been somewhat ignored, but they may received renewed interest in studying solid transport or wave-current interactions along a beach profile. Further study of these two contributions could probably provide useful information for current research.

Biesel (1952) also dealt with another topical subject, that of second-order irregular wave equations. The advantages of second-order equations in studying natural wave propagation on a beach no longer need to be stressed (Hamm et al. 1993) and the equations produced by Biesel have since been proven correct in every detail (mainly in Eulerian variables). The point to be made here is that, according to Biesel, "second-order equations are not just a mathematical curiosity or a simple laboratory tool but a source of new knowledge that may have considerable practical importance, thus justifying the much more complicated calculations that they entail." This attitude was characteristic of Biesel, Danel and their colleagues at the LDH at that period.

Lastly, mention should also be made of the work carried out by Serre (1953) on a non-linear long wave theory in which vertical velocity is no longer neglected but is assumed to be linear between the bed and surface. It is thus possible to deduce a non-hydrostatic pressure that accurately reflects what happens in a wave just prior to breaking. The first numerical application of this theory was presented by the LNH (Hauguel 1980). This was the first non-linear "Boussinesq" type model, which was followed by major developments during the 1990s (Hamm et al. 1993), in particular at the IMG.

After 1952, Biesel's theoretical work was geared more to the requirements of physical and then numerical modelling.

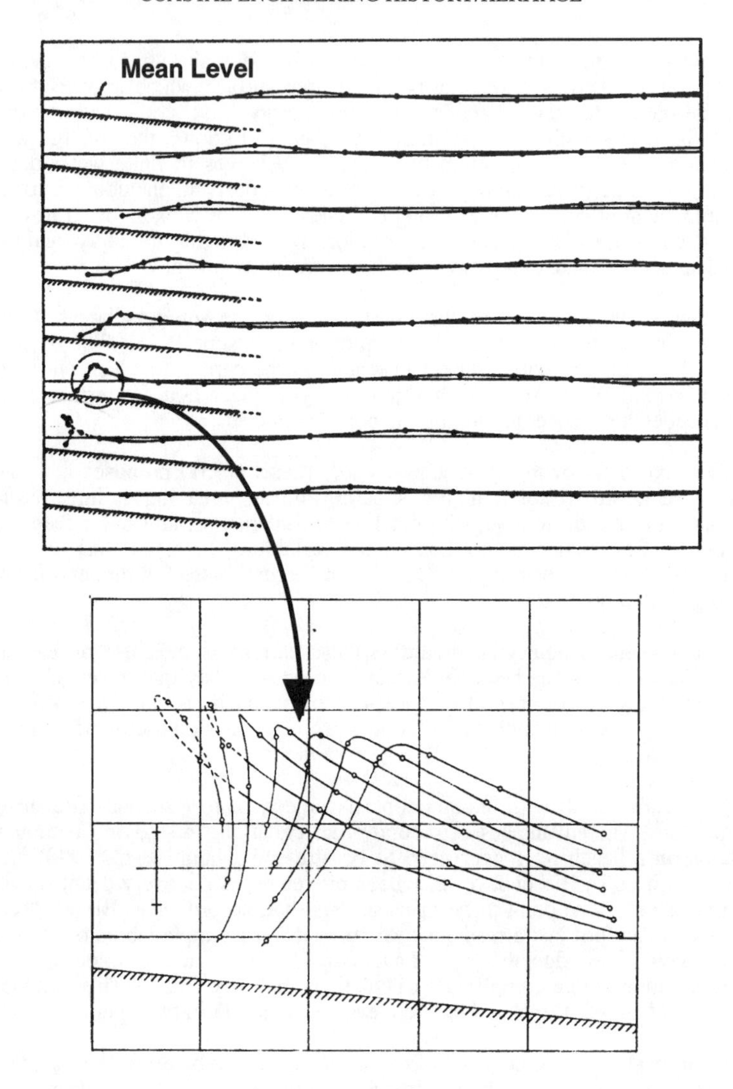

Figure 3 - Propagation and breaking of regular waves on a flat beach (Biesel 1951)

Physical aspects

In 1951-52, Biesel and Suquet produced an exhaustive series of articles on laboratory wave generators. These were translated into English at the St. Antony Falls Laboratory (University of Minnesota) and had a considerable impact from both the theoretical (transfer function) and practical standpoints, though the use of a complex set of mechanical rods has now been replaced by computer-controlled hydraulic jacks or electric motors. However, the use of a 25 m long segmented wave generator consisting of 63 elements was remarkable for the period. A detailed description of this device in English is given by Biesel (1954a). There was also a concern at the time to eliminate spurious waves connected with first-order generation in finite depths by second-order generation. The theory behind this was also established at the University of Grenoble by Fontanet (1961).

The reproduction of harbour seiches in the laboratory also received much attention at the period, along with possible theoretical solutions for simple but highly varied geometries. The theoretical bases for this type of model and the practical results obtained were presented at the 5th ICCE (Biesel 1954b and Le Méhauté 1954) and then published in *La Houille Blanche* (Le Méhauté and Biesel 1956).

Numerical aspects

The most significant contributions in this area concerned wave diffraction modelling. At the IAHR Congress in Dubrovnik, Biesel and Ranson (1961) described a method for computing the wave disturbance in a basin of constant depth. The idea was to solve the equations using an integral method in which wave disturbance at a point inside the basin is taken to be the sum of all constituent disturbances from source points situated on the edge of the basin. This method was described in detail by Barailler and Gaillard (1964) and validated by Montaz (1964) through comparison with results obtained on scale models. The method was developed extensively at Sogreah to include the effects of partial or total reflection on beds of varying depth, multidirectional incident wave spectra, and finally the effect of wave breaking and the computation of radiation stresses generating longshore currents behind a breakwater. Relevant references on these topics can be found in Hamm (1991).

However, this original approach to wave disturbance calculations nevertheless received much less attention than the Berkhoff method (the mild slope equation). Yet there is one particular case in which it is particularly recommended, and that is the numerical simulation of a multidirectional wave tank (Isaacson 1989, Benoît 1995).

Biesel also made an important contribution to the modelling of moderate diffraction. It was he who published the first refraction model including lateral diffraction, thus enabling the problem of caustics to be solved (Biesel 1964, 1972). However, his method was never exploited and was to be superseded by the development of parabolic models.

He also dealt with second-order radiating phenomena in wave propagation. These are free harmonics with half the fundamental period that become visible and of practical importance in certain particular cases of diffraction behind a breakwater or caustics (Biesel 1963, 1966). Experimental verification of the existence of these waves is described in the publications cited, as are the results obtained with a numerical model. Figure 4 shows these waves observed behind a breakwater.

Here again, the numerical method presented by Biesel was not followed up and most of the numerical models of wave disturbance inside harbours are restricted to the linear theory, even though phenomena such as those described have often been observed in the laboratory by the author of this article. This is another field that deserves further exploration.

Figure 4 - Second-order diffracted waves (Biesel 1963)

Tides

A considerable amount of research in France since 1950 has been devoted to the study and modelling of tides and similar long-order waves (for example flood waves). Such studies focused first of all on the English Channel, with a view to meeting the requirements of various projects, including one to stabilise the navigation channel in the estuary of the river Seine (Parthiot 1981) and in particular the design of tidal power stations in the Normandy-Brittany bay area, where tides are exceptionally large (14 m during spring tides at Mont-Saint-Michel). The first physical scale model tested at the LNH in 1955 showed that it was absolutely essential to reproduce the effect of the Coriolis force, which is the only possible explanation for the differences in tidal range observed on the French and English coasts (Michon and Bonnefille 1956). A working group referred to as the "Coriolis Pool" was therefore set up, comprising engineers from the various institutes in Paris and Grenoble that have been described earlier. The conclusions reached by this group led to the building of a test facility consisting of a 14 m diameter, 250 t platform (Figure 5) that could be rotated to reproduce the effects observed in nature, and also to the construction of various mathematical models at the LNH and Sogreah (*La Houille Blanche* No. 5, August 1959).

The rotating platform was built at the University of Grenoble, starting in 1959, and the first results were published three years later (Chabert d'Hières 1962). This model was subsequently operated with a view to publishing an atlas of water levels and currents in the Channel, including contributions from about 20 harmonic components (Chabert d'Hières and Le Provost 1978). This extremely accurate atlas is still in use today, providing boundary conditions for mathematical models of local coastal phenomena.

Figure 5 - "Coriolis" rotating platform at the University of Grenoble
(Chabert d'Hières 1962)

At the same time, beginning in 1956, the first numerical models began to be built at the LNH and Sogreah. The first results were published by Gohin (1960, 1961), and Bonnefille and Voyer (1961). The first attempts to build ocean models at Sogreah date from this period (Gohin 1961).

These early models ignored convection terms and use explicit numerical schemes. An implicit numerical scheme was then programmed at Sogreah in 1967 (Abbott and Cunge, 1975). Work to include convection terms was carried out jointly at the LNH and Sogreah (Benqué et al. 1982) and this led to a third-generation finite-difference computation code. More recently, the LNH developed a finite-element code dedicated to coastal and river applications (Galland et al. 1991), while the University of Grenoble was finalising a global oceanic model (Le Provost et al. 1994), thus completing work that began with the first tests performed thirty years earlier. The rotating platform continues to be used for studying oceanic flows in stratified media. This short overview emphasises the continuous efforts that have been made over the past 40 years in the field of tide propagation modelling.

Modelling of coastal and estuarine morphodynamics

In this area, there has been considerable development in mobile-bed scale modelling, based on a specifically French approach to questions of similitude in such models. The sand that forms beaches is generally simulated with plastic materials that have a similar grain size but a much lower relative density. The vertical scale is generally distorted in comparison with the horizontal scale. Much attention has also been paid to the use of irregular waves and the appropriate sequencing of calms and storms (Migniot et al. 1975). Such models were used for many years to study the changes occurring along beaches, with or without coastal structures. The figures below show an example of such a model that reproduces the erosion observed on Anglet beach to the north of Biarritz.

Figure 6 - Destruction of the seafront at Anglet
Top: model
Bottom: site
(Migniot 1979)

Nowadays this type of study is usually carried out with a mathematical model that simulates the changes in coastline on the basis of the Pelnard-Considère theory (1956). Such a numerical model was first built at Sogreah 1963 to study the development of the north-western sand spit forming at the mouth of the river Rhône, shown in figure 2. This model contains interesting features which have still not been included in more recent models. Scale models are still used, however, for studying three-dimensional phenomena such as, for example, the development of sandy spits or local erosion around various structures.

Models of estuaries with sandy beds call for a greater degree of distortion in view of the areas being simulated, and hence involve using very low-density materials such as sawdust. A model of this kind reproducing the Seine estuary was described by Parthiot (1981). Studies on this model began at Sogreah in 1951 and were completed five years later. The calibration phase was long and extremely meticulous, but it meant that the model could be validated over a period of 84 years between 1869 and 1953 by comparison with the many bathymetric surveys carried out on site. Tests on the proposed new development works were then conducted for a period of 40 years. Because the forecasts obtained with this model were so reliable, it was rebuilt in 1966 for a series of tests that was completed in 1977. Here again, accurate forecasts were obtained, and the model has just been rebuilt yet again in 1995 to study further development works. It is clear then that, unlike fixed bed tidal models, which disappeared in the 1970s with the advent of numerical modelling, mobile bed models are still the only way of performing sufficiently reliable long-term simulations of changes in estuaries.

Physical scale models have also been used to simulate the movement of mud in estuaries, following a procedure developed at the LCHF in the 1960s. Extensive work was carried out by the laboratory at the time concerning the mechanical properties of muds. This work is described in detail by Migniot (1968). By applying the resultant knowledge to the design and operation of a scale model of the Loire estuary, it was possible to validate the model and produce reliable forecasts of future changes after the implementation of development works (Caillat and Migniot 1981).

Coastal works and structures

Artificial beaches

Among the many development works built along the French coasts between 1950 and 1970, artificial beaches were a major innovation, and forerunners of the present trend towards beach nourishment. At that time the term was used indifferently to cover the widening of existing beaches or the creation of new ones in areas where none had existed previously. Tourmen (1962) describes the nourishment work carried out on La Croisette beach in Cannes following the widening of the seafront boulevard. 250 000 m^3 of sand with a median diameter of 0.8 mm (similar to that of the existing sand) were deposited over a linear distance of 1040 m and stabilised by the construction of three 60 m long groynes. This beach has remained remarkably stable up to the present time, as only 5000 m^3 of sand have been required to maintain it over the past 30 years. The construction of a completely artificial beach in Monaco between 1965 and 1967 is also a reference of major historical importance (Tourmen, 1968). This beach was built on a rocky coast over a linear distance of 400 m. 80 000 m^3 of 3-8 mm crushed gravel were used and stabilised by three breakwaters built in water 6-10 m deep (Figure 7).

Here again, little consolidation has been required to maintain the stability of the beach (5000 m^3 of material over the past 30 years). On the other hand, crushed gravel has certain drawbacks. Firstly, the equilibrium slope, which is of the order of 10%, is considered too steep by bathers, particularly for children. In addition, the crushed stone had rough edges in the early years, making it uncomfortable to walk on with bare feet. Wave action has subsequently rounded the edges of the gravel. Lastly, as gravel was blown off the beach during winter storms, shop windows along the seafront needed to be adequately protected.

Figure 7 - Larvotto artificial beach, Monaco
(Tourmen 1968)

Artifical armour blocks

Until the Second World War, the techniques of maritime breakwater protection involved using either natural rocks or concrete blocks of parallelepiped form, whenever the technical and economic conditions of the site made it impossible to use natural rock. Indeed, for small structures, a protective armour with natural rocks is often suitable. However, as the scale of the works increases, this solution is soon limited by the rock extraction potential of nearby quarries and by the poor stability of rock under wave attack. Under such conditions, use was first made of concrete blocks of parallelepiped form. These blocks are placed either at random or in an orderly pattern. In the latter case, however, they are sensitive to underpressure phenomena when the waves recede, which may lift the blocks and disorganise the structure. The slopes are gentle, the volumes substantial and the unit weight of the blocks very high. It was in this context that, in 1950, the LDH introduced the first artificial block, the Tetrapod (Danel 1953), which satisfactorily solved the problems of protecting rubble mound breakwaters from wave attack. The LDH authorized another company called Sotramer (Société d'Exploitation de Brevets pour Travaux à la Mer) to issue licences for the use of the Tetrapod.

Figure 8 - Tetrapod blocks

In fact, although this block was mostly used in coastal works, its special hydraulic properties made its use equally well adapted to other types of works such as dam facings, downstream faces of spillways and bank protection for rivers. Its invention revolutionised the design of maritime structures: the slopes became steeper, close to the natural bank slope, the volume of materials was reduced and the unit weight of the blocks significantly lowered. Placed in two layers on a rubble bank, it forms a facing with a 50% void fraction, which ensures effective dissipation of wave energy while avoiding the problem of underpressure.

The first users, who were well acquainted with the work behind the invention, became confident about the advantages offered. At the end of its patent life, after some 20 years of service, the Tetrapod had been used on more than 300 structures, representing a total of 8 million tons of concrete! And 46 years later, this block is still being used in some countries. After a certain period of cautious reserve, project authorities, contractors and consultants soon realized the qualities of the Tetrapod, starting the widespread use of artificial armour units throughout the world.

In his paper, presented at the 8th ICCE in Mexico in 1962, Pierre Danel stated: "With its special shape, the Tetrapod lends itself to the construction of very rough facings which are capable of standing up to waves of a very considerable height, and is comparatively much lighter than a lump of rock fulfilling the same purpose." He concluded, however, that a careful analysis of the design and comprehensive experimental investigations provide the only means of ensuring structure reliability.

Emphasis was then placed on technical aspects rather than a substantial reduction in the cost of the structures, amounting to as much as 30% in some cases.

Both the shape of the block was and the way the two layers interact were revolutionary: The first layer is laid down so that three legs rest on the mound with the fourth pointing outwards.The second layer is then laid down pointing in the oppposite direction to the first. Tetrapods on the second layer have a natural tendency to settle in this manner, the keying being encouraged by the conical shape of the legs.

Indeed, the Tetrapod opened the way to the first generation of artificial blocks. The tendency at this stage was to seek more effective interlocking between the blocks, thus ensuring better overall stability of the facing. This research led to increasingly complex forms, at the expense of greater fragility of the individual blocks.

The accident which occurred at the Sines oil terminal in Portugal in February, 1978, put a stop to this type of development. The Sines breakwater was subsequently rebuilt using grooved cubic blocks, another French invention (Orgeron et al. 1982). This technique, developed in its final form for the French port of Antifer in 1973, is based on traditional well-tried methods involving the use of classical cubic blocks. An improvement was made, however, by providing the blocks with grooves and giving the sides a slight incline. This ensures a higher void fraction and results in a certain amount of keying between the blocks.

The second-generation type of block, the Accropode®, was developed by Sogreah in 1979, some 30 years after the creation of the Tetrapod. The major difference between the Accropode®) and all its predecessors is that it is designed to be placed in a single layer. In addition, its compact form provides a good balance between strength and stability under wave attack.

It was for this reason that the strength of the Accropode® block was given particularly close attention. The final compact shape gives it excellent mechanical resistance. Stress calculations and rupture tests on full-size blocks have shown it to be stronger than more slender blocks, whose shapes highlight their fragile nature.

Figure 9 - the Accropode® block

Why look for another new shape of block at that time? Mainly because Sogreah's practical experience on work sites using the Tetrapod had shown up a number of difficulties or drawbacks with the earlier block, mostly relating to the particular way of interlocking the upper and lower layers, each to be placed with a specific attitude - hence the idea of a block with a shape that would allow it to be placed in a single layer, with no particular attitude to be respected.

This is the first of the new generation of blocks with a shape specially designed for single-layer armouring. Placing these new types of blocks in one layer makes the armour more flexible and homogeneous. Moreover, because of its interlocking capability, there is an effective group effect activated by the waves, which tends to increase armour compactness while reducing movements of the few units which may not have settled properly on placing. Also, the block's individual strength makes it less vulnerable to rocking (Kobayashi and Kaihatsu 1994). Almost 100 structures have now been built or are under construction without any failure due to unstable armouring.

Environmental aspects

Environmental aspects first began to be considered in France around 1970, owing to the quantity of untreated wastewater and industrial effluent being discharged into the sea. This problem was particularly obvious in the Mediterranean, where the currents are weak and eutrophication occurs during the summer, at the height of the tourist season. This stimulated research at Sogreah with a view to computing wind-induced currents in the Mediterranean and the resultant turbulent diffusion that is the prime cause of pollutant dispersal. Measurement campaigns were also undertaken on site. These demonstrated the effects of thermal stratification that have to be taken into account during modelling. A methodology for designing sea outfalls was defined during this period by the French Ministry of the Environment and was translated into English by Quetin and de Rouville (1986).

Then in the 1980s, it became apparent that certain coastal defence works were failing to fulfil their aims, and that erosion problems were simply being shifted further along the coast or in some cases even aggravated. This type of failure had already been observed by de Rouville (1954); according to him, it was often caused by individual defence works intended to protect apartment blocks that had been built on the dunes or even on the beach itself without any coastal impact studies being carried out. This led to the abandonment of traditional works of this kind made from rockfill and a search for "softer" protection methods. One of these, beach nourishment, is still not very widely used in France. In contrast, it is still too early to draw any conclusions regarding the experiments now being carried out around the country on certain innovatory systems, though some noteworthy failures have occurred.

CONCLUSIONS

This rapid overview of the beginnings of coastal engineering in France between 1930, the year in which the PIANC Congress was held in Venice, and 1972, when politicians and the public became aware of the need to protect the shore, has attempted to emphasise two major themes:

- firstly, the origin of the various institutions involved in the growth of this field, paying homage to the most outstanding figures during the period,

- secondly, the most significant contributions made by France in certain specific areas, namely the modelling of wave and tide phenomena, mobile bed scale models, artificial breakwater armouring blocks and the construction of artificial beaches.

It has also aimed at drawing attention to certain works that were published in French, a fact that has proved over the years to be a major handicap in getting them more widely known.

ACKNOWLEDGEMENTS

This article is the result of major bibliographical research that makes up for the author's lack of any personal memories concerning the events described herein, most of which occurred before he was born. The author wishes to express his deepest thanks to Marie-Christine Vasseur, Sogreah's librarian and archivist, for preserving the documentary heritage of the Laboratoire Dauphinois d'Hydraulique and Sogreah. There is much material there for further specialised historical articles. He should also like to thank Mrs Florence Doux, Head of the Contemporary and Historical Documents Centre at the Ecole Nationale des Ponts et Chaussées and Mrs Caroline Lévy, who provided him with information on the Corps of Engineers. Cyrille-Claude Plénet, author of a thesis on the history of informatics in Grenoble, was of great assistance with this subject and in particular the crucial role played by Francis Biesel. Claude Migniot and Bernard Bellessort provided historical information on the LCHF, Professor André Temperville collected information and biodata on Professor Kravtchenko, and Michel Denéchère, Head of Sogreah's Accropode® Division, wrote the section on armouring blocks invented in France between 1950 and 1980.

REFERENCES

Abbott, M.B. and Cunge, J.A. (1975). "Two-dimensional modelling of tidal deltas and estuaries in unsteady flow in open channels", vol. II, K. Mahmood and V. Yevgevich ed., Fort Collins, Colorado, pp. 795-808.

Abbott, M.B. and Cunge, J.A., editors (1981). "Engineering applications of computational hydraulics. Vol. 1. Homage to Alexander Preissmann" *Pitman Advanced Publishing Program*, Boston, USA.

Barailler, L. and Gaillard, P. (1964). "Exemples de réalisations de modèles mathématiques à Sogreah pour les études de propagation de la houle" *Proc. 9th ICCE*, ASCE, Lisbon, pp. 41-54.

Benoît, M. (1995). "Quelques aspects du développement d'un bassin numérique à houle multidirectionnelle" *Proc. 5èmes Journées de l'Hydrodynamique*, 23-25 March 1995, Rouen, France, pp. 73-86.

Benqué, J.P., Cunge, J.A., Feuillet, J., Hauguel, A. and Holly, F.M. (1982). "A new method for tidal current computation" *J. of the Waterway, Port, Coastal and Ocean Div.*, ASCE, vol. 108, WW3, pp. 396-417.

Biesel, F. (1951). "Study of wave propagation in water of gradually varying depth" *Proc. Semicentennial Symposium on Gravity Waves*, National Bureau of Standards, Circular 521, pp. 243-253, Washington D.C., June 18-20, 1951.

Biesel, F. (1952). "Equations générales au second ordre de la houle irrégulière" *La Houille Blanche*, no. 3, pp. 372-376.

Biesel, F. (1954a). "Wave machines" *Proc. 1st Conf. on Ships and Waves*, Hoboken, New Jersey, Oct. 1954, Council on Wave Research, pp. 288-304.

Biesel, F. (1954b). "The similitude of scale models for the study of seiches in harbours" *Proc. 5th ICCE*, ASCE, Grenoble, pp. 95-118.

Biesel, F. (1959). "Applications de l'IBM 650 dans le domaine de l'hydraulique" *Ingénieurs et Techniciens*, no. 121, pp. 63-79.

Biesel, F. and Ranson, B. (1961). "Calculs de diffraction de la houle" *Proc. 9th IAHR Congress*, Dubrovnik, pp. 688-699.

Biesel, F. (1963). "Radiating second-order phenomena in gravity waves" *Proc. IAHR Congress*, London, pp. 197-203.

Biesel, F. (1964). "Equations approchées de la réfraction de la houle" *Proc. 9th ICCE*, ASCE, Lisbon, pp. 55-69

Biesel, F. (1966). "Les phénomènes du second ordre rayonnants dans les ondes de gravité" *La Houille Blanche*, no. 4, 403-420.

Biesel, F. (1972). "Réfraction de la houle avec diffraction modérée" *Proc. 12th ICCE*, ASCE, Vancouver, Canada, pp. 491-501.

Biesel, F. and Suquet, F. (1951-1952). "Les appareils générateurs de houle en laboratoire" *La Houille Blanche*, vols. 2, 4 and 5 (1951) and vol. 6 (1952).

Blosset, M. (1951). "Théorie et pratique des travaux à la mer" *Editions Eyrolle*, Paris, France.

Bonnefille, R. and Voyer, F. (1961). "Calcul de la propagation de la marée en zone côtière" *Proc. 9th IAHR Congress*, Dubrovnik, pp. 832-853.

Bouvard, M. (1994). "De l'hydroélectricité à la mécanique des fluides "tous azimuths" : Evolution des activités scientifiques et industrielles de la mécanique des fluides - hydraulique à Grenoble" *La Houille Blanche* nos. 5-6, pp. 131-159.

Caillat, J.M. and Migniot, C. (1981). "Modélisation de la vase et de la salinité dans un estuaire" *Proc. 19th IAHR Congress,* New Delhi, vol. V, pp. 403-411.

Chabert d'Hières, G. (1962). "Réglage et exploitation de la plaque tournante de Grenoble" *La Houille Blanche*, supplement no. 1, pp. 130-140.

Chabert d'Hières, G. and Le Provost, C. (1978). "Atlas des composants harmoniques de la marée en Manche" *Annales hydrographiques*, 5th series, vol. 6, fasc 3, pp. 5-36.

Dagallier, H. editor (1970). "Hommage à Pierre Danel, 1902-1966" *La Houille Blanche* no. 6, pp. 509-541.

Danel, P. (1953). "Tetrapods" *Proc. 4th ICCE*, ASCE, Chicago, pp. 390-398.

Danel, P. and Greslou, L. (1962). "The Tetrapod" *Proc. 8th ICCE*, ASCE, Mexico City, pp. 469-481.

Danel, P., Chapus, F.E. and Dhaille, R. (1960). "Tetrapods and other precast blocks for breakwaters" *J. of Waterways and Harbor Div.*, ASCE, vol. 86, WW3, pp. 1-14.

De Rouville, A. (1950). "Renseignements et réflexions sur les ouvrages de défense des côtes" *Annales des Ponts et Chaussées*, no. 21, pp. 489-528.

De Rouville, A. (1954). "Remarques générales sur les mémoires présentées pour la défense des côtes" *Proc. 5th ICCE*, ASCE, Grenoble, pp. 432-440.

Fontanet, P. (1961). "Théorie de la génération de la houle cylindrique par un batteur plan" *La Houille Blanche*, no. 1 pp. 3-31 & no. 2, pp. 174-197.

Galland, J.C., Gontal, N. and Hervouet, J.M. (1991). "TELEMAC: A new numerical model for solving shallow water equations" *Adv. Water Resources*, vol. 14, no. 3, pp. 138-148.

Gerlier, P. (1992). "Le public et son littoral. Travaux de l'Institut Français de la Mer" *La Revue Maritime*, no. 428, pp. 115-162.

Gohin, F. (1960). "Détermination des dénivellations et des courants de marée" *Proc. 7th ICCE*, ASCE, The Hague, pp. 485-509.

Gohin, F. (1961). "Etude des dénivellations et des courants dus à la marée. Emploi de machines à calculer" *Proc. 9th IAHR Congress*, Dubrovnik, pp. 680-687.

Grousson, R. (1975). "La vie et l'oeuvre de l'ingénieur hydrographe général André Gougenheim" *Annales hydrographiques*, 5ème série, vol. 3, fasc. 1, no. 742, pp. 3-5.

Hamm, L. (1991). "Discussion of 'Waves in a harbour with partially reflecting structures' by M. Isaacson and S. Qu" *Coastal Engineering*, 15, pp. 305-308.

Hamm, L., Madsen, P.A. and Peregrine, D.H. (1993). "Wave transformation in the nearshore zone: a review" *Coastal Engineering*, 21, pp. 3-39.

Hauguel, A. (1980). "A numerical model of storm waves in shallow water" *Proc. 17th ICCE*, ASCE, Melbourne, Australia, pp. 746-762.

Isaacson, M. (1989). "Prediction of directional waves due to a segmented wave generator" *Proc. 23rd IAHR Congress*, Ottawa, Canada, Vol. C, pp. 435-447.

Kobayashi, M. and Kaihatsu, S. (1994). "Hydraulic characteristics and field experience of new wave dissipating concrete blocks (Accropode®)" *Proc. 24th ICCE*, ASCE, Kobe, Japan.

Le Méhauté, B. and Biesel, F. (1956). "Mouvements de résonance à deux dimensions dans une enceinte sous l'action d'ondes incidentes" *La Houille Blanche*, no. 3, pp. 348-374.

Le Méhauté, B. (1954). "Two-dimensional seiche in a basin subjected to incident waves" *Proc. 5th ICCE*, ASCE, Grenoble, pp. 119-150.

Le Provost, C., Genco, M.L., Lyard, F., Vincent, P. and Canceil, P. (1994). "Spectroscopy of the world ocean tides from a finite-element hydrodynamic model" *J. of Geophysical Research*, vol. 99, no. C12, pp. 24,777-24,797.

Massoud, Z. and Piboubès, R. (1994). "L'atlas du littoral de France" *Editions Jean-Pierre de Monza*, Paris, 332 pages.

Miche (1944). "Mouvements ondulatoires de la mer en profondeur constante et décroissante" *Annales des Ponts et Chaussées*, Part 1, no. 2, pp. 25-78, Part 2, no. 7, pp. 131-164, Part 3, pp. 270-292, Part 4, pp. 369-406.

Michon, X. and Bonnefille, R. (1956). "La marée dans la Manche. Construction et réglage d'un modèle réduit" *Proc. 4èmes Journées de l'Hydraulique - Les énergies de la mer*, Paris, 13-15 June 1956, *La Houille Blanche*, pp. 344-350.

Migniot, C. (1968). "Etudes des propriétés physiques de différents sédiments très fins et de leur comportement sous des actions hydrodynamiques" *La Houille Blanche*, no. 7, pp. 591-620.

Migniot, C. (1979). "Utilisation des modèles réduits sédimentologiques pour prévoir l'influence d'un ouvrage maritime sur l'évolution du littoral" *La Houille Blanche*, no. 4/5, pp. 291-300.

Migniot, C., Orgeron, C. and Biesel, F. (1975). "LCHF coastal sediment modeling techniques" *Proc. Symp. on Modeling Techniques*, ASCE, San Francisco, Sept 3-5, pp. 1638-1657.

Ministère de la Qualité de la Vie (1977). "Quelques données sur le littoral". 3 vols. published by the Service d'étude et d'aménagement touristique du littoral - Secrétariat d'Etat au Tourisme.

Montaz, J.P. (1964). "Etude expérimentale systématique en vue de l'utilisation des modèles mathématiques pour l'étude de la diffraction pure de la houle" *La Houille Blanche*, no. 7, pp. 785-791.

Neyrpic (1951). "Etudes sur modèle réduit pour le passage du Rhin, Nov. 1944-Avril 1945". Brochure published by Neyrpic/Laboratoire Dauphinois d'Hydraulique.

Orgeron, C., Larras, J., Paolelle, G., Bellipanni, R. and Couprie, P. (1982). "Accident et réparation de la digue de Sines, Portugal" *Travaux*, April 1982, pp. 1-19.

Parthiot, F. (1981). "Development of the river Seine estuary: case study" *J. of Hyd. Div.*, ASCE, vol. 107, HY11, pp. 1283-1301.

Pelnard-Considère, R. (1956). "Essai de théorie de l'évolution des formes de rivage en plages de sable et de galets" *Proc. 4èmes Journées de l'Hydraulique - Les énergies de la mer*, Paris, 13-15 June 1956, *La Houille Blanche*, pp. 289-298.

PIANC (1979). Obituary of André-Gervais de Rouville, vol. II, no. 33, pp. 63-64.

Plénet, C-C. (1996). "Histoire de l'informatique à Grenoble et de ses apports à l'industrie". PhD thesis, University of Grenoble II, France.

Quetin, B. and de Rouville, M. (1986). "Submarine sewer outfalls" *Marine Pollution Bulletin*, vol. 17, no. 4, pp. 133-183.

Rouse, H. (1970). "Pierre Danel's influence on American hydraulics" *La Houille Blanche* no. 6, pp. 526-527.

Serre, F. (1953). “Contribution à l'étude des écoulements permanents et variables dans les canaux” *La Houille Blanche*, no. 3, pp. 374-388 & no. 6, pp. 830-872.

Smith, A.W. and Jackson, L.A. (1995). “The development and impact of harmonic reformed “Miche” wavelets upon a natural beach” *J. of Coastal Research*, vol. 11, no. 4, pp. 1346-1353.

Tourmen, L. (1962). “Artificial beach building on the Croisette waterfront at Cannes” *Proc. 8th ICCE,* ASCE, Mexico, pp. 793-809.

Tourmen, L. (1968). “The creation of an artificial beach in Larvotto bay, Monte Carlo, Principality of Monaco” *Proc. 11th ICCE*, London, pp. 558-569.

Valembois, J. (1954). “Les appareils réalisés à Chatou pour la mesure de la houle naturelle” *Proc. 5th ICCE*, ASCE, Grenoble, pp. 170-176.

HISTORY AND HERITAGE OF GERMAN COASTAL ENGINEERING

Hanz D. Niemeyer[1], Hartmut Eiben[2], Hans Rohde[3]

ABSTRACT: Coastal engineering in Germany has a long tradition basing on elementary requirements of coastal inhabitants for survival, safety of goods and earning of living. Initial purely empirical gained knowledge evolved into a system providing a technical and scientific basis for engineering measures. In respect of distinct geographical boundary conditions, coastal engineering at the North and the Baltic Sea coasts developed a fairly autonomous behavior as well in coastal protection and waterway and harbor engineering. Emphasis in this paper has been laid on highlighting those kinds of pioneering in German coastal engineering which delivered a basis that is still valuable for present work.

INTRODUCTION

The Roman historian Pliny visited the German North Sea coast in the middle of the first century A. D. He reported about a landscape being flooded twice within 24 hours which could be as well part of the sea as of the land. He was concerned about the inhabitants living on earth hills adjusted to the flood level by experience. Pliny must have visited this area after a severe storm surge during tides with a still remarkable set-up [WOEBCKEN 1924]. This is the first known document of human constructions called 'Warft' in Frisian (Fig. 1). If the coastal areas are flooded due to a storm surge, these hills remained

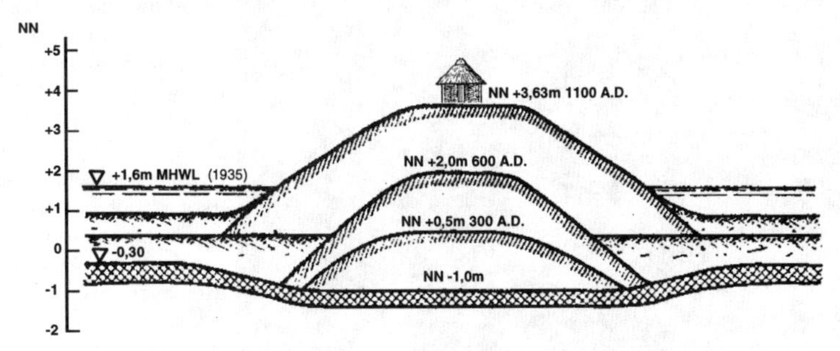

Figure 1. Scheme of a 'warft' with a single building and its adaptions to higher storm surge levels between 300 and 1100 A.D.; adapted from KRÜGER [1938]

---

1) Coastal Research Station of the Lower Saxonian Central State Board for Ecology, Fledderweg 25, 26506 Norddeich / East Frisia, Germany, email: niemeyer.crs@t-online.de
2) State Ministry for Food, Agriculture and Forests of Schleswig-Holstein.
3) formerly Coastal Branch, Federal Institute for Waterway Engineering.

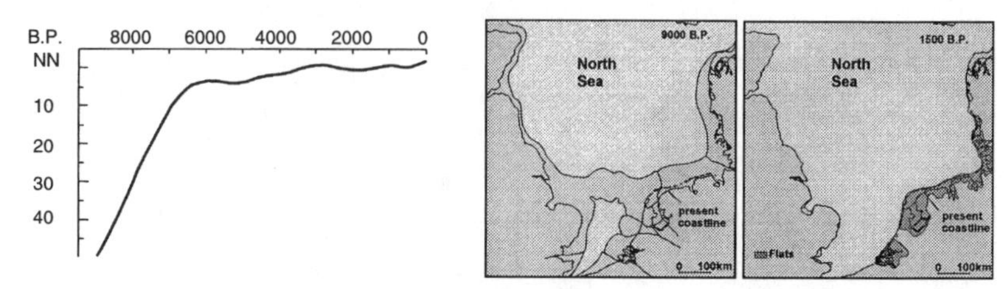

Figure 2. Sea level rise at the coast of Schleswig-Holstein [STREIF & KÖSTER 1978] between 9000 B.P. and present and coastal retreat at the southern North Sea between 9000 and 1500 B.P. [VEENSTRA 1976]

dry: safe havens for the coastal inhabitants, their housing, cattle, and goods. It makes the first engineering response against the sea after millenniums of human retreat enforced by regressive coastlines due to a rising sea level evident (Fig. 2). In modern terms the strategy had changed from retreat to accommodation. It was worthwhile for the inhabitants to take the burden to erect these earth hills and to bear this hostile environment because the accidentally flooded coastal marshes became very fertile due to accompanying sedimentation of silt. In some areas people joined their efforts and erected 'Warften' for whole villages. Partly they have been preserved as monuments of historical coastal zone management (Fig. 3). Though no longer a common standard of coastal protection, the report of Pliny documented the first stage of coastal engineering in Germany still in use today on small islets at the German North Sea coast without flood prevention such as dykes as an adequate and economic solution to protect life and property of the inhabitants.

These human settlements became possible since the coastlines no longer changed as rapidly as in the course of the preceding millenniums after the ice ages due to a deceleration of sea level rise (Fig. 2). In most of the coastal areas it was possible to remain by adjustment of the 'Warften' for the next thousand years (Fig. 1). After that period the construction of small dykes, a marked change in strategy started again: in spite of defending the settlements on the 'Warften' as singular points, whole areas enclosed by the dykes became to a certain extent protected against flooding allowing now more intensive agricultural use. Therefore, the Frisians called their dykes in the middle ages 'Golden Ring' [WOEBCKEN 1924]. In modern terms of coastal management strategy, the adaptive response has again changed: now from accommodation to protection. Moreover this was the first act of human intervention into the morphodynamics of the coastal areas at the southern North Sea. The coast had no longer an autonomous behavior, purely driven by natural forces. That kind of human intervention into the existing dynamical equilibrium caused responses by the sea of which a

Figure 3. Aerial photograph of the village of Rysum in East Frisia. The ring road marks the border of the 'Warft'; the church at the top functioned as a save haven as well during storm surges as against hostile invaders.

remarkable number developed to disastrous lessons for the coastal inhabitants. The chosen strategy of protection required more than the empirical adjustment to the highest flood level for being sufficiently successful. Intervention into the coastal processes forced them to learn how to deal with them. Motivation for the efforts was the prospect of safety and economical welfare being endangered by the sea which was a threat for both the goods of the coastal inhabitants and even their life. The construction of dykes, their failures, and accompanying catastrophic events could be regarded as the fertilized soil for the roots of coastal engineering at the Wadden Sea coasts leading directly to first solutions of problems which are today still of importance.

For centuries the aims of coastal protection have been the driving forces for the efforts to gain further knowledge on coastal hydro- and morphodynamics. But step by step the requirements of sea trade and particularly the demands and efforts of existing ports to keep access to the sea stimulated additionally the processes of learning and the establishment of planning based on increasing knowledge. The aim of this contribution is therefore not to cover the whole field of problems in German coastal engineering but to focus on those subjects of coastal engineering for which a historical heritage exists being of potential interest to an international audience.

THE GERMAN COASTS

Germany borders the North Sea as the Baltic Sea which are mostly separated by the Cimbrian peninsula consisting nowadays of Denmark and the German Federal State of Schleswig-Holstein and intersected by Skagerrak and Kattegat (Fig. 4 + 6). In the North Sea, tidal motion is triggered by the Atlantic tides via the English channel and particularly via the northern approaches between Scotland and Iceland. Due to high damping in its intersection to the North Sea, particularly by the straits between Jutland and Sweden, there are only microtidal oscillations in the Baltic Sea. Both coasts have been shaped after the ice ages: The available material is predominantly clastic being moved and worked out by the huge glaciers covering these areas before. The present coastline were firstly touched by the sea about 9000 years ago in the North Sea area [STREIF & KÖSTER 1978] and about 6000 years ago in the Baltic Sea [NIEDERMEYER et al. 1987].

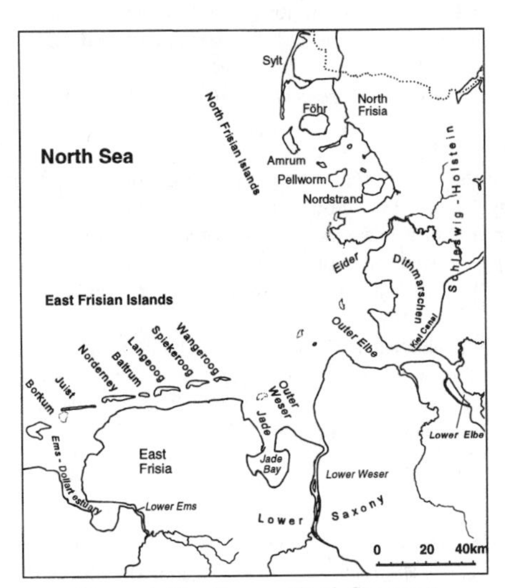

Figure 4. The German North Sea coast

The North Sea coast

The southeastern part of the North Sea is in the German Bight boardered by the German coast. It consists from west to east subsequently counterclockwise of the following parts: Ems-Dollart estuary, East Frisian Islands and Coast, Jade Bay and inlet, Weser estuary, Wursten coast and Elbe estuary. From south to north follow Dithmarschen coast, Eider estuary and North Frisian Islands and Coast (Fig. 4). The present state of the coast has been shaped by holocene transgression [STREIF & KÖSTER 1978] starting about 9000 years ago when the sea-level was about 45 m lower then today (Fig. 2). Its rise was rapid until 5000 years B. P. and decelerated afterwards remarkably with superimposed oscillations. Distinctions in the existing coastal shape and the amount of unconsolidated material caused interactions and specific regional variations. An example of sea-level rise for the coast of Schleswig-Holstein (Fig. 2) highlights the enor-

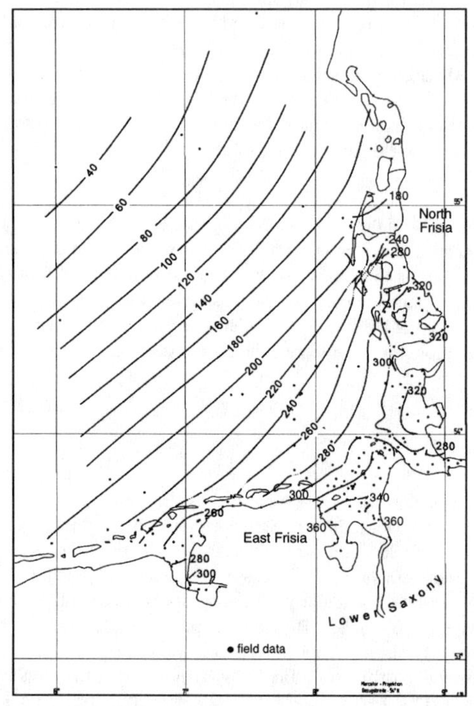

Figure 5. Mean tidal range [cm] at the German North Sea coast [LASSEN & SIEFERT 1991]

mous changes in hydrodynamical boundary conditions for coastal development. The sediments are clastic and have been transferred into the present coastal areas by the glaciers during the ice ages. They have been reworked by currents, waves, and wind causing as well erosion as accretion of dunes, flats and marshlands. Only the island of Heligoland in the North Sea and the island of Rügen in the Baltic Sea have rockey coastlines. In the past storm surges created enormous land losses, particularly in areas with easily erodible soil such as peat for example. The coastal shape changed: Bays were enlarged or even newly created and in North Frisia remnants of former mainland became islands. Afterwards a reshaping of the coast took place driven both by sea and human intervention.

Nowadays mean tidal range increases from 2.2 m in the entrance of the Ems-Dollart estuary to nearly 3 m in the offshore area of Jade, Weser and Elbe estuaries and decreases to about 1.8 m in the offshore area of North Frisia (Fig. 5). The highest tidal range at the German North Sea coast occurs close to the landward border of the Lower Weser estuary in Bremen with about 4 m on the average. The yearly offshore mean significant wave height is about 1 m. Applying the classification of HAYES [1979] the German North Sea coast is high meso- to low macrotidal and in respect of wave climate tide dominated. The set up of storm surges has been measured with maximum values in the range between about 3 m at the islands and 3.7 m in the estuaries. In the offshore areas waves with maximum heights of about 10 m have been measured, on the shoreface of barrier islands of about 5 m and in front of mainland dykes of about 2 m.

At the Wadden Sea coast of the southern North Sea the prevailing longshore drift has an eastward direction bypassing the estuaries and tidal inlets with a remarkable offshore offset due to ebb-dominated tidal currents. They perform in dynamical balance with waves entering from offshore aggregations of sediments as shoals which are interrupted by channels. After the bypassing of the estuarine or inlet areas the ebb delta shoals have at their downdrift a landfall at large intertidal areas or barrier islands.

Most parts of the mainland coasts consist of low lying marshlands being performed by deposition of fine sediments during and after holocene transgression. Only in a few small areas remnants of the pleistocene perform the shore. The whole German North Sea coast is part of the Wadden Sea ranging from Den Helder in the Netherlands across the German Bight to Ribe in Denmark.

The Baltic coast

The northwestern part of the German Baltic coast is characterized by bays and firths, the eastern one by beach berms, spits and shallow lagoons, and islands. The shorelines are partly beach berm with shallow areas backward or sandy bluffs. The morphology has been partly shaped by glaciers during the ice ages

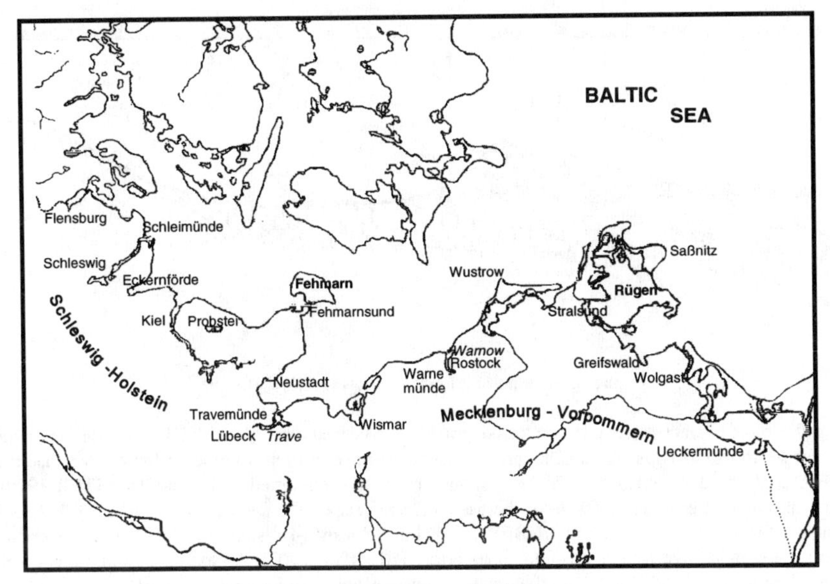

Figure 6. The German Baltic Coast

and later by holocene hydrodynamics. Evidently the present coast has been shaped in the last 6000 years. Basic materials were stones, shingle, sand, and till. The shores can be differentiated in bluffs, lowlands (Fig. 7) and lagoon coasts (Fig. 8). The lowland coasts have beach berms or spits at the shoreline. The beach berms are often created by storm surges. The bluffs have locally different sediment balances. Most changes occur due to storm surges which was already mentioned by HAGEN [1863].

The mean sea level of the Baltic Sea corresponds generally with that one of the North Sea via the intersections. Therefore variations of the North Sea mean level propagate into the Baltic Sea. Tidal oscillations decrease from west to east: An amplitude of about 0.2 m has been analyzed for Travemünde and of about 1 inch for Stolpmünde at the Pomeranian coast [HAGEN 1863]. Of higher importance are seiches and particularly basin oscillations which are triggered by wind action. The mean water level increases from west to east which is an effect of the predominant western wind directions. Eroding bluffs provide neighboring shallow coast with material via seaward bars which act both for transport and sorting. In interaction with local hydrodynamics the sediments create bars or beach berms or spits. In some cases areas with low or even no erodibility act stabilizing [HAGEN 1863].

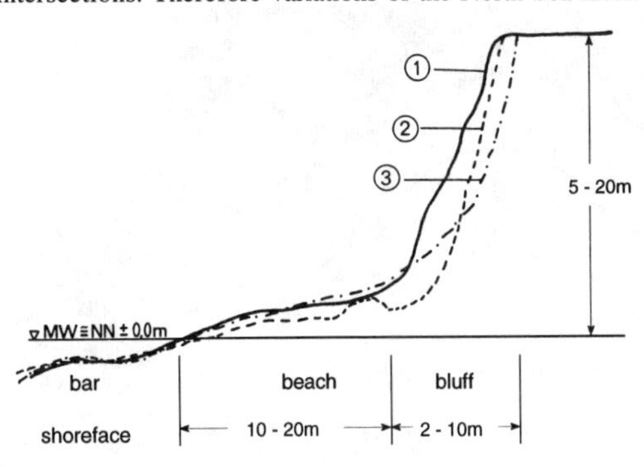

Figure 7. Cross-section of a bluff coast [EIBEN 1992b]
NN : German geodetic datum ≈ MSL

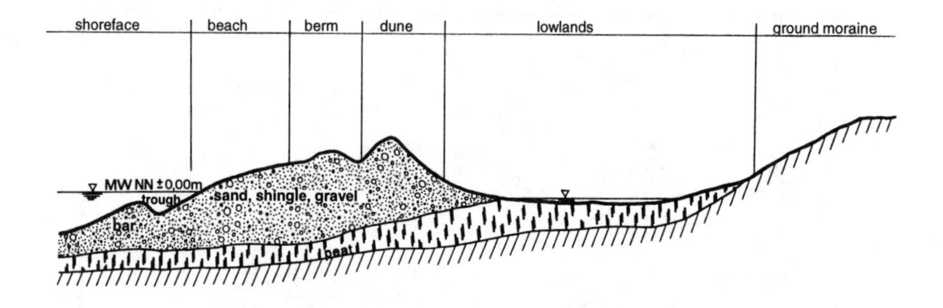

Figure 8. Scheme of a lowland coast [EIBEN 1992b]

Together these distinct shores perform 'physiographic or balanced units' [WYRTKI 1953] for which the longshore transport is balanced. This process has been made evident by tracing the eroded material which could be distinguished by higher variance from the well-sorted native one [KÖSTER 1979]. Coastal retreat at eroding bluffs varies between 0,25 m/year on the average [PETERSEN 1952] and 1m/year for exposed ones [KANNENBERG 1951] in Schleswig-Holstein. In Mecklenburg-Vorpommern the comparative figures are four times larger [GEINITZ 1903]. An extreme example is documented by KOLP [1955] for a coastal area at Darß where since 1827 a total retreat of 150 m occurred. The present state of knowledge allows to determine balanced units for the German Baltic coast (Fig. 9).

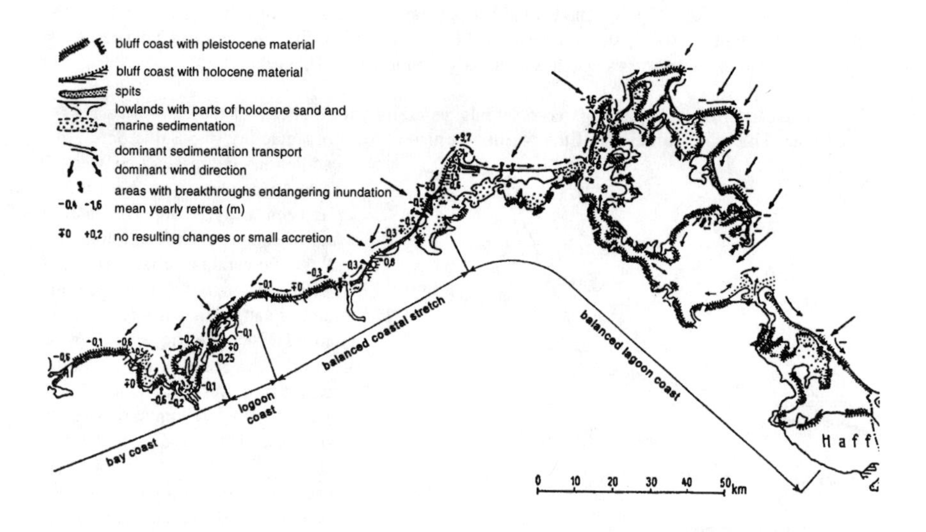

Figure 9. Example of balanced units at the Southern Baltic coast [WEISS 1992]

COASTAL PROTECTION AT THE NORTH SEA COAST

Basic hydro- and morphodynamics

Relative sea-level rise. BRAHMS [1754, 1757], who is nowadays regarded as the pioneer of modern coastal engineering in Germany was a Frisian 'dyke judge': elected head of one the numerous self-ruling communities at the Wadden Sea coast being responsible for coastal protection of its area. He recommended to check regularly the crest height of dykes in respect of mean high tide level which he recommended as a permanent basis for dimensioning dykes and relating them to their environment. Though not knowing or even considering a sea-level rise a thorough execution of that procedure would have highlighted at least the consequences of the sea-level rise for coastal protection. About 150 years later that effect was seriously mentioned by SCHÜTTE [1908], who evaluated a relative sea-level rise of about 33 cm per century being orientated at mean high water level and based on geological and archaeological investigations in coastal areas. KRÜGER [1922] evaluated a relative sea-level rise of about 20 cm per century by using the data of the gauges in Wilhelmshaven, Bremerhaven, and Cuxhaven taking the average of mean high and mean low tidal water level into consideration. Later he gave figures of centennial rises of 24 cm for high, 19 cm for low and 21.5 cm for half tide based on time series of 83 years at the gauge Wilhelmshaven [KRÜGER 1938]. Both still interpreted that effect as long-term subsidence of the coast being interrupted by periods of rising (Fig. 10). Their postulation stimulated an intensive discussion as well among coastal researchers as within the coastal community in their times and triggered the first of up to now three 'coastal levellings' in 1926: The repeated measurement of heights in coastal areas with reference to fixed points inside of the country regarded as vertically stable. Evidently there was then a lack of transatlantic communication among scientists: The explanation of that phenomenon by an eustatic sea-level rise was not taken into consideration though that idea was already introduced by MC LAREN [1842]. Since the 1950s of this century the effect of relative sea-level rise is taken into consideration for the design of coastal structures in Germany, particularly for the dykes at the coasts and tidal estuaries [HUNDT 1954]. Background for the introduction of that additional measure was that in the course of the last centuries the trend of the mean high water level had been nearly the same as that one of the envelope of the highest storm surges during that period (Fig. 11) [LÜDERS 1971; ROHDE 1977]. In respect of an anticipated change of global climate the question of relative sea-level rise is again a subject of the discussion concerning future coastal protection strategy, particularly for the lowlands. Basing on the data and established knowledge a comprehensive analysis on sea-level rise at the German North Sea coast has started [LOHRBERG 1989; LASSEN & SIEFERT 1991] partly also in cooperation with neighbor countries [JENSEN et al. 1993].

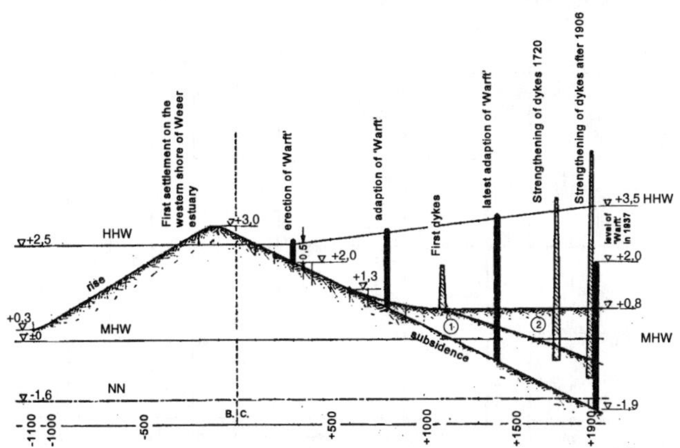

Figure 10. Interpretation of relative sea-level rise as coastal subsidence by KRÜGER [1938]; ①: marshland growth and its stop after impoldering; ②: salt marsh growth; MHW: Mean high tide; HHW: Highest storm surge level; NN: German Geodatic Datum

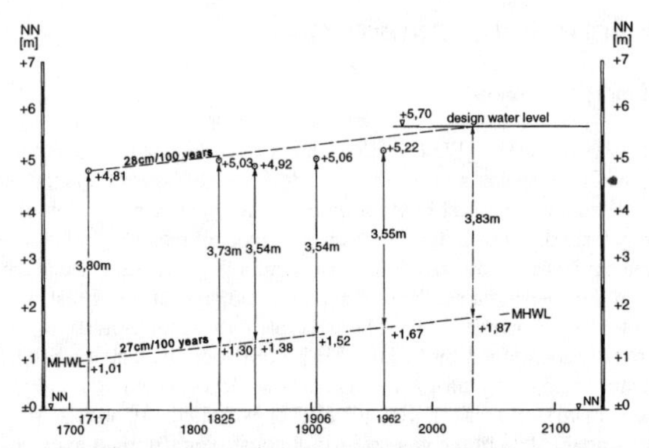

Figure 11. Reconstructed and measured levels of mean high tide and storm surge peaks at the Jade Bay since 1717 [LÜDERS 1971]

Tides and storm surges levels. The oldest known documentation of an ordinary tide at the German North Sea coast (Fig. 12) is inherited from BRAHMS [1754, 1757]. Later navigation purposes led to the installation of the first permanent gauges at the German North Sea coast [ROHDE 1975]. BRAHMS started furthermore the establishment of benchmarks of storm surges in the Jade area after the disastrous storm surge of Christmas 1717 with its high rate of death toll in the coastal areas. They were founded on a stable pleistocene basis in an area being sufficiently sheltered against waves. These benchmarks as well as those ones being measured later have been transferred to a new dyke after the enclosure of the original coastline acting as a monument of coastal engineering (Fig. 13). HUNDT [1954] used historical data at the western coast of Schleswig-Holstein under consideration of the long-term rise of the mean high tide in order to enlarge his data set for determination of return periods of extreme storm surge water levels. Since the coastal structures were rather vulnerable against frequent hydrodynamical impacts BRAHMS [1754, 1757] developed a specific tidal gauge recording the peak of storm surge still water levels (Fig. 14). He used these data to establish two kind of classifications for storm surges: one based on exceeding levels orientated at the distinct parts of the dykes and their seaward apron; the other at their return period. Both approaches are still used in German coastal engineering [LÜDERS 1956] though mostly in respect of the return periods at coastal areas [NIEMEYER 1987] delivering nowadays also basis for judgements on a possible change of storminess and increasing frequency of storm surges at the southern North Sea coast.

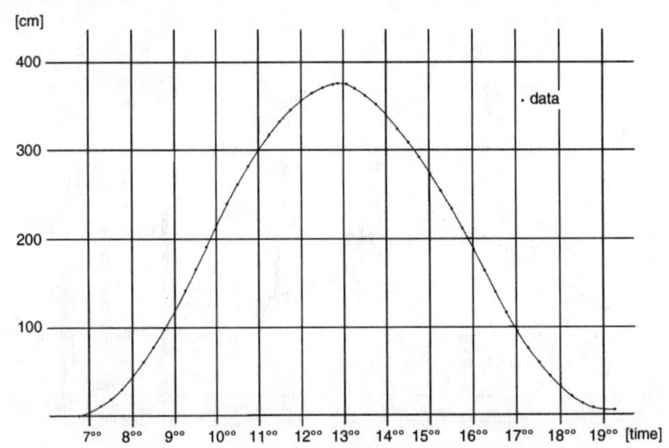

Figure 12. Cycle of an ordinary tide due to measurements of BRAHMS [1754, 1757]; reconstructed by LUCK & NIEMEYER [1980]

Figure 13. Benchmarks (↑) of historical storm surge peaks at Dangast, Jade Bay

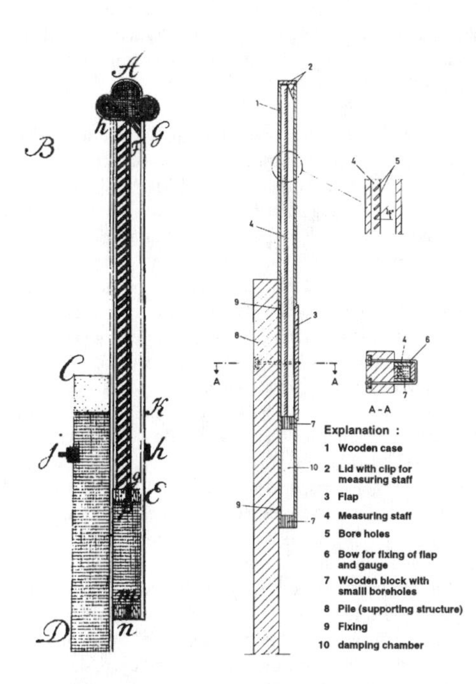

Figure 14. Gauge for the measurement of storm surge peak levels by BRAHMS [1754, 1757]; original drawing left, reconstruction by LUCK & NIEMEYER [1980] right

Waves. In order to consider wave attack for proper dyke design BRAHMS [1754, 1757] observed the nearshore wave climate on tidal flats and supratidal salt marshes close to the coastline. He related wave heights and water depths by

$$H = \sqrt{h}$$

in order to have a general basis for the derivation of local wave heights along all the coastal stretches where dykes had to be designed. It is nowadays well established knowledge that wave heights at Wadden Sea coasts correlate strongly with local water depths [FÜHRBÖTER 1974; SIEFERT 1974; NIEMEYER 1979, 1983]: the basic approach by BRAHMS [1754, 1757] has been modified, but is still not far beyond reality (Fig. 15).

The interactions of waves and coastal morphology in Wadden Sea areas has been quantitatively evaluated by KRÜGER [1911]: The width of intertidal areas is essentially depending on wave energy dissipation. Therefore those in the shelter of barrier islands are remarkably smaller than those positioned in estuarine areas unsheltered against offshore waves. At coasts with island sheltered flats, waves establish a gradient from the tidal inlets where the waves from offshore enter along

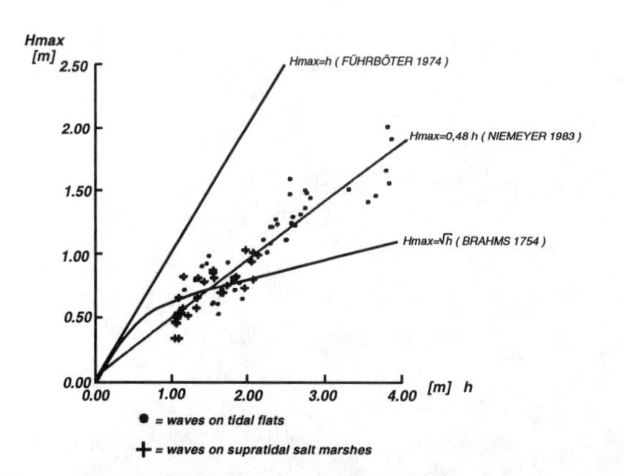

Figure 15. Field data of wave height/water depth relation on East Frisian tidal flats and salt marshes [NIEMEYER 1983] in comparison with the relation developed by BRAHMS [1754] from visual observation

their travel distance: increasing flat levels until the growth of supratidal salt marshes at marshes at sheltered coastal stretches and a tendency from coarser sand to fine silt with significant organic content. KRÜGER [1911] stressed the consequences for mainland protection: strongest wave attack in the vicinity of the inlets and the importance of islands for coastal protection on the sheltered coastal stretches of the mainland. He opposed with his arguments against a recommendation of FÜLSCHER [1905], a high ranking civil servant in the responsible ministry in Berlin, who argued islands were of no benefit for mainland protection and any efforts for their protection should be limited to those few ones which had already then sufficient economical values on their own. KRÜGER's sound description of these complex interactive processes were only based on observations with the human eye and the analysis of mid-term morphological changes. They all have meanwhile been verified by intensive field measurements [NIEMEYER 1979, 1983], evaluation of historical morphological situations [NIEMEYER 1995] and mathematical modeling [NIEMEYER et al. 1995a].

Wave-induced currents. The wave-induced current system on beaches is of high importance for local sediment balance. Intensive knowledge about this driving forces for transport processes on beaches has been gained since the end of World War II, particularly pioneered by the reports of MUNK [1949] on wave set-up, of PUTNAM et al. [1949] on longshore currents, being stimulated by observations of that phenomena made during exercises for allied landing operations of World War II, and of SHEPARD & INMAN [1951] on nearshore circulation. The transport processes due to wave-induced currents have considerable importance for the stability of beaches at the German North Sea coast and particularly on those of the East and North Frisian barrier islands which are partly seriously suffering from structural erosion. Therefore, the scientific findings of their American colleagues were eagerly adapted by German coastal engineers like e. g. LAMPRECHT [1955] who did intensive investigations on beach processes on the North Frisian island of Sylt. He as well as numerous contemporaries and successors were not aware that these phenomena had already been detected and published by predecessors in their own country: HAGEN [1863] described the set-up and undertow which he observed at the Baltic coast which will be discussed in the following chapter. KRÜGER [1911] was beside other obligations also in charge of the shoreline protection of the East Frisian island of Wangeroog and made enormous and successful efforts to understand the processes leading to beach erosion. One of the numerous substantial results is a report which highlights impressively longshore currents on the beaches of the island of Wangeroog: "Due to surfing, the waves drive the water alongshore and generate close to the shore a current which moves the sand being suspended by the breaking waves. During northwestern wind of Beaufort 6 during flood I observed between breaker line and beach a downward racing current in a water depth of 60 cm alongshore in eastern direction. It was impossible for me to keep kneeling on the spot, I was driven away. About 100 m offshore was another strong alongshore current. I did not go further offshore. It is reasonable that the current's strength will decrease further offshore. This assumption is verified by experience gained by landing on the beach across the surf zone. Offshore off the surf zone the current is the same as during calm weather. Between breaker line and beach the boat is driven into the same direction as that one of the incedenting waves. If there is more than one breaker zone a strong current appears at least landward of the central one." Though this observation does not deliver a sound basis of transferrable knowledge according to present standards it does convey an enormous capability of intuitive feeling for natural processes and the target-oriented pragmatic approach to realized problems becomes evident. In recent years more knowledge has become available, particular by intensive prototype investigations. Pioneering efforts in this field have been carried out in recent years on the beaches of the island of Sylt [DETTE & FÜHRBÖTER 1975] stimulating a couple of successful site investigations on the subject of wave-induced nearshore circulation since then.

Ebb delta migration and sediment balance. The bypassing of the tidal inlets by longshore drift in the formations of ebb deltas is of high importance for the sediment balance of the beaches of the downdrift barrier island [GAYE & WALTHER 1935]. Updrift of the landfall of the ebb delta's shoals occurs structural erosion requiring engineering countermeasures. Furthermore, the supply with sediment in the landfall area and downdrift depends on the successive welding of shoals. Therefore intermediate erosion may take place in the periods between and particularly immediately before the landfall [KRÜGER 1911, 1937]. This knowledge is very useful for the evaluation of the necessity of countermeasures or of their appropriateness. A thoroughful analysis and forecasting of these processes have been carried out by HOMEIER & LUCK [1971] and performed the basis for the abandonment of the implementation of costly solid structures (Fig. 16).

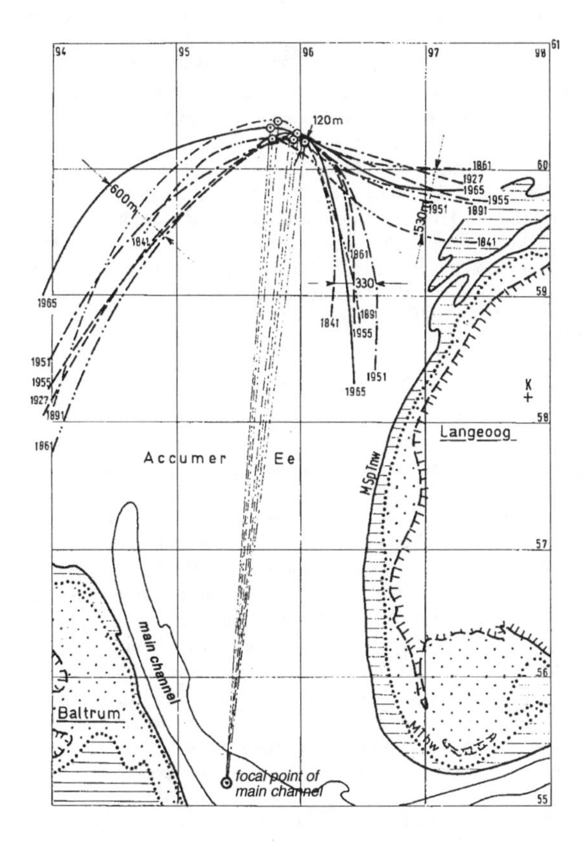

Figure 16. Mean pathways of migration of ebb delta shoals of the tidal inlet Accumer Ee and resulting sediment transport due to distinct landfall positions between 1841 and 1965 [HOMEIER & LUCK 1971]

Erosion and regeneration of beaches and dunes. Storm surges cause often severe erosion of dunes and also of beaches, particularly in front of revetments and seawalls due to scouring. The impression as well as beach profiles leveled immediately after such an event has provoked often countermeasures like beach fills in areas with structural erosion. The remaining buffer was regarded as insufficient with respect to safety. After a series of severe storm surges in 1973 and 1976 HOMEIER [1976] analyzed levelings of dunes and beaches suffering from structural erosion on East Frisian islands gained before, immediately after, and again weeks after the storm surges. He concluded that after a few weeks a regeneration of the beaches had taken places due to hydrodynamical transport. Dunes regenerated if the width of the dry beach is sufficient for wind transport capacity within months. In fact, the well-known phenomenon of winter and summer profiles of beaches and dunes is also valid for shorelines with structural erosion. It has been proven for a number of later storm surges and could also be explained for beaches by the erosional effects of undertows and the regenerating onshore transport due to nonlinear drift effects of steep waves [NIEMEYER 1991a].

Long-term morphological development. Historical maps - mostly for nautical purposes - give an impression of the changes the coast has experienced in the course of the last centuries. Additionally descriptions of the coastal ways called 'sailing handbooks' deliver information on earlier stages of coastal development. Though in nearly all cases not to the same extent reliable as present sources they could perform a basis for deeper insights into long-term processes still governing the present morphological development of the coasts, particularly if a composition of these sources can be justified by still available verification points like then used landmarks as e. g. churches. KRÜGER [1911, 1922] and PLATE [1927] based their impacts for the correction of the Jade and the Outer Weser: they took the anticipated morphological development of the regime into consideration in order to keep their impacts in tune with the tendency of continuing morphodynamical development. The reconstruction of former coastlines, their retreat and subsequent reclamation at the North Sea coast of Schleswig-Holstein has been reconstructed for the period between 1643 and 1648 and compared with the situation of 1888 by GEERZ [KREY 1918]. Particularly for the purposes of coastal protection all available sources for the Lower Saxonian North Sea coast have been evaluated by HOMEIER [1962, 1965, 1969] after World War II. He reconstructed the data from historical sea and land charts and many kinds of additional sources and related them to temporary geographical grids. Very valuable were particularly in the cases of medieval storm surge bays and their reclamation the geological investigations of SCHÜTTE [1908] and WILDVANG [1915]. As a still very useful spin-off HOMEIER [1962] elaborated 15 maps with documentary supplements describing his sources and their use covering the whole coast of Lower Saxony for the situation of 1650, 1750, 1860 and 1960. Each map incorporates the mean high and low water line, positions of the dune foot, border of supratidal salt marshes and dykes. The products of HOMEIER's work of about 30 years has, for example, formed a basis for numerous coastal engineering case studies at the coast of Lower Saxony and furthermore for specific morphodynamical investigations such as the model of tidal inlet migration by LUCK [1977] or the parametrization of long-term morphodynamical development by NIEMEYER [1995].

PROTECTION OF THE MAINLAND COASTS

The lowlands at the German North Sea coast are mainly marshlands developed as deposits of marine sediments settling during phases of flooding from the sea. Beside a few small stretches consisting of high-leveled pleistocene material the coastal areas are therefore below storm surge level and sometimes even below ordinary high tide. Since about 1000 A. D. dykes perform the protection of these areas against inundation. Their dimensions have increased since then (Fig. 17). Early adaptions have carried out due to disastrous experience by the coastal community but since BRAHMS [1754, 1757] an engineering approach has replaced more and more the simple response on catastrophic events in order to prevent another failure of the dykes by planning in advance. BRAHMS was elected as 'dyke judge' one year after the disastrous storm surge of 1717 with thousands of victims and remained in charge until 1752. The state of the art he documented in his 'basic art of dyke construction and hydraulic engineering' [1754, 1757] was unique and far ahead of his contemporaries. In comparison with the present state of knowledge it is evident that he already recognized nearly all key problems and delivered solutions which still must be regarded as breakpathing. To a certain extent this might be an explanation for the large gap between his findings and the continuation in modern German coastal engineering.

Design crest height for sea dykes. The handbook of BRAHMS [1754, 1757] is the oldest known document containing a guide line for the quantitative determination of the crest height of a dyke. His approach to summarize design water level and design wave run-up seems to be simple referring to present knowledge. But he gives further recommendations how to evaluate or estimate these parameters and how to increase the database for their determination.

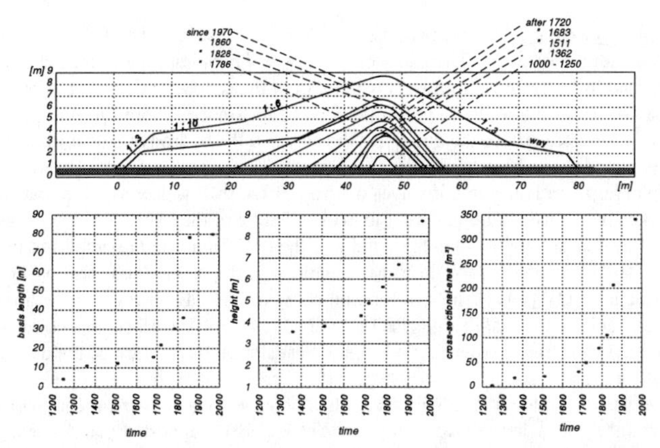

Figure 17. Development of sea dyke profiles at the west coast of the Butjadingen peninsula between the beginning about 1000 and 1972 A. D.; cross-sections adapted from NIEDERSÄCHSISCHE HAUPTDEICHVERBÄNDE [1988]

The determination of the design water level by BRAHMS [1754, 1757] was still based on already observed storm surge levels; a future increase was not anticipated. The major improvements were first his recommendation to gather the necessary basic data by measurements in all coastal areas by aid of his gauge (Fig. 14) and second his instruction to consider the subsidence of the construction in respect of the soft soil beneath the dyke. This approach remained state of the art until as well the discussion about coastal subsidence or respectively sea-level rise as the lessons taken from the disastrous storm surge of 1953 in the neighbored Netherlands led to procedures including a determined measure for future safety in respect of sea-level rise [HUNDT 1954; PETERSEN 1955; LÜDERS 1957] which was improved after the storm surge of February 1962 [LÜDERS et al. 1962; METZKES 1966] with a death toll of more than 300 people. At the North Sea coasts the design water level for sea dykes is still determined on the basis of these procedures. The specific demands in tidal estuaries have been met by the product of an interstate working group on behalf of the Elbe estuary: the design water level is evaluated by superposition of mean high tide, secular rise and the time series of surge set-up at the estuarine entrance [SIEFERT et al. 1988].

BRAHMS must have been aware that his database and theoretical basis were inadequate for derivating a generally applicable formula for the determination of wave run-up. Therefore he recommended first to level benchmarks of flotsam after storm surges marking the highest wave run-up above the still water level. About 200 years later a group of experts repeated his suggestion for getting a sound database [BOTHMANN et al. 1955]. Unfortunately these recommendations were only carried out rarely or the data are lost. Systematic measurements of benchmarks of flotsam are only available from recent storm surges since the 1970s and only a few ones from earlier times. Analysis of these data is carried out in the framework of a research project of the German Committee on Coastal Engineering Research (KFKI) in order to deliver information on the variation of wave run-up along the coastline and furthermore to provide a basis for the derivation of design wave run-up by extrapolation [NIEMEYER et al. 1995b]. Furthermore BRAHMS [1754] determined for specific boundary conditions the dimensions of a dyke. Implicitly these figures include also a relation between wave height and wave run-up [LUCK & NIEMEYER 1980]:

$$R = c \cdot H \cdot \tan \alpha$$

with a constant value of 10,33. The formula turns out to be rather similar to that one published by WASSING [1957] which is well known as the DELFT formula with a constant value of c = 8.

Functional design of sea dykes. The dykes at the Wadden Sea coasts consist of clay and nowadays of a sand core covered with a clay layer. The quality of the clay is of high importance for the capability of the structure to withstand hydrodynamical impacts without or with acceptable damages in respect of safety. But even for clay of good quality there are limitations of resistance. Therefore, the shaping of dykes should take into consideration both the quality of the clay and the hydrodynamical impacts. In order to keep the overtopping of waves he recommended to flatten the inner slopes up to 1:2 and more in respect of clay quality in order to remain stable if waves overtop. BRAHMS was convinced that wave overtopping had to be taken into consideration in order to avoid a failure of the dyke. This basic design philosophy was accepted by coastal engineers in the middle of the last century but rejected in the beginning of this century [PETERSEN 1954]. Basing on investigations in the Netherlands and particularly on the research results of HUNDT [1954] the overtopping of waves was again regarded as inevitable to be responded by a suitable functional design [PETERSEN 1954]. BRAHMS suggested as well a convex shape of the outer dyke profile in order to place the flattest slopes in the place of the design water level and accompanying wave attack (Fig. 18). HENSEN [1955] carried out hydraulic model tests leading him to a general recommendation to replace the then applied concave by a convex profile in order to reduce wave run-up for higher storm surge levels. FÜHRBÖTER [1991] supported that suggestion because a mild slope would decrease the wave impacts on the structure.

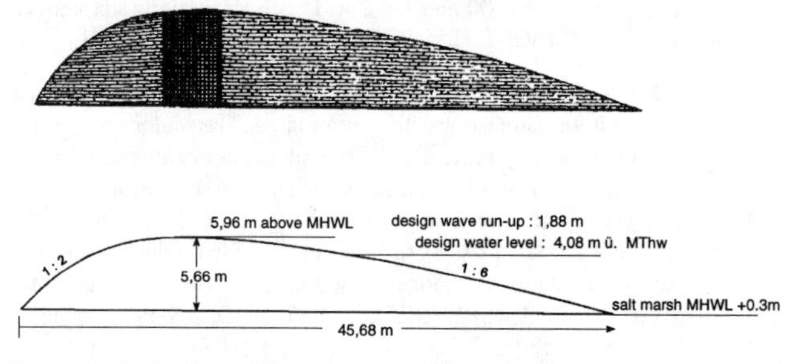

Figure 18. Sea dyke design by BRAHMS [1754]; original drawing above, reconstruction by LUCK & NIEMEYER [1980] down

Cost-benefit analysis. The members of the self-ruling communities being responsible for coastal protection had as well to raise money as to invest working capacity both for maintenance and construction of dykes. Therefore, they were reluctant to follow the ideas of BRAHMS [1754, 1757] which were revolutionary and required enormous higher efforts than the conventional state of the art. The argument of increased safety against disastrous flooding seemed to be insufficient to convince the members of his community though the lessons of the storm surge of 1717 with a death toll of more than 6800 people in the coastal areas between Ems and Weser [ARENDS 1833] were still very recent. BRAHMS [1754, 1757] evaluated first a comparison of different types for the dykes which had to be newly constructed after total damages of the storm surge of 1717 and the succeeding ones. His aim was to prove by detailed figures and costs that the set up of dykes with flatter slopes would not only increase safety but also decrease enormously efforts for maintenance. This kind of evaluation in respect of both economical and safety targets is nowadays called 'optimization'. A very recent example for distinct types of modern sea dykes has been carried out by KRAMER [1977]. The members of the community suffered particularly from the economical losses due to the disastrous effects of the flooding after the storm surges and tried therefore to keep the burden of dyke maintenance as low as possible and were not enthusiastic about the strengthening of existing dykes. BRAHMS [1754, 1757] compared losses due to flooding with the earnings from a successfully protected landscape and the costs of dyke construction which is in terms of our time a typical cost-benefit analysis. Obviously BRAHMS was confronted with a problem being evident for all generation of coastal engineers, particularly in periods of social shortage of money. A later example is given by KREY [1918] who argued that the costs of coastal protection could not only be related to the taxes paid by the landowners of the coastal areas; the economical value of the coastal areas and furthermore the economical spin off for trade and employment should be taken into consideration. Both authors try to convince their audience by rational argumentation appealing to economical aims of human beings. Their approach implies the admission that coastal protection could not be regarded as a purpose on its own but does need social acceptance. Though the figures of BRAHMS are estimated the method is comparable to modern studies dealing with the economic efficiency of coastal protection in respect of the guarded values like that one of KLAUS & SCHMIDTKE [1990] carried out recently for a landscape at the German North Sea coast as a model case.

Land reclamation. The medieval storm surges caused enormous land losses at the Wadden Sea coasts. They enforced an enormous coastal retreat and created and enlarged numerous storm surge bays between Den Helder and the peninsula of Eiderstedt [WOEBCKEN 1924]. In North Frisia continued their impacts the creation of islands from remnants of the former mainland [KREY 1918; HEISER 1933]. In most cases their erosive forces had been very efficient, because the then existing soil of e. g. peat had only low resistance. Particularly the developing large bays got geometrical extensions which were unbalanced with the acting hydrodynamical boundary conditions leading particularly close to the shorelines to a subsequent silting up [NIEMEYER 1991b]. The coastal inhabitants supported this process by the land reclamation works in order to accelerate the growth of salt marshes which were afterwards impoldered. Step by step the coastlines were moved back seaward and the areas of most the storm surge bays were reduced dramatically and only remnants of a few have remained like Dollart, Ley and Jade Bay [HOMEIER 1969] and Dithmarschen Bay [HEISER 1933]. In Schleswig-Holstein the coastline was also moved back towards the sea and in North Frisia even the remaining islands have been enlarged by land reclamation (Fig. 19) [KREY 1918; HEISER 1933]. The reclaimed areas had improved enormously in comparison with the eroded soil from an agricultural point of view. The marine sediments performing the salt marshes delivered after impoldering farmland of very high quality. Numerous court documents of the last centuries report of conflicts between coastal inhabitants on behalf of the rights to reclaim land or the ownership. At the Frisian coast those conflicts initialized even civil wars between the then ruling chiefs. In respect of coastal zone management land reclamation had higher priority than e. g. inland drainage and navigation enforced the erection of numerous new drainage sluices and the cutting off from the access to the sea of historic ports at the Wadden Sea [NIEMEYER 1991b]. Since those days the Wadden Sea coast has experienced a dramatic change of its shape by both natural processes and human interference. There is an aphorism in Frisia referring to selfconfidence of the coastal community: ‘God created the sea, the Frisians the coast.’ The heritage of our ancestors’ efforts are not only the reclaimed landscapes but also unforeseen consequences: The partial enclosures of reclaimed salt marshes have partly created long-term morphodynamical processes being still of importance for present coastal engineering. E.g. the subcompartments of the intake area of the tidal inlet Osterems in East Frisia have changed both their areal extensions and tidal prisms due to the silting up and land reclamation in the Ley Bay. In consequence the watersheds

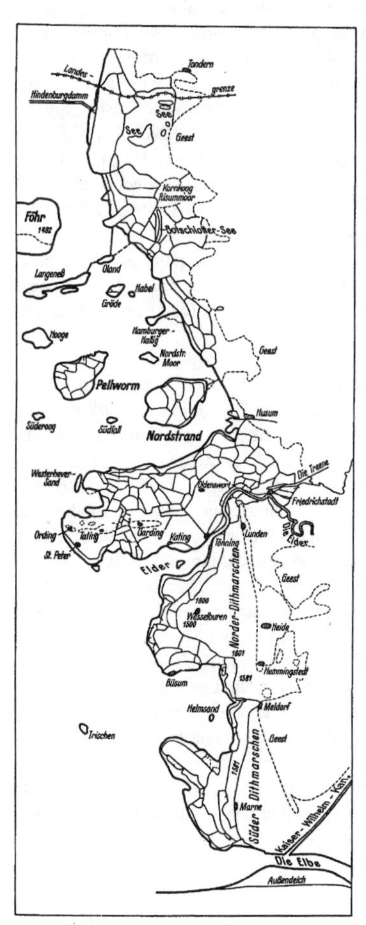

Figure 19. Land reclamation at the west coast of Schleswig-Holstein since 897 A.D. [HEISER 1933]

shifted and effected first a migration of the main channel of the tidal inlet counterdirectional to prevailing drift leading to erosion at the eastern beaches and dunes of the updrift island of Borkum. Second the tidal inlets migrated in response to the updrift reduction of their intake areas downdrift provoking structural erosion on the beaches of the downdrift islands [NIEMEYER 1995]. The traditional coastal management policy to enclose former storm surge bays has been totally changed in recent years: the preservation of this ecologically unique areas has nowadays priority: the planned enclosure of the Ley Bay in East Frisia as a coastal protection measure was abandoned and replaced by an alternative solving the problems of coastal protection, inland drainage, navigability and nature preservation [NIEMEYER 1991b].

The purposes of land reclamation at the Wadden Sea coasts have changed in the course of the last decades. Referring to national economy the gaining of additional farmland has become not only less attractive, it is counterproductive in respect of a total balance. But still land reclamation has been continued to a lesser extent. First the aim was to gain salt marshes serving coastal protection purposes as already explained by BRAHMS [1754, 1757]: dykes with a supratidal salt marsh in their apron do not need heavy solid revetments in their lower parts because the regular interaction with ordinary tides does not take place [HEISER 1933]. After a failure of the dyke repair will also be easier: there is no in- and outflow during ordinary tides. The salt marshes perform a reservoir of clay and saltwater-resistant grass sods being useful for repair purposes in the case of damages and emergency [ERCHINGER 1970]. The assumption wave attack on dykes landward of salt marshes would be reduced due to damping or even breaking [BRAHMS 1754, 1757; ERCHINGER 1970] has been discovered as a misunderstanding of interactions between waves and morphology in Wadden Sea areas [NIEMEYER 1979, 1983, 1995]. The dedication of salt marshes as part of the coastal protection system allowed still their agricultural use. Nowadays these areas are regarded as ecologically unique areas: they became parts of the recently performed National Parks Wadden Sea which implies changes in comparison to traditional use of these areas: prohibition or significant reduction of agricultural use and the limitation of reclamation structures for purposes of salt marsh preservation.

Storm surge barriers. Problems of inland drainage and coastal protection in the upper region of the Eider estuary led to the solution of a tidal barrier which was erected in 1936. The aims were first to avoid further strengthening of dykes on a weak soil which was regarded as difficult and costly. Second the inland drainage in the upper region was independent from tidal water level variations which particularly during storm surges provoked often inland flooding because no or only insufficient discharge due to hydraulic gradients could take place. But the barrier caused also a remarkable reduction of the tidal volume leading to an enormous downstream sedimentation which again hampered inland drainage [WEINHOLDT & BAHR 1952]. The unsolved problems required another solution [ROHDE & TIMON 1967] leading to the construction of storm surge barrier at the estuarine mouth after 1972. The experience with the Eider tidal barrier led to a new concept of a storm surge barrier. The catchment area of ordinary tides was not reduced, only tides with a set-up exceeding a threshold level were cut off by closing the gates. This concept was firstly realized on a small scale at the Lühe, a tributary of the Elbe estuary, in 1939 and second on a larger scale for the storm surge barrier of the Leda in East Frisia, a tributary of the Ems-Dollart estuary in 1952 [LIESE 1956]. Particularly after the storm surge of February 1962 the erection of storm surge barriers at the entrance of tidal rivers was regarded both economical and technical advantageous in comparison to strengthening all the dykes along the embankment of the rivers: the soil was often very weak requiring technically complicated and costly measures to create stable dykes. In total more than twenty have been built, the largest is that one at the mouth of the Eider estuary. An overview on the then existing storm surge barriers at the German North Sea coast was elaborated by GÄTJEN [1979].

Shore protection on islands

The islands at the German North Sea coast have experienced changes in their shape and partly also in their position: The sea borne East Frisian Islands have been reshaped by large-scale morphodynamical processes, particularly inlet and ebb delta migration [HOMEIER 1962; NIEMEYER 1995]. The North Frisian Islands, as remnants of former mainland, have kept their position but experienced enormous land losses which could only gained back partly by reclamation [KREY 1918; HEISER 1933]. In this chapter shoreline protection on islands or parts of islands consisting of lowland areas with cohesive holocene marine sediments is not discussed because the protection against flooding is carried out there by construction of dykes following the same rules as on the mainland. As well the specific situation on the small marshy islets, the 'Halligen', where the protection against flooding is gained by the adaption of artificial earth hills, the 'Warften', will not be part of this contribution. We focus on shoreline protection of the sandy coasts of the East Frisian Islands and those North Frisian Islands being positioned on a pleistocene core.

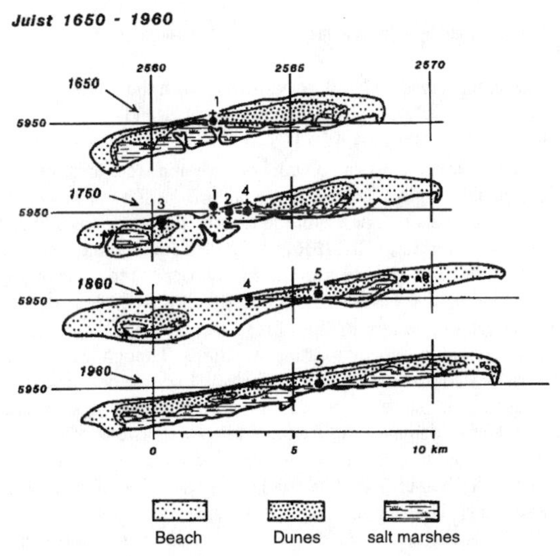

Figure 20. Adaption of settlements on islands to shoreline migration: Island of Juist/East Frisia; the numbers 1 - 4 mark churches built between 1300 and 1717 and abandoned between 1651 and 1779, no. 5 has been built 1779 in the center of the present main settlement on the island [HOMEIER 1962; STREIF 1990]

Looking back the abandoned position of, e. g., churches on the islands make evident that the inhabitants had no possibility beside retreat from areas endangered from the sea after severe structural erosion of beaches and subsequently dunes (Fig. 20). There were no tools available to fight structural erosion on sandy coasts and even if the substance of human goods would have been insufficient to justify the necessary costly impacts. Tourism which increased significantly step by step on all islands after the establishment of the island of Norderney as an official holiday resort in 1798 led to investments and properties of much higher values than before on the islands. Both the owners and the government(s) had as well a motivation as a justification for interfering into the processes endangering infrastructure and properties being worth to preserve even by significant costs. In the middle of the last century continuing downdrift inlet migration caused structural erosion of beaches and dunes at the western spit of the island of Norderney resulting in remarkable shoreline retreat [THILO & KURZAK 1949; HOMEIER 1962; LUCK 1977] which endangered the place, being then also the summer residence of the kingdom of Hannover. In 1858 the chief engineer for hydraulic engineering of the Province East Frisia TOLLE [1864] started the realization of his plans to keep the island's shoreline: the construction of dune revetments and beach groynes first as solid structures at sandy German North Sea coasts (Fig. 21). The introduction of these new construction material and method must have been initialized by the failure of the then traditional protection measures: Bush fences, wooden pile walls and groynes consisting of fascines fixed by wooden piles. After the 2nd World War the responsible administrations for the East Frisian Islands authorized experienced coastal engineers to perform critical state of the art reviews on the shoreline protection on the islands [LUCK 1975]: KATTENBUSCH (Borkum); THILO (Juist, Norderney, Baltrum, Langeoog, Spiekeroog); LÜDERS & WILLECKE (Wangerooge). Unfortunately these reports have remained unpublished though they served as a very valuable basis for both conserving gained knowledge and further improvement of engineering measures. The products of these reports have been used in the framework of published reports to which is here referred to.

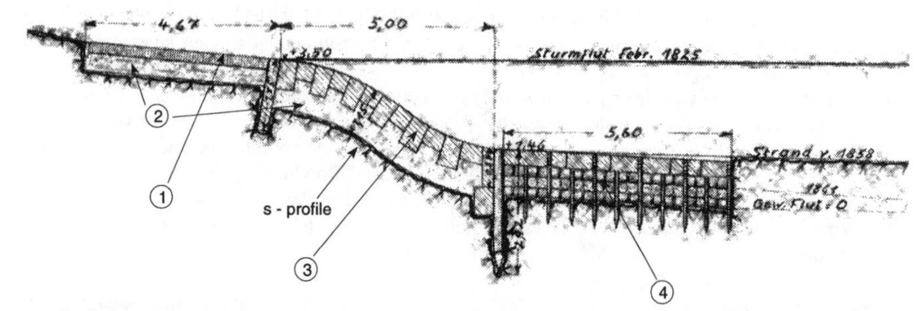

Figure 21. Early dune revetment on the island of Norderney in 1858 with the levels of MHW and of the storm surge of February 1825; beach levels (Strand) of the year of implementation and of 1861 [TOLLE 1864]

Revetments and seawalls. After their introduction revetments came widely in use and were also built on other islands with eroding beaches. The design philosophy changed. The s-shaped profile introduced by TOLLE [1864] was replaced by structures with steeper slopes with the aim to 'push the waves back' [FÜLSCHER 1905]. Already KREY [1906] as well as later HEISER [1932], who favored the s-shaped profile, argued against that opinion. At least a flatter slope was introduced in order to reduce scouring at the toe. In 1936 the first revetment with a slope of 1:4 was built on the island of Sylt being still the prototype of revetments built since then. Design and dimensioning of revetments and seawalls is still widely empirical orientated at damage and survival of already existing structures. Though these structures are very costly the demand for a design procedure has not been very strict, because new constructions of revetments and seawalls have only taken place since the 50s of this century if existing structures must be replaced, in most cases immediately after damages did occur. Further extensions of existing structures for the adaption to eroding beaches are no longer carried out since the more economical tool of artificial beach nourishment is available [GAYE et al. 1952]. Nowadays the question of acceptable wave loads on existing revetments and seawalls is again actual at regularly nourished beaches. In order to determine the necessity of return periods of artificial shoreface or beach nourishments a correlation with the limitation of lowering of beach levels and resulting wave action in front of the structures is necessary if the major purpose of the beach fill is a limitation of wave attack on the structures and of resulting damages [NIEMEYER et al. 1995c]. The revetments fulfilled in nearly all cases their major aim but could not stop beach erosion. In cases of structural erosions extension became necessary in order to prevent damages or even the collapse of the structures. Toe protection and continuing adaption to the

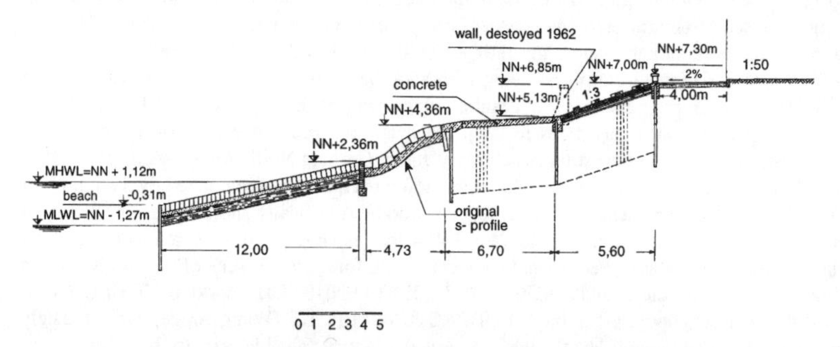

Figure 22. Present Seawall on the island of Norderney extended from the early dune revetment with its s-shaped free stone quader core

lowering levels of the eroding beaches must be carried out. The propagation of higher waves due to enlarged water depths in front of the revetment required also extensions in the upper part including a wide berm [HEISER 1932] in order to prevent to large overtopping volumes creating a return flow beneath the structure leading to damages or even to a total loss of stability [RAGUTZKI 1976]. The initial revetments migrated therefore in the course of the time to seawalls in which the initial structure was still as a small part existing if not destroyed by wave action (Fig. 22). The existing revetments and seawalls have fulfilled their major aim to keep the shoreline and to prevent a flooding of the hinterland. In cases of structural erosion further extensions are no longer economical in comparison with artificial nourishments [GAYE et al. 1952]. But up to now no revetment or seawall has been abandoned in favor of a nourishment: they are maintained and have even been reconstructed if necessary.

Groynes. The first groynes had the aim to prevent erosion of beaches. Already in the 19th century their use was discussed though nobody worked on that problem basically beside HAGEN [1863], who did hydraulic model tests in order to study their effectiveness before implementation. In cases of structural erosion groynes had limited success: At the most they decelerated beach erosion due to longshore currents but failed to stop coastal retreat. The lowering of the beaches made them more vulnerable against hydrodynamical impacts. There were more frequent damages and in some cases their stability was endangered. In order to prevent their collapse numerous groynes have been adapted to the changing morphological boundary conditions (Fig. 23). Most variations in groyne design were targeted at construction methods and the longevity of material but not at their functional purposes [LORENZEN 1954]. The demand of PETERSEN [1961] to carry out investigations delivering more detailed information is still valid. A remarkable approach has been carried out by KRAATZ [1966] who tried to shape groynes in respect of dominant wave climate, particular to reduce beach erosion at the central shore of the island of Sylt. His design was not carried out because the control of beach erosion by an artificial beach nourishment shaped like a groyne was regarded as the most effective tool for that area [FÜHRBÖTER et al. 1974]. As well as the existing revetments and seawalls most of the groynes are maintained and a few ones have been added to the existing scheme, e. g. on the island of Borkum [ASTER et al. 1990]. Particular in the vicinity of migrating tidal inlets the groynes on the beach had no effect to decelerate or even to stop that process. Particularly on the island of Norderney the existing protection structures and the infrastructure in the western part was endangered. Intensive hydrodynamical preinvestigations [GARSCHINA & PANSE 1898] made evident that the existing current

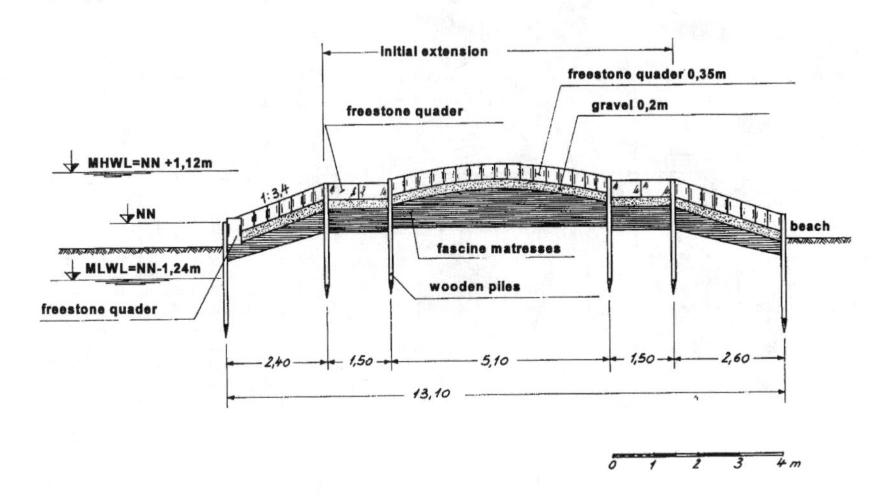

Figure 23. Cross-section of a beach groyne, island of Norderney; initial extension and adaption to lowering beach level

regime of the tidal basin generated a further downdrift migration of the tidal inlet. The chosen response to that development were extensions of the groynes on the beaches close to the inlet's main channel into its steep slope which was covered and successfully fixed. This solution was repeated at other places with comparable boundary conditions in order to stop there also tidal inlet migration and successing shore erosion on the downdrift island. Four of the six East Frisian tidal inlets have been fixed by structures of this type which are called 'stream groynes' in contrast to 'beach groynes' in German professional usage. LUCK [1975] complained that TOLLE had not obtained permission to erect stream groynes at the tidal inlet Norderneyer Seegat. A successful implementation of structures of that type would have provided the next generation of coastal engineers with the heritage of a wider beach than existing nowadays: a more convenient platform for effective beach nourishments. Already KREY [1906] had pointed out that, if necessary, stream groynes should be built as early as possible in order to avoid more costly construction works on steepening slopes.

Large scale groynes. A specific version of the groyne-type has been carried out in the East Frisian tidal inlet Harle: a large-scale structure with a length of about 1460 m. Its construction has not been finished in the course of the 2nd World War. Therefore, the crest height did not reach the design level fairly above mean high tide but remained at mean low tide and is nowadays partly even lower. Though never finished, its effectiveness was already proved a few years later [LÜDERS 1952]. This structure was not only aimed at prevention of further downdrift migration of the inlet's deep channel. It should and has also closed a newly generating secondary channel which was expected to take over the function of the inlet's main channel without any intervention. Furthermore the establishment of the structure should restore the morphodynamical boundary conditions of nearly 100 years ago which were more favorable in respect of sediment supply of the northwestern beaches of the island of Wangeroog (Fig. 24). A similar construction was also discussed after the war to solve the structural problems of the western and northwestern beaches of the island of Norderney but was abandoned because of the high costs and the high technical risks related to crossing the inlet's main channel [KURZAK et al. 1949; THILO & KURZAK 1952]. A beach nourishment was chosen in favor of such a large-scale solution [GAYE et al. 1952].

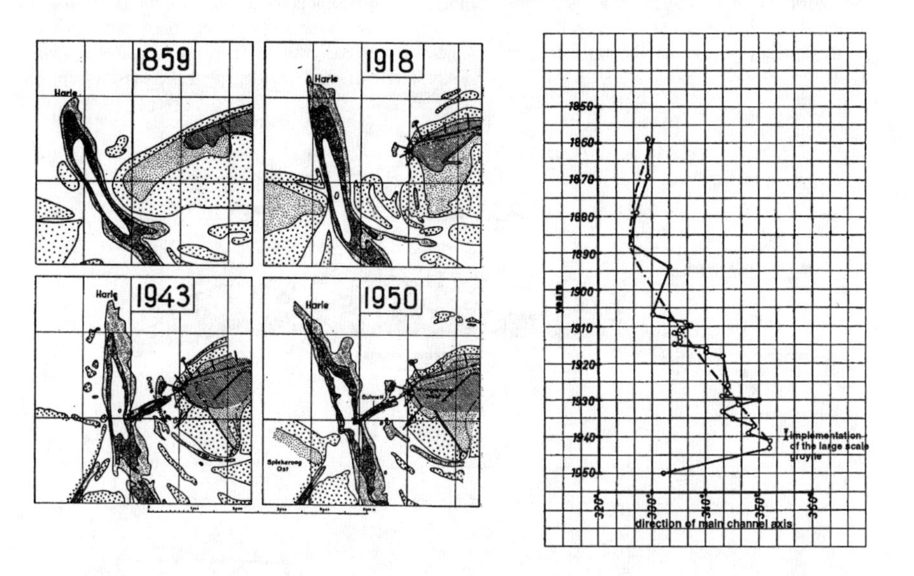

Figure 24. Clockwise migration of the deep channel of the tidal inlet Harle and its shifting backward after the implementation of the large scale groyne H [LÜDERS 1952]

Artificial beach and shoreface nourishments. The first beach nourishment at the German coasts was been carried out on the island of Norderney following recommendation of an expert group [GAYE et al. 1952] on the basis of intensive investigations carried out previously by the Coastal Research Station [KURZAK et al. 1949; THILO & KURZAK 1952]. Since then a remarkable number of nourishment projects have been carried out, particularly on the islands of Norderney [KUNZ 1991] and Sylt [ALW 1984], but also in other places. A very successful application has taken place on the island of Langeoog. Basing on thoroughful investigations on cyclic changes in the welding of shoals at distinct beach areas of the island (Fig. 16) [HOMEIER & LUCK 1971] the planned very costly implementation of revetments and groynes with probable necessary extensions due to their interference into the beach processes could be avoided by the combined application of a beach nourishment being armored by rubber pipes (Fig. 25) [LÜDERS et al. 1972]. Specific solutions of nourishments in respect of local boundary conditions have been discussed and sometimes also carried out: E. g. KRAMER [1960] recommended to counterbalance the structural erosion on the island of Norderney by continuous beach fill instead of repeated concentrated nourishments with return periods of a couple of years; on the island of Sylt a groyne shaped nourishment was carried out in order to feeder the adjacent beaches [FÜHRBÖTER et al. 1974]; later there the beach fills were carried out as deposits in front of the eroding dunes (Fig. 26) [ALW 1984]; on the island of Langeoog the welding process of shoals was accelerated by the closure of the gully in front of the beach by a fill [ERCHINGER 1985]. Already GAYE & WALTHER [1935] mentioned the positive effects on beach morphology on the island of Norderney achieved by dumping of suitable dredging material in their vicinity in the years between 1899 and 1909. This must have been the first nourishment of the shoreface at the German North Sea coast. Nearly one century later again the idea of a shoreface nourishment was renewed: A combined shoreface and beach nourishment was carried out on the island of Norderney. The major aim was to use the deposit on the shoreface as a feeder berm for the intertidal beach [NIEMEYER 1991a]. In order to gain a deeper insight into the governing morphodynamical processes a joint European project was established looking also after the effectiveness of shoreface nourishments in the neighbor countries Denmark and the Netherlands [NIEMEYER et al. 1995c].

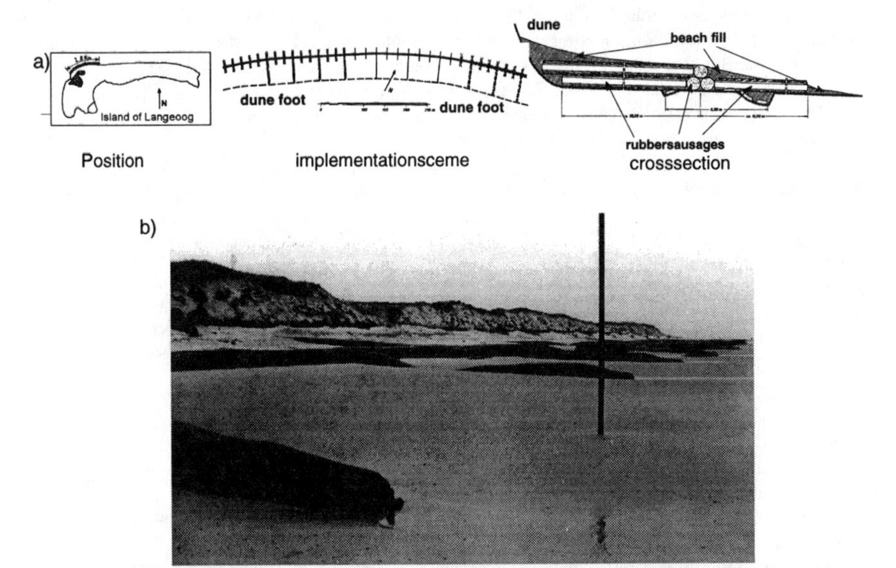

Figure 25. a) Combination of beach nourishment and rubber tubes for stabilization as temporary dune protection on the island of Langeoog
b)Beach nourishment and rubber tubes after implementation on the island of Langeoog

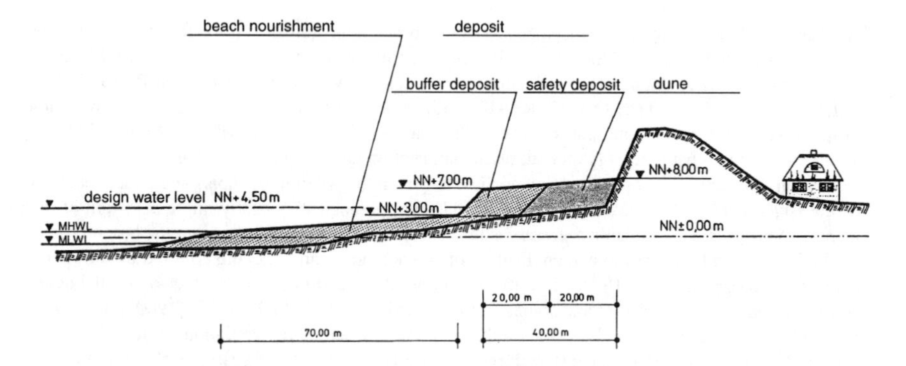

Figure 26. Dune reinforcement by combination of beach nourishment (wave dissipator) and a deposit consisting of a buffer and safety deposit an the island of Sylt. After erosion of the buffer deposit succeeding nourishment will start whereas the safety deposit guarantees that the dune can still withstand storm surges [ALW 1984].

COASTAL PROTECTION AT THE BALTIC SEA

Systematic investigations of water level providing a sound basis for the design of coastal structures has been available since the beginning of the 19th century. The Prussian Hydraulic Engineering Administration erected the first water level gauges in 1810 within the framework of preinvestigations for the construction of ports in Swinemünde, Kolberg, and Memel which belong nowadays to Poland and Lithuania. Therefore the heritage on long-term data sets is shared with coastal engineers in these neighbor countries. In the middle of the 19th century Gotthilf HAGEN [1863] published his encyclopedic manual on hydraulic engineering in which beside other subjects the state of the art for coastal protection at the Baltic coast and its hydro- and morphodynamical basics were described. The manual of HAGEN served for decades as the guideline for coastal engineering at the Baltic coast. The storm surge of 1872 with its disastrous damage stimulated further investigations, particularly on the meteorological boundary conditions and the hydrodynamical effects.

Basic hydrodynamics

Mean water level. In 1845 HAGEN introduced a procedure delivering comparable mean values of water levels for the whole Prussian Baltic coast: A daily value taken at noon. PETERSEN [1952] concluded that these daily water level data provide a sufficient basis for investigations in a medium time scale though a random value. In order to continue the existing time series these values are still documented and additionally the daily maximum and minimum. HAGEN [1863] used these data in order to evaluate the yearly variation of the mean water level by monthly means for distinct areas at the Baltic coast. This approach

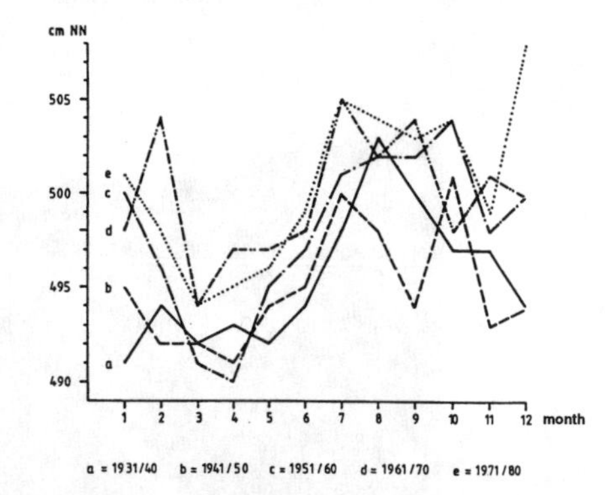

Figure 27. Seasonal changes of mean water level (10 year average) between 1931 and 1980

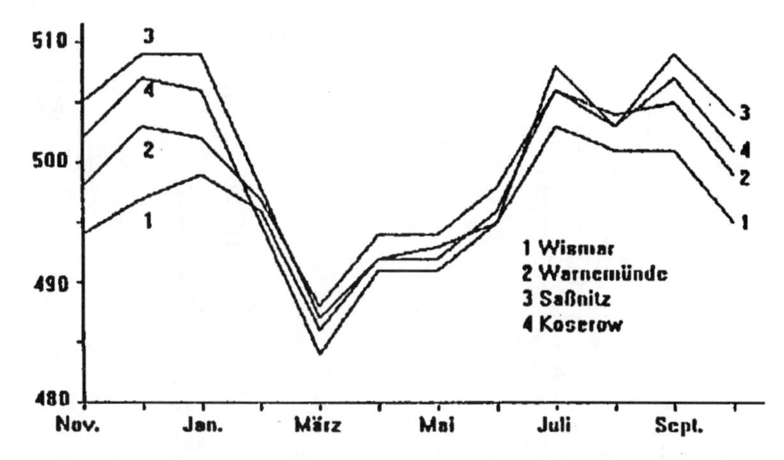

Figure 28. Monthly means (10 year average 1976-85) at four gauges at the southern Baltic Sea

was refined by WITTING for time series between 1898 and 1912 [EIBEN 1992a] which perform still an example for present analysis (Fig. 27 + 28). GAYE [1951] used the existing data sets and hints from HAGEN [1863] and MEYER [1913] for the determination of relative sea-level rise at the Baltic coast: a rate of 2,56 mm per year on the average for all considered gauges and 1,5 mm per year for the southern coast. WEISE [1962] analyzed the time series for the gauge Swinemünde and got for longer runs a rise of about 1 mm per year but got a similar result as GAYE [1951] if using the same data set.

Tidal oscillations. Already HAGEN [1863] delivered figures for the microtidal range of the Baltic Sea decreasing from west to east: At Travemünde 9-10 inches and at Stolpmünche 1 inch. MEYER [1913] evaluated tidal constants for the Baltic Sea and concluded the following results: small tidal range in the whole Baltic Sea, changing harmonic constants and decreasing tidal amplitude in eastward direction. MEYER [1913] succeeded furthermore to detect a tidal oscillation from the record of a gauge at the Imperial Dockyard in Kiel (Fig. 29). PETERSEN [1952] proved a mean tidal range of 19 cm for Travemünde.

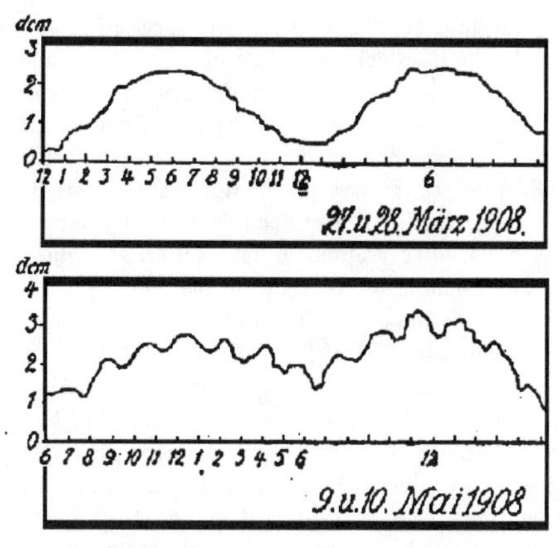

Figure 29. Tidal oscillations at Kiel [MEYER 1913]

Basin oscillations and Seiches. Wind fields induce in the Baltic Sea not only a set-up or set-down as they do, e.g., in the North Sea. They also trigger oscillations of water levels in the whole basin with duration of more than a day. A typical example for a basin oscillation has been recorded at the gauge Kiel in 1978 (Fig. 30): An initial set-down of about 0,6 m in the western part of the Baltic Sea and a set-up in its eastern parts occurred due to western wind. The decrease of the wind intensity rendered possible a basin oscillation with a maximum set-up of about 1.2 m and a succeeding maximum set-down of about 1.0 m in the western part after 36 hours. It is obvious that a superposition of such a basin oscillation with a change in storm direction could be

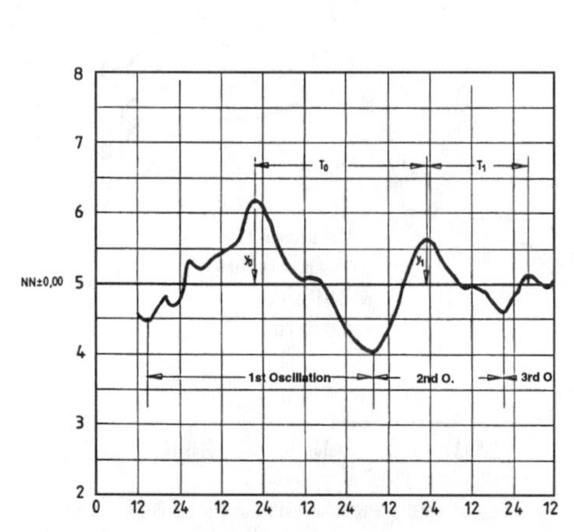

Figure 30. Example of damped basin oscillations, western Baltic Sea [EIBEN 1992a]

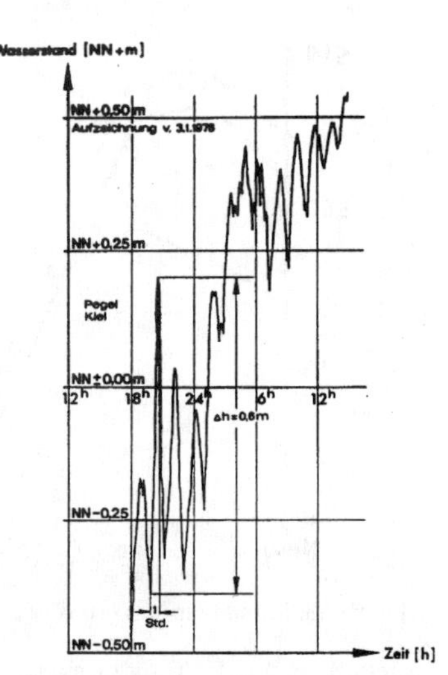

Figure 31. Registration of seiches at the gauge Kiel [EIBEN 1992a]

dangerous for the coast. Furthermore, seiches occur with periods in the range of hours; they appear often in coincidence with the rise of water level set-ups. Already MEYER [1913] had detected those seiches in the bay of Kiel both by measurements and by computation. Later theoretical analysis delivered that the typical seiches in the Baltic Sea are oscillations with one or two knots and duration of 19 or 28 hours [NEUMANN 1941; GEYER 1962]. Those with a duration of about 19 hours is more frequent. PETERSEN [1952] reported a seiches for Travemünde with an amplitude of 0,28 m and a period of 20 minutes. A seiche with an amplitude of 0,6 m and a period of 75 minutes (Fig. 31) has been reported by EIBEN [1992a] for the area of Kiel. The occurrence also of basin oscillations as seiches has often complicated the assessment of the actual development of storm surges, particularly if leading to temporary set-downs.

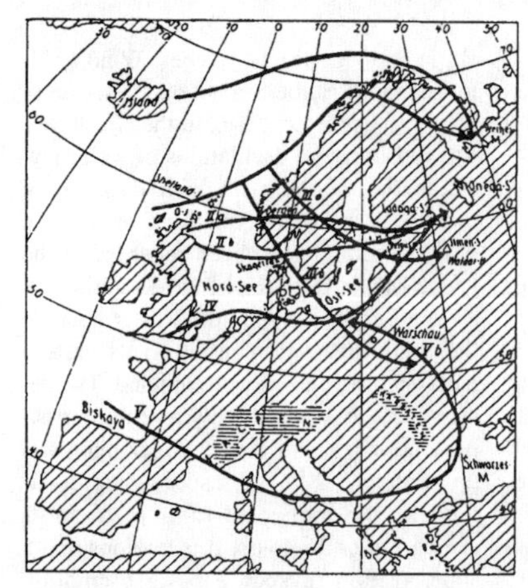

Figure 32. Main tracks of cyclones creating storm surges at the Baltic Coast [EIBEN 1992a]

Storm surge generation. The storm surge of 1872 with its disastrous effects on the German Baltic coast stimulated intensive research. Detailed data inventories on storm surge water levels and meteorological boundary conditions were carried out by BAENSCH [1875] and PRALLE [1875] covering the whole area of the Baltic Sea. Their efforts have been successfully supported by the Danish meteorologist COLDING who provided valuable information on the triggering of the set-up by wind action. The evaluation of main tracks of cyclones by BEBBER [KOLP 1955] and the analysis of resulting water level changes by KRÜGER [1910], REINHARD [1949], STARK [1952], and KANNENBERG [1955] allowed a generalized determination of three typical types of weather conditions resulting in storm surges. Type A: High pressure above North Scandinavia, cyclone tracing from the North Atlantic quickly across the western and southern Baltic Sea (track III or IV, Fig. 32) with initial

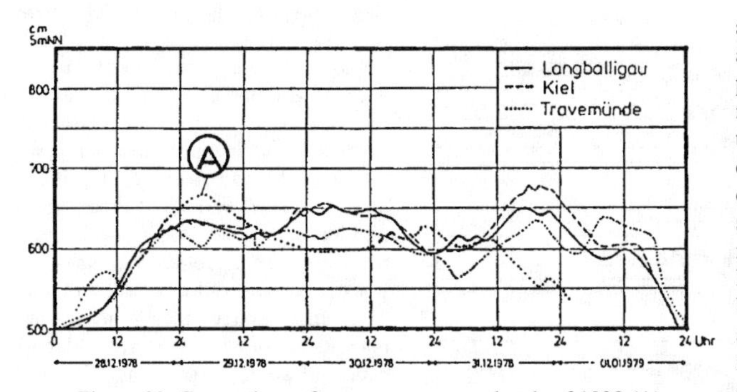

Figure 33. Comparison of storm surge water levels of 1898 (A) and 1978/79 [EIBEN 1992a]

storm directions west and southwest and after passing of the cyclone northeast. According to RODLOFF [1972] conditions of this type create about 50 % of storm surges at the German Baltic coast. Type B: High pressure above Scandinavia, low above the Mediterranean. Slow movement of both to East Germany (track Vb, Fig. 32). Storm from northeast to east with long duration, often turning to southeast and decreasing. According to KANNENBERG [1955] about 30 % of all storm surges are effected by weather conditions of this kind. Type C: Low pressure above North Scandinavia and high pressure in the southwest to west. Northeastern storm at the backside of the cyclone, a rare event. Type A represents both small surges as extreme events such as the one of 1872 due to combination with basin oscillations. The very unfavorable convergence of 1872 occurred due to a succession of three situations [BAENSCH 1875]:

1. Western storms with long duration generating a gradient of about 2 m between the intersection of North and Baltic Sea and a set-up in the eastern part of the Baltic Sea.

2. Gradual turning of the storm direction to northeast and increase of velocities and oscillating back of the set-up from the eastern part.

3. Further growth of the storm intensity up to Beaufort 11. Increase of the set-up from 2.0 to 3.2 m with rising velocities of the water level of up to 0.2 m per hour.

The storm surges occurring due to weather conditions of Type B are characterized by a wind-generated set-up due to onshore directed storms. They often have no significant peaks but set-ups remaining for longer periods on high levels as e. g. the storm surge of 1978 (Fig. 33). Investigations on the frequency of storm surges have been carried out due to German standard classification by STARK [1952] which have been recently updated by EIBEN [1989]. This procedure considers only water level peaks, but not duration and wave action. It is therefore an indicator but not a scale for the endangering of coasts and coastal structures.

Design water levels. The master plans of the Federal States at the Baltic Sea contain design water levels on the basis of the storm surge of 1872 with a set-up of about 3,5 m still following the recommendations of the Prussian Hydraulic Engineering Administration. Recently the probabilistic investigations by JENSEN & TÖPPE [1990] stimulated discussions about lower design levels for an event with a return period of 100 years and a set-up of 2,5m whereas the return period of the storm surge of 1872 with its set-up of about 3,0 m was estimated of about 1000 years. In spite of these suggestion the design water level will be taken in consideration of the set-up of the storm surge of 1872 and a relative sea level rise of 2.5 mm per year for a lifetime of 100 years. Considering uncertainties of the possible maximum of basin oscillation effects and an acceleration of sea-level rise a reduction of the design water level is not reasonable [WIEDECKE et. al 1979]. Furthermore there exist hints on two storm surges between 1320 and 1872 with a set-up of the same order of magnitude as that one of the storm surge of 1872 [KRÜGER 1910]. Considering that information the return period of a set-up of the storm surge of 1872 would be reduced to 180 to 200 years.

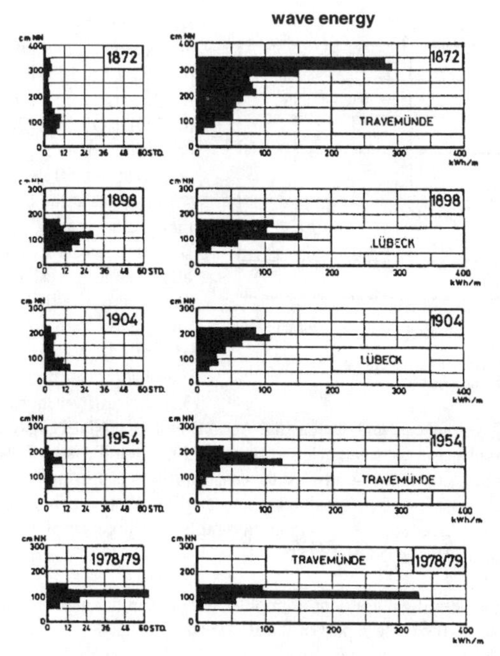

Figure 34. Examples of duration of exceedance of threshold water levels and wave energy for distinct storm surges [EIBEN 1992a]

Storm surge classification. The classification of storm surges has the major aim to get a measure for its severeness. The orientation at water level peaks at meso- or macro tidal coasts is quite reasonable because the variation of storm surge water levels due to tides is still significant and is in shallow coastal areas also a dominant boundary condition for wave action. But at a microtidal coast the duration of high storm surge levels is in some cases much longer than at meso- and macrotidal coasts. Therefore, a classification of storm surges limited to peak levels is insufficient for microtidal coasts. A number of water level thresholds and the duration of their exceedance has been introduced and furthermore an estimation of accompanying wave energy [FÜHRBÖTER 1979]. Though the execution of wave energy determination is rather questionable the basic approach is a very suitable tool for storm surge classification at microtidal coasts. A comparison of distinct storm surges at the Baltic coast makes the improvement evident which has been gained by this approach (Fig. 34).

Storm surge prediction. Severe storm surges occur at the Baltic coast rarer than e. g. at the North Sea coast but sometimes more surprisingly. In order to organize necessary countermeasures at vulnerable or endangered coastal sections a suitable storm surge prediction is urgently required. But it took quite a long time before first tools were introduced. In Mecklenburg-Vorpommern an empirical forecasting was introduced for the estimation of set-up and set-down due to direct wind action [MIEHLCKE 1956]. This prediction method did not consider the effect of basin oscillations and lacked by the very limited information on wind velocities and direction from the northern Baltic Sea. Both effects caused deficits in respect of reliability of prediction. In Schleswig-Holstein the method of KEULEGAN was applied by EIBEN [1983] in order to predict the set-up caused by direct wind action. The prediction delivered results with an accuracy of about 0,25 m if the application was limited on small coastal stretches. After the surprising storm surge of summer 1988 the German Hydrographic Service made great efforts to implement an operational storm surge prediction for the Baltic coast which was introduced in 1992: a numerical model for water levels and currents using predictions of wind action and air pressures and tidal computations. Up to now the accuracy of storm surge prediction has been proven as well sufficient as the phase shift to the occurrence of the event itself.

Wave set-up and undertow. In his manual on the arts of hydraulic engineering HAGEN [1863] mentioned observations of wave-induced set-up of water-levels at the coast during storms with heights of 0.9 to 1.2 m due to the water volume of breaking waves. Furthermore he explained the generation of a near-bottom current driven by the gradient which has been caused by the nearshore set-up. The observation of HAGEN [1863] became lost in the German coastal engineering profession and was brought back to the public more than hundred years later when field measurements had proven his observation [NIEMEYER 1991a].

Coastal protection by natural means

Dune stabilization: Relatively large parts of the German Baltic Sea coast, particularly before World War I and World War II, experienced a natural protection against the sea by foredunes. Therefore dune stabilization was a major concern. Initially its aim was primarily to prevent dune migration in order to avoid the filling of backside lagoons or waterways or the covering of agricultural or forestry areas. Human activities with the aim of dune stabilization at the Baltic coast are already known from the 13th century [WEISS 1992]: Fences have been erected and later pines and willow trees were planted. Knowing about the importance of dune stabilization the Society for Natural Sciences of the City of Danzig donated in 1768 an award for improvements in this field. The winner TITIUS recommended the erection of fences in combination with the planting of vegetation like bushes being replaced successively by trees. This scheme based on experience gained in Denmark by RÖEHL after 1738 by the combined implementation of fences and grass. Obviously a number of tools for dune stabilization have been known and were applied though still a lack of systematic knowledge is evident. A pragmatic approach to reduce these deficits was carried out by BJÖRN in the vicinity of Danzig after 1795: Based on existing knowledge he established test fields of schemes drawn up from sea landward and consisting of distinct kinds of fences in combination with a variety of vegetation (Fig. 35). Within a few years he gained remarkable

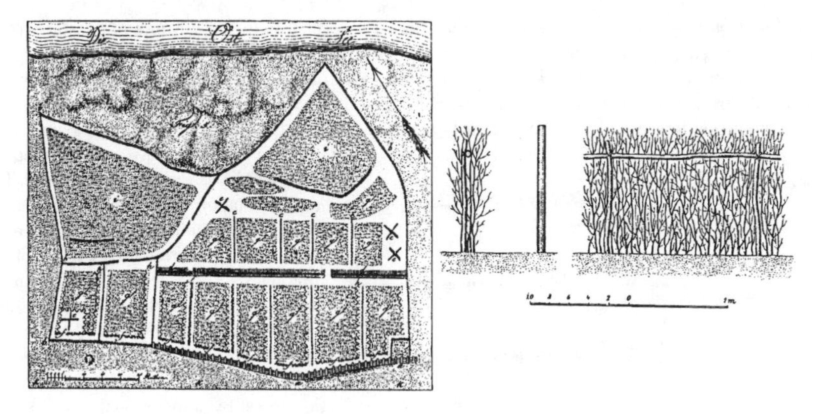

Figure 35. Test fields for dune stabilization by BJÖRN [GERHARDT 1900]

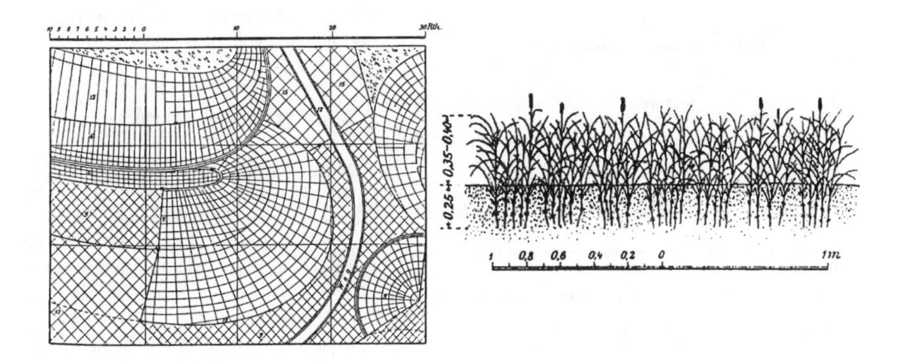

Figure 36. Implementation of sand reed in flexible grids by KRAUSE [GERHARDT 1900]

success allowing a further extension of his fields. After the Napoleon Wars which caused an interruption of these efforts KRAUSE [GERHARDT 1900] improved the scheme of BJÖRN [GERHARDT 1900] by implementation of rows of sand reed in flexible grids varying between 1,25 m and 5,65 m in dependence of dune steepness and wind action (Fig. 36). KRAUSE started furthermore the stabilization of foredunes as a direct coastal protection measure though still not realizing their function as part of the aeolian on- and offshore transport processes. Getting an insight into these processes HAGEN [1863] realized the necessity to equalize the cross-sections of the foredunes. His position as a high-ranking civil servant in the Prussian Ministry for Public Works allowed him to transfer his knowledge into practical application leading to dune reinforcements including in some cases even artificial nourishments and later the planting of sand reed. The motivation of HAGEN [1863] resulted from his insight in the crucial role of stable foredunes for the safety of the hinterland against inundation. But he did not recognize that this aim is only achievable at balanced coastal stretches. This fact was later realized by GERMELMANN [1904] describing the fate of foredunes in areas with structural erosion 'to become always a victim of the sea' which can only be avoided by an organized retreat. The state of the art then available on the basis of developments starting already in the 18th century serves still as the basis for present strategy and practice in dune stabilization at the Baltic coast into which additional materials and tools like synthetic materials and machinery equipment have been introduced.

Coastal protection forests. A unique application of natural vegetation for the purpose of coastal protection is the implementation of forest drawn up as coastal defense schemes (Fig. 37). Coastal protection forests have the function to dissipate wave energy in front of a dyke performing the most landward positioned part of the defense system for a lowland coast. It is expected though not yet proved that the coastal protection forests will absorb wave energy to such an extent that the dykes will experience neatly no wave attack and have only to keep the still water level during a storm surge. The implementation of the coastal protection forests is particularly an additional measure in areas where the capability of the foredunes is regarded as insufficient to guarantee its stability if a storm surge with long duration occurs. It is obvious that it is difficult to estimate or even to prove the effectiveness of coastal protection forest by conventional tools like hydraulic model tests or computations. Therefore, its capability can only tested in the case of a severe storm surge creating breakthroughs of the seaward parts of the system.

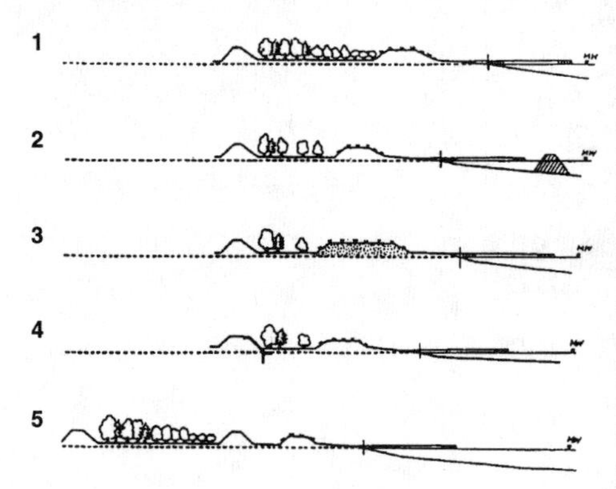

1 : Dyke with grass layer, wide coastal protection forest, dune at a balanced coast with or without groynes

2 : Dyke with grass layer, wide coastal protection forest, with minimum width, weak dune with offshore breakwaters and/or groynes at an eroding coast

3 : Dyke with grass layer, remnants of coastal protection forest, reinforced dune (repeated nourishments) at an eroding coast with groynes

4 : Exposed dyke with revetment; remnants of coastal protection forest, and dune at an eroding coast with groynes

5 : Retreat of the combined coastal defence system dyke - coastal protection forest - dune at an eroding coast with groynes. Dune and old dyke will merge by further coastal retreat

Figure 37. Distinct coastal defense systems with a coastal protection forest [WEISS 1992]

Coastal protection structures

Dykes. The lowlands at the Baltic Sea had in most cases a natural protection by foredunes or high beach berms. Therefore only a few dykes have been erected in the past. The first known construction at the German Baltic coast dates from 1581 at the Gelting Bay [KANNENBERG 1955]. Shape and position of this dyke are unknown, it was destroyed during storm surges in the 17th century. In this area

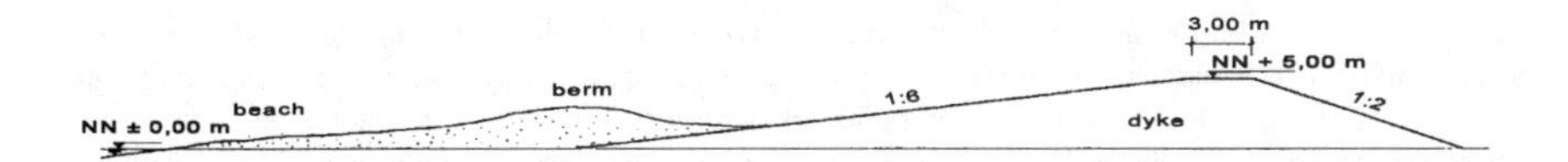

Figure 38. Cross-section of the Prussian Baltic Sea dyke of 1874 [EIBEN 1992b]

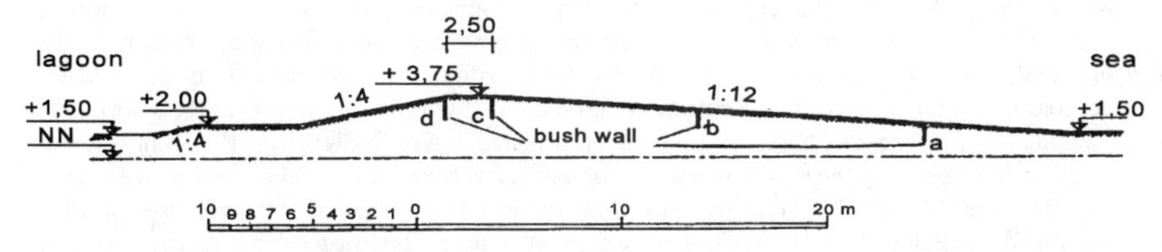

Figure 39. Cross-section of the sand dyke on the peninsula of Wustrow [WEISS 1992]

dykes have also been erected as protection of lowlands in the course of the 18th century. Further dyke construction started in other coastal areas of Schleswig-Holstein in the beginning of the 19th century. The disastrous storm surge of 1872 destroyed numerous foredunes and beach berms leading to the flooding of large areas and causing hundreds of victims. As well in the then Prussian provinces Schleswig-Holstein and Pomerania as in the Grand Duchy of Mecklenburg additional measures were regarded as necessary to keep the lowlands safe. Already 19 days after the event the Prussian government gave orders to erect dykes as additional protection of lowlands landward of foredunes or beach berms which should remained untouched but were regarded as too weak to withstand a very severe storm surge. The shape of the dykes was similar to present ones with a crest height of 5 m above mean sea level, a crest width of 3 to 4 m, an outer slope of 1:6 and an inner one of 1:2 (Fig. 38). If the position close to the shore was inevitable due to insufficient space landward of foredune or beach berm the dyke should be armoured by stone on a shingle layer and a toe protection of piles. The difference to the situations at the North Sea coast makes also the fact evident that at the Baltic coast no self-ruling communities for coastal protection existed. The Prussian government stimulated the coastal landowners to found such communities after the storm surge of 1872. In Mecklenburg dykes were unknown until the storm surge of 1872. But the lesson of this event lead also to the additional implementation of dykes in the coastal protection system. A remarkable construction was carried out in the framework of that programme on the peninsula of Wustrow in order to prevent a breakthrough: a sand dyke with an outer slope of 1:12 to withstand wave attack and an inner slope of 1:4 to reduce the sensitivity against erosion by overtopping (Fig. 39). It reflects a deep insight in the interactions of wave attack on structures

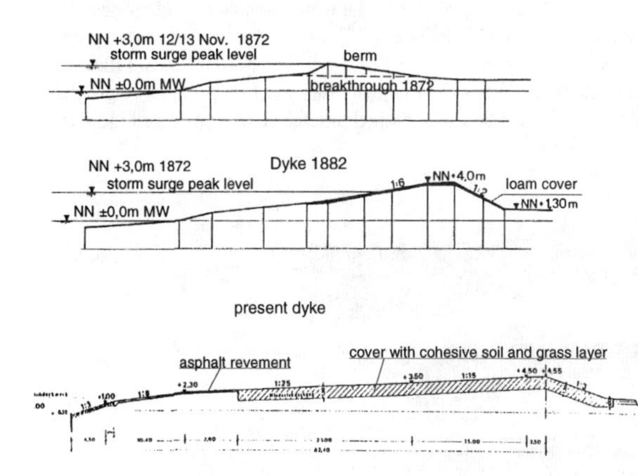

Figure 40. Initial beach berm, dyke cross-sections of 1882 and present at the Probstei coast, Schleswig-Holstein [EIBEN 1992b]

the capability of used material. Generally the design of dykes at the Baltic coast incorporated already all elements of present rules: design water level according to the highest known storm surge and empirical wave run-up in order to avoid overtopping and destruction of the steep inner slope. As well in the Prussian provinces of Schleswig-Holstein and Pomerania as in Mecklenburg in most cases the new dykes were positioned landward of the beach berm or foredune in distances between 100 and 200 m from the shoreline where wave energy was expected to be remarkably or even totally dissipated. This empirical approach proved itself as appropriate considering the fate of those dykes which had been erected close to the shoreline. The dyke in the Probstei was erected in the position of the beach berm after the storm surge of 1872 (Fig. 40). The dyke experienced a number of damages due to storm surges and needed additional armoring by revetments and toe protection before its replacement by a new construction in the beginning of the 80s of this century (Fig. 40) [EIBEN 1992b]. Already in 1898 the notes of a meeting of the local coastal protection community refers to that problem: 'Obviously the danger for the dyke due to direct interaction with the shoreface has been underestimated.' [KANNENBERG 1955]. In order to adress this problem recently the tool of beach nourishments has been introduced to reduce direct wave attack on the dyke. Nevertheless in general the programmes initialized after the storm surge of 1872 were successful, the Baltic coast experienced since then no comparable disaster. Honesty requires us to admit also that no comparable event had occurred in the meantime. Due to more recent assessments many of them would not have been able to withstand a storm surge like that one of 1872. After the World War II and particularly due to the experiences gained from the storm surges of 1954 and 1978 and 1979 reinforcements of existing or replacements by new constructions have increased the safety of lowland coasts at the Baltic Sea in the shelter of dykes.

Groynes. The first application of groynes at the German Baltic coast had the aim of reducing sedimentation at harbor entrances, e. g. at Pillau in 1811 and at Warnemünde in 1850 [GERHARDT 1900]. Later as well HAGEN [1863] as GERHARDT [1900] defined the purpose of groynes to preserve the shoreface and the beach. HAGEN [1863] regarded groynes as a part of coastal protection against storm surges: Keeping the beach as a wave energy dissipator by reducing or even minimizing erosion. He documented also the distinct construction methods of groynes in Mecklenburg and Pomerania being built in the first decades of the 19th century with lengths between about 18 and 37 m and a shore-parallel distance of about 23 m. At first groynes had only been implemented at bluff coasts but HAGEN [1863] recommended also their application at sandy lowland coasts. The design of the first groynes was purely empirical, the construction material was in the beginning fascines, later wooden piles and stones. The first known design criteria have been evaluated by GERHARDT [1900], unfortunately without explanation of their background: relation of length to distance between 1 and 1,5. In 1874 in Prussia the crest height was chosen to 0.2 m above mean water level [WEISS 1992]. In 1887 the first permeable groynes were implemented by spacing the wooden piles by 1/7 to 1/4 of their diameter. This completed the basic types of groynes still in use at the Baltic coast (Fig. 41)

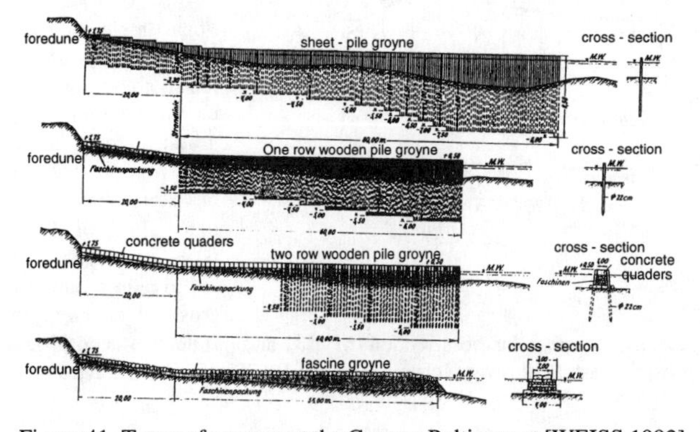

Figure 41. Types of groynes at the German Baltic coast [WEISS 1992]

COASTAL WATERWAY AND HARBOR ENGINEERING

The beginnings until the 19th century

Already in the first century A. D. Roman ships entered the estuaries of the Ems, Weser, and Elbe rivers for trading purposes. Sea trade at a larger scale started after the migration of peoples in Europe had finished in the middle of the first millennium A. D. Frisian sea traders built up settlements and trading places eastward of the Rhine along the North Sea coast but also entering the Baltic coast. In their times vessels were sufficiently small to find access to suitable coastal places by use of existing natural water depths. In the course of the 12th century A. D. by lead of the cities of Lübeck at the Baltic, of Hamburg and of Bremen close to the North Sea coast the Hanseatic League was established: a system of alliances between cities in the Northern hemisphere of Europe establishing an extended exchange of goods, particularly by seagoing vessels with a typical draft of 3,5 m. Those vessels could still get access to the large harbors by use of the existing natural waterways. In order to improve their safe traveling waterways were firstly marked by buoys, beacons and flares in the course of the 13th century.

Beside these efforts engineering activities were only necessary to install quays and similar facilities in the harbors enabling the ships to land and to take goods as easy and quick as possible. Before the 19th century engineering works in order to improve or to control waterways are only known for a few cases. The harbor of Emden had lost its access to the deep waterway in the Ems-Dollart estuary after a storm surge caused a break-through of a new channel in the 16th century and a silting up of the existing waterway. In order to give the seagoing vessels again access to the harbor the closure of this channel was at least tried in vain by erecting a wall of piles. First dredging (Weser estuary and Trave river) and implementation of groynes (Weser and Elbe estuary) is reported from the 18th century.

The operation of waterways and harbors demands for a certain standard of knowledge about available water depths and water level fluctuations. In order to meet these requirements as well survey and hydrographic mapping as water level measurements were established in the 16th and 17th century. A tidal gauge must have been available in the harbor of Hamburg in the 17th century, because since then the phase lag between high tide at London Bridge and this place is known for which a number of measurements is inevitably necessary [ROHDE 1975]. Unfortunately these data have been lost. Systematic tidal water level measurements in the harbor of Hamburg and their analysis had been initiated in 1786 by REINKE [1787] who considers them as needed for tidal compensation of soundings, information on extreme fluctuations and as design basis for engineering constructions. Other tidal gauges had been erected at the end of the 18th century at different places along the Elbe estuary, e.g., that one in Cuxhaven by WOLTMANN in 1784 who invented also an impeller for the measurement of current velocities in rivers [WOLTMANN 1790].

From the 19th century to present

In the course of the 19th century after the end of the Napoleon Wars sea trade with other continents gradually increased, particularly with Northern America which was also destination of an increasing number of emigrants from central Europe. Moreover Germany developed to an industrialized country importing raw material necessary for production and exporting goods. The use of steel instead of wood and the propulsion by steam instead of wind allowed the construction of larger vessels with larger draft demanding deeper waterways as access to the existing harbors [ROHDE 1970]. Thus in the middle of the 19th century modern waterway engineering got its driving force in Germany. The dimension of interference into the existing natural systems increased enormously and asked for a much sounder basis of process knowledge than ever before. As the theoretical knowledge was far beyond our present standard the common approach was empircal: a mixture of mostly regionally bounded observation or measurement in combination with basic process knowledge. According to that fact the history of German coastal waterway engineering will be discussed on the basis of the six most important coastal waterways: the Ems, Jade, Weser, and Elbe on the North Sea coast and Trave and Warnow on the Baltic coast.

EMS

After the already described break through of a new channel in the Ems river far away from the harbor, futile efforts for its closure and the subsequently silting up of its access channel the harbor of Emden lost its importance and position as one of the leading ports in Europe. But, during the last quarter of the 19th century, the import of ore for the industry of the Ruhr area and its coal export got shipped via a newly constructed canal to the port of Emden requiring a sufficient offshore access to the harbor which had to be established step by step. First off all a new harbor entrance was dredged and the harbor was closed off from tidal action and sedimentation by a lock acting moreover as a drainage sluice being used for concentrated outflows in order to maintain the entrance channel's cross-sections. The storm surge bay Dollart being enormously favorable with its large tidal prism for keeping cross-sections in the outer estuary stable was separated by a training wall from the waterway leading to the harbor entrance of Emden and further upstream in order to avoid import of sedimentation by the tidal ebb flow with high turbidity. Following the same purpose at least large tidal flats were enclosed and a new lock was established close to the waterway [SCHUBERT 1970]. In spite of further improvements by construction of new and enlargement of existing training walls the situation of the waterway to the harbor of Emden remained unsatisfactory in respect of the large quantities of maintenance dredging. In the 50s and 60s of this century another - at least again futile - approach to solve the problem was started: Interfering into the system by training walls and groynes in order to use the transport capacity of the flood tide for avoiding sedimentation in the Emden waterway and shifting it upstream where it would be less disadvantageous in respect of needed navigable water depth. In the 70s of this century a plan was introduced to close the Emden waterway by a new lock at both its seaward and upstream end, arranging new safe harbor areas on both shores being protected by a new dyke (Fig. 42). The river Ems should be passed through the storm surge bay Dollart [CARSJENS & CLASMEIER 1986]. Major aim of this plan was to avoid maintenance dredging in the Emden waterway and getting a smoother transition from the harbor entrance to the deeper parts of the estuary in which cross-sectional maintenance is supported by the tidal prism of the Dollart bay. Though never described this was an adaption on the basics of the concept which had been developed earlier and applied with remarkable success by KRÜGER in the Jade area. But the realization had to be postponed for years due to difficult negotiations with the neighbor country, the Netherlands. In the end, the project was abandoned for both ecological and economical reasons.

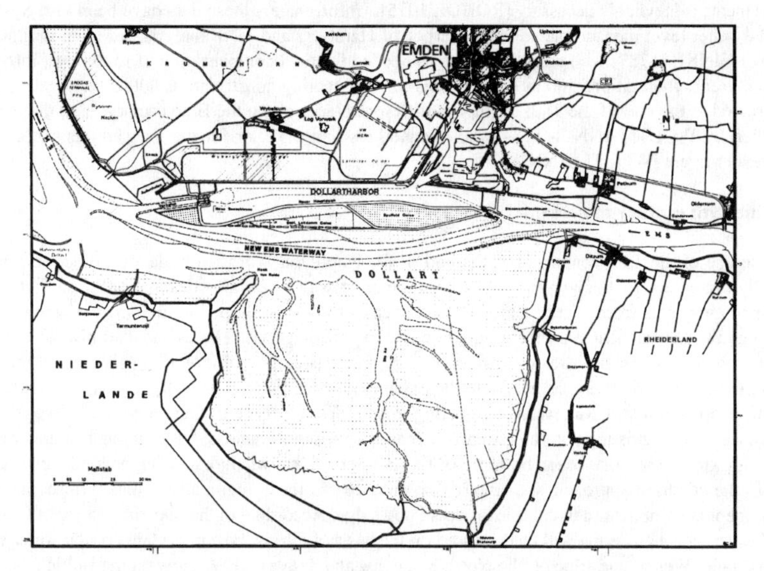

Figure 42. Dollart harbor (plan) [CARSJENS & CLASMEIER 1986]

JADE

The Jade inlet and bay were created by the erosive forces of storm surges after consecutive dyke breaks between the 13th and 16th centuries leading temporarily even to a link to the Lower Weser. The large tidal prism of the bay maintains large, stable cross-sections with remarkable water depths in the Jade inlet. Prussia though in possession of the harbor of Emden at the North Sea coast bought an area for the erection of a naval port called Wilhelmshaven at the Jade from the Grand Duchy of Oldenburg in order to use the natural advantages which became later more evident when the German Imperial Navy operated from that place large battleships of the 'Dreadnought' type.

The analysis of this area was carried out by HAGEN [1856], one of the best educated and most experienced German coastal engineers of his time. His heritage to our time beside others is the second historical record of an ordinary tide in the Jade area after that one inherited from BRAHMS about a century before [LUCK & NIEMEYER 1980]. Furthermore he combined his observations of tidal water levels with the fluctuation of silt content in the water column (Fig. 43). Later the migration of channels in the Jade and resulting occurrence of bars and shallows motivated the Imperial Navy to ask an experienced engineer of the harbor authority of Hamburg for an expertise. LENTZ [1899, 1903] recommended areal dredging in the Jade bay in order to increase the tidal prism and consequently its capacity of maintaining sufficiently large and deep cross-sections in the Jade inlet. The basics of that kind of indirect interference into the coastal processes is similar to the relation of tidal volume and channel cross-section which has become worldwide popular due to O'BRIEN [1931, 1967]. The knowledge about the important role of the bay's tidal prism for the stabilty of the Jade waterway was credited by a legal act in 1883 which forbade any land reclamation or other impacts in the Jade area leading to a reduction of the local tidal prism.

After the invention of the 'Dreadnought' battleships with enormously increasing dimensions the concept of LENTZ was no longer regarded as suitable to deliver sufficiently stable cross-sections in the Jade waterway, particularly in its offshore area. There the updrift banks were fixed by training walls and large groynes creating a small artificial island: Minseneroog. Moreover regular dredging created a stable waterway requiring only limited maintenance [KRÜGER 1922]. Before interfering into the system KRÜGER [1911] had carried out intensive investigations on the acting hydrodynamical forces and the long-term morphological development in this area in order to get a sound basis for engineering measures. He improved his concept in the course of the following decades and even after retirement until

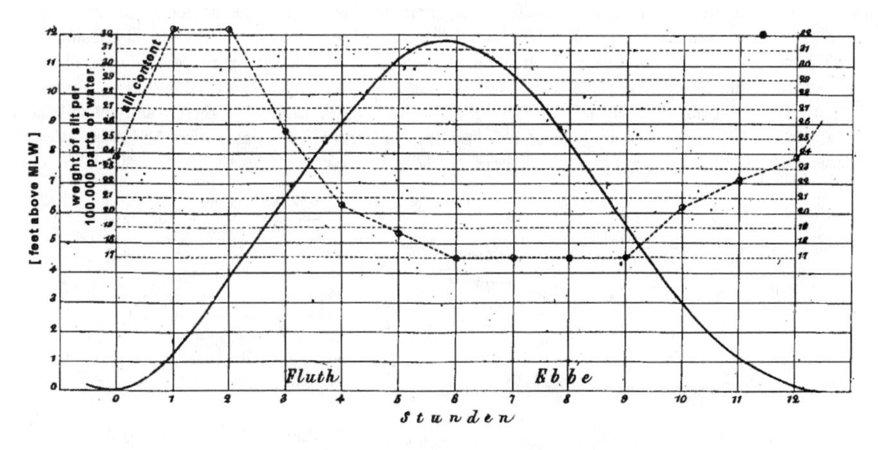

Figure 43. Tidal water level and silt content; Jade area (Fluth: flood; Ebbe: ebb; Wasserstand: waterlevel; Schlickgehalt: silt content) [HAGEN 1856]

his death incorporating an enormous number of scientists from coastal engineering and related disciplines. KRÜGER must be regarded as the initatior of interdisciplinary coastal research in Germany. Though the allied powers did not allow any maintenance of the Jade waterway after the World War II the correction of the Jade mainly inspired by KRÜGER was sufficiently successful to deliver a suitable basis for establishing at the Jade a deep water harbor for large tankers and bulk carriers step by step between 1958 and 1974. Vessels up to 250000 tdw and a maximal draft of 20 m are enabled to enter the piers at Wilhelmshaven [BRAUN & WITTE 1979].

WESER

The Weser estuary with the Outer and Lower Weser is the access for seagoing vessels to the port of Bremen. Since the 16th century the nautical conditions worsened more and more, particularly close downstream of the harbor of Bremen. In order to continue the profitable sea trade first harbor facilities were erected downstream of the existing harbor and at least in 1827 the port of Bremerhaven was founded in the transition area of Outer and Lower Weser by VAN RONZELEN [1857]. From there goods were transferred to Bremen by smaller vessels and land vehicles. For improving the safe access from the North Sea on a flat area bordering the Outer Weser a lighthouse was erected [VAN RONZELEN 1857] which is still in use as a basis for a radar based pilot system.

But still regaining an access by a sufficiently designed waterway in the Lower Weser guaranteeing sea trade as a major economical basis of the city of Bremen had political priority. The appointment of L. FRANZIUS, a very successful member of a well-known dynasty of East Frisian coastal engineers, as chief engineer of the harbor and waterway authority provided that aim with a feasible technical background: First of all the concept of FRANZIUS was contradictory to earlier local impacts the whole Lower Weser by creating a system of continuously increasing cross-sections in downstream direction with the major aim to minimize sedimentation and to achieve cross-sectional stability in the waterway requiring low maintenance efforts. As far as achievable, tidal flow was concentrated in the main channel by closing secondary ones and erecting groynes. The prospected effect was evaluated by estimated tidal curves which were expected to occur after the correction of the Lower Weser and used for computations of the tidal volumes. Applying the law of continuity an adaption of suitable cross-sections was carried out for getting nearly the same tidal current velocities along the estuary. Due to the impact of upstream freshwater a small dominance of ebb current was anticipated provoking a net seaward sediment transport [FRANZIUS 1888]. The approach of FRANZIUS was successfully carried out between 1883 and 1895 allowing ships with a draft of up to 5 m to get access to the harbor of Bremen (Fig. 44) by 'riding on the tidal wave' from Bremerhaven to Bremen. Remarkable also was that the expected effect of natural transport capacity was achieved: Only 50% of the necessary 50 million m^3 of sediment had been taken by dredging, the rest was evacuated by tidal flow. In order to stop the lowering of tidal water levels occurring after the correction of the Lower Weser further upstream in 1905 a tidal barrier was erected upstream of the harbor of Bremen [FLÜGEL 1988] reflecting the tidal wave there totally. The basic concept of FRANZIUS has been applied for a number of consecutive adaption of the Lower Weser waterway until the last one performed from 1974 to 1977 which enables ships with a draft of 12.5 m and

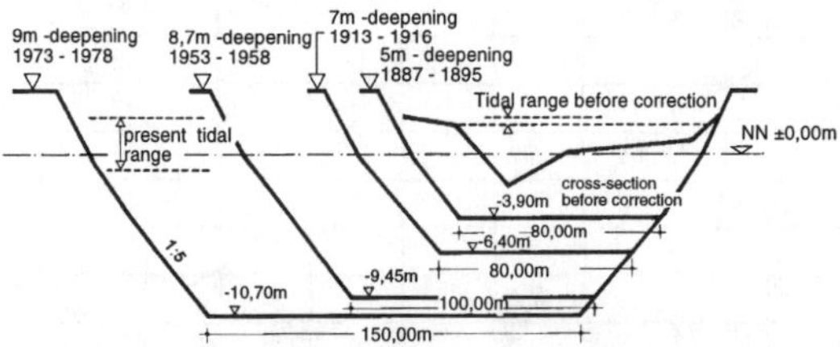

Figure 44. Cross-sections in the Lower Weser close downstream of Bremen since 1885 [WETZEL 1988]

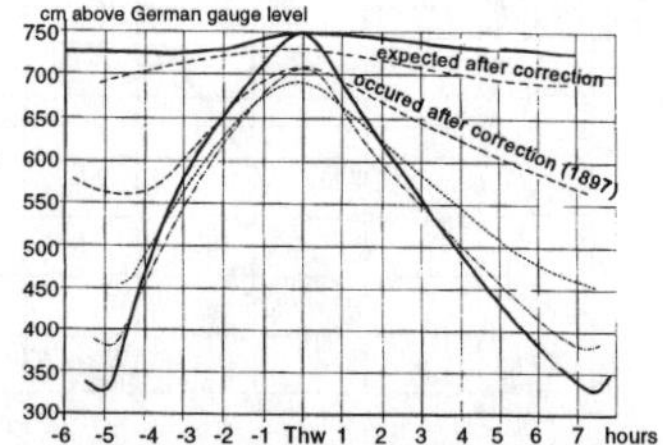

Figure 45. Tidal curves of the Lower Weser in Bremen between 1885 and 1978 [ROHDE 1970, 1980]

about 30.000 tdw to go upstream to Bremen harbor by adapting their traveling to the higher part of the tidal wave (Fig. 44). When FRANZIUS started his correction the tidal range in Bremen was about 0.3 m. As a result of the five successive deepenings tidal range has increased to about 4 m at Bremen (Fig. 45). Honesty requires us to admit that nevertheless the number of ships traveling to Bremen has been reduced in recent years and ecologists complain about the effects of the subsequent corrections demanding a restructuring of the Lower Weser to a more natural river [BUSCH et al. 1989].

Already in 1890 the growing passenger ships from the North Atlantic routes required engineering impacts in the Outer Weser which was carried out also by FRANZIUS [1895]. But this measure was less successful than the correction of the Lower Weser. The maintenance of the existing waterway required increasing efforts without gaining sufficient stability. The multiple channel system remained migrating with a tendency of changing cross-sections in its branches due to variations of local tidal volumes.

Already in 1825 the waterway had been shifted from the most westward situated channel to the eastern one. Another shift to the then deepening one was regarded by FRANZIUS as favorable, but not carried out in respect of the necessary high efforts. In 1921 PLATE [1927] started to shift the waterway from the 'Wurster Arm' in the eastern part to the central one 'Fedderwarder Arm' (Fig. 46) and fixed the then implemented system by groynes and training walls. His work was the basis for later improvements of the navigability of the Outer Weser (Fig. 46) [HOVERS 1975; WETZEL 1988] and still continues in present days for an adaption to the requirements for container vessels of the fourth generation for which a specific analysis of necessary waterway dimensions in respect of the design vessel was carried out [DIETZE 1990].

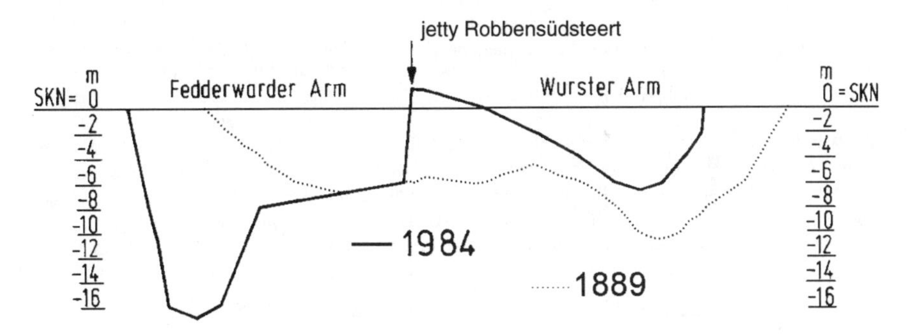

Figure 46. Cross-sections of the Outer Weser waterway in 1889 and 1984 [WETZEL 1988]

ELBE

In the beginning of the 19th century the increasing size of seagoing vessels led to difficulties in the Elbe waterway downstream of Hamburg harbor. In order to achieve a sound basis for engineering impacts HÜBBE [1842] started systematic hydrodynamical investigations. He installed e. g. permanent measuring tidal gauges in Cuxhaven at the estuarine mouth and in Hamburg between 1841 and 1843 [ROHDE 1975] which data highlighted the necessity of levelings in order to get a sound reference for the measured tidal water levels. This requirement was later fulfilled by LENTZ. River sections being critical in respect of navigability were streamlined by training walls and groynes in order to concentrate tidal flow in the navigation channel and to keep it away from tributary ones. Additionally steam dredgers were used to achieve the goal [HÜBBE 1853]. HÜBBE [1861] investigated also intensively the morphological development of the river he had to deal with; remarkable are his observations on bedforms which he already distinguished in the four classes: ripples, dunes, tidal ridges and shoals. Additionally he used tracers in order to study the migration of bedforms (Fig. 47).

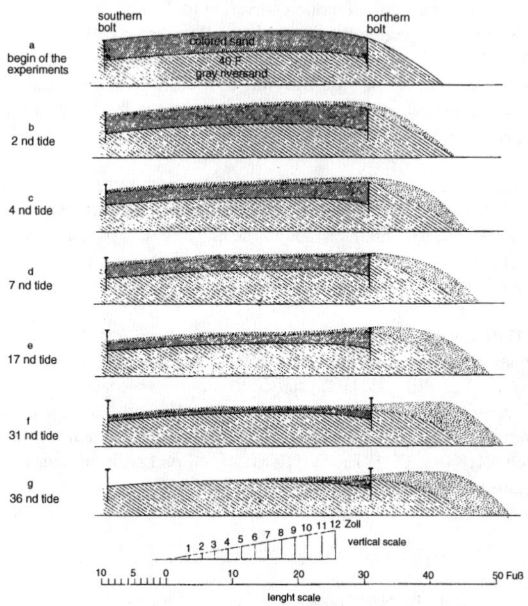

Figure 47. Results of a field experiment on bedform migration by tracer on an intertidal shoal [HÜBBE 1861]

His successor DALMANN [1856] studied the contemporary experience in coastal waterway engineering available in the neighbor countries England, France and the Netherlands. His consequence for further improvement of navigability and maintenance in the Elbe river was focussing on dredging and restriction of groyne and training wall construction on specific sections.

The increasing draft of seagoing vessels required at least dredging in the Elbe estuary, even seaward of the city of Cuxhaven [ROHDE 1971]. For the stabilization of that estuarine section between 1948 and 1966 a training wall with a length of 9.2 km was erected at the western edge of the waterway seaward of Cuxhaven due to a recommendation of HENSEN [1941]. After the 2nd World War four subsequent deepenings of the Elbe waterway have taken place. The planning for a fifth one in respect of the demands of large container vessels is presently finished [SCHLÜTER 1993]. A comparison of present and planned navigational water depths in the Elbe estuary with historical ones [ROHDE 1971] highlights the enormous changes the regime has experienced by coastal engineering interference (Fig. 48). In the meantime there had been plans to erect a deep water port at the estuarine entrance seaward of Cuxhaven. This prospects have been abandoned for economical and ecological reasons [LAUCHT 1982]. The intensive preinvestigations for that project delivered a remarkable amount of information being generally valuable for coastal engineering problems.

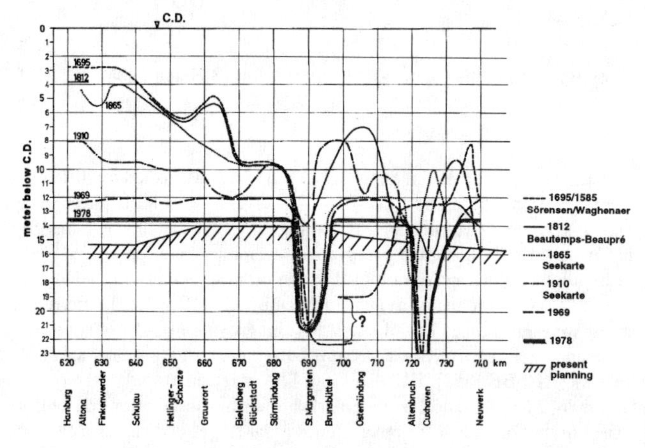

Figure 48. Change of navigational depth in the Elbe estuary and planned deepening; earlier situations adapted from ROHDE [1971]

Table 1. Coastal waterways at the German Coasts

Waterway						1994	
	between	length	depth below C. D.	maxi-mum vessel	mainte-nance dredg-ing volume	number of ves-sels	cargo volume
		km	m	tdw	10^6 m^3		10^6 t
Elbe	North Sea-Brunsbüttel	70	13.5	110000	13.0	58450	81.5
	Brunsbüttel-Hamburg	64	13.5	100000			
Weser	North Sea-Bremerhaven	60	12.0	80000			16.2
	North Sea-Brake	85	10.0	45000	1.3	29600	6.6
	Bremerhaven-Bremen	60	9.0	35000			14.7
Jade	North Sea-Wilhelms-haven	55	18.5	250000	13.4	2760	34.5
Ems	North Sea-Emden	70	8.5	40000	10.6	1970	2.0
Trave	Baltic Sea - Lübeck (Stadt)	25	9.5	14000	0.02	23500	20.3
Warnow	Baltic Sea-Rostock	11	13.0	60000	0.06	23150	15.8

TRAVE

At the beginning of the 19th century sedimentation in the Lower Trave had led to a reduction of water depth allowing only ships with a draft of about 2 m traveling to the harbor of Lübeck. The necessary information for navigation was delivered by a gauge the data of which since 1826 are still available [JENSEN & TÖPPE 1986]. In order to improve navigability in the Lower Trave in 1835 dredging was started and in 1840 a channel passing through its barrier was established. The first river correction between 1850 and 1854 enabled vessels with a draft of about 4 m to enter the harbor of Lübeck. During the periods from 1879 to 1883 and from 1899 to 1907 the second and third correction were carried out due to plans of REHDER [1898], who was then in charge of waterway and harbor engineering. Major means were: reduction of river length by cutting bows, erection of groynes and deepening by dredging. Afterwards the waterway was navigable for vessels until Travemünde at the mouth with a draft of 8.5 m and until Lübeck with a draft of 7.5 m. After the fourth (1908 - 1961) and the fifth (1961 - 1982) correction the harbor of Lübeck is accessible by ships with a draft of 9.5m.

WARNOW

The harbor of Rostock was until the World War II of minor importance in comparison with other German harbors at the Baltic coast such as Lübeck and particularly Stettin. Nevertheless already in the course of the 19th century the waterway was step by step dredged up to a water depth of about 5 m. In 1903 the railway connection between Berlin and Copenhagen was closed by a railway ferry service between Warnemünde at the mouth and the Danish island of Falster. Therefore a new access channel to the Baltic Sea was established being fixed by new jetties. The navigable draft for vessels increased due to those and accompanying impacts afterwards to 7 m until Warnemünde and 6 m until Rostock. After the World War II the former G.D.R. decided to establish a basis for its sea trade in Rostock. At the southern shore of a bay connected with the Lower Warnow new harbor facilities were erected. The seaward access channel was shifted eastward of the existing jetty system with a depth of about 10,5 m and a width increasing from 60 m to 120 m at its seaward end. On its eastern bank a new breakwater with a length of 655 m was additionally established. In 1977 the water depth was increased to 13 m. Nowadays a further deepening for bulk carriers is planned including deepening to a water depth of 14,5 m and a reshaping of the existing breakwater and jetty system.

Present situation

The efforts in the past have created an effective system of harbors and waterways for sea trade being an important contribution to nation's welfare. They provide furthermore a sound basis for future development in respect of changing requirements of modern cargo vessels and new techniques of sea transport. Table 1 makes evident the navigable water depths on the major German waterways and the accessability of its ports. A major problem is already the deposition of dredged material, particularly in those areas where it has been contaminated by human waste water. Moreover major ports such as Hamburg and Bremen are located far upstream from the sea: as well the economical as the ecological limitations in respect of further adaptions to increasing seagoing vessels have been be achieved or will be soon.

EDUCATION, RESEARCH AND COOPERATION

The manuals on coastal engineering by BRAHMS [1754, 1757] and HAGEN [1863] highlighted the impressive level then available for further education. Though the predecessors of the later Universities of Technology were mostly founded in the first decades of the 19th century, no profiled school of coastal engineering was established before this century. Particularly between World War I and II Otto FRANZIUS at Hanover, WINKEL at Danzig, and DE THIERRY and AGATZ in Berlin established a widely accepted education of coastal engineering. After World War II the FRANZIUS-Institute for Hydraulic Engineering at the University of Technology in Hanover became under the directorate of HENSEN the most prominent school of coastal engineering in Germany attracting most of those students with specific interest in coastal engineering and providing German coastal engineering with a respectable number of splendid graduates. Nowadays in Germany two specific institutes for coastal engineering exist: at the Universities of Hannover and Brunswick. Furthermore coastal engineering is part of the education in hydraulic engineering at the Universities of Technology at Berlin, Darmstadt, Dresden and Rostock-Wismar.

Research has a respectable tradition at German universities, particularly in hydraulic modeling. For example, the FRANZIUS-Institute has carried out hundreds of model tests dealing with coastal problems in Germany or abroad. But also others operate in this field. For example model tests for the closure works of the Zuiderzee in the Netherlands were carried out at the Theodor-Rehbock Laboratory of the University of Karlsruhe. German coastal engineering has been provided with a unique tool by the German Research Foundation (DFG) in the 1980s: the Large Wave Flume at Hanover, a joint institution of the Universities of Hanover and Brunswick.

Traditionally, applied research has been carried out also by specific institutes belonging to administrations being responsible for coastal engineering. Before World War II the most prominent ones have been the Naval Waterway Engineering Department at Wilhelmshaven, the Prussian Hydraulic Laboratory in Berlin and the Prussian Coastal Research Stations at Büsum, Husum and Norderney. The Coastal Research Stations focused on field investigations, the Naval Waterway Department carried out both hydraulic modeling and field research. Nowadays there are still two governmental agencies for consultancy and research in the field of coastal engineering existing: the Coastal Branch of the Federal Institute for Waterway Engineering at Hamburg and the Coastal Research Station of the Lower Saxonian Central State Board for Ecology at Norderney/East Frisia.

The Naval Waterway Department became particularly under the directorate of KRÜGER the center of German coastal engineering research at the North Sea coast. KRÜGER pursued not only cooperation among coastal engineers but also with other disciplines such as geology and biology leading to integrated coastal research. He initiated as a platform for scientific discussion on coastal research the 'Annual meetings of Northwestern German Geologists' with contributions of all fields of coastal research with coastal engineering in a - still existing - prominent position. In order to establish exchange of information he initiated with colleagues from neighboring countries the International Association of Hydraulic Research (IAHR), its first congress took place in Berlin 1937.

After World War II the division of the country enforced also a separation of the coastal engineering community. In West Germany coastal researchers of all disciplines founded the Coastal Council North and Baltic Sea (Küstenausschuß Nord- und Ostsee) which served as a platform both for exchange of information and cooperation. In 1973 it was replaced by the German Committee on Coastal Engineering Research (KFKI: Kuratorium für Forschung im Küsteningenieurwesen) which has been established by a formal agreement between the Federal Government and the coastal states being both represented in the Committee which coordinates the applied research in German coastal engineering. The Committee is supported by a research coordinator and an advisory group which might be regarded as an institution in which existing knowledge is transferred between generations of coastal engineers.

The first contribution of a German colleague at an International Coastal Engineering Conference took place in 1960 [KRAMER 1960]. In 1970 for the first time a group of German coastal engineers attended the 12th ICCE at Washington/DC and presented 11 contributions. The chairman of the German group suggested celebrating the 16th ICCE in Germany in 1978. The suggestion was appreciated though not already fixed [RAMACHER 1971]. The decision was made by the Coastal Engineering Research Council at the following 13th ICCE at Vancouver/Canada on the basis of a formal invitation by the Coastal Council North and Baltic Sea, the German Harbor Engineering Association and the City of Hamburg [PARTENSCKY 1973]. In 1978 the 16th ICCE took place in Hamburg. That location enabled a large number of German coastal engineers to attend, particularly the younger ones. Both information and even more the impressions initiated in the German coastal engineering society increasing research efforts and interest in international cooperation. A lot of those who attended then for the first time an ICCE have been motivated to participate in the unique exchange of information being offered at the following ones until today.

SUMMARY AND OUTLOOK

The tracing back of the contemporary state of the art in German coastal engineering makes evident that beside a historical development a remarkable amount of knowledge has been inherited from our professional predecessors. Furthermore, many of their projects provide still a sound basis for further developments. Taking present tools and their effectiveness into consideration, a comparison with the performance of former coastal engineers reduces the scaling of present products significantly requiring moderation. Emphasis should be laid on the message that the looking back in a profession with a remarkable empirical component is mostly worthwhile and more effective than the 'reinventing of the wheel' which often ends up as repeating a blunder. Particularly, but not only, our younger colleagues are encouraged to make more intensive use of the heritage of our professional history.

LITERATURE

ALW [1984]: Masterplan coastal protection island of Sylt. Amt f. Land- u. Wasserwirtschaft Husum (in German)

ARENDS, F. [1833]: Description of the storm surges between February 3rd and 5th 1825. Comission W.Kaiser, Bremen (in German)

ASTER, D.; JÜRGENS, H.H.; WEITZEL, H. [1990]: Construction of groynes on the island of Borkum. Jb. Hafenbautechn. Ges. 1989, 43 (in German)

BAENSCH, J. [1875]: The storm surges at November 12th and 13th 1872 at the Baltic coasts of Prussia. Zeitschr. f. Bauwesen 25, Berlin (in German)

BRAHMS, A. [1754]: Basic arts in dyke construction and hydraulic engineering, part I, Verl. H. Tapper, Aurich/Ostfriesland (in German)

BRAHMS, A. [1757]: Basic arts in dyke construction and hydraulic engineering, part II. Verl. H. Tapper, Aurich/Ostfriesland (in German)

BRAUN, J.; WITTE, H.H. [1979]: Enlargement and marking of the Jade waterway. Jb. Hafenbautechn. Ges. 1977/78, 36 (in German)

BOTHMANN, W.; KATTENBUSCH, E,; LORENZEN, J.M.; LÜDERS, K.; SCHAUBERGER, H.; SNUIS, H.; THILO, R. (Arbeitsgruppe Küstenschutz im Küstenausschuß Nord- und Ostsee) [1955]: General recommendations for coastal protection in Germany. Die Küste, 4 (in German)

BUSCH, D.; SCHIRMER, M.; SCHUCHARDT, B.; ULLRICH, P. [1989]: Historical changes of the river Weser. in: G.E. Petts (ed.): Historical changes of large alluvial rivers: Western Europe. J. Wiley & Sons, London

CARSJENS, R.; CLASMEIER, H.D. [1986]: Project Dollart Harbor: technical presentation. Jb. Hafenbautechn. Ges. 1985/86, 41 (in German)

DALMANN, J. [1856]: On river corrections in tidal areas. Hamburg (reprint: Die Küste, 46, 1988) (in German)

DETTE, H.H.; FÜHRBÖTER, A. [1975]: Field investigations in surf zones. Proc. 14th Int. Conf. Coast. Eng. Copenhagen/Denmark, ASCE, New York

DIETZE, W. [1990]: Expertise on the dimensions of the Outer Weser waterway for tide-independent access of container vessels. Wasser- u. Schiffahrtsdirektion Nordwest, Aurich/Ostfriesland (unpubl. in German)

EIBEN, H. [1983]: Specific hydrological conditions at the southwestern Baltic Coast. Amt f. Land- u. Wasserwirtschaft Kiel (unpubl. In German)

EIBEN, H. [1989]: Wind, water levels and waves at the Baltic coast of Schleswig-Holstein during the stormy periods in winter 1986/87. Die Küste, 50 (in German)

EIBEN, H. [1992a]: Specific hydrological conditions at the Baltic Coast. in DVWK: Historical coastal protection. Wittwer-Verl. Stuttgart (in German)

EIBEN, H. [1992b]: Coastal protection at the Baltic Coast of Schleswig-Holstein. in DVWK: Historical coastal protection. Wittwer-Verl. Stuttgart (in German)

ERCHINGER, H.F. [1970]: Coastal protection by salt marsh reclamation, dyke construction and maintenance in East Frisia. Die Küste, 19 (in German)

ERCHINGER, H.F. [1985]: Beach fill by turning the course of sandbars. Proc. 19th Int. Conf. Coast. Eng. Houston/Tx., USA, ASCE, New York

FLÜGEL, H. [1988]: Correction of the Lower Weser and harbour extension in Bremen during the last hundred years. Jb. Hafenbautechn. Ges. 1987, 42 (in German)

FRANZIUS, L. [1888]: The correction of the Lower Weser estuary. Bremen (reprint: Die Küste, 51, 1991) (in German)

FRANZIUS, L. [1895]: The correction of the Lower Weser and a project for the correction of the Outer Weser estuary. Leipzig (in German)

FÜHRBÖTER, A. [1974]: Some results from field investigations in surf zones. Mitt. Leichtweiß-Inst., 40 (in German)

FÜHRBÖTER, A. [1979]: Duration of wave energy impacts. Mitt. Leichtweiß-Inst., 65 (in German)

FÜHRBÖTER, A. [1992]: Wave impacts on dyke- and revetment-slopes. Jb. Hafenbautechn. Ges. 1991, 46 (in German)

FÜHRBÖTER, A.; KÖSTER, R.; KRAMER, J.; SCHWITTERS, J.; SINDERN, J. [1974]: A sand groyne for beach preservation on the island of Sylt. Die Küste, 23 (in German)

FÜLSCHER, J. [1905]: On protection structures for the preservation of the East and North Frisian Islands. Zeitschr. f. Bauwesen, Verl. Ernst & Sohn, Berlin (in German)

GÄTJEN, B. [1979]: Storm surge barriers at the German North Sea coast. Tasks, planning and construction. Jb. Hafenbautechn. Ges. 1977/78, 36 (in German)

GARSCHINA; PANSE [1898]: Hydrological investigations in the area of the East Frisian Islands. Part 1: Tidal inlet of Norderney. Wasserbauinspection Norden/Ostfriesland (unpubl. in German)

GAYE, J. [1951]: Changes of water levels in the Baltic and the North Sea during the last hundred years. Die Wasserwirtschaft, Sonderheft Vorträge gewässerkundl. Tag. (in German)

GAYE, J.; HENSEN, W.; LORENZEN, J.M.; LÜDERS, K.; PLATE, L. ;ROLLMANN, A.; SCHUMACHER,W.; WALTHER, F.: (Küstenausschuß Nord- und Ostsee, Arbeitsgruppe Norderney) [1952]: Expertise on the investigations concerning beach erosion at the western and northwestern shores of the island of Norderney and the recommended engineering measures for the protection of the island. Die Küste 1, 1 (in German)

GAYE, J.; WALTHER, F. [1935]: Migration of ebb delta shoals along the East Frisian Islands. Die Bautechnik, 13, 41 (in German)

GEINITZ, E. [1903]: Land losses at the coast of Mecklenburg. Mitt. Großherzogl. Geol. Landesanst. 15 (in German)

GERHARDT, P. [1900]: Manual on dune stabilization in Germany. Berlin (in German)

GERMELMANN [1904]: Expertise on necessary protection measures at the bluff coasts of Alt-Graatz, Heiligenhafen, Graal, Müritz and Wustrow. Berlin (unpubl. in German)

GEYER[1962]: Basin oscillations and water exchange in the Bay of Eckernförde under specific consideration of the storm at December 5th and 6th, 1961. Kieler Meeresforsch., 20 (in German)

GÖHREN, H. [1975]: Dynamics and morphology of sand banks in the surf zone of outer tidal flats. Proc. 14th Int. Conf. Coast. Eng. Copenhagen/Denmark, ASCE, New York

HAGEN, G. [1856]: On the tides and soil of the Prussian Jade area. Monatsber. Königl. Akad. Wiss., Berlin (reprint: Die Küste, 51, 1991) (in German)

HAGEN, G. [1863]: Marine and harbor engineering. in: Manual of hydraulic engineering. 3rd Part: The sea, Vol. 1. Verl. v. Ernst & Korn, Berlin (in German)

HAYES, M.O. [1979]: Barrier island morphology as a function of tidal and wave regime. in: S.P. Leatherman: Barrier islands. Academic Press, New York

HEISER, H. [1932]: Shore protection at coasts with and without dominant sediment transport. Die Bautechnik, 10, 40 (in German)

HEISER, H. [1933]: Coastal preservation and land reclamation at the German North Sea coast. Die Bautechnik, 11, 13 (in German)

HENSEN, W. [1941]: The development of navigability in the Outer Elbe estuary. Jb. Hafenbautechn. Ges. 1940/41, 18 (in German)

HENSEN, W. [1954]: Model tests on the impact of dyke shaping on wave run-up. Mitt. Franzius-Inst., 7 (in German)

HOMEIER, H. [1962]: Reconstructed historical morphology 1:50000 of the Lower Saxonian coast. Jber. 1961 Forsch.-Stelle f. Insel- u. Küstenschutz, 13 (in German)

HOMEIER, H. [1965]: Historical-morphological investigations at the Coastal Research Station on long-term developments at the Lower Saxonian coast. Jber. 1964 Forsch.-Stelle f. Insel- u. Küstenschutz, 16 (in German)

HOMEIER, H. [1969]: Changing shape of the East Frisian Coast in the course of centuries. in: J. Ohling (ed.): East Frisia in the shelter of the dyke, 2, Deichacht Krummhörn, Pewsum /Ostfriesland (in German)

HOMEIER, H. [1976]: Effects of severe storm surges on beaches and dunes on East Frisian Islands. Jber. 1975 Forsch.-Stelle f. Insel- u. Küstenschutz, 27 (in German)

HOMEIER, H.; LUCK, G. [1971]: Investigations on the morphological development of the tidal inlet Accumer Ee as a basis for the future behavior of beaches and dunes on the western and northwestern shore of the island of Langeoog. Jber. 1970 Forsch.-Stelle f. Insel- u. Küstenschutz, 22 (in German)

HOVERS, G. [1975]: Morphological changes in a fine sand tidal estuary after measures of river improvement. Proc. 14th Int. Conf. Coast. Eng. Copenhagen/Denmark, ASCE, New York

HÜBBE, H. [1842]: Some observations of water levels in the tidal area of the Elbe river. Hamburg (reprint: Die Küste, 46, 1988) (in German)

HÜBBE, H. [1853]: Experience and observations in the field of river engineering, part I. Hamburg (reprint: Die Küste, 46, 1988) (in German)

HÜBBE, H. [1861]: On the state and on the behavior of sand. Zeitschr. F. Bauwesen, Verl. Ernst & Sohn, Berlin (reprint: Die Küste, 46, 1988) (in German)

HUNDT, C. [1954]: Design storm surge levels for dimensioning of dykes at the western coast of Schleswig-Holstein. Die Küste, 3, 1/2 (in German)

JENSEN, J.; HOFSTEDE, J.L.A.; KUNZ, H.; DE RONDE, J.; HEINEN, P.F.; SIEFERT, W. [1993]: Long term water level observations and variations. in: R. Hillen & H.J. Verhagen: Coastlines of the southern North Sea. Proc. Coastal Zone '93

JENSEN, J.; TÖPPE, A. [1986]: Compilation and evaluation of measurements at the gauge Travemünde. Deutsche Gewässerk. Mitt., 30, 4 (in German)

JENSEN, J.; TÖPPE, A. [1990]: Investigations on storm surges of the Baltic Sea under specific consideration of the gauge Travemünde. Deutsche Gewässerk. Mitt., 34, 1/2 (in German)

KANNENBERG, E.G. [1955]: The surge from January 1st 1954 at the German Belt Coast. Urania, 18, 1, Leipzig/Jena (in German)

KLAUS, J.; SCMIDTKE, R.F. [1990]: Valuation expertise for dyke construction planning at the mainland coast -model area Wesermarsch -.. Bundesmin. Ernähr., Landw. u. Forsten, Bonn (in German)

KÖSTER, R. [1979]: Sediments at the Probstei coast. Mitt. Leichtweiß-Inst. 65

KOLP, O. [1955]: Endangering of the German coast between the rivers Trave and Swine by storm surges. Seehydrograph. Dienst der DDR (in German)

KRAATZ, D. [1966]: Beach and shoreline changes at the west coast of the island of Sylt and the impact of technical measures. Marschenbauamt Husum (unpubl. in German)

KRAMER, J. [1960]: Beach rehabilitation by use of beach fills and further plans for the protection of the island of Norderney. Proc. 7th Conf. Coast. Eng. Berkeley/Ca., USA, ASCE, New York

KRAMER, J. [1977]: Safety of sea dykes against storm surges. Die Küste, 31 (in German)

KREY, H. [1906]: On protection structures for the preservation of the East and North Frisian Islands. Zentralbl. Bauverw., 26, 54 (in German)

KREY, H. [1918]: Wadden Sea, marschlands and marshy islets at the North Sea coast of Schleswig-Holstein. Zentralbl. Bauverw., 38, 89, 93, 95, 96 (in German)

KRÜGER, G. [1910]: On storm surges at the German coasts in the western part of the Baltic Sea under specific consideration of the storm surge at December 30th and 31st 1904. Jber. Geograph. Ges. Greifswald (in German)

KRÜGER, W. [1911]: Sea and coast in the area of the island of Wangeroog and the forces acting on their shaping. Zeitschr. f. Bauwesen, Verl. Ernst & Sohn, Berlin (reprint: Die Küste, 51, 1991) (in German)

KRÜGER, W. [1922]: The Jade inlet, waterway of Wilhelmshaven, its genesis and state. Jb. Hafenbautechn. Ges. 1921, 4 (in German)

KRÜGER, W. [1937]: Migration of ebb delta shoals close to the island of Wangeroog. Abh. Naturw. Ver. Bremen, 30, 1/2 (in German)

KRÜGER, W. [1938]: Coastal subsidence in the Jade area. Der Bauingenieur, 19, 7/8 (in German)

KUNZ, H. [1991]: Artificial beach nourishment on Norderney, a case study. Proc. 22nd Int. Conf. Coast. Eng. Delft/The Netherlands, ASCE, New York

KURZAK, G.; LINKE, O.; DECHEND, W.; KRAUSE, H.; THILO, R. [1949]: Causes for erosion of the western and northwestern shores of the island of Norderney. Jber. 1949 Forsch.-Stelle f. Insel- u. Küstenschutz, 1 (in German)

LAMPRECHT, H.O. [1955]: Surf and shore variations at the west coast of the island of Sylt. Mitt. Franzius-Inst., 8 (in German)

LASSEN, H.; SIEFERT, W. [1991]: Mean tidal water levels in the southeastern North Sea and secular trend. Die Küste, 52 (in German)

LAUCHT, H. [1982]: Harbor project Scharhörn. Planning in reflection of time spirit. Aumühle (unpubl. in German)

LENTZ, H. [1899]: Expertise on the conservation of the Jade waterway. Unpubl. (reprint: Die Küste, 51, 1991) (in German)

LENTZ, H. [1903]: Second expertise on the Jade Waterway. Unpubl. (reprint: Die Küste, 51, 1991) (in German)

LIESE, R. [1956]: First experience with the management of the storm surge barrier of the Leda river close to the city of Leer. Wasser & Boden, 8, 6 (in German)

LOHRBERG, W. [1989]: Changes of mean tidal water levels at the North Sea coast. Deutsche Gewässerk. Mitt., 33, 5,6 (in German)

LORENZEN, J. M. [1954]: Hundred years coastal protection at the North Sea. Die Küste, 3 (in German)

LUCK, G. [1975]: Impact of the protection structures on the East Frisian Islands on the morphological processes in the area of the tidal inlets and basins. Mitt. Leichtweiß-Inst., 47 (in German)

LUCK, G. [1977]: Inlet changes of the East Frisian Islands. Proc. 15th Int. Conf. Coast. Eng. Honolulu, Hawaii/USA, ASCE, New York

LUCK, G.; NIEMEYER, H.D. [1980]: Albert Brahms and the storm surge of 1717. Die Küste, 35 (in German)

LÜDERS, K. [1952]: Effect of the groyne H at the western spit of the island of Wangerooge on the tidal inlet Harle. Die Küste, 1 (in German)

LÜDERS, K. [1956]: What is a storm surge? Wasser & Boden, 8, 1 (in German)

LÜDERS, K. [1957]: The reestablishment of dyke safety at the German North Sea coast between the Dutch border and the Elbe estuary. Wasser & Boden, 9, 2 (in German)

LÜDERS, K. [1971]: On the valid time of the design water level for sea dykes at the Lower Saxonian North Sea coast. Jber. 1969 Forsch.-Stelle f. Insel- u. Küstenschutz, 21 (in German)

LÜDERS, K.; FÜHRBÖTER, A.; RODLOFF, W. [1972]: New dune and beach protection method on the island of Langeoog. Die Küste, 23 (in German)

LÜDERS, K.; LORENZEN, J.M.; RODLOFF, W.; FREISTADT, H.; TRAEGER, G.; KRAMER, J. (Küstenausschuß Nord- und Ostsee, Arbeitsgruppe Küstenschutzwerke) [1962]: Recommendations for hinterland protection by dykes after the storm surge of February 1962. Die Küste, 10, 1 (in German)

MC LAREN, C. [1842]: The glacial theory of Professor Agassiz. Am. J. Sci. & Arts, 42

METZKES, E. [1966]: Report on dyke construction and coastal protection in Lower Saxony after the storm surge of February 16th/17th 1962. Die Küste, 14, 1 (in German)

MEYER, K. [1913]: Water level variations in the Firth of Kiel. Heider Anzeiger, Heide (in German)

MIEHLCKE; O. [1956]: On the water levels at the coast of the GDR in respect of the storm surge at January 3rd and 4th 1954. Ann. Hydrographie, H. 5/6 (in German)

MUNK,W.H. [1949]: The solitary wave theory and its application to surf problems. New York Acad. O. Sc., 51

NEUMANN, G. [1941]: Basin oscillations of the Baltic Sea. Arch. dt. Seewarte & Marineobserv., 61, 4 (in German)

NIEDERMEYER, R.O.; KLIEWE, H.; JANKE, W. [1987]: The Baltic coast between Boltenhagen and Ahlbeck. Gotha (in German)

NIEDERSÄCHSISCHE HAUPTDEICHVERBÄNDE [1988]: No dyke, no land, no existence. Leaflet of the Union of the Lower Saxonian Coastal Protection Commuties (in German)

NIEMEYER, H.D. [1979]: Wave climate study in the region of the East Frisian Islands and Coast. Proc. 16th Int. Conf. Coast. Eng. Hamburg/Germany, ASCE, New York

NIEMEYER, H.D. [1983]: On the wave climate at island sheltered Wadden Sea coasts. BMFT-Forschungsber. MF 0203-83 (in German)

NIEMEYER, H.D. [1987]: On the classification and frequency of storm surges. Jber. 1986 Forsch.-Stelle Küste, 38 (in German)

NIEMEYER, H.D. [1991a]: Field measurements and analysis of wave-induced nearshore currents. Proc. 22nd Int. Conf. Coast. Eng. Delft/The Netherlands, ASCE, New York

NIEMEYER, H.D. [1991b]: Case study Ley Bay: an alternative to traditional enclosure. Proc. 3rd Conf. Coast. & Port Eng. i. Devel. Countr., Mombasa/Kenya

NIEMEYER, H.D. [1995]: Long-term morphodynamical development of the East Frisian Islands and Coast. Proc. 24th Int. Conf. Coast. Eng. Kobe/Japan, ASCE, New York

NIEMEYER, H.D.; KAISER, R.; DEN ADEL, J.D. [1995a]: Application of the mathematical wave model HISWA on Wadden Sea areas. Die Küste, 57 (in German)

NIEMEYER, H.D.; GÄRTNER, J.; KAISER, R.; PETERS, K.H.; SCHNEIDER, O. [1995b]: The estimation of design wave run-up on sea dykes in consideration of overtopping security by using benchmarks of flotsam. Proc. 4th Conf. Coast. & Port Eng. i. Devel. Countr., Rio de Janeiro/Brazil

NIEMEYER, H.D.; BIEGEL, E.; KAISER, R.; KNAACK, H.; LAUSTRUP, C.; MULDER, J.P.M.; SPANNHOF, R.; TOXVIG, H. [1995c]: General aims of the NOURTEC-project -Effectiveness and execution of beach and shore face nourishments-. Proc. 4th Conf. Coast. & Port Eng. i. Devel. Countr., Rio de Janeiro/Brazil

O'BRIEN, M.P. [1931]: Estuary tidal prisms related to entrance areas. Civ. Eng., 1, no.8, ASCE, New York

O'BRIEN, M.P. [1967]: Equilibrium flow areas of tidal inlets on sandy coasts. Proc. 10th Int. Conf. Coast. Eng., ASCE, New York

PARTENSCKY, H.W. [1973]: 13th International Conference on Coastal Engineering in Vancouver/ Canada from July 10th to 14th 1972. Die Küste, 24 (in German)

PETERSEN, M. [1952]: Erosion and protection of the bluff coasts at the eastern coast of Schleswig-Holstein. Die Küste, 1, 2 (in German)

PETERSEN, M. [1954]: On the basics for the design of dykes in Schleswig-Holstein. Die Küste, 3, 1/2 (in German)

PETERSEN, M. [1961]: German literature about groynes at sandy coasts. Die Küste, 9 (in German)

PLATE, L. [1927]: Deepening of the Outer Weser waterway by adaption of the gully Fedderwarder Arm. Jb. Hafenbautechn. Ges. 1926, 9 (in German)

PRALLE [1875]: Observations of the Baltic Sea storm surge from November 13th 1872. Zeitschr. Arch. & Ing. Vereins Kngr. Hannover, Neue Folge d. Not-Bl., 21, 4 (in German)

PUTNAM, J.A.; MUNK, W.H.; TRAYLOR, M.A. [1949]: Prediction of long shore currents. Trans. Am. Geophys. Un., 30

RAMACHER, H. [1971]: The Coastal Engineering Conference 1970 at Washington. Die Küste, 21 (in German)

RAGUTZKI, G. [1976]: Effects of the storm surges of January 1976 on the protection structures on the island of Norderney. Jber. 1975 Forsch.-Stelle f. Insel- u. Küstenschutz, 27 (in German)

REHDER, P. [1898]: Report for the Engineering Council of the City of Lübeck on the deepening of the Trave river to 7.5 m water depth. Lübeck (unpubl. in German)

REINHARD, H. [1949]: The storm surge at the Baltic coast of Mecklenburg on March 1st and 2nd 1949. Zeitschr. Meteorol. 3, 7, Potsdam (in German)

REINKE, J.T. [1787]: On observations of ebb and flood tide in the Elbe river. Hamb. Adr.-Comptoir-Nachr. (reprint: Die Küste, 46, 1988) (in German)

RODLOFF, W. [1972]: Hydrological analysis of the Baltic Sea storm surge at November 13th 1872. Deutsche Gewässerk. Mitt., 16, 6 (in German)

ROHDE, H. [1970]: Development of waterways at the German North Sea coast. Die Küste, 20 (in German)

ROHDE, H. [1971]: On the development of the Elbe tidal river as a waterway. Mitt. Franzius-Inst., 36 (in German)

ROHDE, H. [1975]: Water level measurements at the German North Sea coast before the middle of the 19th century. Die Küste, 28 (in German)

ROHDE, H. [1977]: Storm surge levels and secular rise of water levels at the German North Sea coast. Die Küste, 30 (in German)

ROHDE, H. [1980]: Problems concerning enlargements of tidal river waterways. in: DVWK: Course enlargements of waters. Goslar (in German)

ROHDE, H.; TIMON, A. [1967]: Preinvestigations for a solution of the problems in the Eider estuary. Wasserwirtschaft, 57, 5 (in German)

SCHUBERT; K. [1970]: Ems and Jade. Die Küste, 19 (in German)

SCHLÜTER, K. [1993]: Basics of the adaption of the Elbe waterway. Hansa, 130, 4 (in German)

SCHÜTTE, H. [1908]: Holocene subsidence effects at our North Sea coast. Jb. Oldenburger Ver. Altertumskde. Landesgeschichte, 16 (in German)

SHEPARD, F.P.; INMAN, D.L. [1951]: Nearshore circulation. Beach Eros. Board, T.M. 26.

SIEFERT, W. [1974]: On waves in shallow water. Mitt. Leichtweiß-Inst., 40 (in German)

SIEFERT, W.; KRAUSE, G.; PROBST, B.; SCHERENBERG, R. [1988]: Design water levels in the Elbe estuary. Die Küste, 47 (in German)

STARK, E. [1952]: High water levels in the Lübeck Bight between 1885 and 1949. Die Küste, 1, 2 (in German)

STREIF, H. [1990]: The coastal areas of East Frisia - North Sea, islands, tidal flats and marshlands. Set of geological guides, 57, Gebr. Bornträger, Berlin/Stuttgart (in German)

STREIF, H. ; KÖSTER, R. [1978]: The geology of the German North Sea coast. Die Küste, 32

THILO, R.; KURZAK, G. [1952]: The causes for the erosion of the western and northwestern beaches of the island of Norderney. Die Küste, 1,1 (in German)

TOLLE, A. [1864]: Beach protection structures on the island of Norderney. Zeitschr. Arch. & Ing. Vereins Kngr. Hannover, Neue Folge d. Not-Bl., 10, 1 + 2 (in German)

VAN RONZELEN, J.J. [1857]: Description of the construction of the Bremen lighthouse replacing the Bremen beacon at the entrance of the Weser estuary. Commission L. v. Vangerow, Bremerhaven (reprint: Die Küste, 51, 1991) (in German)

VEENSTRA, H. [1976]: Structure and dynamics of the tidal area. in: J. Abrahamse et al.: Wadden Sea. K. Wacholtz, Neumünster (in German)

WALTHER, F. [1934]: Die Gezeiten und Meeresströmungen im Norderneyer Seegat. Die Bautechnik, 12, 13 (in German)

WASSING, F. [1957]: Model investigations of wave run-up on dikes carried out in the Netherlands during the past twenty years. Proc. 6th Conf. Coast. Eng.

WEINHOLDT, E.; BAHR, M. [1952]: The sedimentation in the Eider estuary - causes and countermeasures. Wasserwirtschaft, 42, 8 (in German)

WEISS, D. (1992): Coastal protection in Mecklenburg-Vorpommern. in DVWK: Historical coastal protection. Wittwer-Verl. Stuttgart (in German)

WETZEL, V. [1988]: Improvement of the Weser waterway between 1921 and today. Jb. Hafenbautechn. Ges. 1987, 42 (in German)

WIEDECKE, W.; EIBEN, H.; DETLEFSEN, G. [1979]: History of the protection of the Probstei lowland against Baltic Sea surges. Mitt. Leichtweiß-Inst., 65 (in German)

WILDVANG, D. [1915]: The alluvial age between Ley and northern Dollart coast. Aurich/Ostfriesland (in German)

WOEBCKEN, C. [1924]: Dykes and storm surges at the German North Sea coast. Friesen-Verl., Bremen-Wilhelmshaven (in German)

WOLTMANN, R. [1790]: Theory and use of the hydrometric impeller, or a reliable method to observe wind and water velocities. Hoffmann, Hamburg (in German)

WYRTKI, C. [1953]: Long shore transport balance in the surfzone. Deutsche Hydrogr. Zeitschr. 6 (in German)

HISTORY OF COASTAL ENGINEERING IN GREAT BRITAIN

Edited by Rendel Palmer and Tritton Limited
Development and Engineering Consultants

There has been a long tradition of coastal estuary engineering in Great Britain (England, Wales and Scotland), dating back to Roman times. As a result Britain has often been at the forefront in the application of coastal defence techniques and engineering methods, and the development of coastal process theories. This paper provides a summary of these developments. The British coast is vulnerable to tidal flooding of low lying areas and the recession of soft cliffs. In the past the most significant problems have occurred in the developed parts of the east and south coasts of England and in North Wales. This has been reflected in the concentration of coastal engineering works and studies in these areas. By contrast, Scotland (which has its own legislative framework) has had less need for coastal defences of the scale found in parts of England and Wales. A brief review of the long history of coastal defences and engineering works is provided in Section 2 and 5.2. Section 3 outlines the nature of the coastline and natural processes. The legal and administrative framework of responsibility for defences, leading up to the strategy formalised in the last decade, is then described. Section 5 dealing with the development of theory and practice, first looks at research organisation. Following an outline of types of defences in relation to needs, it goes on to review the history of research on waves and of physical and computer based modelling; leading on to environmental considerations and the impact of 'soft' engineering on coastal planning and management. Seven case studies then follow. These illustrate the increasing need for technology to be compatible with the now more articulated environmental and social aims of society. Finally, the future of coastal engineering work is briefly considered with emphasis on the relationship of private and public financing and the continuing need for strategic research and long term statistical information.

1. INTRODUCTION

The coastline of Great Britain is a dynamic environment, subject to waves, the rise and fall of the tides and the movement of currents and storms. Due to the nature of the coastal geology and landforms, these processes give rise to natural hazards which can cause considerable losses in vulnerable locations. Large areas of the coast and low lying land adjacent to tidal estuaries are vulnerable to flooding, indeed, it is estimated that over 5% of the population of 60 million live in areas below the 5m contour (around 8,000 km^2) and, hence, are at risk from tidal flooding, whilst others live on or close to eroding cliffs (Figure 1).

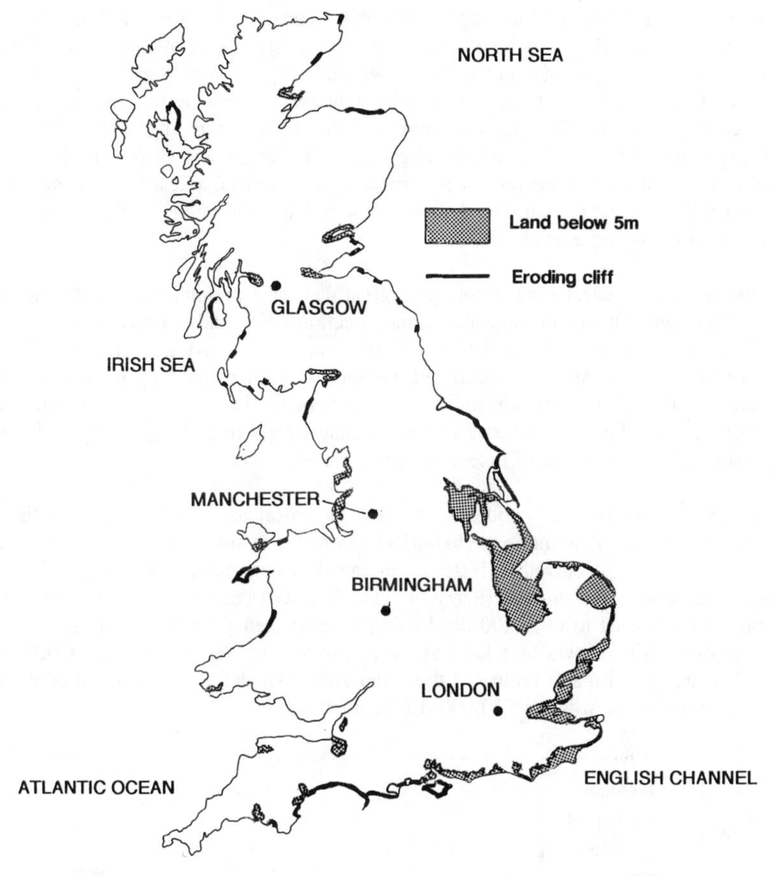

Figure 1 Land subject to coastal erosion and to flooding

Figure 2 presents the frequency of reported coastal erosion and flood events for each decade since 1700. The pattern reveals an almost exponential increase in frequency of events up to the 1950s after which the number of events has remained fairly constant (Rendel Geotechnics, 1995a). The factors influencing this trend are likely to be highly complex, reflecting, amongst other things, the spread of development into vulnerable locations. The downturn in incidence in the 1960s was undoubtedly due to the substantial investment in the reconstruction and renovation of defences following the extensive damage caused by the tidal surge in the North Sea in 1953. This coincided with the growth of seaside towns, especially during the Victorian era and the expansion of ports, heavy industry and energy generating facilities in estuaries. In Britain today, there is a growing tendency for people to move away from the major conurbations to live in the country; the fastest growing populations are in Cambridgeshire, Dorset and the Isle of Wight, with the largest increases near the coast. As a result, more and more development is being sought in coastal areas, both for new housing and to meet the varying recreational demands of a mobile population, resulting in an increase in the risk of damaging events.

Development on the coast has, of course, a great deal of economic gain. Settlements were essential in order to maintain trade and communication links, and to derive a living from the sea. Coastal land is often flat and highly fertile and, hence, easy to cultivate. The varied nature of the coastline provides significant benefits of shelter and deep water for ports and harbours, locational benefits for industry and opportunities for recreation and tourism. However, our occupation of the coastal zone has had a price in building substantial seawalls, breakwaters and other forms of defence to protect these assets.

Flooding is a more serious and widespread risk than coastal erosion. Examples of distressing and costly events can be found in the historical record for many parts of Britain. Perhaps the most severe flooding occurred in 1606 in Severnside when about 2,000 people drowned as the sea defence were overtopped (Perry, 1981). The east coast floods of January 31st and February 1st 1953 led to over 300 deaths and an estimated £900M of damages (at current prices) as over 800 km² was inundated as a result of a storm surge, (Grieve 1959; Summers 1978) It is estimated that a failure of the Thames Barrier defence in London could lead to property losses in the order of £10,000M (Clement, 1995).

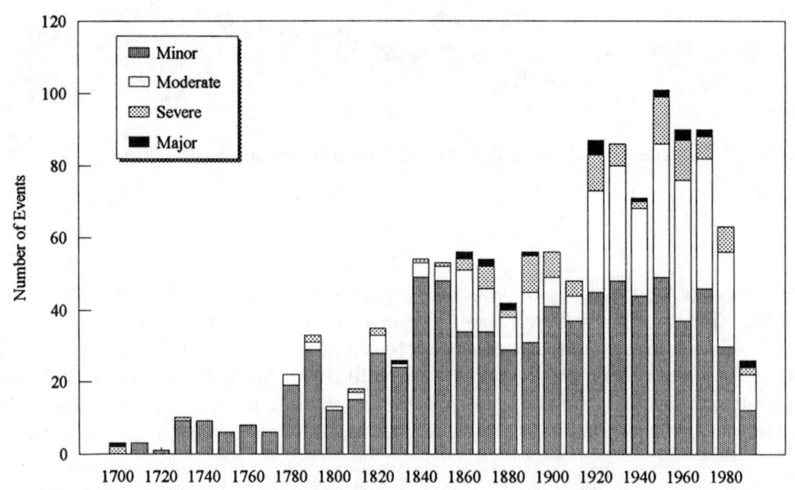

Figure 2 **The frequency of coastal erosion and flooding events of different magnitudes, from 1700 to 1993 (Rendel Geotechnics 1995a)**

Some stretches of coastline are prone to rapid cliff erosion (Figure 1). Along the Holderness coastline, for example, over 8000ha of land has been lost over the last 1,000 years (average annual recession rate around 1.8m) including at least 26 villages listed in the Doomsday survey of 1086. Major coastal landslides are a feature on many coastlines and affect many communities including Herne Bay, Sandgate, Barton-on-sea, Bournemouth, Swanage, Charmouth, Lyme Regis and Torbay, (Jones and Lee, 1994). At Scarborough, a sudden cliff failure in 1993 resulted in an overnight loss of 60m of cliff and the destruction of the Holbeck Hall Hotel (Clark and Guest 1994). The most extensive coastal landslide problem is at Ventnor on the Isle of Wight, where the town of 6000 people has been built on an ancient landslide complex (Lee and Moore, 1991). Contemporary ground movements within the town have been slight; however, because movement occurs in an urban area the cumulative damage to roads, buildings and services has been substantial.

2. BACKGROUND

The earliest coastal engineering works in Britain probably dated from the time of the Roman occupation. On the Medway, for example, embankments built as sea defence works by the Romans remained until the 18th century. By the Middle Ages, the Church had become instrumental in reclaiming and protecting many coastal marshes. The monks of Furness reclaimed the Walney marshes, Cockersands Abbey reclaimed parts of the Fylde and the Bishop of Durham instigated extensive drainage and flood defence works along the northern shores of the Humber estuary. Following the dissolution of the monasteries in the 1530s, the impetus for sea defence works shifted to the courts of sewers (a "sewer" was a drainage ditch) or individual landlords. Over the next 350 years or so, extensive drainage and flood embankment works were undertaken as part of ambitious land reclamation schemes in wetland areas such as the Fens, the Somerset Levels and Romney Marsh (see Godsin, 1978; Williams, 1970: Brooks, 1981).

Much early coastal engineering was also undertaken for ports and fishing harbours, often involving the construction of breakwaters, docks or flood embankments. The scale of the endeavour, especially between the late 18th century and the end of the 19th century was considerable; the following examples can only serve to illustrate this unprecedented period of maritime development. Brixham harbour, Devon was established in 1799, but the fishing fleet quickly outgrew the capacity of the port. Following an Act of Parliament in 1837, over 400m of breakwaters were constructed from local limestone. In the 1830s James Meadow Rendel designed the first port on an open beach at Par in Cornwall to serve the china clay industry, using a 150m pier containing an outer basin linked to an inner floating basin and protected by a 400m breakwater (Lane, 1989). The naval dockyards at Portland, Dorset were developed between 1850 and 1872 as a coaling and watering place between Plymouth and Portsmouth; it involves a breakwater built of Portland Stone with the workforce provided by 900 inmates from Portland Jail.

During the latter half of the 19th century and early 20th century, many local authorities constructed seawalls (often as a way of relieving unemployment) which combined the functions of coastal defence and promenades or drives for the expanding tourist industry. In Scarborough, North Yorkshire, seawalls were constructed in the 1880s and 1890s to provide protection against flooding and, along with landscaping and drainage works to stabilise the eroding coastal cliffs. These walls have provided an invaluable asset to the development of the town as a popular seaside tourism centre. At Bognor Regis, Sussex, a 1 mile long concrete seawall and promenade was constructed in the 1890s. The Undercliff Drive promenade and seawall in Bournemouth was opened by 1911, with 29 pre-cast concrete groynes installed to attempt to maintain beach levels between 1987 and 1939 (Lelliott, 1989).

By the end of World War II, (WW2), Britain's coastal defence were in a fairly neglected state. However, the severe winter storms in 1945 and the east coast floods of 1953 led to a major programme of defence improvements and reconstruction. Some 1200 beaches had to be repaired and hundreds of kilometres of defence made good, coupled with the raising of the defence heights to withstand a recurrence of the flood level.

A Storm Tide Warning Service (STWS) was established after the 1953 flood to provide warnings of high surge tides on the east coast. It now also covers the south and west coasts.

The distribution of surges is heavily biased towards the winter months when Equinox Spring Tides occur, with little surge activity experienced in the summer months. The STWS, therefore operates on a seasonal basis with 24 hour a day manning from 1st September to the end of April, and with a general overview of the situation during the summer. The STWS is operated by the Meteorological Office on behalf of Ministry of Agriculture Fisheries and Food (MAFF), who are responsible for ensuring that adequate warning procedures exist.

Following recent national surveys of the state of the defence (NRA, 1992; MAFF,1994), there are currently some 1,000km of sea defence and 860km of coast protection in England, representing around 50% of the 3763km coastline of England; of these defences it is estimated that around 13% have elements with a residual life of 5 years or less. There is, however, a firm government commitment to carry out improvements to defences to provide effective protection for coastal communities and property (MAFF, 1993a; NRA, 1993).

Considerable impetus for an ongoing programme of coastal engineering has come from the need to improve coastal water quality and improve derelict and despoiled areas in industrial estuaries. The European Community (EC) Bathing Waters Directive (76/160/EEC) has the dual objective of raising and maintaining bathing water quality off tourist beaches and protecting public health. The major sources of pollution of coastal water is sewage effluent which enters the sea through a variety of outfalls. It has been estimated that 17% of Britain's sewage is disposed of at sea (DoE, 1990), the vast majority of which is untreated; much has been discharged close inshore leading to the fouling of recreational beaches.

In recognition of the unsatisfactory state of many beaches, the Government, in 1989, announced a £1.4 billion capital investment programme to bring all identified bathing waters into compliance with the directive (DoE, 1989). In 1990 the Government announced that all significant discharges of sewage of coastal and estuarine waters should receive treatment leading to the investment programme being increased by £1.5 billion. It is estimated that the EC Urban Waste Water Treatment Directive (91/271/EEC) which requires member states to improve sewage treatment facilities to particular standards would add a further £2.2 billion to the cost of the necessary improvements (DoE, 1991). A fair proportion of this investment will involve coastal engineering works, with schemes such as the Lyme Regis Environmental Improvements taking the opportunity to tackle pollution control and coastal defence problems in joint schemes.

In the 1980s the Government announced an urban regeneration initiative, creating Urban Development Corporations to encourage investment in inner cities and former industrial areas. As part of their regeneration programmes both Teeside and Cardiff have commissioned estuarine tidal barrages to provide a wide range of planning and environmental benefits to river front land. The Tees Barrage (see Section 6.6) was completed in 1994 at a cost of £54M and has transformed 22km ot tidal river upstream into an attractive freshwater corridor (Norgrove, 1995).

The coastline has, however, an inheritance of what were (in modern terms) poorly planned and designed coastal defence structures, many built over a hundred years ago. Seaside promenades and vertical seawalls built to protect fashionable Victorian resorts, such as at Llandudno and New Brighton on the Wirral have promoted beach erosion in front of the seawall. Many such walls had no or inadequate sheet piling at the toe and have had to be extended seaward to prevent undermining by erosion of the beach. Many artificial harbours such as the Cobb at Lyme Regis have had a significant impact in disrupting natural sediment transport around the coast, leading to problems of accelerated erosion and beach depletion. Groyne fields have had a similar effect in places; apparent increases in cliff erosion at Fairlight, East Sussex during the 1980s are believed to have been related to a diminished supply of shingle arriving at the base of the cliff, partly as a result of the groynes at nearby Hastings (Penning-Rowsell et al 1992). Coast protection works to prevent cliff recession have prevented the supply of sediment and led to the depletion of the beaches that front many defences, increasing the risk of undermining and failure, a situation that has occurred on parts of the East Anglian coast (Wakelin 1989; Clayton 1990) and at Bournemouth (Leliott, 1989). Although the construction of seawalls and other cliff foot structures has generally reduced recession rates, they have not always prevented coastal landslides. The Holbeck Hall landslide, Scarborough, of June 1993 occurred on a protected cliff and resulted in the loss of 60m of cliff top land in a single failure (Clark and Guest 1994); the lesson of this event and other less dramatic failures has been that protection against marine erosion has to take account of the stabilisation of the slopes behind the defence.

In estuaries, the progressive constriction of the channel through the construction of embankments and channel improvements can lead to a reduction in the storage capacity of the estuarine lowlands and marshes. (Horner 1978), for example, has demonstrated that high tides on the Thames at London Bridge have risen dramatically over the last 150 years (Figure 3), this can only be partly attributed to rising sea levels and settlement in the London area caused by groundwater abstraction. A similar trend has been identified on the Humber estuary (Humberside County Council, 1994). These changes can act to pass the peak of the flood tide upstream, increasing the potential flood problems above the reaches that have been defended. Intertidal areas have been reduced by the construction of sea defence "coastal squeeze" resulting in a loss of habitat and the valuable contribution that saltmarshes and mudflats can make to flood protection (Davidson et al 1991).

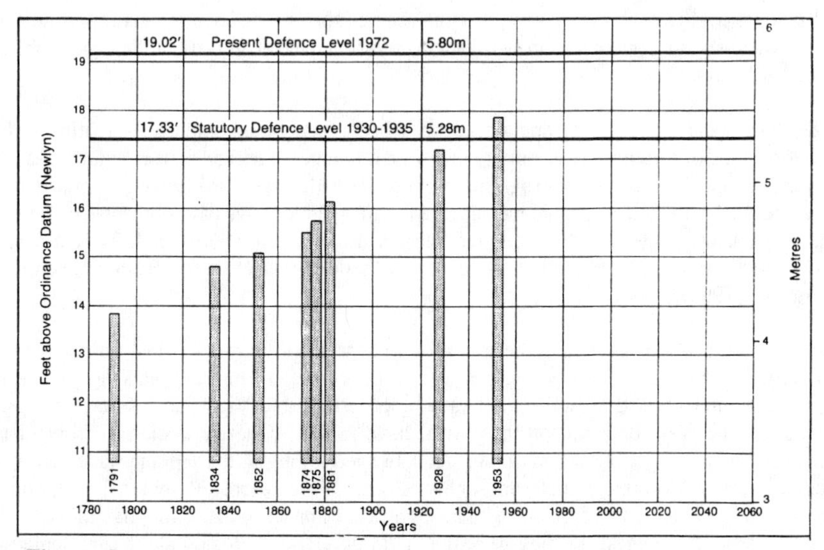

Figure 3 **Increasing tidal surge levels in the Thames Estuary (Horner 1979)**

In the light of these problems, attention has increasingly been turned towards adapting and supplementing natural processes, with the aim of creating more environmentally acceptable and sustainable coastal defences. A developing understanding of physical processes, the availability of computer-based modelling and changes in attitude towards the environment have all reinforced this trend towards "soft engineering" and a more strategic view of the planning of defences.

3. THE NATURE OF THE COASTLINE

3.1 INTRODUCTION

The British coast is noted for its diversity of shoreline conditions and landforms, reflecting considerable variability in rock type, topography, supply of sediments, wave energy and tidal range around these islands. The most erodible bedrocks tend to be the sedimentary rocks of Lowland Britain although the trend of a general increase in erodibility towards the south and east is complicated by the presence of weak superficial deposits. The nature and composition of these deposits is extremely varied, including spreads of "boulder clay" (glacial till) or "head" (periglacial solifluction deposits) mantling cliff tops or forming the whole cliff profile, marine or river alluvium fringing estuaries and coastal spits of sand and gravel. The superficial deposits tend to be highly erodible, with the highest rates of cliff recession occurring on the boulder clay cliffs of Holderness and East Anglia.

The potential for erosion and sediment transport is provided by wind-generated waves and tides. Figure 4 indicates the 50 year recurrence interval maximum wave height and highlights the exposed nature of much of the coastline, especially the west coast. Tidal range (Figure

5) exerts an important control on coastal evolution particularly where the shoreline material is readily eroded. Where there is a small tidal range, e.g. of about 2m, as on the Norfolk coast, the wave action is concentrated over a narrow range. This makes the waves more effective in shaping the shoreline and features such as sandy beaches and spits dominate. In areas with larger tidal ranges, as in Lincolnshire and the Bristol channel, tide-related landforms dominate with features such as saltmarshes and mudflats.

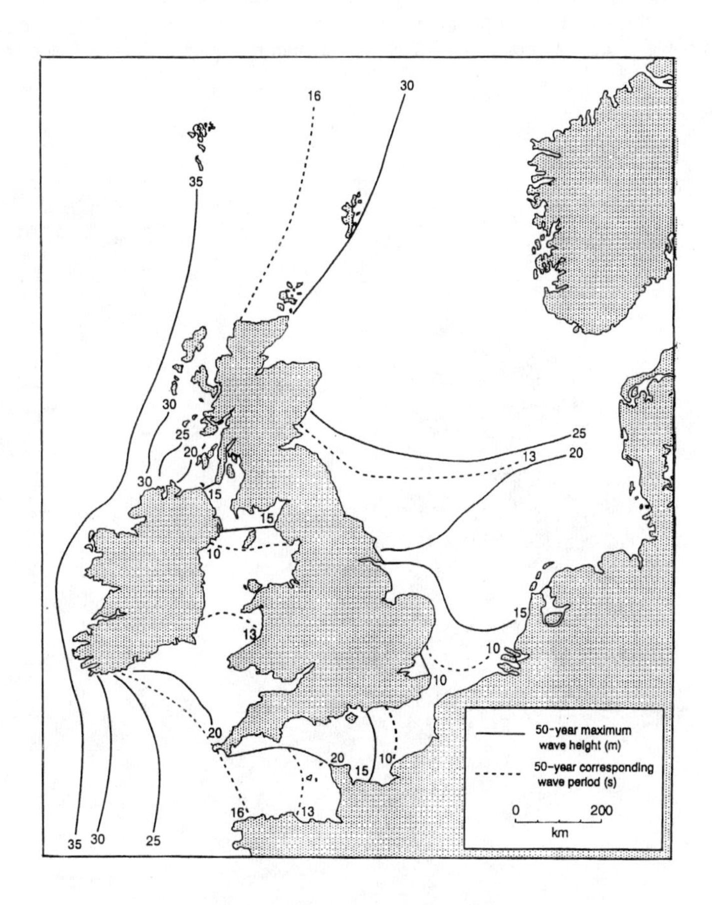

Figure 4 **Wave characteristics around the British Isles (NERC 1991)**

The tidal range can be severely modified by surges when the actual water height can be considerably in excess of the expected level. Such surges are, in part, the tidal response to changes in barometric pressure (water level will rise by 0.01m for every millibar fall in pressure), In shallow coastal water the shoaling transformation of these waves can cause a surge with a height of 2m of more. The North Sea is particularly susceptible to storm surges from the North Atlantic, or generated by storms crossing the sea itself, and there has been a long history of damaging surge events, most notably 1825, 1894, 1897, 1906, 1916, 1921, 1928, 1936, 1942, 1943, 1949, 1953, 1969, 1976 and 1978. All appear to have been associated with strong north westerly winds accompanying the passage of deep depressions moving across the sea to the Norwegian coast, (Lamb, 1991). The 1953 surge was the most memorable event and, as described later, had a major impact on coastal engineering in Britain.

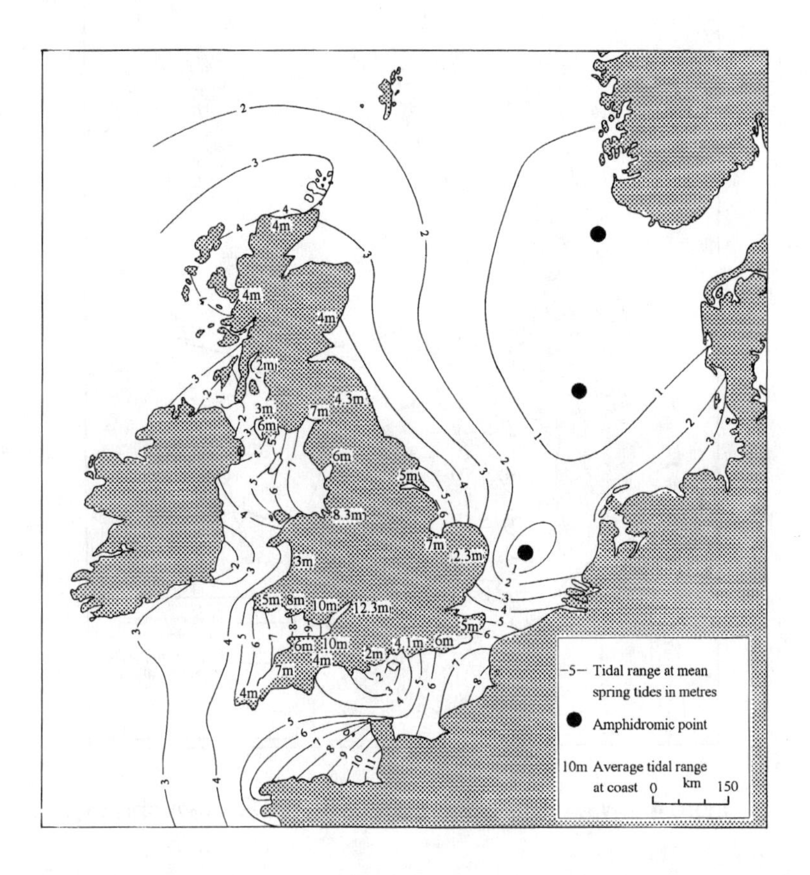

Figure 5 Tidal range at mean spring tides (NERC 1991,Pethick 1984)

3.2 SEDIMENT TRANSPORT SYSTEMS

Waves and tides generate currents which can transport sediments from source areas (eroding cliffs and shore platforms, rivers and the sea bed) to temporary sediment stores or permanent sediment sinks, in a series of linked sediment transport systems (littoral cells), Along a particular stretch of coast there may be a series of inter-linked systems, often operating at different scales. Suspended fine sediments may be carried many thousands of miles indeed, some of the cohesive material supplied from erosion of the Holderness cliffs is believed to be carried across the North Sea to continental Europe (Pethick 1992). Deposition of suspended sediments generally occurs in calm waters of sheltered bays and in estuaries where mudflats and saltmarshes develop. Granular materials are generally transported on the sea bed, either in a longshore or cross-shore direction depending on the pattern of currents and dominant wave action. Coarse sediments are not carried as far as some suspended sediments and, occasionally, may be confined to a single bay. Although headlands can be identified which appear to mark the limits of coarse sediment transport they are often not permanent boundaries as material may be moved around these sediment divides in severe storm conditions. Despite these difficulties, a series of littoral cells have been identified around the British Coast (Figure 6 ; HR Wallingford 1993, 1995); these represent a practical sub-division of the coastline into broad management units.

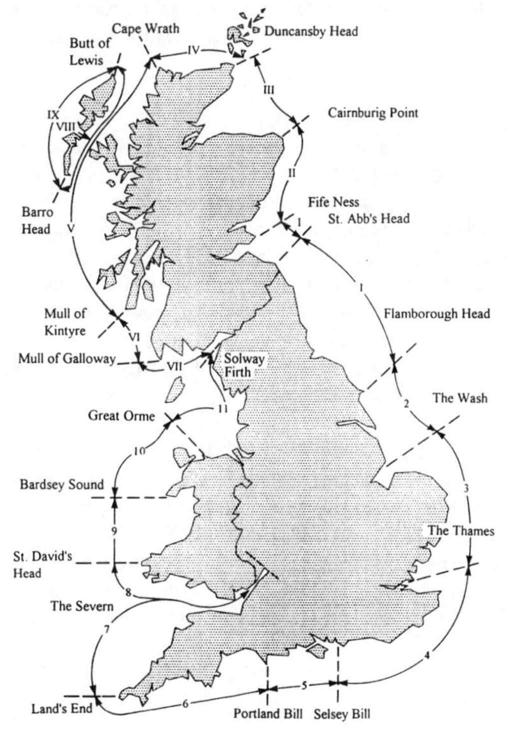

Figure 6 **Littoral cells around the coast of Great Britain (HR Wallingford 1993a)**

The dynamic nature of many coastal landforms and their dependence on continued sediment supply is viewed as a critical factor in effective coastal management, both from a coastal defence and conservation perspective (eg MAFF 1993a; Bray and Hooke, 1995; Rendel Geotechnics 1995b). Changes in one part of a littoral cell can lead to adjustments in other parts, often causing an increase in flood risk or accelerated erosion. This has important implications for any development or works on the shoreline which interferes with the movement of sediment around the coast (eg harbour breakwaters, groynes etc).

3.3 NATURAL COASTAL DEFENCES

Many coastal landforms offer a degree of protection against coastal flooding. Sand dunes serve as a natural barrier against high water levels and form effective coastal defences for may communities around the coast. Beaches and shingle ridges absorb as much as 90% of the wave energy arriving at the coast by continuously adjusting their form (Brampton, 1992a), providing an important component of sea defences either alone or where they front embankments and sea walls. Saltmarshes and mudflats are also effective in dissipating wave energy. (Brampton 1992b). An example of the efficiency of saltmarshes in floodwave attenuation can be found on the Gwent Levels, where there are two types of defence. Where defences are fronted by healthy saltmarsh they comprise an earth embankment with crest heights around 8.8-9.5mAOD; where there is no marsh, the defence levels are at 10.5mAOD and involve a near vertical wall capped by a wave return wall and fronted by an 8m high rock slope (Green, 1984).

3.4 RELATIVE SEA LEVEL RISE

There has been much debate in recent years of the global warming/rising sea level issue and the possible effects on the coast. The Intergovernmental Panel on Climate Changes has identified the extent to which rising sea levels are in evidence around the world, and the extent to which further rises may be expected (Houghton et al 1990). The Second World Climate Conference (Jager and Ferguson, 1991) reached similar conclusions, which in the case of the British Isles suggest that there could be a rise of between 50 and 79cm over the next 100 years. However, it is clear that this rise would not be the same in all parts of Britain since long-term vertical land movements are still taking place in some areas, and the more northerly parts (e.g. the coastal margins of the Highlands of Scotland) may continue to see a relative drop in sea level (Figure 7).

As yet, British tidal gauge records show no clear evidence of an acceleration in the rate of relative sea level rise (Woodworth, 1990; Woodworth et al, 1991). However, even if there is no acceleration, mean sea level is predicted to rise by as much as 100mm over the next 20 years on parts of the south coast (Bray et al, 1992). Allowances given by MAFF for the design or adaption of coastal defences with an effective life beyond 2030 range from 6mm per year (south east and southern England) to 4mm per year (north west and north east England) and 5mm per year (the remainder of England and Wales).

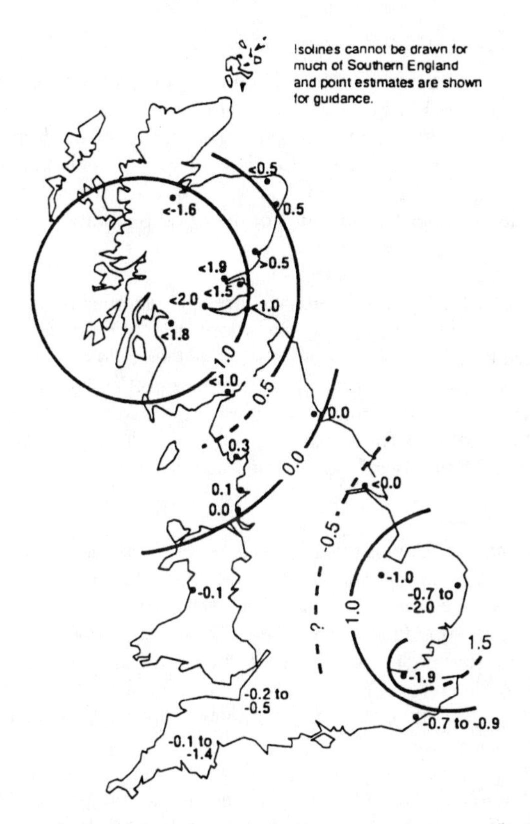

Figure 7 Estimated current rates of crustal movement (mm/yr) in Great Britain (Shennon 1989)

4. LEGISLATIVE AND ADMINISTRATIVE FRAMEWORK

The primary responsibility for dealing with coastal flooding and erosion problems rests with the landowner, under common law, rather than the state. However, the scale of many problems and the need to carry out works for the common good has led to the development of statutory law, vesting powers in specific local authorities and bodies or state agencies. These powers are permissive; the authorities are not required to undertake works and are only expected to promote schemes which are of benefit to the community. Under statutory law, a clear distinction has been made between sea defence (ie. protection against tidal flooding) and coast protection (i.e. protection against coastal erosion). Legislation relevant to sea defence has always been linked to the drainage of agricultural land and land reclamation, and has its origins in the "commissions of sewers" which were first established in the 13th century (the first was set up in Lincolnshire in 1288). By contrast coast protection legislation has a much shorter history, dating from a 1906 Royal Commission on Coastal Erosion and Afforestation and the 1939 Coast Protection Act.

4.1 SEA DEFENCE

The early Commissions of Sewers were required to report on flooding problems throughout England and Wales. In 1390, for example, a commission was appointed to inspect and repair flood banks and dykes on the Thames Marshes between Greenwich and Woolwich. The powers of these bodies were set out in a succession of legislation, culminating in Henry VIII's Statute of Sewers in 1552. These commissions (or "courts of sewers") were to survive until the 1930s; the levying of rates by the Lords of the Level of Romney Marsh carried on until 1932, after 7 centuries of enforcement.

The Land Drainage Act 1930 reformed land drainage administration, following a Royal Commission review (the Bledisoloe Commission) which reported in 1927. The Act specifically included defence against tidal flooding and set up 49 catchment boards to tackle the more urgent problems arising from the previous neglect of both river and sea defences. Relatively minor changes were introduced in the River Boards Act 1948 which reorganised the Catchment Boards into 34 River Boards. However, the Boards were only allowed to undertake works on "main rivers", as designated by the Minister (in general, the larger watercourses).

As a direct consequence of the east coast floods of 1953 and at the recommendation of an interdepartmental committee on coastal flooding (the Waverley Committee), the Land Drainage Act 1961 extended the River Boards powers to construct sea defences wherever the needs arose. The Waverley Committee also recommended that flood warning systems should be set up. The Water Resource Act 1963 reduced the number of River Boards to 27 and re-named them River Authorities. Ten Regional Water Authorities were established under the Water Act 1973 which introduced the concept of integrated river basin management, including flood and sea defence.

Established under the Water Act 1989, the National Rivers Authority (NRA; as of April 1995 part of the Environment Agency) has a statutory obligation to exercise a general supervision over all matters relating to flood and tidal defence in England and Wales, in accordance with the Water Resources Act 1991. The authority can promote its own schemes and regulate defence works by other bodies. These responsibilities are undertaken through 10 Regional Flood Defence Committees. Under the 1991 Act the NRA has the power to provide and operate flood warnings, and is required to carry out surveys to identify flood risk areas.

Under the Land Drainage Act 1991, Internal Drainage Boards (responsible for administering areas with special drainage needs such as the Fens) and local authorities have powers to protect land against flooding.

Thus two, or sometimes three authorities have powers to carry out sea defence works along a stretch of coastline. The division of responsibility between the NRA and local authorities has traditionally been resolved at a local level, with the NRA generally accepting responsibility for defences which protect large areas of urban or agricultural land where failure could lead to disastrous flooding. The local authorities are generally responsible for defence of coastal resorts, especially where the defences have an amenity function.

A separate legal system operates in Scotland, where schemes to protect non-agricultural land are promoted by the Island and Regional Councils under the provisions of the Flood Prevention (Scotland) Act 1961. In many instances sea defence works are undertaken through other legislation. For example, the Roads (Scotland) Act 1984 makes provision for the protection of roads from natural hazards. Flood defence of agricultural land is the responsibility of the owner.

4.2 COAST PROTECTION

Local authorities had no specific statutory powers to construct works prior to 1949, although many provided defences under general powers or by means of Private Acts. A Royal Commission appointed in 1906 reviewed the extent of coastal erosion problems and recommended, amongst other things, that local authoritie's works should be brought under central supervision and that there should be controls on the removal of beach materials. (Royal Commission on Coastal Erosion and Afforestation, 1911). The Coast Protection Act 1949 included provisions for the latter and the Board of Trade was empowered to make orders restricting the removal of such materials wherever there was thought to be danger of erosion if the beach was reduced. This legislation extended earlier powers dating back to the Harbours Acts of 1814 which empowered the Admiralty to prohibit the removal of shingle and ballast from the shores or banks of any port, harbour or haven.

Following severe winter storms in 1945 investigation were undertaken to assess the extent to which existing defences had deteriorated in World War II. The survey revealed serious and urgent problems and identified the need for financial assistance and coordinated planning of future works (Ministry of Health, 1949). The resulting legislation, the Coast Protection Act 1949, gives local authorities in England, Wales and Scotland powers to carry out works to protect land from erosion or encroachment by the sea, but it does not impose a duty to protect all land from erosion.

4.3 STRATEGY FOR COASTAL DEFENCE

Since 1985 a single government department, the Ministry of Agriculture, Fisheries and Food (MAFF), - the Department of the Environment had previously administered the coast protection legislation - is responsible for policy and providing financial support to the various operating authorities in England. In Scotland the Secretary of State for Works has an equivalent responsibility. A comprehensive policy framework was published in 1993 (MAFF, 1993a), the aims and objectives of which are:

- to encourage the provision of adequate and cost effective flood warning systems;
- to encourage the provision of adequate, technically, environmentally and economically sound and sustainable flood and coastal defence measures;
- To discourage inappropriate development in areas at risk from flooding or coastal erosion.

Emphasis is placed on the safeguarding of life and minimising stress and disruption to communities. As a result the priorities for financial support through grant-aid are, in descending order:

- avoidance warning systems;
- urban coastal defence;
- rural coastal defence.

Grant-aid for capital works for 1993/94 was around £70M which supported around 250 new schemes. However, it is recognised that to attempt to protect every inch of coastline would not only be uneconomic, but could lead to significant degradation of the natural environment or result in the disruption of coastal process to the detriment of defences elsewhere. Defence measures must, therefore, contribute to wider social, economic and environmental objectives. To qualify for grant-aid schemes must meet technical, environmental and economic criteria, as set out in various guidance notes (MAFF, 1991b,1993b) and described below:

(i) technical soundness; in addition to being technically sound, schemes should be sustainable insofar as they take account of the interrelationships with other defences, developments or processes operating around the coast.

(ii) environmental acceptability; environmental issues place considerable constrains on the choice of coastal defence strategy. Large stretches of the open coast and estuaries are of national or international importance and are protected by a wide variety of statutory or non-statutory conservation designations (Rendel Geotechnics, 1993).

Environmental assessments are required to be carried out and submitted in support of any application for grant-aid, involving extensive and early consultation with both statutory and voluntary environmental bodies. The potential impact on habitats and the operation of coastal processes are key considerations. Indeed, there is a presumption that natural coastal processes should not be disrupted except where life or important man-made or natural assets are at risk (MAFF 1993a). Grant aid is only provided for schemes which are judged environmentally acceptable.

(iii) Economic viability and cost-effectiveness; as financial support for coastal defence involves significant sums of public expenditure, schemes should be cost-effective and have a benefit to cost ratio of at least unity. To this end, MAFF have commissioned a series of manuals which deal with the quantification of relevant costs and benefits, including so called "intangibles" (Middlesex University Flood Hazard Research Centre, 1977, 1985 and 1992). Indicative target standards of protection have been published which provide a guide to the appropriate levels of sea defence for different categories of land use (MAFF 1993b) and table below:

CURRENT LAND USE	INDICATIVE STANDARDS OF PROTECTION (RETURN PERIOD IN YEARS)	
	TIDAL	NON-TIDAL
High density urban containing significant amount of both residential and non-residential property	200	100
Medium density urban. Lower density than above, may also include some agricultural land	150	75
Low density or rural communities with limited number of properties at risk. Highly productive agricultural land	50	25
Generally arable farming with isolated properties. Medium productivity agricultural land	20	10
Predominately extensive grass with very few properties at risk. Low productivity agricultural land	5	1

Environmental considerations are dealt with in detail in Section 5.6 hereinafter.

5. DEVELOPMENT OF THEORY AND PRACTICE

5.1 RESEARCH ORGANISATION

Until the 19th century, theory and model testing in maritime matters was largely confined to ship design and interaction with the sea.

Much of the work on the theory and practice of coastal engineering, started last century, was carried out in universities. Then engineers working on port schemes became involved and the use of models was steadily developed, increasing in scope and size.

The development of governmental laboratories followed. These initially worked on government funded basic research. In due course they started to carry out studies for specific projects, generally on a 'cost plus' basis.

Private and public sector laboratories developed after WW2 for studies on specific topics and projects. Universities expanded their ability to undertake external work. That post-war period was a time of considerable activity both in the UK and overseas. There was ample work for the majority of organisations.

In recent years the work load and the activities of non-governmental laboratories have reduced. In addition, the government has rationalised, and has sought to 'privatise', the national laboratories or at least make them operate on a more commercial basis. Some national laboratories, are now linked to universities, and much specific research is funded from external sources. In the context of this paper the following brief account of some national organisations illustrates the current trend.

The Proudman Oceanographic Laboratory (POL) is the oldest marine laboratory in the UK. It started life as the Liverpool Observatory in 1845, transferring to Bidston in 1866. The Liverpool Tidal Institute started in 1919 relocated to Bidston in 1924, creating the Liverpool Observatory and Tidal Institute. In 1969 it became part of the National Environmental Research Council (NERC), and was renamed the Institute of Oceanographic Sciences (IOS) in 1973. It became the POL in 1987. It reports via the Centre for Coastal and Marine Sciences (CCMS) through which it is linked to other NERC and CCMS institutes. All research work is organised in time-limited Strategic Research Projects and Technology projects.

The future of POL and other NERC institutes is currently under review.

The National Physical Laboratory set up in 1909 was the main UK laboratory for the study of ship design and performance. Its Ship and Marine Science Divisions became the National Maritime Institute (NMI) in 1976. A change to NMI Ltd followed in 1992, to sharpen the commercial aspect. A subsidiary company for coastal engineering work CEEMAID (Coast and Estuary Engineering Management for the Acquisition and Interpretation of Data) was set up. In 1985 a further change to British Maritime Technology Ltd took the company further into the private sector. Some coastal engineering work continued, but BMT concentrates on its main ship related activities. It is an international source of wave statistics (Dacunha et al, 1984).

After WW2, UK engineers pressed strongly for the creation of a national laboratory for the study of loose boundary hydraulics. At that time the NPL was the only national laboratory carrying out tidal model tests - a model of the Forth Estuary had been built there. The Hydraulics Research Station (HRS) was set up in 1947, Sir Clause Inglis, former Director of the Central Water and Power Research Station at Poona, being the first Director. It moved to Wallingford in 1952. Following a study on possible privatisation in 1979, a company was formed in 1982. In 1984 privatisation was completed and HR Wallingford Ltd (HRL came into being. HRL (as is BMT) is limited by guarantee. In 1981 government funded strategic research was 80% of turnover: in 1994 it was only 30%. Collaboration with universities in the UK and overseas is growing.

5.2 SEA DEFENCES

About 1600km of the coastline of the UK is protected with some form of artificial defence. Each scheme is different and in the confines of this paper it is not intended to review the types in detail. Typical designs are well reviewed in (CIRIA 1986). For many years sea wall cope levels were set on the basis of Mean High Water Spring tides plus between 9 and 12 feet depending on the degree of exposure to surges and wave attack and the risks associated with overtopping. The advent of computer aided design methods has enabled a wider range of coastal structures to be considered beyond simple walls. However, it is important to stress that many design practices still rely on empirical methods and engineering judgement.

The ‘hard’ defences were first built to protect coastal communities. Land was locally valuable and so the defences were vertical or near vertical walls. As the theory for the hydraulic design of seawalls was not well developed it was not until the last few decades that flatter front slopes were used. Such walls often fronted roads or promenades. Often severe problems resulted from scour in front of vertical sea walls and many beaches were lost or severely depleted as a result of early engineering works. Experience has shown that the flatter the seawall the less energy will be reflected, and the preferred slope in most locations in between 1 in 4 and 1 in 3. Further details of the use and design of seawalls can be found in CIRIA (1986) and Thomas and Hall (1992).

Timber construction used initially gave way to masonry and then concrete and steel sheet piling. Until such piling became available protecting the toe against undermining was difficult. Where beaches were wider, or the land was less valuable, these ‘hard’ walls could take the form of a wave or splash wall at the top of a sloping stepped construction. When beach levels were progressively lowered by erosion additional aprons, flat or stepped, were constructed: again with toe piling.

Some protection to the toe of a wall could be provided by masonry or concrete blocks, but the effect of these were limited and they could be removed by severe storms. Such blocks became more useful as machinery was improved so as to handle larger blocks and to be fully mobile in tidal areas.

However a beach in front of the wall was still the best protection, and groynes were a valuable adjunct to hold beach in place to absorb the wave energy and thereby reduce damage to the wall and undermining.

These systems were particularly effective where the protected frontage was contained between headlands so that the beach material could circulate within the system. They were also effective on more open coastlines where the littoral drift was weak or there was continuous supply of material provided by the drift. However, the trapping of the drift sediment usually resulted in erosion downdrift.

In recent years there have been considerable advances in the design and use of rock armoured beach control structures. The main types that have been used include; near-shore detached breakwaters, low crest or reef breakwaters, submerged breakwaters or sills, rock groynes and rubble revetments. Typical configurations of such structures have been described in the CIRIA/CUR manual on rock armour structures. (CIRIA/CUR, 1991).

Communities with trading or fishing activities built breakwaters to provide harbours. Figure 8. records that these endeavours were not always successful (Illustrated London News, c 1850). Training banks were also provided for navigable river estuaries. Whilst these provided protection to the frontage on the lee side, that benefit was often offset by the interception of the littoral drift leading to denudation of the lee side beach.

Away from built up areas, natural salt marshes or sand dunes would generally provide stability, although there were sections of coastline, such as Holderness on the East coast, where steady erosion has continued for centuries. Where the agricultural land behind such areas was valuable, or drift was intercepted by other defences, or development progressed, protection was initially provided by earth bunds and/or timber revetments, sometimes retaining rock fill. As the frontagers became able to afford better defences, the bunds were faced with stone, concrete blocks or bituminous material - solid or permeable. Toe piling was a further improvement.

Again groynes were able to play an essential part in retaining beach in front of the defences.

These 'less hard' defences are particularly vulnerable to being breached by storms and surges accompanying them.

As most locations needing protection already have defences of some type, the construction of completely new major works in the future will be infrequent. Renovation and reconstruction will be the main activity. Where the defences are 'hard' scope for significant new designs will be limited. However developments in construction plant will enable more robust works to be provided, and, as already mentioned, larger concrete blocks, such a Tetrapods or rock armour may be placed in front of existing structures.

Environmental and amenity issues have now become more important. 'Soft' defences that cause less disturbance to shoreline processes are increasingly being studied or used. Integration of activities along a section of coast that can be considered to be a more or less a self-contained cell of littoral processes is growing. This integration is highlighting the responsibility for the costs involved, between present and future generations and between frontagers - public and private - and the wider community that wishes to use the coast.

5.3 WAVE RESEARCH

The history of wave research divides itself rather neatly into two periods. In the first, up to 1944, waves were treated deterministically. After 1944, the randomness of wind-generated waves was taken into account. The first period will be treated only briefly, with the main emphasis on the period after World War Two.

There was a great deal of interest in waves in the nineteenth century, much of it in connection with naval architecture, and many of the classic names are still frequently on our lips. Sir G.B. Airy solved the linear equations for a regular periodic long-crested wave train, and hence this solution has become the Airy Wave: the reference usually given is to a long and comprehensive review article (Airy 1845), though it is not clear how much of this is new and how much was already known. Soon afterwards, Sir G.G. Stokes (1847) developed the method of expansion which enabled him to calculate the properties of a finite-amplitude (and hence non-linear) regular wave, and his method, with only minor variations, is still in use. With regard to ships, Lord Kelvin wrote a classic paper on the pattern of waves produced by a moving ship (Kelvin 1887), and Professor. T.H. Havelock wrote so many excellent papers that the US Office of Naval research has recently issued a volume of his collected

CAPT. TAYLORS BREAKWATER, OFF BRIGHTON.

Our artist at Brighton has sketched the annexed representation of the experimental portion of Captain Taylor's Breakwater, consisting of three sections, moored off Hove, (one mile from Brighton,) at the distance of a mile and a quarter from the shore, and nearly parallel to it about East and West. It was placed there in December last, under Capt. Taylor's superintendence, in her Majesty's tug, *Monkey*, the three sections being arranged thus ___ ‾‾‾ ___; and two men were stationed on it, to keep a light to warn vessels off. Although up to the time of our informant writing, (March 27) there had been several smart breezes since the Breakwater had been stationed, there had been nothing like the gales to which the Brighton coast is subject, to testify the utility of the invention.

The Wreck of Captain Taylor's Breakwater on the beach of Brighton

Brighton has no natural harbour and, until the construction of the Marina, no successful artificial one. Several schemes were moored during the nineteenth century, including one based on an invention by Captain J. N Taylor, R. N. , C. B. Captain Taylor's Floating Breakwater consisted of large timber sections moored in formation, their open-work construction designed to absorb rather than to resist the force of the waves. The experiment was not a success, the sections broke their moorings and the timber which fetched up on Brighton Beach was eventually sold for £148.

Figure 8 Captain Taylor's breakwater

works (Havelock 1963). Some of these are relevant to forces on offshore structures. R.E. Froude (1876) developed the theory and practice of model testing, and invented the Froude Number to allow correct scaling. Lord Rayleigh published many papers on water waves. For example, in Rayleigh (1877) he worked out the energy transport in waves. Finally in this brief survey of pre-war work, Sir Horace Lamb must be mentioned. His famous book "Hydrodynamics" is still on our shelves and much used. It was first published in 1879 with the sixth edition (remarkably, edited by himself) published 53 years later in 1932.

Modern wave research in the UK. started when George Deacon, FRS (later Sir George Deacon) set up the Wave Group at the Admiralty Research Laboratory (ARL), Teddington in 1944. It was realised right from the start that the development of the theory of ocean waves and the development of instruments to measure them must go hand-in-hand, and the group included several engineers. In 1949 this group was brought together with a group of whale biologists to form the independent National Institute of Oceanography (NIO), which had terms of reference covering the whole of oceanography. This later became a constituent laboratory of the Natural Environment Research Council (NERC) when this was set up in 1965: then in 1973 it became the headquarters and largest laboratory of the Institute of Oceanographic Sciences (IOS). A decade later control of its remaining satellite laboratory was taken from it and it was renamed the IOS Deacon Laboratory (IOSDL), finally losing its identity in 1995 when the remaining staff were moved to become part of the Southampton Oceanography Centre.

The key to the early work at the ARL was the realisation, that study of the wave spectrum held the key to advancing our understanding of waves.

Early wave analysers were optical-mechanical (Barber et al 1946), since digital computers of sufficient power for spectral analysis did not become available till the mid 1950s; and for routine analysis, not for another 10 to 15 years after that.

A spectacular early achievement of spectral analysis using this analyser was the identification of swell arriving on UK beaches which had been generated in the Southern Ocean 10 days or so previously (Barber and Ursell 1948). The time during which the storm was in a position to generate waves travelling in the correct direction to reach the UK was comparatively short and could be considered to be effectively instantaneous. Thus, the first waves to arrive were the lowest frequency ones, followed by progressively higher frequency waves. We are used to considering the modulation of the amplitude of waves by tidal effects, but Barber and Ursell (1948) found that the *period* of such swell was also modulated by the tidal streams over the continental shelf.

The requirement in 1944 to forecast wave conditions for the D-Day Landings on the Normandy beaches was the trigger which started modern wave research in the UK. However, after WW2, estimating wave climate for coastal engineering design soon became an important application, and still is. J. Darbyshire (1952) published the first study of wave generation based on spectral methods, and this became the basis of a forecasting and hindcasting system which remained in use by coastal engineers in particular (with improvements as better data became available) until computers were powerful enough to allow numerical gridded models to be used. M. Darbyshire and L. Draper (1963) gave graphs for wave forecasting and hindcasting which are still widely used.

The disastrous flooding in 1953 of the coastal areas of East Anglia and the Netherlands led to the development of a flood warning service, and to the development of surge prediction models. For many years now a tide and surge prediction model has been run in real time using data from the Meteorological Office weather prediction model. However, as well as tides and surges, waves contribute to the overtopping of sea walls, and an important function of the wave prediction model referred to above is as an input to the flood prediction service. Tides, surges and waves all interact, so a combined prediction model which takes account of these interactions has been developed by the Proudman Oceanographic Laboratory (Wu and Flather 1992).

The late 1950's and early 60's saw a major programme of research into ship motion. This was a cooperative programme with the British Ship Research Association, the Ship Division of the National Physical Laboratory (these organisations later joined to become the National Maritime Institute, or NMI), the Royal Navy and NIO taking part. Some results are given by Canham et al (1962). Although this programme failed to fulfil the initial high hopes in terms of improved ships, it did result in the development of improved directional wave measuring buoys (Cartwright and Smith, 1964).

For coastal and offshore engineers the determination of wave climate is extremely important. The first record of systematic wave measurements is that by Robert Stevenson (1818-1887) who came from the famous Scottish family of lighthouse and harbour engineers and who was the father of Robert Louis Stevenson (Townson 1976). He was building a small harbour of refuge at Lybster in Caithness, Scotland, and instructed his resident engineer to measure the waveheight every day by observing the movement of the water surface up and down a calibrated pole. From these measurements he produced the earliest known wave prediction formula, which applies to storm waves (Townson 1981) : $H\ (\text{ft}) = 1.5\sqrt{F}\ (\text{miles})$.

A remarkable series of wave measurements is reported by Hagger (1979). The Port of Tyne Authority measured waves once per day using a theodolite to observe the motion of buoys 100 yards off the end of the Tyne Piers. Measurements were made from 1908 to 1929 with a break during the 1914-18 war.

About 25 years ago the determination of wave climate for the design of offshore platforms for oil and gas production, and the prediction of wave conditions for offshore operations took over as the main practical applications of wave research and remain so, though still followed closely by Coast Protection.

The problem of predicting the properties of the "100 year wave", that is, the waveheight exceeded on average once in 100 years, and its associated period and direction, is a difficult one. Pioneering work was done by Draper (1963), who later produced charts of design wave height and period covering UK waters which were incorporated into the Department of Energy's "Guidance on the design and construction of offshore installations" (Department of Energy 1977, Third Edition 1984). These have stood the test of time surprisingly well.

In more recent times Carter and Challenor have made major contributions in this field, and perhaps their work is best illustrated by their background document written for a later edition of the Guidance Notes (Carter and Challenor 1990).

At the end of WW2, the only satisfactory wave sensors available were pressure recorders mounted on the sea bed. Inverted echo-sounders mounted on the sea-bed had been tried, but failed in storm conditions because of aeration of the water produced by breaking waves. The pressure sensors could only be used effectively in depths of less than about 12 m, and there was an urgent requirement to measure waves in the open sea. This problem was solved by the development of the Shipborne Wave Recorder (SBWR) (Tucker 1952 and 1956).

An important step forward in the UK. was the development of directional wave buoys already mentioned, in particular the pitch-roll-heave buoys in which a flat discus buoy follows the slope and displacement of the sea-surface. Analysis of the pitch, roll and heave acceleration of such a buoy allows the mean wave direction and two measures of the directional spread about this mean direction to be obtained (Longuet-Higgins et al 1963). This line of development was largely taken over by Datawell in the Netherlands, who produced the widely-used Waverider and Wavec buoys. However, early in the 1970's IOS obtained backing from the UK Dept of Trade and Industry to develop a robust buoy (designated DB1) for routine recording of metocean data in deeper waters up to perhaps 200 m (Rusby et al 1978). Following successful trials of this it was taken over by the UK Offshore Operators Association and successfully used to record metocean data (including directional wave spectra) in the Western Approaches to the English Channel. In the light of experience with this buoy, UKOOA commissioned the design of two improved buoys, DB2 and DB3 (Woollen 1981), and these have produced some invaluable data sets in various exposed areas of interest to the oil companies.

An interesting early attempt at remote sensing was made in 1946-48 by Deacon's Group at the Admiralty Research Laboratory. They used a narrow-beam radar altimeter mounted in a Lancaster bomber and flown at an altitude of 25-30 m over the Irish Sea, to obtain a footprint of about 2 m (Deacon et al 1949). Flying so low in winds up to force 7 was hazardous, but some indications of variations in waveheight and period with fetch were obtained. The combination of high cost and operational difficulties caused the programme to be curtailed.

Other land-based remote sensing methods have been tried with limited success, the most intensively studied being the use of the Doppler spectrum produced when HF radar is backscattered from the sea surface due to diffraction by the sea-waves. Out of many papers on the subject, a representative one is probably Shearman and Wyatt (1987)

In recent years, precision altimeters carried on satellites have been able to measure significant waveheight. However, the altimeter does not give satisfactory measurements close to shore, partly because of the comparatively large footprint, partly because when moving offshore it takes some time to settle down, and partly because the measurements are sparse in both space and time, so it will not solve the coastal engineer's need for nearshore measurements.

Neil Hogben (1988). and his colleagues used the huge mass of worldwide visual wave observations from ships to generate a Global Wave Climate. The basis of the study was that while individual observations are not reliable, the averages ought at least to be consistent. By using known relationships and calibrating these averages against instrumental measurements where these were available, it was possible to estimate the wave climate in most areas of the world's oceans. This climate included wave period and direction, which is not obtainable at present from the satellite data. The Global Wave Climate data have been widely used by offshore operators and others.

Breaking waves are the main enemy of hard coastal defence structures, and their study is therefore important. Brigadier Bagnold, famous for his work on sediment transport, seems to have been the first to study this by modern methods (Bagnold 1939). More recently, Howell Peregrine and his co-workers have studied the kinematics of breaking waves since the early 1980s by numerical simulation and have been able to reproduce the high transient pressures observed under certain conditions of impact (Cooker and Peregrine 1990 and 1992). Equally importantly, they have shown how the pressure gradients generated at the sea bed are capable of moving large stones or rocks away from the foot of the wall. On the more analytical front, although considerable progress has been made with the understanding of the breaking of regular waves (Longuet-Higgins 1980), little is known about the spectral composition of the energy loss due to wave breaking in a complex sea. This is important, since it is the only factor in the energy-balance equation for waves travelling into shoaling water towards a coast which is not really understood at all.

It is not possible in this brief history to do full justice to the work of Michael Longuet-Higgins. In Longuet-Higgins (1953) he showed theoretically that the interaction of waves with the bottom boundary layer produced a very significant forward drift in this, accounting for the field observation that swell builds up a beach. Together with Bob Stewart he worked out the theory and effects of radiation stress in waves: for example, in producing "set-up" on beaches (Longuet-Higgins and Stewart, 1964). In Longuet-Higgins (1970) he applied this theory to calculate the longshore currents generated by obliquely incident sea waves.

5.4 PHYSICAL MODELLING OF ESTUARIES

BACKGROUND - THE 18TH AND 19TH CENTURIES

Beginning in the mid 18th century, industry based on cast and wrought iron, fed by plentiful supplies of coal, put great demands on the infrastructure of the United Kingdom. Steam power led to rapid development of manufacturing industry and the opening of markets throughout the UK and then of the world. This was the era of canal construction for the movement of coal and other raw materials and manufactured goods, connecting sources of goods to outlets in the rapidly developing cities. Coastal traffic increased dramatically and, as the population grew exponentially, so did the demand for food - much of it imported, paid for by manufactured goods. Sea fishing also expanded greatly, with major fishing ports such as Grimsby, Lowestoft and Aberdeen on the east coast each becoming the base for hundreds of trawlers. There were similar developments round the whole coast of the United Kingdom. This expansion reached its height by the mid-19th century and continued well into the 20th century.

This was also the time of population movements from the over-crowded European cities to the developing world - the United States, Canada, Australia, New Zealand, South Africa. Ports rivalled each other to provide facilities for emigrants and had to fight to retain their positions as leading passenger terminals.

The British Empire had been established to ensure sources of raw materials and markets for manufactured goods. It had to be defended, and by the end of the 19th century, the British Navy was the most powerful in the world, just as the Merchant Navy had by far the greatest capacity for world trade. Shipbuilding was on a tremendous scale, with major yards on the Severn, Mersey, Clyde, Tyne, Tees with other important yards in Ireland and round the whole British coast.

Port development was an essential part of this ferment and it had a profound effect on coastal evolution. Most ports were on the banks of estuaries and access to them was inevitably through the channels at the sea-face. Many had to be protected from the elements - ports on the east coast were particularly vulnerable to wave action and alongshore drift of sediment, so breakwater construction was adopted at many of them. One example of many is the Tees estuary which had become the site of major industry based on steel. The estuary itself had been changed from a freely-meandering tidal river at the beginning of the 19th century to a comprehensively re-aligned and trained system by its end (Taylor 1864, Wheeler 1893), extensive use being made of blast-furnace slag in construction of training banks and breakwaters. Between 1855 and 1877, 20 miles of training walls were constructed using 1,500,000 tons of slag, at a cost of £50,000. Between 1862 and 1888 a curved southern breakwater 2 miles long was built into the sea at a cost of £308,853 of which £56,671 was received from Ironmasters for using their slag! A northern breakwater 6,000ft long was constructed later.

There were few constraints on these developments. Free enterprise ruled the day and there was little real concern for the environment. Parliamentary approval had to be obtained, nevertheless, and there grew up a coterie of professional engineers who built up expertise based on intelligent observation. Some were remarkably successful, but even the best of them found it hard to convince others that their ideas were sound. The data and tools necessary for comparison of different schemes of development did not exist until the end of the 19th century.

It would be wrong to get the impression that rapid development led to careless work. Most schemes were carefully thought out and constructed. However, it was not until late in the 19th century that hydraulic model testing of tidal waters was demonstrated and used for studies of the Mersey and Severn estuaries and of the Seine in France. Before that, reliance had to be placed on experience. There were few people with the vision to take in the complexity of tidal and river flows, sediment transport and wave action and to predict how estuaries would respond to works. There were many who would profess to do so....

The work of three able men is worthy of mention. The first of these is L.F. Vernon Harcourt, an engineering consultant who specialised in development of rivers and estuaries; the second is Osborne Reynolds, Professor of Engineering at the University of Manchester;

the third is W.H. Wheeler, whose book Tidal Rivers (Wheeler 1893) is an invaluable record of developments in the British Isles up to then.

L.F. Vernon Harcourt built up a reputation as a coastal and tidal engineer based on careful observation and analysis of tidal and wave behaviour. He was employed as consultant by many port authorities to advise on protection of facilities and improvement of navigation channels. He toured European coasts and ports, wrote papers on development of ports and estuaries on sandy coasts (Vernon-Harcourt 1882), advised on development of the Mersey and Ribble estuaries and on the Seine in France. He also visited Calcutta, India to advise on development of the approaches to the port up the tidal river Hooghly and wrote a masterly paper based on his report to the Commissioners for the Port of Calcutta (Vernon-Harcourt 1905).

All this was solely based on his experience, without the aid of hydraulic models; but he followed closely Osborne Reynolds' work on the use of tidal models and later applied them to the Seine and Mersey.

The limitation of relying on experience were well illustrated in the development of the Port of Preston at the beginning of the century. Proposal by Sir John Coode, reported and commented on by Wheeler (1893), were disagreed with by Vernon-Harcourt acting for Southport Corporation. A compromise layout was recommended by the Commissioners of the Board of Trade. The works constructed included training banks extending 22kms seaward of Preston. By 1965, despite extensive dredging the channel was effectively closed and adjacent beach areas and foreshore were significantly and detrimentally changed. The opinions of Vernon-Harcourt and Wheeler were vindicated. A Parliamentary Bill in 1991 empowered the Borough of Preston to close the port.

The experience at Preston and elsewhere showed that important decisions were having to be made concerning the environmental effects of proposed works, but the tools that would enable sensible comparisons to be made did not exist. That changed dramatically once physical modelling had been introduced by Osborne Reynolds at the end of the 19th century.

TIDAL HYDRAULIC MODELS

Osborne Reynolds was appointed to the first Chair of Engineering at Owens College, Manchester in 1868. His experiments on turbulence which led to the Reynolds number were reported to the British Association for the Advancement of Science (Reynolds 1883, 1884). He became renowned as a scientific observer of natural phenomena and produced explanations for many of them (McDowell and Jackson 1970). He had noticed that in the Mersey estuary, tidal currents reversed in some parts of the estuary before others. To study this, he had a model of the Mersey estuary built to a horizontal scale of 1:31,800 and a vertical scale of 1:960. Tides were generated by a hinged tray at the seaward end of the model. Tests were done with sand on the bed, initially laid flat. Reynolds found that only one tidal period in the model - of about 40 seconds - gave a correct imitation of the tidal phenomena that had been observed in the Mersey; *"...a result that might have been forseen from the theory of wave motions, since the scale of velocities varies as the square root of the wave heights, so that the velocities in the model which would correspond to the velocities in the channel would be as the square root of the vertical scales - about 1:33 - and the ratios of the periods would be in the ratio of horizontal scales divided by this ratio of velocities....* ". He then operated the hinged tray by means of a continuously-driven shaft and found that the sand, originally placed in the model to arrange the correct mean high-water level, was shifted by the tides so as to form the principal features of the natural estuary. He also noted " *And what is as important, the causes of these as well as all minor features could be distinctly seen in the model.* "

Vernon-Harcourt viewed these experiments with interest. He was studying the river Seine at the time, and built a very small-scale model of the tidal river - horizontal scale 1:40,000, vertical 1:400 - in which he tried various schemes of regulation that had been proposed (Vernon-Harcourt 1889). As a result of this, French engineers built a model with horizontal scale 1:3,000, which confirmed the preliminary findings from the Vernon-Harcourt model.

In 1885, Reynolds built a larger model of the inner Mersey to a horizontal scale of 1:10,560 and a vertical scale of 1:396. After running this model for 6,000 tides, a survey of the model showed a remarkable resemblance to the general features of charts of the actual Mersey for the years 1861, 1871 and 1881. He observed that the model survey showed as great a resemblance to any one of these charts as they did to each other. This model was used to assist with planning of the Manchester Ship Canal and was instrumental in locating it along the southern bank rather than through the centre of the estuary.

Vernon-Harcourt also built a model of the Mersey, to scales of 1: 30,000 and 1:500, including the outer estuary to its interface with the Irish Sea. It extended from near Warrington at its head, to the open sea beyond the bar (Vernon-Harcourt 1890). The bed was formed in Bagshot sand, chosen after testing many alternatives. He tried various schemes that had been proposed for training the inner Mersey and the channels in Liverpool Bay. His conclusions were that training the inner Mersey would be injurious because of serious accretion, whereas training the sea approaches, combined with dredging, would offer the best prospect of forming a direct stable, and deepened channel across the bar.

Reynolds presented his results to the British Association in an historic communication (Reynolds 1887). As a result of this, a committee was appointed by the British Association to investigate the Action of Waves and Currents on the Beds and Foreshores of Estuaries by means of Working Models. Membership of this committee included Professor O. Reynolds and A.H.Wheeler. A programme of tests was carried out in the Whitworth Engineering Laboratory of Owens College, Manchester in two specially-constructed flumes; one as big as the available space would allow, and another having half its linear dimensions. The first experiments were "*directed to determine in what respects, and to what extent, the distribution of sand in the beds of model estuaries of similar lateral configuration is affected by the horizontal and vertical dimensions, and the relation which these bear to one another and to the tide period so as to place the laws of similarity on which the practical applications of the method depend, on as firm an experimental base as possible.*"

The large tank was 11'-10" long and 3' 9" wide (3.62m by 1.16m). Tides were generated by a hinged tray rocked via a system of gears driven by an hydraulic motor. The bed was laid initially with fine sand. After operation of many tides, when the bed had reached a stable form, it was carefully surveyed.

Osborne Reynolds reported on the experiments in Reports to the British Association (Reynolds 1889, 90, 91). In these reports he noted that the two tanks had been continuously occupied in these investigations, when not stopped for surveying or arranging fresh experiments. They were thus run about five-sixths of the time day and night. In this way, in 1890 the large model had run 500,000 tides in ten experiments, equivalent to 700 years. The smaller model ran more tides, distributed over fourteen experiments. In 1891 the models had run 600,000 tides, corresponding to 840 years. These tides had been distributed over six experiments in the large tank, and four in the small tank, in number from 50,000 to 250,000 per experiment.

The outcome of this was that there then existed for the first time a firm basis for the laws of similarity, based on comprehensive tests on models with mobile bed, varying tides and fresh water flows. Reynolds devoted much time to finding a criterion that would ensure similarity of friction in models with exaggerated vertical scale. His proposal was not always used in his own experiments and was shown by Jack Allen to be much too conservative (Allen 1947).

It was only after 1920 that tidal hydraulic modelling was used frequently by British investigators, some of it overseas. In 1926, A.H. Gibson began work on a model of the Severn estuary, intended to assist with plans for a barrage which would have been used for generation of tidal electric power (Gibson 1933). The model, built at the Unversity of Manchester, had scales of 1:8,500H and 1:100V, later changed to 1:200V, to reduce local instability of flow, due to the very high velocities in the estuary (Allen 1947, Ippen 1970). It was followed by many other models built at Manchester and elsewhere.

Another advance was made by consultants Sir Alexander Gibb and Partners in 1931, when they built and operated a model of Rangoon harbour in which they had to represent the movement of fine silt (Elsden 1938). That model had scales of 1:8,060H and 1:192V.

Sir Bruce White, Wolfe-Barry and Partners operated two models of the Prai Wharves in the Straits of Penang, Malaya in 1948/49. These had several innovations, some of which became common practice (Rolfe and McDowell 1953). The problem concerned maintenance of depths at the wharves, which had not been used since their construction in 1915. Fine silt discharged from the river Muda, north of the river Prai, was carried down to the Prai by the tidal currents which were out of phase with the tidal rise and fall. One model of the whole width of Penang Straits was built, while another larger-scale model covered the river mouth with part of the adjacent coast. Both models were controlled with the aid of servo-motors, which produced the desired tidal levels and flow speeds at monitored points in the models. The variations of levels and flows for a whole year were recorded on paper charts which were read by optical curve-followers. Measurements of currents in the Straits model were used to assist with calibration of the Prai river model. Both models had mobile beds, pumice being used in the Straits model and a carefully chosen mixture of lightweight plastic grains in the Prai model. The finest sediment was represented by polystyrene, specific gravity 1.05. The grains of different densities were of different colours, which was of great help in locating the destinations of fine silts. The lower density of fresh water flowing down the river Prai compared with that in the Straits was represented by heating the water fed into the tidal head of the river. The models gave good qualitative results and experience with them led to a programme of research into the behaviour of sediments under oscillating flows.

Systematic research into the fundamentals of behaviour of estuaries and coasts got underway when HRS was created in 1947.

In addition to model experiments in the laboratory, field studies into the behaviour of estuaries were undertaken. These included field measurements in the Thames estuary (Inglis and Allen 1957) which resulted in a fundamental change in disposal of dredged material; a comprehensive study of the evolution of an estuary that had been trained in the mid-19th century (the Lune) with a neighbouring one that had been more free to develop naturally (the Wyre) (Inglis and Kestner 1958); and a study of short-term changes in distribution of fine sediments in estuaries, based on systematic bed sampling in the Wash (eastern England) (Kestner 1961). The Lune and Wyre are in a region of high-energy flows - spring tidal ranges being of the order of 10m. This work demonstrated clearly that training of the Lune had resulted in massive sedimentation and loss in tidal volume.

By the 1950s, digital computers were coming into use. From then onwards, numerical modelling took over more and more from physical modelling, but physical modelling still has an important role in the study of complex three-dimensional flows.

MODELS OF WAVE ACTION

Reynolds made provision for generation of short-period waves in his larger tank, but he did not mention its use in his reports to the BA. However, E.R. Matthews, Professor of Municipal Engineering at London University (Matthews 1913), described experiments conducted in a very small tank, 4ft 6in x 2ft 6in x 3in deep (1.37m x 0.76m x 0.076m) in which he reproduced the effects of wave action on coasts in the presence of an alongshore current. He tried various configurations of coastal structures, including a representation of a harbour between two breakwaters, based on Madras Harbour. The model reproduced the redistribution of beach sand around such structures. Unfortunately no details of model scale or wave periods were given in his book.

Coastal modelling is dependent on representation of wave climate, tidal and wave-induced currents and the movement of sediment by them. In the mid-1930s there was much interest in the UK in the action of waves on structures such as breakwaters and quay walls. R.A. Bagnold had made a classic study of the movement of wind-blown sands, based on observations in deserts while on leave from the army (e.g. Bagnold 1936). On retirement from the army, he was employed at Imperial College, London in Professor C.M. White's department. He used a wave tank to conduct experiments on the impact forces generated by waves on vertical structures (Bagnold 1939), jointly sponsored by the Research Committee of the Institution of Civil Engineers and the Buildings Research Station, and on beach formation by waves (Bagnold 1940). After the war, on his second retirement from the army,

Bagnold continued his research into sediment transport and beach processes (Bagnold 1988). He played a leading part in setting up the Hydraulics Research Station, where major work on modelling of wave action on coasts continues.

5.5 COMPUTER MODELLING

While the design of coastal engineering works has involved and, indeed still involves, the use of physical models, it was not until the early 1960's that computer models became available to aid and refine designs. Development of such tools has largely been controlled by the availability and power of computer hardware. In the 1950's and early 1960's only Government Research Laboratories and a few universities had their own computers, which themselves were limited in execution speed and information storage. The development of North Sea oil fields in the 1960's/1970's together with the production of portable/desk-top computers widened accessibility of computer models to large commercial companies both in the civil engineering and oil industries. The continued miniaturisation of hardware in the 1980's/1990's with the production of networked work-stations as well as mini-super-computers has enabled both large and small engineering companies to make use of computer models and also assist with their development.

Early models in the 1960's were concerned with the prediction of tide/surge currents and water levels. They were needed to assist with the design of flood prevention works in the Thames Estuary (Thames Barrier): a problem highlighted by the extreme North Sea surge of 1953 which damaged coastal defences and flooded areas of Southern England. The early models (Otter and Day, 1960; Rossiter and Lennon, 1963) were one-dimensional and drew on even earlier numerical, theoretical and field studies of many authors including Taylor (1921), Lamb (1932), Grace (1936), Proudman (1940, 1952), Bowden and Fairbairn (1952) and Doodson (1956).

The North Sea oil developments and the 1953 surge event in the North Sea also led to the production at the Proudman Oceanographic Laboratory (POL), of two and three dimensional models for tide/surge prediction in coastal seas, see Heaps (1969, 1972a), which also included the effects of density currents, Heaps (1972b). Such models have proved invaluable in investigating extreme surge events (Flather, 1984). The effects of internal tides and correct reproduction of density effects emerged somewhat later (Chen and Falconer, 1992). Similar developments also occurred in the early 1970's at HRS for use by the civil engineering industry, (Odd et al., 1976; Weare T.J., 1976; Odd et al., 1985) The pioneering work at POL on tide and surge modelling in two-dimensions now forms part of the routine Storm Warning Service operated by the UK's Met. Office. A useful review of tides and modelling is provided by Prandle (1991), and Davies et al. (1996).

As regards coastal wave climate, methods for offshore (20-200 m water depth) prediction were developed in response to the needs of WW2 and prediction charts were in regular use by the early 1960's (Darbyshire and Draper, 1963) with help being provided by national agencies such as NERC'S MIAS (Marine Information and Advisory Service, Institute of Oceanographic Sciences, Wormley). However, it was not until the mid 1970's that generalised prediction of wind and wave spectra became possible with work at the UK's Met.

Office on a 50 Km-grid of the European Shelf (Golding, 1983). Such methods form the basis of present-day Met. Office forecasts and are integrated with atmospheric forecast models and tide/surge models. While such models are useful at planning stage of projects and provide the only means of providing long term records, it is more usual to use statistical extrapolation of measured wave climate for detailed coastal designs.

The first elementary ray refraction models appeared in the mid 70's, Heaf (1974), and made use of American experience, Griswold (1963). Similar developments, including spectral refraction, seem to have occurred about the same time at HRS (Abernathy and Gilbert, 1975; Brampton, 1977) although theoretical work on spectra had been completed much earlier (Longuet-Higgins, 1957). The diffracting effect of structures was included in models using the analytical solutions of Penney and Price (1952), while the refracting effect of currents on waves was in use by the mid 70's (Fleming and Hunt 1976), although theoretical work was done much earlier (Longuet-Higgins and Stewart, 1960). The numerical solution of diffraction equations using finite element techniques had already been demonstrated in the late 1960's (Arlett et al., 1968) and was used to study harbour oscillation problems (Taylor et al., 1969). Combined refraction, diffraction models using finite element solutions of the mild-slope equation were also emerging by the late 1970's (Bettess and Zienkiewicz, 1977) and were based generally on earlier theoretical work (Gilbert, 1968). Finite difference approaches for complex bed contours appeared in 1980 (Williams et al., 1980). At the same time, work on bore and swash zone motions was also in progress (Hibberd and Peregrine, 1979) . However, it was not until the mid 1980's that general inshore wave climate models emerged, which were able to deal with combined refraction, diffraction, wave-current interaction, inter-active bed friction, wave breaking and surf zone mixing based on turbulence model (k, k-ε) closures (Yoo and O'Connor, 1986, 1988 and O'Connor and Yoo, 1988). Such models made use of earlier theoretical work of Longuett-Higgins and Stewart (1962), Hedges (1976) and Battjes (1968, 1975), see Peregrine (1976) for a technical review. Inclusion of wave reflection using intra-wave period approaches also emerged about the same time (Copeland, 1985) and with wave/current interactions, interactive bed friction, random waves and k-ε closures in the late 1980's and early 1990's (Anastasiou et al., 1987; Yoo and O'Connor, 1988; Yoo, Hedges and O'Connor, 1989; Li and Anastasiou, 1992). However, reflection problems within harbours using ray techniques and physical models still provided data for design (Southgate, 1984). The more complex wave climate models have also been simplified by modifying diffraction processes so that they can be used on a routine basis (Dodd, 1991). Specification of wave climate for non-linear waves using a Boussinesq approximation have emerged more recently and proved useful to study harbour resonance and vessel motions due to wave set-down effects (Smallman and Cooper, 1989), although the basic equations were produced much earlier (Peregrine, 1967). Specification of the three-dimensional surf zone currents is also becoming possible in the 1990's as computing power and storage increase.

Computer prediction of shoreline and nearshore seabed erosion and accretion, including longshore sediment transport disruption by structures has been possible since the 1970's using simple one-line sediment models coupled to wave ray climate models (Price et al, 1972; Ozasa and Brampton, 1979). About the same time, general area models were produced which were able to predict wave climate, longshore currents and seabed changes (Fleming and Hunt, 1976, and O'Connor et al., 1981), although analytical work on currents had been completed earlier (Longuet-Higgins, 1970). The inclusion of onshore-offshore transport due to under-tow currents and long waves in such models was possible for beach profile changes by the late 1980's (O'Connor and Nicholson, 1989; Southgate H.N. and Nairn R.B., 1993; Hardisty et al., 1993), in three-dimensions for harbour siltation studies (O'Connor et al., 1992) and with two dimensional wave-current-seabed interaction for tombolo formation behind breakwaters by the 1990's (O'Connor et al., 1995; Price et al., 1995). Further details are provided in reviews by Yoo et al., (1989); O'Connor (1991a, b); Lawson and Gunn (1996); and Cooper (1996).

In UK estuaries, siltation problems are generally associated with muddy material. Simple one-dimensional two-layer models and generalised solution concepts for full two and three-dimensional sediment models existed in the early 1970's (Odd and Owen, 1970; O'Connor, 1971) along with theoretical ideas for zero-dimensional models (Harrison and Owen, 1971) although generalised methods for dock entrances and local regions did not appear until later (O'Connor and Tuxford, 1980). Both one-dimensional two-layer and laterally-integrated two-dimensional models of mud movement were in routine use by the late 1970's (Odd and Rodger, 1978). Generalised three-dimensional models appeared somewhat later (Nicholson and O'Connor, 1986, O'Connor and Nicholson, 1988) along with simple models for long term morphological changes and models for simulation of fluid mud flow down underwater slopes and within navigation channels (Odd, 1986; Odd and Cooper, 1989; O'Connor et al., 1991). Models for sand movements in estuaries appeared by the mid 1970's, often combined with physical models (Odd et al., 1976), and for navigation channel siltation studies about the same time (O'Connor and Lean, 1977), often with k-ε closures (Smith and O'Connor, 1977) based on earlier turbulence modelling work in the field of mechanical engineering (Launder and Spalding, 1974). Two and three-dimensional sand transport models were in routine use at HR Ltd both in estuaries and coastal zones, by the early 1990's (Odd and Metcalfe, 1991). Models for the description of sand wave influence on navigation depths appeared by the late 1980's and early 1990's (Johns et al., 1990; O'Connor, 1992) along with the concept of integrated system models so that one, two or three dimensional models of tides, surges, sediment, salt and pollutant transport could be used to suit particular problems (Hydraulic Research Ltd., Tideway System, Thorn and Huntingdon, 1989).

Modelling of salt intrusion in estuaries started in the mid 1960's (Bowden and Gilligan, 1971) using finite difference methods and with finite element techniques in the late 1960's and early 1970's (Bebbia, 1972, Smith et al., 1973), using earlier theoretical work of Pritchard (1952) and Bowden (1967). Prior to this time, simulation techniques were in regular use for both salt and pollutant dispersal studies (Preddy, 1954, Downing, 1971). Further details are provided by Gilligan (1972) and McDowell and O'Connor (1979). By the mid 1980's, pollutant problems and their impact on dissolved oxygen levels in estuaries were routinely being studied with finite difference models (Maskell, J.M., 1985). A similar level of expertise existed with coastal models in the early 1990's (Odd and Murphy, 1992).

Computer models have also been used in recent time to study fluid loading and foundation stability for both coastal breakwaters and offshore structures as well as the stability of armour layers. A useful review is provided by Abbott and Price (1994).

For the future, there is no doubt that the use of three-dimensional models will increase, although over the next decade, such models are still likely to be limited to local areas, which are nested dynamically within larger area, two-dimensional models. Computer concept packages are also likely to increase in number and reduce in cost. The continued development of hardware is likely to make sophisticated software directly accessible to the design engineer with a greater ability to test "what-if" scenarios. Virtual reality technology is also likely to be increasingly used to aid visualization of data and results.

5.6 ENVIRONMENTAL CONSIDERATIONS

Over the last 15-20 years, conservation issues have had an increasing influence on coastal defence strategies and the design of schemes. Previously, few concerns had been expressed over the impact of sea defence improvements on wetland habitats after, for example, the 1953 floods, or as coast protection works were constructed on eroding clifflines with the loss of important geological exposures and fossil assemblages. However, the Wildlife and Countryside Act 1981 reflected on the growing national awareness of the importance of protecting and enhancing the environment. This Act introduced the SSSI designation (site of special scientific interest) which has proved to be the cornerstone of conservation in Great Britain by providing protection to sites of national importance, including:

- outstanding examples of semi-natural habitats;
- areas supporting large populations of birds and other animals;
- areas supporting endangered plant and animal species;
- features of national geological or physiographic interest.

This designation has become a significant constraint for coastal engineers, as much of the coastline has been afforded protection as SSSI. The 1981 Act also introduced the following statutory obligations for water authorities (predecessors of the NRA), Internal Drainage Boards and local authorities in the exercise of their sea defence functions:

- to further the conservation and enhancement of natural beauty;
- to further the conservation of flora, fauna and geological or physiogaphic features of special interest;
- to have regard to the cosirability of protecting buildings or other objects of archarological, architectural or historic interest.

These obligations have been re-enacted in the Water Resources Act 1991 and the Land Drainage Act 1991 which also introduced specific reactional duties:

- to have regard to the desirability of preserving for the public any freedom of access to areas of woodland, mountains, moor, heath, down, cliff or foreshore and other places of natural beauty;

- to have regard to the desirability of maintaining the availability to the public of any facility for visiting or inspecting any building, site or object of archaeological, architectural or historic interest; and

- to take into account any effect which the proposals would have on any such freedom of access or on the availability of any such facility.

The Harbours Act 1964 was amended in 1992 to include similar environmental duties for harbour authorities.

Conservation guidelines for drainage authorities have been prepared by the Government (MAFF 1991b) which recommend consultation with a wide range of nature conservation agencies (ie. English Nature, Countryside Council for Wales, Scottish Natural Heritage) and grouped in the early stages of preparing proposals. Although these conservation obligations do not apply to coast protection works, MAFF ensure that consultation with the relevant conservation agencies does take place.

The importance of protecting heritage sites is set out in the Ancient Monuments and Archaeological Areas Act 1979, which extends to monuments on or under the sea bed. Under this Act, the consent of the Secretary of State is required before any work is carried out that could destroy or damage a heritage site.

The importance of environmental issues in scheme design has been further enhanced by the introduction of the EC Environmental Assessment Directive in 1985. This directive requires an environmental assessment (EA) to be carried out before development consent is granted for certain types of major project, listed in 2 Annexes to the directive. For Annex I projects EA is mandatory. For Annex II projects EA is required if there are likely to be significant environmental effects. Where EA is required, the developer must prepare and submit an environmental statement setting our their own assessment of the likely environmental effects of the proposed development.

The land use planning system is one of the main instruments for taking account of an EA in the decision making process, under the Town and Country Planning (Assessment of Environmental Effects) Regulations 1988. The regulations apply to certain projects that require planning permission under the Town and Country Planning Act 1990, including new sea defence works and, since 1994, coast protection works. The requirement for EA for the sea defence improvements is covered by the Land Drainage Improvement Works (Assessment of Environmental Effects) Regulations 1988.

Despite the increasing awareness of the environment by coastal engineers, the impact of coastal defences on conservation interests remains a major concern for conservation agencies. English Nature, for example, are particularly concerned about the problems that have arisen as a consequence of disruption of natural sediment transport systems and "coastal squeeze" fears that are central to the overall objective of their "Campaign for a Living Coast".

"English Nature will seek to halt and reverse the loss of coastal habitats and natural features resulting from coastal squeeze and from the disruption of natural sedimentary systems. We shall try to establish a principle tht new or replacement sea defence, coast protection or similar works should not exacerbate coastal squeeze or disruption of systems and should reverse these whenever possible, so as to maintain habitats and natural features at least equivalent to their present distribution (1992) and in sustainable condition". (English Nature, 1992).

The profile of environmental issues continues to rise. The EC Habitats Directive requires measures to be taken to maintain or restore natural habitats and wild species. Landscape conservation, visual amenity and architectural design considerations are becoming important factors in scheme design, moving coastal engineering away from structures designed solely to protect property towards structures that complement or enhance the local heritage.

Indeed, recent sea defence improvements at Lyme Regis has won awards for architectural and engineering merit. Engineers have traditionally considered schemes in terms of the technical options, followed by the cost/benefit assessment and a check that the environment will not be adversely affected. There has, however, been a recent shift in emphasis, giving priority to environmental considerations before technical and economic matters (Hutchison and Leafe, 1995).

5.7 SOFT ENGINEERING

This rise in importance of environmental issues has been matched by a notable trend towards "soft engineering" on the coast. Such engineering reflects a desire to work with nature by manipulating natural systems to the benefit of environmental interests as well as protecting property (Pethick and Burd 1993).

In reality most schemes will be a mixture of "soft" and traditional "hard" techniques, but it is important to stress that the approach engenders a different design philosophy (Richardson 1995). In traditional engineering, safety margins can be added to design criteria; in soft engineering, failure can occur from over-design as well as under-design. A detached breakwater and beach scheme, for example, will have failed if it does not provide the required standard of protection. As a "soft" scheme it will also have failed if it absorbs too much energy and allows the formation of a tombolo in its lee which disrupts sediment transport to downdrift beaches. This approach inevitably depends on a broad scale view of coastal systems and a detailed knowledge of physical processes, a requirement that should be well-served by the shoreline management plan process described in Section 5.8.

Beach management is probably the most obvious form of "soft engineering", with beaches now recognised as one of the most effective forms of sea defence (Brampton 1992a). However, in many areas there appears to have been a long term decline in beach levels, often reflecting a loss of sediment source areas and disruption of sediment transport (eg Lelliott

1989; Wakelin 1989). The use of groynes to control or retain a beach has been a long-standing practice with early examples found in front of many south coast tourist resorts such as Bournemouth, Hastings and Brighton (CIRIA 1990). Beach recharge is becoming increasingly popular, not least because beaches are important tourism and amenity assets. The method involves placement of sand or shingle of specific grain size on a designed beach profile, after using hydraulic pumping from a dredger. It is suited to three main situations (Simm et al, 1995):

(i) in areas experiencing losses of beach due to depletion of sediment inputs;

(ii) on beaches acting as a self-contained coastal cell but where redistribution of material is occurring with localised areas of erosion and accretion;

(iii) to enhance the recreational value by creating a wider beach.

The recharge material is usually from offshore sources, although there is increasing competition with the construction industry for marine aggregates. As a result alternative sources, such as dredgings from ports and harbours are now being considered.

Recharge alone is unlikely to provide a long term solution to declining beach levels, as the coastal processes that gave rise to the need for positive management will continue after the scheme is implemented. This problem can be addressed either by introducing beach control structures (e.g. groynes, detached breakwaters, shore connected breakwaters, beach drainage systems etc.) or by undertaking substantial maintenance recharges at regular intervals. Practical guidance on beach management has recently been prepared by CIRIA (1996), and provides the engineering, economic and environmental information, and technical methods required to effectively manage British beaches.

Saltmarsh restoration and managed retreat are now important policy options within many low lying rural areas because of the difficulty in justifying economically the continued protection of agricultural land. In many areas relative sea level rise, mainly on the East coast has been accompanied by the erosion of saltmarshes that form an integral part of existing sea defences, increasing the probability of overtopping and breaching to the extent that maintenance and repair becomes uneconomic. At Orplands in the Blackwater Estuary, Essex it was estimated that the reinstatement defence cost was £350,000 with a further £250,000 to prevent overtopping and flood damage over the next 20 years (i.e. a total cost of £600,000). The defences currently protect some 38 hectares, at an average cost of £16,000/ha, although the market value is only £3,700/ha (Dixon and Wright, 1995). An alternative solution being considered is relocating the line of defences further inland and encouraging the development of saltmarsh habitats between the old and new defence lines (NRA 1995a, b). Experimental managed retreat and restoration schemes are currently being developed and monitored at Orplands, Abbot's Hill and Tollesbury in Essex.

On eroding cliffs, soft engineering methods have been promoted to achieve a balance between coast protection and conservation interests, with the aim of reducing erosion sufficiently to protect property whilst allowing a small degree of erosion to maintain geological exposures.

This approach has been promoted by the nature conservation agencies who published guidance on the selection of scheme options for geological SSSIs (HR Wallingford, 1991).

The need for environmental guidance in selecting appropriate coastal defence solutions was identified by MAFF in 1991. As a result they commissioned the Institute of Estuarine and Coastal Studies (Hull University) to prepare guidance to coastal engineers, planners, managers and environmental scientists. The manual (Pethick & Burd, 1993) provides the background to environmental issues in coastal defence, the natural physical and ecological processes involved and the engineering techniques available. It embraces the principle of **multiple sustainable use** of the coast which attempts to achieve the solution of coastal management problems through the use of environmentally sensitive schemes involving the use of natural coastal systems.

5.8 COASTAL PLANNING AND MANAGEMENT

Following the 1953 east coast floods, the Waverley Committee commented that much of the damage had been *"... the result of sporadic and ill-considered development near the coast which has led to unnecessary expense ... by way of additional expenditure on restoration and improvement works"*. The committee also expressed the view that *"all possible steps should be taken to prevent further undesirable development"*.

The planning system controls the development and use of land (originally set up through the Town and Country Planning Act 1947). However, one of the most serious failings of coastal management in the past has been the lack of coordination between land use planning and decisions over coastal defence strategy. Many parts of the UK coast suffer from an inheritance of unplanned communities and developments, built before planning control was established in 1947. Numerous properties now at risk also have been permitted under the planning system such as on the coastal lowlands of north Wales (the scene of the Towyn flood disaster of February 1990; Parker 1992), within the area affected by the 1953 East Coast floods where thousands of new properties have been built and on numerous eroding cliff tops (Rendel Geotechnics 1993, 1995a, b).

Since the mid 1980s there has been a notable change in planner's attitudes to coastal hazards, as directed by recent Department of the Environment (DoE) advice:

(i) PPG 14 Development on Unstable Land (DoE, 1990a) advises local authorities to consider a presumption against development in areas of coastal landslides or rapid coastal erosion;

(ii) DoE Circular 30/92 Development and Flood Risk (DoE, 1992a) advises local authorities to guide development away from areas that may be affected by flooding. Development should be avoided within set-back lines identified by the NRA;

(iii) PPG 20 Coastal Planning (DoE 1992b) advises local authorities to avoid putting further development at risk from erosion or flooding. Development should not be allowed to take place on coastal cliffs where erosion is likely to occur during the lifetime of the building.

It is envisaged that those changes in the operation of the planning system will ensure that an additional burden of responsibility is not placed on future generations by increasing unnecessarily the number of areas that need coastal defences. This is an important principle of MAFF's aim for sustainable coastal defence strategies (MAFF 1993a) and should set the framework for integrated coastal planning and management (Rendel Geotechnics 1993).

The change of attitude has lead to the setting up of informal coastal defence groups as a mechanism for dealing with issues related to, for example, the disruption of sediment transport. These groups, comprising local authorities, the NRA and conservation agencies provide a forum for strategic planning within a Coastal Process Unit - a discrete system largely independent from processes operating along adjacent sections of coastline. Much of the pioneering work in this field has been undertaken by Canterbury City Council who, in the 1980s, developed a management system for the London Clay Cliffs of the North Kent (Roberts & McGown, 1978).

In response to the establishment of the groups, MAFF and the Welsh Office provide financial support for the preparation of shoreline management plans (MAFF, 1993a, 1995). Such plans are intended to set out a strategy taking account of natural coastal processes and human and other environmental influences and needs. The principal objectives in preparing a plan are to:

* assess a range of strategic coastal defence options and agree a preferred approach;

* outline future requirements for monitoring, a management of data and research;

* inform the statutory planning process and related coastal zone planning;

* identity options for maintaining and enhancing the environment, taking account of any specific targets set by legislation or locally;

* set out arrangements for continued consultation with interested parties.

Once a preferred option has been identified, on the basis of a preliminary appraisal of costs and benefits, consultation with all interested organisations should take place to ensure that the preferred option is acceptable and can be adopted by the operating authority.

As of early 1996 there are 39 plans in progress or about to commence (see, for example, Ash et al 1995; Pos et al 1995). Expectations on their ability to deliver sustainable defence strategies are high, although this first generation of plans may be hindered by the relative limited amount of data on shoreline processes. However, they should identify research and monitoring needs.

6. CASE STUDIES

With the increasing emphasis on environmental and amenity aspects of both sea defences and coast protection, authorities are having to take a broader view of the type of hard and soft works to be used. The following case studies illustrate a range of such schemes.

6.1 LONDON: THE THAMES TIDAL DEFENCES AND THE THAMES FLOOD BARRIER

Some 14,000 ha of Greater London lies below high water level in the Thames and is at risk from tidal flooding. At least seven floods have had major impacts since the 13th century.

In the past, London has been protected by river walls in the built-up areas and embankments extending down the Thames Estuary in front of the low lying costal areas. These defences were repeatedly raised and strengthened, as, a result of passing of the Thames Flood Act 1879, and following floods in 1928 when 14 people were drowned.

In 1953 a surge tide rose some 1.6m above the level of a high spring tide at London Bridge. the defences in the main part of Central London held. Downstream the defences were breached and overtopped thus preventing the full effect of the surge reaching Central London. If it had the defences would have been overtopped and there would have been extensive dislocation of the heart of the capital city. In the event some 65,000 ha of farmland were flooded, there was extensive industrial damage and 300 people were drowned.

The Thames Technical Panel was set up to look into the problem confirmed that there had been a progressive absolute and relative rise in tide levels in the Estuary at least since the mid-18th century (Figure 3) of 0.75m per 100 years. This was caused by a combination of crustal movement, land subsidence and river training. The Panel decided that the defences in Central London needed to be raised by 1.8m to provide a 1 in 1000 year protection. This would have meant unacceptably high defences which would have hidden the river from view. Accordingly it was decided to construct a Barrier at Woolwich 31km from London Bridge. Downstream of the Barrier the walls and banks were to be raised to the extent necessary to provide the same level of protection. Smaller barriers were to be provided on tributaries downstream.

The basic requirement of the Barrier (Figure 9) were that it could be operated with complete reliability during any state of the tide or river flow and that it would allow shipping to pass without restrictions during non-surge conditions.

In 1971 the Greater London Council commissioned Rendel Palmer & Tritton to design and supervise the construction of the Barrier.

Drop-gates suspended from high towers were ruled out on environmental grounds. A rising sector gate barrier was chosen. The barrier is 520 metres wide with ten gates four of which have a clear navigational span of 61 metres. When not in use the gates lie in recesses in concrete sills at or below bed level.

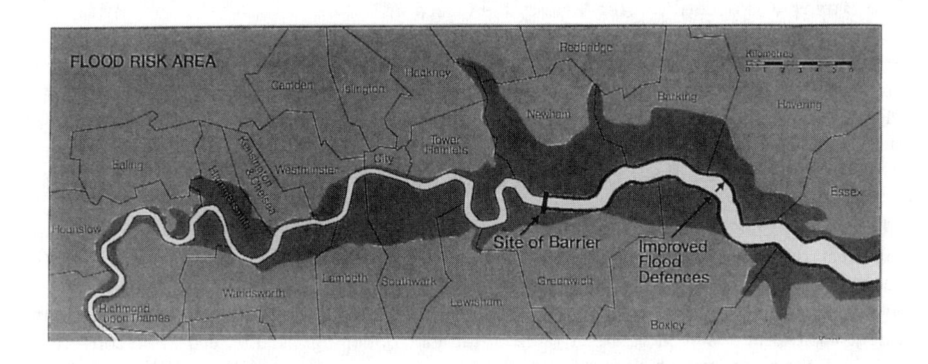

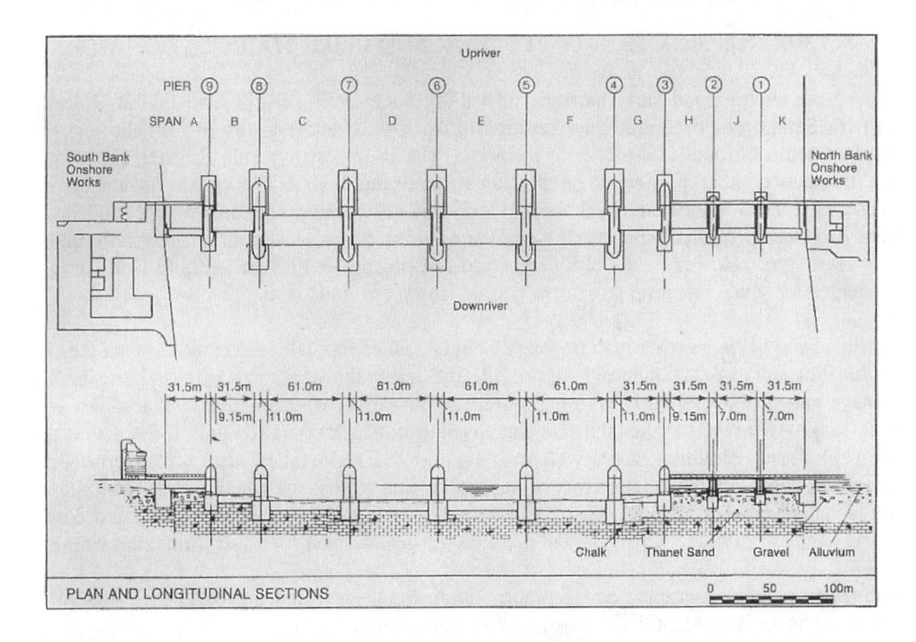

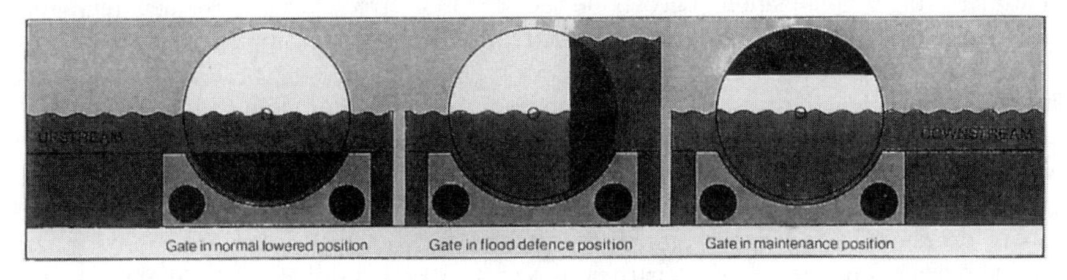

Figure 9 The Thames Barrier

The Barrier was opened by Her Majesty the Queen in 1984.

Whilst the Barrier has not yet had to be closed against a maximum surge it has been used to stop smaller surges which otherwise would have caused minor damage. It has also been closed at low tide when there has been a high river run-off to provide a reservoir upstream into which the upland flow could discharge without being held back by the rising high tide. This has reduced local flooding upstream.

The design of the Barrier is such that it could be used to maintain water levels upstream in the tidal reaches at, any, half tide level to cover the mud banks and provide a stable recreational water area. However, studies would be needed to determine the effects on land drainage, ground water levels and the stability of the river walls when full tidal conditions might need to be resumed. One or more of the gated openings would need to be converted to a lock for shipping, which could not wait for free passage during the top half of the tide.

6.2 SCARBOROUGH: COAST PROTECTION AND CLIFF STABILISATION WORKS

During the evening and early morning of 3rd/4th June 1993 a major landslide at Holbeck cliff, in Scarborough's South Bay, resulted in the loss of around 70m of cliff top and the displacement of around 1M tonnes of material. The slide destroyed the Holbeck Hall Hotel and threatened other properties on the top of the 60m high cliffs formed in glacial till overlying Jurassic mudstones and sandstones. The cliff, along with similar cliffs in South Bay, had been protected from marine erosion by mass concrete seawalls, constructed about 100 years ago. However, the cliffs remained oversteep (30-40°) and despite landscaping, drainage and small retaining structures had a history of minor instability.

The local authority, Scarborough Borough Council, asked Rendel Geotechnics to visit the site on 4th June and evaluate the problem, provide and design the works necessary to limit further damage and provide an early warning system to determine whether further evacuation was necessary. As a result of the landslide a peninsular of debris extended some 135m across the shore platform, providing a toe weight to support the main failed mass. Concerns were expressed about the potential effects of marine erosion of this toe weight on the stability of the landslide and, ultimately, the safety of the remaining cliff top assets (Clark and Guest 1994, Clements 1994). This problem dictated the urgent need for coast protection works.

Following rapid topographic and geomorphological surveys, and a ground investigation a preferred option was identified. This involved immediate coast protection of the landslide toe peninsular using a combination of geotextile bags and rock armour with subsequent regrading and drainage of the slipped mass and geotextile reinforcement with some cliff top reprofiling.

The final conceptual design was derived through a process of elimination of potential alternative details and it incorporated design features which have been successfully used elsewhere by the Consultants. The total costs associated with the scheme were estimated to be £2 million, (including construction costs of the order £1.5M) which gave a cost benefit ratio in excess of 1.5. The nature of the ground dictated that a flexible approach was required during construction, especially in the earthworks and drainage operations. A total of 22,000 tonnes of rock armour (in 5-10 tonne blocks) was used, imported by barge from Norway to nearby Teesside and transported to site on flat bed lorries.

The emergency works nature of the project required a fast track response from initial problem identification through to remedial works construction. In three months from the date of the landslide, contracts for survey and site investigation work were awarded and completed; engineers reports were prepared and submitted to MAFF for scheme approval; conceptual and then detailed design of the remedial works were carried out and a tender document issued and contract awarded for the construction of the works. The 40 week Contract for the work was undertaken by Tarmac Civil Engineering Ltd and work started during the first week of September 1993 and was completed by the 20th May 1994.

6.3 LYME REGIS, DORSET: COAST PROTECTION AND SEWAGE TREATMENT WORKS

Lyme Regis is a small coastal town on the Dorset coast, made famous by The Cobb (a harbour breakwater dating back to the 12th century) and now a popular tourist centre, with sandy beaches an important attraction. However, by the 1980s pollution from the Victorian sewage system, which discharged untreated sewage directly into the sea through two short outfalls, had led to the pollution of the tourist beaches and failure to meet the EC Bathing Waters Directive Water Quality Standards. This situation promoted the water company, South West Water, to upgrade the system, pumping sewage to a treatment works 1.5km inland and discharging the effluent to a short sea outfall by means of a gravity pipeline (Cole and Jay 1994). the discharge consent from the Department of the Environment required 1500m^3 of storm water storage to be provided before the system became operational; thus it was necessary to provide storm water tanks and a pumping station on the foreshore.

Around the same time concerns began to be expressed about the condition of the seawalls fronting the town. The walls date from the beginning of the 18th century and were constructed of limestone quarried from the foreshore. The slopes behind the walls are unstable and there is a history of a wall damage caused by the ground movement. By the 1980s some sections comprised a mixture of the original limestone blockwork and hessian bags filled with concrete. The frequency and degree of failure of the walls increased in 1989 and several emergency repairs were undertaken in 1990. As a result of these problems, West Dorset District Council proposed to upgrade and improve the coastal defences.

In 1989, with both West Dorset District Council and South West Water proposing to undertake engineering work on the foreshore, it was agreed to promote the coast protection and the sewage treatment proposals as a joint scheme.

Following extensive consultation and various public meetings a scheme was developed by Rendel Geotechnics which took account of the various environmental issues. Of particular importance was the need to preserve the appearance and maintain the view of the existing sea walls, and to ensure that the new walls were designed and constructed in a sympathetic nature to the existing walls.

It was decided to incorporate the storm water tank and sewer inlet to the pumping station within the seawall structure, with an access road for maintenance vehicles constructed over the top. One of the benefits in combining the facilities is that it reduces costs and provides a new promenade which improves the amenity of this area, with the above-ground works modelled on a castle complete with studded timber doors. The scheme was commissioned in 1995 and was awarded the Secretary of State for the Environment's Special Commendation for Environmental Excellence in the British Construction Industry Awards.

6.4 BOURNEMOUTH : EROSION CONTROL AND BEACH MANAGEMENT

Coast protection measures to prevent the recession of the 30m high cliffs comprising Tertiary Sands at Bournemouth were constructed between 1907 - 1911. The defences involved a promenade and seawall extending over 3km along the sea front. However, beach levels quickly fell as the natural supply of sediment from the cliffs had ceased, resulting in the need to construct a series of 29 concrete groynes by 1939. A further 19 groynes were built in the 1950s as beach loss continued to be a problem for a town boasting over 5km of "golden beaches" (Lelliott, 1989).

By the late 1950s the lower apron and toe of the older seawalls had begun to deteriorate with resultant breaches and damage to the promenades and associated infrastructure. These problems prompted a programme of research between 1960-1972 by Bournemouth Borough Council and the Hydraulics Research Station, including seabed surveys, drifter and surface float experiments, fluorescent tracer studies, construction of experimental concrete permeable groynes and a pilot beach recharge scheme. As a result of these studies, a "soft engineering" strategy was adopted in 1972, involving groyne building and beach recharge.

The groyne building programme commenced in 1974 and has added a further 46 groynes to the frontage. Beach recharge was carried out in 1974-75 and involved:

(i) placing 106,000m^3 of material 200m offshore west of Bournemouth pier over a 1800m length;
(ii) pumping directly onto the beaches, over a 8.5km frontage, some 654,200m^3 of marine dredged sand.

The was the largest scheme of its kind in Britain (May 1990) and remained effective for about 13 years. The recharge increased the total net beach volume from a low of around 6Mm^3 in 1975 to a peak of 7.7Mm^3 in 1979 (initial losses in the placement process gradually moved back onshore to further nourish the beaches). Beach volume fell to 6.9Mm^3 by 1988, resulting in damage to the seawalls (Harlow and Cooper 1995).

A further recharge scheme was undertaken from 1988-1990 involving 1Mm^3 of dredged material from nearby Poole Harbour. This material was pumped onshore above mean high water and allowed to form its own profile, increasing the beach volumes to 8Mm^3.

Detailed monitoring of beach profiles after each recharge scheme has revealed a similar pattern of rapid initial fall followed by a gradual decline in beach volume (Figure 10). This pattern has been used to predict the timing of a future recharge scheme; assuming that the

1988 volume of 6.9Mm³ is a critical volume and that sea level will rise at 6mm/year, it can be anticipated that the next scheme will be needed before the year 2003 (Harlow and Cooper 1995). Continued monitoring should identify whether this prediction proves to be reliable.

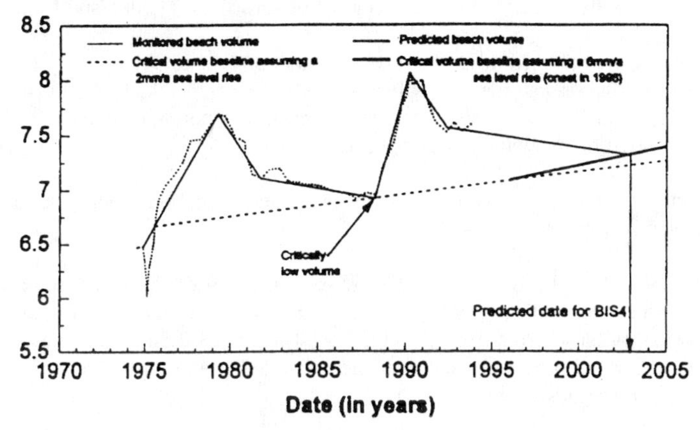

Figure 10 Bournemouth beach recharge

6.5 EAST NORFOLK : OFFSHORE REEFS

The sea defences on the east coast of Norfolk which protect several villages and 6,000ha of low lying land from tidal flooding have progressively deteriorated since their construction after the 1953 floods. The steel sheet piled toe of the reinforced concrete seawall fronting sand dunes has become the defences (Gardner and Runcie, 1995). These problems are believed to be related to the disruption of sediment supply to the beaches fronting the defences from the eroding cliffs to the north west. It has been suggested that coast protection works, mainly since 1945, have reduced sediment inputs from these cliffs by 25% (Clayton 1988) with beach levels now lower than at any time in the previous 5,000 years.

Since 1991 around 100,000 tonnes of rock toe protection has been placed to reinforce the sea defences at critical locations. This rock armour revetment has prevented undermining of the seawalls but has not controlled the wave induced beach volatility, and sudden losses of up to 2m of sand are experienced. Loss of sand exposes the erodible clay shore platform and promotes undermining of the revetment.

The key problem facing the NRA (the operating authority on this stretch of coast) is one of developing a sea defence solution that is sustainable. The scheme should reduce the flood risk minimise beach volatility and shore platform lowering whilst minimising the environmental effects elsewhere on the coast. A series of groynes would meet the technical aims but would interrupt longshore drift and promote downdrift erosion. Beach recharge would not be successful because of the exposed nature of the coast; it is estimated that the wave and current climate has the potential to remove $1Mm^3$ of sand a year. The preferred option was to construct a series of offshore reefs to reduce wave energy whilst maintaining sediment transport, albeit at a reduced rate. This programme would be followed by beach recharge to protect the shore platform and control beach losses (Garden and Runcie 1995).

It is intended that the reefs will lead to the development of a series of cuspate spits or salients which will protect the seawall toes. Reef location will be such that the formation of tombolos attached to the reefs will be prevented allowing sediment transport to continue. Mathematical modelling was used to check the design features.

Physical model tests were carried out to examine the structural stability and wave transmission of the proposed reef cross sections.

To date four reefs have been constructed between 200-300m offshore and aligned parallel with the coastline. the reefs are formed of 1-3 tonne rocks covered by 8-16 tonnes armour rocks, imported by barge from Sweden with a crest height of 2.8m above mean sea level. A further 12 reefs will be constructed over the next 15-20 years, together with a series of beach recharge operations.

6.6 TEESSIDE : THE RIVER TEES BARRAGE

The Teeside Development Corporation was established in 1987 to oversee the regeneration of a derelict river and docks area by using public sector resources to stimulate new private sector investment. Teesdale (a 100 ha site) in Stockton-on-Tees was seen as having the potential to provide the residential, commercial cultural and leisure facilities which were so much needed on Teesside, not only to create new employment opportunities (some 8000 jobs) and financial investment (some £500 million), but to provide new homes for the existing community and offer accommodation for the incoming employees. (Hall et al 1995)

The site had two fundamental problems:

(a) the quality of the environment given the exposed mudbanks and the heavily polluted tidal waters of the River Tees estuary

(b) the lack of a transport system capable of sustaining the massive development potential

The presence of mudbanks, exposed at low tide, and heavily polluted waters carried in on the high tide were simply not attractive to new investment (Norgrove 1995). It was originally proposed, therefore to construct a weir across the estuary, creating a waterside setting which would clean the river and create the environment suitable for new high quality development. However, in order to maximise the waterside setting, provide flood defence and safeguard against pollution and siltation, Teesside Development Corporation found that a barrage would be the best solution. The barrage acts as a bridge and provides vehicular access to the Teesside site which, together with infrastructure improvement elsewhere, has enabled an efficient transport system to be developed. The architectural concept for the bridge was drawn from the early ironmasters.

The barrage essentially separates the fresh and saline waters, but has sluiceways so that any saline water accumulating upstream can be discharged. It is designed to retain the upstream levels at an amenity range such that the mud flats are kept covered.

The structure is supported on a 5m thick concrete raft 70m wide and 35m long. Four buoyant fish-belly gates each 13.5m long, 8.1m high, 2.0m thick and weighing 50 tons each, close off the river. Alongside there is a navigation lock, a Denil ladder fish pass (primarily for salmon), an elver pass for the passage upstream of juvenile eels, and a 320m long white water slalom course (for use when the tidal water level is below the upstream level).

The operation of the gates provides for the following circumstances:

- At times of low river flow the gates are raised to retain the upstream level and prevent the ingress of saline water at high tide.
- At times of medium river flow the gates are partially lowered to discharge surplus water over the gates; they are raised to prevent tidal ingress.
- when a high river flow coincides with a high tide surge the gates are lowered to minimise the upstream level.

The scheme obtained the necessary approvals through a Private Act of Parliament, the Tees Barrage and Crossing Act 1990. The design and construct package with Tarmac Construction Ltd as the contractor was completed in 1995 at a cost of £54m.

6.7 SEAFORD : SHINGLE BEACH RECHARGE

Shingle is a generic term used in the UK for beaches which contain a significant proportion of gravel from granules to cobbles. The steep seaward face of shingle beaches may extend seaward below the low tide level (e.g. Chesil Beach) or may form the upper portion of an otherwise sandy foreshore. They may be found along eroding shorelines, as ridges along the seaward edge of low lying marshes or as spits extending across inlets and estuaries. They are common in areas of the world subjected to Quaternary glaciation, but are also found in the lower latitudes, usually as pocket beaches between eroding cliffs. Research on shingle beaches has been most intense in the UK due to their importance along the heavily populated south east coast.

The first major shingle beach recharge scheme in the UK was undertaken at Seaford in the East Sussex coast in 1987 (Chester FT, 1988). This scheme was among the first to be designed using predictive modelling techniques for determining the volume of material required, the expected beach profile and the long term maintenance programme. The design process involved both physical and numerical modelling. The design and modelling procedures for Seaford continue to be used to the present.

Although a range of beach control options, including groynes and offshore detached breakwaters, were considered during the design and modelling of the Seaford scheme, the eventual solution relied on an 'open' beach managed by periodic recycling of material from the down-drift to up-drift boundaries. Subsequent monitoring of the beach has confirmed this to be the best option for this particular site, and also indicates that the beach volume at Seaford has increased over the years, through natural recharge, at a rate of around 40,000m^3 per annum.

The physical modelling was undertaken by HRS in a wave basin equipped with a mobile random wave generator and a mobile beach. The scaling relationships used to determine the beach material were based on previous research funded by the MAFF and recognised the need to reproduce correctly on-offshore transport, beach slope and the threshold of motion. A compromise was required between the model sediment size and density to ensure that the fall velocity and beach permeability were representative of the actual beach material. Anthracite coal was selected as an appropriate bed material.

The physical model was supplemented by a numerical model of longshore shingle transport. This model, (Brampton and Ozasa), 1980 was based on the familiar CERC equation developed for sand transport but with a modified transport coefficient. Field verification of the model equations was not available, but continued beach monitoring at Seaford has shown the original model to be reliable.

Following the success of the Seaford physical model, a research programme was undertaken to develop a numerical model for the prediction of shingle beach profiles under a range of conditions and with a range of sediment gradings, (Powell 1990). This work culminated in the widely used SHINGLE computer model. Subsequent research has used further physical models to investigate shingle beach response in front of vertical seawalls, within groyne fields and in the lee of detached breakwaters. Beach management operations now make use of these research results to improve long term success rates.

Following on from Seaford many other shingle beach replenishments schemes have been undertaken successfully in the UK. The future of this technique is, however, somewhat clouded by depletion of reserves of suitable material. Thus future emphasis may need to be more on careful management of existing beaches rather than large scale replenishments.

Numerical models of shingle transport and shingle beach development have also continued to evolve although the lead in this work is usually provided by sand beach research; present models concentrate on cross-shore and longshore wave induced shear stresses and have proved to provide good results when verified against the Seaford database.

The main reason for the lack of initiative on shingle transport modelling is the shortage of quality field data. A new field research programme involving HRL and four south coast Universities is underway and will begin to address the lack of short term beach response information for open and groyned beaches. The programme involves the use of electronic tracers, a new generation of sediment traps, directional wave monitoring, real time beach profile monitoring and traditional beach surveys. The results of this programme will further improve the success of shingle beach management.

Although relatively few innovative techniques have so far been adopted for engineering shingle beaches, a scheme undertaken at Chesilton, at the extreme eastern end of the Chesil Beach shingle ridge, is of particular interest for its use of gabion mattresses and leeside drains to aid in the stabilisation of the beach crest, (Riddell and West 1988). So far this scheme appears to have worked well and opens the way for further developments to supplement traditional groyne fields and detached breakwater as beach control structures.

7. THE FUTURE OF COASTAL ENGINEERING

The effects of global climate change and relative sea level rise present a range of challenges even over the next 50 years. It has been suggested, for example, that the present 1 in 100 year flood event could become as frequent as 1 in 10 years (Pethick and Burd 1993).

However, it seems unlikely that the response to this challenge will involve automatic replacement or improvements to existing defences (MAFF 1993a), it being recognised that a balance needs to be struck between risk, reducing the financial burden placed on future generations, and environmental interests. Areas of current high investment will undoubtedly continue to be protected. This may involve raising embankments or seawall heights or design of replacement schemes using a combination of 'hard' and 'soft' engineering. In high value areas at risk from cliff recession, it seems likely that greater emphasis will be placed on schemes which stabilise the slopes as well as prevent marine erosion.

Elsewhere there will be difficult decisions to be made. Abandonment of existing defences in some less developed areas may be an economic reality, but will be resisted by affected property owners and communities. Here, the long-standing view of the need to 'win the war' against the sea, runs very deep. Environmental issues are likely to gain influence on the nature of coastal defence strategies and the design of schemes. In many instances, these have already gained precedence over and above technical issues, a trend that has seen the rise of 'soft' engineering solutions and managed retreat. Nevertheless, such solutions are already being criticised and viewed by many with scepticism and concern (Child 1995).

But history teaches an important lesson: the risks of coastal flooding are ever present and when events such as the 1953 East Coast floods occur they can shape the politics of coastal defence for many decades afterwards.

Shoreline management plans are a central component of the new approach to coastal engineering. They mark an important shift in focus from 'engineer' to 'strategic shoreline manager' (Hutchinson and Leafe 1995) and towards greater public participation in developing coastal defence strategies. However, it seems inevitable that some provision to allow compensation of afflicted interests will be necessary to achieve the objectives of 'sustainable' coastal defence. Such money will have to come from the public sector and it may be that schemes, desirable in the general public interest may be held up through shortage of funding.

The consolidation and privatisation of national research organisations in recent years raises questions. Primarily, there needs to be an assurance that strategic and basic research, carried out by both universities and research organisations will continue to be adequately funded. Because this research cannot usually be directly related to specific projects it has to be funded nationally. Indeed, MAFF have an important role in this respect though its Research and Development Programme of strategic research. Such research is not only needed for direct national benefit but to enable the UK to play its full part in dealing with such problems internationally. Secondly, it is important that scientific excellence, initiative and impartiality are not compromised.

In spite of the advances made in recent years, understanding of coastal processes is in many areas still an inexact science. Whilst data on wave, tides, current, sediment movement, performance of schemes etc., has been collected at many sites and over many years, the data is usually piecemeal and does not cover a sufficiently long period to allow full interpretation, especially of the joint probabilities of waves and water levels. Coastal defence engineers working on specific projects can obtain only as much information as the client is willing to pay for. The other essential information about long term and external factors vital to making the right recommendations can only come from strategic research and statistics. (Payne et al, 1996).

Conferences such as the annual MAFF Conference of River and Coastal Engineers in Great Britain and this International Conference on Coastal Engineering provide necessary opportunities for ensuring that research advances and technological developments are effectively disseminated to the practising coastal engineers who will have to meet the future challenges.

8. ACKNOWLEDGEMENTS

Rendel Palmer and Tritton Ltd., were invited to contribute towards and to coordinate the production of this review. They wish to express their gratitude to the individuals who have provided written contributions, namely: P A Cox, S Guest, D W Hookway, E M Lee, D M McDowell, B A O'Connor, R L Soulsby, and M J Tucker.

Thanks are also extended to those who have assisted with their advice, information and comments.

9. REFERENCES

Abbot, M.B. and Price, W.A. (1994) "Coastal estuarial and harbour engineers' reference book", Spon. UK.

Abernethy, C.L. and Gilbert, G. (1975), "Refraction of wave spectra", Rep. No. 117, Hydraulic Research Station, Wallingford, Berks.

Airy, Sir G.B. 1845. "Tides and waves" *Encyclopedia . Metropolitana*

Allen, J. 1947. *Scale Models in Hydraulic Enginnering,* (Longmans, Green)

Allen, J. 1970. The Life and Work of Osborne Reynolds, (In McDowell and Jackson 1970)

Allsop, N.W.H. 1986. Seawalls - a literature review. HR Wallingford Research Report.

Anastasiou, K., Dong, P. and Walker, D.J. (1987) "Turbulence modelling and the effect of directional random waves on computations of nearshore circulation", Numeta '87, Numerical Techniques for Eng. Analysis and Design, G. N. Pande and J. Middleton (Eds.), Martinus Nijhoff Pubs, Holland.

Arlett, P.L., Bahrani, A.K., and Zienkiewicz, O.C., (1968) "Application of finite elements to the solution of Helmholz's equation", Proc. IEE., 115, 1762-6.

Ash, J.R.V., Nunn, R. & Lawton, P.A.J. 1995. Shoreline management plans; a case study for the North Norfolk Coast. In Coastal Management 95 Putting Policy into practice. Thomas Telford.

Bagnold, R.A. 1939. "Interim report on wave pressure research" *J. Inst. Civil Engnrs.*, Vol 12, pp. 201-226.

Bagnold, R.A. 1940. Beach formation by waves; some model-experiments in a wave tank,*Journal of the Institution of Cvil Engineers*, Paper 5237, 27-52

Bagnold, R.A. 1988. *The Physics of Sediment Transport by Wind and Water: a collection of hallmark papers* (Ed: C.R. Thorne, R.C. MacArthur and J.B. Bradley). American Society of Civil Engineers.

Barber, N.F., Ursell, F., Darbyshire, J. and Tucker, M.J. 1946 "A frequency analyser used in the study of ocean waves" *Nature,* Vol 158, p.329 (7 Sept 1946).

Barber, N.F. and Ursell, F. 1948. "The generation and propagation of ocean waves and swell: 1 Wave periods and velocities." *Phil. Trans. Roy. Soc. A,* Vol. 240, pp. 527-560

Battjes, J.A., (1968) "Refraction of water waves", J. WHCOD, ASCE, 94, WW4, 437-451.

Battjes, J.A., (1975) "Turbulence in the surf zone" Proc. Modelling Tech., ASCE, 1050-1061.

Bettess, P. and Zienkiewicz, O.C. (1977) "Diffraction and refraction of surface waves using finite and infinite elements", Int. J. for Numerical Math. in Eng., II, 1271-1290.

Bell, A.O. 1972. "North Sea wave spectra" Note for the North Sea Environmental Study Group (Issued by British Petroleum Ltd.)

Bowden, K.F. and Fairbairn, L.A. (1952) "A determination of the frictional forces in a tidal current" Proc. Roy. Soc. (A), 214, 371.

Bowden, K.F. and Gilligan, R.M. (1971), "Characteristic features of estuarine circulation as represented in the Mersy estuary", Lim. and Ocean, 16, 3, 490-502.

Bowden, K.F. (1967) "Circulation and diffusion", Estuaries, G.H. Lauff (Ed), AAAS Pub., 83, 15-36.

Brampton, A.H. (1977). *"A computer method for wave refraction"* Rep. IT 172, IIRS Wallingford, Berks.

Brampton, A. 1992a Beaches - The natural way to coastal defence. In MG Barrett(ed) Coastal Zone Planning and Management, pp 221 - 229. Thomas Telford.

Brampton, A. 1992b. Engineering significance of British Saltmarshes. In JRL Allen and K

Brampton, A.H. and Ozasa, H. Mathematical modelling of beaches backed by Seawalls. Coastal Engineering, Vol 4, No.1, 1980. Pye (eds) Saltmarshes: morphodynamics, conservations and engineering significance pp 115 - 122. Cambridge University Press.

Bray, M.J. and Hooke, J.M. 1995 Strategies for conserving dynamic coastal landforms. In MG Healy and JP Doody (eds) Directions in European Coastal Management, PP 275 - 290. Samara Publishing.

Bray, M.J, Carter, D.J. and Hooke J.M. 1991. Coastal Sediment Transport Study. Report to SCOPAC (5 Volumes).

Bray, M.J. Carter, D.J. and Hooke, J.M. 1992. Sea-level rise and global warming; scenarios, physical impacts and policies. Report to SCOPAC.

Brebbin, C.A. (1972) *"Some applications of finite elements for flow problems"*, Int. Conf. on Variational Methods in Eng., Dept. of Civil Eng., Southampton University, 5, 25-29.

Brooks, N.P. 1981. Romney Marsh in the Early Middle Ages In: The Evaluation of Marshland Landscapes Oxford University Department for External Studies.

Butters, K. and Lane, J.J. 1975. Flood alleviation on some River Thames tributaries Journal of the Institution of Water Engineers, 29 pp 67-84.

Canham, H.J.S., Cartwright, D.E., Goodrich, G.H. and Hogben, N. 1962 "Seakeeping trials on OWS 'Weather Reporter" *Trans. Roy. Inst. Nav. Archit.*, Vol. 104, pp. 447-492 Carter, D.J.T. and Draper, L. 1988 "Has the north-east Atlantic become rougher?" *Nature*, Vol. 332, p. 494.

Carter, D.J.T., Foale, S. and Webb, D.J. 1991 "Variations in global wave climate throughout the year" *Int. J. Remote Sensing*, Vol. 12, pp. 1687-1697.

Carter, D.J.T. and Challenor, P.G. 1990 Part 2 in "Metocean Parameters-wave parameters *"Department of Energy,"* Offshore Technology Report No. OTH 89, Her Majesty's Stationery Office, London.

Cartwright, D.E. and Longuet-Higgins, M.S. 1956 "The statistical distribution of the maxima of a random function" *Proc. Roy. Soc. A*, Vol. 237, pp. 212-232.

Cartwright, D.E. 1958 "On estimating the mean energy of sea waves from the highest wave in a record." *Proc. Roy. Soc. A*, Vol. 247, pp. 22-48.

Cartwright, D.E. and Smith, N.D. 1964 "Buoy techniques for obtaining directional wave spectra" *Trans. of the 1964 Buoy Technology Symposium*, Marine Technology Soc., Washington, USA. pp. 112-121.

Chen, Y. and Falconer, R.A. (1992) "Advection-diffusion modelling using the modified QUICK scheme" Int. J. Num. Meths. in Fluids, 15, 1171-1196.

Chester, F.T. The Seaford sea defence scheme. MAFF Conference of River and Coastal Engineers, Loughborough, UK. July 1988.

Child, M. 1995. Taking plans forward through Consultation and Participation : are Plant Sustainable. In Coastal Management '95. Thomas Telford.

CIRIA Technical Note 125, 1986. Sea Walls. Survey of performance and design practice.

CIRIA/CUR 1991. Manual on the use of rock in coastal and shoreline engineering.

CIRIA special Publication 83.

Clark, A.R. and Guest, S. 1994. Holbeck Hall landside: Coast protection and cliff stabilisation. Proceedings of the MAFF Conference of River and Coastal Engineers, Loughborough.

Clayton, K.M. 1989 Sediment input from the Norfolk Cliffs, eastern England: a century of coast protection and its effect. Journal of Coastal Research, 5,pp 433-442.

Clayton, K.M. 1990 Sea-level rise and coastal defences in the UK. Quaterly Journal of Engineering Geology, 23, pp 283-287.

Clement, D. 1995. Property Insurance and flood risk. Proceedings of the MAFF Conference of River and Coastal Engineers, Keele.

Clement, M. 1995. The Scarborough experience - Holbeck Landslide 3-4th June 1993. Proceedings of the Institution of Civil Engineers: Municipal Engineer, 103, pp63-70.

Cole, K. and Joy,D. 1994. Lyme Regis environmental improvements. Proceedings of the MAFF Conference of River and Coastal Engineers, Loughborough.

Cooker, M.J. and Peregrine, D.H. 1990 "Computations of violent motion due to waves breaking against a wall" *Proc. 22nd Int. Conf. Coastal Engineering*, Delft, A.S.C.E. Vol 1, pp. 164-176.

Cooker, M.J. and Peregrine, D.H. 1992 "Wave impact pressure and its effect upon bodies lying on the bed" *Coastal Engineering,* Vol. 18, pp. 205-229.

Cooper, A.J. (1996) " Guidelines in the use of computational models for coastal and estuarial studies - wave transformation and wave disturbance models"., HR Wallingford, Rep. SR. 456.

Copeland, G.J.M. (1985) "A practical alternative to the mild-slope equation", Coastal Eng., 9, 125-149.

Count, B. (Ed.) 1980 "Power from sea waves" *Proc. Conf. on Power From Sea Waves,* organised by the Institute for Mathematics and its Applications, Edinburgh 26-28 June, 1979. Published by Academic Press.

Construction Industry Research and Information Association (CIRIA) 1990. Guide to the use of groynes in coastal engineering.

Construction Industry Research and Information Association (CIRIA) 1996. Beac Management Manual.

Crabb, J.A. 1984 "Assessment of wave power available at key United Kingdom sites" IOS Report No. 186. 113 pp. Available from the Southampton Oceanography Centre, Southampton, England, SO14 3ZH.

Dacunha, NMC, Hogben, N, Andrews, KC. *"Ocean Wave Statistics - a new book".* Oceonological Int. Conference, Brighton 1984.

Darbyshire, J. 1952 "The generation of waves by wind" *Proc. Roy. Soc. A,* Vol.. 215, pp. 299-328.

Darbyshire, M. and Draper, L. 1963 "Forecasting wind-generated sea waves" *Engineering,* London, Vol. 195, pp. 482-484.

Davidson, N.C., d'A Laffoley, D., Doody, J.P., Way, L,S., Gordon, J., Drake, C.M., Pienkowski, M.W., Mitchell, R. and Duff, K.L. 1991. Nature conservation and estuaries in Great Britain. Nature Conservancy Council.

Davies, A.M., Jones, J.E. and Xing J. (1996) "A review of recent developments in tidal hydraulic modelling, Pt. 1 Spectral modelling, Pt. 2 Turbulence energy modelling" Journal of Hydraulic Engineering, ASCE, New York (in press).

Deacon, G.E.R., Darbyshire, J. and Smith, N.D. 1949 "Use of airborne sea and swell recorder to measure changes in the wave spectrum from West to East across the Irish Sea." *Admiralty Research Laboratory,* Report ARL/R1/103-18/W Dept. of Energy 1977: (Third Edition 1984) "Guidance on the design and construction of offshore structures" *Her Majesty's Stationery Office*, London.

Department of the Environment (DoE) 1989. The Quality of UK bathing water improves. Environmental news release.

Department of the Environment (DoE) 1990a. Planning Policy Guidance note 14 Development on Unstable Land. HMSO.

Department of the Environment (DoE) 1990b. Guidance note on ministerial declaration 3rd International Conference on the Protection of the North Sea.

Department of the Environment (DoE) 1991. Environmental news release No. 183.

Department of the Environment (DoE) 1992a. Development and Flood Risk. Circular 30/92. HMSO.

Department of the Environment 1992b. Planning Policy Guidance note 20 Coastal Planning HMSO.

Derbyshire, J. and Draper, L., (1963) "Forecasting wind-generated sea waves", Engineering, 195, 482-484.

Dixon, A.M. and Weight, R.S. 1995. Managing Coastal realignment. NRA Saltmarsh Research Seminar.

Dodd, N. (1991) "Reflective properties of parabolic approximations in shallow water wave propagation" SIAM J. App. Maths., 51, 635-657.

Doodson, A.T. (1956) "Tides and storm surges in a long uniform gulf" Proc. Roy. Soc (A), 237, 325-343.

Downing, A.L. (1971) "Forecasting the effects of polluting discharges on natural waters, II, Int. J. Envir. Studies, 2, Part II.

Draper, L. 1963 "Development of a 'design wave' from instrumental records of sea waves" *Proc. Inst. Civil Engrs,*. Vol. 26, pp. 291-304.

Draper, L. 1991 "Wave Climate Atlas of the British Isles" *Dept. of Energy* Offshore Technology Report no. OTH 89 303, Her Majesty's Stationery Office, London.

Elsden, Oscar, 1938. Investigation of the Outer Approach-Channels to the Port of Rangoon by means of a Tidal Model.

English Nature 1992. Campaign for a living coast. Peterborough.

Flather, R.A. (1984) "A numerical modelling investigation of the storm surge of 31st January and 1st February 1953 in the North Sea", Quart. Jnl. Roy. Meterological Soc. Vl 110, 591-612.

Fleming, C. A. and Hunt, J. N. (1976) "Application of a sediment transport model". Proc. 15th Coastal Eng. Conf., 1184-1202.

Forcott, S.E., Hewling, M. and Burt, N. 1995. River Tees Barrage and Arcillary Works Part 2 Design of the Barrage. Proceedings of the Institution of Civil Engineers: Municipal Engineer, 109 pp 153-167.

Froude, R.E. 1894 "On ship resistance" *Papers of the Greenock Phil. Soc.,* 19 Jan 1894

Garoner, J. and Runcia, R. 1995. Planning and construction of four offshore reafs in Norfolk. Proceedings of the MAFF Conference of River and Coastal Engineers, Keele.

Gibson, A.H. 1933. *Construction and Opertion of a Tidal Model of the Severn Estuary,* H.M Stationery Office, London.

Gilbert, G. (1968) "Mild-slope equation for nearshore waves", HRS Annual Report, Wallingford, Berks.

Gilligan, R.M. (1972) "Forecasting and effects of polluting discharge on estuaries", Pt. II, Chem. and Ind., Dec., 19-35.

Godsin, H. 1978. Fenland: its ancient past and uncertain future. Cambridge University Press.

Golding, B. (1983) "A wave prediction system for real-time sea state forecasting", Quart. J. Roy. Met. Soc., 109, 393-416.

Grace, S.F. (1936), "Friction in the tidal currents of the Bristol Channel", Mon. Not. Roy. Astron. Soc. (Geophys. Suppl.), 3, 388.

Green, C. 1984. Saltings and sea defence on the Gwent Levels Proceedings of the MAFF Conference of River Engineers, Cranfield.

Grieve, H. 1959. The Great Tide. Chelmsford, Essex County Council.

Golding, B. 1983 "A wave prediction system for real-time sea-state forecasting" *Quart J. Roy. Met. Soc.*, Vol. 109, pp.393-416.

Griswold, G.M. (1963) *"Numerical calculation of wave refraction"* J. Geo. Res., 68, 6, 1715-1723.

Hagger, P. 1979 "Waveheight at the Tyne Piers" Report by the Port of Tyne Authority. Hall D, Firth SJ and Odd NVM 1995. River Tees Barrage and Ancilary Works Part 1 Planning. Proceedings of the Institution of Civil Engineers: Municipal Engineer, 109, pp 131-152.

Hardisty, J., Davison, M., Russell, P., Huntley, D. and Hoad, J.P. (1993) "Numerical experiments with gravity and infra-gravity waves on a macro-tidal beach profile, J. Coast. R., Sp. Iss. 15, 198-214.

Harlow, D.A. and Cooper, N.J. 1995. Bournemouth beach monitoring: the first twenty years. In Coastal Management '95 Putting policy into practice. Thomas Telford.

Harrison, A.J.M. and Owen, M.W. (1971) *"Siltation of fine sediments in estuaries"* Proc. 14th IAHR Conf., D1, 4, 1-8.

Hasselmann, K., Barnett, T.P., Bouws, E., Carlson, H., Cartwright, D.E., Enke, K., Ewing, J.A., Gienapp, H., Hasselmann, D.E., Kruseman, P., Meerburgh, A., Muller, P., Olbers, D.J., Richter, K., Sell, W. and Walden, H., 1973 "Measurement of wind-wave growth and swell decay during the Joint North Sea Wave Project (JONSWAP)" *Deutsches Hydrographisches Zeitschrift*, Reihe A(8°).

Havelock, Prof. Sir Thomas "The collected papers of Sir Thomas Havelock on hydrodynamics" Published by the Office of Naval Research (US Navy) 1963, ED. W.C.S. Wigley.

Heaf, N.J. (1974) "Wave refraction", Ph.D Thesis, Department of Civil Engineering, University of Liverpool, UK.

Heaps, N.S. (1972a) "Estimation of density currents in the Liverpool Bay area of the Irish Sea", Geophys. J. of the Roc. Ast. Soc., 30, 415-432.

Heaps, N.S. (1972b) "On the numerical solution of the three-dimensional equations for tides and storm surges", Mem. Soc. Royale des Sciences de Liège, 6, 2, 143-180.

Heaps, N.S. (1969) "A two-dimensional numerical sea model" Phil. Trans. Roy. Soc. (A), 265. 93-137.

Hedges, T.S.H., (1976) "An empirical modification to linear wave theory", Proc. ICE, 2, 61,75-579.

Hibbered, S. and Peregrine, D.H. (1979) *"Surf and run-up on a beach"*, JFM, 95, 323-345.

Hinde, B.J. and Gaunt, D.I. 1967 "Microseisms" *Contemporary Physics* Vol. 8, pp. 267-283.

Hogben, N. 1988 "Experience from compilation of global wave statistics" *Ocean Engng.*, Vol 15, pp. 1-31

Horner, R.W. 1978. Thames tidal flood works in the London excluded area. Journal at the Institution of Public Health Engineers 6, pp 16-24.

Houghton, J.T., Jenkins., G.J. and Ephraums, (eds) 1990. Climate change: The IPCC Scientific Assessment Cambridge University Press.

Humberside County Council 1994. Humber Estuary and Coast.

Hutchison, J. and Leafe, R.N. 1995. Shoreline management a view of the way ahead. In Coastal Management '95 Putting policy into pactice Thomas Telford.

HR Wallingford 1991. A guide to the slection of appropriate coast protection works for geological SSSIs. Nature Conservancy Council.

Inglis, C.C. and Allen, F.H. 1957. The regimen of the Tahmes estuary as affected by currents, salinities and tidal flow, *Proceedings of the Institution of Civil Engineers,* 7.

Inglis, C.C. and Kestner, F.J.T. 1958. The long-term effects of training walls, reclamation and dredging on an estuary, *Proceedings of the Institution of Civil Engineers,* 9.

Interdepartmental Committee on Coastal Flooding (the Waverley Committee) 1954 Report HMSO.

Ippen, A.T. 1970. Hydraulic Scale Models, (in McDowell and Jackson 1970).

Jager, J. and Ferguson, H.L. (eds) 1991. Climate changeL Science impacts and policy. Cambridge University Press.

James, I.D. 1986 "A note on the theoretical comparison of wave staffs and Waverider buoys in steep gravity waves" *Ocean Engng.*, Vol. 13, pp. 209-214

Jensen, H.A.P. 1953 Tidal inundations past and present. Wheather, 8, pp 85-89, 99 108-112.

Johns, B., Soulsby, R.L. and Chesher, T.J. (1990) "The modelling of sandwave evolution resulting from suspended and bed load transport of sediment". J. Hyd. Res., 28, 3.

Jones, D.K.C. and Lee, E.M. 1994. Landsliding in Great Britian: a review HMSO Kelvin, Lord (Sir W. Thomson) 1887 "On the waves produced by a single impulse in water of any depth or in a dispersive medium" *Proc. Roy. Soc. A,* Vol. 42, pp. 80-85.

Kestner, F.J.T. 1961. Short term changes in the distribution of fine sediments in estuaries, *Proceedings of the Institution of Civil Engineers,* 19.Lamb H. (1932) "Hydrodynamics", Dover, New York.

Lamb, H.H. 1991. Historic Storms of the North Sea, British Isles and Northwest Europe. Cambridge University Press.

Lamb, Sir Horace (1932) "Hydrodynamics" Sixth Edition, Cambridge University Press. Reissued by Dover Publications in 1945. (First edition published in 1879).

Lane, M.R. 1989. The Rendel Connection: A Dynasty of Engineers. Quiller Press.

Launder, B.E. and Spalding, D.B. (1974) "The numerical computation of turbulent flows", Comp. Meth. in Applied Mech. and Eng., North Holland Pub. Co., 3, 269-289.

Lawson, J. and Gunn, I.J. (1996) "Guidelines in the use of computational models for coastal and estuarial studies - wave transformation and wave disturbance models"., HR Wallingford, Rep. SR. 450.

Lee, E.M. and Moore, R. 1991. Coastal landslip potential assessment; Ventnor, Isle of Wight. Report to the DoE.

Lelliott, R.E.L. 1989. Evaluation of the Bournemouth defences In Coastal Management pp 263-277. Thomas Telford.

Li, B. and Anastasiou, K. (1992) "Efficient elliptic solvers for the mild-slope equation using the multigrid technique", Coastal Eng., 16, 245-266.

Longuet-Higgins, M.S. 1952 "On the statistical distribution of the heights of sea waves" *J. Marine Res.,* Vol. 11, pp. 245-266.

Longuet-Higgins, M.S. 1953 "Mass transport in water waves" *Phil. Trans. Roy. Soc. A,* Vol. 245, pp. 535-581.

Longuet-Higgins, M.S. 1957 "The statistical analysis of a random moving surface" *Phil Trans. Roy Soc. A,* Vol. 249, pp. 321-387.

Longuet-Higgins M.S. and Stewart, R.W. (1960) "Changes in the form of short gravity waves on long waves and tidal currents" J.F.M., 8, 565-583.

Longuet-Higgins, M.S. 1962 "The statistical geometry of random moving surfaces" in *Hydrodynamic Stability: Proc. 13 Symp. Appl. Math.*, pp. 105-144, American Mathematical Society, Providence, Rhode Island..

Longuet-Higgins M.S. and Stewart, R.W. (1962) "Radiation stress and mass transport in gravity waves with application to surf beats", JFM, 13, 481-504.

Longuet-Higgins, M.S., Cartwright, D.E. and Smith, N.D. 1963 "Observations of the directional spectrum of sea waves using the motions of a floating buoy" *Ocean Wave Spectra*, Prentice Hall, Englewood Cliffs, New Jersey, USA, pp. 111-132.

Longuet-Higgins, M.S., and Stewart, R.W. 1964 "Radiation stress in water waves: a physical discussion with applications" *Deep-Sea Research*, Vol. 11, pp. 529-562.

Longuet-Higgins, M.S., 1970 "Longshore currents generated by obliquely incident sea waves", *J. Geophys. Res.*, Vol. 75, pp. 6778-6789

Longuet-Higgins M.S., (1970) "Longshore currents generated by obliquely incident sea waves", 1-2, J. Geophys. R., 75, 33, 6778-6801.

Longuet-Higgins, M.S. 1980 "The unsolved problem of breaking waves" *Proc. 17th Int. Conf. Coastal Engineering*, pp. 1-28, Am. Soc. Civil Engrs.

Maskell, J.M. (1985) "The effect of particulate BOD on the oxygen balance of a muddy estuary", Est. Manage. and Quality Assessment, J.G. Wilson and W. Halcrow (Eds.), Plenum Press, New York.

Mathews, E.R. 1934. *Coast Erosion and Protection*, (Griffin).

May, V.J. 1990. Replenishment of the resort beaches at Bournemouth and Christchurch, England. Journal of Coastal Research SI(6) pp 11-15.

McDowell, D.M. and Jackson, J.D. (Eds.), 1970 *Osborne Reynolds and Engineering Science Today.* (Manchester University Press; Barnes and Noble Inc.)

McDowell, D.M. and O'Connor, B.A. (1977) "Hydraulic behaviour of estuaries",MacMillan Press, London.

Miles, J.W. 1957 "On the generation of surface waves by shear flow" *J. Fluid Mech.*, Vol. 3, pp. 185-204.

Ministry of Agriculture, Fisheries and Food (MAFF) 1991a. Advice on allowance for sea level rise. November 1991.

Ministry of Agriculture, Fisheries and Food (MAFF) 1991b. Conservation Guildlines for Drainage Authorities. MAFF Publications.

Ministry of Agruculture, Fisheries and Food (MAFF) 1993a. A Stratgy for Flood and Coastal Defence in England and Wales. MAFF Publications.

Ministry of Agriculture, Fisheries and Food (MAFF) 1993b. Project Appraisal Guidance notes for flood and coastal defence. MAFF Publications.

Ministry of Agriculture, Fisheries and Food (MAFF) 1994. Coast Protection Survey of England.

Ministry of Agriculture, Fisheries and Food (MAFF) 1995. Shoreline Management Plans: A guide for coastal defence authorities. MAFF Publications.

Ministry of Health 1949. Coast Protection Survey *Minutes of the Proceedings of the Institution of Civil Engineers,* Paper No. 5100 *Minutes of the Proceedings of the Institution of Civil Engineers,* 24 NERC 1991. United Kingdom ..... marine atlas. NERC, Swindon.

National Rivers Authority (NRA) 1991 Bathing Water Quality in England and Wales - 1990. Water Quality Series No.3.

National Rivers Authority (NRA) 1992. Sea Defence Survey. NRA Bristol.

National Rivers Authority (NRA) 1993. Food Defence Strategy. NRA Bristol. National Rivers Authority (NRA) 1995a. A guide to the understanding and management of saltmarshes. R and D Note 324.

National Rivers Authority (NRA) 1995b. Maintenance and enhancement of saltmarshes. R and D Note 567.

Nicholson, J. and O'Connor, B.A. (1986) "Cohesive sediment transport model", J. Hyd. Eng. ASCE, 112, 7, 621-640.

Norgrove, W.B. 1995. River Tees Barrage - a catalyst for urban regeneration. Proceedings of the Institution of Civil Engineers: Civil Engineering, 108, pp 98-110.

Odd, N.V.M. and Rodger, J.G. (1978) "Vertical mixing in stratified tidal flows", J. Hyd. Div., ASCE, 104, HY3, 337-351.

Odd, N.V., Miles, G.V. and Mann, K. (1976) "Mathematical and physical model studies for the wash water storage scheme", Proc. ICE/CWPU Sym. on the Wash Storage Scheme, Hyd. Res., Wallingford.

Odd, N.V. and Owen, M.V. (1970) "A two-layer model of mud transport in the Thames Estuary", Proc. ICE, Sup., 175-205.

Odd, N.V.M. and Murphy, D.G. "Experience and recommended practice for the construction and objective calibration of coastal pollution models", Hydraulic and Environmental Modelling of Coastal Waters, R.A. Falconer, S.N. Chandler-Wilde, S.Q. Liu (Eds), Ashgate Pub. USA, 291-239.

Odd, N.V.M., Wolfe-Barry, J.N. and Ali Berrahim (1985) "Hydraulics of a tidal lagoon at Benghazi", Int. Conf. on Hyd. Model of Ports and Harbours, BHRA, Bedford, UK., 43-53.

Odd, N.V.M. and Metcalfe, G.A. (1991) "Hydraulic studies for the development of Port Klang", 3rd Int. Conf. on Coast. and Port Eng. in Developing Countries, Copedec, Lanka Hyd. Inst., Sri Lanka, 1036-1041.

Odd, N.V.M. and Cooper, A.J. (1989) "A two-dimensional model of the movement of fluid mud in a high energy estuary", J. Coast. Res., 5, 185-193.

Odd, N.V.M. (1986) "Mathematical modelling of mud transport in estuaries" Int. Sym. Physical Processes in Estuaries., J. Dronkers and W. van Leussen (Eds), Springer-Verlag, Berlin, 503-531.

O'Connor, B.A. and Lean, G. (1972) *"Estimation of siltation in dredged channels in open situations"* Proc. 24th PIANC Congress, 52, 12, 163-177.

O'Connor, B.A. and Tuxford, C. (1980) *"Modelling siltation at dock entrances"* Proc. 3rd Int. Conf. on Dredging Tech., BHRA, Bedford, England, 359-372.

O'Connor, B.S. Marcos Fanos, A. and Cathers B., (1981) "Simulation of coastal sediment movements by computer models", Proc. 2nd Int. Conf. Eng. Software, R.A. Adey (Ed), 554-568

O'Connor, B.A. (1991a), "Suspended sediment transport in the coastal zone" Proc. Int. Symp. on the Transport of Suspended Sediments and its Math. Modelling, IAHR/University Di Firenze, Italy, 17-64.

O'Connor, B.A. (1992) *"Prediction of seabed sand waves"*, Computer Modelling of Seas and Coastal Regions, P.W. Partridge (Ed), Computational Mchanics Pubs., Southampton, UK, 321-338.

O'Connor, B.A., Kim, H. and Yum, D. (1992) "Siltation at Chukpyan Harbour", Computer Modelling of Seas and Coastal Regions, P.W. Partridge (Ed)., Computational Mechanics Pubs., Southampton, 380-397.

O'Connor, B.A. (1971) "Mathematical model for sediment distribution", Proc. XIV IAHR Cong., Paris, D, 1-8.

O'Connor, B.A. (1991b) "Coastal Modelling", IEM/ICE Joint Conference on Coastal Eng. in National Development, Kuala Lumpur, Malaysia, 1-19.

O'Connor, B.A. and Nicholson, J. (1988) "A three-dimensional model of suspended particulate sediment transport", Coastal Eng., 12, 157-174.

O'Connor, B.A. and Yoo, D.H. (1988) "Mean bed friction of combined wave/current flow", Coastal Eng., 12, 1-21.

O'Connor, B.A., Nicholson, J., Rayner, R., (1991) "Estuary geometry as a function of tidal range", Proc. 22nd Coastal Eng. Conf., B. Edge (Ed), ASCE, New York, 3050-3062.

O'Connor, B.A., Nicholson, J.N. and MacDonald, N.J. (1995) "Modelling morphodynamic changes associated with an offshore breakwater", Computer Modelling of Seas and Coastal Regions II, C.A. Brebbin, L. Traversani and L.C. Wrobel (Eds), Computational Mechanics Pubs, 265-272.

O'Connor, B.A. and Nicolson, J., (1989) "Modelling changes in coastal morphology" Proc. Int. Sym. Sediment Trans. Modelling, S.Y. Wang (Ed), ASCE, New York, 160-165.

Otter, J.R.H. and Day, A.S. (1960). *"Tidal Flow comutations"*, Engineer, 209, 177 182.

Ozasa, H. and Brampton, A.H. (1979) *"Models for predicting the shoreline evolution of beaches backed by seawalls"*, Rep. IT 191, HRS, Wallingord, Berks.

Parker, (Ed). J. Wiley and Sons, New York, 125-152.

Parker, D.J. 1992. Flood disasters in Britain: lessons from flood hazard research. Disaster Prevention and Management, 1, pp 8-25.

Parker, D.J., Green, C.H. and Thompson, P.M. 1987. Urban flood protection benefits: a project appraisal guide. G..... Technical Press.

Patrick, J. and Burd, F. 1993. Coastal defence and the environment: a good practice guide. MAFF Publications.

Payne, H., Rayner, R., Cramp, A., Barber, P., McCue, J. Coastal process measurement - what should you be doing? Proceedings Inst. Civ. Engineers, June 1996.

Penney, W. G. and Price, A.T., (1952) "Diffraction of sea waves by breakwaters and the shelter afforded by breakwaters" Phil. Trans. Roy. Soc., (A), 244 (882), 236-253.

Penning-Rowsell, E.C. and Chatterton, J.B. 1977. The benefits of flood alleviation: a manual of assessment techniques Saxon House.

Penning-Rowsell, E,C., Green, C.H., Thompson, P.M., Coker, A.M., Tunstall, S.M., Richards, C. and Parker, D.J. 1992. The economics of coastal management - a manual of benefits assessment techniques. Belhaven Press.

Peregrine, D.H. (1976) "Interaction of water waves and currents" Adv. in App. Mech., 16, 9-117.

Peregrine, D.H. (1967) "Long waves on a beach", JFM, 27 815-827.

Perry, A.H. 1981. Environmental hazards in the Britian Isles. George Allen and Unwin.

Pethick, J. and Burd, F. 1993. Coastal defence and the environment : a good practice guide MAFF Publications.

Phillips, O.M. 1957 "On the generation of waves by turbulent wind" *J. Fluid Mech.*, Vol. 2, pp. 417-445.

Phillips, O.M. 1958 "Spectral and statistical properties of the equilibrium range in wind-generated gravity waves" *J. Fluid Mech.*, vol. 156, pp. 505-531.

Pierson, W.J. and Moskowitz, L. 1964 "A proposed spectral form for fully developed wind seas based on the similarity theory of S.A. Kitaigorodskii" *J. Geophysical Res.*, Vol. 69, pp. 5181-5203.

Powell, K.A. Prediciting short term profile response for shingle beaches. HR Wallingford Report SR 219, February 1990.

Prandle, D. (1991). *"Tides in estuaries and embayments"*, Tidal Hydrodynamics, B.B.

Preddy, W.S. (1954), "The mixing and movement of water in the estuary of the Thames", J. Mar. Biol. Ass. UK., 33.

Price, D.M., Chesher, T.J. and Southgate, H.N. (1995) "PISCES, a morphological coastal area model. Final Report", HR Wallingford, Rep. SR 411.

Price W.A., Tomlinson, K.W. and Willis, D.H. (1972) "Predicting changes in the plan shape of beaches", Proc. 13th Coastal Eng. Conf., 1321-1329.

Pritchard, D.W. (1952), "Estuarine hydrology", Ad. Geophys., 1, 243-280.

Proudman, J., "Dynamical Oceanography", Methuen and Co. Ltd., London, 1952.

Proudman, J. (1940), "The effect of coastal friction on the tides", Mon, Not. Roy. Astron. Soc. (Geophys. Suppl.), 5, 23.

Purnell, R.G. 1995. Shoreline management plans - national objectives and implementation. In Coastal management '95 Putting policy into practice. Thomas Telford.

Pos, J.D., Young, S.W. and West, M.S. 1995. Lizard Point to Land's End Shoreline Management Plan. In Coastal Management '95 Putting Policy into practice. Thomas Telford.

Rayleigh, Lord 1877 "On progressive waves" *Proc. Lond. Math. Soc.* (1), 9, 21.

Reynolds, Osborne, 1887. "On Certain Laws relating to the Regime of Rivers and Estuaries, and on the Possibility of Experiments on a Small Scale", *British Asociation Report.*

Reynolds, Osborne, 1889, 1890, 1891. Reports 'On Model Estuaries' *Annual reports of the British Association.*

Rice, S.O. 1944-45 "The mathematical analysis of random noise" *Bell System Technical Journal,* Vol. 23, pp. 282-332 and vol. 24, pp. 46-156.

Riddell, K. and West, M. Chesil sea defence scheme. MAFF Conference of River and Coastal Engineers, Loughborough, UK, July 1988.

Rendel Geotechnics 1993. Coastal Planning and Management: A Review HMSO.

Rendel Geotechnics 1995a. The Occurrence and Significance of Erosion, Deposition and Flooding in Great Britain. HMSO.

Rendel Geotechnics 1995b. Coastal Planning and Management: A Review of Earth Science Information Needs. HMSO.

Rolfe, J.A.S. and McDowell, D.M. 1953. A New Technique in Tidal Models. *Transactions of the Liverpool Engineering Society* LXXIV, 18-44.

Rossiter, J.R. and Lennon, G.W. (1965) *"Computation of tidal conditions in the Thames Estuary by the initial value method"* Proc. ICE, 31, 25-56.

Rusby, J.S.M., Hunter, C.A., Kelly, R.F., Wall, J. and Butcher, J. 1978 "The construction and offshore testing of the UK Data Buoy (DB1 Project)" *Proc. Oceanology Int. 1978,* pp. 64-80 (Society for Underwater Technology, London) Richardson, D.B. 1995. Soft engineering on the coast-where to now? In Coastal Management '95 Putting policy into practice. Thomas Telford. Roberts, A.G. and McGown, A. 1978. A coastal area management system as developed for Seasalter-Reculver, North Kent. Proceedings of the Institution of Civil Engineers 1,85 pp 777-797. Royal Commission on Coast Erosion and Afforestation 1911. Third ad Final Report.

Simm, J.D., Beech, N.W. and Hohn, S. 1995. A manual for beach management. In Coastal Management '95 Putting policy into practice. Thomas Telford.

Smallman, J.V. and Cooper, A.J. (1989) "A mathematical model for set-down in harbours", Coastal Eng., 13, 247-261.

Smith, I.M., Farraday, R.V. and O'Connor, B.A. (1973) "Rayleigh-Ritz and Galerkin finite elements for diffusion-convection problems", Water Res. Research, 9, 3, 1-15.

Smith, T.J. and O'Connor, B.A. (1977). *"A two-dimensional model for suspended sediment transport"* Oric, 17th IAHR Congress, A11, 1, 79-86.

Southgate, H. (1984) "Techniques of ray averaging" Int. J. Num. Math. in Fluids, 4, 725-747.

Southgate, H.N. and Nairn, R.B. (1993) "Deterministic profile modelling of nearshore processes, Pt. 1,Waves and currents", Coastal Eng., 19, 27-56.

Southgate, H.N. and Nairn, R.B. (1993) "Deterministic profile modelling of nearshore processes, Pt. 2, Sediment transport and beach profile development", Coastal Eng., 19, 57-96.

Stevens, D. W., Carstairs, N.A. and Carter, E. 1995. The Tees Barrage and Ancillary Works Part 3 Design of the Bridge. Proceedings of the Institution of Civil Engineers: Municipal Engineer, 109, pp 168-176.

Stokes, Sir G.G. 1847 "On the theory of oscillating waves" *Camb. Trans.*, Vol. 9, p. 441. See also *Mathematical and Physical Papers, Vol.* 1, pp. 227 and 314, Camb. Univ. Press, 1888.

Summers, D. 1978. The East Coast Floods. David and Charles Wakeline MJ 1989. The deteriaration of a coastline. In Coastal Management pp 135 - 152. Thomas Telford.

Taylor, C., Patil, B. S. and Zienkiewicz, O.C., (1969) "Harbour oscillation: a numerical treatment for unclamped natural modes", Proc. ICE, 43, 141-156.

Taylor, G.I., (1921), "Tides in the Bristol Channel", Proc. Cambridge Philosophical Soc., Math. and Phys. Sciences, 20, 320-325.

Taylor, J. 1864. "Description of the River Tees, andof the works upon it connected with the Navigation".

Thomas, R.S. and Hall, B. 1992. Seawall design. CIRiA/Butterworths.

Thorn, M.F.C. and Huntington, S.W. (1989) "Hydraulic aspects of port and harbour design", Hyd. Res., Wallingford.

Townson, J.M. 1976. *"Piorneers in coastal studies III Thomas Stevenson"* Shore and Beach, vol. 44. No.2.

Townson, J.M. 1981 "The Stevenson formula for predicting wave height" *Proc. Inst. Civil Engrs.*, Part 2 (R&D). Vol. 71, pp. 907-909

Tucker, M.J. 1952 "A wave recorder for use in ships" *Nature*, Vol. 170, pp. 657-661.

Tucker, M.J. 1956 "A shipborne wave recorder" *Trans. Inst. Nav. Archit.*, Vol. 48, pp. 236-250.

Tucker, M.J. 1963 "Analysis of records of sea waves" *Proc. Inst. Civil Engrs.*, Vol. 26, pp. 305 316.

Vernon-Harcourt, L.F. 1882. "Harbours and Estuaries on Sandy Coasts", *Minutes of the Proceedings of the Institution of Civil Engineers,* Paper No. 1772

Vernon-Harcourt, L.F. 1886. on the trainingof the Seine estuary, *Minutes of the Proceedings of the Institution of Civil Engineers,* 84, 241-252 .

Vernon-Harcourt, L.F. 1894. "The Trainingof Rivers", *Minutes of the Proceedings of the Institution of Civil Engineers,* 98, 3-48; also Partiot, H.L. "Estuaries", 49-79. Discussion of both papers, 80-202.

Vernon-Harcourt, L.F. 1889. "Principles of training Rivers through Estuaries", *Proc. Roy. Soc. London,* 45, 504-524.

Vernon-Harcourt, L.F. 1890. "Effects of Training Walls in an Estuary like the Mersey", *Proc. Roy. Soc. London,* 47, 142-144.

Wakelin, M.J. 1989. The deteriaration of a costline. In Coastal Manatement pp 135- 152. Thomas Telford.

Weare, T.J. (1976) "Finite element or finite difference methods for three-dimensional shallow water equations" Comp. Math in Appl. Mech. and Eng., 7, 351-357.

Wheeler, W.H. 1893, "Tidal Rivers".

Williams, M. 1970. The draining of the Somerset Levels. Cambridge Univeristy Press.

Williams, R.G., Darbyshire, J. and Holmes, P. (1980) "Wave refraction and diffraction in a caustic region. Numerical solution and experimental validation", Proc. ICE, 2, 69, 635-649.

Woodworth, P.L. 1990. A search for accelerations in records of European mean sea level. International Journal of Climatology, 10, pp 129-143.

Woodworth, P.L., Shaw, S.M. and Blackman, D.L. 1991 Secular Friends in mean tidal range around the British Isles and along the adjacent European Coastline. Geophys. J. Internat. 104, pp 593-609.

Woollen, W.B. 1981 "The role of a large data buoy in satisfying environmental data requirements in a new UK exploration area". Pages 405 to 435 in *Petromar 80: Proceedings of a conference on petroleum in the marine environment,* held in Monaco, May 1980. London 1981, Graham and Trotman Ltd. for Eurocean.

Wu, X. and Flather, R.A. 1992 "Hindcasting waves using a coupled wave-tide-surge model" pp. 159-170 in *Third Int. Workshop on Wave Hindcasting and Forecasting*, Montreal, Quebec, 19-22 May 1992. Published by Environment Canada, Ontario, Canada.

Yoo, D.H. and O'Connor, B.A. (1988) "Numerical modelling of waves and wave induced currents on a groyned beach" Proc. IAHR Int. Sym. Math. Modelling of Sed. Transport in the Coastal Zone, Copenhagen, 127-136.

Yoo, D.H. and O'Connor, B.A. (1988) "Computer modelling in ocean engineering" B.A. Schrefler and O.C. Zienkiewicz (Eds). Balkema, Rotterdam, 151-158.

Yoo, D.H. and O'Connor, B.A. (1986) "Mathematical modelling of wave-induced nearshore circulations" Proc. 20th Coastal Eng. Conf., ASCE, 1667-1682.

Yoo, D.H., Hedges, T.S.H. and O'Connor, B.A. (1989) "Numerical modelling of reflective waves on slowly varying currents" Advances in Water Modelling and Measurement, M.H. Palmer (Ed), BHRA, 2, 16, 231-244.

Yoo, D.H., O'Connor, B.A. and McDowell, D.M., (1989) "Mathematical models of wave climate for port design", Proc. ICE, 1, 86, 513-530.

HISTORY OF COASTAL ENGINEERING IN ITALY

Leopoldo Franco [1]

ABSTRACT: The paper first describes the main relevant geographical, morphological and meteoceanographic characteristics of the Italian seas and coasts. A broad overview is then given of the long historical evolution of coastal and harbour engineering in Italy, with emphasis on the ancient developments and particularly on the advanced technological achievements of the Romans. The historical perspective includes the "architectonical" approaches of the Renaissance age, the fascinating hydraulic developments of Leonardo da Vinci and the valuable experience of the Venetian and Genoese engineers. The more recent and innovative technical and scientific contributions are also highlighted, with special reference to the typical tradition of composite caisson breakwaters. In the end the present status and prospects of both research and practical applications of coastal engineering in Italy are summarized and referenced.

INTRODUCTION

The Italian peninsula is like a giant *pier* jutting out in the middle of the Mediterranean Sea, the world closest large continental sea. Due to this unique strategic geographical location at the convergence of three continents, with a total coastline extension of 7500 km and over 3000-year long history of human civilization (with difficult inland communications, for both geographical and political divisions), it is easy to understand that Coastal Engineering has played here a relevant role since the ancient times.

Coastal Engineering is a newly defined specialist branch of Civil Engineering which developed in the last decades with the growing human pressure in the coastal zones and the increasing beach erosion problems. As young modern scientific discipline it generally deals with the hydrodynamics of the nearshore zone and with its interactions with sediments and structures, mainly related to harbour and shore protection problems. As old technical field of practical engineering it includes the design and

---

Associate Professor of Coastal Engineering, Terza Università degli Studi di Roma, Dipartimento di Scienze dell' Ingegneria Civile, Via C. Segre 60, 00146 ROMA, Italy

construction of any work at sea (traditionally related to maritime navigation) and it may be considered as a complex "art", strongly based upon experience. Today Coastal Engineering can broadly encompass many other aspects, such as environmental control, survey and planning, water pollution, economics and management.

Up to 170 ports are now classified in Italy (77 of national interest) which need upgrading and maintenance, and new activities are related to offshore platforms and marina developments. However a greater attention is paid to the protection of the coastal areas and to the understanding of the complex nearshore processes. The present importance attached to these problems in Italy is also demonstrated by its high population density (some 56 million people mostly concentrated along the coastal zone, due to the largely mountainous interior land) and by the tourist summer pressure upon its beaches: over 100 million people are coming just to the northern Adriatic shores from all over Europe. It is claimed that the present capital value of a square meter of beach, as derived from all revenues from tourist-related activities, can reach US $ 3,500 ! (Brambati 1993).

Given the special long-term Italian heritage and the fundamental contributions in practical construction at sea, this historical review is mostly addressed to the development of harbour technology and hydraulic design since antiquity. In fact it can be noted that only Italy and a few neighbouring Mediterranean countries retain the unique valuable heritage of ancient harbour engineering. Since no other country from the south-east Mediterranean region is represented in this volume, the detailed review of the ancient heritage has been given a wider Mediterranean scale, thus covering somewhat the global history of early harbour engineering.

Emphasis is then on the remote history, since the related bibliography is of more difficult access for coastal engineers, being mostly produced by archaeologists, historians and geographers. A humanistic and historical background is believed useful to modern coastal engineers, who can always learn from others'experience, also because the present abundance of computational approaches may overshade the fundamental importance of the old natural observation of the physical processes.

On the other end, the review of the "contemporary age" is rather short, in relation to the present much larger amount of available information, since it is also too close to allow a proper historical perspective. However a final quick overview of the present state-of-the-art of both research and applications is included to complete the overall picture and provide useful references to projects and organizations of international interest. Preference is given to review papers and to publications in English.

In fact the aim of this review is to document the Italian coastal setting, facts, circumstances, locations and personalities that have contributed to the progress of Coastal Engineering and indirectly to the national economic development. It may be observed that most advancements are due to individuals, rather than to institutions, as typical of the character of the Italians. It is quite difficult to summarize in a relatively short paper the enormous amount of information. Attention is therefore focussed on the main original and innovative *Italian* achievements. One typical example is the old and new tradition of monolithic breakwaters, which has been favoured by the necessity to build deep sea defence structures along open coasts in nearly tideless conditions.

It can be noted that our rich history of water and coastal sciences has hitherto been largely neglected, being documented to a very small extent if compared to other scientific and technological disciplines (eg.architecture, medicine). In fact no previous historical review of the scattered information on maritime engineering could be found to help the preparation of this paper. This is actually a concise compact report of a long exciting "adventure" in the "ocean" of precious old manuscripts and books (often in Latin or arcaic Italian) which are preserved in many not easily accessible libraries. Due to page and time limitations, this history cannot be fully complete. The content reflects the personal interests and experience of the author, who apologizes for any overlooked fact, personality or reference publication that could have been unwillingly omitted.

Before the historical review, a paragraph is introduced, which gives a brief synoptic description of the Italian coastal environment, with general figures of the relevant geographical, morphological and meteoceanographical characteristics of the Italian coasts and seas. It is reasonable to believe that these conditions have not changed since the ancient times, although there is some evidence that the sea-level has risen by over 1.5 m in the last 2500 years. Most old survived coastal structures are infact underwater, but coastal tectonics and erosion/deposition effects have locally altered the coastline position, so that ancient port sites may be found totally landlocked (Ostia) or deeply submerged (Mysenum).

ITALIAN COASTAL ENVIRONMENT

Geographic, Geomorphological and Administrative Conditions

Italy is located in southern Europe in the middle of the Mediterranean Sea, approximately within latitude 36-46 N and longitude 8-18 E. The country total surface is just above 300,000 Km^2 and the coastline extension, including the islands, is about 7,500 km (40 km^2/km or 0.025 km/km^2). Therefore its "insularity index" (defined as the ratio of shoreline length to the circumference of the circle of equivalent continental area) is pretty high, since it nearly reaches the value of 4. The maximum land distance from the sea is only about 200 km. Maps are shown in figs.1,2,3.

Coastal Geomorphology. The coasts are for some 55% of the high rocky type, while 45% are beaches, mostly shallow sandy beaches, particularly along the northern Adriatic Sea. Steep gravel beaches are common in the regions of Liguria, Calabria, Sicily and Sardinia, where small "pocket-beaches" are often found between rocky headlands. The 32% of Italian beaches is suffering erosion problems, only 5% is accreting and 63% is stable (Caputo et al.1991), also with the help of protection works, which globally defend some 700 km or 20% of beaches. An overall "radiography" of the morphological asset and evolutive trends of some 2000 km of

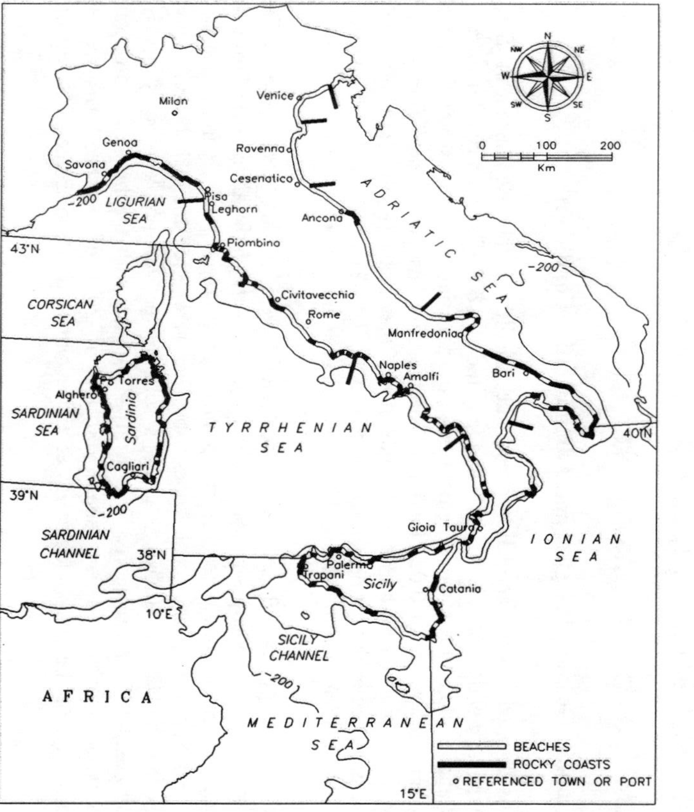

Fig.1. Geographical subdivision of Italian seas and coasts

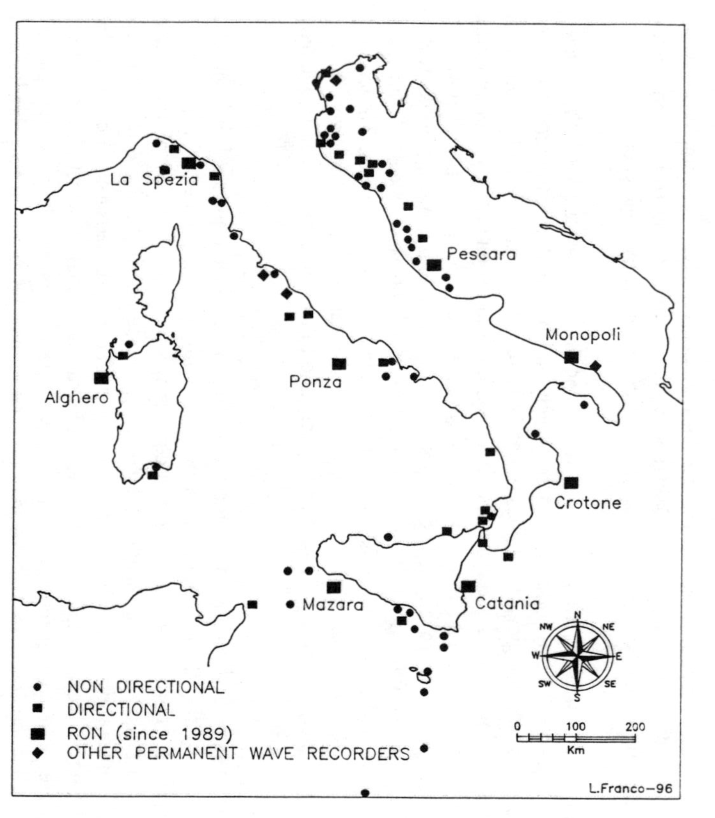

Fig.2. Distribution of wave recording stations off the Italian coasts

littorals is given in the 60 maps at 1:100,000 scale of CNR's Atlas of Italian Beaches (1984). Reference is also made to Zunica 1976, 1985 and to the vast CNR's bibliography.

Italy is a "geologically young" region, but the variety and ancient age of rock formations indicate a very complex geological history. Volcanic and tectonic activities are still present. The articulated Italian orography has kept rising and, despite the small catchment basins of the fragmented hydrographic network, the numerous streams and rivers have been feeding the beaches with sediments eroded from the steep slopes. This material built up wide coastal plains which have stopped marine ingression: morphological deposition forms can be observed, such as deltas, lagoons and "*tombolos*" (a widely used Italian word, because typically formed by wave diffraction behind nearshore islands in almost tideless seas). The coasts have been modelled by the sea level variations induced by the glaciations (with final rise of 120 m between 15000 and 5000 years ago) and "fossil" beaches can be found either on the mountains of Calabria or in the sea floor of the south-central Tyrrhenian Sea and northern Adriatic (the latter is now quarried for beach nourishment of Venice lidos).

Quaternary deposits are made with marine and continental rocks of various kinds. Sediments are predominantly quartz but a variety of mineralogic types (siliceous, volcanic) can be found, with the exception of the carbonate sands from marine biota. Italian beaches have been accreting until the end of last century, but they are now generally eroding, especially after the reduction of river sediment supply (mainly due to human activities such as damming and aggregate excavation).

Sea Level Rise. Measurements in the Tyrrhenian sea have been carried out since 1884 in Genoa. The computed rate was 1.2 mm/year for the period 1897-1942 and 1.54 mm/year in 1931-1971, while the present mean rate of sea level rise is 1.6 mm/year (Mosetti,1989). However this worldwide issue has not been so far a problem on fashion for Italian scientists and coastal engineers, with few exceptions such as the recent studies related to the design of the Venice flood barriers (CVN) and an application of a new Bruun-type model to predict shoreline retreat along the Tuscany coast (Pranzini et al.1995). However high rise eustathic scenarios would have dramatic consequences, especially for the shallow Adriatic beaches. Here, also coastal subsidence (due to the compaction of sediment deposits favoured by groundwater abstraction) has played a negative role with a land settlement of some 0.15 m in the first half of this century until recent deceleration after proper countermeasures.

Administrative Regulations. It can shortly be said that all Italian coasts belong to the State, although the jurisdiction for coastal protection works is now shared with the Regions (numbering 15 with sea shores out of the total 20) and new laws are now just being enforced which define coast classifications according to the importance of the river catchment basin. The technical authority for design and supervision of coastal works is the *Genio Civile Opere Marittime* of the Ministry of Public Works, subdivided in 11 geographical compartments as shown in fig.1.

Meteoceanographical Conditions

Seas and shelf. The very special shape of the 1200 km long Italian peninsula and the near large islands of Sardinia, Sicily and Corsica defines a number of "narrow" seas or basins (named in fig.1) with somewhat different meteoceanographical features. The largest depth of the Mediterranean occurs in the Ionian Sea with more than 5000 m, while the deepest point of the Tyrrhenian Sea is at -3700 m. The continental shelf is quite small (typically 10 to 40 km wide) all along the western and southern coastline, which is often bordered by nearshore deep waters. Only the narrow Adriatic basin has a relatively shallow seabed with the shelf extending for over 500 km from the northern shores to the area off the big Gargano promontory to the southeast.

Tides and Surges. This long shelf, together with the special semi-closed geometry of the Adriatic basin is responsible for the well known storm surges in Venice and neighbouring coastal areas, due to the setup of southeasterly winds and 22-hours seiching. The north Adriatic is also the only sea area in Italy where the astronomical tidal range can reach 1 m at springs. In most other Mediterranean coastal seas the semidiurnal micro-tides have an amplitude of just 0.4 m at springs and 0.2 m at neaps. In fact chart elevations are often simply referred to the Mean Sea Level Datum.

This peculiar feature has obvious important consequences in the design of harbours and shore protection works: port developments could not generally take advantage of sheltered estuaries as in oceanic locations and thus often needed costly breakwaters in deep wave-exposed waters; with regards to coastal defence works, the modesty of tides promoted the diffusion of detached breakwaters with tombolo formation and now the boom of submerged structures, which are more "environmentally friendly".

Currents Either density, tidal and wind-driven currents are generally weak (less than 1m/s), with the exceptions of the Messina Straits (up to 3 m/s due to tidal phase shift between the Ionian and Tyrrhenian Sea) and the Bonifacio's (between Corsica and Sardinia) and a few tidal inlets, such as those of Venice lagoon. A permanent weak density current runs in the anticlockwise direction all along the coastal water surface. The full water exchange of the Med Sea takes about 100 years, thus producing little autodepuration capacity and consequent ecological fragility, also due to the increasing human pressure along the shores and to pollution. Average seawater salinity is 3.8%.

Wind. Wind conditions are generally moderate: average speeds range between 5 to 7 m/s onshore at ground level, with strongest and more persistent winds occurring in northern Sardinia under prevailing north-westerly directions (Troen et al. 1989). Peak velocities associated with the fall/winter low pressure systems tracking across the Med Sea from west to east may reach 30 m/s. Useful overwater wind data have been recorded at some 60 stations located near the shore and on small islands by the Hydrographic Naval Institute of Genoa from 1927 to 1960 (IIMM 1984) and by the Air Force Meteorological Institute (AM) since 1951. Statistics from the latter data had been produced together with ENEL, National Electricity Board (AM-ENEL 1980).

Global wind forecasts are issued by AM, based on the advanced model results from the European Centre for Medium-range Weather Forecast located in Reading (UK).

Waves. Wind waves are the main hydraulic parameter affecting the response of Italian beaches and coastal structures. Storm waves are mostly generated in geographically limited fetches, thus producing short steep 3-D seas and modest swell activity. Directional and extreme wave statistics were typically gained from offshore shipboard observations and with simple hindcasting methods, but more accurate and reliable data are now obtained from advanced 3rd generation wave models and from systematic instrumental measurements. An updated inventory (Franco 1993) showed nearly 100 wave recording stations installed since 1974 along the Italian coasts (fig.2) by some twenty different institutions (including ENEL, oil companies, universities), after the pioneering observations off Genoa breakwater in the early 30's (described later).

A remarkable achievement is represented by the integrated permanent National Wave Measurement Network (RON), operational since 1989 with an average data acquisition rate above 90%. It is managed by the Italian Hydrographic and Tidal Service and includes 8 directional Wavec buoys moored in 100 m depths (remote-controlled by the Argos satellite) and the corresponding onshore receiving stations for real-time data transmission (De Boni et al.1992). New deepwater wave statistics from these 6-year directional records can be found in Archetti et al.1995.

The wave climates are more consistent, unidirectional (westerly) and energetic along the western shores, while reduced and more variable wave activity affects the Adriatic and Ionian coasts (with prevailing northerlies). In these eastern basins the average specific gross wave power is smallest (3-4 kW/m), while it is largest at the Sardinian Sea coasts (16 kW/m), but yet not enough for a convenient production of wave energy with the present technology.

Typical storm durations are 2-4 days, with spectral peak periods in the range of 5-13 s and 1-year return significant wave heights of 3-4 m and 50-year return H_s of 7-8 m. Wave energy spectra at storm peaks are typically well fitted by the unimodal JONSWAP shape, though the mean value of the peak enhancement factor γ is found to be 1.9 (instead of 3.3 as for the North Sea). The directional energy spread is well described by the cosine-power function with an exponent between 2 and 4.

The most severe sea states occur off the western coasts of Sardinia and Sicily, under mistral wind storms (from WNW): the highest sea state was instrumentally recorded by the RON buoy off Alghero on 1.1.1994 with H_s=9.2 m, T_p=12.5 s and H_{dmax}=15.0 m. However, severe damages to various port and coastal structures along the eastern Tyrrhenian shores were caused by another exceptional storm reaching H_s=11 m off west Sicily on 11 January 1987, as hindcasted with a refined wave spectral model covering the whole Mediterranean Sea. In fact either 2nd and 3rd generation wave forecasting models are being used with good results. The advanced WAM model is now operational for the Med Sea upon a computational grid of just 25 km mesh size (Cavaleri et al.1991).

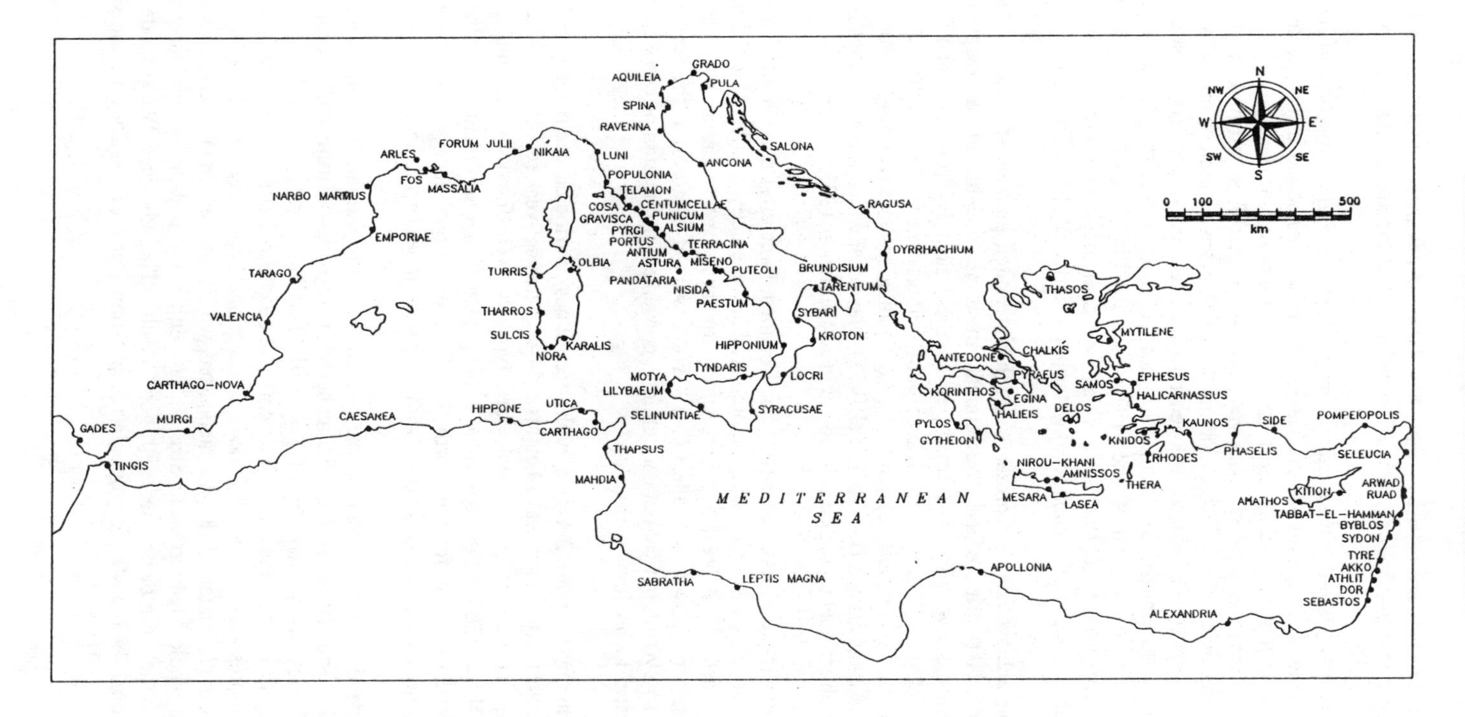

Fig.3. Location map of main ancient Mediterranean harbours, generally called with their original names (mostly with visible remains)

HISTORICAL DEVELOPMENTS

Coastal Engineering developed very early in the history of human civilization, particularly in the Mediterranean basin, together with the origin of maritime traffic, which has always been a key factor in the economic and political growth of nations. In fact past efforts were mostly devoted to port structures, with the exception of a few places where life has been dependent on the coastline protection. One such case is Venice and its lagoon, where sea defences (hydraulic and military) were vital for the survival of the thin coastal strips: the old impressive shore protection works built by the Venetians are still admired today.

The developments in the ancient classic times were related to the different dominations and cultures which followed especially in the eastern and central Mediterranean basin: Egyptians, Minoans, Phoenicians, Carthaginians, Greeks, Etruscans and Romans. After the Roman age nearly no technical evolution took place until Napoleon times! Therefore it is believed worthwhile to give a broader review of this unique valuable antique heritage, which is still visible in many surprisingly advanced and intact coastal structures. The study of these "monuments" can give to present coastal engineers and managers a useful humanistic background and even some good ideas for new designs, especially in view of the *gentle* environmental harmony of the early natural harbours. Moreover the conservation, restoration and valorization of these remains in suitable "archaeological coastal parks", or even re-use for modern marinas, could enhance the touristic-cultural offer of many Mediterranean countries (Franco, 1996).

In a simple historical classification coastal/harbour engineering might be subdivided in a "naturalistic" *antique age* (2000 BC - 500 AD), a "technological" *modern age* (500-1950 AD) and a "scientific" *contemporary age* (after the first ICCE in 1950). However these chronological boundaries, as usual in the history of sciences, are inevitably conventional and somewhat artificial: they are not be considered as a rigid frame, but as a simple marking criterion for the organization of this paper.

Antique Age

For a general review of ancient harbour archaeology reference is made to Shaw (1974) and Blackman (1982) with an extensive bibliography. Two monographical books on ancient Italian harbours were also edited by the Navy (Marina Militare 1905). A more technical overview has been recently given by De la Pena et al.1994. A reference map of the main ancient harbour sites in the Mediterranean Sea is proposed in fig.3.

It is expected that more information on the outstanding achievements of ancient coastal engineers is still to be gained after future discoveries from modern underwater archaeology and more sophisticated survey techniques. In fact very few written reports on the ancient methods for design and construction of coastal structures are available. The only technical handbook is the one of Vitruvius (27 BC) mainly related to the Roman engineering experience. Further useful descriptions are given in the classic Greek and Latin literature by Herodotus, Josephus, Suetonius, Pliny, Appian,

Polibius, Strabo and others. They also show the ancients' capability to understand and handle various complex physical phenomena without supporting data or computational tools, such as the tidal phenomenon, the Mediterranean currents and wind patterns and the wind-wave cause-effect link. The Romans first introduced the wind roses too.

Pre-Roman harbours. The so-called *proto*-harbours were mainly used for refuge, unloading of goods and freshwater supply for the shallow-draught wooden vessels cruising along the "inside routes" of the Mediterranean Sea (only during the good season). These early harbours were "natural", typically located in favourable geographical conditions, such as sheltered bays near capes or peninsulas, behind coastal islands, at river mouths, inside lagoons or deep coves, where short breakwaters were often sufficient to supplement the natural protection. Harbours were generally close to high coastal mountains easily visible from the sea in the distance and they were possibly spaced at 40-50 km intervals to allow safe day by day tranfer to the vessels sailing coastwise at a speed of 3-5 knots. Ports were necessarily built on the numerous islands and along the mainland coasts to serve a large hinterland and they were often closely linked with city sites. In fact the harbour basin was often enclosed with fortifications, even closable from the sea, and separated from the city for reasons of security (military harbours) or control of goods and passengers.

Probably the oldest man-made seaport was the first harbour of Alexandria (Egypt), built by the **Minoans** to the west of Pharos island around 1800 BC. The main basin was 2340 m long, 300 m wide and 6-10 m deep to accomodate 400 ships of 35 m length; the numerous breakwaters and docks (14 m wide) were made with rock blocks (5 m long). Remains of early Minoan coastal structures are also visible at the harbour of Nirou Khani, Crete (1500 BC), which still shows two rectangular slips (43 x 6 m) cut into the dune rock and divided by a 0.8 m wide masonry rock wall (Inman 1974). The earliest harbour works on the Italian coasts were probably made later by the expert **Phoenicians** who set up the maritime trade with various southern coastal towns since the 7th century BC. However the best known Phoenician harbour structures are found on the eastern Mediterranean shores of today's Lebanon and Israel (e.g. Arwad, Byblos, Tyr, Sidon, Dor, Akko, Athlit) (fig.4).

The design of these early harbours was mainly dictated by nautical and military constraints, such as providing safe access even in hard weather through usually narrow entrances, which could be easily controlled and even closed with chains. Two or more entrances were sometimes provided, either to ease navigation under variable winds, and to separate different port routes and traffic (commercial, military and fishery) and to favour water flow in the relatively tideless basins, in order to keep silt in suspension and avoid harbour siltation. Prevention of siltation was infact a fundamental issue, since mechanical dredging was not feasible. Not only multiple port entrances were provided, but also underwater channels and tunnels through the moles (eg. Syracuse). River flow, sometimes diverted in settling-tanks, was also used to prevent siltation. The flushing channels, often controlled by sluice gates, would be open to the sea above MSL at the side where the seabed was rocky and shallow enough for the waves to break and release most suspended sediment before reaching the breakwater. Even ramps were constructed to allow the wave crests to sweep over the breakwater and collect sand-free water in a tank at an higher elevation for periodical release into the harbour.. This "wave pump" system is now proposed for flushing Mediterranean marinas with the novel scope to improve the quality of polluted water and openings through the breakwaters (also with the help of pumps) are again being created for the same reason.

The natural de-silting concept was still in use in the imperial Roman harbour of Caesarea Maritima. The recent discovery of a dump off the entrance seems also to show that the outflowing current not only prevented the sand from entering the port, but even kept the floor of the closed basin free of jettisoned garbage. Moreover the funnel-like topography of the entrance would channel the outgoing current through the whole opening. (Raban 1988).

In order to reduce large wave overtopping at the vertical rocky cliffs and breakwaters the Phoenicians excavated hollows and trenches in the rock. These "wave catchers" would channel and drain the water off back to the sea and were also used as an immediate source of building stone. This concept was also later applied by the Romans at the harbour of Pandataria island (now Ventotene) (fig.5) and a modern artificial version of this crownwall modification was recently patented by Sogreah under the name of "Overspill Basin" breakwater and it was used at Fontvieille, Monaco (on top of precast cellular caissons towed from Genoa in 1968) (Borzani 1973).

Another typical characteristic of the earliest harbour works cut out of rock reefs or islands was that the rock mass was flattened on its sheltered landward side to make a quay, leaving a protective rock wave-wall on the sea side (Frost 1972). A wonderful example of the natural “carved breakwater” is again represented by the imperial Roman harbour of Pandataria (fig.5). This remarkable “coastal structure” is cut from the bed-rock by creating a suitable wave-absorbing seaward profile, with a deep parabolic shape and a mild grooved absorbing slope near the waterline.

A high level of technology for submarine construction was already achieved, as shown by the regular placement of small stones to build smooth walls which are still in place. An innovative technical feature was the use of ashlar header quaywalls: the slim blocks were laid in tight courses with their long joints in a very close fit to give maximum drag and thus solving the problem of the vacuum effect of the retreating water from the vertical wall (Raban 1988). The Phoenician engineers were also able to prevent the undertrenching of walls laid on sand by paving it with a layer of large pebbles. External breakwater slopes were mostly armoured with randomly placed rock.

The western phoenician colonies later assumed the naval preminence in the Mediterranean under the guide of Carthage (750-146 BC). The characteristic novel feature of **Carthaginian** port layouts was the so-called *cothon*, an internal artificial basin, excavated behind the coast and joined by one or more channels to the sea. The cothon could be used for the home fleet (leaving the outer harbour for foreign ships), or for the military fleet, or just for ship repair works. Beautiful examples are still

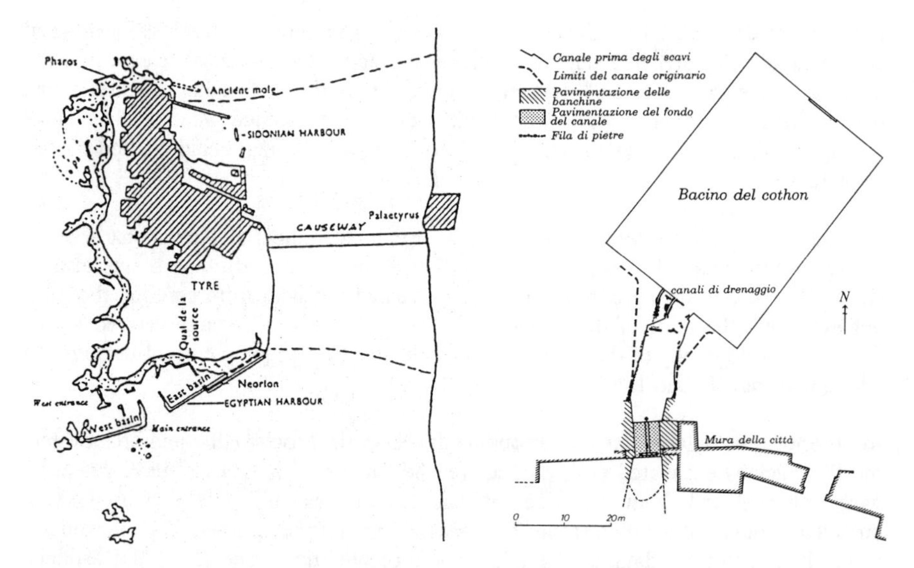

Fig.4. The multiple harbours of Tyr

Fig. 6 Plan of the Phoenician harbour of Motya (Shaw)

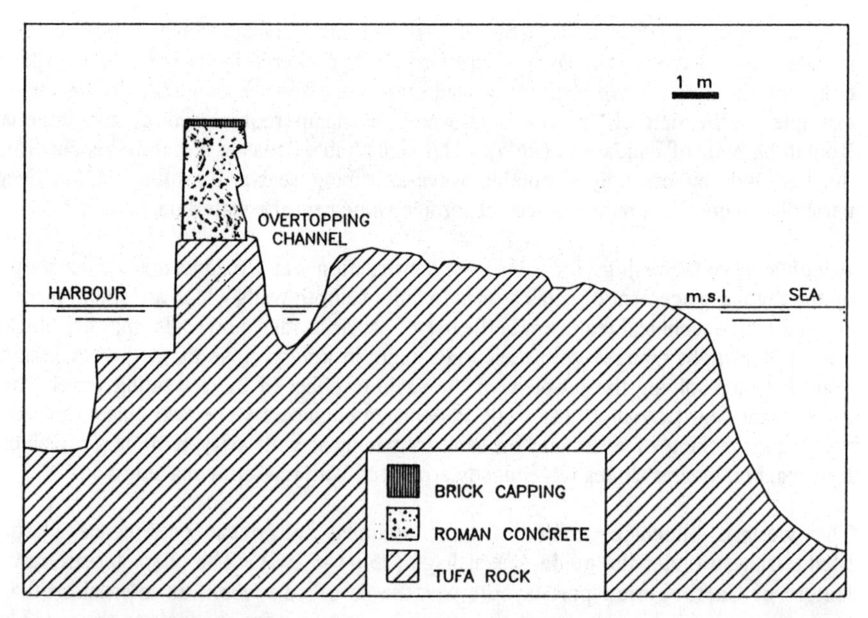

Fig.5. Cross-section of the Roman "carved rock breakwater" with overspill stilling channel at Ventotene

visible at Motya (Sicily) and Carthage (Tunisia). At Motya the cothon might have been closable and drainable to be a dry-dock (fig.6).

The famous harbour of Carthage, recently studied under a UNESCO safeguard program, shows two large basins excavated inland, a rectangular one probably devoted to commercial traffic and the inner ring-shaped refuge basin for the navy, invisible behind a long high wall. Right in the centre of the inner basin the 125 m wide circular Admiralty island was used as a shipyard for up to 200 war-ships and dominated by a control tower. It seems that radial wooden finger piers were also provided along the circular perimeter for an easier sideways berthing, like in modern marinas. Columns were also used as bollards instead of mooring rings and covered berths were arranged within monumental arcades. Other Punic harbour ruins are visible in Sardinia (Nora, Tharros) and in Sicily (Lylibeum), besides the north-african coast.

The **Greeks** (VI-III c. BC) also took advantage of narrow peninsulae for building safe "multiple harbours" (Cnidos had four). The breakwaters were mostly built with cut rocks regularly placed without mortar on rubble mounds. Herodotus reports of a breakwater at Samos built on 530 BC in water depths up to 35 m ! The large Athenian triple harbour of Piraeus is famous for his 372 covered slipways ("shipsheds" 37 m long and 6 m wide with a slope of 1:10). Alexandria was also a well known monumental Greek export port, actually built by Alexander the Great in later Ellenistic times (III-I c. BC) behind the coastal island of Pharos. A 1.5 km long breakwater, with two openings, joined the island to the mainland dividing two basins with an area of 368 hectars and 15 km of quay front . The port also became famous for its impressive 130 m high lighthouse tower, to guide ships from a distance of about 50 km towards the port in a sea without landmarks. The multi-storey building, one of the Wonders of the Ancient World, eventually collapsed due to earthquakes in the middle ages, though it was made with solid blocks cemented with melted lead and lined with white stones, which are just recently being finally recovered underwater by archaeologists.

Another peculiar feature of the Graeco/Ellenistic harbours was the use of colossal statues to mark the entrance. The most famous application reported by historians was the 30 m high Colossus of Rhodes, standing on top of the breakwater heads (but not yet discovered). It is interesting to observe that three ancient windmill towers are still surviving upon Rhodes breakwater, impressive precursors of today's wind turbines for energy production. Greek ports in Italy were created in the southern regions (*Magna Grecia*). Typically they are still part of the town (even closed within the walls like Delos), whereas in the Roman Empire they will become an independent infrastructure, with own buildings and goods storage deposits (*horrea*).

In the same times the western central part of the Italian peninsula was ruled by the **Etruscans,** who also constructed new harbours (often in coastal lagoons), later used and upgraded by the Romans. Several marks still exist along the Tyrrhenian coast north of Rome. An interesting example is the *colonia maritima* of Pyrgi (now S.Severa), where an inland basin with quaywalls is connected to the sea where submerged offshore rubble breakwaters still exist (Protani et al.1989). Shore protection works have recently been built here also to defend a superimposed nice medieval castle, deteriorated by the wave activity focused on the shallow promontory.

The Etruscans were also expert in controlling water circulation against siltation by linking the lagoon to the sea with one or more canals, as still visible at Ansedonia. Here the excavations of the ancient harbour of Cosa (further developed by the **Romans** in the II century BC) were carried out by the American Academy in Rome since 1948 (Lewis 1973, Brown 1980). The harbour represents a transition between the natural anchorages of the early Etruscans and the elaborate artificial harbours of the later Romans (fig. 7). It was composed by a lagoon and an outer basin sheltered by the limestone promontory and by breakwaters (now submerged) made with 2 t rock blocks directly quarried from the adjacent cliff, piled on the seabed and later cemented with natural sand concretion, hydraulic cement and addition of broken pottery to increase the bond. The rocks are now worn to an oval shape and reduced in size due to sand abrasion and animal borings over 2000 years. A few tufa-and-mortar eroded piers (docks?) are still standing out of the water and some detached breakwater extensions are visible underwater near the 50 m wide entrance: their staggered arrangement was probably intended to provide the usual scouring de-silting currents. Existing spectacular features are the gigantic natural sluiceways formed by two nearly parallel cuttings along the adjacent rocky cliff, the natural crevasse Spacco della Regina (260 m long, 30 m deep and 1 to 6 m wide, after suitable wall scarping and bed clearing) and the artificial Tagliata (70 m long and 4 to 5 m wide, partly tunneled), which link the deep sea with the inner harbour and lagoon (fig. 8). Vertical rockcut slots are clearly visible on opposite sides of the channel which were surely used for sliding boards as sluice-gates to control the water flow (probably also the fish flow to and from the lagoon) according to wind conditions and tidal cycle.

<u>Roman harbours.</u> The first harbours of the golden imperial times (since I century BC) are Forum Julii (Frejus), Mysenum (Miseno) and Puteoli (Pozzuoli). Other marks of well known Roman harbours on the Italian shores are still visible at Nisida, Terracina, Antium (Anzio), Portus (Ostia-Roma), Ancona, Centumcellae (Civitavecchia), Astura, typically protected by monolithic concrete breakwaters.

The revolutionary innovation in harbour engineering was infact introduced by the Romans, who learned to build walls underwater and therefore managed to construct solid breakwaters to protect fully "external" harbours with free planshape, even curvilinear (to enhance de-silting currents) along a coastline without many natural protection. They learned the use of metal joints and clamps to fasten neighboring blocks and discovered the hydraulic cement made with pozzolanic ash (from the volcanic region around Naples), which hardens underwater. Therefore the Romans replaced the traditional Greek rubble mound breakwaters with vertical and composite concrete walls (*opus pilarum*), even for the rehabilitation of the old Greek harbours. In fact on deep seabeds they also used to lay a rock foundation up to -6 or -7 m MSL and then construct the vertical wall. Monolithic coastal structures could be built rapidly and needed little maintenance: in fact many works have survived to sea attacks for over 2000 years. Another possible reason for the Roman preference for vertical

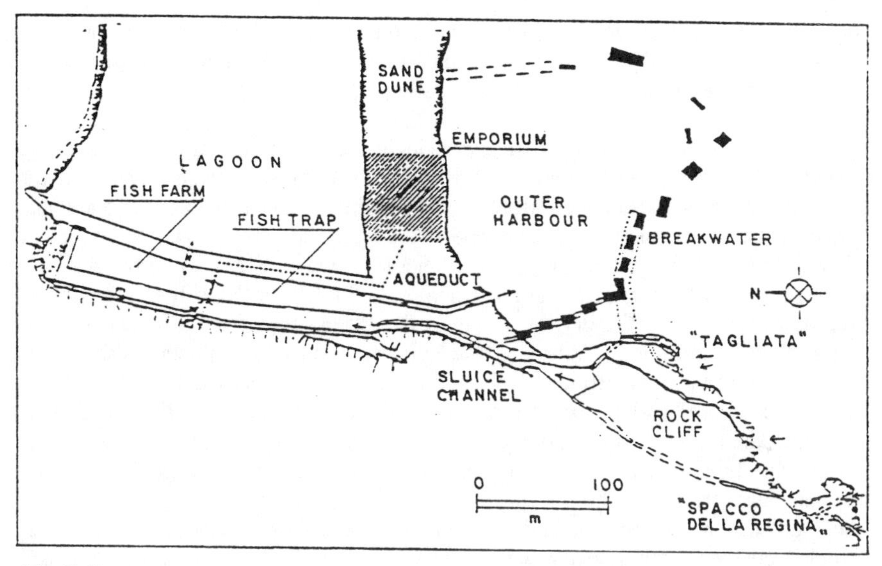

Fig.7. Reconstruction of the Etruscan and Roman harbor layout at Cosa (Brown)

Fig.8. View of the ruins of Cosa harbour from an ultralight plane, also showing the Etruscan artificial rock-cut channel named *Tagliata* (photo by Franco 1991)

coastal structures was the potential of wave reflection for desilting. Anyway either rubble or bronze slabs were placed by the Romans in front of the vertical walls to protect the seabed from toe scour (Oleson 1988).

However the Romans did not just follow a single codified tradition, but properly used a variety of design concepts and construction techniques at different coastal sites to suit the local hydraulic and geomorphological conditions, and availability of materials. Geotechnical conditions were analysed in order to choose a suitable foundation method. In hard bottom soils only a superficial layer was dredged to be replaced by a smooth rock bedding layer. If the bottom was sand a trench wider than the wall was excavated (typically in the dry, as shown in fig. 9) and then filled. In case of high layers of mud, according to Vitruvius, the Romans used to drive numerous short piles, (with 0.45 m square section) made of wood from olive tree, black poplar or holm-oak with scorched tip, and put coal rock between the pile heads. Infact the Romans developed the mechanical technology of cranes and pile drivers (Julius Caesar 55 BC). For pile driving in water the crane was placed on a raft or barge, and guides, hoist and pile-hammer were used. The iron pile caps were covered with lead against corrosion. The fig.8 also shows the method for building a rock-block vertical structure after pumping out water and leveling the seabed within two 1.5 m wide cofferdams made with sheetpiles (wooden posts set closely together) filled up with clay bags (Prada et al. 1995), probably strengthened by cross tie-rods (ropes).

Instead, the typical breakwater construction technique in sheltered sites consisted in pouring a mix of cement, pozzolan and brick pieces (impermeable mass concrete) within immersed wooden forms (*arcae*) supported by driven piles and tie-rods (*catenae*) and later casting a concrete emerged superstructure, about 6 m wide, covered by bricks or joined squared rock slabs (fig.10) (Clementi 1981). Figure 10 also shows the characteristic block with a hole (*dactylium*) used for ship mooring on the rear side: the hole axis could be either horizontal for mooring lines or vertical for tie piles, as still visible at the river port quaywall of Aquileia (fig.11).

Several prints (and even some remains) of a more complex reinforcing "skeleton" of oak pillars and cross-bars can be observed in the moles of the Neronian port of Antium. The pillars inside the formwork (*destinae*) could have worked as anchors, while the horizontal cross-bars (at +1 m MSL) could have also usefully supported a working floor. The modular frames are repeated with 2.5 m spacings. The joints show that half mole width (6 m) was probably cast against the quickly hardened solid wall edge of the other half. The "staggered"construction sequence thus allowed saving the forms for the adjoining portions. The perimetral wooden forms could also be "sunk" in a first fresh cast of mortar to increase the frame stability (Felici 1993).

Sometimes, instead of the forms, old ship hulls were sunk and filled with concrete, saving time and material. A well known example is the main west breakwater of Portus (Rome) built under Claudius (around 50 AD) partly by sinking the 7400 t DWT, 104 m long Caligula's ship (probably the largest wooden hull ever), which had transported the Vatican obelisk from Egypt, in order to create a solid foundation for the big

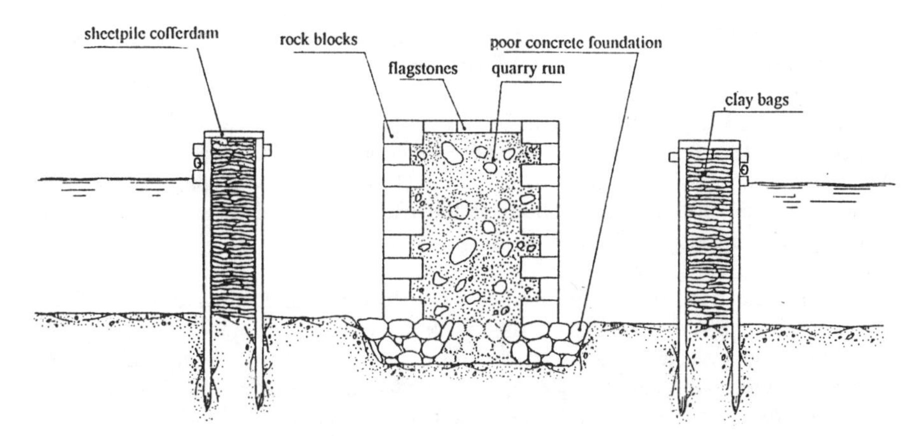

Fig.9. Roman method of construction of a rock block vertical breakwater in good seabed conditions at exposed sites according to Vitruvius (Prada et al.1995)

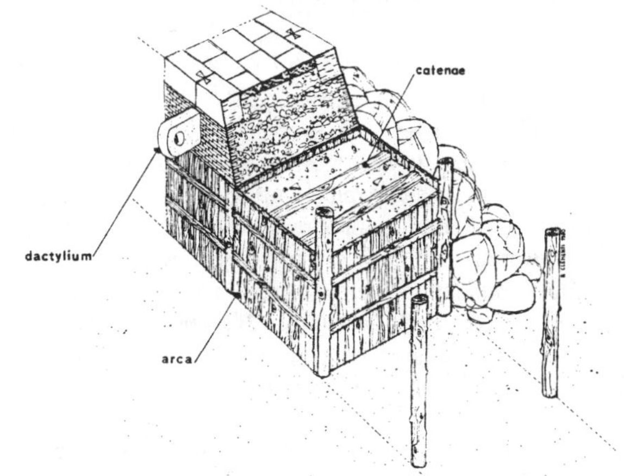

Fig.10. Reconstruction of the Roman construction system of a vertical breakwater with cast-in-situ concrete within wooden forms and tie-rods (Clementi)

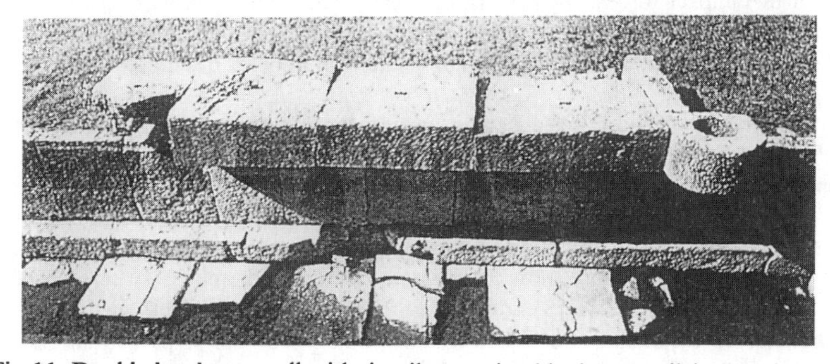

Fig.11. Double level quaywall with tie-pile mooring block at Aquileia river harbor

lighthouse (fig.12) (Testaguzza 1970). The great breakwater is still partly visible, abandoned within the grass grounds of Fiumicino airport.

In wave-exposed locations a different construction method could be used according to Vitruvius. Work progressed from the shore by dumping a submerged rubble mound. A perimetral sheet-piling was then filled with sand and topped with concrete. After the opening of little doors in the cofferdam the natural removal of sand allowed the concrete block settlement (Prada et al.1995).

Another advanced technique was invented by the Romans for deep water applications: the watertight floating cellular caissons, precursors of the modern widespread technology of monolithic breakwaters. Double-walled wood forms constructed nearshore were towed into position over a foundation of boulders on sandy bottoms and waterproof mortar packed between the double walls to sink the form. Concrete was then poured into the water-filled frames by lowering baskets. This system was used by Herod the Great's engineers in 18 BC to build the 60 m wide breakwaters of Sebastos (Caesarea) harbour (Hohfelder 1987).

The recent large excavation project at Caesarea (CAHEP) also revealed a subsidiary parallel breakwater to reduce wave impacts onto the main walls (fig.13). The crest of this small "tandem" rubble barrier was probably at the sea level, some 15-30 m seaward of the main breakwater. It was interrupted with gaps to provide an exit for rip currents and prevent the wave setup piling up in the "stilling basin"(Raban 1988). Another unique feature of the main breakwater at Caesarea is the "natural" construction technique used for building up the very wide core: the block walls or caissons on the rubble foundation were framing hollow "trap"compartments which would be filled by the wave-carried sand within 2-3 years and then paved and built-on above water level (fig.14). The large width of the imperial port breakwaters allowed the innovative location of various installations (eg.warehouses) upon the crown.

An original vertical wall breakwater was built at Thapsus (today Rass Dimas, Tunisia), whose 259 m long and 12 m wide impressive remains were still surviving in 1869. The peculiar feature of this monolithic mole was the presence of vents through the wall to reduce wave impact forces (fig.15) (Tasco et al. 1965). This idea was resumed some 30 years ago by Jarlan to provide absorbing chambers in perforated caisson breakwaters and has numerous modern applications in Italy (as described later).

Another particular version of the Roman vertical-type breakwaters was the system of detached piers joined by arches. Remains of "arched moles" have been found in Nisida, Mysenum and Puteoli (all in Naples Gulf). As depicted in a famous fresco of Stabiae (fig.16c), the ancient monumental superstructure of Pozzuoli's mole was adorned with arches and columns to be an important social walkway. It was 372 m long and rested on 15 piers each 16 m square (now incorporated in the modern breakwater) as shown in the surveys of De Fazio in 1814 (fig.16 a, b). The large water depth at the last pier (-16 m) can be surprising due to the shallow draught of the ships at that time: it demonstrates that not only nautical but also hydrographic reasons (wave disturbance,

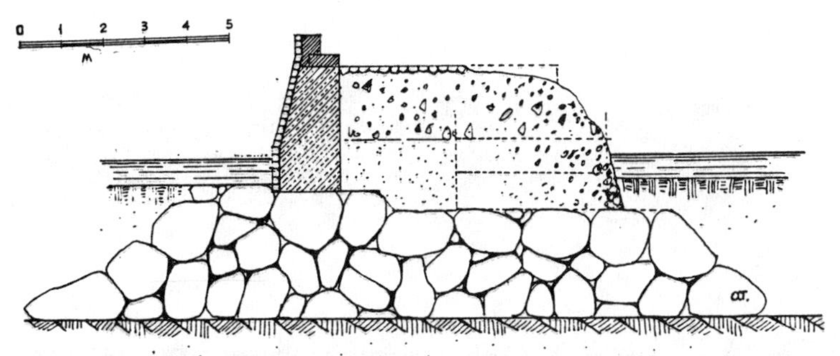

Fig.12. Cross-section of the main breakwater of Claudius Port (Rome): the concrete superstructure was cast after sinking ship hulls as lost forms (Testaguzza 1970)

Fig.13. Conceptual depiction of the ancient harbour of Sebastos based on recent archaeological data (Center for Maritime Studies, University of Haifa, Israel)

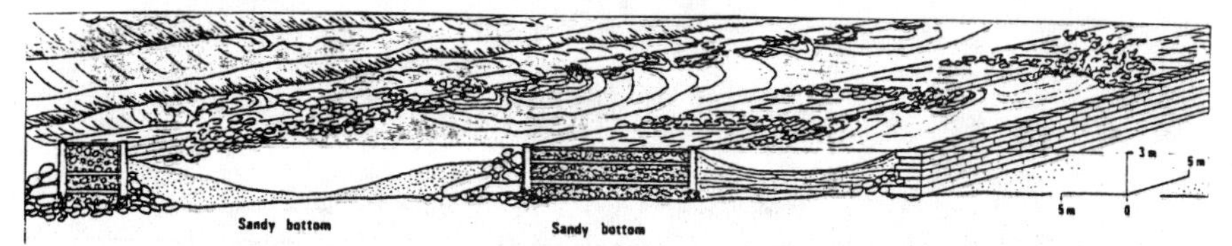

Fig.14. Schematic block diagram across Sebastos main breakwater during the initial phases of construction (C..M.S.- University of Haifa, Raban 1988)

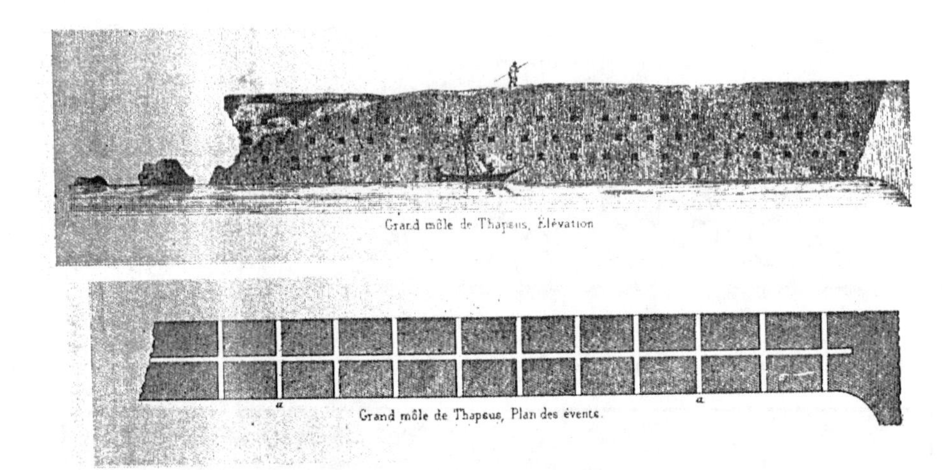

Fig.15. The perforated vertical breakwater at Tapsus in a drawing of the XIX century

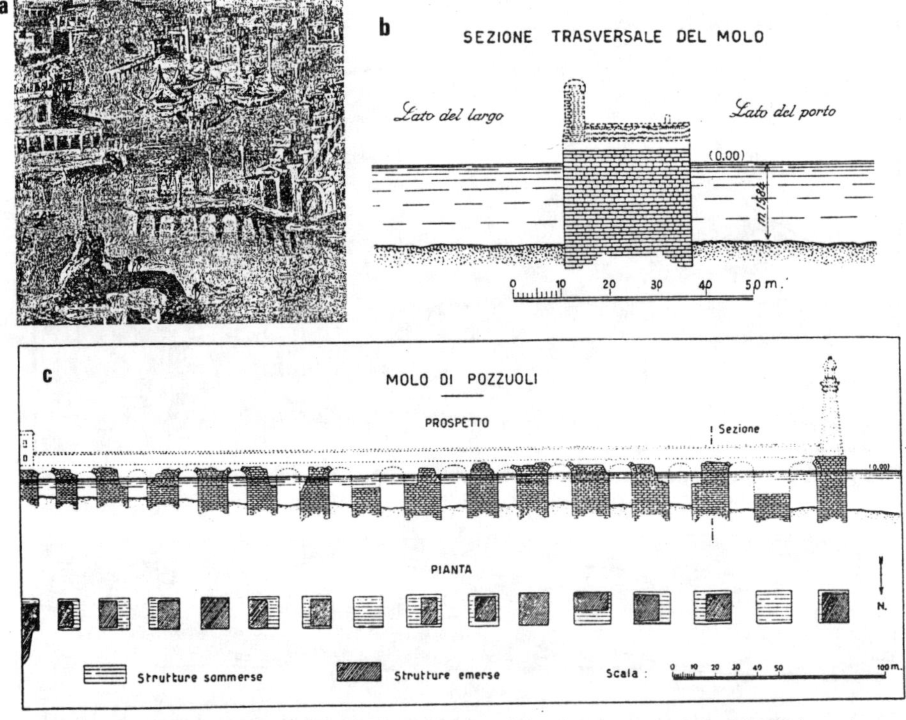

Fig.16. The arched mole of Puteoli as depicted in the fresco of Stabiae (a) and as surveyed by De Fazio in 1814: b) typical pier section; c) view and plan

port siltation) were taken into account in the design of harbour protection structures. At Mysenum the breakwater was even made with a double row of arches in an off-set position to reduce wave penetration. The technical reasons for these unconventional arched breakwaters may be: control water circulation against siltation; reduce wave reflection which affects coastal navigation; save material in deep water; borrow the successful construction techniques and aesthetical views of the famous Roman aqueducts and bridges. This kind of "open" breakwater was still favoured by harbour engineers of the XIX century (De Fazio, 1814), despite its obvious uneffectiveness due to the actual sedimentation and unacceptable wave disturbance in the sheltered basin (at Astura they appear partially closed at a later stage). However, they were probably only used in sheltered locations as outer port protection, and the arcades could have been equipped with sliding gates for temporary closure during storms.

In antiquity the largest artificial harbour complex (1.3Mm^2) was the imperial port of Rome: the maritime town at the Tiber mouth was infact named Portus (=*The* Port) (Lugli et al.1935, Testaguzza 1970). It is now some 4 km from the sea, partly buried under Rome-Fiumicino airport (the outer port of Claudius) and partly within a private estate (the inner hexagonal basin later built by Trajan with sides of 360 m and a depth of 5 m) (fig.17). Despite its importance for the supply to the empire capital (over 300,000 t/year of wheat from Egypt and France), the port always suffered from river siltation, but this is also the reason for its conservation in modern times, though not yet fully accessible to the public. Numbered columns have been found, set back from the edge, on the quays of Trajan harbour, which meant to identify each berth.

Trajan (around 100 AD) also built the ports of Terracina and Centumcellae (Civitavecchia). The former one was excavated at a river mouth and the mooring quays and columns are well visible along the nice circular perimeter (Schmiedt 1975). The harbour of Centumcellae was built just to serve his villa in a site chosen on purpose for the favourable rocky morphology, but after the decline of Portus it became (and still is today) the port of Rome and remained unchanged for over 1000 years. The inner Roman Basin, presently still in use, was dredged in the rock (200,000 m^3), which was reused for the construction of the composite type breakwaters. The main structure has been reshaped over the centuries to reach an efficient mild-sloping profile (1:10 down to -2 m, 1:8 until -7.5 m, and 1:2 down to the bottom around -15 m MSL), which is considered a reference for modern rubble mound breakwater design (fig.18).

The harbour layout shows a characteristic Roman scheme with an island breakwater which supplemented the two main converging arms in order to reduce wave penetration through the gap, thus providing a double entrance for manouvrable vessels and possibly avoiding siltation at the entrance. The offshore breakwater generally supported a large fire-lighthouse (fig.17a). A similar planshape is observed in the small well preserved port at Astura point near Antium, again serving a seaside villa (where Cicero had stayed) and fishing ponds (Clementi 1981).

However, the most modern and efficient harbour planshape can be observed in the beautiful intact Roman harbour built under Augustus in the exposed small prison

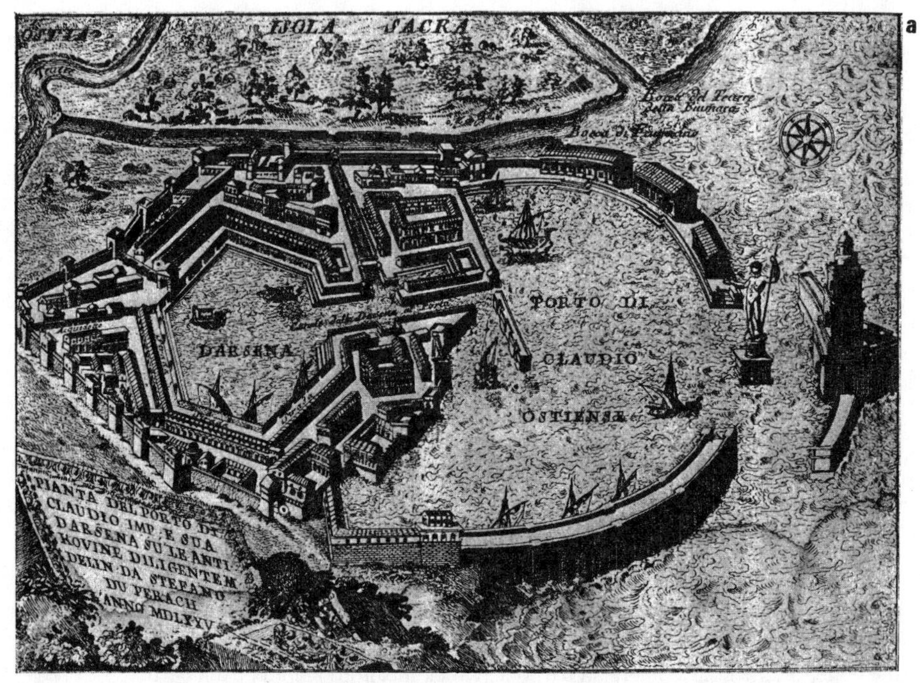

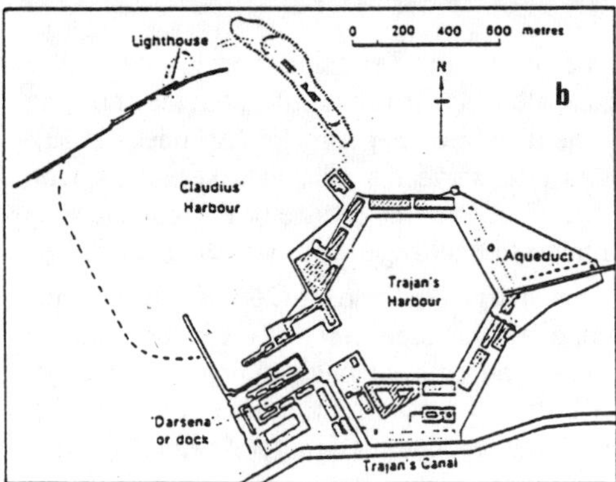

Fig.17. The layout of Portus Ostiae, the great port complex of Rome:
a) in a drawing of Du Perac in 1595,
b) new correct reconstruction (Lugli)

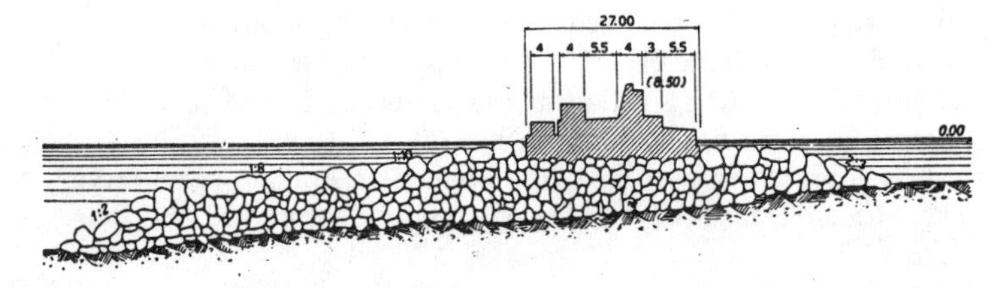

Fig.18. Modern section of the Trajan island breakwater of Civitavecchia harbour

-island of Pandataria (now Ventotene): a bay with spending beach (used as slipway) is located straight after the entrance and a lateral mooring basin is still preferred for its tranquillity to the modern port by the fishing and tourist fleet (fig.19). This layout is very similar to the most modern industrial harbour of Gioia Tauro (fig.20). Ventotene harbour is actually a colossal sculpture, fully excavated into the dark tufa-rock (60,000 m^3 with an average cut depth of 9 m) to create artificially a "natural" basin of 7,000 m^2 with 3 m depth, a "carved breakwater" (see fig.5), quays and grotto-storerooms, carved within deteriorated arcades resembling petrified elephants. Other apparent ancient port facilities are two aqueduct outlets, a number of large rock-cut bollards and a cave at the roundhead to contain the chain for closing the entrance. Again a small secondary opening was still documented by drawings of the 18th century, but is now obstructed despite the need of harbour flushing (De Rossi 1993).

Well preserved monumental structures are also existing at Leptis Magna (Libya), partly covered by red quick-sands. Here early Punicum and Neronian harbours created at the mouth of a wadi had rapidly silted up. A new basin of 400 m diameter was later excavated in the dry behind breakwaters under the emperor S.Severo (210-216 AD) and the river was diverted upstream with a dam to avoid siltation, which occurred anyway due to the littoral drift.

The Roman engineers also made elaborate harbour works in northern Europe, often near river mouths and along the main waterways of the Rhine and Danube, or even in lakes (Geneva). Recent archaeological discoveries have revealed that they were the first dredgers in the Netherlands in order to maintain the use of a river harbour at Velsen on 15-30 AD (van Rijn 1995). Siltation problems at Velsen were finally solved by building new "open" piled jetties to replace the solid piers which had previously been made by filling the drained space between sheet-piles with locally available material, such as wicker work and clay in reed mats. Another nice example of inland river port is the early harbour of Aquileia built at various stages since the 2nd century BC. The well preserved landlocked remnants of the Istrian stone quaywalls show two loading levels to cope with high and low water levels. Also visible at some 35 m spacing are the mooring blocks made with three cut stones morticed together to form a vertical round hole of 43 cm diameter along the upper crown, and horizontal 35 cm mooring rings in indented blocks of the lower step (fig.11).

Moreover it seems that the Roman officer T.Abudio Vero (19 AD) first built a detached breakwater just for the sea defence of the town of Parenzo (D'Arrigo, 1940).

Modern age

Early Middle Ages. After the decline of the Roman empire (V century AD) due to the northern barbaric invasions, a long "blackout" occurred to the development of human civilization in the medieval age, till about 1000 AD. In general this historical period is the least studied and very little is reported about civil engineering achievements. The danger of attacks from the sea by the Saracens caused the abandonment of the coastal zones and their ports which rapidly silted up (also due to the continuous

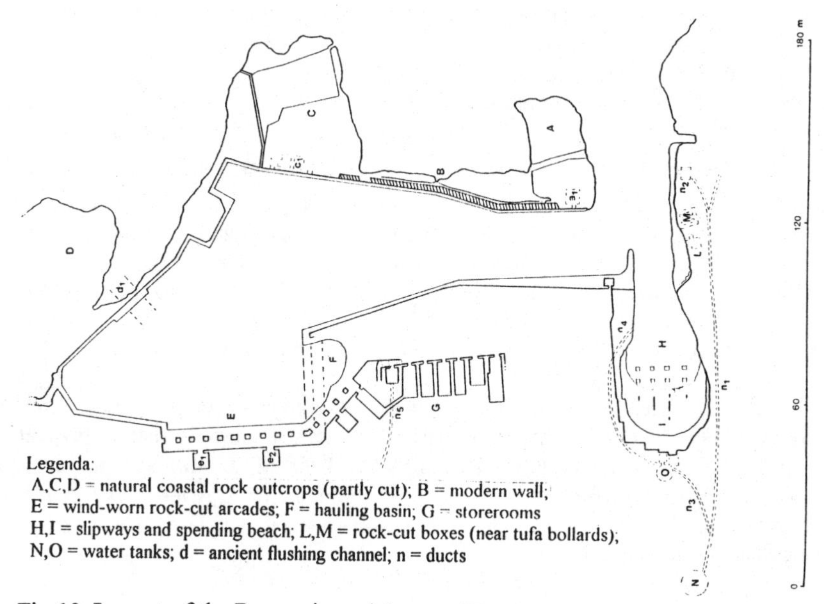

Fig.19. Layout of the Roman imperial port of Pandataria-Ventotene (De Rossi)

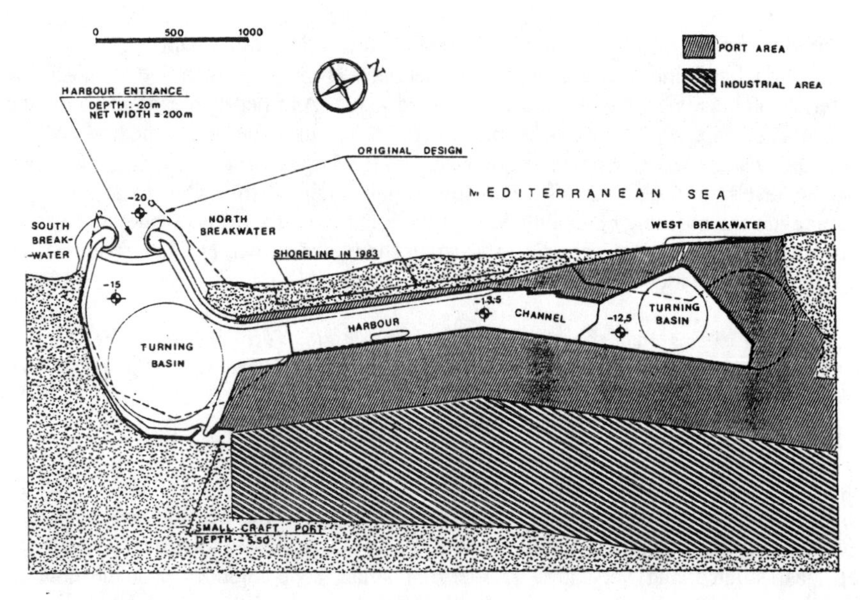

Fig.20. Layout of the modern industrial port of Gioa Tauro (Grimaldi et al.1984)

shoreline advance, like at Portus near Rome) and often became unhealthy swampy places. Other harbours declined for natural reasons, like Puteoli which sank deeply due to a well known local bradeysism.

However many little harbours were surviving along the Italian shores, as described in the famous *Tabula Peutingeriana* and reviewed by Schmiedt (1978), although the only important port was the one of Ravenna, keeping strong commercial links with the East. This harbour was created in a coastal lagoon and the two inlet jetties were made with waterproof caissons filled with concrete according to the Roman tradition. The shoreline here has now advanced by some 7 km into the Adriatic sea.

One of the few inhabited coastal areas was the Venice lagoon, where an autonomous "land-detached" community had settled upon island marshes to defend from barbarians. Written reports of local shore protection works date back as early as 537 AD: wicker faggots were used to hold the earth dikes reinforcing the sandy dunes formed from river supply, wind and wave action. The elastic but fragile willow branches for bank protection were then supplemented by the use of timber piles and stones, often combined in a sort of cribwork. The protection from the sea was so vital for the Venetian lidos since the early times that strict "environmental" regulations had been issued to preserve the littoral defences. Law documents of 1282 to 1339 state the prohibition to cut or burn trees from coastal woods; to pick out mussels from the rock revetments; to make cattle transit upon the dikes; to remove sand or vegetation from the beaches and dunes; to export materials used for shore protection (Grillo 1989).

Late Middle Ages. The Maritime Republics. The dawn of modern times appeared at the end of XIIth century when from the coastal village of Amalfi (near Naples) new sea routes and trades were opened. The new political and economic power was then held in succession by the four "Maritime Republics": Amalfi, Pisa, Genoa and Venice. In particular during the Crusader times the Genoese and Venetians constructed some remarkable ports also in the eastern Mediterranean and a brief resurgence in harbour development took place. A new integration of ports and cities is then observed, also for military defense purposes. Important reference documents of these times (1275) are the Mediterranean Chart (called *Pisana*) and the *Compasso de Navigare*, which describe the status of all coastal harbours, mainly from a nautical point of view.

Amalfi rapidly declined, also due to the disastrous tsunamis of January 1270 and November 1343, which destroyed the harbour structures and a portion of coastline. Many new artificial harbours developed, such as Naples, Leghorn, Palermo, but it may be just worth recalling here the construction of the most important one: Genoa. In its exposed coast the old east breakwater was built with a massive walled superstructure to serve as fortification too. This coastal structure was so important for the city life that in 1245 it was declared "Pious Work" (*Opera Pia*), which means that each citizen had to leave a legacy in his last will to support the breakwater maintenance.

Renaissance Technology. The Renaissance actually took place in the XV-XVIth centuries, when Italy was again playing a leading role in science and technology, including the field of coastal engineering, although the standards for design and construction of works at sea were still those codified by the Romans (eg.for monolithic concrete moles). However a great technological progress was favoured by the development of mechanical equipment, such as for driving piles, dumping blocks and placing forms with pontoons. One of the innovative techniques was the impermeabilization of the forms by means of pitch or with hemp curtains laid over the caisson bottom, typically floated into position (fig.21a) (Di Giorgio Martini 1470).

Even primitive techniques for breathing and working underwater were developed. Moreover the first dredging machines were developed: in 1413 a Genoese inventor was hired in the port of Marseille. In the *Codicetto* by Di Giorgio Martini the first dredger is drawn with two boats holding a wheel equipped with four arms ending with sharp buckets (fig.21b). Advanced dredgers designs were later made by Leonardo and Veranzio 1595 (fig. 21c) and others. However seabed levelling and cleaning was generally carried out by closing and drying out the harbour basin. Siltation was still the greatest problem of harbours. Openings through the moles (generally equipped with mobile gates) were still a favoured solution, but even fixed sheetpiles or mobile barriers made with wooden plates hinged to the seabottom at the harbour entrance were experimented by Tibaldi (a forerunner of the present floodgates for Venice...). These innovative devices were also introduced for military defense purposes (together with other artificial submerged obstacles). Ingenious rhomboidal wooden elements, also in inclined position, were conceived to strengthen current velocity at the entrance.

Renaissance Architectonics. Ideal city planning began in Renaissance Italy during the 15th century by preserving a balance between beauty and usefulness. This humanistic vision of symmetry and proportion is also evident in harbour design. Moreover the Italian engineers introduce a new design procedure, in which several alternatives are studied and the chosen solution is finally visualized by means of small-scale wooden models. The famous architect L.B.Alberti in his global handbook *De re aedificatoria* (1452) follows Vitruvius' technical recommendations with a nature-wise approach, even in the hydraulic design of coastal structures, such as harbour breakwaters and navigational channels. He suggests to have a mild seaward slope for rubble mounds "to allow a smooth run-down to restrain the assault of next waves" and avoid toe scour. A high crownwall is also recommended to provide shelter from wind too. He also deals with siltation and dredging problems and with historical shoreline variations and beach equilibrium profiles.

In his first general description of the artificial harbour layout Alberti, as well as other famous engineers of those times, strongly favours narrow entrances with curvilinear or even circular basin planshapes, to satisfy the ideal aesthetical/phylosophical (as well as the hydraulic/functional/military) standards in the "closed" vision of the fortified city-port (pirates' raids in fact continued until about 1570). It is even a symbolic reference to the elementary spherical shape of water according to Leonardo. The "perfect" monocentral circular shape is also the one which encloses the largest basin area. A semicircular port layout would even resemble to a theatre since it offers an attractive "show" (fictitius ship battles have been actually performed in harbour basins...).

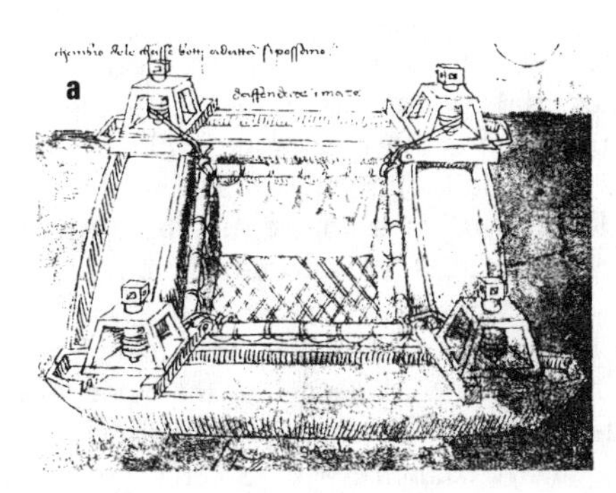

Fig.21. Technological progress in the Renaissance Age:

a) floating equipment for dumping foundations in shallow water (Di Giorgio Martini 1475)

b) Lagoon bucket dredger (Di Giorgio Martini 1475)

c) A dredger (Veranzio 1595)

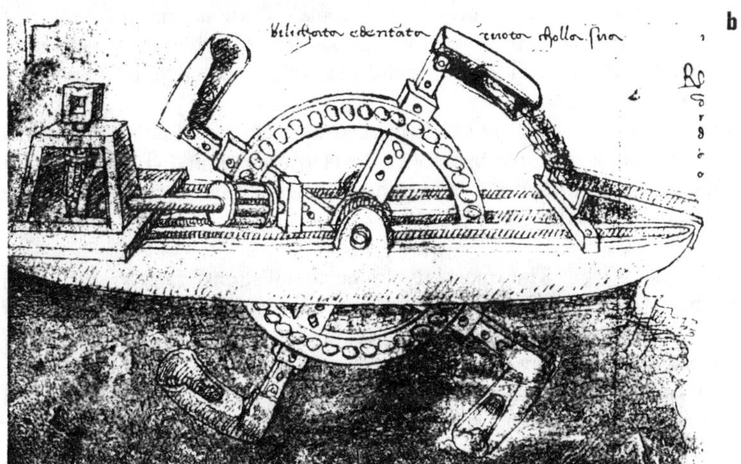

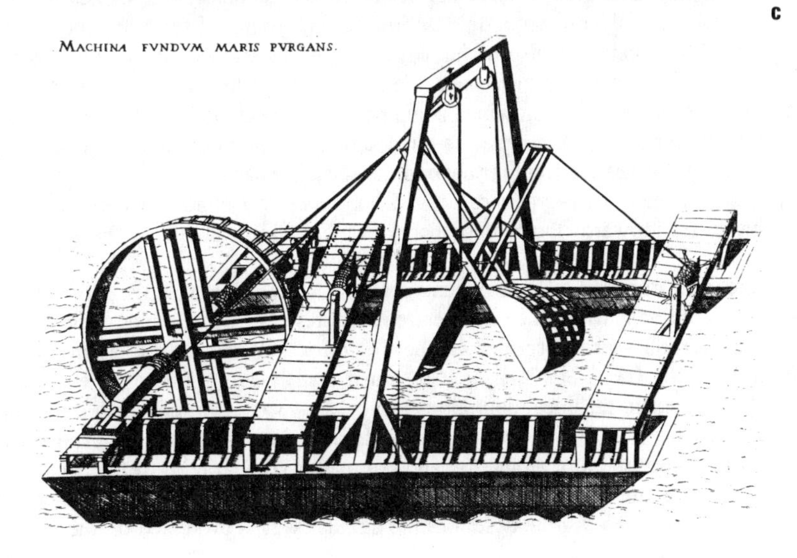

Di Giorgio Martini (1480) even fixed the "golden measures" of the ideal harbour layout: a convex-shaped island breakwater, 100 m long, to shelter the 70 m wide central entrance between two moles 70 m behind and a breakwater width of 25 m. However he also proposed a different layout with a long mole parallel to the coast, which would ease the defense of the fortified town, or a double nearshore entrance with a horse-shoe shaped breakwater with closable gaps for water circulation (fig.22).

Therefore the classic Roman layout of Centumcellae harbour was considered in the Renaissance, also by Leonardo (fig.24c) and later by Michelangel, as the model of the "ideal city-port". The oval planshape of its converging arms was even considered as a symbolic reference in the design of the famous columnates of S.Peter's square in Rome: "the Port of Salvation of the Roman Catholicism devoted to the Saint fisherman and true safe pilot". Other famous Italian architects worked in the port of Civitavecchia and its fortifications, such as Sangallo, Bramante, Bernini and Vanvitelli.

However the Renaissance period is marked by the important birth of the hydraulic sciences, including maritime hydraulics. In fact it was noted (Rouse et al, 1957) that "The Italian school of hydraulics was the first to be formed and the only one to exist before the middle of the 17th century".

Leonardo da Vinci. The most important innovations can be attributed to the extraordinary eclectic genius of Leonardo da Vinci (1465-1519). By means of his well known experimental method, based on the systematic observation of natural phenomena supported by an acute intellectual reasoning and a passionate creative intuition, Leonardo was really the precursor of the modern coastal engineering science, often anticipating ideas and solutions by more than three centuries. Unfortunately, such anticipation was to remain unfruitful, because for centuries his results continued to be practically unknown to both the scientific and technical world. His notebooks exhibit some astonishingly accurate descriptions and graphical representations even of complicated flow patterns, also taking advantage of his extraordinary skill in painting. Some descriptions of water movement are essentially qualitative, but often so correct and acute, that some of his drawings could be usefully included in a modern treatise of coastal hydrodynamics. Of course the quantitative interpretation and the mathematical formulation of the results were far beyond the scientific capabilities of those times. The jump from kinematics to dynamics proved impossible for almost two centuries, until some basic steps towards the correct theory of gravitation were taken (Fassò 1987).

The variety of problems of hydrokinematics dealt with in Leonardo's notebooks is so vast that would be hard even to enumerate in this review. Most phenomena related to maritime hydraulics are described in the 36 *folios* (sheets) of the beautiful Codex Leicester (1510) (ex C.Hammer,now renamed by B.Gates), where he even gives a glossary of the main technical definitions (folio 12 *verso*). A full English translation is given by Richter (1970). Leonardo clearly defines the progressive nature of surface waves with transmission of "impetus" and not matter comparing it to the oscillation of wheat in a field under the impulsion of wind (36 *verso*) ("a *tremor* rather than a motion"). In folio 14 *verso* he describes the circular waves produced on the still water surface when one throws a pebble into it and explains that “floating bodies do not change position”. The observation of this phenomenon is rather trivial but Leonardo tried to go deeply into the mechanics of waves by studying either the intersections of the families of circular waves produced by the impact of three pebbles at a time and the wave pattern produced by the impact of a triangular and bar-shaped objects (rapidly assuming a circular shape). In the same page he also anticipates the concept of relative wave celerity by writing “An object thrown into flowing water will make an oval undulation, and this will extend little in the direction opposite to the flow and much in the direction following it”. Leonardo also describes the nearshore wave shoaling (12 *recto*), wave breaking (4 *verso*), air entrapment (28 *verso*) and wave reflection and impacts at walls (25 *verso*), for which he suggests a concave parabolic profile “to make the wave percussion falling onto itself...”. Some nice drawings of sea waves (also breaking at walls) are shown in fig.23.

Leonardo beautifully illustrates the water movement around obstacles (14 *recto*), the techniques for measuring current velocities (13 *verso*) and the mixing of fresh and salt waters (21 *verso*). .Moreover he correctly defines the current circulation pattern in the Mediterranean Sea and explains the astronomical tides, also giving their exact amplitudes at various coastal locations (6 *verso*, 27 *recto*). Furthermore Leonardo studied sediment transport, even using small-scale mobile model tests with sand (15 *recto*), dealing with sand ripples (23 *recto*), bed-load (20 *verso*), cross-shore transport and grain size distribution (“the coastal silt is transported offshore by storm sea waves...”, 35 *recto*). He even suggests measures to reduce erosion effects, such as with a counter-jet against the erosive toe current (13 *verso*). Well ahead of times he attributes to the sea waves the prevailing role in the morphological evolution of beaches (2 *verso*, 6 *verso*) and defines the sediment equilibrium line and the seabed depth at which the orbital motion of the waves is still present (9 *verso*), three centuries before the experiments of the Weber brothers.

As far as we know, Leonardo was also the first to describe and probably to test several experimental techniques now employed in all hydraulic laboratories, such as flow visualization by means of both suspended particles and dyes; glass-walled tanks; use of movable bed models, both in water and in air. A synthetic review of such techniques was recently made by Levi 1982, who pointed out in particular that Leonardo was aware of both the limitations and possibilities of the experimental method and noted the scale effects when testing on small models.

Leonardo often used simple suggestive similitudes to clarify complex physical processes in his unitary vision of nature.(“pulsating sea motion alike the blood flow in the human body...”, 34 *recto*). The blood circulation system analogy is resumed again to describe the tidal flushing of lagoons and “canal-ports”. As a matter of fact he also gave a significant contribution to harbour engineering, particularly as related to the so-called “canal-ports” which are typical of the northern Adriatic coast, taking advantage of both river and ebb tidal flows to keep the inlet navigable. In his famous drawing of Cesenatico harbour in 1502 he shows a canal with a wider section at the bend to ease navigation and a narrower section at the mouth to increase flushing current velocities

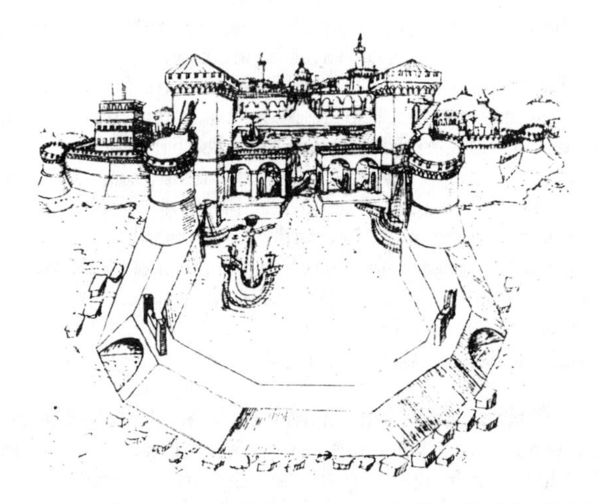

Fig.22. Drawing of an ideal harbour (Di Giorgio Martini 1475)

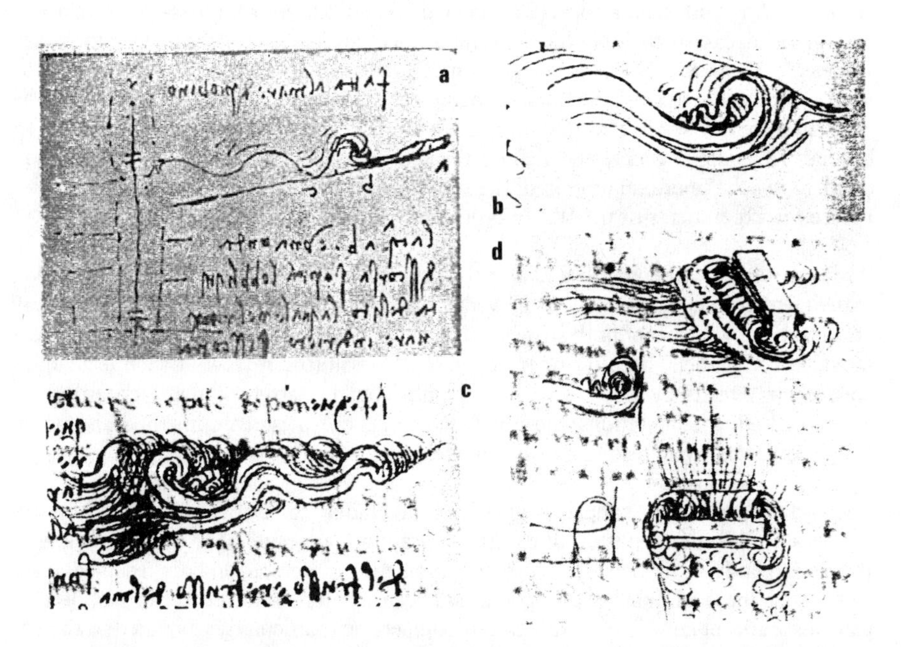

Fig.23. Leonardo's drawings of breaking sea waves (1510): a) Codex L, folio 6r; b) c) d) Codex Leicester, f. 4v, 26v, 25v

(fig.24a) (D'Arrigo 1940). He also introduces the favourable effect of an inland large pond or artificial basin linked with the canal to increase the ebb flow and remove the inlet siltation. The upstream bridge had a wooden mobile gate which could also release water at low tide to clean the entrance. This system was then favoured by the Venetian hydraulic engineers until today (with their motto "a large lagoon makes a good port"), also with the occasional artificial help of sliding gates. In the quoted plan the shoreline position at the updrift side of the inlet jetty is about 100 m from the tip.

The interest of Leonardo in harbour engineering is also reflected in the Codex Madrid, where he represents a design of a harbour at Piombino with a triangular planshaped island breakwater (to reduce wave forces) and two other small port designs (one without the island breakwater) (fig.24b). Leonardo is the first harbour designer to understand that the above breakwater layouts would not only reduce wave penetration but also the siltation at the entrance, which was to be caused by the wave action. His port designs are inspired to the ancient ones but he first performs a detailed survey according to his new scientific approach. In the Codex Atlanticus, Leonardo reports the mentioned detailed survey of the port of Centumcellae (folio 271 *recto*) (fig.24c) and its vertical block east breakwater (63 *verso*). In folio 4 he also examines the damage to the "Old Mole" superstructure of Genoa harbour occurred during a great storm in February 1498. An innovative harbour layout is then presented in folio 285 *recto* (fig.24d): the special spiral shape should provide a good shelter and recalls the wave form itself, as typically depicted by the artist, or the interior curl of a sea shell.

Leonardo da Vinci was also involved in the design and construction of locks and is credited with having introduced the idea of locks in France, where he spent the last years of his life from 1516 to 1519: still in use are the well known *Vincian gates*.

Locks. However the first navigational lock had been already constructed in 1438-39 at Viarenna (Milan) by Fioravante da Bologna and Filippo degli Organi in order to ease the inland waterway traffic and particularly the transport of material for the construction of Milan's Cathedral. Most historians regard the Viarenna lock as the first one in Europe, though it was preceded by the Vreesvijk lock in Holland (1373) (Fassò 1987). But the two engineers were surely unaware of the existence of previous applications and must be credited for the independent invention of this useful device. This simple but revolutionary idea of the short canal between two gates had a tremendous impact on the economic and social progress of entire countries and was later used also for tidal ports. Viarenna lock is 38 m long and 6.2 m wide and could rise boats by about 3 m. It was dried in 1936 after 500 years of regular operation.

Port and Breakwater Developments (1600-1850). A general decline of harbour activities occurred in Italy after the development of oceanic navigation at the end of XVIth century . The center of scientific and technological development began to move from Italy to northern Europe. Even harbour design concepts started to change. Since then the technical aspects have been dissociated from the formal architectonic ones and the new professional category of the hydraulic engineers took over this ever more specialized technical field, despite the lack of a sound "harbour theory", losing any

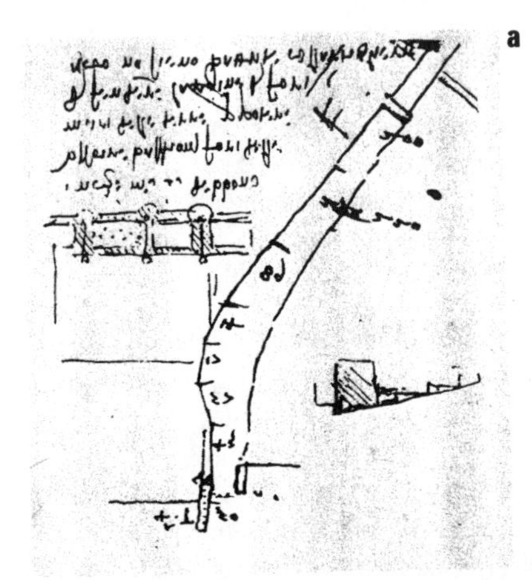

Fig.24 Leonardo's drawings of harbours:
a) Cesenatico (Codex L, folio 66 *verso*)
b) Piombino (Codex Madrid II, f.88 v)
c) Civitavecchia (Codex Atlanticus, f. 271r)
d) Spiral planshape (C. Atlanticus, f.285 r)

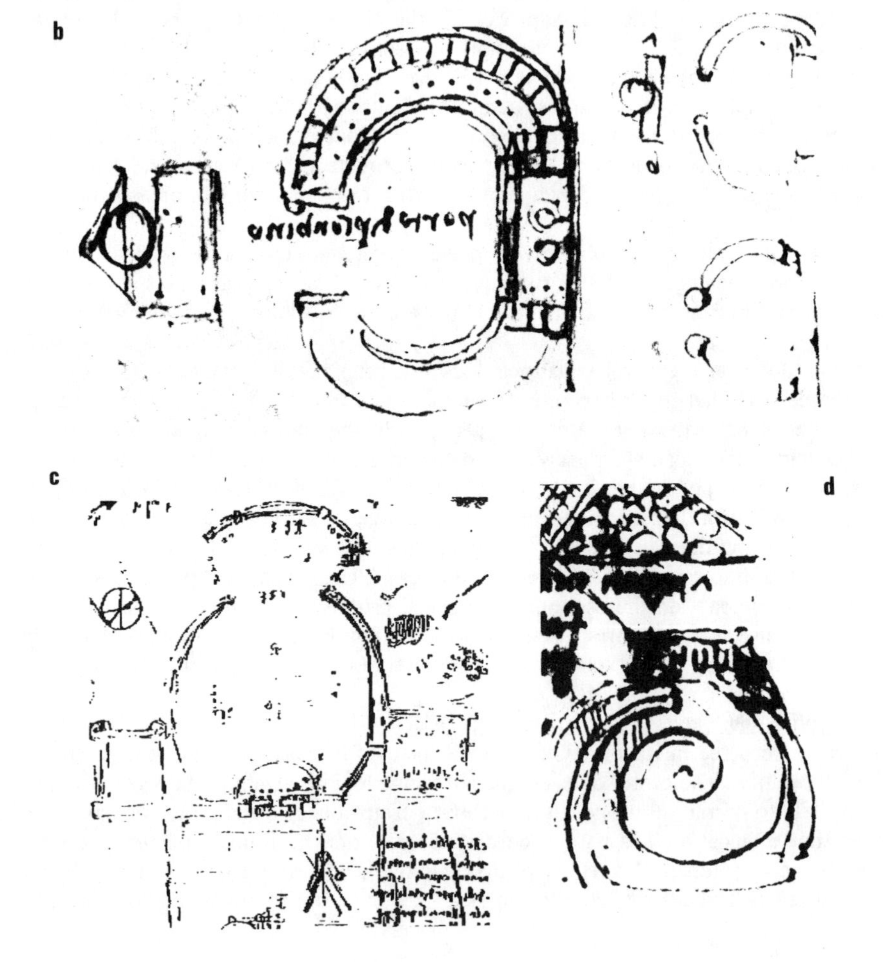

interest for the formal and functional aspects of ports. The unitary approach of the Renaissance was replaced by the merely decorative aggregation of building elements in formal figurative baroque stile (Simoncini 1993). The harbour planshape became polycentric: Vasari in 1598 proposed elliptic basin layouts. However circular shapes were still attractive as shown by De Marchi (1545) (fig.25) and Gallaccini (1603), who resumed Leonardo's idea to produce a number of fascinating, though unrealistic, spiral-harbour designs (fig.26a,b). In his beautifully illustrated treatise on sea harbours Gallaccini (1603) also gave a decalogue of prescriptions for a good harbour design and described breakwater construction methods, advanced floating equipment and original mechanical devices to work underwater and even walk overwater ! (fig.27a,b,c).

In fact in the XVIIth century the technology of coastal structures was again receiving attention. The eternal debate about the choice between rubble mound or vertical breakwaters was still active. Based on his experience in the restoration of Civitavecchia harbour, Crescentio (1607) recommended the use of irregular rock mounds (so-called "lost stones" system...) with a pozzolanic concrete crown above sea level. Quite modernly he was suggesting not to use small stones to fill the porosity of the large rock armour and to let the sea waves naturally shape the rubble mound profile. Scamozzi (1615) proposed to use this system only in deep waters and on weak soils, while heavy squared blocks could be regularly placed on hard bottoms. In shallow calm waters the Roman system of cast-in-situ monolithic walls was preferred. Actually the caisson technique kept being widely used in Italian harbours according to the literature (in Naples on 1302 and 1470, in Palermo on 1575, in Savona on 1626)

Long discussions on this matter also occurred for the necessary protection of Genoa harbour, which was suffering severe damaging storms in 1613, 1630, 1636, also due to high reflections inside the basin. As described by Faina (1969), interesting design solutions for the new breakwater were submitted by L.Bianco (1637), who proposed the use of tronco-pyramidal caissons for a better connection (fig.28a). Finally an innovative composite-type design by De Mari was approved in 1638, which had a larger and lower rubble foundation up to - 4 m MSL and a monolithic superstructure cast-in-situ within caissons or ship hulls to be sunk exactly over the leveled foundation (fig.28b). Another new feature was the use of both small and large stones to obtain a more compact foundation. Important merits of the composite caisson breakwater are the reduced costs and time of construction, which particularly reduce the risk of storm damage during the execution of works. De Mari also stated some general principles (only partly correct), such as: "the wave force strongly reduces with depth and it is almost zero at -5 m MSL in the Mediterranean; in fact 5-10 t rocks are stable at this depth; the optimum rubble mound profile has a 1:2 seaward slope and 1:1 landward".

Despite the scientific limitations the methods of De Mari had a large international resonance, as stated by Clark (1948): "The English were indebted, in the building of the Tangiers Mole,1663-83, *the largest engineering work till then undertaken by their nation*, to the advice of Genoese engineers". He also admitted that "Italy in 1600-1700 was still the most prolific country for technical and scientific developments and its superiority lasted longer than supposed". The success of the composite caisson

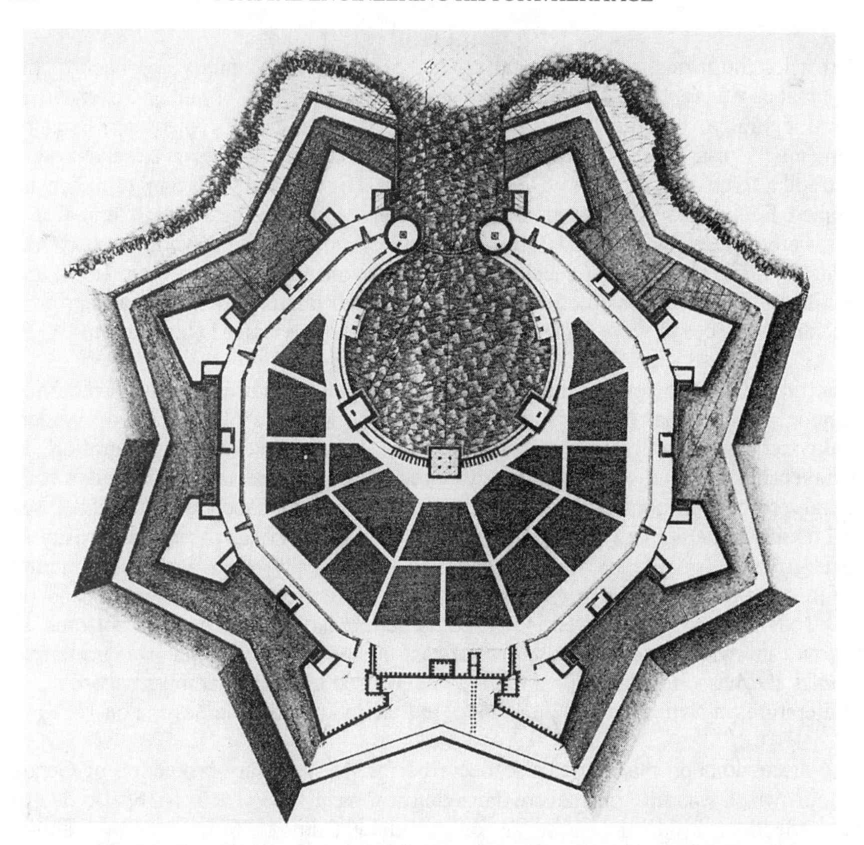

Fig.25. Fortified sea town with circular harbour (De Marchi 1545)

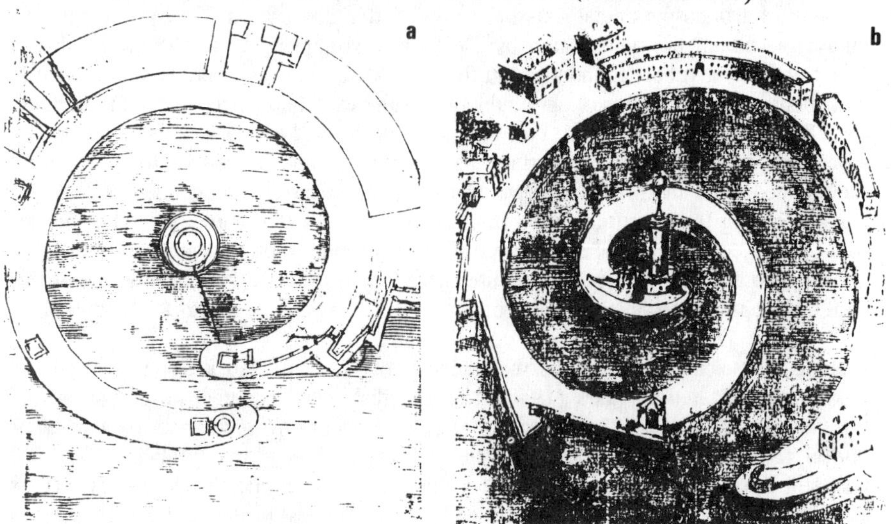

Fig.26. Drawings of circular (a) and spiral (b) harbours by Gallaccini (1603)

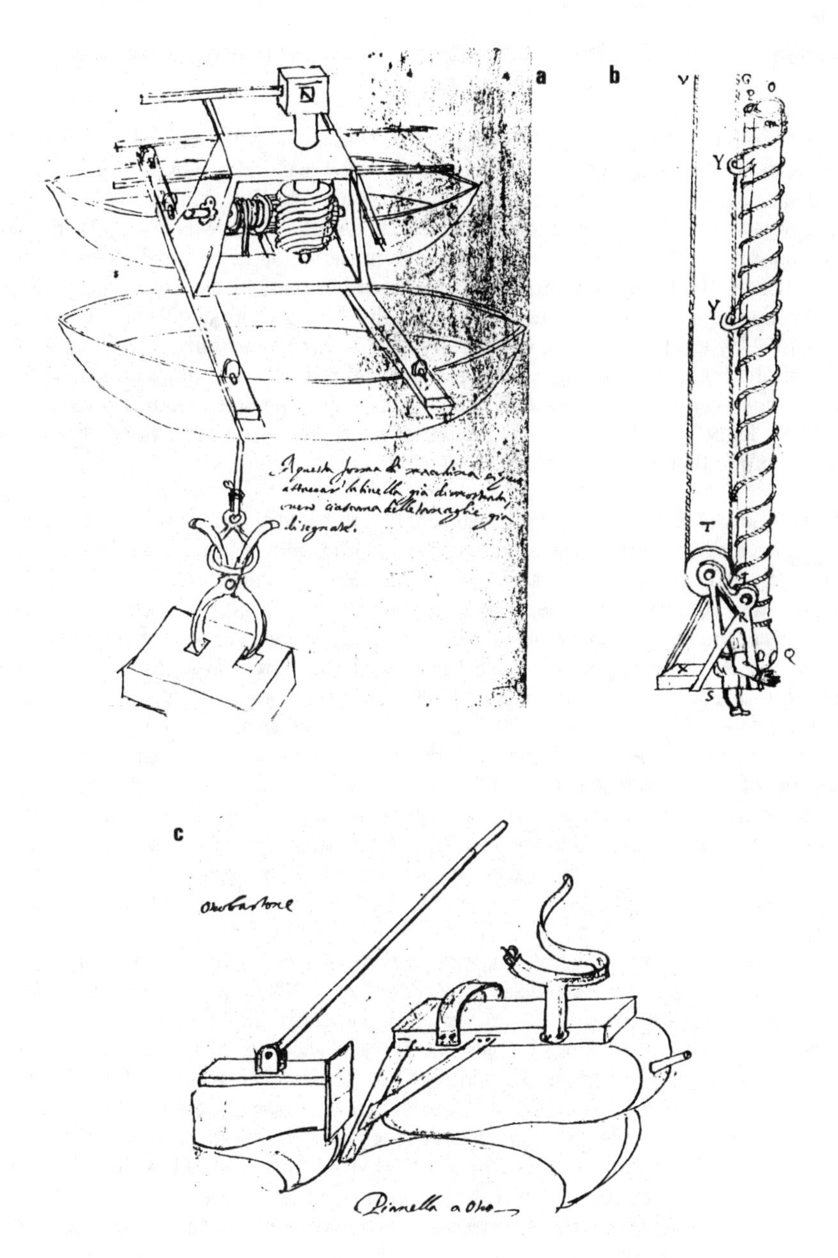

Fig.27 Advanced mechanical devices for maritime works (Gallaccini 1603):
a) placement of a breakwater armour block from a pontoon
b) a new device to allow underwater work
c) a system to walk overwater

breakwater built in Tangiers by Shere then demonstrated a more general applicability of this Italian solution even in high tidal ranges.

Venice - Shore Protection (1500-1800). In the XIV-XVIII centuries it was the independent Venetian Republic to become the main maritime power, the bridge of communication and commerce between the West and the East, and also the leader in the progress of hydraulic and coastal engineering (Adami 1992, Marchi 1992) (fig.29). The importance of the problems related to the preservation of the tidal lagoon, to the protection of the littoral barriers dividing it from the sea and to navigation, required the creation of a special absolute Water Authority ("Magistrato alle Acque") managed since 1501 by elected hydraulic experts. A careful management and severe control of its delicate hydro-morphological system and related engineering works has been carried out ever since in Venice. The works included the river mouths diversion to avoid the lagoon siltation, the construction of inlets jetties to ease navigation and the defence of the thin barrier islands from wave-induced erosion.

The earliest shore protection works before the 18th century were typically revetments and groins made with timber piles in multiple longitudinal rows to contain layered rock slopes (fig.30). The forerunners of modern groins were about 100 m long and their purpose was actually to intercept the littoral drift to avoid both erosion and siltation problems at the navigable tidal inlets. However maintenance costs were very high, due to the limited durability of wood (5 years) and the long travel distance for rock transport. Frequent repair was needed after damaging storms, sometimes afforded by sinking barges filled with silt, reed and stones. Therefore the Authority seeked for innovative designs: various technical solutions were proposed and experimented at the own risk of consultants and contractors (who were paid only after the effectiveness of the work had been proved...), such as: riprap revetments; gabions; smooth marble blocks linked with mortar and steel; flexible steel strips; stepped limestone blocks in regular pattern; various artificial elements to increase the roughness of the revetment slope; and even beach nourishment with sand dredged from offshore ! (Grillo, 1989).

Finally the project of a durable massive seawall presented by a group of experts headed by the State Mathematician (B.Zendrini et al.1743) was approved. The so called "murazzi" are composed by a smooth heavy flagstone revetment supported at toe and crest with massive block walls; the average width is 12 m and crest elevation at +5 m MSL (see original drawing in fig.31a). The innovative technology (actually based upon an earlier idea of Abbot Coronelli in 1706) was represented by the use of pozzolanic cement as effective bond between the rock-cut blocks, instead of the timber pile fencing. This seawall was built in about 40 years for some 20 km along the lidos and has lasted up to the present time, becoming a significant Venetian monument and a historical landmark of coastal engineering. However repairs have been necessary, particularly after the exceptional storms of 1825 and 1966: the reinforcement has been made with a rubble mound at the toe and recently with anchor piles and jet-grouting diaphragms (fig.31b). The main impermeable screen has been constructed along the lagoon side (down to a clay layer at -20 m MSL) to prevent siphoning risk, while the seaward screen was to reduce the uplift pressures enhanced by the main diaphragm.

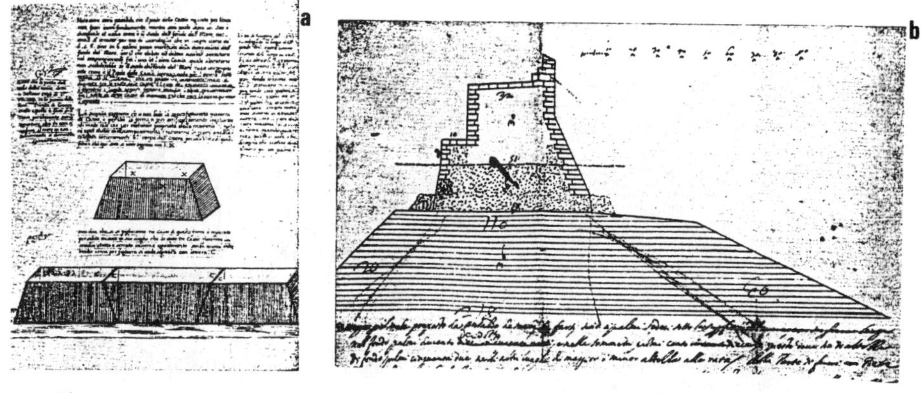

Fig.28 Designs for the new breakwater of Genoa harbour in XVII century:
a) Models of pyramidal caissons by Bianco (1637)
b) The new composite-type design by De Mari (1638)

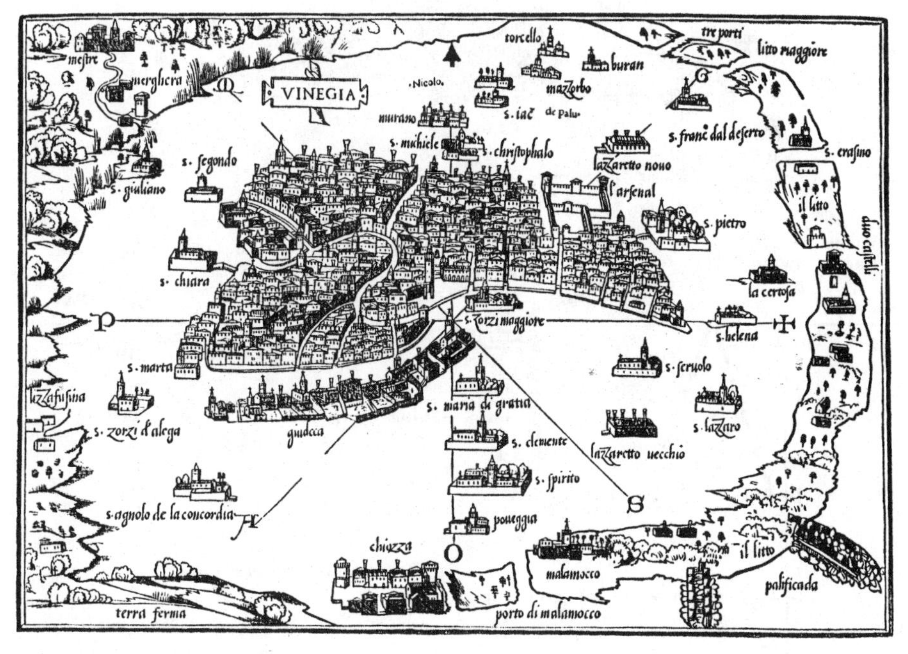

Fig.29. Map of Venice lagoon in 1528 (B.Bordone) with piled groins for littoral defence (CVN 1995)

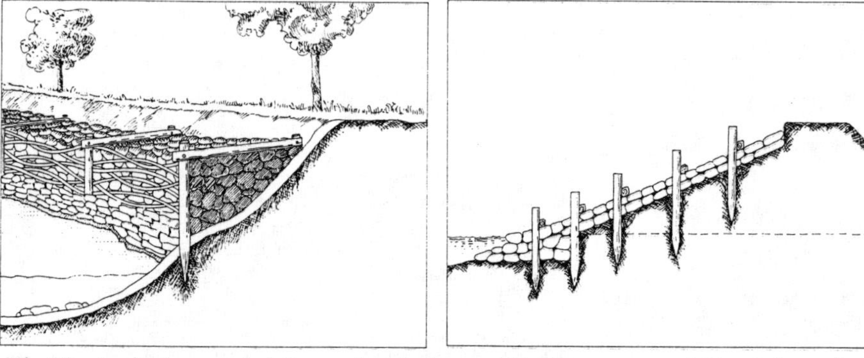

Fig.30: Ancient coastal defences in Venice: rubble revetments within timber piled fences (6 to 17th century)

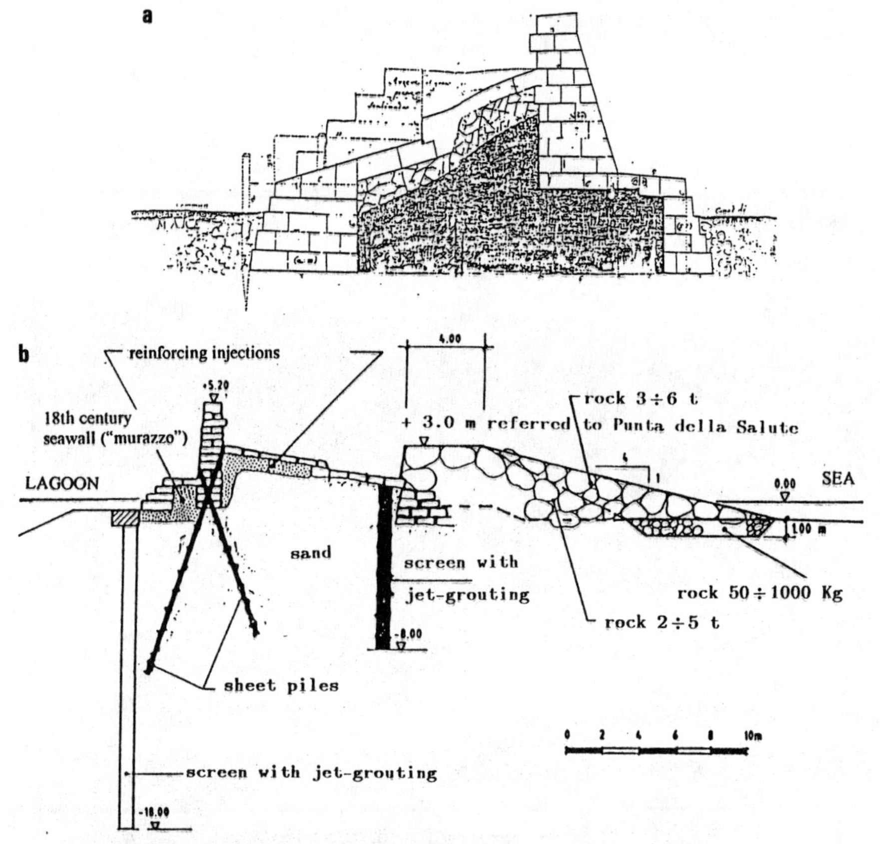

Fig.31: a) Original design of the "murazzi" seawall by Zendrini in 1743
b) Modern reinforcements to the 250-year old *murazzo* at Caroman

Other important Venetian maritime works were related to navigation in and out the lagoon, which was dangerous for the variable shallow waters. In fact a special pilot company had been set up long since, navigational aids were installed and channels dredged. In 1725 a new 4.5 m deep, 18 m wide shipping channel was excavated in order to link Venice to the main (deepest) inlet of Malamocco. However, the old groins could not prevent progressive siltation of the three tidal inlets and the need for longer training jetties was envisaged by the Venetian engineers. The first big project to develop the Port of Venice was commissioned by Napoleon in 1807 to a French group of experts, who followed the proposal of the local naval officer Salvini to build two parallel breakwaters normal to the shore with the longer north jetty overpassing the bank off the Malamocco mouth and deviating the littoral currents. After many interruptions the 2 km north jetty was finally completed in 1845 after the brilliant redesign and supervision of P.Paleocapa (one of the engineers who worked at the Suez Canal project), who avoided any dredging to rely upon the tidal flow for the natural deepening of the access channel (Noli 1990), which in fact rapidly happened (today reaching -15 m MSL). The construction of the south jetty was completed in 1872 and the successful results stimulated similar works at the other two inlets of Lido (1882-1910) and Chioggia (1911-34). The rubble mounds were built with advanced criteria, but they are presently undergoing substantial rehabilitation works. Moreover they have assured safe navigability even to the larger vessels which have been later entering the lagoon to reach the large modern (XX century) industrial port of Marghera. However the jetties have also stopped the littoral sediment transport, thus contributing to the recent erosion problems of the central lidos, which are now been tackled with new protection systems, as described further on.

Advances in Maritime Hydraulics and Coastal Dynamics.(1700-1950). As far as the study of sea and beach dynamics is concerned, valuable contributions in the XVIII-XIX centuries were given by a number of scientists, like Marsili, Montanari, Cialdi, Rovereto, Uzielli, Chelussi, Clerici, Cornaglia, Boscovich and others. L.F.Marsili was a precursor of modern oceanography, since he made in 1715 detailed investigations on submarine sand banks off the Adriatic shores.

A large interesting collection of shipboard observations of storm waves and their effects in deep sea-bottoms was produced by captain Cialdi (1866). He was convinced that wind waves could induce strong water movements down to the seabed ("even at 200 m depth in the ocean"), which he called "current-wave" (*flutto-corrente).* He correctly understood that the wind waves were the main cause of sediment transport along beaches and felt the need for a general theory describing both beach evolution and harbour siltation. His ideas were opposed by other experts of his times, such as Paleocapa, who believed that the main transport agent was the general littoral density current (which was first measured in the Mediterranean Sea by Montanari). Cialdi also produced a treatise on harbour construction and a book on Leonardo's wave theories.

However, the most original theoretical studies are due to P.Cornaglia (1891) who proposed his famous theory of the "bottom wave"(*flutto di fondo*) to explain either coastal sediment.transport and wave forces on breakwaters. He demonstrated that the elliptical orbital motion induced by sinusoidal surface waves degenerates into alternate horizontal particle movements at the sloping seabed and a high momentum is then transmitted near the bed with wave propagation. Cornaglia then defined a *neutral line* on the seabed where the shoreward and seaward components are balanced : the bodies are then pushed onshore within the nearshore zone and are directed offshore if located off the neutral line He also tried to determine the wave pressures on walls from the elevation of water jets at the impact. Cornaglia's theories are partly contradicted by the evidence of real observation (eg. the too high speed of the "wave current" approaching the shore in relatively deep water), but they were among the few theoretical references available to coastal engineers for decades.

Later on an important contribution to the theoretical analysis of wave kinematics was given by the mathematician Levi-Civita (1925) who found the rigorous exact solution of the theory of irrotational oscillatory waves on infinite depth, which had been given with approximate solutions by Stokes and Raileigh.

In the early decades of the XXth century detailed studies were also produced on the Italian shoreline variations (Toniolo 1910, Marinelli, Merciai, Mori) and systematic surveys were promoted already in the 30's by the National Research Council (CNR).

Further Evolution of Breakwaters (1850-1950). Despite the little financial support given by the newly unified kingdom of Italy (1861) to port development in order to keep pace with the rapid evolution of shipping (also due to its historical heritage of fragmented local policies without global planning), the most important developments are again observed in the design and construction technology of harbour breakwaters. The state-of-the-art at the start of the new century was summarized by Luiggi (1907). An updated general overview is given by Borzani (1995) and, with special reference to the typical composite structures, by Franco (1994) and by Lamberti et al. (1994).

With regards to rubble mound breakwaters, which were more popular in the XIX century, the historical experience (e.g. Civitavecchia, fig.18) had shown the effectiveness of a mild slope in the critical zone near the waterline (Lo Gatto 1925). But in deeper more exposed waters the need to save material and to use heavier blocks, favoured by the rapid increase of capacity of the lifting equipment, had led to the modern type section with steep slopes and large *precast* armour blocks (30 t in the first application of Marseille offshore breakwater in 20 to 35 m depth). The first example in Italy was the S.Vincenzo mole in Naples, built in 35 m depth since 1850 and followed by the Duca di Galliera mole in Genoa built in 1877-88 and reaching 29 m depth (Coen-Cagli et al.1905) (fig.32). Both breakwaters suffered soon severe damages and settlements, thus requiring reinforcement, typically achieved with heavier armour blocks, even placed in an original steep stepped profile (called *Parodi* system from the engineer who invented it). This new armour arrangement had many applications until the beginning of the XX century, but proved to be vulnerable to mound settlements with consequent fragile blocks displacements. Another famous storm in 1898 caused serious damages to the Galliera breakwater in Genoa, as described by Bernardini (1901) (fig.33).

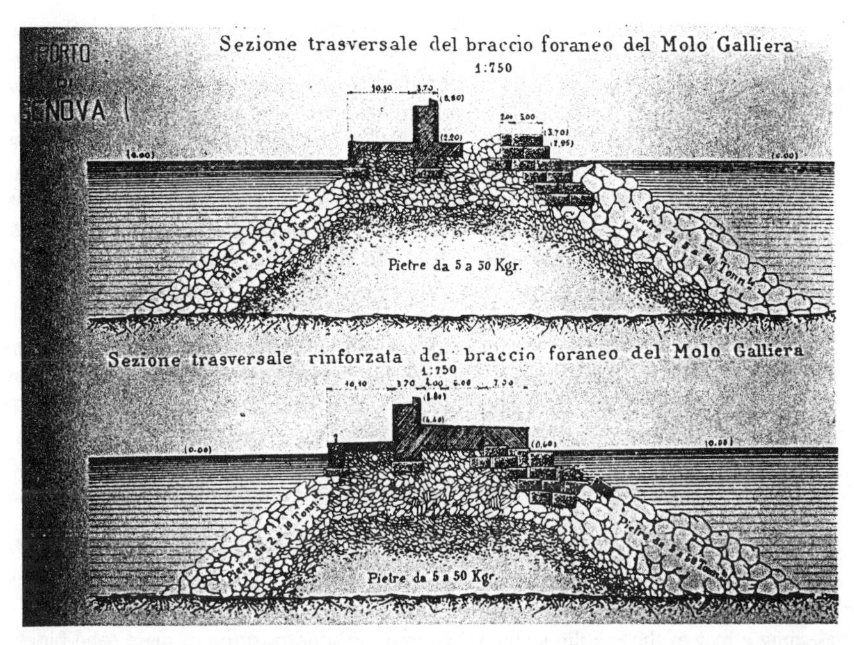

Fig. 32. The Galliera breakwater in Genoa with stepped block armour, before and after reinforcing

Fig.33. View of Genoa breakwater after Storm of 1898 (Cunningham 1908)

These failures then promoted the revival of the classic vertically composite structure, which was actually reintroduced in Italy by Coen-Cagli after his visit to some English vertical breakwaters. In fact in 1896 he designed the southern extension of the Trajan offshore breakwater in Civitavecchia, which however was soon transformed in a horizontally-composite structure with a block mound protection. The actual first applications of the modern walls, made with 50 t concrete blocks arranged in horizontal layers with staggered joints upon a rubble foundation ("masonry blockwork"), took place in Trapani harbour in 1896 (Colombaia mole) and in Naples harbour in 1905 (offshore breakwater Thaon de Revel, fig.34a) (Coen-Cagli 1936, Greco 1953).

Soon after, both the Galliera mole in Genoa and the offshore breakwater in Naples were extended with a slight different solution for the vertical structure (proposed by the engineer Inglese): a single or double column of weakly reinforced 220 t *cellular* blocks was filled in-situ with mass concrete to join the elements in one monolith (fig.34b). However the filling concrete did not bond well with the walls of the precast blocks under the wave dynamic action and repair works were needed.

Therefore in the 20's the bigger pontoon capacity promoted the use of full-width solid *cyclopean* precast blocks (up to 450 t) with vertical pits to allow the in-situ solidarization with concrete and reinforcing rails or bars (fig.34c). The fear about the weakening effect of these hollows then favoured the superposition of plain solid blocks in single columns without vertical connections, as made in the breakwaters of Genoa, Palermo and Catania (1928-31). The latter structure soon became famous, because it failed dramatically during a storm in 1933, shortly followed by the failure of Mustapha jetty in Algiers in 1934. The main failure reasons were the high wave breaking loads due to the shallow toe depth, the lack of wall monolithicy and toe scour.The rehabilitation of Catania breakwater led to a sloping section with a "natural" parabolic profile (fig.34d). These disasters opened new discussions and international guidelines were given at the next PIANC congress in Bruxelles in 1935. The new design criteria were summarized by Coen-Cagli (1936) and Ferro (1936).

Vertical breakwaters became again out of fashion (mainly outside Italy), but a new construction technology allowed to overcome the above drawbacks: the large monolithic cellular r.c.caissons. Constructed in yards the caissons can be floated into position and then sunk with water ballast and filled with mass concrete (as in the earliest applications at Alghero in 1915 and at Civitavecchia in 1931-36) (fig.35) or with incoherent material (as at Genoa airport breakwater in 1938). The construction of these latter caissons was made with a special fixed platform, which was the first industrial prefabrication yard for maritime structures and judged "worth of attention" by Schorn 1959 (fig.36). The caisson technology later evolved with advanced floating plants and deeper-water applications. In the 70's a 42 m high caisson was designed and built by an Italian contractor in Sines, Portugal (CSPP 1976).

The ingenious creativity of the Italian engineers from either contractors and the *Genio Civile* was especially active in finding more efficient and economic solutions for the

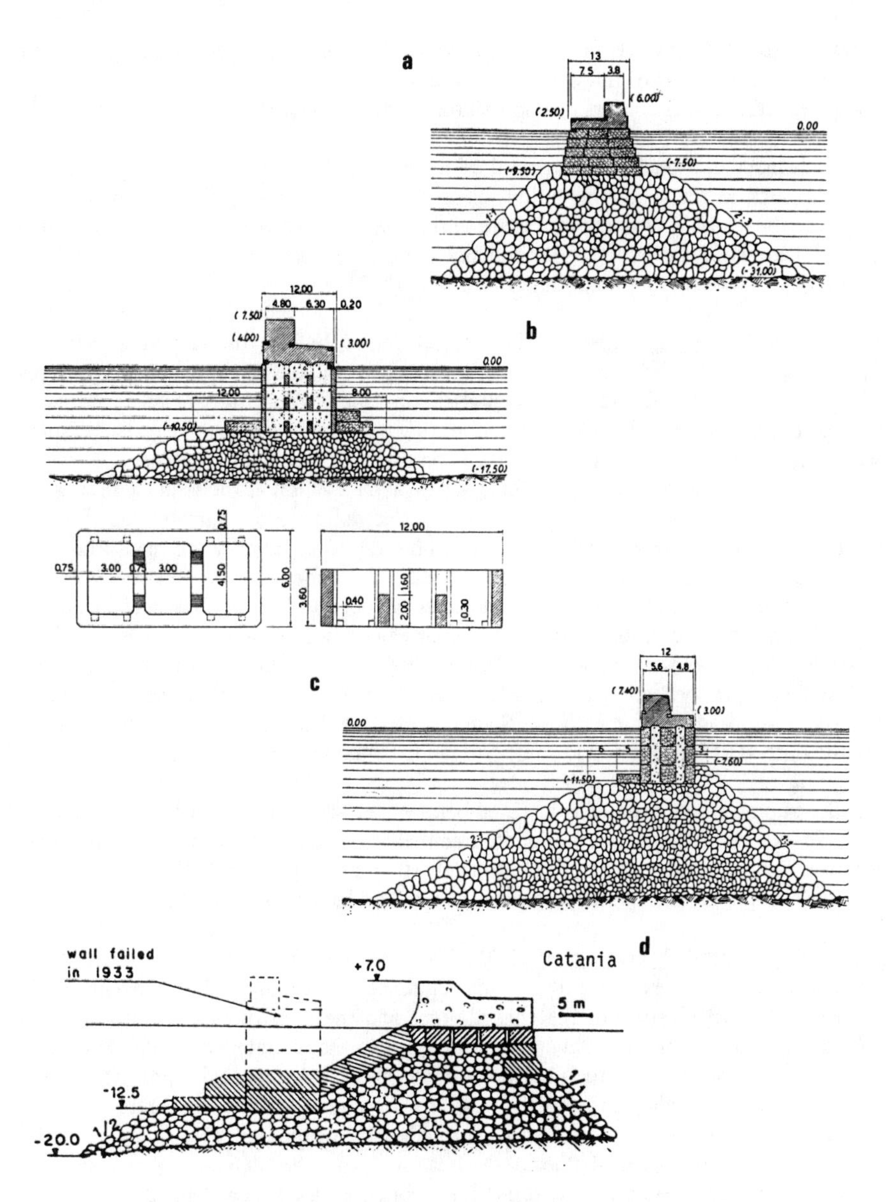

Fig.34. Developments of vertical breakwaters between 1900 and 1940:
a) Early block-type composite breakwater at Naples (1905)
b) Cellular block-type composite breakwater at Genoa (1912)
c) Cyclopean block-type breakwater at Naples "Duca degli Abruzzi" (1929)
d) The failed cyclopean block breakwater at Catania harbour, repaired with a sloping profile

construction of coastal structures, although without giving much publicity of their experience (nor patents). New ideas for artificial armour units were applied, such as the cubes with cut-off corners to improve the work compactness (Mariani,1935).

Early advanced approaches were also used for breakwater geotechnical designs on soft seabeds, such as the sand foundation fill on the silty seabed of La Spezia harbour to support the rubble mound offshore breakwater. Geotechnical problems also conditioned the design of the horizontally composite breakwaters which were later built at Ravenna and Sibari harbours (Matteotti 1993).

Early Prototype Measurements. The interest raised by the mentioned breakwater failures also promoted the modern scientific approach in coastal engineering, by means of advanced theoretical and experimental studies and even prototype measurements. One of the first ever used full-scale monitoring system of waves and wall pressures was infact installed on a section of the Genoa offshore breakwater extension in water 15 m deep (Albertazzi 1932, Levi 1933). The wave elevation at the wall (with crest at +7.5 m MSL) was measured by a series of 19 electric contact terminals spaced at 0.5 m intervals between -2 and +7 m. Wave heights on 20 m depth were measured by a 2.5 m spherical buoy bearing a vertical cardanic suspended spar collimated from the shore. Wave length and direction was obtained by observation of some smaller buoys, placed 1 km offshore up to 30 m depth, outside the range of reflected waves, at fixed distances along two orthogonal lines. Two sets of 7 pressure cells were mounted in the wall to measure wave-induced forces. The results of these early prototype experiments showed a good agreement with the Sainflou pressure distribution of standing waves: the pressure at -15 m was found to be half the maximum value at the waterline.

Another monitoring system was installed before the last war in the Duca degli Abruzzi mole in Naples, where a stereophotogrammetric technique was used to measure incident and offshore waves. Unfortunately the transducers did not give reliable results. The station was damaged during the war and later rehabilitated (Greco 1955).

Progress in Coastal Protection (1900-1950). Attention to coast defense engineering started to receive more attention since the beginning of the XXth century, even outside Venice. The urbanization of the coastal areas and the construction of littoral roads and railways grew along the Italian peninsula (due to the mountaneous hinterland) and sea protection works were required on exposed shores. A new legislation was issued in 1907 (only now being updated!) to regulate the intervention of the State, which is the actual "owner" of the coastal strip (normally 75% of the budget is financed by the state and 25% by local Authorities), but only for the defence of endangered infrastructures, typically with "hard passive" works, such as seawalls and revetments. As far as technical innovations are concerned, an original solution was produced by the engineer G.Lenzi: he was probably the first advocate of the detached parallel breakwater system, which he successfully designed for the defense of the endangered marine promenade of Salerno in 1905. However, yet in the 1930 edition of the largest Italian encyclopaedia "Treccani", the term Coastal Defence was still only referred to its military meaning !

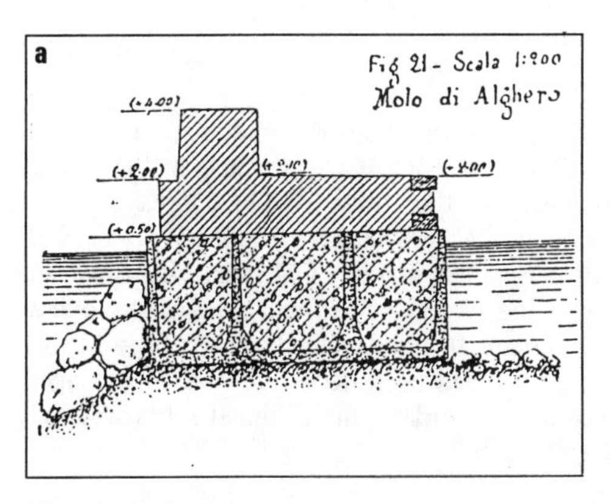

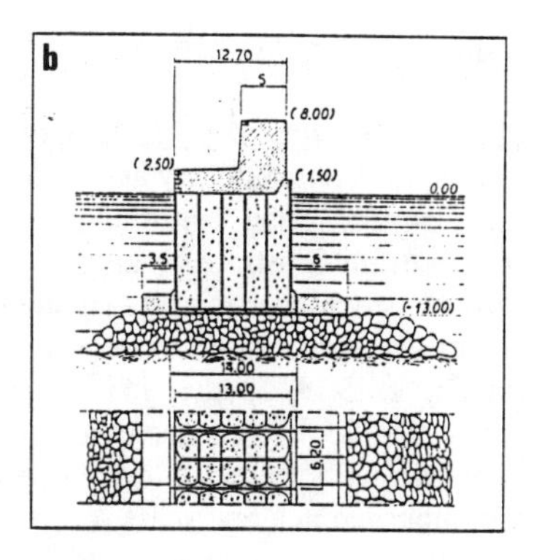

Fig.35. Earliest applications of the floating caisson technology at Alghero in 1915 (a) and Civitavecchia harbour extension in 1935 (b)

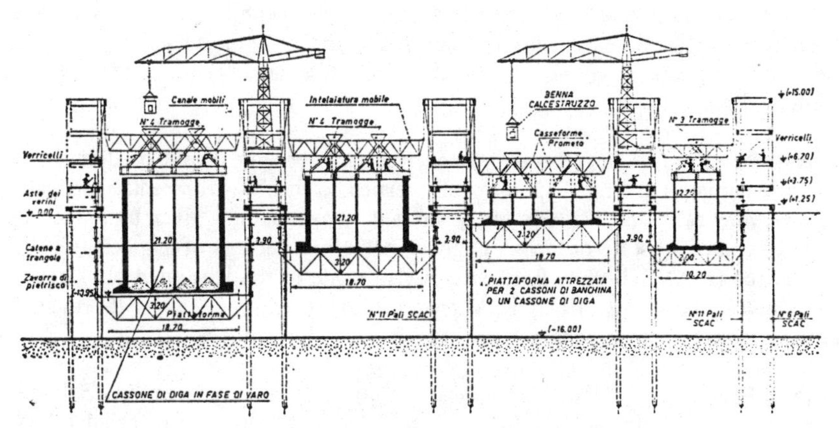

Fig.36 The first fixed caisson construction yard at Ponte Canepa, Genoa

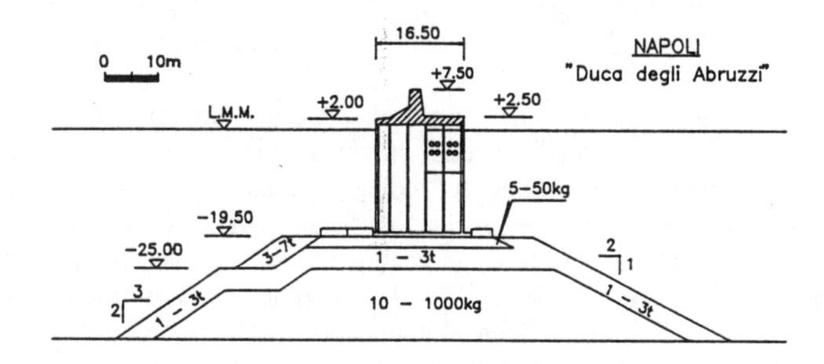

Fig.37. Section of a modern caisson breakwater at Naples

Contemporary Age

Harbours Reconstruction. As mentioned before we could conventionally define the start of the contemporary age after WW2 when the first ICCE was held in 1950. Italy was just recovering from the destructive war. At the 18th PIANC Congress in Rome an interesting illustrated book prepared by the Italian Ministry of Public Works (1953) was handed to participants. It contains a detailed survey of all port structures and facilities before and after the war. Out of the 91 km of harbour breakwaters overall existing in 1939, 20 km had been destroyed and 10 km damaged, while of the 140 km quaywalls (33 km deeper than 9 m) some 51 had been destroyed and 43 damaged. However with a great effort the reconstruction work of the National Ports was fully accomplished in a few years with modern criteria.

Beside the necessary upgrading of the historical harbours, including deeper breakwater extensions, a few fully new ports have been constructed in the last 50 years. The most modern is probably the industrial harbour of Gioia Tauro, which was designed with advanced criteria (Grimaldi et al.1984). Its plan layout with long internally excavated basin and normally oriented outer harbour is very similar to the ancient Roman port of Ventotene (fig.20). The two deepwater breakwaters have been armoured with 30 t reinforced dolosse.

New developments in port design have been recently observed in the modern sector of pleasure harbours and marinas, especially related to their environmental implications. An overview of the new criteria for design and construction is given by Franco et al. 1993. An updated state-of-the-art is given in the proceedings of a specialty national conference (PIANC 1995). Reference is also made to the illustrated book given to delegates at the ICCE (1992) in Venice, which generally describes the main new Italian projects of coastal engineering.

Vertical Breakwaters. The technological evolution again progressed, especially in the familiar field of composite breakwaters, with the systematic use of cellular floating caissons, with larger size and more complex geometries (CSPP 1976, Tosi 1980). Caissons with semicylindrical external walls were prefabricated for the first time by an Italian contractor to be towed from Sicily to the west breakwater of Marsa el Brega (Libya) in 1967. An even longer journey (2500 km from the yard in Genoa) was undertaken in 1975 by the 5000 t traditional square caissons used for the east mole.

A modern (1988) application of the semicircular planshaped front is observed in the new extension of the Duca degli Abruzzi breakwater in Naples, whose cells on the harbour side are partially absorbing to reduce internal wave disturbance (fig.37). Other new caisson designs include seaward perforated walls and absorbing chambers (to reduce wave reflection, toe scour, as well as forces and overtopping) and sloping curved and set-back parapet walls (mainly to reduce wave forces), sometimes combined with the perforations. Original applications of these designs can be found in the new breakwaters at Civitavecchia, Sorrento, Bagnara and Porto Torres (Noli et al.1995).

A unique example of "pure vertical" breakwater is represented by the steel-piled structure with concrete screen (fig.38), built in the early 70's to shelter an island-harbour at Manfredonia in 11 m of water on a shallow sloping beach. This unusual solution was chosen due to the soft clayey silt foundation soil (Benassai et al.1974).

Better designs have then been favoured by the revolutionary progress in the knowledge of random wave statistics and development of laboratory facilities, which occurred after WW2. Further lessons have been learned from analysis of new failures occurred to old vertical block walls, such as the famous one of Genoa breakwater in 1955 (Grimaldi 1955) and even an old caisson case in Naples slid in 1987 (Franco et al.1992). Again prototype measurements of wave pressures on vertical walls have been carried out on Genoa new offshore breakwater (Scarsi et al.1974, Marchi 1977, Boccotti 1984), which also demonstrated the possibility of a nearly rectangular distribution of uplift pressures on the caisson base. More recently a double measurement station operated on both plain and perforated neighboring caissons of Porto Torres industrial harbour breakwater (De Girolamo et al.1995) (fig.39).

Rubble mound breakwaters. It should be remembered that the long valuable experience of the Italian contractors in the field of maritime engineering has been revealed also abroad. In particular in the last 20 years they have built important rubble mound harbour breakwaters, sometimes in very deep and exposed waters and often with innovative technical solutions, in Sines (Portugal), Bandar Abbas (Iran), Mohammedia (Morocco), Homs (Libya), Djen-Djen (Algeria), Bosaso (Somalia), Ras Laffan (Qatar) (Noli 1993, Burcharth et al.1991, Franco et al 1995).

Beach Dynamics and Coastal Protection. Substantial progresses are observed in the new expanding sector of beach morphodynamics and shoreline defense, due to the dramatic increase of erosion processes in the last 40 years (mainly related to human activities, such as river damming and quarrying, coastal urbanisation, sporadic maritime works, subsidence etc.). The growing socio-economic importance of beaches for recreation and the new environmental concerns called for new territorial policies with more systematic strategies. The earliest Regional and Interregional Plans for coastal defence started at the end of the seventies in the north-central Adriatic areas (Moretti et al.1984).

Research in the field of coastal geomorphology and dynamics had a remarkable boost in the 70's thanks to the "Special Programme for Soil Conservation" promoted by the C.N.R.. A series of interdisciplinary research projects were carried out by sedimentologists, mineralists, geologists, geographers and hydraulic engineers, who produced a large scientific production, listed and summarized by A.V.1985. These projects also led to the edition of the mentioned "Atlas of Italian Beaches" and to a reference technical manual for shore protection works (Cortemiglia et al.1981).

An overview of the present state-of-the-art can be found in the proceedings of a national specialty conference, where engineers and geologists managed to bring together once again to report on the main Italian case studies (Aminti et al.1993).

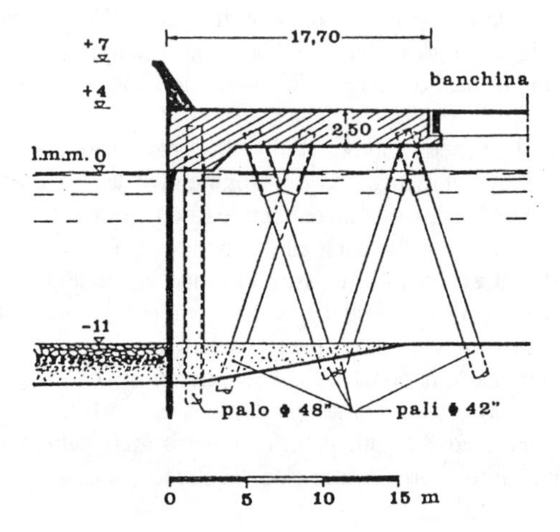

Fig.38. The piled breakwater with vertical screen of Manfredonia island-port

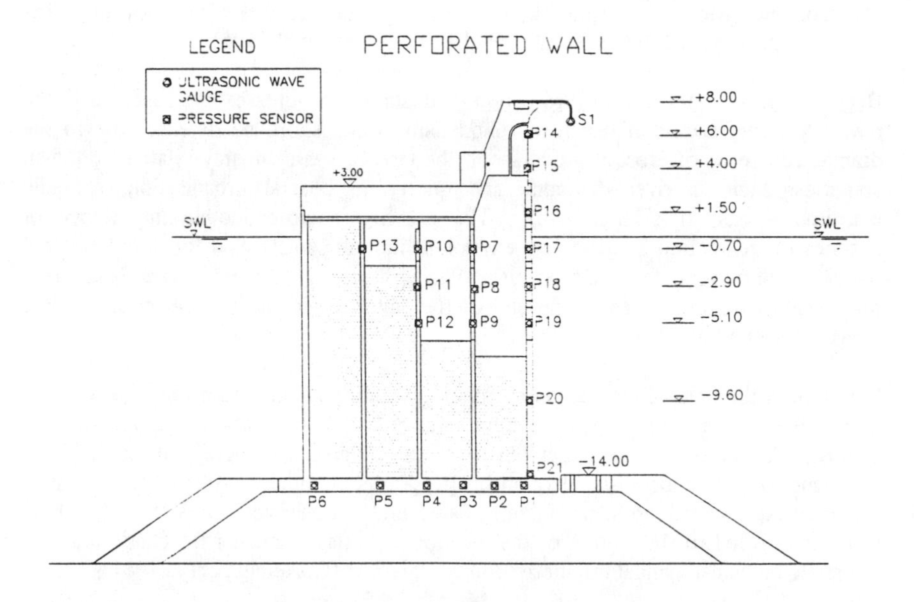

Fig.39. The new instrumented perforated caisson breakwater at Porto Torres

Further references are found in the proceedings of the first edition of the new biennial "Italian Days of Coastal Engineering" promoted by the Italian PIANC section in 1993. Moreover a general inventory of 113 studies related to coastal protection (including bathymetric, sedimentologic, meteoceanographical surveys and modelling), which were carried out in Italy in the last 15 years, has been published by ESTRAMED 1995.

Again a vast empirical knowledge has been gained after many practical field applications, mostly with partially successful "hard" protection structures such as seawalls, groins and the typical rock detached breakwaters : more than 500 barriers of the latter type have been inventoried along 300 km of central Adriatic shoreline. Further information on Italian detached barriers is given by Liberatore (1992).

Sometimes original designs of shore protection structures or systems have been applied. A new original Italian structure is represented by the permeable piled barriers named *Ferran* (fig.40a), which have been used off two sandy beaches near Ancona in 1981-85 with a partial success (Vitale et al.1985). The star-shaped r.c.piles can promote beach accretion and allow water circulation, but they have problems of stability, durability, aesthetics and recreational safety. Other patented concrete elements (perforated or articulated blocks) are used within submerged modular structures also to enhance fish survival (e.g.*Nettuno* fig.40b, *Monobar*).

In fact the traditional emerged discontinuous detached breakwater systems are now being replaced by longitudinal even continuous submerged barriers for their favourable aesthetical and hygienic effects. However the present trend for shore protection works in Italy (as in most countries) is toward natural sand nourishment, although typically combined with retaining structures to reduce sediment losses and maintenance work. The scarcity of offshore sources of suitable borrow material generally forces to use more costly land quarries. Actually early "soft" solutions had been already used in the 50's, when one of the first fixed *bypassing* plants was installed at Viareggio harbour (fig.41). The bypass system was later modified and it is still operational with more flexible floating equipment (Fiorentino et al.1985).

An original perched beach scheme has been applied in 1990 along 3 km of shoreline at Ostia Lido near Rome (Ferrante et al.1992), featuring a sand fill with an underlayer of coarser poorly sorted material and a longitudinal rock sill with crest now at about -2 m MSL (fig.42). The present monitoring shows a predicted beachline rotation due to the longshore sediment transport gradient and to the lack of groins, and the accumulation of gravel near the waterline. Another "protected nourishment" scheme has been used to defend the coastal railway along the exposed steep western Calabrian shores: it combines a gravel beach nourishment with large T-shaped groins (Guiducci et al.1993) (fig.43).

Various shore protection projects made in the last decade along the northern Adriatic beaches are often characterized by sand containment within cells made with groins (only partly emerged) and a submerged longitudinal barrier, sometimes made with sand bags (fig.44). A similar system is now also being constructed for the new protection of

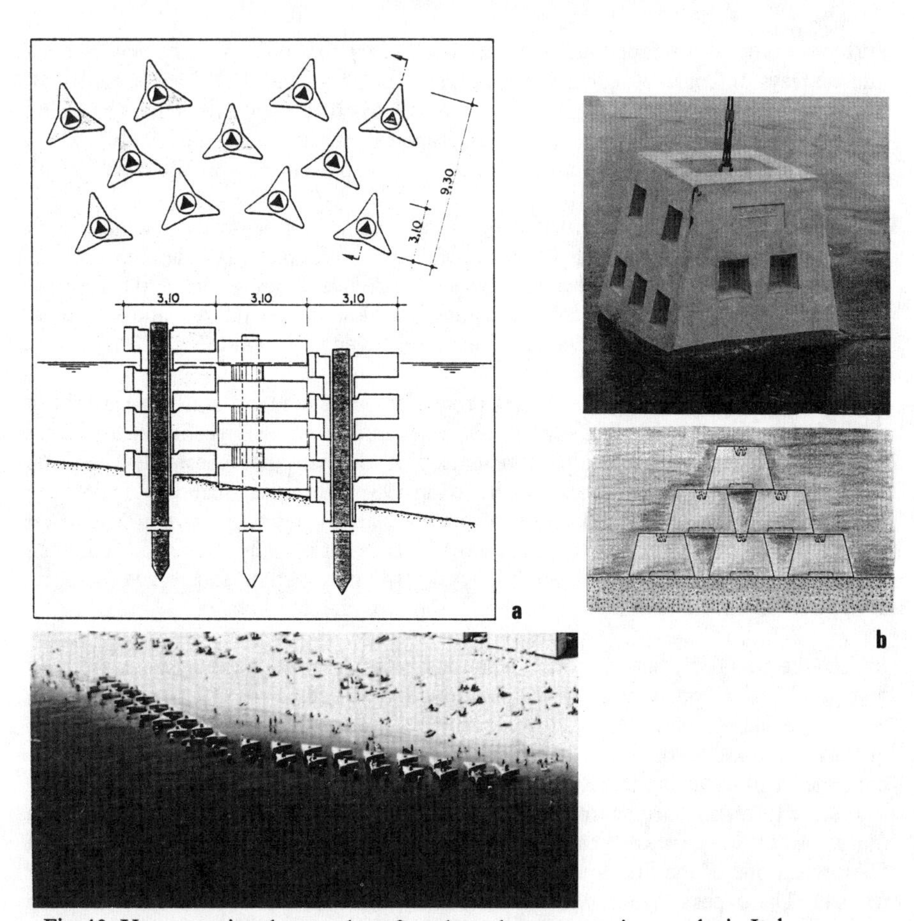

Fig.40. Unconventional examples of modern shore protection works in Italy:
a) *Ferran* piled barriers near Ancona
b) Perforated concrete blocks *Nettuno* for submerged barriers and to enhance fish habitat

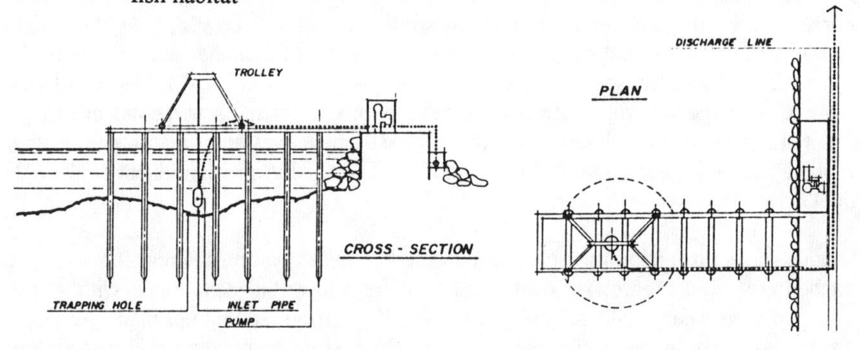

Fig.41. First fixed bypassing plant at Viareggio (1956)

protection of the Venetian littoral of Pellestrina, where, for the first time in Italy, the borrow fill is dredged from 20 m deep offshore "fossil beaches". The same material is used to nourish the near beach of Cavallino, where new groins are only built after the uneffectiveness of the shore-parallel barrier has been proved with advanced integrated modelling techniques (Noli et al.1993, CVN 1995). This shore protection works are the modern development of the historical seawalls previously described and they are part of the integrated Venice Safeguard Project. This project also includes the storm surge mobile barriers at the three lagoon inlets (now designed as shown in fig.45 and undergoing the E.I.A.) and represents today the largest ongoing project of coastal engineering in Italy (Marchi 1992).

Offshore. Significant Italian developments in the last half century may also be mentioned in the new "neighbouring" field of offshore engineering. Actually the first offshore oil well in Europe was drilled in the Sicily Channel in 1959 (by AGIP) and the first European discovery of offshore gas was made in 1960 near the Italian coasts of the North Adriatic Sea. Since then over 100 piled jackets have been installed, mainly in these two areas, together with important submarine pipelines (eg. for gas transport across the 600 m deep Sicily Channel from North-Africa). Offshore related engineering and research activities have then been in the forefront: innovative designs of steel gravity platforms (by Tecnomare)and technology for lying deep sealines (by Saipem) have had important applications even in the North Sea.
As elsewhere the fast progress in the offshore hydrodynamics and practical applications has produced benifits also to the *coastal* community. The engineers of the main oil companies have even joined the researchers and contractors in the traditional civil-hydraulic fields of coasts and harbours to form an Association of Offshore and Marine Engineering (AIOM), which since 1985 is organizing national congresses, meetings, courses and other activities.

Status of Scientific Research. In Italy the research activity in the field of coastal hydraulics and engineering has been quite scattered and isolated in the last century as for other similar scientific disciplines. This is due to various reasons such as: the late diffusion among Italian scholars of the English language (which has replaced Latin and French as official scientific language) and especially the fact that hydraulic research in Italy is run by Governmental Agencies (for 90% State Universities, which are not commercially oriented) with the inherent deficiencies of such institutions. In particular coastal engineering research is mainly carried out by many small separated academic institutes and laboratories within Hydraulic and Civil Engineering departments (over 20 distributed all across the country), which had somewhat neglected it in favour of fluvial and damming problems. Only after 1985 3-year PhD courses are offered each year by Italian Universities: a thesis on coastal engineering subjects can be carried out within three consortiums with administrative seats at the Hydraulics Institutes of the Universities of Padua, Naples and Milan Polytechnic. Admission is open to home and foreign students under public competition.

Another drawback is represented by the absence of a large national hydraulic laboratory for coastal scale models. Advanced experimental facilities are available

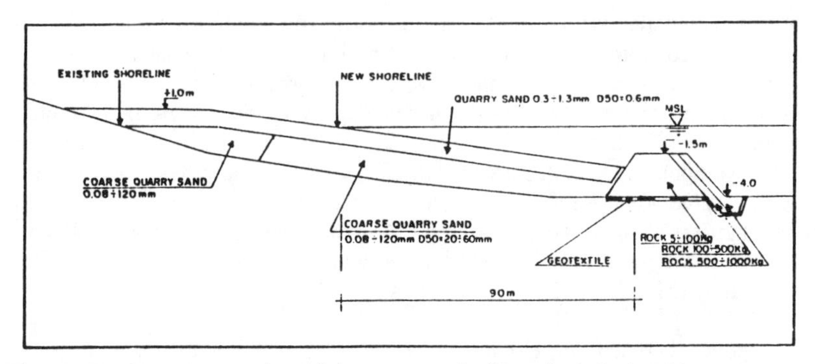

Fig.42. Design cross section of the new perched beach at Ostia Lido (Ferrante 1992)

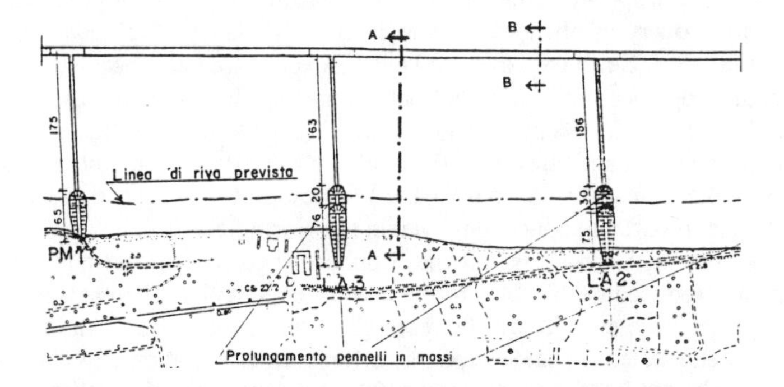

Fig.44 Example of new protected nourishment scheme along the north Adriatic coast

Fig.43. Photo of the beach nourishment within T-shaped barriers to protect the littoral railway at Paola-S.Lucido in Calabria (Guiducci et al.1993)

today at ENEL-CRIS in Milan and at the Center of Voltabarozzo, Padua, which belongs to the *Magistrato alle Acque* (mainly devoted to the problems of Venice), and a new lab is under construction in Bari.

However, despite the traditional lack of coordinated organization and of financial support, either theoretical, experimental and field research work in coastal engineering have produced valuable contributions, especially in the last years. This is partly due to the diffusion of mathematical modelling with inexpensive computational facilities and also by taking advantage of new European programs, such as the various MAST (Marine Science and Technology) projects carried out since 1989 on Coastal Morphodynamics and on Coastal Structures with the participation of some Italian Universities (Rome, Milan, Naples, Bologna, Genoa, Florence) and private companies (Snamprogetti, Tecnomare, ENEL). For the complete list of related subjects and publications the reader is referred to the Commission of the European Community 1995. Another EU-funded program which is worth mentioning is the Large Installation Plan (LIP): it allows young researchers from any European country to carry out selected experimental projects by using the advanced expensive facilities of a few specialized hydraulic laboratories (e.g. DH, HR, DHI). An example of Italian participation is given by a novel 3-D physical model study on the hydraulic performance (wave forces and overtopping) of caisson breakwaters under multidirectional seas (Franco C.et al.1996). Research projects of each university also receive limited annual funds at the national level by the Ministry of University and Scientific Technological Research (MURST).

Most publications of Italian authors on maritime hydraulics and coastal engineering can be found within the national congresses of Hydraulics and the AIOM's ones, and more and more frequently in the international proceedings of the ICCE, IAHR, PIANC and ICE Breakwater conferences, besides various specialized journals. The present production of papers is of the order of hundreds per year and it is obviously impossible to give a full account of them here. However, though uncomplete, a short list of the main research topics is provided in order to outline the global picture and give some useful references of international interest.

The first and only ICCE held in Italy (Venice, 1992) has witnessed - together with a properly organized high surge (...!) - the presentation of twenty papers by Italian authors (out of 263). Nine of these papers were devoted to hydraulic and coastal engineering problems related to Venice. The others dealt with wave forecasting models (Cavaleri et al.), wave transformation over submerged bars (Liberatore et al., Petti et al.), mechanics of wave groups (Boccotti et al.), sand ripple formation (Foti et al), shallow water wave theories (Brocchini et al., Mattioli), sediment suspension (Longo), breakwater monitoring (Muraca et al.), beach profile evolution (Chiaia et al.) and case histories of new shore protection projects. At the next most recent ICCE's in Kobe (1994) and Orlando (1996) the number of papers by Italian authors and co-authors is still around ten, generally dealing with the same above subjects, and also covering 2-D and 3-D model tests of wave overtopping and loads on breakwaters, the stability of berm breakwaters and the behaviour of submerged barriers.

Generally, relevant Italian contributions are given in the definition of the characteristics of deep and shallow water wave spectra and their interaction with coastal structures, based on theoretical approaches, field measurements and on numerical and physical models. Remarkable is the new theory of "quasi determinism of the highest waves" (Boccotti 1988), which predicts the characteristics of the highest wave groups in a sea-state. Advanced theoretical studies on bedload transport and ripples formation are performed by Blondeaux et al.1990. An analytical computation of random wave direction is given by De Girolamo 1995, who also treated harbour resonance problems (1996). The statistical properties of random waves and groups are also investigated with numerical simulations (Rebaudengo Landò et al. 1989). A numerical study has been addressed to the flow interaction with sinusoidal bed dunes (Sammarco et al.1994). New efforts are devoted to wave propagation in shallow waters, including simulation of nonlinear steep waves at vertical walls (Passoni 1995). Hydraulic model studies on coastal structures have been addressed to wave overtopping response (Aminti et al.1988, Franco et al.1994), toe stability and reshaping breakwaters (Lamberti et al.1994). Original small scale experiments on the real seas action against a vertical wall have been performed at Reggio Calabria (Boccotti 1992). New field/lab tests on the structural integrity of prototype armour blocks confirmed the high strength and durability of Italian marine concretes (Franco et al.1995). Among the few studies on beach dynamics, 2-D mobile bed model tests have been conducted at the University of Bari (Chiaia et al.1992) and Naples (Di Natale et al.1992).

National papers in coastal hydraulics are also internationally distributed through the annual issues of the journal EXCERPTA, published by GNI (Gruppo Nazionale Idraulica) which reviews the Italian academic contributions to the whole field of hydraulic engineering.

Finally a new study area is represented by the modelling and monitoring of the natural coastal ecosystems, including the hydrodynamic transport of surface pollutants, beach dune stabilization and the ecological impacts of coastal works. An example reference on the detailed studies of the vegetation of the Venetian sandy littorals is given in CVN 1995. New protection works at Cavallino beach have included transplanting of local *Ammophila*, sheltered by wooden fences, to "armour" the littoral dunes degradated by erosion and high touristic use.

CONCLUSION AND OUTLOOK

This condensed historical review has shown how the ingenious individual contributions of Italian hydraulic engineers, architects, mathematicians, oceanographers, geologists and even naval officers have helped the development of Coastal Engineering in a time span of over 2000 years. It is in fact acknowledged that Italy has been the leader country in the technological progress since the II century BC until the XVIIIth AD.

The scarcity of natural harbours along the Italian tideless coasts has forced the Romans and later the Italians to build exposed deepwater breakwaters (in particular monolithic composite breakwaters) and hydraulically efficient harbour layouts, developing most modern design solutions well ahead of times. Earliest shore protection works were also experimented by the Venetians. This bimillennial field experience can also provide a valuable knowledge of the long-term response of coastal structures and morphology.

Despite the lack of large organized research centers and laboratories, Italian individuals have also contributed to the scientific development of this discipline, since the earliest advanced intuitions of Leonardo da Vinci to the present revival of both theoretical and experimental research. The growing interest in Italy about the fascinating problems of coastal engineering is also revealed by the number of the PIANC memberships (now the Italian delegation is the world largest together with USA's). In the wake of the ICCE'92 in Venice a new series of biennal national congresses, specifically devoted to coastal engineering, are taking place under the auspices of the Italian PIANC delegation.

It is expected that future coastal engineering works will be mainly related to the restoration of sandy beaches and of old harbour structures, possibly converted to different functions, especially for recreational use. New marinas will be built, but with greater environmental concern. Soft shore protection systems will be preferred.

The understanding of the complex nearshore processes and the reliability of coastal structures design and construction are likely to be improved in the near future by the increasing development of the computational models and remote sensing techniques for field monitoring. Modelling the coastal systems will be account for either physical, ecological and socio-economic aspects. The growing pressure on the Mediterranean shores and on the delicate ecosystem of this semiclosed basin will in fact require a more integrated and careful Coastal Management (Vallega 1993). The necessary coordination of various "expertises" could also include underwater archaeologists, in order to enhance the discovery, preservation, revaluation and even musealization of ancient harbour structures, which are still a neglected valuable heritage.

It is clear that the present trends in Coastal Engineering are again strongly addressed towards harmony with the environment. Therefore it is worth remembering Leonardo's statement in Latin *"Ne coneris contra ictum fluctus: fluctus obsequio blandiuntur"*, which means that "Nature should not be faced squarely and opposed, but wisely circumvented".

ACKNOWLEDGEMENTS

The author wishes to thank all the persons who provided useful information and historical references for the preparation of this paper. In particular the cooperation and advice of prof. G.Borzani (University of Genoa), prof. A.Noli and prof. G.B. La Monica (University of Rome), dott.L.Cipriani (University of Florence), ing. M. Gentilomo (Consorzio Venezia Nuova), ing.A.Ferrante (PIANC-Italy), ing. C.Franco, ing. A.Triola and dr G.Passoni (Politecnico di Milano) are gratefully acknowledged.

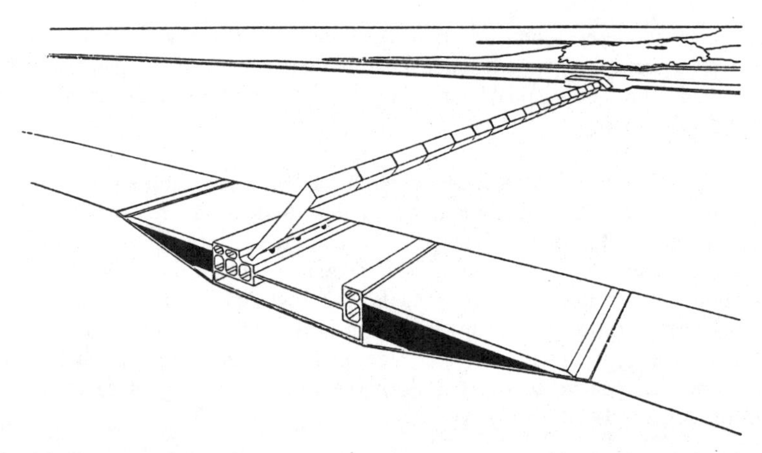

Fig.45 Conceptual drawing of the buoyant flap gates for Venice surge barrier

REFERENCES

Adami A., 1992. "Sea defence works in Venice", *Proc.of ICCE'92 Short Course on Design and Reliability of Coastal Structures,* Venice, pp.397-414

Albertazzi A., 1932. "Recent experiences on dynamic action of waves against breakwaters", Annali Lavori Pubblici, Roma , Feb 1932, p.134 (in Italian)

Alberti L.B., 1465. "De re aedificatoria", G.Simoncini Editor, De Luca , Roma

AM-ENEL, 1980. "Diffusive characteristics of the low layers of the atmosphere" 16 volumes (in Italian)

Aminti P. and Franco L. 1988 "Wave overtopping on rubble mound breakwaters" *Proc. ICCE,* Malaga, ASCE NY, pp.770-781

Aminti P. and Pranzini E. (Editors), 1993. "The littoral defense in Italy" Ed.Autonomie n.34, Roma (in Italian)

Archetti, R., and Franco, L. 1995. "New analysis of wave measurements in the Italian seas", *Proc. AIPCN 2nd Italian Days of Coastal Engineering,* Ravenna (in Italian)

AV1985 "Italian Research on Physical Geography and Geomorphology: an overview", *1st Int. Conf. On Geomorphology,* Manchester, Tecnoprint, Bologna, pp.17-25

Benassai E., Grimaldi F., Scotti A., 1974, "Interaction between waves and the vertical piled breakwater of the new Manfredonia harbour", *Proc. XIV Convegno di Idraulica e Costruzioni Idrauliche,* Napoli (in Italian)

Bernardini O.,1901. "On the damage due to the storm of 27 Nov.1898 and on the repair of Genoa Breakwater" *Giornale del Genio Civile* pp.665-692 (in Italian)

Bianco L., 1637. "On Genoa harbour rehabilitation", ed. FINCOSIT 1955 (in Italian)

Blackman D.J., 1982. "Ancient harbours in the Mediterranean", *Int.Journal of Nautical Archaeology and Underwater Exploration,* part 1 in 11.2 pp.79-104; part 2 in 11.3 pp.185-211

Blondeaux P., Vittori G., 1990. "Sand ripples under sea waves", *J.Fluid Mech.* vol.218, pp.1-39

Boccotti P., 1984. "New wave pressure measurements on the caisson breakwater of Genoa-Cornigliano", *Proc.XIX Congresso di Idraulica e Costruzioni Idrauliche*, Pavia (in Italian)

Boccotti P., 1989. "Quasi-determinism of sea wave groups", *Meccanica* vol.24, n.1

Boccotti P., 1992. "The behaviour of a very light vertical breakwater", *Proc.XXIII Congresso di Idraulica e Costruzioni Idrauliche*, Firenze, (in Italian)

Borzani G., 1973 "On a vertical breakwater for reclamation defence at Fontvieille, Monaco", *L'Industria Italiana del Cemento* n.7-8, pp.481-512 (in Italian)

Borzani G., 1995. "Evolution of design criteria for harbour breakwaters", PIANC Italian Section (in Italian)

Brambati A., 1993. "Preliminary investigations on the physiographic unit", *Proc. ICCE92-PIANC 1st Italian Days of Coastal Engin.*, vol.A, pp.218-237 (in Italian)

Brown F.E., 1980. "Cosa, the making of a Roman town", Michigan University Press

Burcharth H.F., Toschi, P.B., Turrio E., Balestra T., Noli A., Franco L., Betti A., and Mezzedimi S., 1991. "Bosaso harbour, Somalia, a new hot-climate port development", *Proc. 3rd COPEDEC Conference on Coastal Engineering in Developing Countries*, Mombasa, pp. 649-667,

Caesar Julius, 55 BC, "De Bello Gallico" IV,17

Caputo C., D'Alessandro L., La Monica G.B., Landini B., Lupia Palmieri E. , 1991. "Present erosion and dynamics of Italian beaches", *Z.Geomorph.N.F.* Supp.Bd.81, pp.31-39

Cavaleri L., Bertotti L, 1992. "The Mediterranean Sea Wave Forecasting System," *Proceedings 23rd Coastal Engineering Conference*, ASCE, NY, pp.116-128

Chiaia,G.,Damiani,L.,and Petrillo, A. 1992."Evolution of a beach with and without a submerged breakwater:experimental investigation", *Proc.23rd ICCE, Venice, ASCE,NY, pp.1959-1972*

Cialdi A., 1866. "On Sea Waves and currents, especially near the littorals", Roma (in Italian)

Clark 1948. "Science and Social Welfare in the Age of Newton", Oxford

Clementi R.,1981. "The Roman port of Astura". *L'Universo* n.6, I.G.M. (in Italian)

CNR (National Research Council), 1984. "Atlas of the Italian Beaches" SELCA, Florence

Commission of the European Communities,1995. "Marine Sciences and Technologies" *Proc.2nd MAST days and Euromar market, Sorrento*, Project reports, Bruxelles

Coen-Cagli E. and Bernardini O.,1905. "News on Italian maritime ports" Ministry of Public Works, Pirola, Milano (in Italian)

Coen-Cagli E., 1936. "Italian Docks and Harbours" *Journal of the Inst.of Civ.Eng.*, London

CSPP (Centro Studi Problemi Portuali) 1976. "Technological progress and maritime works" *Serie Documenti e Ricerche* n.16-17 (in Italian)

CVN (Consorzio Venezia Nuova) 1995. "QT- Quarterly Notebooks" n.2-3 (in Italian)

Cornaglia P.,1891. "On beach regime and harbour regulation" Stamperia Reale , G.B.Paravia, Torino (in Italian)

Cortemiglia,G.C., Lamberti,A., Liberatore, G.,Lupia Palmieri, E., Stura,S., and Tomasicchio, U.1981. "Technical Recommendations for Coastal Protection" CNR, P.F."Soil Conservation", S:P."Littoral Dynamics" (in Italian)

Crescentio B., 1607. "Nautica Mediterranea", vol.5, Ed.Bonfaldino, Roma (in Italian)
D'Arrigo A., 1940. "Leonardo da Vinci and the regime of Cesenatico beach. Researches on the origins of canal-ports in the Renaissance" Annali Lavori Pubblici n.4, pp.293-325 (in Italian)
De Boni M., Cavaleri L., Rusconi A., 1992. "The Italian waves measurement network", *Proc. 23rd Coastal Engineering Conference*, ASCE, NY, pp.1840-1850
De la Pena J.M., Prada E.J.M., Redondo M.C., 1994. "Mediterranean Ports in Ancient Times". PIANC Bulletin n.83/84, Bruxelles, pp.227-236
De Girolamo P., 1995. "Computation of sea wave direction of propagation of random waves", *Journ.Wat.Port.Coast.Ocean Eng.*, ASCE NY, vol 121 n.4
De Girolamo,P., Noli A.,and Spina D., 1995 "Field measurements of loads acting on smooth and perforated vertical walls" *Proc. ICE Conf.on Breakwaters and Coastal Structures*, T.Telford, London
De Girolamo P. 1996 "An experiment on harbour resonance induced by incident regular and irregular short waves" *Coastal Engineering* (pending publication)
De Fazio G., 1814. "On the best harbour construction method". Napoli (in Italian)
De Marchi F., 1545 "Military Architecture" in Simoncini 1993 (in Italian)
De Rossi G.M., 1993. "Ventotene e Santo Stefano", Rome, (in Italian)
Di Giorgio Martini F., 1475. "Treatise of Architecture, Engineering and Military Art", Ed Il Polifilo, Milano 1967 (in Italian)
Di Natale M., Ragone A., 1992. "Profiles of beach models under regular and irregular wave attack", *EXCERPTA* vol.6, pp.55-96
ESTRAMED (Paolella,G.,Bolatti Guzzo,L.,Corsini,S.,and Schintu C.), 1995. "Survey of the studies performed for shore protection in Italy", Maggioli Editore, 129
Faina G., 1969. "Genoese harbour engineering in 1600" Istituto Italiano per la storia della tecnica, Giunti, Milano (in Italian)
Fassò C.A., 1987. "Birth of hydraulics during the Renaissance period" in "Hydraulics and Hydraulic Research: a Historical Review", *IAHR*, G.Garbrecht Editor, Balkema pp.55-79
Ferro G., 1936. "On wave action on vertical breakwaters", *Annali dei Lavori Pubblici* pp.764-781, 935-946 (in Italian)
Felici E., 1993. "Observations on the Neronian port of Antium and the Roman technology of concrete harbour works", in Archeologia Subacquea, IPZS, Roma (in Italian)
Ferrante A., Franco L. and Boer S., 1992. "Modelling and monitoring of a perched beach at Lido di Ostia", *Proc. 23rd Coastal Eng. Conf.* ASCE, NY, pp.1879-1895
Fiorentino A., Franco L. and Noli A., 1985. "Sand bypassing plant at Viareggio, Italy", *Proc. Australasian Conference on Coastal and Ocean Engineering*, Christchurch (NZ), Vol.1, pp. 441-451
Franco C., Van der Meer J.W. and Franco L., 1996. "Multidirectional wave loads on vertical breakwaters" *Book of Abstracts, ICCE'96*, Orlando, ASCE, NY
Franco L. and Passoni G., 1992 "The failure of the caisson breakwater Duca D'Aosta in Naples harbour during the storm of 11 January 1987", *Proc.2nd MAST G6-S workshop* (also in Bauingenieur 69, 1994, in German)
Franco L., Marconi R., 1993. "Marina design and construction" in Marina Developments. Computational Mechanics Publications, Southampton, pp.143-213

Franco L., 1993. "Wave records in the Italian Seas: results of a new inventory and statistics of extreme storms," *Proc. Italian Days of Coastal Engineering, ICCE'92-AIPCN*, vol.D, Genova, pp.146-181 (in Italian).

Franco L., 1994. "Vertical Breakwaters: the Italian experience", *Coastal Engineering*, 22, pp.31-55

Franco L., De Gerloni M. and Van der Meer J.W., 1994. "Wave overtopping on vertical and composite breakwaters", *Proc.24th ICCE Conference on Coastal Engineering*, Kobe, ASCE, NY, pp.1030-1045

Franco L., Gentilomo M., Noli A. and Passacantando G., 1995. "Constructional aspects of some large rubble mound breakwaters in northern Africa", *Proc.ICE Conf.on Breakwater and Coastal Structures*, London, T.Telford

Franco L., Noli A., De Girolamo P. and Ercolani M., 1995. " Concrete strength and durability of prototype tetrapods and dolosse: results of field and laboratory tests", *Proc. final workshop MAST2-RMBFM project, Sorrento*

Franco L., 1996. "Ancient Mediterranean Harbours: a Heritage to Preserve", *Journal of Ocean and Coastal Management*, Special Issue on the Mediterranean, Elsevier

Frost H., 1972. "Ancient harbours and anchorages in the Eastern Mediterranean". *Proc. UNESCO Congress "Underwater archaeology: a nascent discipline"*, Paris, pp.95-114

Greco L., 1953 "On port technology- Half century of engineering" L'ingegnere n.3, Milano (in Italian)

Greco L., 1955. "On wave theories and necessity of experimental research", *Giornale del Genio Civile*, pp.507-513 (in Italian)

Grillo S.,1989 "Venice Sea Defences" Arsenale Editrice, Venezia (in Italian)

Grimaldi M., 1955. "The disastous cyclone damage at Genoa Breakwater" The Dock and Harbour Authority , vol.36, n.418 pp.117-120

Grimaldi F. and Fontana F., 1984. "Redesign of the main breakwater of Gioia Tauro (Italy)", *Proc.Int.Symp.on Maritime Struct. in the Med Sea,* Athens, pp.2.153-172

Guiducci F., LoPresti F. and Scalzo M.,1993 "Beach nourishment at Paola-S.Lucido", in La difesa dei litorali in Italia, editors P.Aminti and E.Pranzini, pp.195-214

Hohlfelder R.L., 1987. "Caesarea Maritima". *National Geographic*, vol.171, n.2

ICCE, 1992. "Coasts, harbours, lagoons protection works".(under patronage of the Italian Ministry of Public Works, Rome) *23rd Coastal Engineering Conference*, Venice

IIMM (Navy Hydrographic Institute), 1984. "Winds and waves along the Italian coasts" Genova (in Italian)

Inman D., 1974. "Ancient and modern harbours: a repeating phylogeny". *Proc. 15th ICCE,* ASCE NY, pp.2049-2067

Italian Ministry of Public Works, 1953. "The reconstruction of National Maritime Ports" addendum to the *Proc. XVIII PIANC Congress* Rome (in French)

Lamberti A., Franco L., 1994. "Italian experience on upright breakwaters" *Proc. Int. workshop on Wave Barriers in deepwaters*, PHRI Yokosuka, Japan, pp.25-72

Lamberti A., Tomasicchio G.R. and Guiducci F.,1994 "Reshaping breakwaters in deep and shallow water conditions" *Proc. 24th ICCE*, Kobe, ASCE NY

Leonardo da Vinci, edited 1826 "On water motion and measurements", Raccolta d'Autori idraulici italiani, Opuscoli idraulici, Vol.X. Bologna Cardinali e Frulli

Levi S., 1934. "Wave pressures on vertical breakwaters" L'Ingegnere, August issue (in Italian)

Levi E., 1982. "Leonardo precursor of the hydraulic science" in Leonardo e l'età della ragione. Milan: Scientia (399) (in Italian)

Levi-Civita T., 1925. "Rigorous determination of permanent waves of finite amplitude", Mathematische Annalen, XCIII, pp.264-314 (in French)

Lewis J.D., 1973. "Cosa: an early Roman harbour". Marine Archaeology, Colston Papers 23, London

Liberatore G., 1992. "Detached breakwater and their use in Italy", *Proc.of ICCE'92 Short Course on Design and Reliability of Coastal Structures,* Venice, pp.373-395

Lo Gatto D., 1925. " Treatise on Maritime Works" Anonima Editoriale Romana, Roma (in Italian)

Lugli G. and Filibeck G., 1935. "The port of Imperial Rome". Roma (in Italian)

Luiggi L., 1907. "Maritime constructions at the beginning of the XXth century" Tip.Genio Civile, Roma

Marchi E., 1992. "Coastal Engineering in Venice", *Proceedings 23rd Coastal Engineering Conference*, ASCE, NY, pp.4-39

Marchi,E., 1977. "Problems of vertical wall breakwater design", *Proc.17th IAHR Congress,* Baden Baden, pp.337-349

Mariani,E., 1935. "A new type of breakwater armour unit" Annali dei Lavori Pubblici, n.4, pp.319-320 (in Italian)

Migliardi Tasco A., D'Arrigo A., 1965. "Italian Report at the XXI PIANC Congress" S.I , II-6 (in Italian)

Marina Militare, 1905. "A historical monograph on ancient Italian harbours", (in Italian)

Matteotti G., 1993. "Stability of soil-breakwater complex", *Proc. ICCE92-AIPCN 1st Italian Days of Coastal Engineering,* vol.A, pp. 202-217 (in Italian)

Moretti M., Pedone F., 1984. "Defence plan of the central Adriatic coast, Italy" *Proc.Int.Symp.on Maritime Structures in the Mediterranean Sea, Athens,*pp 4.133

Mosetti F.,1989. "Sea level variations and related hypoteses", *Boll.Ocean.Teor.Appl.*, vol.VII, n.4, pp.273-284

Noli,A., 1990. "The activity of Paleocapa in maritime engineering and the project of Malamocco inlet", *Proc. conf. "Ingegneria e politica nell'Italia dell'Ottocento: P. Paleocapa"* Ist.Veneto di Sci.Lett.Arti, Venezia (in Italian)

Noli A., 1993. "Constructional problems", *Proc. ICCE92-AIPCN 1st Italian Days of Coastal Engineering,* vol.A, pp.218-237 (in Italian)

Noli A., Galante F. and Silva P.,1993 "The Venice littoral nourishment project" *Proc. ICCE92-AIPCN 1st Italian Days of Coastal Eng.,* vol.A, pp.74-89 (in Italian)

Noli A., Franco L., Tomassi S., Verni R. and Mirri F., 1995. "The new breakwater and ore-carrier quay of the industrial harbour of Porto Torres (Italy)", *Civil Engineering,* London, February issue, vol.108, pp.17-27.

Oleson J.P., 1988. "The technology of Roman Harbours", IJNA 17.2, pp.147-157

Passoni,G. 1995. "Coupling nonlinear wave models for the simulation of steep waves at vertical walls", *Proc.Int.Conf.on Finite Elements in Fluids,* Venice

Prada E.J.M., De la Pena J.M., 1995. "Maritime engineering during the Roman Republic and the early empire", *Proc. Conf. MEDCOAST'95,*E.Ozhan Ed., pp.305

Pranzini E., Rossi L., 1995. "A new Bruun-Rule based model: an application to the Tuscany coast, Italy", *Proc. MEDCOAST'95 Conf.*, E.Ozhan Editor, Tarragona

Preti M., 1993 "Coastal defence in Emilia Romagna" in La difesa dei litorali in Italia, editors P.Aminti and E.Pranzini, pp.283-296

Raban A., 1988. "Coastal Processes and Ancient Harbour Engineering". Proc. 1st Int. Symp. "Cities on the Sea- Past and Present", BAR Intern.Series 404, pp.185-261.

Rebaudengo Landò L., Scarsi G., Taramasso A.C., 1989. "Random waves in directional seas", *EXCERPTA* vol.4, pp.77-116

Richter, J.P., 1970. "The notebooks of Leonardo da Vinci", Dover Publ., New York

Rouse H. and Ince S. 1963. "History of Hydraulics", Dover N.Y.

Sammarco P., Mei C., Trulsen K., 1994. "Nonlinear resonance of free surface waves in a current over a sinusoidal bottom: a numerical study", *J.Fluid Mech.*, vol.279, pp.377-405

Scamozzi V., 1615. "The idea of universal architecture", Venezia p.333 (in Italian)

Schmiedt G., 1975. "Ancient Italian Ports", L'Universo 45/2,46/2,47/1. I.G.M., Firenze (in Italian)

Schmiedt G., 1978. "Italian ports in the early middle ages", Rome

Schorn E., 1959. "Construction of large r.c.floating bodies". Deutscher Beton Verein E.v., Vortrage auf dem Betontag, Munchen pp.216-252 (in German)

Shaw J., 1974. "Greek and Roman harbour works", from Navi e Civiltà, Milan

Simoncini G., 1993. "On sea ports: The treatise of Teofilo Gallacini and the architectural concept of the harbour since the Renaissance to the Restoration", L.S.Olschki Editor

Testaguzza O., 1970. "Portus". Ed.Julia, Roma (in Italian)

Toniolo A.R., 1910. "On beach variations at Arno's mouth since the end of the 18th century", Pisa, Tipografia Municipale (in Italian)

Tosi R., 1980. "Progress in maritime works" L'Industria Italiana del Cemento n.11, pp.1059-1094 (in Italian)

Troen I. and Petersen E.L., 1989. "European Wind Atlas", ISBN 87-550-1482-8. Riso National Laboratory, Roskilde DK

Vallega, A., 1993. "From the Action Plan to the Mediterranean Agenda 21", *Proc.1st Conf. MEDCOAST'93*, Antalya, E.Ozhan Editor

Van Rijn P., 1995. "The Roman Harbour of Velsen", Terra et Aqua n.61, pp.25-28

Veranzio F.1595."Machinae Novae" (New Machines) Ed.Manunzio,Venezia (in Latin)

Vitale A., Mancinelli A., Cipriani M., 1985. "Structures in prefabricated elements for the defence of beaches: experimental result obtained employing the breakwater in Porto Recanati sandy shore and considerations about the plan of such structures" *Proc. XXVI PIANC Congress*, Bruxelles, pp.85-93

Vitruvius M.L., 27 BC., "De Architectura", vol II -6; vol V-13 (in Latin) (in English in Morgan M.H. 1914, Cambridge or in Loeb Classical Library 1983, London)

Zunica M., 1976. "Coastal changes in Italy during the past century", Italian Contributions to the 23rd Int.Geog.Congr.

Zunica M., 1985. "57.Italy" in *The World's Coastline,* Van Nostrand Reinhold Cmp, New York

HISTORY OF COASTAL ENGINEERING IN JAPAN

Kiyoshi Horikawa[1], F.ASCE

ABSTRACT : This paper describes the history and heritage of coastal engineering in Japan. The storm surge damage at Ise Bay in 1953 opened the dawn of coastal engineering in Japan. This storm occurred three years after the year of 1950 when the First Conference on Coastal Engineering was held in Long Beach, California, U.S.A. Before that, coastal related works in Japan were executed as a branch of river engineering and/or harbor engineering by government agencies, and the related research activities in this field were in an early stage of development. Therefore, the contents of this paper are mainly related to achievements in the period after 1953. This paper consists of four parts; that is, Introduction, Chronological Review of Coastal Engineering in Japan, Coastal Engineering Research Activities, and Dissemination of Coastal Engineering Knowledge. The Introduction describes the characteristic features of coastal conditions in Japan. The third part of this paper covers the subjects of wave mechanics, long-period wave mechanics, nearshore current, coastal processes and sediment transport, coastal works and structures, and coastal environment protection.

---

1) President, Saitama University, 255 Shimo-Okubo, Urawa, Saitama 338, Japan,and Professor Emeritus, The University of Tokyo, Japan.

INTRODUCTION

Geographical Conditions of Japan

Japan consists of four main islands, namely Hokkaido, Honshu, Shikoku and Kyushu from the north to the south, and about 4,000 other small islands. The total length of the coastline is about 34,360 km, whereas the total land area is 372,000 km^2. Hence, the coastline per unit land area is 92 m/km^2, which is about 1.8 and 42 times as large as that of the U.K. and U.S.A.,respectively.

The mountainous region including hilly area covers 71% of the total land, and the remaining area is divided into the foot of mountains, the terrace, and the low-lying land with the ratio of 4%, 12%, and 13%, respectively. The low-lying land is mainly situated along the coast, particularly at the head of relatively large bays such as Tokyo Bay, Ise Bay, and Osaka Bay. These areas are densely populated and keep the nation as one of the most advanced societies in the world from social as well as economical perspectives.

The Japanese archipelago, however, is situated in the Circum Pacific Earthquake Zone and in the zonal route of typhoons. As it is located on the fringe of the Asian Continent, strong low pressures migrate across the Sea of Japan in winter. Therefore, we have frequently suffered from numerous disasters caused by huge waves, storm surges, tsunamis[1)], and earthquakes.

Utilization of Coastal Zone

Because Japan is an island country, the people have traditionally utilized the coastal zone covering the seaward and landward areas from the shoreline for their daily life. The nearshore area has been widely used for fishing and marine transportation, whereas the coastal landward area is used for agricultural and industrial production,

---

1) The Japanese word "tsunami" has been accepted internationally as a technical term to express seismic sea waves possibly since the 1896 Sanriku Tsunami. The words "tsu" and "nami" mean anchorage basin and waves, respectively. Hence "tsunami" meant originally dangerous big waves appeared in a harbor basin or in a bay. Sometime in the period of Edo (1603 - 1867) the word "tsunami" was used in order to record in literature disasters causted by seismic sea waves.

dwelling, and recreation. Due to enormous expansion of human activities, tremendous amount of engineering works has been executed along the coast during the last hundred years. These works cover construction of ports and harbors as well as of fishery harbors, land reclamation for agricutural purposes, reclamation works for industrial lots and urban area, shore protection works, and various facilities for recreation.

It is well known that Japan is a daughter of Chinese civilization (Reichauer, 1981). In ancient times, Japan was heavily isolated from the Asian Continent, where one of the oldest civilization in the world had developed. This fact caused by the great distance of open sea from Japan to the neighboring countries, Korea and China, was the very big barrier for the primitive arts of navigation. Even though the Japanese people imported Chinese cultures to Japan since the time more than 1000 years ago, as indicated by the first record of the Japanese mission to China in 52 AD. On the other hand, domestic trade was actively done in the medieval ages through coastwise sea routes, such as the Seto-Inland Sea route first and then the routes of the Sea of Japan as well the as the Pacific Ocean. These activities were accomplished by the construction of the numerous harbors along the Japanese coast. One of recorded harbor construction works is Ohwada Domari (harbor) initiated by Kiyomori Taira in 1172. His idea was to built an artificial island named Kyoga-shima with the area of about 30 ha by filling with stones and earth of about 1.4 million m^3 in order to provide a sheltered anchorage basin in cooperation with the Cape of Wada. The harbor construction work was succeeded by Buddhist Priest Chogen. Thus the completed harbor played an important role in the international trade business with China and other countries in the southeast Asia in the 15th Century. This is a story for the initiation of the present Port of Kobe.

At present, there are about 4,000 harbors, which consist of about 1,000 commercial and about 3,000 fishery harbors. From this fact it can be realized that how densely the coastal zone of the Japanese Islands has been utilized for marine transportation and fishing activity. The entire nearshore area is completely covered by fishing rights held by Fishermen Associations. Therefore, it has been a difficult task to negotiate with the Fishermen Associations before initiating any kind of works in the nearshore area even for public purposes. It is also true, however, that fishing grounds have become narrower and narrower during the last thirty five years due to the rapid development of the coastal zone.

Active reclamation works have been done during the last hundred years, in particular in the period of 1956 to 1970, amounting about 35,000 ha as shown on Table 1(a). About 80 percent of the above reclaimed land was for industrial lots, and the remaining was for urban area. In the following period of 1970 to 1988, the reclamation works were still active as shown on Table 1(b), but the spot for the works has been moving from shallower area to a little offshore. That is to say, a number of artificial islands have been constructed in this period. Typical examples are the Kobe Port Island (436 ha) completed in 1981 and the Rokko Island in Kobe (580 ha) completed in 1990. These artificial islands suffered partly the earthquake damage in 1995 in the form of liquifaction. Another example is the New Kansai International Airport (1,200 ha) constructed at the site 5 km offshore in Osaka Bay. The airport opened for operation in September, 1994.

Table1. Reclaimed land area on seashore and offshore (Horikawa, 1991a)

(a) 1956-1970

Year	Area(ha)	Year	Area(ha)
1956	244	1964	3,061
1957	736	1965	3,084
1958	1,149	1966	2,289
1959	1,271	1967	2,549
1960	1,483	1968	3,390
1961	2,198	1969	4,076
1962	2,733	1970	4,136
1963	2,875	Total	35,274

(b) 1970-1988

Year	Area(ha)
1970-1974	17,111
1975-1979	9,114
1980-1984	5,182
1985-1988	2,930
Total	34,337

Two other big national construction projects were completed in 1988. The first is the Seikan Underwater Tunnel crossing the Tsugaru Strait to connect Hokkaido with Honshu by railway. The total length of this tunnel is about 54 km, and the maximum water depth along the route is about 140 m. Nearly 25 years were needed to complete this outstanding project. The second is the Seto-Ohashi consisting eleven bridges (about 12 km long in total), among which five are the majors to cross the Seto-Inland Sea, connecting Shikoku Island to Honshu by railway as well as highway. As results, Hokkaido, Shikoku,and Kyushu have now been connected to Honshu. The construction of the second route via Awaji Island between Honshu and Shikoku will be completed in the near future by construction of the Akashi Bridge which will be the longest suspention bridge in the world.

Environmental Problems in Coastal Zone

In parellel to the extensive utilization of the coastal zone for industry, Japanese people have encountered various painful problems particularly those related to the coastal environment. In the initial stage of coastal preservation in Japan, beach erosion was recognized as one of the biggest natural disasters. It is true that beach erosion is caused by natural forces, such as shallow water waves and nearshore currents, acting on coastal sediment. However, various artificial causes, such as construction of dams for flood control, water resources and electric power generation; mining of sand from river bed; and construction of coastal structures including harbor breakwaters, have given strong impact on natural coast, inducing severe beach erosion as a result of a sediment budget inbalance. Considering the present strong demand and shortage of sandy beaches for recreation, it is recognized at present that the beach erosion is one of the most serious environmental problems in Japan in a broad sense.

In addition to the beach preservation requirement, water quality in the nearshore area has been a serious concern since the 1970s. The reason is that the coastal water was once badly contaminated by the sewage discharged from urban area as well as industrial plants. In Minamate Bay the organic mercury discharged from a plant deposited on the bed inside the bay dissolved gradually into water. Through the food chain of marine growthes a certain number of dwellers near the bay suffered severly from the poisonous material. These serious problems have been treated for many years, say more than fifteen years, by the efforts of government officials as well as inhabitants including fishermen, resulting in a slightly improved situation at present.

In recent years, the Japanese people have been realizing strongly the importance of human dignity in their daily life. Hence, the concept of seeking a healthier lifestyle has been introduced gradually into the utilization of coastal zone. The main target of the above approach is to establish a comfortable environment in the coastal zone.

CHRONOLOGICAL REVIEW OF COASTAL ENGINEERING ACTIVITIES IN JAPAN

Introductory Remarks

The Japanese coast has been recognized as a public property for several hundred years. Based on the above traditional acceptance, the Japanese coast has been controlled by the government agencies for the past hundred years or more. Since the end of the World War II, the Japanese coast has in practice been seperately governed by four government agencies, namely Ministry of Construction, Ministry of Transport, Ministry of Agriculture, Forestry and Fisheries, and Fishery Agency, for particular coastal regions depending on the designation and usage of the region.

In September, 1953, Typhoon No.13 crossed over Ise Bay from the southwest to the northeast and generated a storm surge which produced tremendous damage in the area, with a death toll of 393, missing of 85, wounded of 2,559, destroyed houses of 26,071, and inundated houses of 455,872. The great scale of this disaster was a strong impact or shock to politicians as well as government officials. That is why the Sea Coast Act was newly approved by the Japanese Diet and issued to be effective on 12 May, 1956.

The above situation was an important epoch in the coastal engineering history in Japan. According to the Sea Coast Act, Governors of Prefectural Governments have strong responsibility to maintain and preserve the coast in their regions. The Governor should designate any necessary coast as an endangered coastal region and keep that region at the acceptable level. However, the four central government agencies exert strong influence in their related coastal regions through subsidies supplied from the national budget for coastal preservation works.

Another important article contained in the Sea Coast Act established the Standard for Design and Execution of Coastal Structures. Coastal structures are those to be built to preserve and protect the coast against tsunamis, storm surges, stormy sea waves, ground subsidence, and sea level change. These structures such as sea walls, sea dykes, jetties, groins, and detached breakwaters had previously been designed independently by each of the government agencies without any unified design standard. Due to the bitter experience at the 1953 Typhoon Storm Surge, government engineers strongly realized the necessity of a common standard for the design and execution of coastal structures jointly authorized by the related agencies.

Under the above circumstances, the Committee on Coastal Engineering was organized in the Japan Society of Civil Engineers (JSCE) in April, 1955. The main task of the Committee was initially to edit a Design Manual for Coastal Structures, which was published in August, 1957 through JSCE. For the preparation of the Design Manual, the following two materials were very important references. One was "Shore Protection, Planning and Design" published by Beach Erosion Board, US Army Corps of Engineers in 1954. The second was the manual "Hydraulics Formulae" edited by JSCE. Based on the above Design Manual, the Standard for Design and Execution of Coastal Structures was issued by the Japanese Government in December 1958. The Standard has since been revised twice, in 1969 and 1987, to incorporate newly developed knowledge and technology in it.

From the facts stated above, it is clear that the coastal engineering in Japan started substantially in 1953. However, in the following discussion, a chronological review of coastal engineering activities in Japan will be made in each decade starting from 1950 to the present.

The First Decade (1950-1960)

The main concern of coastal engineers in this decade was coastal disaster prevention as described in the following. In this decade Japan was anxious to recover as a nation from the miserable destruction caused by World War Ⅱ.

As stated previously, in September 1953, Typhoon No.13 induced a big storm surge inside Ise Bay facing the Pacific Ocean, and thus caused tremendus disaster along the bay shore. Due to this natural disaster the importance of coastal preservation became a serious concern of the Japanese people.

Just after the end of the war, serious beach erosıon at the Niigata, Kaike, and Toyama coasts, all facing the Sea of Japan, was recognized from the view point of national land conservation. Thus, government agencies organized an investigative committee for each locality in cooperation with researchers in the fields of oceanography and civil engineering in order to determine the causes of the beach erosion and to establish suitable prevention measures.

Under such circumstances, the Sea Coast Act of 1956 was issued by the Japanese Government as mentioned previously. In parallel to this event, the Committee on Coastsl Engineering JSCE was founded by Masashi Hom-ma with the aim of promoting research activities of coastal engineering in Japan. Since that time, an annual conference on coastal engineering has been organized by that committee up to the present. The Proceedings of the Japanese Coastal Engineering Conference are written in Japanese and an English Journal, "Coastal Engineering in Japan," has been issued under the editorship of the Committee on Coastal Engineering since 1955 and 1958, respectively.

In September, 1954, the Tōyamaru Typhoon took a route along the Sea of Japan and generated heavy waves near Hokkaido causing miserable disaster. The Tōyamaru, a ferry boat between Hakodate in Hokkaido and Aomori in Honshu, sank with all her passengers and crew. This unfortunate accident was a motivation for the underwater tunnel construction mentiond in the Introduction. This huge Seikan Underwater Tunnel construction project was completed in 1988 after overcoming numerous technological difficulties.

Following the above typhoon disasters, Ise Bay storm surge damage caused by Typhoon No.15 in 1959, and Chilean Tsunami damage in 1960 successively happened. These natural disasters raised again strong interest in coastal preservation among the Japanese coastal engineers for maintaining the coastal land area as a basis of social and economical development in Japan.

The Second Decade (1960-1970)

This decade corresponds to the period of recovery of the nation from exhaustion induced by the war and for lifting up the national economy toward development. However, natural disasters occurred still continuously for a while, for example, the Niigata Earthquake in 1964 caused heavy earthquake damage on port facilities at

Niigata Harbor, and a tsunami hazard in the Niigata City area. Another serious problem in Niigata was land subsidence induced by pumping up ground water from which subsoil natural gas was separated for production at plants.

In the meantime coastal development works such as land reclamation for industrial sites were very active as seen in the record of 27,738 ha of reclaimed land within this decade. Hence, coastal disaster prevention was still kept as a top priority in coastal works. The keen themes in coastal engineering researches were, for example, 1) to clarify the characteristics of ocean waves in the nearshore zone as well as in deeper water, 2) to evaluate the wave action on coastal structures, 3) to predict storm surge behavior by numerical simulation, 4) to evaluate tsunami deformation inside a bay and tsunami run-up on beaches, and 5) to clarify the mechanism of beach erosion (Proceedings Japanese Conferences on Coastal Engineering). Reflecting these activities, this period was highlighted by active coastal development. In addition to the above, environmental problems in the nearshore area existed actually in the shadow of coastal development.

Here it should be mentioned that the Tenth International Conference on Coastal Engineering (ICCE) in 1966 was held in Tokyo, in which M. Hom-ma took the leading role as Chairman of the Local Organizing Committee. The conference was the first ICCE held in Asia and gave young reserchers in particular not only in Japan but also in other Asian countries strong motivation for promoting coastal engineering research intensively.

Prior to this conference, a US-Japan Seminar on Coastal Engineering was organized in 1965 by the joint effort of J. W. Johnson (University of California) and M. Hom-ma (The University of Tokyo) under the financial support of the National Science Foundation (NSF) in U.S.A. and the Japan Society for Promtion of Seience (JSPS) in Japan. The members visited typical sites of coastal engineering works in Japan and gave seminars at various locations on different topics. It was a kind of Pre-ICCE and offered a very good opportunity to exchange the views on common interests among the members and also Japanese observers.

The Third Decade (1970-1980)

In the previous period, the Japanese people aimed eagerly at economic development at a high rate and ran day by day with little consideration on natural environment including ecological aspects. The people could not afford the time, nor the money, to take coastal environmental change into consideration. As a result contaminated areas spread over from rivers to bays, to nearshore areas and finally to the ocean. Thus, we are seriously concerned about marine pollution from not only the domestic but also international view points. Based on the stated circumstances, numerous environment protection acts were successively issued in 1967. Therefore, field measurements of pollutant concentration in sea water were conducted very actively in various locations to investigate its real state, and then numerical model studies were carried out by coastal engineers to predict the behavior of contaminated sea surface areas.

On the other hand, a number of power plants have been constructed along the coast in Japan since the 1960s. As the capacity of each unit of power station increased very much, the discharge of heated water also increased rapidly. At the initial stage of power development, the engineer's main interest was how to take cooling water effectively for a power plant from the sea without interference to the heated water discharged from the plant. However, as environmental concerns, the engineer's interest has changed from the above subject to prediction of sea surface temperature rise caused by the heated water discharged from the plant. However, it was unfortunate for us for preserving our good environment that basic investigations on the negative or positive effects of heated water on marine growth were almost ignored in order to expedite agreement between electric companies and the related Fishermen Associations.

Hence, research subjects related to water pollution, sea bed material pollution, and chain reactions, and consequences to marine growth have become an additional field in coastal engineering. The coastal sediment subject also has been treated as one of the coastal environment preservation problems.

Here it should be noted that the first oil crisis happened in 1973, owing to the Fourth Middle East War, resulting in a high rise in oil prices and a high rate of oil production cut. Due to this unpredictable event, people have really recognized that limitations exist in natural resources as well as in economical development, as was pointed out in the report issued by the Club of Rome in 1972. On the other hand, due to tremendous advancement of space technology, it has become a daily experience for us to look at global images came from the satellites. These satellite images assist people in realizing that the size of the globe is not infinite, but limited. Hence, the people can now understand various environmental phenomena in a global scale manner.

The Fourth Decade (1980-1990)

In the meantime, the Third United Nations Conference on the Law of the Sea continued for many years since 1973, and a UN treaty related to that subject was compiled in 1982. Based on this treaty the Japanese Government enacted a law, in which the territorial seas were defined by a line drawn at a distance of 12 miles seaward from the shoreline. Reflecting the above worldwide trend, resources including not only natural resources such as petroleum but also fishery resources for protein have been a great interest to the Japanese people. Thus, the research subject related to fishery resources has become one of the keen interests of researchers. Hence, exchange of information between coastal engineers and fishery scientists has become more active in these days.

The third decade can be characterized as a time for recovering the marine and coastal environment from the contaminated to a better state for human life. As a result of painful effort, the quality of sea water improved very much and reached to an acceptable level. On the other hand, due to rapid economic development, people are wealthy enough to enjoy their leisure time. Thus, the demand for recreational facilities have become very large. Even though the tempo of seashore industrial development is slowing down, development of coastal area has been still active owing to increasing demand of urban development in the coastal zone. Adequate adjustment between conservation and development in the coastal zone has become one of the important research subjects.

It should be noted that the Japanese Government initiated a new project entitled "Coastal Environment Improvement Work" in 1984 with the aim of establishment of better coastal environment. Even though the initiation of the above work seems to be too late, the important role of natural beaches in absorbing or reducing the impact of wave energy has been publicly or officially recognized by government engineers.

It was in May, 1983, that tsunamis generated by a submarine earthquake with its epicenter in the middle part of the Sea of Japan coastal area killed a number of people including school children who were taking their lunch on the beach. Differing from the past major tsunamis in Japan, this tsunami happened nearly at noon; hence, many local people took tsunami records by using their video cameras. These tapes have been valuable for looking at the actual behavior of tsunami waves in the nearshore area. Up to that time scientists had tried to do numerical simulations of past tsunamis by using fault models, and improved its technique very much for practical use. The numerical simulation result of the 1983 tsunami was displayed on TV showing successively the generation and deformation of tsunamis in the whole region of the Sea of Japan. Video recording and the display of the tsunami simulation result were televised frequently, and they were quite instructive to the people in understanding clearly the importance of evacuation from low-lying land to higher area and of engineering technology advancement.

The Fifth Decade (1990-present)

During the last four decades, coastal protection works have been carried out intensively by the Japanese Government, thus occurrence frequency and magnitude of coastal disaster have been reduced greatly. At the same time, the Japanese people are gradually losing the appreciation of these efforts. Looking at this tendency and considering the following factors such as magnification of earthquake activity, extraordinary meteorological phenomena including typhoon behavior, and global mean sea level rise due to global warming, we have to recall always the prevention measures against coastal disasters as an eternal theme of our research. These subjects have been treated and clarified by specialists in each field; however, the people have very little knowledge on the facts which are very common among specialists. Therefore it should be the duty of specialists to provide a campaign of disaster prevention education for the benefit of the people.

Global warming was taken as one of the topics at the UN Human Environment Conference held in Stockholm in 1972. This kind of topic, particularly mean sea level rise due to global warming, has been a serious concern by Government Agencies around the world. It is needless to say that the mean sea level rise should give strong impacts on human activities at the low-lying land. For countries occupying coral reefs and atolls, mean sea level rise is becoming a matter of great importance for life or death. According to a report issued in 1990 by the Intergovernmental Panel on Climate Change (IPCC) mean sea level will rise in the amount 30-110cm in 2100 (Warrick et al., 1990) . Reliability of this prediction should be checked by future investigations. As a remarkable response to the above prediction, two workshops were successively held in 1993 in the U.S.A. and in Japan for the western and eastern hemispheres, respectively. These workshops were preparatory to the World Coast Conference held in the Netherlands in 1993 with the aim of preparing input to the IPCC Second Assessment Report (Mimura et al.,1993). Such effort might be valuable to form habitable and stabilized coastal environment in the world.

In 1994, the Twenty-fourth ICCE was held at the Port-Island in Kobe as the second ICCE in Japan. This conference was successfully held by the great effort of T. Sawaragi (Osaka University) and Y. Tsuchiya (Meijyo University, formally Kyoto University) who took the role of Co-Chairmen of the Exective Committee of the Local Organizing Committee. None of the participants expected that the conference site, Kobe, would suffer the tremendous earthquake damage on January 17, 1995, as reported worldwide.

COASTAL ENGINEERING RESEARCH ACTIVITIES IN JAPAN

Introductory Remarks

As stated in the previous part, the dawn of coastal engineering in Japan opened a new era of research activities in 1953. Even before, however, a certain number of research papers related to the present coastal engineering discipline were published by engineering scientists. These were distributed through JSCE to civil engineers. It was quite natural that these papers were directly connected to the public works at that time, particularly to harbor construction. Therefore, at the beginning of this chapter, we will review those important contributions to the advancement of civil engineering society in Japan. Following the above pre-dawn period, a description will be made to review the remarkable advancement of coastal engineering research activities in Japan after 1953 up to the present by using statistical summaries.

In the following chapters the trend and the main achievements in various subjects will be discussed. These subjects cover the fields of wave mechanics, long-period wave mechanics, nearshore current, coastal processes and sediment transport, coastal works and coastal structures, and coastal environment protection.

Pre-dawn period. Civil engineers encountered numerous difficulties in the coastal zone during their construction practices. In order to solve the problems they learned essentials from introductory articles that appeared in the Journal of JSCE. One example is a paper on seiches written by T. Okada in 1918. He was a prominent geophysicist and introduced in this paper his study on the mechanism of seiche that occurred inside a bay and/or a harbor. This kind of interaction between engineers and scientists seemed to be quite effective to allow civil engineers to understand natural phenomena.

I. Hiroi, a well-known harbor engineer, presented two papers in 1919 in the Journal of the College of Engineering ,Tokyo Imperial University. In the first paper he proposed a formula to evaluate pressure intensity of breaking waves acting on a vertical breakwater. In this paper he also described the wave gage meter which he developed to record the maximum wave force and was applied in the field to obtain data. In the secound paper, he discussed the importance of wave power utilization. His contribution to the design practice of breakwaters in Japan was outstanding. The so-called Hiroi formula is still mentiuned in the Technical Standards for Port and Harbor Facilities, even if some modifications have been introduced into the original form as explained in the related chapter. The original formula is expressed in such a simple form as $p = 1.5 \rho g H$ where p is wave pressure intensity, ρ is sea water density, g is acceleration of gravity, and H is the incident wave height at the water depth where the breakwater is expected to be constructed. The total wave force per unit length of breakwater was evaluated by the product of the stated wave pressure intensity and the total height of breakwater as shown in Figure 1. Japanese harbor engineers believed for many years that the Hiroi method was effective as a practice to estimate the wave forces in order to maintain the stability of breakwaters. It has been realized that the above rule was essentially correct because the crown height of breakwaters at that time was normally not high enough to prevent wave overtopping. Since then numerous engineers have discussed repeatedly wave forces in connection with the stability of vertical type and/or composite type breakwaters, because the design of breakwaters is the main interest of harbor engineers.

In 1933, a submarine earthquake occurred near the continental fringe of the Pacific Ocean and generated a disastrous tsunami in the northeastern part of Honshu called Sanriku in March, 1933. Along the Sanriku Coast there are a number of bays with various sizes and configurations. The tsunami reached the maximum record of more than 20 m above the mean sea level at the head of Ryori Bay, and the residents felt a strong shock from the result. With the aim of explaining the stated phenomena, M. Hom-ma (1933) extended his analytical treatment to the transformation of long-period waves inside a bay with a linearly decreasing width toward its bay head and a

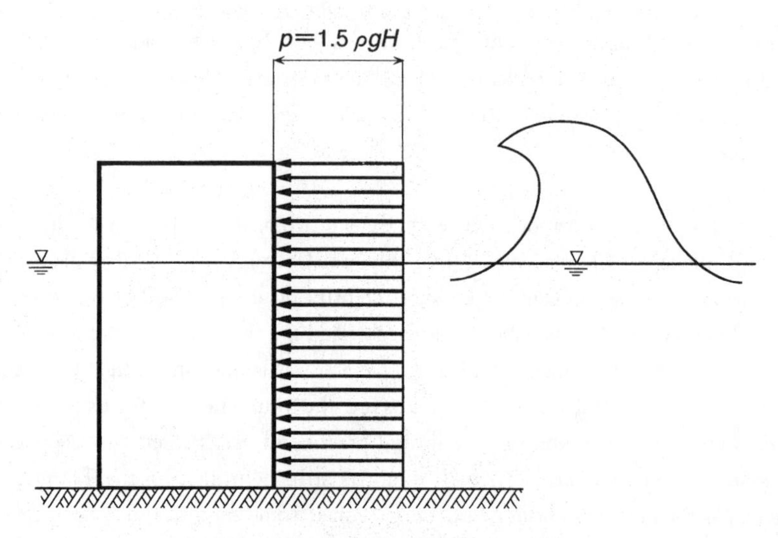

Figure 1. Basis for Hiroi's formula

uniform bottom slope (including a horizontal bed) and noticed that the tsunami wave height could reach a height such as recorded.

Post-dawn period. In 1954 a speciality meeting on coastal engineering was held in Kobe by the efforts of T. Ishihara (Kyoto University) under the sponsorship of the Kansai Branch of JSCE. The purpose of this conference was to enlighten Japanese engineers; hence, introductory lectures on different subjects closely related to coastal engineering were given by sixteen Japanese scholars. The conference style was quite similar to the First Conference on Coastal Engineering held in Long Beach, California, U.S.A., in 1950. As the Committee on Coastal Engineering was organized in JSCE in 1955, the Japanese Conference on Coastal Engineering has been decided to be held annually by the Committee and the above speciality meeting was named the First Japanese Conference on Coastal Engineering.

Figure 2 indicates the trend of increasing number of papers presented at the International Conference on Coastal Engineering and at the Japanese Conference on Coastal Engineering. These two curves show that the coastal engineering research activities in Japan, as well as in the world, have been almost in parallel and been expanding remarkably since the 1960s. In the same figure the number of papers appeared in "Coastal Engineering in Japan," JSCE English edition since 1958, are also shown. Considering the fact that the papers presented at the Japanese Conferences on Coastal Engineering were basically written in Japanese, we can realize that about five to fifteen percent of Japanese papers have only been published in English. It is rather unfortunate that most of the Japanese contributions have been unknown in other countries owing to language difficulty. However, an English abstract of the papers presented at the Japanese Conferences on Coastal Engineering are contained in the Coastal Engineering in Japan, and provides available information on the Japanese contributions to foreign researchers.

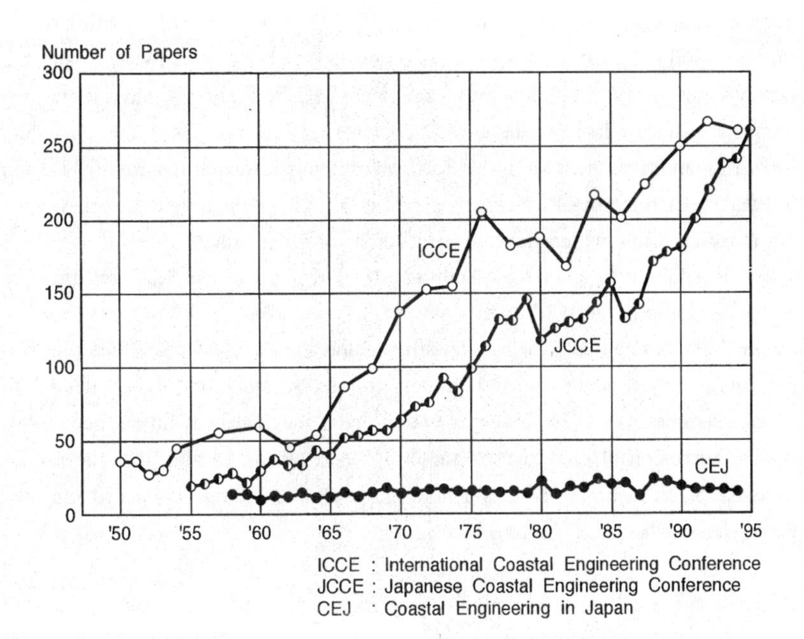

Figure 2. Number of published papers

Coastal engineering research activities in Japan are principally conducted by the researchers at universities and research institutes. However, we have kept a definite policy in the management of the Committee on Coastal Engineering, JSCE, since its foundation. That is to say, close interaction between research scientists and practicing engineers should always be maintained. Hence, we have tried to encourage practicing engineers to present results of their field investigations conducted for their practical works at the annual conferences. The intentional policy has taught us a great deal about the actual phenomena appearing in the nearshore region and to know real problems with which the coastal engineers are encountering in practice.

There are many research institutions in the field of coastal engineering in Japan, and these institutions are categorized into three groups, namely governmental research institutes, universities, and private research institutes.

As mentioned in the previous part, the Japanese coast is controlled by four governmental agencies. These are Ministry of Construction (MC), Ministry of Transport (MT), Ministry of Agriculture, Forestry and Fisheries (MAFF) and the Fishery Agency (FA). Among these agencies, MC, MT and FA have their own research institutes related to the coastal engineering subjects. These are the Public Works Research Institute (MC), the Port and Harbour Research Institute (MT), and the Fishery Engineering Research Institute (FA). These institutes have their own laboratory facilities which have been utilized for basic studies as well as model studies. In addition to laboratory facilities, the Port and Harbour Research Institute has operated a long pier with total length of 427 m and the width of 3.3 m at the Hazaki Oceanographical Research Facility facing the Pacific Ocean. The pier has been intensively used for the field investigations of nearshore waves, nearshore currents, sediment transport, beach evolution under rough sea conditions, ecology of bivalve, corrosion of construction materials, and others. In addition, the Public Works Research Institute operates a field observation station at Ajigaura facing the Pacific Ocean where a 200 m long pier exists.

In Japan there are about 100 national universities, 50 public universities, and 400 private universities. Among these universities, at about 50 universities coastal engineering research activities have been actively carried out. However, at the universities in Japan, the number of staff members in each speciality is quite small in general. Under such a condition, the Disaster Prevention Research Institute of Kyoto University and the Applied Mechanics Research Institute of Kyushu University should be mentioned as unique research institutes attached to the national universities. At these research institutes, a substantail number of staff members engage in the research works by using well-equipped laboratory facilities as well as field observation facilities including observation towers and an observation pier.

In addition to the above research institutes, the Central Research Institute for Electric Power Industry (CRIEPI) and a number of Technical Research Institutes attached to general contractors and consulting firms should be mentioned to indicate potentiality of coastal engineering research in the private sector in Japan. As a typical example, an outline of CRIEPI will be introduced in the following. This institute is financially supported by nine electric companies which supply electric power to their related regions in Japan. Thus the institute has responsibility to respond to the practical needs raised by these electric companies. In 1984 a big wave flume was installed there to investigate various coastal engineering subjects by generating waves with height of 1 to 2 meters and period of 5 to 10 seconds. The flume is 205 m long, 6 m deep and 3.4 m wide. Nowadays this facility has been utilized to carry out a cooperative research project with the aim of developing simulation model of two-dimensional beach profile evolution.

Research subjects in Japan. The research subjects in Japan are closely related to the practical problems realized in each decade stated in the previous part. In order to observe the general trend of coastal engineering research activities in Japan, Table 2 was prepared, in which the papers presented at the Fifth to the Fortieth Japanese Coastal Engineering Conferences held at five-year intervals were classified into nine categories. These are 1) wind waves and swell, 2) long-period waves such as storm surges, tsunamis, and harbor oscillation, 3) nearshore currents, 4) tidal currents, 5) coastal sediment and beach evolution, 6) wave actions on coastal structures, 7) coastal water quality, 8) coastal and marine recreation, and 9) miscellaneous. The number of papers in each category and its rate among the total number of papers presented at each conference are listed on this table. From the table, the outlook for future prospect of coastal engineering research in Japan can be introduced as shown in the following items.

1) Researchers have strong interest in wave characteristics of deep water and wave transformation in shallow water. This fact reflects the enormous advance of technology in measuring the physical quantities of waves in the field as well as in the laboratory, and in collecting and processing tremendous amount of data. In addition

Table2. Subjects of papers presented at Japanese Coastal Enginerring Conferences

Subjects	5th (1958)	10th (1963)	15th (1968)	20th (1973)	25th (1978)	30th (1983)	35th (1988)	40t (1993)
1) Waves	1 (4)	10 (31)	13 (23)	23 (25)	30 (23)	36 (27)	51 (30)	39 (16.5)
2) Long waves	1 (4)	7 (22)	5 (9)	4 (4)	10 (7)	10 (7)	10 (6)	10 (4)
3) Nearshore currents	1 (4)	0 (0)	0 (0)	2 (2)	4 (5)	8 (6)	1 (0.5)	6 (2.5)
4) Tidal current	1 (4)	1 (3)	1 (2)	4 (4)	12 (9)	8 (6)	8 (4.5)	6 (2.5)
5) Coastal sediment	11 (47)	4 (13)	10 (17)	15 (16)	18 (14)	21 (16)	41 (24)	59 (25)
6) Wave action on coastal structures	7 (29)	8 (25)	16 (28)	31 (34)	29 (22)	32 (24)	39 (23)	59 (25)
7) Coastal water quality	0 (0)	0 (0)	4 (7)	9 (10)	14 (10)	8 (6)	12 (7)	28 (12)
8) Recreation	0 (0)	0 (0)	0 (0)	2 (2)	0 (0)	1 (1)	1 (0.5)	4 (2)
9) Miscellaneous	2 (8)	2 (6)	8 (14)	3 (3)	13 (10)	9 (7)	8 (4.5)	25 (10.5)
Total	24 (100)	32 (100)	57 (100)	93 (100)	130 (100)	133 (100)	171 (100)	236 (100)

upper: number of papers

lower in parentheses: %

to the above, the techniques of computational analysis have been developed in conjunction with the remarkable advancement of computation technology.
2) Coastal sediment movement and its related problems have been one of the most important subjects in which coastal engineers have been deeply concerned throughout the whole period. This fact indicates the actual situation in the coastal engineering community; that is to say, beach evolution and siltation problems are still one of the most difficult subjects in order to solve the practical problems confronted and to produce a better and pleasent coastal environment.
3) The interaction between waves and coastal structures has been a great interest of coastal engineers. In recent years new types of coastal structures with low reflection coefficient have been developed to match with numerous needs in producing favorable coastal environment with safety, effectiveness, and pleasantness for human beings.
4) The subjects related to coastal environment protection have been treated during the last thirty five years in order to evaluate the influence of human activities on environment, particularly coastal water quality, and to reduce the effect or to produce more favorable coastal environment at the questioned site.
5) Each of the numerous subjects other than the above has been investigated in response to the needs of our society. During the past five years the expected mean sea level rise due to global warming has drawn a great interest of coastal engineers in the sense of global environment.

In the following sections, the achievement of research activities in each specified field will be discussed (Coastal Engineering Committee, 1994a; Sawaragi, 1995).

Wave Mechanics

The characteristics of wind waves and swells have been recognized by harbor engineers and coastal engineers for many years as key parameters in designing harbor and coastal structures, particularly breakwaters and operating harbor facilities. However at the early stage of coastal engineering development, our knowledge on wave characteristics was very limited due to the lack of wave data. Owing to the stated circumstances, coastal engineers devoted their great efforts to several subjects related with wave mechanics. These are 1) development of wave measuring devices, 2) analytical treatment of nonlinear water waves, 3) statistical and spectral analysis of wind waves, 4) prediction of wave transformation in shallow water, and 5) prediction of directional wave spectra.

Development of wave measuring devices. In order to obtain real data on waves, wave measuring devices were developed from the early stage of coastal engineering researches, particularly at the Port and Harbour Research Institute (formally Kurihama Branch of Transportation Technical Research Institute). Pressure-type wave gages were commonly utilized for many years with numerous difficulties to be overcome. Later several types of wave gages have been successively developed and utilized for the practical use. These are wave gages of step-type, resistance-type, capacitance-type, and ultrasonic-type. Among these gages, the ultrasonic-type wave gages are most commonly used to obtain incident wave information, whereas capacitance type wave gages are occasionally employed for field observations in the surf zone.

With the increase of interest in the irregularity of wave characteristics and its spectrum, electro-magnetic current meters with wave pressure gages were proposed to be used. On the other hand, stereophotographic techniques from ground and air respectively were applied to look at the spatial and temporal variations of nearshore waves, including breaking waves.

Analytical treatment of nonlinear water waves. Studies on wave mechanics began with reviewing previous researches; for example, a survey of sea wave research, the engineering approach to wave refraction and diffraction, and known characteristics of nearshore waves were introduced at the First Japanese Coastal Engineering Conference in 1954. At an early stage of the research, primary effort was devoted by reseaschers to finding analytical solutions for liner refraction or diffraction problems. However, as the waves in the nearshore zone have been of particular interest by coastal engineers, various nonlinear wave theories with lower orders were derived on the basis of perturbation method. These are in the category of Stokes waves progressing with a constant form over a horizontal bottom in shallow water. On the other hand it was well-known that the cnoidal wave theory should be applied to describe the wave forms in shallower water beyond a certain limit. From that view point the applicability criterion between the Stokes wave theory and the cnoidal wave theory was determined. In order to reduce the mathematical difficulty in the cnoidal wave theory, the so-called hyperbolic wave theory was introduced as an apporoximation of the cnoidal wave theory (Iwagaki, 1972). Many years later, the extremely higher order solutions of the Stokes and the cnoidal waves were treated by analytically systematic manner and the validity range of various finite amplitude wave theories was defined clearly on the basis of above theories as well as the stream function method (Nishimura et al., 1977; Isobe et al.,1982). Figure 3 shows the validity ranges of finite ampituade wave theories, where H is the wave height, h is the water depth , L_0 is the wave length in deep water, g is the acceleration of gravity, T is the wave period, and $\tan\beta$ is the beach slope. The term $Ur = gHT^2/h^2$ is the Ursell parameter, and S-5, C-3, SFM19 indicate the 5th-order Stokes wave theory, the 3rd-order cnoidal wave theory, and the 19th-order Stream Function Method (Dean, 1965), respectively.

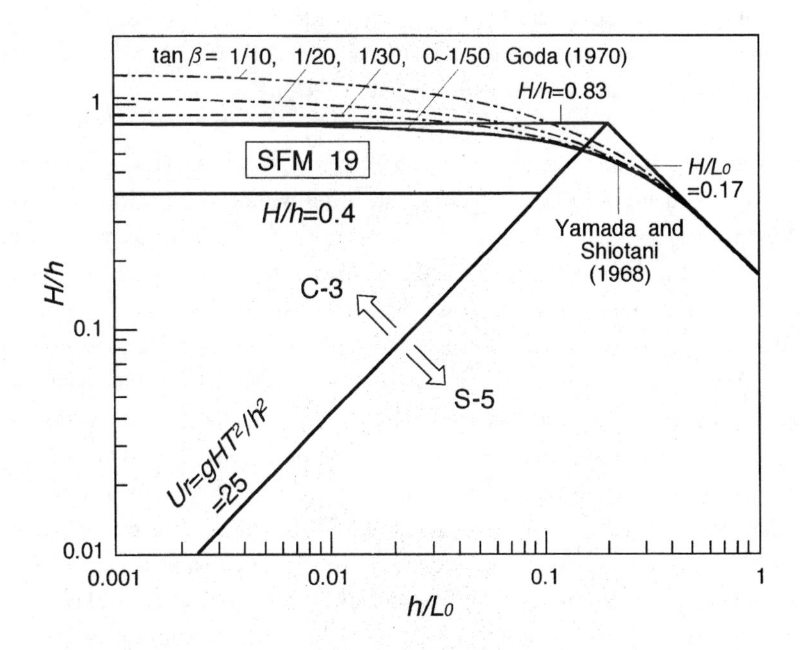

Figure 3. Validity range of finite amplitude wave theories (Horikawa, 1988)

Statistical and spectral analysis of wind waves. In conjunction with the accumulation of wave data in the field, the statistical characteristics of wave height and wave period in shallow water and the joint distribution of wave height and wave period were discussed in detail by referring to the theoretical treatments by Longuet-Higgins (1975). In 1964, several of introductory papers on the wave spectrum of wind waves were presented at the Eleventh Japanese Conference on Coastal Engineering. Since then, coastal engineers introduced the wave spectrum concept in understanding the coastal wave phenomena. On the other hand extensive field observations were made by using the clover-leaf buoy type wave gauges developed at the Applied Mechanics Research Institute of Kyushu University, under various physical conditions. Based on abundant reliable wave data some number of standard wave spectra under the various conditions of sea state were proposed (Goda, 1985; Mitsuyasu, 1995). In wind-wave hindcasting and forecasting, spectral models have been developed as well as simple models such as SMB method.

<u>Prediction of wave transformation in shallow water.</u> It is well-known that the waves in shallow water change suddenly their transformation at the point where waves are breaking. To define clearly the breaking wave, a criterion was proposed in the form of empirical curves with a parameter of beach slope by using voluminous data obtained mainly in laboratory experiments (Goda, 1970).

Prediction of wave transformation in shallow water began with employing analytical solution in such a simple situations as shoaling without/with refraction on a uniformly sloping beach and diffraction due to semi-infinite breakewaters. In recent years, however, numerical simulations in more realistic and complicated situations have been drawing primary attention. There are many factors, such as shoaling, refraction, diffraction, reflection and dissipation of wave energy, to be considered in the modeling. In most models, linear periodic waves are usually assumed, but nonlinearity and irregularity should be taken into account in reality. The mild-slope equation and its extension to irregularity of waves named time-dependent mild-slope equation can be applied to predict the trasformation of linear random waves in shallow water. It is impossible to cite all of the contributions in Japan on this subject due to enormous amount of papers presented at the Japanese Conferences on Coastal Engineering. Therefore, some of the representative papers will only be cited here. Ito et al. (1971) presented a pioneering paper to propose a new method of numerical analysis for wave propagation. In order to improve primarily the treatment of boundary conditions of the mild-slope eqation proposed by Berkhoff (1972), two types of so-called time dependent mild-slope equations were proposed independently (Tanimoto et al, 1975; Nishimura et al., 1983; Watanabe et al., 1986a). The nonlinear mild-slope equations were also presented.

Another important aspect on this subject is the wave transformation inside the surf zone. This subject has repeatedly been treated by many researchers. The main point is how to evaluate the wave energy dissipation in this region. A simple idea was proposed; that is the waves inside the surf zone were assumed to keep always the critical wave height defined by the breaking wave condition. The paper appeared in 1956 and was the first one which discussed the variation of breaking waves on the basis of the stereographic data taken from ground (Ijima et al., 1956). In 1966, calculated results of wave height variation after breaking on a uniformly sloping bottom were presented based on the wave energy equation (Horikawa et al., 1966). In this treatment, the wave energy dissipation in the surf zone was evaluated under appropriate assumptions. Since then many treatments on this subject have been made repeatedly to model the real phenomena more appropriately (Horikawa, 1988).

Prediction of directional wave spectra. With increase of our understanding on wave characteristics, in the engineering treatment random waves have been practically used for the planning and design of coastal structures in particular harbor structures. Following the above tendency, directional wave spectrum have been deeply interested by researchers (Nagata, 1964). Hence a certain amount of field data have been accumulated in these days and then various theories have been proposed for estimating directional wave spectra. These include the Maximum Lihelihood, Maximum Entropy Principle, and Bayesian Directional Spectrum Estimation Method (Isobe, 1988).

Long-Period Wave Mechanics

The study on long-period waves was greatly motivated by disasters due to storm surges after typhoons in 1953 and 1959, and also after the Chilean tsunami in 1960. The 1960's was the initial period in Japan for predicting numerically storm surge behavior by using electronic computers on the basis of a typhoon model. A typical example in the early stage was that at the Meteorological Agency in Japan (Unoki et al, 1962). On the other hand, tsunami behavior was initially treated as a transformation and run-up of long-period waves on a sloping beach (Watanabe, 1964). As another aspect of tsunami behavior, bay oscillation induced by incident long-period waves was investigated for particular bays and for a simple L-type bay in the 1960s.

Study on long-period waves was motivated by seiche motion appearing in harbors (Honda et al., 1908), after that resonance oscillations in harbors have been investigated analytically and numerically.

After the 1968 Tokachi-oki Tsunami, research activity of tsunamis has been accelerated (Aida, 1969). By the advancement of seismology, a fault model was proposed to simulate the earthquake ground movement; thus, numerical simulation has been actively done for the past tsunamis, and its validation has been made by the comparison between the calculated result and the record of tsunami traces (Mano et al., 1976). In recent years, nonlinear models including tsunami run-up have been developed in the numerical simulation covering a wide region from deep water to the shore. Long-period waves in harbors are drawing attention in regards with the ship motion which is critical in designing harbor configurations. Further study will be necessary in focusing on long-period components in incident waves, long- and short-wave interaction and dissipation at the harbor entrance.

Storm surge protection works have been rapidly progressed since the second decade of coastal engineering, that is the 1960s, so that Japan has not suffered serious damage due to storm surges. However, because population and assets concentrate in low-lying, water front areas are more intensively utilized than before, and the technology to predict storm surges should be improved both in theoretical background and numerical scheme to raise the level of safety. Earthquakes and resultant tsunamis cannot be predicted reliably by the present technology and, if once occur, run-up of a tsunami might be as large as experienced in 1993 at Okushiri Island in the Sea of Japan, near the southern part of Hokkaido, so that coastal structures cannot provide full protection. Therefore, in the planning of tsunami disaster prevention, warning system and evacuation should be developed as well as construction of coastal structures (Iida et al., 1983; Tsuchiya et al., 1995).

Nearshore Current

The longshore current was initially considered as an important factor controling beach erosion in the 1950s on Katase and Kamakura beaches and at the Toban Coast. However, due to the difficulty in measuring longshore current velocity, study on nearshore current was stalled for about a decade.

Study on longshore current velocity started again in connection with sediment transport rate and beach evolution in the laboratory in the 1960s. Prior to the above investigation the cross-shore velocity distribution of longshore current on a fixed uniformly sloping bed was measured by using miniature propeller-type current meters and a rough picture had been obtained. However, the mean velocity of the longshore current was practically correlated with incident wave characteristics.

Bowen (1969) and Longuet-Higgins (1970), treated analytically the cross-shore distribution of longshore current velocity. The latter compared his theoretical curves with laboratory data of Galvin et al. (1965). On the other hand, Thornton (1970) treated the same subject on a natural beach and compared his result with field observation data of Ingle. Thus the study of the longshore current was extended from the laboratory to field investigation of nearshore current. In order to cover an appropriate range of coastal area and to obtain spatial distribution of nearshore current velocity, a balloon-camera system was developed (Horikawa, 1988). For the development of the above system the assistance offered by Sonu (1969) was very much appreciated. Figure 4 indicates the development process of the balloon-camera system, the last of which is for taking stereo photographs. Applying the above systems, various observation data were accumulated (Sasaki et al.,1976). These were quite effective to look at the real nearshore phenomena, not only nearshore current but also spatial and temporal variations of waves and water surface elevation.

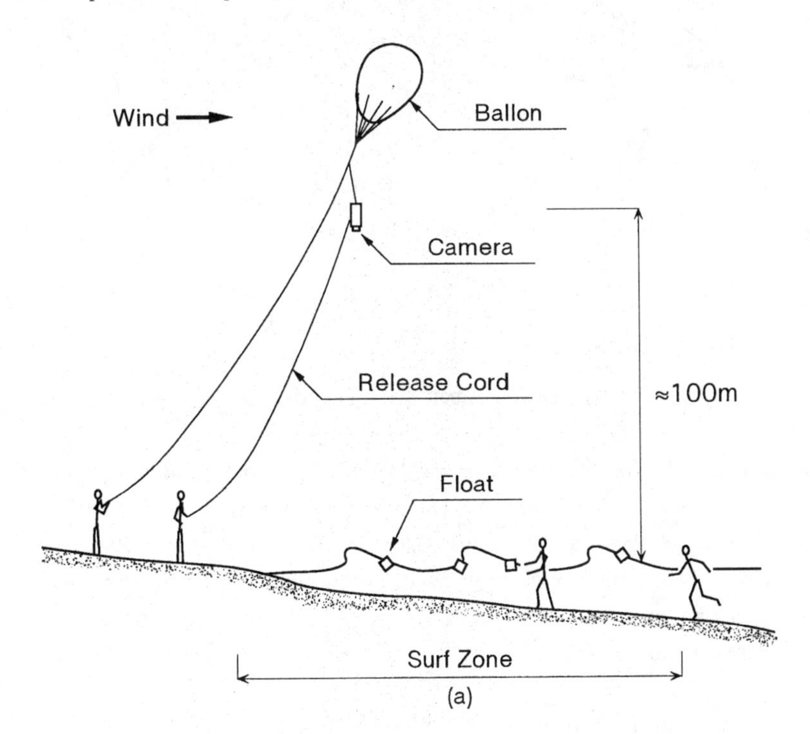

(a)

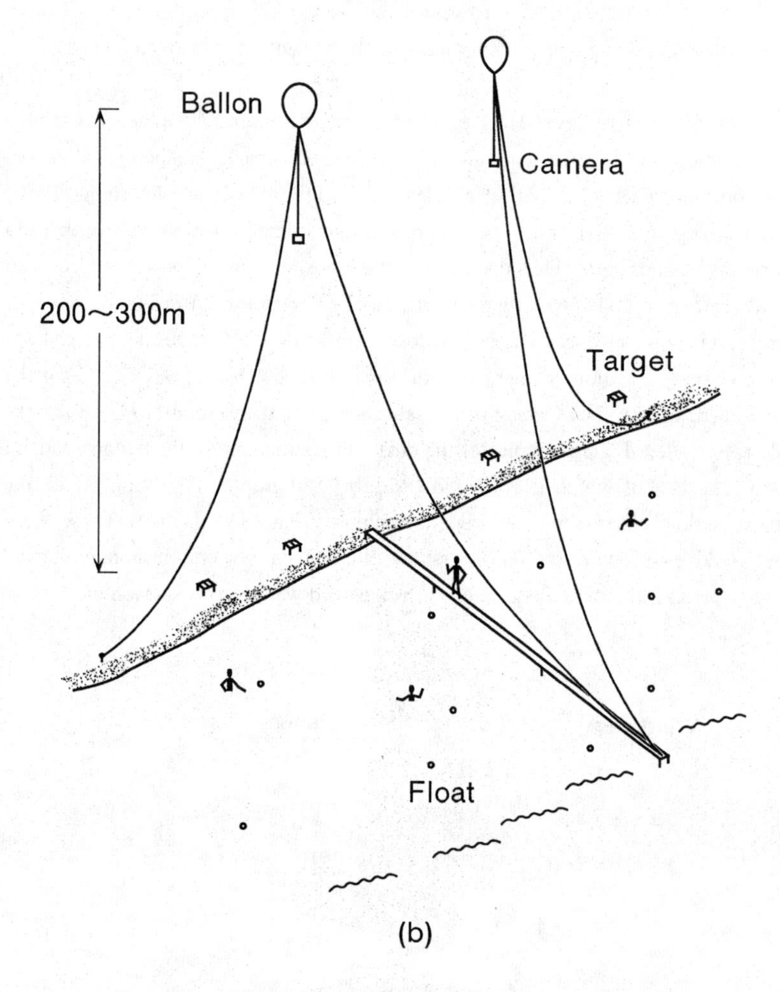

Figure 4. Ballon camera system (Horikawa, 1978)

In parallel to the field observations, numerical analysis of nearshore current was carried out under the real conditions corresponding to those in field observations. The aim was to investigate the availability of numerical simulation method and of obtaining effective information on the methods to evaluate the friction term and the lateral mixing term, both of which are included in the fundamental equations of nearshore current.

In order to close the numerical calculation of nearshore current, various wave transformation models are used to calculate the distribution of the radiation stress contained in the basic equations as external force. Recently, being the second-order quantity of the wave amplitude, the nearshore current is directly obtained by calculating wave transformation by Boussinesq equations which are second-order nonlinear equations. Thus, in the future, nonlinear wave equations incorporated with breaking wave models will be developed and used more frequently, yielding both wave and current fields from numerical solution.

Coastal Processes and Sediment Transport

Coastal processes, especially sediment transport, have been one of the major fields of coastal engineering research, and great numbers of papers on these subjects have constantly been published since the First Japanese Conference on Coastal Engineering. Coastal process and sediment transport are closely related to coast preservation such as topography change and beach erosion.

Understanding of coastal processes began with field surveys of coastal topography, which has often revealed serious beach erosion appeared in many places during the last fifty years. Up to the present, a great amount of data has been accumulated on bottom topography change of various beaches. In order to treat quantitatively these coastal phenomena, many devices have been proposed and tried to measure the sediment transport rate directly, but none of them has been completely successful (Horikawa, 1988).

Treatment to prevents closure of river mouths have been done at numerous places for more than a half century, and the problem has been realized as difficult. Many field and model studies have been performed for many river mouths. However, the river mouth treatment is at present in the stage of trial and error, hence much more study is required.

The coastal sediment phenomenon is very complex; that is, the interaction among waves, nearshore currents, and bottom topography should be taken into account in the analysis of coastal sediment transport. Therefore, it is believed that three approaches, namely micro-scale, meso-scale, and macro-scale approaches, should be done in parallel and the gaps among these approaches be filled step by step in order to clarify the whole aspect of coastal sediment transport.

In the micro-scale approach, theoretical treatments on coastal sediment transport mechanism started with studies on the oscillatory bottom boundary layer flow characteristics and on the inception of sediment movement. Recently, turbulence models have been applied to oscillatory boundary layer flow to simulate the flow above rippled beds and succeeded in reproducing vortices generated near ripple crests.

In the meso-scale approach, evaluation of sediment transport rate is of great importance in treating quantitatively the various engineering problems. From the practical view point, treatment of coastal sediment transport is conventionally separated into two parts, namely longshore sediment transport and cross-shore sediment transport. As for the longshore sediment transport rate, frequent field observations have been carried out on various coast by injecting colored sand with three kinds of fluorescent material into the bed and then by sampling bed material. These data were effectively used to propose a formula of longshore sediment transport rate. On the other hand, cross-shore sediment transport has been treated in various ways as stated in the following.

As it is known well, type of sediment movement is classified into suspended sediment movement, bed load movement and sheet flow movement, and the predominant type among these is defined mainly by the action of wave motion on sea bed. Careful observations of suspended sediment particles were made to propose a sediment transport model above a rippled bed. Repeated beach profile measurements were actively done to evaluate the direction and magnitude of the rate of overall cross-shore bed load transport. For sheet flow, intensive measurements of sand particle movement were done in oscillatory flow tank and formulas of sediment transport rate under sheet flow condition were presented. Based on these formulas, cross-shore bottom topography change was simulated numerically, and the calculated results were compared with the beach profiles measured in small and/or large scale laboratory flumes and in field in order to check the validity of models and to improve the simulation techniques (Horikawa, 1988).

Detached breakwaters were proposed as a type of coastal structures to prevent beach erosion and to store a certain amount of sand behind the breakwaters as sand spits or a tombolo (Toyoshima, 1974). According to the statistics of Japanese coast compiled by the Ministry of Construction, detached breakwaters with the length of 572 km in total have been constructed up to 1992. In recent years artificial reefs consisting of submerged breakwaters with wide crown width have often been substitute for detached breakwaters and other coastal structures from an aesthetic view point.

A three-dimensional model of beach topography change has been developed in recent time (Watanabe et al., 1986b). This model includes a wave transformation model based on the time-dependent mild-slope equations, a nearshore current model, and sediment transport rate formula due to waves and currents. The model has been used for the last decade to predict beach evolution in field. Due mainly to the progress in wave transformation models and sediment transport rate formulas, the model is being improved for more accurate prediction of beach evolution.

Study on wind blown sand has significantly progressed by doing frequently large scale field observation tests and by conducting wind flume tests in laboratory (Sherman et al., 1990; Horikawa, 1991b). The problem originated in protecting farms from wind blown sand. Nowadays, wind blown sand has been treated as one of the causses of beach topography change.

In the above treatment, movement of sand particles was only treated for the sake of simplicity. However, transport of cohesive sediment has drawn much attention rather recently in the practical cases. Therefore, in recent years, study on behavior of cohesive sediment has been intensively done by researchers. Rheological properties of cohesive sediment have been examined by using shear meters, and numerical models of cohesive sediment transport have been proposed in order to predict variation of muddy beach profile.

In refererence to the process stated above and future prospects on the subjects, some comments will be given in the following. Application of turbulence models to oscillatory boundary layer flow is thought to have potential to reproduce generation of sand ripple over sea bottom and inception of sheet flow in the future. This kind of effort will give clear understanding of the sediment transport mechanism. Improvement of sediment transport rate formulas is necessary especially for irregular waves and combined waves with currents at an arbitrary crossing angle. Along with the progress in the modeling of the wave transformation and three-dimensional nearshore current, beach topography change will be predicted more reliably.

Coastal Works and Coastal Structures

Study of coastal structures began with field and laboratory investigations on the effects of various types of coastal structures on sediment transport. The structures cited here include groins, detached breakwaters, submerged breakwaters, pneumatic breakwaters, and others. In addition, wave forces on structures (including impulsive wave forces), wave run-up, and wave overtopping over sea walls have been intensively investigated.

As an example of wave forces acting on vertical or composite type breakwaters, the historical change of design procedures of breakwaters will be explained briefly. As described before, the Hiroi formula of breaking wave forces had widely been used to evaluate wave forces for many years. At the early stage of development, irregularity of wave characteristics could not be introduced into the formula. Therefore the definition of wave height used in the formula was unclear. In 1967, the Design Standards of Port and Harbor Structures defined it as the significant wave height of incident waves at the breakwater site, and a uniform distribution of the above wave force intensity was applied up to the height of 1.25 $H_{1/3}$ as shown in Figure 5. The concept of wave irregularity was first introduced into the design criterion of breakwater by Ito in 1971 in the form of expected sliding distance of a caisson placed on rubble mound due to standing and breaking wave forces. Later on, Goda (1974) proposed a new formula to predict wave forces acting on a vertical face caisson of a composite type breakwater on the basis of analytical as well as empirical considerations (Figure 6). A typical merit of this formula is to cover the whole region of water depth at the site of the breakwater, being independent of whether breaking wave forces or standing wave forces act . The readers are requested to refer the original paper (Goda, 1974; Goda, 1985) in order to define the terms p_1, p_2, p_3, p_u, and η^*.

During the past two decades, many new types of breakwaters have been proposed for use in relatively deep water and rough seas. One example is a dual cylindrical caisson breakwater (Tanimoto et al., 1994). The development of these structures was made on the basis of laboratory experiments, and several structures were constructed in field to test their functions. Thus, their full performance for practical use will be evaluated in the future.

Various types of detached breakwaters for shore protection were newly developed and tested in laboratories in the program Marine Multi Zone established by the Ministry of Construction. Some of these structures are installed in the field to

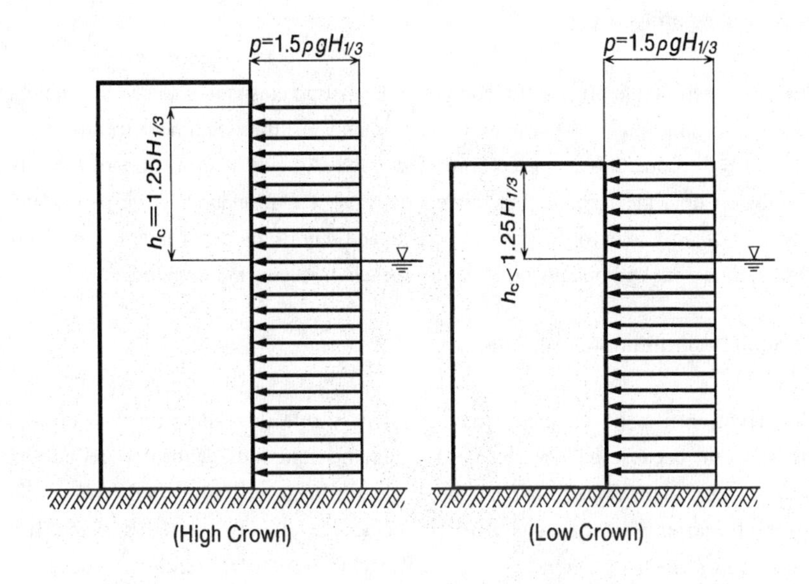

Figuer 5. Breaking wave force in the revised Hiroi formula

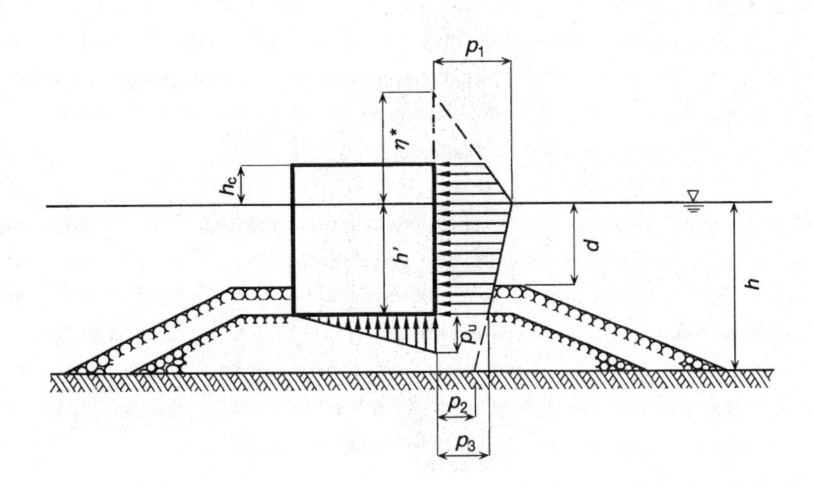

Figure 6. Wave pressure distribution against a vertical wall in Goda's formula (Goda, 1985)

examine their efficiency.

Up to the present, the primary function of coastal structures is to change wave height distribution through which nearshore current and resulting sediment transport are controlled to promote beach accretion effectively. As indicated in some of recently established non-linear simulation models of wave transformation, wave period and direction as well as wave height can be changed due to nonlinear interaction. This kind of treatment will be utilized to control beach processes in the future.

Coastal Environment Protection

Conservation of sandy beaches plays a primary role in maintaining the coastal environment because beaches have advantageous functions of disaster reduction, water quality improvement, habitat supply, recreational utilization, and others. All research subjects in the coastal engineering are closely related with coastal environment in broad sense. In the following, recent research activities are introduced for selected subjects.

At the initial period, numerical models for diffusion of heated water discharged from power plants situated along open sea coast were developed to predict the thermal impact on the ambient water. In parallel to the above procedure, field observations were intensively done on occasion to obtain information about the surface water temperature pattern in wide area including the vicinity of the outlet. These data have been utilized to verify the applicability of the models and to improve them. Therefore, hydrodynamics of contaminated water and water quality in a closed bay have been treated by numerical simulation.

In recent years, the effect of coastal structures on the ecological environment in the nearshore area has been studied. The population change of marine plants and animals due to construction of coastal structures has been examined through joint efforts of biologists and coastal engineers. Based their experience, a predictive model has been proposed. On the other hand, the importance of tidal flats on the ecological environment has gradually been recognized by the public, hence technology development has been made to create artificial tidal flats.

Sea level rise due to global warming is expected to accelerate in the next century, which will cause serious impacts on the coastal region. A research subcommittee was organized in the Coastal Engineering Committee of JSCE to study various issues related to the global environment including the Intergovernmental Panel on Climate Change (IPCC) Report in 1990 (Warrick et al., 1990). In the committee, climate change of Japan in the past was reviewed carefully, impacts of sea level rise on natural and developed coastlines were predicted, and strategies for coping with the expected circumstance were discussed at the subcommittee level (Coastal Engineering Commitee, 1994b).

The coastal environment includes a variety of phenomena, such as physical, chemical, biological and ecological processes. In order to obtain better understanding of the coastal environment, the importance of field observation cannot be exaggerated. Then, each unit process included should be developed to improve the quality of the coastal environment. For example, models of water quality have already been proposed by considering physical and chemical processes. When biological processes are included in these models, the combined model will give more important and useful information for use by society.

DISSEMINATION OF COASTAL ENGINEERING KNOWLEDGE

Coastal engineering research activities and coastal engineering works have progressed at a high rate during the past forty years. In the above progress, it is believed that the following important events contributed very much to promote the coastal engineering in Japan as a whole. These events are the Tenth International Conference on Coastal Engineering in Tokyo in 1966 and the Twenty-Fourth International Conference on Coastal Engineering in Kobe in 1994. The former conference can be regarded as the first international conference in Asia in our field, which gave strong influence in encouraging young coastal engineers in Japan and in promoting their research activities. The recent conference contributed very much in introducing the present state of coastal engineering in Japan to coastal engineers from various countries.

In addition to the above, it should be mentioned that the Nearshore Environment Research Center (NERC) Program was conducted in the period of 1975 to 1980 as a multi-institutional research program. A detailed description on this program can be found elsewhere (Horikawa et al., 1987). Many field observation data of nearshore waves, nearshore currents, and sediment transport were presented in conjunction with the papers by members of the NERC Program. The book entitled "Nearshore

Dynamics and Coastal Procesess" is the most important outcome of this program (Horikawa, 1988.)

Dissemination of available knowledge on coastal engineering discipline has effectively been done in Japan through the annual Japanese Conferences on Coastal Engineering and the International Conferences on Coastal Engineering. In the case of the NERC Program, international exchange of information was made with Nearshore Sediment Transport Study (NSTS) Program in the U.S.A. (Seymour, 1989) in order to encourage beneficial interaction between these two programs during the course of study.

CONCLUDING REMARKS

In this paper, the history and heritage of coastal engineering in Japan were described briefly. As stated previously, the dawn of coastal engineering in Japan opened a new era in 1953. Hence the period under discussion was divided into two, namely pre-dawn and post-dawn periods. For the pre-dawn period, limited number of contributions were described with their historical background. In contrast, in the post-dawn period, many papers have appeared particularly in the conference proceedings, and it was beyond the author's capacity to review all of them. Hence the subjects and related contents in this paper are based on the author's personal selection.

ACKNOWLEDGEMENTS

The author would like to express his sincere appreciation to Dr. M. Isobe, Professor at The University of Tokyo, and Dr. K. Tanimoto, Professor at Saitama University. They offered their kind assistance in various ways in the process of manuscript preparations. The author is also indebted to Dr. N.C. Kraus, Director of Conrad Blucher Institute for Surveying and Science at Texas A&M University - Corpus Christi, and former member of NERC, for his valuable comments on the structure and phraseology of this paper.

REFERENCES

Aida, I., 1969. "Numerical experiments for the tsunami propagation - the 1964 Niigata Tsunami and 1968 Tokachi-oki Tsunami," Bulletin Earthquake Research

Institute, Vol. 47, pp.673-700.

ASCE. "Proceedings International Conferences on Coastal Engineering."

Berkhoff, J. C. W., 1972. "Computation of combined refraction-diffraction," Proceedings 13th International Conference on Coastal Engineering, ASCE, pp. 471-490.

Bowen, A. J., 1969. "The generation of longshore currents on a plane beach," Journal Marine Research, Vol. 27, pp. 206-215.

Bureau of Port and Harbour, and Port and Harbour Research Institute, Ministry of Transport (ed.), 1991. "Technical Standard for Port and Harbour Facilities," Overseas Coastal Area Development Institute of Japan, 438pp.

Coastal Engineering Committee. "Proceedings Japanese Conferences on Coastal Engineering," JSCE. (in Japanese)

Coastal Engineering Committee. "Coastal Engineering in Japan," JSCE.

Coastal Engineering Committee, 1994a. "Nearshore Waves - Analysis of Interaction among Waves, Structures and Sea Bottom," JSCE, 520pp. (in Japanese)

Coastal Engineering Committee, 1994b. "Effect of Global Warming on Coastal Region - Real State, Effect and Strategy of Sea Level Rise and Meteorological Change," JSCE, 221pp. (in Japanese)

Dean, R. G., 1965. "Stream function representation of non-linear ocean waves," Journal Geophysical Research, Vol. 70, pp. 4561-4572.

Galvin, C. J., and Eagleson, P. S., 1965. "Experimental study of longshore current on a plane beach," U. S. Army, Coastal Engineering Research Center, Technical Memorandum No. 10, 80pp.

Goda, Y., 1970. "A synthesis of breaker indices," Proceedings JSCE, No. 180, pp.39-49. (in Japanese)

Goda, Y., 1974. "New wave pressure formula for composite breakwaters," Proceedings 14th International Conference on Coastal Engineering, ASCE, pp. 1702-1720.

Goda, Y., 1985. "Random Sea and Design of Maritime Structures," University of Tokyo Press, 323pp.

Hiroi, I., 1919. "On a method of estimating the force of waves," Journal College of Engineering, Tokyo Imperial University, Vol. X, No.1, pp. 1-20.

Hiroi, I., 1919. "An experimental determination and utilization of wave power," Journal College of Engineering, Tokyo Imperial University, Vol. X, No.1, pp. 21-37.

Hom-ma, M., 1933. "Deformation of long waves," Journal JSCE, Vol. 19, No.9, pp.741-763. (in Japanese)

Honda, K., Terada, T., and Ishitani, D., 1908. "On the secondary undulations of ocean tides," Philosophical Magazine, Vol. 6, No. 15, pp.88-126.

Horikawa, K., 1978. "Coastal Engineering, An Introduction to Ocean Engineering," University of Tokyo Press, 402pp.

Horikawa, K., (ed.), 1988. "Nearshore Dynamics and Coastal Processes - Theory, Measurement, and Predictive Models," University of Tokyo Press, 522pp.

Horikawa, K., 1991a. "Introductory remarks on coastal protection in Japan," Research Report Department of Foundation Engineering & Construction Engineering, Saitama University, Vol. 21, pp. 1-14.

Horikawa, K., 1991b. "Chapter 14 : Sand transport by wind," in Handbook of Coastal and Ocean Engineering, Vol. 2, edited by J. B. Herbich, Gulf Publishing Company, pp. 771-798.

Horikawa, K., and Kuo, C. T., 1966. "A study on wave transformation inside surf zone," Proceedings 10th International Conference on Coastal Engineering, ASCE, pp. 217-233.

Horikawa, K., and Hattori, M., 1987. "Accomplishments of the Nearshore Environment Research Center Project," Proceedings Coastal Sediment '87, ASCE, pp. 756-771.

Hydraulics Committee, 1985. "Hydraulics Formulas, Revised," JSCE, pp.479-598. (in Japanese)

Iida, K., and Iwasaki, T., 1983. "Tsunamis : Their Science and Engineering ," Terra Scientific Publishing Company, D. Reidel Publishing Company, 563pp.

Ijima, T., Takahashi, T., and Nakamura, K., 1956. "Waves in the surf zone observed by a photographic method," Proceedings 3rd Japanese Conference on Coastal Engineering, JSCE, pp. 99-116. (in Japanese)

Isobe, M., 1988. "Chapter 3 : Measurement of wave direction," in Part V Field Observations, Nearshore Dynamics and Coastal Processes, edited by K. Horikawa, University of Tokyo Press, pp. 407-422.

Isobe, M., Nishimura, H., and Horikawa, K., 1982. "Theoretical considerations on perturbation solutions for waves of permanent type," Bulletin Faculty of Engineering, Yokohama National University, Vol. 31, pp. 29-57.

Ito, Y., 1971. "Stability of mixed-type breakwater - A method of probable sliding distance -," Coastal Engineering in Japan, Vol. 14, pp.53-61.

Ito, Y., and Tanimoto, K., 1971. "A new method of numerical analysis of wave propagation and its application to wave height distribution along structures," Proceedings 18th Japanese Coference on Coastal Engineering, pp. 87-90. (in Japanese)

Iwagaki, Y., 1972. "Practical utilization of cnoidal wave theory," Course B, Summer Seminar on Hydraulic Engineering, JSCE. (in Japanese)

Longuet-Higgins, M.S., 1970. "Longshore currents generated by obliquely incident sea waves 1 & 2," Journal Geophysical Research, Vol. 75, pp.6778-6801.

Longuet-Higgins, M.S., 1975. "On the joint distribution of the periods and amplitudes of sea waves," Journal Geophysical Research, Vol.80, pp. 2688-2694.

Mano, A., and Iwasaki, T., 1976. "Hindcast of the Sanriku Tsunami by using the fault model of earthquakes occurred off the Sanriku Coast," Proceedings 23rd Japanese Conference on Coastal Engineering, JSCE, pp443-447. (in Japanese)

Mimura, N., and Morvell, G. (ed.), 1993. "Proceedings of the IPCC Eastern Hemisphere Workshop, Vulnerability Assessment to Sea-Level Rise and Coastal Zone Management," IPCC, 429pp.

Mitsuyasu, H., 1995. "Physics of Ocean Waves," Iwanami-Shoten, 210pp. (in Japanese)

Nagata, Y., 1964. "The statistical properties of orbital wave motions and their application for the measurement of directional wave spectra," Journal Oceanographic Society Japan, Vol.19, pp.169-181.

Nishimura, H., Isobe, M., and Horikawa, K., 1977. "Higher order solution of the Stokes and the cnoidal waves," Journal Faculty of Engineering, University of Tokyo, B-34, No. 2, pp.267-293.

Nishimura, H., Maruyama, K., and Hiraguchi, H., 1983. "Wave field analysis by finite difference method," Proceedings 30th Japanese Conference on Coastal Engineering, JSCE, pp.123-127. (in Japanese)

Okada, T., 1918. "Study on seiche motion," Journal JSCE, Vol. 4, No. 4, pp. 753-766. (in Japanese)

Reichauer, E.O. 1981. "Japan, The Story of Nation," Charles E. Tuttle Co., 428pp.

Sawaragi, T. (ed.), 1995. "Coastal Engineering - Waves, Beaches, Wave-Structure Interactions," Elsevier, 479pp.

Sasaki, T., Horikawa, K., and Hotta, S., 1976. "Nearshore current on a gently sloping beach," Proceedings 15th International Conference on Coastal Engineering, ASCE, pp. 624-644.

Seymour, R.J. (ed.), 1989. "Nearshore Sediment Transport," Plenum Press, 418pp.

Sherman, D.J., and Hotta, S., 1990. "Chapter Two : Aeolian sediment transport : theory and measurement -." in Coastal Dunes, Form and Process, edited by K.Nordstorm, N. Psuty. and B. Carter, John Wiley & Sons, pp.17-55.

Sonu, C.J., 1969. "Tethered balloon for study of coastal dynamics," Proceedings Symposium on Earth Observations from Balloons, American Society of

Photogrammetry, Technical Report No.66, pp.91-103.

Tanimoto, K., and Kobune, K., 1975. "Computation of waves in a harbor basin by a numerical wave analysis method," Proceedings 22nd Japanese Conference or Coastal Engineering, JSCE, pp.249-253. (in Japanese)

Tanimoto, K., and Takahashi, S., 1994. "Design and construction of caisson breakwaters - the Japanese experience -," Coastal Engineering, Vol.22, Elsevier, pp.57-77.

Thornton, E.B., 1970. "Variation of longshore current across the surf zone ," Proceedings 12th International Conference on Coastal Engineering, ASCE, pp.291-308.

Toyoshima, O., 1974. "Design of detached breakwater system," Proceedings 14th International Conference on Coastal Engineering, ASCE, pp.1419-1431.

Tsuchiya, T., and Shuto, N. (ed.), 1995. "Tsunami: Progress in Prediction, Disaster Prevention and Warning," Kluwer Academic Publishers, 336pp.

Unoki, S., and Isozaki, I., 1962. "Several results of numerical experiments on storm surges," Proceedings 9th Japanese Conference on Coastal Engineering, JSCE, pp.1-6. (in Japanese)

Warrick, R.A., and Oerlemans, H., 1990. "Sea level rise," in Climate Change - The IPCC Scientific Assessment, edited by J.T. Houghton, G.J. Jenkins, and J.J. Ephraus, pp.257-282.

Watanabe, A., and Maruyama, K., 1986a. "Numerical modeling of nearshore wave field under combined refraction, diffraction and breaking," Coastal Engineering in Japan, Vol.29, pp.19-39.

Watanabe, A., Maruyama, K., Shimizu, T., and Sakakiyama, T., 1986b. "Numerical prediction model of three-dimensional beach deformation around a structure," Coastal Engineering in Japan, Vol.29, pp.179-194.

Watanabe, H., 1964. "Studies on the tsunamis on the Sanriku Coast of the Northeastern Honshu in Japan," Geophysics Magazine, Vol.32, pp.1-65.

HISTORY OF COASTAL ENGINEERING IN MEXICO

J. Antonio Maza*, Rodolfo Silva† & Carlos Sánchez‡

ABSTRACT: Coastal engineering has a relatively short history in Mexico beginning with the arrival of sea-farers from Europe in 1518. In this article we trace the growth in importance of port engineering through the colonial era and the struggles in continuing this development for an independent Mexico beset by political, social and economic difficulties. Ambitious projects at the turn of this century were largely the result of foreign investment and many of them doomed in Revolutionary and post Revolutionary turmoil. In the last four decades Mexico's economic advances have been coupled with advances in research and investigation in coastal engineering as well as physical developments along the coast to accommodate the advances in national industries, the most important being the petrol industry and tourism. With a developing economy and new trade agreements with other nations Mexico looks to the sea to enhance her progress in economic and social development.

INTRODUCTION

The Mexican seaboard, of some 11,500 km in length, is mainly Pacific (72%), the Gulf of Mexico and the Caribbean bordering what is generally the eastern side of the country. Territorial waters total almost 3 million km^2. Along these coasts are around 130 lagoons and inlets, covering 15,000 km^2. Surprisingly, with a coastline of such length, Mexico counts on only 45 ports of any size and of them only 21 are able to accommodate ocean going traffic. Notwithstanding, 80% of Mexico's exports pass through these ports, Table 1.

Mexico´s petrol industry is largely based at sea. Three quarters of her reserves are found below the sea and a high percentage of the extraction is often via mooring buoys in the open sea.

From the 1960s Mexico has invested in the field of coastal engineering through various development programmes and in research. In 1960 the Secretaria de Marina founded the first laboratory for the study of maritime works. In 1977 this laboratory, one of the finest in Latin America, came under the control of the Secretaria de

* Manager of Civil Engineering Studies, CFE, Oklahoma 85, 03810, México, D.F
† Research Fellow, Instituto de Ingeniería-UNAM, Cd. Universitaria, Apdo. Postal 70-472, 04510, México, D.F.
‡ Head of Oceanography Department, CFE, Oklahoma 85, 03810, México, D.F.

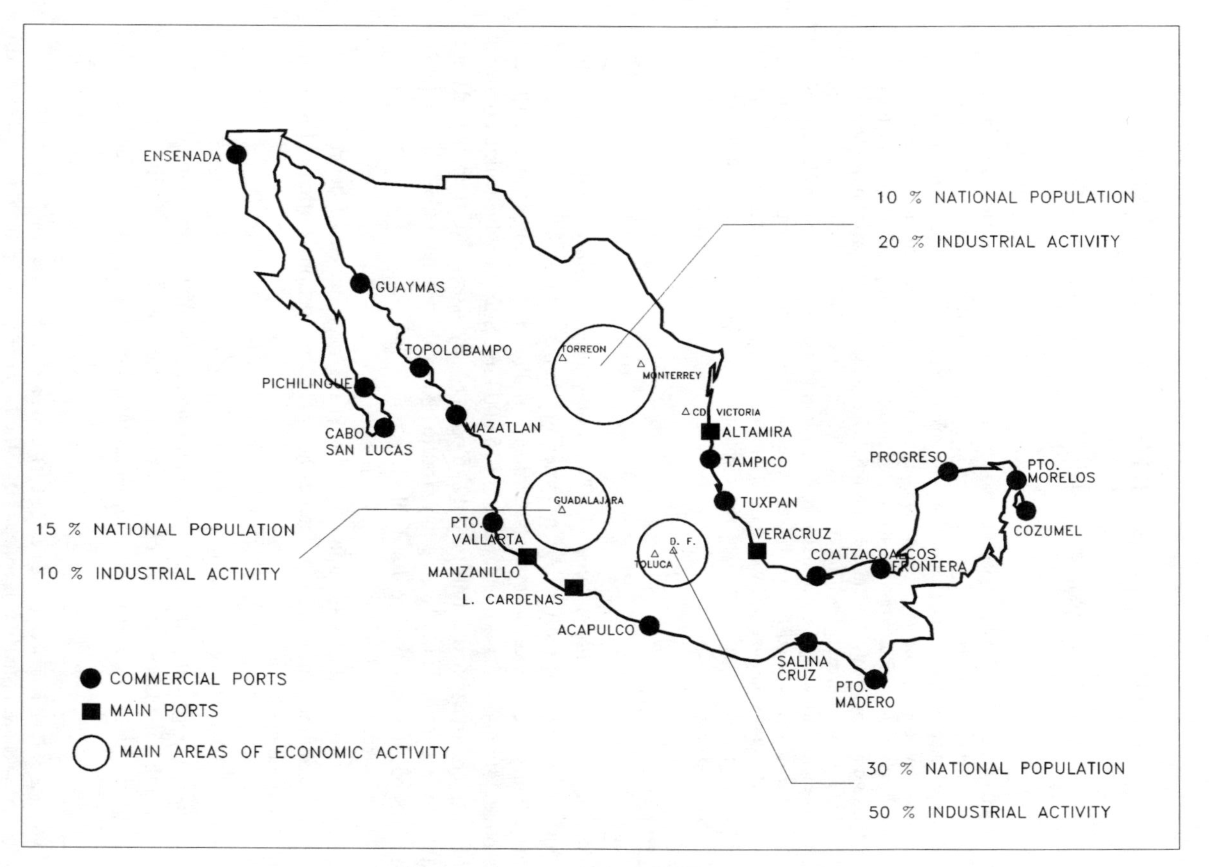

Figure 1 MAIN PORTS

Comunicaciones y Transportes. During the 1980s its main thrust was in the area of beach dynamics and port engineering in conjunction with the Japanese government.

Table 1. Port Activity in 1995 (source Puertos Mexicanos).

PORT	CARGO MOVED IN THOUSANDS OF TONNES
Pacific Coast	
Ensenada, Baja California Norte	826.30
Pichilingue, Baja California Sur	34.40
Cabo San Lucas, Baja California Sur	no data available
Guaymas, Sonora	2,661.70
Topolobampo, Sinaloa	129.30
Mazatlán, Sinaloa	184.80
Pto. Vallarta, Jalisco	no data available
Manzanillo, Colima	4,585.90
Lázaro Cárdenas, Michoacan	10,707.40
Acapulco, Guerrero	39.30
Salina Cruz, Oaxaca	152.00
Puerto Madero, Chiapas	103.9 (in 1994)
Gulf Coast	
Altamira, Tamaulipas	2,591.00
Tampico, Tamaulipas	4,110.30
Tuxpan, Veracruz	271.20
Veracruz, Veracruz	6,480.70
Coatzacoalcos, Veracruz	1,669.80
Frontera, Tabasco	0.20
Progreso, Yucatan	1,098.20
Puerto Morelos, Quintana Roo	391.00
Cozumel, Quintana Roo	364.90

A measure of the growing importance of coastal engineering to Mexico can be seen in her offer to host the 8th ICCE, in 1962, which was held in Mexico City. Shortly after this both the Universidad Nacional Autonoma de Mexico (UNAM) and the Instituto Politecnico Nacional (IPN) opened their first facilities, at the Instituto de Ingeniería (II) and Escuela Superior de Ingeniería y Arquitectura (ESIA), respectively. Elsewhere in the Republic, subsequent years have seen facilities developed at the Universidad Autonoma de Tamaulipas, and in Baja California at the Facultad de Ciencias Marinas and CICESE. Nowadays these are the main centres for the study of coastal engineering at undergraduate and post-graduate level. Smaller laboratories operate in Colima, Sinaloa, Tabasco, Campeche and Yucatán, but the main research centres are at the UNAM, IPN and CICESE along with those of Instituto de Investigaciones Electricas (IIE). From these institutions most of our coastal engineers have emerged and in the last twenty years 'homegrown' authors have also published works in the field, the most important of which are 'Ingeniería Maritima' Roberto Bustamante Ahumada (1976), 'Análisis Hidrodinámico de Estructuras y Vehículos Marítimos' José Luis Sánchez Bribiesca (1981), 'Hidráulica Marítima' José Antonio Maza Alvarez (1983) and "Ingeniería de Costas" A. Frías and G. Moreno (1988).

A HISTORICAL OVERVIEW

Despite the highly developed civilisations of the indigenous peoples of Mexico before the Spanish Conquest, there was no sea-faring culture among the original inhabitants. Although much of the indigenous population lived on the coast, rarely did they undertake more than short trips along the coast in small canoes, while fishing was shoreline activity. Hence the consternation of the coastal population when European vessels were first seen off the Gulf coast. In 1518 one of Moctezumá's subjects on the Gulf Coast went to the Court to inform the Emperor of his sighting of "a range of mountains, or some big hills, floating in the sea." These were in fact the ships of the Spanish explorers and adventurers, so far removed from the canoes of the indigenous peoples of contemporary Mexico that the observer could not hazard a guess as to their real nature. Moctezuma's spies, sent out to investigate further, came back with descriptions of the strange activities of the people who came from these "mountains" in small boats, casting nets into the sea or using rods and hooks; fishing methods unfamiliar to them

Sixteenth to Eighteenth Centuries

The colonial period for Mexico implied the development of commerce and communication with Spain and her other dependent territories. It was vital to have ports linking Europe and America and, using America as a bridge, Europe and Asia. At this time the average draught of sea-going vessels was only 2.5 metres and thus natural bays and river mouths provided adequate depth, cliffs and islands offering natural protection from winds and waves. With little exertion maritime interests were served by a few ports. In 1542 an observation tower, a wharf and a small jetty were constructed at Veracruz. By 1596 it was seen necessary to build a wooden warehouse to store delicate goods. The repeated attacks of pirates decided the population to build a fortress and surrounding wall to protect the port. A little further up the Gulf coast a stone jetty was built at Tantoyoquita thereby facilitating trade with the interior region of Huasteca and the development of Tampico as a port. Shipyards served the merchant fleet at Campeche, near the Yucatán peninsula. Meanwhile on the Pacific coast, Hernán Cortés' wood burning lighthouse continued to serve for decades at "La Ventosa" near Salina Cruz in the South-East (and is still visible today) where wooden ships were built. Acapulco became a hub of the silk trade as shipments from the Philipines unloaded for a cross country journey before going on to Europe. Sadly, there is no visible evidence of this international activity today. Further north, New Spain's first naval base was established at San Blas, Nayarit, with dockyards fronting a bay with a natural depth of 9 metres.

Unfortunately the alluvial deposits from the nearby Rio Santiago delta caused problems of sedimentation and two bays at San Blas gradually closed off and became useless as harbours.

Independent Mexico

Mexico became independent in 1821 and one of her first moves was to transfer naval facilities from San Blas to what is today still an important base, Manzanillo, 300 kilometres further down the Pacific coast.

The introduction of steam power brought larger vessels, of metal instead of wood, to Mexico and, as a result, the need for improved protection for shipping from the elements. Breakwaters were constructed and dredging began to be considered at some ports. The railway too came into its own in the Steam Age. Five ports were seen as the backbone of a planned railway network: Veracruz, Tampico and Manzanillo were linked to the capital, Mexico City, by rail and plans made to link Coatzacoalcos and Salina Cruz across the isthmus of Tehuantepec. These ports began to be considered as part of a national maritime framework and the port itself as a factor in urban development.

By the turn of the century at all five of the above mentioned ports breakwaters protecting the ports from sediment transport had been constructed. At Tampico, Isleta Pérez was joined to the mainland by a series of bridges across the rivers Panuco and Tamesi with wharves and warehouses built on the island. Similarly the Isla de San Juan de Ulúa was connected to Veracruz by a breakwater on the northern shore which continued out to sea from the island, protecting the port from "nortes", the severe winds which bedevil the Gulf coast. Concrete blocks were laid to raise the height and widen the breakwater to accommodate warehouses and moorings. From the south bank of the river a third breakwater defined the entrance to the port. Inside the newly developed harbour another breakwater, "Fisherman's Wall", served to control wave action. New wharehouses of two storeys were also built. Meanwhile in the town itself street lighting and drinking water appeared along with workers' housing and rail links into the hinterland. In 1901 the total tonnage of cargo handled at Veracruz was 457,915 which produced a healthy balance sheet for the goverment through the taxes imposed. The docks at Manzanillo were built at this time as well as the dry dock at Salina Cruz, which is still in operation. Suddenly dredging became all-important as a follow up to such investments.

The Isthmus Dream

As early as 1520, having conquered the dominant people, the Mexica, Cortés set about developing New Spain. He saw the potential value of a trade route between Europe and the Far East which avoided having to circumnavigate Cape Horn, as early as 1520. Following the studies he had commisioned he actually laid out a carriage road across the isthmus of Tehuantepec in the second decade of the sixteenth century.

Shortly after this, the Portuguese navigator Antonio Gavao suggested the building of a canal to link the Atlantic and the Pacific. In 1550 he was still torn between Tehuantepec, Nicaragua, Panama, or Darién. Nothing came of his ideas however, though surely it enjoyed a resurgence 300 years later as prospectors in the 1849 Californian Gold Rush used Cortés' carriage road having started and finished the journey by sea. Perhaps it was this which prompted an American engineer, by the name of Eads, to suggest a broad-gauge railway to carry ships, in their entirety, from the Gulf to the Pacific coast.

Around the turn of this century projects aiming to cross the isthmus were begun at various points in central América and with varying degrees of success. The President of Mexico, Porfirio Díaz, was determined not to fall behind, and enlisted the services of British engineer Sir Weetman Pearson to improve the ports of Salina Cruz on the Pacific and Coatzacoalcos (to be renamed Puerto Mexico) on the Gulf coast as well as to construct a railway linking the two.

When work began in 1902 the route of the railway itself had already been laid out. Five years earlier another firm had constructed what turned out to be a "white elephant" of a railway due to insufficient foresight. Curves had to be straightened, bridges replaced, new rails and ballast laid and concrete culverts built. In total 300 kilometres of railway track covered the distance across the isthmus and in 1906 the first cargo of sugar passed along it, on its way from Hawaii to the United States' East Coast.

The improvement of the port at the mouth of the Coatzacolacos river presented few problems. A channel of ten metres depth was dredged through the delta to the open sea. Two converging jetties of 1300 metres each were built from either side of the river to form a harbour entrance. Finally, further dredging scoured out a deep basin between the jetty confines. Wharves, railway yards and workers' housing transformed the waterfront.

The Pacific port of Salina Cruz was the most troublesome part of the project in that it was necessary to enclose an open bay with breakwaters. High winds, summer and winter, presented problems of sand storms and damage to equipment while a series of minor earthquakes hampered the progress of the work in 1902.

To protect the port from the seasonal winds two harbours were built. The outer harbour is 8.1 hectares in extension with an eastern breakwater of a kilometre reaching towards its partner of 581 metres, leaving a harbour entrance of 200 metres. This outer harbour was 20 metres deep while the calmer inner harbour, with a 30 metre entrance, had a depth of only 10 metres at low tide. Two swing bridges cross the harbour entrance which is itself 1000 metres by 222 metres. Steel wharves and railway tracks run the length of the waterfront and state of the art electrical and steam-driven facilities installed for efficient cargo handling. A dry dock of 180 by 30 metres was built and is still in use at present.

In 1907 the main rival to the Tehuantepec railway, that of Panama, in America hands, being some 2000 kilometres further south-east was less attractive for freight going to or from the US or Canadian West coast, Hawaii or the Far East. That first year 900,000 tons of freight crossed the isthmus on the Mexican railway which ran seven trains daily in both directions - by 1911 that number had doubled.

But the success of this project was short lived. The outbreak of the Mexican Revolution, in 1910, was soon followed by the taking of trains to "serve the cause" thus disrupting the internal network of the railways. Then in 1915 the Panama Canal opened; a far more cost-effective route for shipping in that it cut out the double handling costs at the two ports. The nail in the coffin came in 1918 when Mexican President Carranza took over the running of the railway for Mexico in a thinly disguised appeal to the xenophobia rampant in Mexico following the Revolution. The improverished government had difficulty in maintaining the line and installations and the route soon became even less cost-efficient.

After the Revolution

Mexico suffered grave disruption to civil order and investment from 1910 to the 1920s. No new maritime projects were undertaken and maintenance even of the five aforementioned ports was limited. 1924 saw the only maritime construction work of the decade: breakwaters built at Yavaros in Sonora, a small, relatively unimportant fishing port. Mexico fell further behind in the development of a maritime culture.

In the 1930s, the economic situation began to improve. The port terminal at Salina Cruz, having been so long abandoned to sedimentation and the effects of sand transporting winds, had become an enormous recreation ground with the local residents enjoying baseball games in the docks where so recently they had handled cargoes from all over the world. Fortunately a dredging vessel had been left, forgotten in silt of the harbour, and with minimal investment it was put to work and the inner harbour opened up again. In Yucatán, Progreso enjoyed a period of development and became an important focus for coastal shipping, particularly with Veracruz, as a harbour wall of 2,200 metres was built.

In the early 1940s Mexican companies finally got involved in Mexican coastal engineering. At Veracruz a new breakwater (called Numero 2) was built, as well as the first dry dock on San Juan de Ulúa. Less propitious was the attempt to build a dry dock at Coatzacoalcos, which was a complete disaster, due to the lack of knowledge regarding the ground conditions at this site.

Later in the decade more work was done at Tampico where dredging and breakwaters improved the viability of the port. Acapulco and Mazatlán on the Pacific were modernised, Tuxpan, in Veracruz, and Guaymas, in Sonora, had jetties built and work began in the construction of a new port, Ensenada, in Baja California.

As the country found its feet and the economy improved, the concept of coastal development gained credence. Calls for economic regeneration of the Mexican coast were put forward and the belief that the sea and regions bordering it should be further exploited resulted in more funding for the improvement of fishing ports above all.

From 1959 a series of improvements to facilities and the construction of new wharves and warehouses brought back to Veracruz some of its old life. PEMEX invested in a jetty and a grain storage facility diversified the handling capacity of the port.

Figure 2 Oil platforms in the Gulf of Mexico

The oil industry also produced improved investments and returns. In 1959 the first under sea drillings were took place near the Laguna del Carmen, Campeche. In 1966, in the same area, PEMEX discovered an important oil field 35 kilometres offshore, to be named the Atún field. Oil platforms and rigs were necessary for the exploitation of the oil and by 1973 some eleven platforms were in place with a pipeline leading to Punta Piedras. In the 1970s PEMEX required improved port facilities; Dos Bocas in Tabasco, became the terminus for offshore pipelines in the Gulf and Salina Cruz, in Oaxaca, was given a new lease of life.

The 1960s saw further economic growth resulting in more ports being modernised and the birth of ambitious new projects. The Comision Federal de Electricidad (CFE) began to build a series of thermal power stations, six on the Pacific and three on the Gulf coast. The construction of the first plant was at Rosarito in Baja California Norte where large volumes of sea water are used to cool condensers. A nuclear power station at Laguna Verde in Veracruz was also built to take advantage of sea water for cooling. Together with the nine thermal power station these generate 36% of the country's electricity. CFE continues to encourage research and development in coastal engineering across the Republic.

Figure 3 Thermoelectric Power Plant in Rosarito, Baja California

Recent History

The need for industrial ports was seen to be met by the development of new dock lands at Salina Cruz, which have since been abandoned along with those of Puerto del Ostión, in Veracruz, while the improvements at the ports of Lázaro Cardenas, on the Pacific, and Altamira, on the Gulf coast, have proved more long lasting.

The environmental movements of the 1980s were not ignored in Mexico and during this decade much lip-service, at least, was paid to conserving and improving the coastline, especially the lagoons and rivermouths. However, to date, only 2% of

Mexico's 15,000 km^2 of marshlands are functioning as Nature intended, and almost all of them are used for intensive crustacean farming.

Tourism is Mexico's third industry in economic terms, after petrol and manufacture industries, and the main centres of international tourism are on the coast. Even so, not until the devastation caused to the Caribbean coastline, especially Cancún, by Hurricane Gilbert in 1988, were serious calls made for beach protection and conservation. Cancún is one of 5 integrally planned beach resorts which have been developed by FONATUR (the National Fund for the Development of Tourism), an agency founded in 1974 with the brief of identifying key locations, developing and consolidating infrastructure and real estate there and finally selling and/or leasing land and buildings. The jewel in the crown of FONATUR is undoubtedly Cancún which was a sleepy fishing village in 1969. Nowadays some two million visitors (national and international) stay in Cancún fold annually and the actual population has risen a thousand to 250,000. The FONATUR masterplan also included development at Ixtapa on the Pacific, where the Marina is a centrepiece of Ixtapa's world-class sports-fishing facilities. Further south in Oaxaca, a virgin site of 22 miles of coastline, comprising 9 secluded bays, makes up the resort of Huatulco. Also on the Pacific, in Baja California is the resort of Los Cabos a site of tremendous natural beauty, and Loreto a resort based around watersports, especially fishing.

Figure 4 Port of Salina Cruz

Figura 5 Port of Veracruz

TOWARDS THE FUTURE

Eighty percent of the total volume of Mexico's foreign trade is via the sea and this tendency is likely to increase as markets diversify thanks to trade agreements such as NAFTA (North American Free Trade Agreement). In terms of international commerce and the competitiveness of Mexican exports, four ports have assumed greatest importance. The main objective for future investment is to ensure the efficiency of Veracruz and Altamira on the Gulf and Manzanillo and Lázaro Cardenas on the Pacific. At Lázaro Cárdenas and Altamira container handling facilities are being updated and there are now 5380 m^2 of warehousing at the former while Lázaro Cárdenas has a totally automated agricultural bulk handling system which allows unloading of 600 tons per hour. Another development of great consequence is the privatisation programme of ports, begun in 1989, and the encouragement of foreign investment-projects.

Interest has recently been reborn in the idea of an inland water link between Mexico and the US which would save shipping from the fierce and unpredictable 'nortes', winds in the Gulf capable of destroying ships no matter how close they stick to the coast. It is hoped to extend the canal which runs from Tuxpan to Tampico and link this with the US network of waterways which begins at Brownsville, Texas, thus cutting transport costs substantially.

As regards inshore fishing, the protection of water quality in existing lagoons as well as the preservation of their ability to function as lagoons is of great importance.

The Secretaria de Medio Ambiente, Recursos Naturales y Pesca (Environment, Natural Resources and Fishing) have undertaken a programme for the improvement of lagoons in Nayarit covering some 10,000 hectares. Dredging work and improved drainage should increase the catches and livelihoods of 1000 families there. On a slightly smaller scale in Oaxaca the river mouth of the Oro and its adjoining lagoon are being given a facelift with a view to improving the fishing potential for the locals, raising production to 500 tonnes. Wherever possible the potential for lagoon-to-lagoon communication is to be exploited thus circumventing the need for fragile fishing vessels to enter the open sea as well as opening up new wetlands. Thus the Secretaria is currently dredging and improving the canal which runs from Agua Grande to Tapo Revolution in Sinaloa. Similarly appropiate infrastructure, capable of dealing competetively with the catches of our deep sea fishermen is all important in ensuring the development of coastal regions around the fishing ports and a healthy future for their populations.

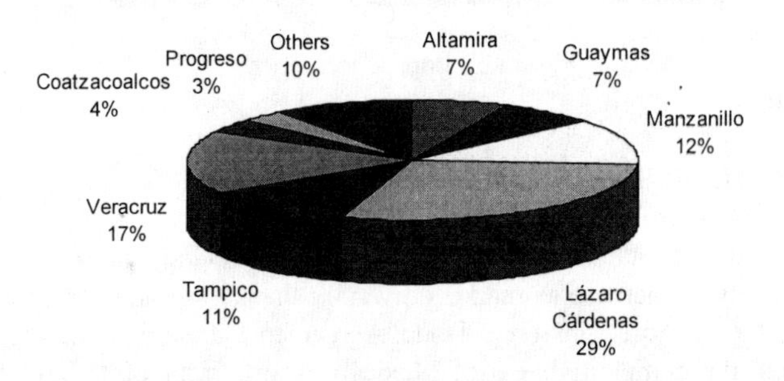

Figure 6 Percentage of Cargo Handled in Principal Mexican Ports, 1995 (source Puertos Mexicanos).

The oil industry is a mainstay of the Mexican economy (2685 barrels were being produced daily in 1995) and as a result the exploitation of offshore oil fields assumes ever increasing importance. However, hand-in-hand with this must go greater care for the environment and a more sensitive approach to exploration, transportation and processing of this "black gold."

In the past few decades the importance of international tourism to the national economy has rocketed. Most visitors from abroad come to the 5 resorts mentioned earlier but FONATUR have highlighted 7 more coastal destinations for further development and instigated a clean-up campaign at Acapulco, Mexico´s most traditional resorts. Scuba-diving and snorkelling off the coast at Cancún, Isla Mujeres and Cozumel coupled with improved access to, and information about, the archaeological sites on the Yucatán peninsula have made the Cancún-Tulum corridor one of the most popular beach destinations in the world.

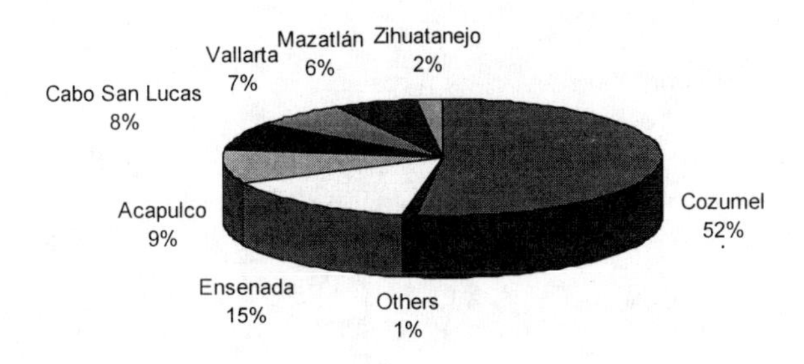

Figure 7. Movement of Cruise Passangers in Mexico, 1995 (source Puertos Mexicanos).

There are ever-more facilities to accommodate cruisers, ferries and yachts in the Republic. Cozumel is the most important tourist terminal in the country with a quay of 344 m^2 for crusiers and ferries and an up to date terminal station to receive passengers. There are 44 marinas operating on our coasts, the majority being situated on the Gulf coast and Caribbean sea whilst in terms of size and importance the Pacific marinas are far more developed.

Beach and shore conservation has begun along with attempts to control the dumping of urban waste at sea. However there is still a long way to go before Mexico's tourist industry can be pronounced "healthy".

Undoubtedly one of the biggest challenges to Mexico is the call of the 1950s, the "Marcha Hacia El Mar" (the march towards the sea), that of developing our coastal regions and encouraging a decentralisation of industry, services and population. Despite the enormous length of her coastline only 13% of the Mexico's current 91 million population live in the coastal zones, which remain backward and inhospitable in many cases. Too much in Mexico happens on the central plateau, far from the sea. Coastal engineers are few in number, material resources are scarce and thus the challenge to improve our coastline is great. There is a growing number of Mexicans studying in schools of excellence elsewhere in the world, returning home with a strong desire to further coastal engineering and contribute to Mexico's progress to First World status.

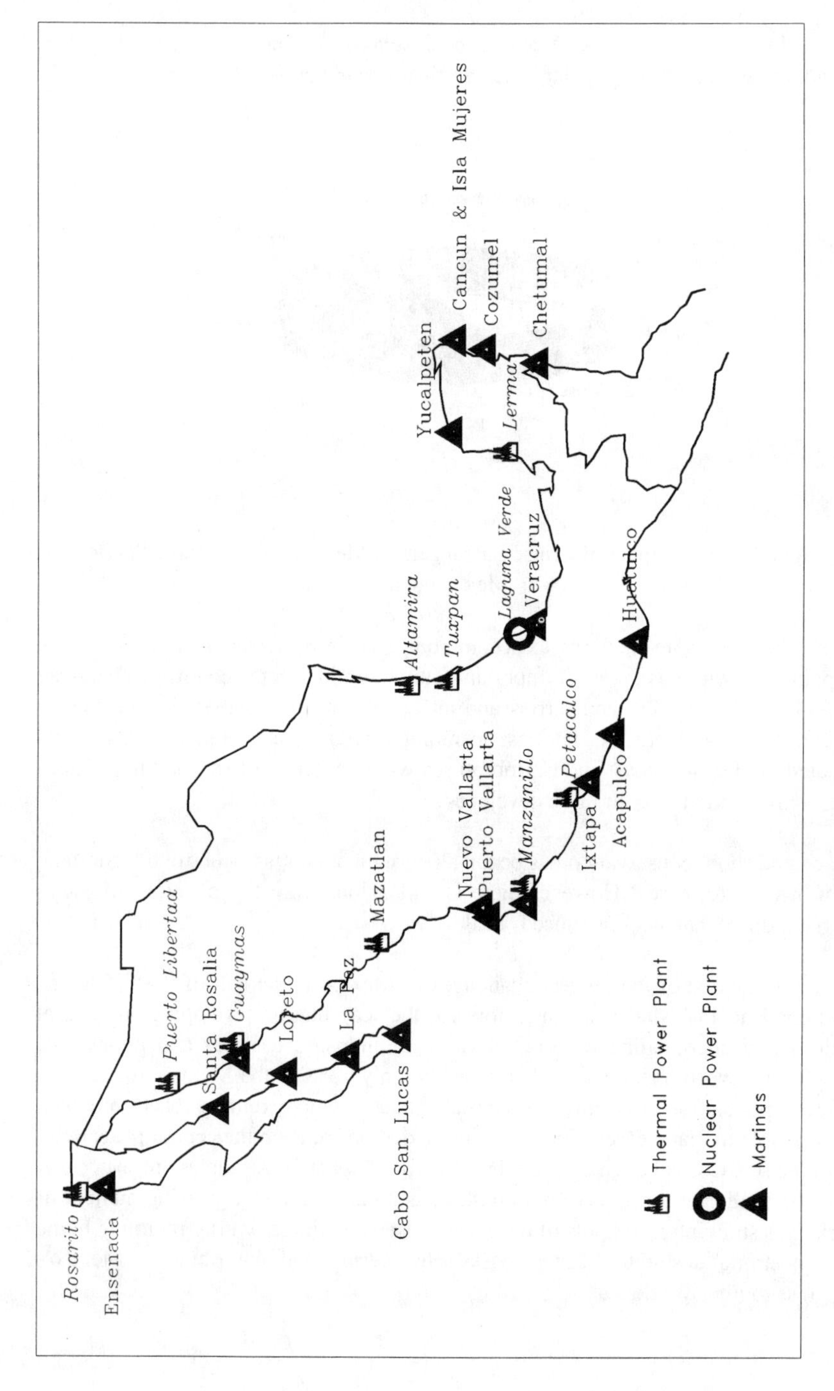

Figure 8 Localization of Power Plants & Marinas

Illustrations

- Figure 2 courtesy of Instituto Mexicano del Petróleo
- Figure 3 courtesy of Comisión Federal de Electricidad
- Figures 4 and 5 courtesy of Dirección de Puertos y Marina Mercante

References

- Bosch, C. 1981. "México Frente al Mar." UNAM, México, 472 pp.
- Bustamante, R., et al. 1976. "Ingeniería Marítima." Obras Marítimas, México, 344 pp.
- Cárdenas, E. 1965. "Urdaneta y el Tornaviaje." Secretaria de Marina, México, 290 pp.
- Comisión Nacional Coordinadora de Puertos, 1976. "La Reforma Portuaria." Comisión Nacional de Puertos, Mexico, 240 pp.
- Fodor, 1995. "Guide to Mexico." New York, 800 pp.
- FONATUR, 1994. "20 Anniversary Report" FONATUR, México.
- Frías, A. and Moreno, G. 1988. "Ingeniería de Costas." Editorial Limusa, México.
- Gerencia de Planeación, 1995. "Estadísticas de Movimiento Portuario 1994-1995." Puertos Mexicanos, México.
- Hernández, A. 1983. "Síntesis Histórica del Puerto de Veracruz." Talleres Gráficos de la Nación, México, 24 pp.
- Maza, J.A. 1983. "Hidráulica Marítima." CFE, Manual de Diseño de Obras Civiles, A.2.1.3, 468 pp.
- MOPU, 1990. "Desarrollo y Medio Ambiente en América Latina y el Caribe." MOPU, España, 232 pp.
- Ortiz, F. 1976. "Los Puertos Mexicanos." Fondo de Cultura Económica, México, 64 pp.
- Pasquel, L. 1980. "San Juan de Ulúa, Fortaleza, Presidio, Residencia Presidencial." Colección Suma Veracruzana, 109 pp.
- Reséndiz-Núñez, D., editor. 1994. "El Sector Eléctrico de México." Comisión Federal de Electricidad, Fondo de Cultura Económica, 772 pp.
- Sánchez, J. 1981. "Análisis Hidrodinámico y Vehículos Marinos." Instituto de Ingeniería, No. 446, 376 pp.
- Secretaría de Comunicaciones y Transportes, 1989. "Catastro Portuario." Vocalía de Planeación de Puertos Mexicanos, 543 pp.
- Secretaría de Comunicaciones y Transportes, 1996, "Reporte Estadístico del Movimiento de Carga y Pasajeros en los Principales Puertos del País." Coordinación General de Puertos y Marina Mercante, 35 pp.
- Secretaría de Marina, 1963. "Canal Intracostero Mexicano Posibilidades de Tráfico y Justificación Económica." Dirección General de Obras Marítimas, 50 pp.
- Secretaría de Marina, 1963. "Sobre el Diagnóstico del Sector Portuario y Algunas Evaluaciones Preliminares." Dirección General de Obras Marítimas, 172 pp.
- Secretaría de Turismo, 1995. "Indicators of Tourist Activity January-June'95" (report), México.
- Thomas, H. 1993. "The Conquest of Mexico." Pimlico, London, Great Britain, 812 pp.
- Young, V., 1988. "An Unlikely Quintet." The British in Mexico, Number 2, The British and Commonwealth Society Mexico City, México, 44pp.
- Young, V., 1993. "The Tehuantepec Railroad, Coatzacoalcos, and Salina Cruz." The British in Mexico, Number 9, The British and Commonwealth Society Mexico City, México, 60pp.
- WWW.inegi.gob.mx

HISTORY AND HERITAGE IN COASTAL ENGINEERING IN THE NETHERLANDS

Eco W. Bijker[1]

ABSTRACT

The history of the struggle of the Dutch people against the sea is described from the beginning of our era until present. With this history also the heritage through the main testators and landmarks of Dutch coastal engineering structures is described. Finally the work of the present generation of coastal engineers and researchers, who possibly are the testators for the next generation, is discussed.

INTRODUCTION

From the title of this paper it will be clear that I have been asked to discuss the History and Heritage of Coastal Engineering in the Netherlands. This means that I will inform you of our history in relation to the water that surrounds our country, which forms the basis for our present day knowledge and is indeed something that we may call our heritage. However, time flies and "today" will be history "tomorrow." I will therefore discuss the developments in coastal engineering in our country up to this very moment. Not only because I think that the initiative of the Coastal Engineering Research Council to compile surveys like this from all over the world will certainly produce a valuable document for our successors and because "today" might prove to be the heritage we leave to the future generation, but also because I feel it is impossible to appreciate the recent history without looking at the latest developments that resulted from it.

A study of the history of the struggle of the Dutch people against their traditional enemy, the sea, shows that many current developments result from events in the past. It also leads to modesty. We Dutch like to quote the statement: "God created the world and the Dutch created the Netherlands." A nice statement, but apart from some mild blasphemy, you might expect such a creation to be based on some sort of plan. Such a plan, however, did not exist at the time when the rough outlines of the Netherlands were shaped. Our ancestors only reacted to the continuous attacks by the sea, and that has been the way in which they shaped their country (Dirkzwager, 1977).

In this paper I will first discuss the Old and Middle Ages and the Renaissance. For these periods I will not distinguish between the various disciplines in Hydraulic Engineering, since no such specializations existed in those days. Only from the 16th century onwards I will try to make such a distinction.

---

[1] Professor Emeritus of Coastal Engineering at the Delft University of Technology.

THE OLD AND MIDDLE AGES

The sand ridges along our coast were already inhabited some 3000 years ago. Quite good information about the inhabitants is available as from the year 600 B.C. (van den Broeke and van Londen, 1995). These early inhabitants were rather wealthy and had a relatively high cultural standard, which probably resulted from the overseas trade from these areas via the various inlets. Because of this trade there was no great pressure to close these inlets to keep the sea out. The people in those days preferred to live on man-made mounds and to be able to sail the coastal waters. It is about this people that the much quoted Roman author Plinius writes. The, partly quoted, report of Plinius reads as follows:

"There the ocean flows with two intervals, at day and at night, in a tremendous flood over an immense land. Because of this everlasting struggle with the course of nature, it is doubtful whether the bottom is land or sea. In the area a miserable people live on high hills, or better on man-made mounds, just above the highest water level known by experience. On these mounds they have built their shacks and when the water is high they are like sailors, but when the water is low they look more like shipwrecked sailors. Then they hunt for the fish that flee with the water around their huts. They don't have cattle, so they cannot feed on milk, like their neighbouring people. Nor can they hunt for game since the sea washes all brushes away. Out of reed and rush they make a sort of string, of which they knot fishing nets. They dry clods of soil (peat!) in the wind rather than in the sun, and burn them to cook their food and to warm their limbs that are chilled by the northern wind."

However, his remarks have to be regarded with some prudence. We know now that he referred to the Chauk, who at that time lived in the area presently known as East Friesland in Germany (around Emden, Norden and Jever). But since the northern coast of the Netherlands was very similar, the quotation of Plinius nevertheless gives an indication of how people not familiar with these circumstances, regarded upon the people living in these areas.

Secondly we have learned from archaeological surveys that they certainly were not as "miserable" as Plinius suggests in his description. They had cattle (sometimes up to 30 or 40) and their handicraft (also made of iron) was of a relatively high level. Even remnants of workshops of bronze, silver and goldsmiths have been found. Their economy was apparently based on trade with the Romans.

In the area behind the coastal ridge a layer of peat was formed, which at a certain point in time, probably around 600, became high enough to enable habitation. However, some protection against the surrounding water was also required and, undoubtedly, rather weak dikes were constructed. Due to the inevitable settling of the peat these dikes had to be raised continuously. At a certain moment these dikes probably also offered some protection against the sea. This not only happened in the south of our country, in the estuary of the rivers, but also in the north around the Wadden Isles. According to archaeological and historical research this must have been around 800. It is worthwhile to observe that when these rather primitive dikes came so close to the actual sea coast and the inlets that the latter were threatened to be closed, there was a clash of interest between the seafarers and traders on the one hand and the people who relied on fishing, hunting and most likely some modest cattle breeding, on the other.

The dikes along the sea coast to protect areas of land from flooding by salt water were mainly made of and situated on marine clays. This development enabled the growing of crops such as wheat, of which the first reports date from around 1030 in Flanders. This seems a likely development since the population in Flanders at that time had grown considerably due to the cloth industry of Bruges and Gent. This cloth industry in its turn had developed on the basis of the available wool from the sheep that grazed the saltings. Incidentally, it is nice to realize that Flanders literally means "flooded land." The sheep could graze the saltings without any dike protection, but this changed when the growing population had to be fed and the land was required to grow crops. All this is known thanks to the existence of monasteries in the area, where the monks duly recorded what happened.

fig. 1. The Netherlands.
The dotted line indicates the area which would be flooded during HW without protection by dikes and dunes.

The first written notice found in the county of Holland is about the so called Zanddijk near Egmond dated in or shortly before 1105.
Since Friesland had a rather large and wealthy population in the early Middle Ages (500-1000), it is certainly possible that the building of sea dikes in that area started around the same time. But since there were no monasteries in these regions, no written evidence is available.

It is most likely that every landowner originally had to maintain the section of the dike that bordered on his land. This probably worked well for the simple and clear maintenance of the dikes in the moor areas. For sluices to discharge the rain water, however, additional arrangements had to be made. In small communities this was not very difficult, but when the areas that were protected as one unit grew larger, more complicated organizations had to be set up. This marks the beginning of the development of the Water Boards. The development of the Water Boards in our country was stimulated by the fact that local rulers acquired increasingly more power and influence. They even had the possibility to enforce decisions by arms. The population made use of this development by submitting their mutual agreements to the ruler to have them checked on legal aspects, and subsequently a written agreement in the form of a charter would be drawn up. When disagreements arose with regard to the contents of such a charter, they had the possibility to turn to the Chancellery of the count to ask for a ruling. These groups can be regarded as the forerunners of the Water Boards. Since such a charter would basically be drawn up after discussions between all the people involved, the Water Boards can be regarded as one of the oldest democratic institutions in our country. One of the oldest, still existing Water Board Charters is that of Rijnland, which is dated in the year 1255.

Another example of the real democratic attitude in the Netherlands at that time is a ruling issued by count Floris V in 1280, which stated that everybody had to pay for the maintenance of the dikes: "the monastery, the knight, the priest, the common man, everybody alike."

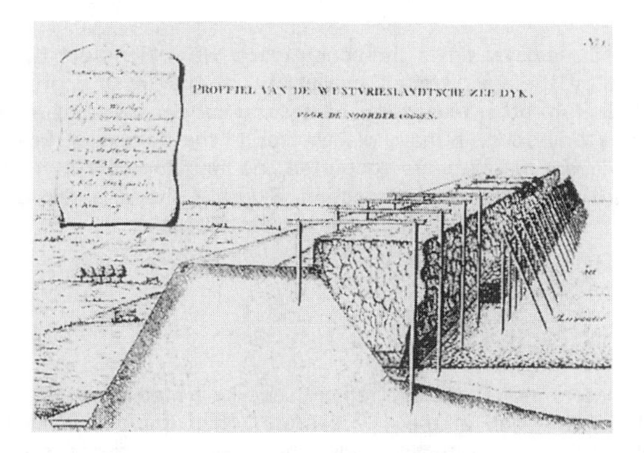

fig. 2. Kelp dike.
The picture shows the construction of the dike. A rectangular shaped mound of kelp, supported by wooden piles. The kelp is moreover partly protected by stones (from Schilstra).

fig. 3. Remnants of a kelp dike.
The wooden support piles came exposed during an excavation.

fig. 4. Remnants of a kelp dike.
This is a part of a still more or less intact kelp dike at the south border of the island Wieringen in the former Zuiderzee (from Lambooy, Getekend land).

In ancient and medieval times the construction work along our coast and rivers was strongly influenced by the very limited availability of the most appropriate construction material: stone. Therefore other reasonable heavy and current-resistant materials were used, such as for instance clay sods. In many places around the Zuiderzee, kelp dikes were built. In these structures layers of kelp were supported and kept together by timber. Small stones dumped in front of it gave additional protection. Figure 2 shows a kelp dike, and Figures 3 and 4 show relics of a kelp dike as found at an excavation site.

RENAISSANCE

Up until the 16th century stone was still rarely used. As a matter of fact this remained valid for Holland until the first half of the 18th century. Then it was mainly used, though not always very effectively, to protect wooden piles against marine borers. Especially in Flanders stone was used to form protective layers on dikes that were attacked by waves. This mainly concerned flat-formed limestone. Sometimes stone was also used to close dike breaches. In that case two stone bunds would be made across the breach, and clay sods would be dumped in between these bunds. Also with the fascine mattresses that were already used at that time, mainly clay sods were used as ballast.

Actually, clay was still the most important material for the closure of the final gaps in the closing dam of the Zuiderzee in 1932. In this case "boulder clay" was used, which is a very heavy clay deposited in the surroundings of the Zuiderzee as an end moraine of the glaciers.

Although not much written evidence of old times is available, there is one very interesting exception: the unfinished manuscripts of Andries Vierlingh. Andries Vierlingh, who lived from 1507 to 1579 and who spent a great part of his working life on planning and building river and coastal structures, decided to spend the last years of his life on summarizing his own experiences for the benefit of his successors. Andries Vierlingh was a well educated patrician and gentleman-farmer who served in many rather high government offices. In this respect his most important position was that of "dike-reeve" (in Dutch *dijkgraaf,*) which is the highest rank in a Water Board. Then and now, the Water Boards in our country are responsible for the protection of the land against the water. Vierlingh was active in the entire field of hydraulic engineering: river works, polders, sea defences and, probably most importantly, the closing of dike breaches that resulted from storm surges. From the text of the manuscript it is clear that Vierlingh intended to publish it. However, he only managed to finish two of the five books he planned and he died half way writing the third. We don't even have the manuscripts in their original form, but only a rather poor copy. In 1895, so more than three centuries later, this copied manuscript emerged at the *Rijksarchief* (State Archives).

In 1920 the Dutch Minister of Education, Culture and Sciences ordered that the finished part of the manuscript should still be published. In 1973 the *Nederlandse Vereniging van Kust- en Oeverwerken* (Netherlands Coastal Works Association) published a facsimile edition of Vierlingh's work, which he named: *Tractaat van Dyckagie* (A discussion on Dikes).

The first book discusses the accretion and reclamation of saltings and tidal marshes. The second book discusses the diking-in of tidal marshes which have grown sufficiently high and the third concerns the maintenance and eventually the repair of dikes. The fourth book was planned to be on river and harbour works and the fifth on inundations.

Although some changes in the techniques did occur in the long period between the writing of the work and its publication in 1920, it is striking to find that these changes were only minor. The changes that occurred in the three quarters of a century after 1920 have been much larger than those in the previous three and a half centuries. However, the latest big leap could never have been made without the heritage built up previously.

From a technical point of view the most important message of Vierlingh is probably the adage "Don't fight the sea with brute force but with soft persuasion" (*niet met fortsigheit maar met soetigheit*), which actually is still characteristic for the coastal defence policy in the Netherlands.

In his books Vierlingh not only discusses technical aspects, but he also describes the social context of the works. He gives a detailed description of the bad habit of the authorities responsible for dikes and coasts, to give high posts to their friends or to men who had become "friends" by handing out valuable presents to the authorities concerned. This often concerned the king himself. In his book Vierlingh addresses king Phillips II and reproaches him for appointing somebody to the office of dike-reeve "who knows just as much about dike construction as a pig knows about eating with a spoon!" For safety's sake he also added the plea "please forgive this grey beard his boldness."

Also the men who actually carried out the work were not the skilled and devoted craftsmen you might expect them to be. These labourers were a bunch of roughnecks who lived rather rough lives and only produced some useful work when they were very strictly supervised. The contractors who had undertaken the work were not much better. Their main aim was to try to cheat both the authorities and their labourers (Vierling, ≈1578). Sometimes you wonder how it was possible at all that they managed to keep the feet of their countrymen dry. A very great difference indeed with the contractors and workers of today!

Another intellectual testator of this time was Simon Stevin (1548-1620), who is also rather well-known outside our country. He was an engineer and a scientist, born in Bruges, Belgium. After working in relatively modest jobs in Antwerp and Bruges he moved to Leiden in 1581, where he registered at the University. Not long after this registration he started publishing on more or less mathematical subjects. During a great part of his life Stevin maintained a close contact with Prince Maurits, first as his teacher, and later probably as his adviser in mathematics and fortress construction. Stevin has been of importance because he partly laid the foundations for the development of science in our country. Perhaps even directly for the field of Hydrodynamics, by his explanation of the forces exerted by water on a submerged body.

Although historians are not too sure about it, he has probably also been a professor in Leiden for some time. In that case he might have been the first professor in hydrodynamics in Europe. Rather important was his demand to be allowed to give his lectures in Dutch instead of in Latin, which was still common use at that time. He made this demand in order to enable common boys - mind you not girls! - who did not understand Latin, to be educated as engineers.

Simon Stevin also had an impact on the development of hydraulic engineering in the Netherlands through his son Hendrik Stevin (1614-1670), who drew up a plan for the enclosure of the Zuiderzee. At that time it was, however, probably not yet possible to execute such a great work.

For the understanding of storm surges also the fixation, in the year 1684, of a reference level for tidal observations in Amsterdam, the AP (*Amsterdamse Peil*), was important. Between 1797 and 1812 this level was transversed to an increasing number of rivers and estuaries. Because this reference level spread all over the country, and, which was even more important, all along the coast, the registration and elaboration of storm surge levels became possible. Moreover, at an early stage coast and dune profiles were already measured by the Water Boards responsible for the coast.

At this point we have reached the more recent history of hydraulic engineering in the Netherlands and it will also be possible to pay some more attention to coastal engineering. First I will discuss the developments directly related to the protection of our country against storm surges, the inevitable dike breaches and their repair. Subsequently I will discuss the developments in knowledge on inlets, estuaries and the lower parts of tidal rivers. Finally, I will return to the "real" coastal engineering subject of the behaviour of sandy coasts.

PROTECTION AGAINST STORM SURGES

Apart from all other concerns our ancestors might have had with the sea, protection against the sea was of primary importance. It literally was a matter of life and death.

From Vierlingh we know how the craftsmanship of dike-construction and repair developed through the centuries. A very great part of this knowledge and experience was stored in the heads of the, by now much more respectful and expert, although still rather rough, workers. For instance, the sinking of the rather large fascine mattresses in tidal waters required a lot of experience. To make sure that the mattress would reach the bottom exactly within the period of slack water, it required a lot of skill to determine the moment at which the actual sinking had to start. The way in which the foreman would establish this moment was almost regarded as magic! Of course this knowledge could only be obtained by long training. It should, however, be realized that this skill was tied to both the place and the circumstances. When the circumstances changed suddenly and strongly a new learning process by trial and error had to start.

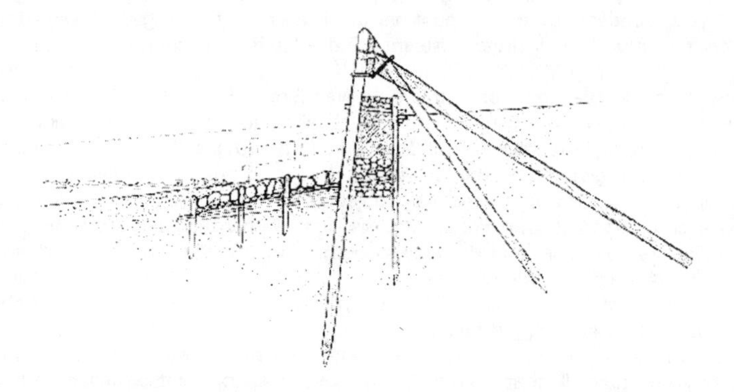

fig. 5. Hondsbossche Seadefence.
Cross section of the dike in the first half of the 19th century. A structure of timber and stones (from Schilstra, 1981).

A good example of the way in which our ancestors protected our country by trial and error is the story of the *Hondsbossche Zeeweering* (Hondsbossche Seadefence) (Schilstra, 1981). Basically this seadefence was nothing more than a profiled and somewhat strengthened dune. The first report of financial support to a coastal defence system in this area dates from 1343. Then there is a report from 1406 which states that one of the two villages behind the defence is lost and flooded by the sea. In 1395 a dike-reeve (*dijkgraaf*) was appointed for the area. In 1421 the dune is again heavily damaged by the, in our country well known, Elisabeth flood in which a lot of people lost their lives. After that the struggle against the sea continued with changing chances, but a more or less adequate solution was not found. In the period from 1506 to 1548 a major effort was made to solve the problem. In front of the dune defended by some toe protection, 22 groynes were constructed, but this also was not the final solution. Fig.5 shows a cross section of the dike constructed with timber and stone. Fig.6 shows the cross section of the present dike. When you compare the old dike of Fig.5, which is roughly indicated with a dotted line in Fig.6, with the present dike, you can only feel sorry for the people that had to live behind it.

Although there was no photography and television to inform the population about disasters in those days, drawings were made to show the people how bad it was. Fig.7 shows a fine example of such a drawing. Just like today it was the intention to raise funds for the victims of the floods.

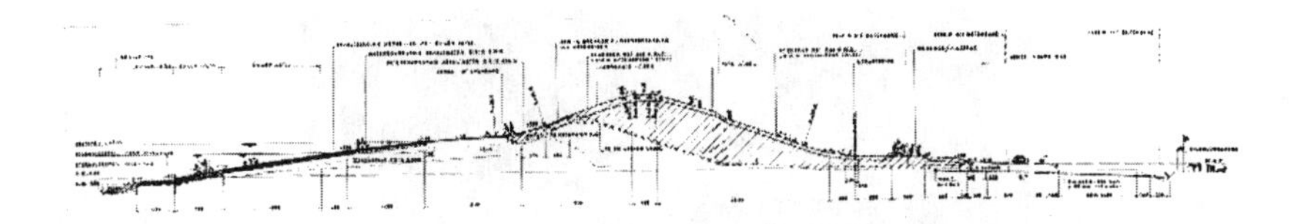

fig. 6. Hondsbossche Seadefence.
Cross section of the present dike. The dotted line indicates roughly the profile of the dike of fig. 5. (from Schilstra, 1981).

In 1792, after another disaster, it was decided to take more or less drastic measures. The sand dike, a profiled dune, was enforced and again groynes made of rock were planned. From then on the situation has been more or less under control, be it that according to today's standards the dike was still not very safe. But in the course of the years the dike was gradually reinforced further so that it stood up against the storm surge of 1953. Since the dike has been maintained at the same place in an eroding coast it now forms a marked protrusion in the coast line as is clearly visible in the aerial photograph of Fig.8.

fig. 7. Print of a flood disaster.
The print showing the disaster and dispair of a flood was printed in 1675. The sub title reads as: **Help us Lord, because we perish.** The print was probably distributed to generate money for the stricken population (from Gedenkboek Noord Hollands Noorderkwartier).

In retrospect, the skills in the field of coastal engineering in the Netherlands until early this century were mainly based on experience. Our craftsmen and engineers learned, indeed in the course of the centuries, how to deal with our coast and how to protect the land behind it that lies below sea level. This is also why until early this century, I made the observation previously, not so very much had changed since Vierlingh's time. Nevertheless in the late 19th and early 20th century some initial studies on the hydrodynamics of coastal and tidal waters were published, but these were mainly qualitative and based on physical feeling.

fig. 8. Hondsbossche Sea defence.
The dike developed into a cape due to the continuous erosion of the adjacent coasts (from Schilstra, 1981).

A sudden break in this more or less gradual development occurred after the storm surge of 1916, which caused heavy damage around the Zuiderzee. This disaster caused a revival of the plan to close off the Zuiderzee from the Waddenzee and, contrary to the time of Stevin, the state of hydraulic engineering had now developed in such a way that it also had become technically possible. The great man in this development was Cornelis Lely (1854-1929), a hydraulic engineer with political interests and skills. During various periods in his life he was a member of parliament and minister of Public Works and Water Management. He was strongly in favour of the enclosure of the Zuiderzee and succeeded to pilot this project through parliament. However, before this project could be realized a great deal of what we would now call basic research, had to be executed. With this research we enter the field of the hydraulics of tidal movement in inlets, because it was realized that closing off the Zuiderzee would have large repercussions with regard to the tidal motion in the Waddenzee. Especially the height of the future storm surges that had to be resisted by the new dam and the coast of Holland and Friesland was of course very important. Although, mainly qualitative, work had already been executed on the basis of physical understanding, the government decided that this problem had to be studied thoroughly.

In 1918 a State Committee was installed to study this problem and Lorentz was asked to be the chairman. Hendrik A. Lorentz (1853-1928) was a theoretical-physicist. In 1878 he was appointed professor in theoretical-physics in Leiden. In 1902 he received the Nobel price for physics together with Zeeman. Lorentz was very bright indeed, but he had no specific experience in the field of hydraulic engineering.

J.Th. Thijsse (1893-1984), a young engineer of the Directorate of the Zuiderzeewerken, was appointed as second secretary. Thijsse once told me that he accompanied Lorentz on a trip on the Waddenzee, because Lorentz should at least once have seen the sea. During the greater part of their trip Lorentz sat down, quietly overlooking the water and mumbling a few times: "It has to be possible to calculate this." And indeed he proved it to be possible. By linearizing the quadratic terms in the tidal equations he developed an amazingly simple and solid method to predict the changes in the tidal movements as result of the construction of the dam. With this method it appeared to be especially possible to predict the changes in the current velocities. One of the calculated results, an increase in the velocities in the inlets to the, now smaller, Waddenzee was first regarded with some suspicion, also by the members of the committee. But when the dam was closed in 1932, the predictions proved to be very good. This mathematical model was the largest operational model at that time (Staats Commissie Lorentz, 1926 and Thijsse, 1922 and 1933). In this model the equations were analytically solved, so it cannot be compared with the numerical models of today.

Also with regard to the design of the structures and the execution of the works the existing experience proved to be insufficient. The final closing-gaps were much larger and had much greater discharges then ever before. This was equally valid for the design of the discharge sluices. Therefore Thijsse was sent to participate in model investigations in Karlsruhe, where a hydraulics laboratory was already in operation at the time. This may well be regarded as the direct introduction of science in hydraulic engineering in our country. Soon after this, in 1927, the Delft Hydraulics Laboratory was established and Thijsse was put in charge.

Back to the closing-off of the Zuiderzee. Although the velocities in the closing gaps of the Zuiderzee dam were higher than they had been until then, they could be closed with basically the same technique as used before. Only the scale changed. Large cranes were used to dump large quantities of qualitatively good clay in the final gap, of which the bottom was protected against scouring by fascine mattresses with a heavy stone cover.

Another leap forward in the development of the technique was made when the dikes of Walcheren, which had been bombed in the last year of World War II, had to be repaired. In the final winter of the war the dikes of Walcheren were bombed at two places in order to silence the heavy guns of the enemy that threatened the access to Antwerp via the Westerschelde. P.Ph. Jansen (1902-1982), an experienced engineer of *Rijkswaterstaat* (the Dutch Department of Public Works), was in charge of the closing of these gaps. For this operation the engineering applied at the closure of the Zuiderzee dam was not adequate. The winter had eroded the gaps to great depth so that vast amounts of material would be required, such as heavy clay and stones. Moreover, the velocities in the gaps during the final phase of the closing would be considerably higher than at the Zuiderzee operation, due to the greater tidal ranges. Therefore Jansen now used an idea he had already thought of during the closing of the Zuiderzee dam, namely the use of big units by which the final gap could be closed practically instantly, before the velocities became too high. At the time the contractors were still rather suspicious of this method. Nevertheless they did the job and the operation was concluded successfully. The epos of this work, of the workers and of the inhabitants of the island, has been described by Anton den Doolaard in his novel *"Het Verjaagde Water"* (The water chased off).

The next event that had a large impact on the engineering knowledge on dikes was the storm surge of February 1953. This storm surge drowned 1834 people, flooded 1340 km^2 of land and more than 750,000 inhabitants were affected. It will be clear that the first concern was to repair the dikes and to reclaim the inundated land. Nevertheless, on 18 February, less than three weeks after the storm surge on 1 February, the Minister of Public Works, Water management and Traffic established the so-called Delta Committee. On 27 February 1954, the Committee published an interim report which included a proposal for the Delta Plan.

To close the gaps in the dikes resulting from the 1953 storm surge, the method used in Walcheren was further developed. The cooperation between engineers in the field and in the laboratory, which already started on Walcheren, was intensified and now also the contractors participated actively in the planning stage of the operations. Fig.9 shows the tests for the manoeuvring of the caissons in the closing gap.

fig. 9. Manoeuvring tests.
In the Delft Hydraulics Laboratory tests for one of the closing operations with a Phoenix caisson are executed (from Dirkzwager, 1977).

When the direct damage of the 1953 storm surge had been repaired, the actual work on the Delta Plan could start. For the further design of the plan a big model of the Delta, which had already been made operational at the Delft Hydraulics Laboratory, could be used. In this model the plans of Joh. van Veen, a Rijkswaterstaat engineer, were studied. Van Veen had come up with the idea to connect the various islands in the South-West Delta of our country by dikes into groups of 3, 4, or 5 islands in order to create bigger units that could be better protected against storm surges. This plan had also been studied immediately after World War II when this same model had been set up in a large army tent. This model enabled the study of the effects of storm surges in the existing situation and in the 3, 4 and 5 islands situation. Because of these studies our engineers were at any rate not completely unprepared when the storm surge actually stroke. Fig.10 shows the rather primitive, but nevertheless reliable way in which the measurements were performed.

fig. 10. Tidal model of the Delta.
All measurements, just as the generation of the tides, were performed by hand and visual observations. Of course not so very accurate but rather reliable (from Dirkzwager, 1977).

It will be clear that the experiences gained during the repair of the damage of the storm surge were now used in the definite closing of the delta. Not only the method of putting big units into place was further refined, also other methods were developed. The onset of these developments is probably hidden in a remark by the engineer in charge of the closing of Schelphoek, the largest gap of 1953, made during the placing and sinking of the final caisson, a Phoenix caisson left over from the Allied landing operations on the French coast at the end of World War II: "Everything works out excellent, but every sinking operation is still an adventure and I don't like adventures." Among the other methods developed was the dumping of rock or concrete blocks of 2.5 tons by cable-way or even by helicopter.

The set-up of the sequence of the various closing operations was such that the Oosterschelde, the by far greatest branch, would be closed last. In this way all experience gained in the previous operations could be used for this final one. It was decided that the Oosterschelde would be closed by dumping concrete blocks by cable-way. This method was applied successfully for the closure of the Grevelingen and the Brouwerhavense Gat.

Then, however, something unforseen happened, although it had been "in the air" for some time. The method of "repressive tolerance" might work to fight the sea, it did not work with regard to the Dutch population. They became increasingly aware of the ecological value of the Oosterschelde as a large and relatively clean salt water tidal basin. They realized that the planned fresh water basin would suffer a lot from the inflow of highly polluted water from the great rivers. Moreover the creation of a fresh water basin for agricultural purposes was no longer essential because of the already existing overproduction of agricultural products. In combination with the desire to save the Oosterschelde on behalf of the oyster and mussel culture, this resulted in a strong social movement that pleaded for consideration of the possibilities to maintain the salt tidal environment. Two options existed. The most drastic option was not to close the Oosterschelde at all, but to only heighten the dikes. This solution was not very popular with the engineers of Rijkswaterstaat and the majority of the population of Zeeland. Another less radical but much more sophisticated solution was to create a storm surge barrage that would allow for a sufficiently large tidal movement into the Oosterschelde, but which would prevent storm surges from entering. The inevitable committee was established (Committee Klaasesz) in August 1973 and in March 1974 they came up with an interesting, although not very realistic solution. Their proposal comprised a porous dam which would allow a sufficient tidal movement in the Oosterschelde under normal tidal conditions. Under storm surge conditions the water levels in the Oosterschelde would be reduced to acceptably low values.

After the presentation of yet another alternative by a group of Dutch contractors, eventually a working group of Rijkswaterstaat, contractors and, last but not least, the Delft Hydraulics Laboratory, developed the eventually accepted plan. This cooperation in the design phase between contractors and engineers of Rijkswaterstaat had by now already become common practice.

Since Rijkswaterstaat realized that the experience obtained in the realization of one project should be used in the next, the so called "learning effect," the traditional practice of public tendering subsequently followed by negotiations "with the knives out" was not used for the large Delta Project. Moreover, the works were so extensive that consortia between contractors were formed to execute the jobs. So Rijkswaterstaat discussed the design and execution of the plan, including the price, with one consortium of contractors.

The final plan for the Oosterschelde comprised a number of large piles with gates that can be closed in case of a storm surge. Due to the decrease of the cross-section the tide in the Oosterschelde was reduced by some 10 percent, which was regarded sufficiently small to maintain the valuable ecological environment.

In the meantime work on the closure according to the original plan had already started. The bottom protection for the closing operation had been constructed and some of the piles for the cable-way were ready. All this had to be removed. Moreover, the engineers faced another problem. As I mentioned previously, the policy had been to lastly close the Oosterschelde, the by far largest branch, in order to be able to use the experience gained in the previous works. This was no longer possible. Now the most difficult job would have to be executed without the help of the learning effect. Although the closing of final gaps by big units was not very much in favour any longer (the dislike of adventures), now an even bigger adventure had to be undertaken: the positioning of the huge piles with equipment that had been especially developed for this purpose. This could only be done thanks to the increasingly developed technology. It is a great compliment to everybody concerned that the job was finished almost flawlessly.

Although the Delta Project was originally based almost exclusively on technical, physical and economical considerations, it also had the effect that coastal engineers realized that the Delta is not only a complex physical system, but also an ecological one. This growing awareness of ecological values was picked up quickly by the coastal engineers. In 1970 H.J. Ferguson, at the time in charge of the Delta Works, established an ecological research group within Rijkswaterstaat of which the status equalled that of the coastal engineering group within the project team of the Delta Works.

Nowadays this environmental awareness is commonly accepted all over the world and has led to the requirement to execute Environmental Impact Assessments for large projects.

In line with these large coastal structures presently the storm surge barrage in the Rotterdam Waterway is being constructed. In this case the design of the structure has been entirely made by a group of contractors after winning a competition between various consortia of contractors. Only after the contract had been awarded, Rijkswaterstaat engineers began to play an active role in the final design.

The protection of our sea dikes also has been subject for further research. J.W. van der Meer carried out research with the aim to develop design criteria for rock slopes and gravel beaches under wave attack (Van der Meer, 1988). Together with the Construction Industry Research and Information Association (CIRIA) and Rijkswaterstaat, the Centre for Civil Engineering Research and Codes (CUR) published manuals on this subject (CUR and CIRIA, 1991 and CUR and Rijkswaterstaat, 1994).

INLETS, ESTUARIES AND TIDAL RIVERS

In our country inlets, or better outlets, from rivers to the sea have always been a very important natural feature. They were and are vital for the approach to the harbours as well as a potential danger for the inland, since apart from being an entrance for ships, it was and is also an entrance for storm surges.

The struggle for the approach of Rotterdam is a good example. Due to the increasing size of the ocean-going ships the approach through the outlet of the river Maas was no longer sufficient. After some attempts to develop new approaches through the Delta, P. Caland (1827-1902), a Rijkswaterstaat engineer, developed an idea, which had originally been put forward by N.S. Cruquius (1678-1754) in 1739, to make a cut through the dunes at Hook of Holland from the Maas to the sea. Caland predicted that the thus created new river outlet would be self-flushing, and would therefore not require maintenance dredging. Although the plan was mildly criticized, in 1862 the government decided to execute it and in 1863 the work started. The new channel was finished in 1868, but proved to be not self-flushing and quite extensive shoaling occurred. In 1877 a committee was installed to find a solution and the result was that the plan for what had by now been called the Nieuwe Waterweg, was not officially accepted. However, other solutions were certainly not attractive because of the long detour the ships would have to make. Since Rotterdam still had confidence in the project and managed to make money available, the new outlet to the sea was maintained. For this maintenance dredging was essential and hopper-dredges with centrifugal pumps were introduced to our country by the English contractors who developed them. Part of the success has certainly been due to Leemans (1841-1929), who was in charge of the maintenance and dredging operations. The moral of this story is that you should not be too afraid to make mistakes. Caland's prediction was proved wrong but due to the decision based on this prediction Rotterdam was enabled to develop into a large harbour and the dredging industry in our country developed.

Of course more work had to be done in order to maintain the main port position of Rotterdam. In the "de Voorst" branch of the Delft Hydraulics Laboratory a large, open air model was built to study the approaches to the new extension of Rotterdam: Europoort. During these investigations a new development became clear. Originally most attention was paid to minimizing the maintenance dredging in the harbour entrance. This changed with the developments in the dredging equipment. Big trailing suction hopper dredges with swell compensated suction pipes, made maintenance dredging cheaper and, even more important, much more reliable. At the same time the size of the ships increased, a development which in its turn increased the manoeuvring problems. I remember Ferguson, at that time in charge of the development of Europoort, telling us, engineers of the Delft Hydraulics Laboratory studying the entrance to Europoort: "Now you concentrate on the nautical problem of how to get the ships in, the dredgers will deal with the maintenance."

Also the influence of the density currents as a result of the interaction between salt and fresh water had to be studied. To this end a large tidal model, operating with salt and fresh water was built in the Delft branch of Delft Hydraulics, as the laboratory was called by now. In this model the effect of the Coriolis force due to the rotation of the earth was reproduced by vertically rotating rods. This device, an idea of H.J. Schoemaker, was a real breakthrough compared to the models on turning tables that had been used for the reproduction of this effect until then.

Gradually numerical models began to take over from "real" hydraulic models. A first step in this development was taken by J.P. Mazure (1899-1990) in his doctoral thesis on the computation of tides and storm surges on tidal rivers. (Mazure, 1937). The elegant and solid method of Lorentz could not be used in this case since it was not possible to deal with upper-water discharge in that method. At the same time and subsequent to Mazure other researchers further developed the computation of tidal motions. Some of the ideas did not prosper, such as, for instance, the Electrical Method developed by Van Veen in 1937. This method was based on the analogy between alternating currents and tidal motion. Although these ideas may not have been suitable for further use, they definitely improved the understanding of the physics of the phenomena.

The basis for modern computational methods in coastal and estuary engineering was laid by J.J. Dronkers (1910-1973), who completed a doctoral thesis in mathematics at Leiden University in 1936. The above mentioned Van Veen recognized the importance of Dronkers' mathematical work in relation to computational hydraulics and so the first academic who lacked an engineering degree entered Rijkswaterstaat. His work was of great importance for many closing operations after the inundation of Walcheren in 1945 and the 1953 storm surge. Dronkers described his experiences in a book titled "Tidal Computations in Rivers and Coastal Waters", which is still the standard work for computational coastal engineering (Dronkers, 1964).

In 1967 one of his pupils, J.J. Leendertse, who worked in the USA for the Rand Corporation at the time, made a further important contribution to the implementation of mathematical models on the computer in his publication: "Aspects of a Computational Model for Long-Period Water Wave Propagation" (Leendertse, 1967).

Nowadays these models have developed to such an extent that it is possible to simulate density currents and silt deposits. However, with some modesty we might refer to a remark of Mazure in his thesis: "The result of a computation is never the solution to a problem. At most it is information that may lead to the solution. The value of this information must be evaluated in relation to validity and the reliability of the computational method" (Mazure, 1937).

For a good understanding of the physical phenomena, field research was also necessary. Partly this research was rather fundamental, such as the study of the transportation of silt by the rivers to the North Sea. This study was performed by adding markers to the silt, which would react differently from the silt when irradiated.

Another part of the field surveys concerned the collecting of essential information on behalf of model studies. Various rather extensive field surveys were executed, in the Netherlands as well as abroad, in order to obtain these data. These studies were performed both by

Rijkswaterstaat and by Delft Hydraulics. In this respect the development of the *Studiedienst* (Research Department) of Rijkswaterstaat should be mentioned. The former chief engineer of the Delft Hydraulics Laboratory, J.B. Schijf (1906-1987), was the driving force behind this development, which was also essential for the development of coastal engineering in our country. In this respect Schijf may well be regarded to be the "father" of modern coastal engineering in the Netherlands.

COASTAL MORPHOLOGY

As mentioned before, the engineering knowledge on how to maintain our sandy coast and dunes was mainly based on experience. However one very interesting fundamental study in the field of coastal morphology has been executed by the previously mentioned Joh.van Veen (1893-1995), entitled: *Onderzoekingen in de Hoofden in verband met de Gesteldheid van de Nederlandse Kust* (Measurements in the Straits of Dover and their relation to the Dutch Coast) (Van Veen, 1936). Since this work was published in Dutch it hardly reached the international engineering community.

With regard to the daily routine, however, we still mainly relied upon experience. Of course this experience was influenced by the more fundamental work of scientific pioneers such as Van Veen, and it was their influence that, at a later stage, enabled us to take the step towards more scientifically based hydraulic and coastal engineering.

Apart from the cut-through at Hook of Holland for the Nieuwe Waterweg, another exception to this more or less gradual development was the construction of the breakwaters at IJmuiden around 1875. They formed the outer harbour to the Noordzeekanaal, which leads to Amsterdam. These breakwaters were constructed by English contractors and it is questionable whether or not our engineers at the time studied the impact of the harbour moles on the adjacent coast.

Also the Delft Hydraulics Laboratory, established in 1927 with Thijsse as its first director, fairly soon started the study of harbours. The first harbour that was studied was the harbour of Zeebrugge in Belgium, in order to investigate the possibility to decrease the siltation in the harbour by guiding a part of the longshore tidal current through the harbour by a gap in the harbour mole. The second project was the harbour of Abidjan, Ivory Coast. In this project the inlet from the ocean into the lagoon at which Abidjan is situated was studied. Shortly after World War II the mouth of the river Volta in Volta and the harbour of Lagos in Nigeria followed. This was the start of an intensive involvement of the laboratory with coastal problems all over the world.

Some research on coastal problems had, however, already been executed before that time. In 1919 a model was built in the garden of the Directorate of the Zuiderzeewerken in Den Haag in order to study the wave uprush against the dikes. This work was performed within the framework of the State Committee Lorentz. Also the wind set-up was investigated during that period.

After the establishment of the Delft Hydraulics Laboratory, a 25 m long wind wave flume was built for the study of wind generated waves. Later this flume was lengthened to 50 m. Partly during World War II this flume was used for further tests. Also the influence of the water depth could be reproduced by changing the scales. The results of these investigations proved to correspond very well with the work of Sverdrup and Munk on the prediction of waves in connection with the landing operations at the end of World War II.

At the Hydraulics Laboratory, fundamental research became more and more important. In the discipline of coastal engineering the work on waves has been very important, as well as the work in the field of describing and measuring the nature and the model of the waves. Work was also done in order to better understand the scale effects in hydraulic models. All this basic research was much stimulated and guided by the man who would later succeed Thijsse as director of the Delft Hydraulics Laboratory, H.J. Schoemaker. Schoemaker gave the Delft Hydraulics Laboratory its scientific base. During this period an attempt was also made to eliminate the scale effects from the morphological models. This work later resulted in the development of a longshore transport formula.

In 1968 the field of coastal engineering got a more formal position at the Delft University of Technology because of the appointment of a professor in coastal engineering. Since then a close and fruitful cooperation between the University and the Delft Hydraulics Laboratory has developed. In 1973 J.A. Battjes was appointed lecturer in wave mechanics. After his appointment as professor in fluid mechanics in 1980, Battjes continued his pioneering work in this field. The two main points in Battjes's research through the years have been the incorporation of random-wave effects in wave models for application in offshore and coastal engineering and in coastal hydrodynamics (Battjes, 1968, 1974, 1975, 1979 and Battjes and Jansen, 1978).

fig. 11. Delta Flume.
In this 100m long, 5m wide and 7m deep flume, waves with a maximum height of 2.5 m can be generated. The wave generator which generates waves of any required spectrum is programmed to suppress waves wich are reflected by the tested structure.

The research by Rijkswaterstaat and the Delft Hydraulics Laboratory into coastal problems gradually intensified during the fifties, but it really boosted after the installation of the TAW, the *Technische Adviescommissie voor de Waterkeringen* (Technical Advisory Committee on Water Defences), subsequent to the 1962 flooding of Tuindorp-Oostzaan, a polder just north of Amsterdam. This committee was installed to advise, which it still does, the minister on all problems concerning water defence systems, which includes all dikes and ancillary works around polders, along canals, rivers and the coast. As expected, the dune coast was a rather important item from the very beginning. Therefore one of first study groups of the Committee concerned the "Erosion of Dunes." This study group investigated the erosion of dunes when attacked by a storm surge. The already known principle of this erosion is that the sand, eroded from the dune during the relatively short duration of the storm surge, stays in the foreshore, practically entirely within the breaker zone. The eroded sand is deposited in a so-called erosion profile and at calm weather after the storm surge it is transported back to the beach and from the beach back to the dune again. So when the volume of a dune is sufficiently large, no harm to the coast is done. A procedure to determine the erosion of dunes on the basis of this principle was formulated in the "*Leidraad voor de beoordeling van de veiligheid van duinen als waterkering*," in short the Guideline on Dune Erosion, which was issued by the TAW in 1984. The basis for this method was laid by J. van de Graaff (Van de Graaff, 1977) and further developed by P. Vellinga (Vellinga, 1986). For the development of the erosion profiles, research was executed in the 300 m long and 7 m deep Delta flume of Delft Hydraulics. Fig.11 gives an impression of this flume in operation.

Of course - since safety was concerned - a lot of rather fundamental research still had to be performed. Eventually a numerical model that runs on a PC was developed, with which the dune erosion under various circumstances could be calculated. However, this model still worked with an assumed, constant water level during the storm surge. Another defect inherent to this model was that fluctuations of the wave height and the bed level of the foreshore during the storm surge were not included. Especially the level of the foreshore is important since a low level foreshore provides more volume between the bed of the foreshore and the erosion profile. In that case more sand is required to form the erosion profile and consequently more dune erosion will occur. Therefore a time dependent model was developed by H.J. Steetzel [Steetzel, 1993], which was as such, however, not yet sufficient. Also the probability of the level and the position of the foreshore, the wave height and the water levels had to be included. So a probabilistic approach was required and applied.

At this moment the basis for the design of our coastal defence system is still that it is designed for a surge level and wave condition occurring with a by the Delta Law prescribed probability. For central Holland this probability is 1:10000. Under these conditions the defence system should not fail. This however, can never be stated with absolute certainty. A lower surge with more severe wave conditions can be moreover just as or even more dangerous. The scientifically right method is to calculate the failure probability under a range of combinations of surge level, wave condition and foreshore situation. The by this method calculated failure probability should have an acceptably low level. Roughly this probability is a factor 10 lower than the before mentioned probability of the desing level of the storm surge. This results for central Holland in a 1:100000 probability for failure of the defence system.

Other phenomena should also be taken into consideration, such as a possible gradient in the longshore transport and the effect of groynes. With regard to the latter the "multi line theory" as developed by W.T. Bakker was used (Bakker, 1968). In this way we have now been provided with a series of numerical models, running on PC's, to predict dune erosion and coastline changes caused by storm surges.

With regard to the defence of our dune coast, in 1990 our government officially decided in favour of the policy to maintain this coast as much as possible with soft measures such as sand suppletion, which is a continuation of the old adage of Vierlingh. It is called "Dynamic Maintenance," which intends to point out that the chosen strategy will not be followed at all costs, but that deviations are possible when required by the circumstances. "Intelligent Maintenance" would probably have been a better name. Annually the Dutch government makes available sixty million Dutch guilders for the execution of this policy, which is described in a report titled Dutch Coastal Policy (Rijkswaterstaat, 1990). The proceedings of the 22nd ICCE (Part IV) in 1990 include a comprehensive report on this subject.

Parallel to these developments within the framework of TAW, Rijkswaterstaat started the Coastal Genesis project, which includes the development of the Dutch coastline, the coast of the delta in the southwest and the Wadden Islands. This study is supposed to provide answers to, among other things, the question how our coast can be maintained at reasonable costs and how it will react to a possible rise of the sea level. Especially the behaviour of the inlets between the islands and the behaviour of the Waddenzee itself call for more fundamental research. This research is now concentrated with the NCK, *Nederlands Centrum voor Kustonderzoek* (Dutch Centre for Coastal Research), a combined effort of the Delft University of Technology, Utrecht University, Delft Hydraulics, Geological Survey and Rijkswaterstaat.

The most innovating aspect introduced by the Coastal Genesis project was the awareness that research has to be approached on a variety of temporal and spatial scales (Fig.12). Moreover, it was realized that these scales are coupled at the points where interactions occur between fluid forces (waves, tides, sea level rise) and bed responses (the physical structure of the bottom). To logically approach this challenge, three sub-projects were defined that distinguish geophysically in temporal and spatial scales. Within these sub-projects important progress was made both by working interdisciplinary and by combining field work, laboratory studies and mathematical-physical modelling. In this way perceptions and research techniques from the fields of engineering, geology and physical and historical geography were combined. In this way reconstructions of the coastline over periods of hundred to thousands year before present could be obtained. This resulted in a much better understanding of the behaviour of our coastline. Simultaneously the results were applied in the implementation of the new coastal defence policy. Although these insights are not completely new, they were rather hypothetical and could not be applied in the context of coastal fringe management prior to this project. A more process-based and quantitative foundation was accomplished by combining several complementary research approaches, aiming at:

- the geological evolution of the Dutch coastal system;
- the exchange processes between coast and tidal inlet systems;
- the sand redistribution along the coastal fringe and the exchange with the sea.

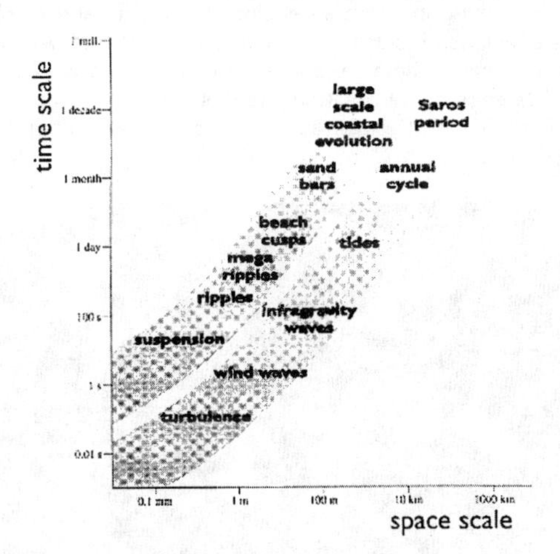

fig. 12. Time and Space Scales.
The relation between the scales in time and space of the various physical phenomena, viz. fluid motion and bed respons along a sandy coast, are shown. (from Stive and de Vriend, 1995)

This work is summarized in three reports about the Morphodynamics of the Netherlands coast on various temporal and spatial scales. (Beets et al, 1992), (Van Rijn, 1994) and (Ribberink and De Vriend, 1994). Of course this work could not have been realized without analysis of the field surveys that have been executed along the Dutch coast (Wijnberg and Terwindt, 1995). It should once more be stated that the Dutch coast is one of the best and longest observed coasts in the world. We have a 100 year long observation of HW and LW lines and a 25 year long observation of the foreshore, beach and dune profiles.

Important lessons were derived from the Coastal Genesis project. Among them is the lesson that the Dutch coast is subject to a structural sand deficit, caused by losses due to dune formation and sea level rise. On the one hand the sand supply from the rivers has decreased after the Roman era, and on the other hand there is a decreased sediment supply from the shelf and shore face towards the coast after the period of younger dune formation. Also a significant movement of sand masses within the coastal fringe has occurred, due to both natural processes and human impact.

In order to be able to understand all this, research was performed on the movement of sand in the boundary layer under influence of waves and currents in the oscillating water tunnel of Delft Hydraulics (Ribberinik and Al-Salem, 1994). In combination with the knowledge on cross-shore currents (Roelvink and Stive, 1989) this made it possible to model the shore face profile evolution (Stive and De Vriend, 1995).

Presently the lessons from the Coastal Genesis project are used in the context of coastal management. They lead to a more sustainable approach of the solution for the erosion problems of the coast and towards substantial savings in maintenance budgets, among other things by better design and location of nourishments and structures.
Coastal Genesis has taught us that local changes in coastline positions have consequences for the coastal evolution on larger scales and longer terms. This coherence in the coastal system is an important aspect in the evaluation of plans for a substantial extension of the coastal zone in the North Sea.

In the defence against the sea the concern about the environment has also awakened. Although safety is still the primary issue, we now realize that we can probably fulfil the safety requirements and at the same time maintain and protect the precious environment of the dune coast. Within the study group "Dune Erosion," now renamed "Sandy Coast" - and in this case there *is* something in a name - ecologists play an active role. Our knowledge of the behaviour of the coast has increased to such an extent that we can give nature a well balanced position in our coastal defence policy. In many cases we can even include natural aspects in the actual defence system. Of course this calls for further research in the field of the ecology of the dunes. Currently this research is performed within the frame work of the study group "Sandy Coast."

fig. 13. Breach in a sand dike.
The picture shows the breach in a sand dike some 10 minutes after the rupture. This test was executed in the fall of 1994 in the "Zwin" an inlet at the North Sea coast at the frontier between Belgium and the Netherlands (from Somers, Meetverslag Zwin proeven, 1995).

Today much of the research is performed by mathematical-physical models. However, sometimes it is still necessary to execute model tests or even full-scale tests. An example is the research performed by the Delft University of Technology and Delft Hydraulics into the development in time of dike breaches. Without further research it was not possible to model the sand transportation by the very high currents in the gap. To investigate this a breach was forced in a sand dike in the Zwin, a tidal gully at the frontier between Belgium and the Netherlands. The development in time of the width and depth of the gap was measured. Fig.13 shows the execution of the test. These full-scale tests were combined with two and three dimensional model tests. The ultimate goal was to develop a model which would be able to predict the development of a breach in a dike, in size as well as in time. This knowledge is required for the development of inundation models for polders.

At this moment research has started into the question whether it is possible to combine "hard" measures such as, for instance, offshore breakwaters, with sand suppletion. Rijkswaterstaat, Delft Hydraulics and consulting engineers and contactors working in this field participate in this research. It is coordinated by the CUR, the Centre for Civil Engineering Research and Codes.

EPILOGUE

Coastal engineering has developed from an ordinary discipline within general hydraulic engineering into an independent specialization. Apart from the fact that it is useless to regret this for mainly nostalgic reasons - it was nice to work on large upper rivers, estuaries, sand and mud coasts as well - it is also not realistic. We simply cannot meet the challenges of our present society without a much improved scientific foundation for our knowledge. Even within the field of Coastal Engineering various specializations have emerged. This is inevitable since it is physically impossible to work at the frontiers of research as well as on armour layers for breakwaters and on the possible response of offshore bars on, for instance, edge waves. There is, however, also a danger attached to this development. At a certain stage in the development of a plan there has to be a man or a woman with a sufficiently broad view, who is able to decide what methods should be used and what the reliability of those methods is in that specific case.

I think it is our mutual responsibility to warrant that these people will also be available in the future. It is most likely that conferences like the ICCE, where "practical" engineers and researchers meet, are invaluable in this respect.

ACKNOWLEDGEMENTS

I would like to thank my friends and colleagues, as listed below, for their invaluable support.

- H.J. Schoemaker, former Director of the Delft Hydraulics Laboratory and Professor Emeritus of Irrigation at the Delft University of Technology, for the extensive information, and his critical and stimulating remarks on all concepts.
- G.J. Borger, Professor of Historical Geography, Faculty of Environmental Sciences at the University of Amsterdam, for the information on the interaction between community, dike building and environmental developments in hydraulic engineering.
- Mrs. S. Jelgersma, formerly of Geological Survey, for the information on dike construction in North Holland in general and kelp dikes in particular.
- J.J. Schilstra, for his information on kelp dikes.
- J. Dronkers of the *Rijks Instituut voor Kust en Zee* (Department for Coast and Sea) of Rijkswaterstaat and part-time professor in the Physics of Coastal Systems at Utrecht University, for his information on the development of mathematical modelling at Rijkswaterstaat.
- J. van de Graaff of the Hydraulic Engineering Group at the Delft University of Technology, for his general assistance and his critical and stimulating remarks.
- M. Stive of Delft Hydraulics and part-time professor of Coastal Morphodynamics at the Delft University of Technology, for his information on the Coastal Genesis project.
- Dronkers, Van de Graaff, and Stive are also members of the Netherlands Centre for Coastal Research.
- Mrs. A.A. Verbeek for editing the manuscript.
- CUR and especially I. van Beek for the assistance and support with the final realization of the paper.

REFERENCES

- Bakker, W.T., 1968. Dynamics of a coast with a groyne system. ICCE 1968.
- Bakker, W.T. and Vriend, H.J.de, 1995. Resonance and morphological stability of tidal basins. Marine Geology, Vol.126 No.1/4.
- Battjes, J.A., 1968. Refraction of water waves. Journ. Waterways Harbors and Coastal Engineering Division, ASCE, 94, WW4, 1968, pp. 437-451.
- Battjes, J.A., 1974. Computation of set-up, longshore currents, run-up and overtopping due to wind-generated waves. Doctoral thesis Delft University of Technology.
- Battjes, J.A., 1975. Modelling of turbulence in the surf zone. Proc. Symp. Modeling techniques, San Francisco, 1975, pp. 1050-1061).
- Battjes, J.A. and Janssen, J.P.F.M., 1978. Energy loss and set-up due to breaking random waves. Proc. 16th ICCE, Hamburg, 1978, pp. 569-587.
- Battjes, J.A., Encounter probability of extreme structural response values based on multiparameter descriptions of the physical environment. Proc. BOSS 1997, Second Int. Conf. on the Behaviour of Offshore Structures, Londen 1979, Vol.3 pp. 609-616
- Beets, D.J., Spek, A.J.F.van der, Valk, L. van der, 1995. Holocene development of the Dutch coast. Rijks Geologische Dienst, report 40 016, Haarlem, The Netherlands
- Broeke, P.W. van den en Londen H. van, 1995. 5000 jaar wonen op veen en klei (in Dutch). Dienst Landinrichting en Beheer, Utrecht.
- Dirkzwager, J.M., 1977. Water, van natuurgebeuren tot dienstbaarheid (in Dutch). Martinus Nijhof, den Haag.
- Dongeren, A.R.van, Vriend, H.J.de, 1994. Geometry prediction for wave-generated bedforms. Coastal Engineernig Vol.22 Nos. 3,4.
- Doolaard, A. den, 1946. Het verjaagde water (in Dutch). Querido Uitgeversmij, Amsterdam.
- Dronkers, J.J., 1964. Tidal computations in rivers and coastal waters. North- Holland Publishing Company, Amsterdam.Wijnberg, K.M.and Terwindt, J.H.J., 1995. Extacting decadal
- Graaff, J.van de, 1977. Dune erosion during a storm surge. Coastal Engineering, Vol.1 no.2.
- Kox, A.J.& Chamalaun, M., redacteuren, 1984. Van Stevin tot Lorentz, Portretten van Nederlandse natuurwetenschappers (in Dutch). Intermediair Biliotheek Amsterdam.
- Leendertse, J.J., 1967. Aspects of a computational model for long-period water wave propagation. Rand Corporation, Santa Monica, Cal., USA.
- Manual on the use of rock in coastal and shoreline engineering, 1991. CUR and CIRIA, CUR report 154, Centre for Civil Engineering and Codes.
- Manual on the use of rock in hydraulic engineering, 1995. CUR and RWS, CUR report 169, Centre for Civil Engineering and Codes.
- Mazure, J.P., 1937. De berekening van getijden en stormvloeden op benedenrivieren (in Dutch). Doctoral thesis Delft University of Technology.
- Meer, J.W.van der, and Pilarczyk, K.W., 1987. Stability of breakwater armour layers-Deterministic and probabilistic design. Delft Hydraulics Communication No 378.
- Meer, J.W.van der, and Pilarczyk, K.W., 1984. Stability of rubble mound slopes under random wave attack. Proc. 20th ICCE Taipei, Taiwan.
- Pilarczyk, K.W.(ed.), 1990 Coastal defences. A.A,Balkema, Rotterdam, 1990.
- Rapport Staatscommissie Lorentz (in Dutch). Staatsdrukkerij, 1926.
- Rapport Deltacommissie (in Dutch). Staatsdrukkerij, 1960.
- Ribberink, J.S. and Al Salem, A.A., 1994. Sediment transport in oscillatory boundary layers of rippled beds and sheet flow. Journal of Geophysical Research, Vol 99, No C6.
- Ribberink, J.S. and Vriend, H.J., 1995. Morphodynamics of a meso-tidal barrier-islands coast. Delft Hydraulics Report H 2129
- Rijn, L.C.van, 1995. Dynamics of the closed coastal systems of Holland. Delft Hydraulics Report H 2129.
- Rijkswaterstaat, 1990. Dutch coastal policy.

- Roelvink, J.A. and Stive, M.J.F., 1989. Bar-generating cross-flow mechanisme on a beach. Journal of Geophysical Research. Vol. 94, No. C4.
- Schalkwijk, W.F., 1947. A contribution to the study of storm surges on the Dutch coast. Staatsdrukkerij.
- Schilstra, J.J., 1981. De Hondsbossche (in Dutch). Hoogheemraadschap Noordhollands Noorderkwartier.
- Steetzel, H.J., 1993. Cross-shore transport during storm surges. Doctoral thesis Delft University of Technology.
- Stive, M.J.F. and Vriend, H.J.de, 1995. Modelling shoreface profile evolution. Marine Geology, Vol.126 No.1/4.
- Thijsse, J.Th., 1922. Berechnung von Gezietenwellen mit beträchtlicher Reibung. Voträge aus dem Gebiet der Hydro- und Aerodynamik, Innsbrück. Springer, Berlin.
- Thijsse, J.Th., 1933. L'influence de la fermeture du zuyderzee sur le régime des marées le long des côtes Néerlandaises. Bulletin de l'Association permanente des congrès de navigation, No.15, 1933.
- Veen, J.van , 1936. Onderzoekingen in de Hoofden, in verband met de gesteldheid der Nederlandse kust (in Dutch). Algmene Landsdrukkerij,
- Veen, J van , 1948, 1962. Dredge drain and reclaim. Martinus Nijhof, Den Haag.
- Vellinga, P., 1986. Beach and dune erosion during storm surges. Doctoral thesis Delft University of Technology.
- Verslag van de stormvloed van 1953 Staatsdrukkerij (in Dutch), 1961
- Vierlingh, A., between 1576 and 1579. Tractaat van dyckagie (in old Dutch) Martinus Nijhof, Den Haag, 1920.
- Visser, P.J., Kraak, A.W., Bakker, W.T., Smit, M.J., Snip, D.W., Steetzel, H.J. and Graaff, J. van de, 1995. Proc. Coastal Dynamics '95 Gdansk, Poland.
- Vriend, H.J.de, Bakker, W.T. and Bilse, D.P., 1994, A morphological behaviour model fot the outer delta of mixed- energy lidal inlets. Coastal Engineering, Vol.23, Nos. 3-4.
- Wang, Z.B., Louters, T. and Vriend. H.J.de, 1995. Morphodynamic modelling for a tidal inlet in the Wadden Sea. Marine Geology, Vol.126 No.1/4.
- Wijnberg, K.M.and Terwindt, J.H.J., 1995. Extacting decadal morphological behaviour from high-resolution, long-term bathymetric surveys along the Holland coast using eigen-function analysis. Marine Geology, Vol.126 No.1/4.

HISTORY OF COASTAL ENGINEERING IN PORTUGAL

F. Vasco Costa[1], F. ASCE, F. Veloso Gomes[2], F. Silveira Ramos[3], Claudino M. Vicente[4]

ABSTRACT: The long coastline and the well-known maritime tradition of Portugal make Coastal Engineering undisputedly an important branch of Portuguese Engineering. Difficult coastal engineering problems arise from frequent simultaneous occurrence of 4-meter tides and 8-meter plus significant wave heights.

In this historical summary, the coast is physically described and many engineering works are mentioned in which artificial blocks of different shapes (tetrapods, cubes and dolosse) with weights of up to 90 tons are currently used. Success and failure of these works are also discussed.

The most recent coastal erosion and rehabilitation problems and the new challenges arising today in Portuguese coastal planning and management, owing to increasing human occupation and activity in water courses and littoral areas, are considered in this paper and some insights offered.

INTRODUCTION

The Portuguese tradition relating to maritime navigation is well known as well as the settlement of Portuguese people along the coast of all continents from the XVth century on. Such tradition was always supported by a strong and innovating technology in ship construction (lighter and better manoeuvring vessels), navigation (introduction of

---

1) Formerly Professor at Technical University of Lisbon. Rua Joaquim A. Aguiar, nº 27, 9º Dt, 1070 Lisboa, Portugal.

2) Professor at Oporto University. Instituto de Hidráulica e Recursos Hídricos, Rua dos Bragas, 4099 Porto Codex, Portugal.

3) Partner, Consulmar. Rua Joaquim A. Aguiar, nº 27, 9º Dt, 1070 Lisboa, Portugal.

4) Senior Research Officer. Laboratório Nacional de Engenharia Civil, Av. do Brasil, 101, 1799 Lisboa Codex, Portugal

the astrolabe and navigation methodologies), and cartography. Thus, for a long time, the development of maritime navigation and technologies was almost exclusively related to the Navy. The survey and recording of the characteristics of marine bottoms, cartography, winds, tides, currents and storms were - and to a great extent still are - accountable to the Navy in Portugal.

In 1836, the course of Hydrographic Engineer and in 1849 the Navy Hydrographic Service were created by royal decree. The Portuguese Navy Hydrographic Institute was established in 1960, and continues that tradition by developing a broad range of activities in the field of Physical Oceanography, including the characterization of sea and swell, tides and sea level monitoring, and physical and dynamical characterization of the ocean and coast of Portugal.

The origins of sea and navigation studies in the city of Oporto go back to 1762, when the "Aula de Náutica" (Nautic Class) was founded. This was replaced in 1779 by the "Aula de Desenho e Debuxo" (Drawing and Design Class) and in 1803 by the "Academia Real da Marinha e Comércio da Cidade do Porto" (Royal and Commerce Academy of the City of Oporto). This, in turn, originated the Polytechnic Academy in 1837, which is considered the first Civil Engineering School in Portugal. In this Academy, the 14th course included three weekly lessons on harbours and lighthouses.

Early this century, Adolpho Loureiro was the author of an extensive and comprehensive publication with more than 3000 pages and 250 figures and drawings, inventorying hydrographic data as well as port designs and structures for all harbours in mainland Portugal and on the Portuguese islands (Loureiro 1904/1910).

From the 1940s on, Portuguese coastal engineering took a great leap forward and extended beyond the national borders. Portuguese coastal engineering technicians started to participate in international meetings: in 1949 they were called upon to organize the congress of the Permanent International Association of Navigation Congresses (PIANC), in 1957 the 7th Congress of the International Association for Hydraulic Research (IAHR), and in 1964 the 9th International Conference on Coastal Engineering (ICCE).

For a number of years, Carlos Abecasis was one of the four international members who appraised and decided on the acceptability of the papers submitted to ICCE; Gervásio Leite presided over the PIANC Commission for the study of acceptance conditions for large ships; Fernando Abecasis was a member of the Maritime Hydraulics Commission of IAHR and one of the four vice-presidents of PIANC; Vasco Costa was an active member of several PIANC commissions, notably those concerned with the study of acceptance criteria for large ships and mooring systems (Abecasis, F. 1990).

Another important milestone in coastal engineering development in Portugal was the creation, in 1949, of the Hydraulics Sector of the National Civil Engineering Laboratory (LNEC), in which research and applied studies have been carried out in the domain of coastal engineering up to now, covering the areas of the hydraulic and structural behaviour of port structures, coastal protection, wave climate, and hydrodynamics and sedimentology of estuaries, lagoons and coast.

The Directorate-General of Ports, a government department recently designated as Directorate-General of Ports, Navigation and Maritime Transports, has developed important activity referring to secondary harbours management, navigation and coastal problems, particularly as concerns control of beach erosion. The main harbour authorities, such as those of Leixões, Aveiro, Lisbon and Sines, have always stood outside the technical, administrative, and financial control of this Directorate-General.

From 1992/93 on, this department's responsibilities as regards coastal management were passed to the Ministry of Environment.

THE PORTUGUESE COAST

The Portuguese coast consists of a continental part located at the western end of Europe and an insular part in the Azores and Madeira archipelagoes (Figure 1). The continental part of the coast comprises two main stretches: one facing west, about 640 km in length, and the other facing south, about 170 km long (Figure 2).

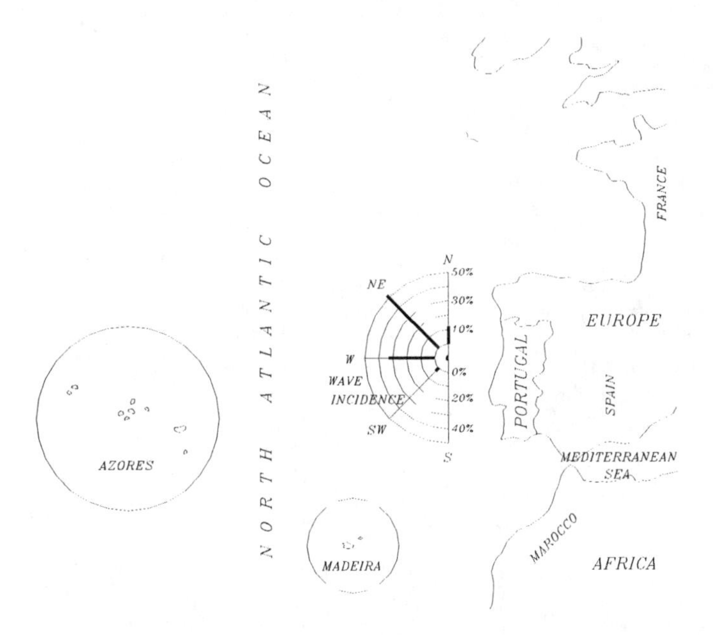

Figure 1. Location Map - Continental part and insular part in the Azores and Madeira archipelagos.

This is mostly a sandy shore, with a large number of siliceous sand beaches and some stretches of rocks and cliffs. The shoreline is cut by the estuaries of a number of rivers of the Iberian Peninsula, of note being the rivers Minho, Douro, Mondego, Tagus, Sado and Guadiana. There are two tidal lagoons of considerable size: the Aveiro lagoon on the western coast, and the Formosa lagoon on the southern coast.

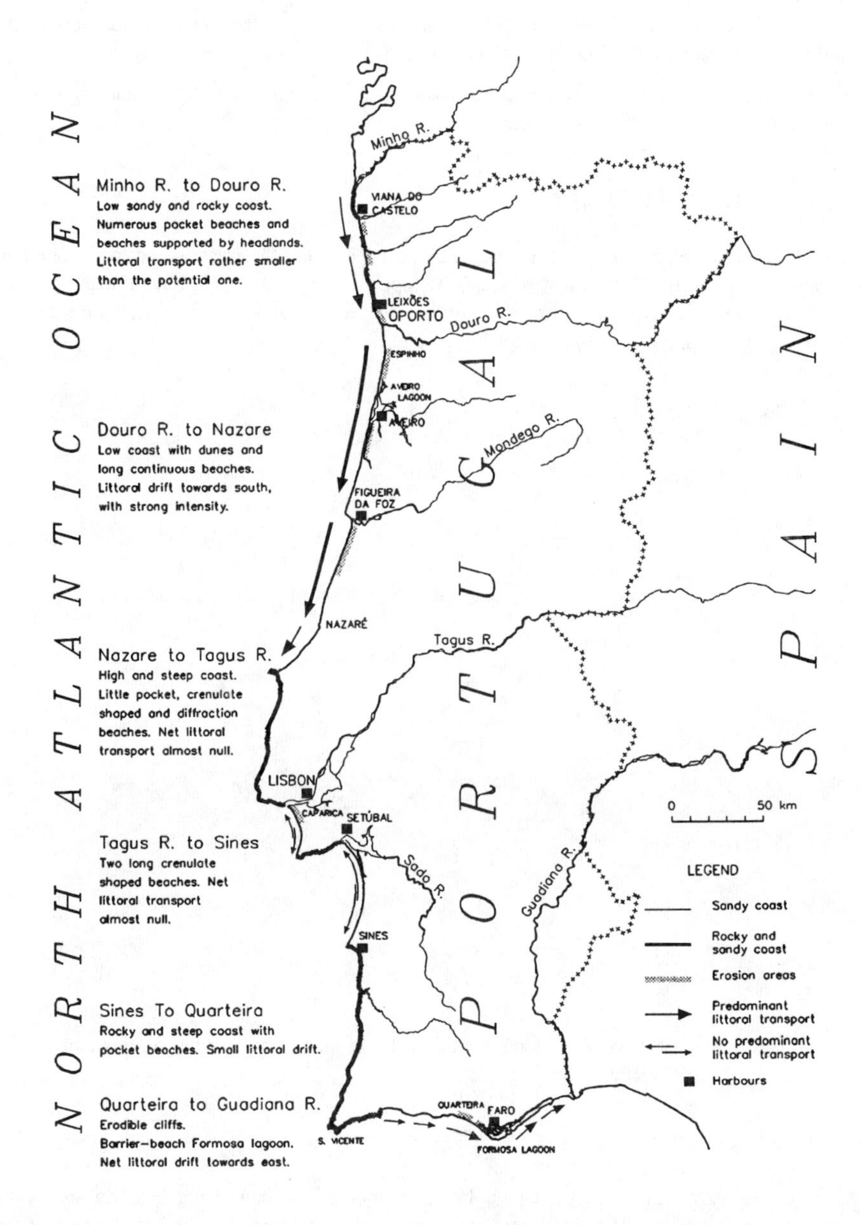

Figure 2. The Portuguese continental coast - Physical characteristics and littoral drift.

On the Portuguese continental coast, tides are of the semi-diurnal type, reaching a range of about 3.8 m for spring tides. Wave characteristics are different in the western coast from those in the southern coast. The western coast has a high-energy wave climate and is struck by storms from the North Atlantic, particularly in the October to March period. Directions between W and NW prevail, with some occurrences from SW. The most frequent periods fall in the 8-12 sec range, and the significant wave heights are in the 1-3 m range. Under storm conditions, however, waves with significant heights above 8 m and periods exceeding 16 sec may occur. The southern coast experiences a milder wave climate, with many calm periods, significant wave heights seldom above 4 m, prevailing directions from SW and SE and periods similar to those in the western coast.

On the western coast, owing to predominance of waves from N-W directions and to the their high energy, net littoral transport usually proceeds to south and reaches values up to about 10^6 m^3/yr. On the southern coast, with a milder wave climate, net littoral drift is eastward and has markedly smaller values (Abecasis, F. 1994). The main sediment sources are the numerous rivers and watercourses. Very often the capacity of these sources has been reduced by the construction of dams and increase of sand extraction for the construction industry.

Geomorphologic conditions and sediment processes make it possible to divide the Portuguese coast in different stretches (Castanho et al 1981), whose characteristics are briefly outlined in Figure 2.

The main harbours in the Portuguese western coast are those of Viana do Castelo, Leixões, Aveiro, Figueira da Foz, Setúbal, Sines, Portimão and Faro. Of these, only Leixões and Sines are open coast ports, whereas the remaining are located inside estuaries or lagoons. Besides, there is a large number of small fishing harbours as well as some marinas.

The archipelagoes of Azores and Madeira are of volcanic nature, with rocky shores where high cliffs predominate. Only a few pocket beaches can be found, usually formed of coarse sand and basalt pebbles.

The wave climate of the Azores islands is somewhat less rough than that of the continental western coast. The Azores islands are in a wave generation zone, and so waves tend to be relatively steep. The most frequent periods fall in the range of 6 to 10 sec. The most frequent directions are between NW and W. During storms, waves tend to turn from NW to SW, their height gradually decreasing. Tides are of the semi-diurnal regular type, with a maximum range of 1.8 m. The main harbour in the archipelago is found in Ponta Delgada.

The wave climate in the Madeira region is mild or moderate; NW-NE directions prevail, except in the southern coast of the island, where SE-SW directions prevail. The local tides are of the semi-diurnal regular type, with a maximum range of 1.3 m. The port of Funchal is located on the southern coast of Madeira.

The collection of data on the Portuguese continental and insular coasts, including estuaries and lagoons, has been the object of a number of field campaigns of topo-hydrographical surveys, tide recording and sedimentological studies, and measurements of waves, sea currents and tides (Abecasis F. et al 1961). These field campaigns have mostly been developed by two institutions: the Navy Hydrographic Institute and the Directorate-General of Ports, Navigation and Maritime Transports.

COASTAL ENGINEERING WORKS

Portuguese Experience at Home

At present the main structures existing on the Portuguese coast consist of groins, revetments, and rubblemound breakwaters.

These structures have been studied, tested, and constructed in Portugal for more than 100 years; many of them were reported or described in several Proceedings of ICCE, PIANC and other international conferences (Abecasis, C. et al 1949, Abecasis, C. et al 1953, Costa et al 1969, Gomes, N. et al 1985, Peixeiro et al 1990, Silva et al 1985).

Most of the works on the Continent use rock of up to 10 tons, easy to obtain from a large number of Portuguese quarries, and concrete blocks with weights varying between 15 and 90 tons. The most frequently used concrete block is the tetrapod, with weights up to 40 tons; dolosse and modified cubes have also been applied. As examples of the use of tetrapods up to 25 tons, we can mention the cases of Viana do Castelo, Póvoa do Varzim, Aveiro, Figueira da Foz, Baleeira, Faro/Olhão and Funchal. Works with heavier blocks were constructed in Leixões (cubes and tetrapods), Sines (dolosse and cubes), Praia da Vitória (tetrapods), Santa Maria (tetrapods and cubes), and Flores (tetrapods and cubes).

For the first time in Portugal, a berm breakwater solution was designed and is now under construction. This structure will protect a land reclamation area for the extension of the Funchal Airport on Madeira island.

The tough natural conditions of the Portuguese coast - rough wave climate, high tidal ranges and intense littoral drift - can easily lead to serious alluvial unbalances at coastal works, making these more complex and very expensive, both for initial investment and subsequent maintenance.

Attempts at sustaining the shoreline, keeping erosion off urban or sensitive zones by trapping littoral drift, led to construction of several groin systems and revetments (Campos et al 1949). The large tidal range and the rough wave climate, particularly on the continental western coast where littoral drift is heavier, led to the use of concrete blocks whenever more than two or three metres depth in low water was reached (Espinho - Figure 3). As a consequence, the majority of works were constructed so as to hardly exceed low water level, using rockfill alone, and are thus not very effective or require considerable maintenance (Vagueira - Figure 4, Cova/Lavos, Caparica, Quarteira).

In addition to these structural works, other actions are worth mentioning, such as dredging of channels and turning basins in most harbours, estuaries and lagoons; the cases of Douro, Aveiro, Figueira da Foz, Lisbon and Setúbal should be singled out owing to the volume or frequency of maintenance dredging required. Of note is also the artificial nourishment of a few beaches, as an indirect result of dredging operations (Rocha, Figueira da Foz, Sesimbra and Setúbal), and three instances in which artificial beach nourishment was the main purpose (Faro, Estoril and Esposende). Nevertheless, most sand dredged for improving navigation conditions was not used for replenishing beaches, but dumped offshore (Lisbon, Leixões, Setúbal) or to reclaim new land areas in some ports (Aveiro, Setúbal), or supplied to the construction industry (Viana do Castelo, Douro, Faro, Aveiro, Figueira da Foz, Alvor).

Figure 3. An element of the groin field protecting the city and beaches of Espinho.

Figure 4. Longitudinal protection at Vagueira - Wave action at high tide.

The majority of these structures and actions was carried out over the past fifty years. With the single exception of Sines, all of them were studied and designed in Portugal by Portuguese professionals. These were public works, launched by the central Administration bodies in charge of the ports sector over those fifty years, especially the Directorate-General of Ports, Navigation and Maritime Transports and the agencies that preceded it, successively under the orientation of engineers Carlos Abecasis, R. Vieira de Campos, M. Fernandes Matias, and F. Munhoz de Oliveira.

Particular mention should be made of LNEC - National Civil Engineering Laboratory - and of its Hydraulics Department, set up in 1949, in which most of those actions were studied and tested under the direction of Fernando Abecasis. LNEC acted as a true "school" for technologies and technicians in coastal engineering in Portugal. A large number of technicians now in the Universities and in the main design offices in this field (Hidrotécnica and Consulmar) came from or were trained in LNEC.

Portuguese Experience Abroad

To fully understand the Portuguese experience in coastal and harbour engineering, the extensive contribution of interventions on the thousands of kilometres of coast of the colonial territories of Angola, Mozambique, Guinea and Cape Verde, up to 1975, must be ackowledged (Abecasis, F. et al 1961, Abecasis, F. et al 1985, Castanho 1971).

Of particular significance were studies and works on the intricate systems of bays, lagoons, sandspits and estuaries in Africa, for which the understanding and study of local natural conditions were decisive, particularly as concerns sediment transport. The success of the measures undertaken in the Lobito and Luanda sandspits, in Angola, which halted the natural trend towards destruction of the sheltered bays where important commercial ports are located, is a good example. The studies, unfortunately discontinued, of the complex hydrodynamics of the large estuaries and lagoons of Mussulo (Angola), Maputo and Beira (Mozambique) are further relevant examples.

Over the last decades the prestige of the Portuguese coastal engineering profession allowed it to cross beyond its European and colonial frontiers. LNEC's studies for establishing or widening new beaches at Flamengo, Botafogo and Copacabana, in Brazil, from 1962 to 1970, and its studies regarding navigational access to the ports of Huelva (Spain), Montevideo (Uruguay) and to lake Maracaibo (Venezuela), already in the seventies and eighties, cleared the path for internationalization of Portuguese coastal engineering, which was continued by the main Portuguese design offices (Abecasis, F. et al 1969, Vera-Cruz 1971).

ENGINEERING PROBLEMS

The main problems with which the Portuguese coastal engineering profession has dealt during the past decades are:

- Stability of maritime structures, especially in the continental west coast and in the Azores;
- Maintenance of access conditions to harbours located inside estuaries and lagoons on the continental coast;

- Coastal erosion, both of the sandy shores of mainland Portugal and of some cliffs in the mainland and islands.

Stability of Maritime Structures

The rough wave climate of the western coast and of the Azores has originated stability problems in breakwaters constructed in those areas. In most instances, heavy storms have caused controlled damage, which is overcome through periodic maintenance and repair. Nevertheless, cases of important destruction have also occurred.

No doubt the most serious accident happened to the Sines breakwater in 1978/79. This accident is well known to the international community, and sea conditions, design and construction are well documented (Dias et al 1994, Edge et al 1982, Ligteringen et al 1994, Magoon et al 1994, Mettam 1976, Oliveira, F. et al 1994).

Other important accidents are: in 1962 the rubblemound breakwater of Praia da Vitória port, Azores, was partially destroyed by a storm with waves of 5 to 6 m of significant height, leading to a new design using concrete blocks; in 1967 the breakwater under construction at Póvoa do Varzim was hit by a storm with significant waves up to 9 m, leading also to changes; in 1979, in Leixões, an accident prompted a study to re-evaluate the incident waves; in 1987, the 25 ton tetrapod armour of a breakwater at the final stage of construction in Santa Maria port, Azores, was destroyed by waves of 6 m of significant height.

In order to detect damages and compare the actual behaviour of the structures with that predicted in the design, the National Laboratory of Civil Engineering, and the Directorate-General of Ports, Navigation and Maritime Transports have for some years undertaken a systematic program of observation of maritime works on the continental coast.

Maintenance of Access Conditions to Harbours

The access to many harbours is through inlets which are subject to sediment inflows from the corresponding rivers and from the adjoining shores. This sediment inflow ranges from moderate values, as in Viana do Castelo and Portimão, to very high volumes, above one million cubic metres per year, as in Aveiro and Figueira da Foz inlets. Heavy sediment inflows give rise to important offshore bars, through which navigation access channels to harbour inlets are established. These access channels usually follow natural ebb channels, which are deepened and maintained by dredging; however, constant sediment transport means access channels are often unstable as to depth and layout, making their maintenance heavy and difficult. To improve navigation conditions and reduce sand inflow, these inlets are often protected with breakwaters (Abecasis, C. 1953).

Examples are: the Douro inlet, where in recent years dredging for access to the estuary amounted to about 300 000 m^3/year; the Aveiro lagoon inlet, giving access to the port of Aveiro, where, despite maintenance dredging of some hundred thousand m^3/year, there are frequent limitations to ships using the port; the Tagus inlet, for which 1×10^6 m^3 maintenance dredging every three years was estimated, to ensure general access with depths at (-16 m CD) (Oliveira, I. 1992); the Sado inlet, for which dredging to deepen the access channel is expected to amount to $14 \times 10^6 m^3$.

Coastal Erosion

With increasing notoriety from the 1950's, the continental coast has been subject to erosion processes (Abecasis, F. et al 1990, Castanho et al 1981, Oliveira, I. et al 1982) . Shoreline recession has affected more seriously stretches of sandy beaches; an exception is the zone of sandy/clayey cliffs east of Quarteira, on the southern coast, which has also receded at a fast rate. The main cause for erosion is the decrease in the amount of sand supplied by the numerous rivers, which constitute the only significant sources of beach nourishment. The sand supply to the coast has been decreasing markedly, mainly due to the construction of dams and to the dredging of sand in estuaries for maintenance of navigation channels or for the construction industry. Two other causes with significant, though more localized, effects are also worth mentioning: breakwaters, constructed to protect ports or improve estuary or lagoon inlets, and maintenance dredging in harbour access channels.

Up to the late 1980s, there was not a perspective of littoral management; shore protection and coastal engineering were understood in terms of the construction of protective structures (groins and revetments) and artificial nourishment of beaches, to control erosion and reduce risks of built-up areas exposure to wave and tidal attack.

Studies of littoral problems for several stretches of the Portuguese coast, ordered by the Directorate-General of Ports in the eighties, constitute very important landmarks for the physiographic characterization and understanding of erosion phenomena. The proposals contained in the General Plans of Works included in these studies pointed to the construction of groin systems in several stretches of the coast.

In the seventies and eighties, the occurrence of several emergency situations near built-up areas led to the construction of several groins and revetments. These groins were already constructed at locations near those indicated in the Plans mentioned; they fulfilled the objectives of localized protection, but accelerated erosion phenomena downdrift.

The main cases of erosion of the Portuguese coast are indicated next (Figure 2):

Minho River to Leixões Port

Erosion has occurred in several zones as a result of the reduction of sediments provided by the Minho and other rivers due to dam construction and sand extraction. Although it is intense, such erosion is not generalized because pocket beaches and beaches between rocky headlands prevail here.

Leixões Port to Douro River

The small pocket beaches that existed along this stretch, 4 km long, suffered strong erosion and completely disappeared following the construction of the breakwaters of Leixões Port, at the end of the XIXth century. The erosion process was triggered by the halting of littoral drift in this stretch, which had a net value of about 150 000 m^3/year southward. The trapping effect of the structures and dredging for maintenance of the access to the port practically cancelled sand nourishment from the north and resulted in the vanishing of the beaches. In the last two years, bypass operations were undertaken for the rehabilitation of those beaches.

Espinho Beaches

The Douro River was the most affected by dam construction and sand extraction (Oliveira, I. et al 1982). It is estimated that the annual volume of sand discharged to the coast has decreased from about one million to less than two hundred thousand cubic metres per year over the last half century. As a result of lack of nourishment, a continuous erosion process has developed south of the river mouth, over some ten kilometres of coastline. The shore in the neighbourhood of Espinho suffered the most serious erosion, leading to the construction of protection works. These works, whose last stage of construction occurred in the eighties, consist of a field of long groins, coupled with a longitudinal protection. Coastline retreat was locally stopped, but erosion proceeded south, prompting the construction of further groins and revetments.

Aveiro lagoon inlet

The inlet to this lagoon, inside which the Aveiro port is located, was trained in the fifties by two breakwaters, about 700 m long, one of which was extended in the eighties (Abecasis C. 1954). It is located on a sandy, straight, shore, where the net littoral drift is southward, and exceeds 1.0×10^6 m^3/year. Littoral drift is disturbed by the breakwaters, originating intense accretion to the north and erosion to the south of the works, affecting the shoreline over about 30 km. The coastline receded by about 150 m, and a groin system had to be built to protect urban areas (Oliveira, I. et al 1982).

Mondego estuary inlet

A similar occurrence took place at the inlet to this river, in which estuary the port of Figueira da Foz is located (Figure 5). The breakwaters constructed here provoked high accretion on Figueira da Foz beach, which lies next to the north. The coastline advanced substantially, up to a maximum of four hundred meters at the north breakwater. South of the inlet a parallel recession of the coast occurred over some ten kilometres, with maximum values of about two hundred metres. To protect threatened infrastructures and urban areas, a few groins had to be constructed (Castanho et al. 1981).

Caparica Beach

Morphological changes on the southern border of the offshore bar at the Tagus inlet caused erosion of the adjacent sandy shoreline. A rockfill groin system was constructed, complemented with a longitudinal revetment.

Quarteira Beach

A rockfill groin system, coupled with a revetment, was also constructed on the southern coast, in the Quarteira area. The beach here was subject to erosion, which was worsened by the construction of breakwaters at the Vilamoura marina. Erosion extends eastward and affects a stretch of erodible cliffs.

Faro Beach

This stretch, at the west end of the Formosa lagoon barrier islands, is undergoing intense erosion. To counter mitigate this, two artificial nourishment operations were carried out in recent years, involving one hundred thousand cubic metres of sand taken from inside the lagoon, (Figure 6).

Figure 5. Mondego estuary inlet, access to Figueira da Foz port.

Figure 6. Wave attack at Faro beach, located on a barrier island limiting Formosa lagoon.

PLANNING AND MANAGEMENT OF THE PORTUGUESE COASTAL ZONE

Over the last decades the management of the Portuguese coastal zone was conducted within a complex juridical and institutional framework, involving different agencies and resulting in a dilution of responsibilities.

Concerning a very limited strip of the shore, the "Public Maritime Domain" (a strip 50 m wide from the line of maximum high water springs), the Directorate-General of Ports carried out an important activity in the fields of physiographic studies and coastal protection.

In 1992/93, jurisdiction over most of the coastline was turned over to the Ministry of Environment, together with enhanced powers in the domains of land use planning and environment. This authority, however, has not yet been provided with financial means, technical staff and internal organization appropriate to its broad responsibilities in the management of coastal areas. The lack of a systematic program for collection of field data, namely topo-hydrographic, and the deficient maintenance of the existing protection structures (groins and revetments) are other serious drawbacks.

In zones markedly exposed to sea attack, as is the case of the Portuguese coast, the existence or the possible construction of protection structures, such as groins and adjoining longitudinal revetments, should not be used as an excuse to allow building in areas of risk. These structures may locally reduce risks of exposure to sea action, but do not eliminate them when time frames identical to those adopted in the design of the buildings are considered (Gomes, F. et al 1994).

Even where groins and revetments exist in front of built-up areas, critical situations have been documented due to the failure of those structures and/or to the inexistence or vanishing of a beach wide enough to avoid wave attack. The progressive extension and structural strengthening of these works may transform the shore into a (conquerable) fortress, change the natural features of the coast, and require a huge financial effort of maintenance.

Over recent decades, in several places subject to erosion, urban areas close to the shore have continued to expand, without a large part of the community, developers, designers and authorities considering that numerous physical aspects of the shoreline had significantly and (with high probability) irretrievably changed, as compared to the past.

NEW CHALLENGES

Coastal engineering must continue to play in Portugal a decisive role in solving problems of coastal protection; however, it must provide for a more thorough integration of objectives and methodologies of land management, taking into account environmental components and natural resources preservation.

The challenges that coastal engineering has to face now in Portugal include: active participation in coastal zone planning studies, particularly as concerns maritime and estuarine developed areas; follow-up of detailed development plans and rehabilitation of ports; participation in the definition of the much needed systematic field campaigns of data collection; study of solutions for artificial by-passing of breakwaters and navigation channels, and artificial nourishment of beaches; search for technical solutions that may minimize the negative aspects of the installation of coastal revetments and groins, or other sand-trapping works; quantification of cost-benefits associated with these actions; quantification of risks and demarcation of risk zones; rehabilitation of dune systems; improvement of prediction capacity through research and modelling; training of new technicians and researchers; improvement of the interdisciplinary dialogue capacity.

Mention should be made, as particular aspects of current research in coastal engineering, of the studies carried out for the installation in the Azores of an experimental plant for harnessing sea wave energy and the construction of artificial reefs for fostering of fishing resources.

In the field of planning and management of the coastal zone, a prevention policy, with all its consequences, must be adopted to avoid new occupation of areas prone to erosion, which would put them under situations of non-quantified risk.

Papers presented by several Portuguese researchers at the Second International Symposium of the European Coastal Zone Association for Science and Technology, LITTORAL'94, characterize present coastal problems and propose a set of management actions for the future.

The urban development of the Portuguese coastal zone, which till recently proceeded without any control and induced several environmental disfunctions, will be governed by a series of regional plans, municipal plans and coastal management plans, all of which require an effort of harmonization, sometimes difficult to achieve.

The coastal zone management plans aim to organize the different uses and activities specific to the coastal border, classifying the beaches and regulating bathing water uses, improving the beaches considered strategic for environmental reasons, directing the development of activities of the coastal zone, preserving and protecting nature. These plans should have been completed in 1995; their preparation, however, was held up by the difficulty of the studies and a relative institutional unsuitability.

Experts in Coastal Engineering must continue to strive for clarification and strengthening of the institutional responsibilities referring to the Portuguese coastal zone, so as to make the appropriate management of uses and resources viable, as well as for the assignment of adequate financial means for these purposes.

REFERENCES

Abecasis, C., 1949. "Le Port de Leixões (Portugal)," *Proceedings 17th International Navigation Congress,* Lisbon, Section II, Communication 4, pp. 173-186.

Abecasis, C., 1953. "Données d'Expérience sur les Côtes du Continent et des Îles Portugaises de l'Atlantique du Nord," *Proceedings 18th International Navigation Congress,* Rome, Section II, S. II - Q.I, pp. 199-216.

Abecasis, C., 1954. "The History of a Tidal Lagoon Inlet and its Improvement - the Case of Aveiro (Portugal)," *Proceedings 5th Coastal Engineering Conference*, Grenoble, pp. 329-363.

Abecasis, F., Matias, M., Carvalho, J., and Vera-Cruz, D., 1961. Paper, *Proceedings 20th International Navigation Congress,* Baltimore, Section II, Subject 5, pp. 149-173.

Abecasis, F., Castanho, J. P., and Matias, M. F., 1969. Paper, *Proceedings 22nd International Navigation Congress,* Paris, Section II, Subject 4, pp. 203-242.

Abecasis, F., Tomé, E. C., Gomes, A.N., Carvalho, J. R., Bicudo, J., Oliveira, I. M., and Elias, N. P., 1977. Paper, *Proceedings 24th International Navigation Congress,* Leningrad, Section II, Subject 2, pp. 107.

Abecasis, F., and Ramos, F. S., 1985. "Aspects of Port Planning, Design and Construction in Developing Countries," *Proceedings 26th International Navigation Congress*, Brussels, Section I, Subject 5, pp. 103-117.

Abecasis, F., 1990. "Aspectos Recentes da Presença da Técnica Costeira e Portuária Portuguesa no Mundo." Academia da Marinha. Lisboa.

Abecasis, F., and Oliveira, I. M., 1990. "Littoral Problems in the Portuguese Coast," *Proceedings 27th International Navigation Congress,* Osaka, Section II, Subject 2, pp. 55-64.

Abecasis, F., 1994. "Geomorphological Characterization of the Portuguese Coast," *Proceedings Littoral'94,* European Coastal Zone Association for Science and Technology (EUROCOAST), pp. 25-29.

Campos, R. V., and Schreck, H., 1949. "Ouvrages de défense pour éviter la destruction de la ville d'Espinho," *Proceedings 17th International Navigation Congress,* Lisbon, Section II, Communication 1, pp. 157-165.

Castanho, J. P., 1971. "Os Problemas da Restinga do Lobito," *3as Jornadas Luso-Brasileiras de Engenharia Civil*, Luanda-Lourenço Marques.

Castanho, J. P., Gomes, N. A., Oliveira, I. B., and Simões, J. P., 1981. "Coastal Erosion Caused by Harbour Works on Portuguese Coast and Corrective Measures," *Proceedings 25th International Navigation Congress*, Edinburgh, Section II - Vol. 5, pp. 877-898.

Costa, F. V., Carvalho, J. R., and Matias, M. F., 1969. "The Probabilistic Approach in the Design of Maritime Structures," *Proceedings 22nd International Navigation Congress,* Paris, Section II, Subject 5, pp. 95-106.

Dias, M. D. C., 1994. "Port of Sines West Breakwater Rehabilitation Strategy Selected," Reconstruction of the West Breakwater at Port Sines, Portugal, A.S.C.E., ISBN 0-7844-0044-X/94, pp.25-34.

Edge, B. L., Baird, W. F., Caldwell, J. M., Fairweather, V., Magoon, O. T., and Treadwell, D. D, 1982. *Failure of the Breakwater at Port Sines, Portugal.* A.S.C.E., ISBN 0-87262-298-3.

Gomes, F. V. and Pinto, F. T., 1994. "Urban Expansion in High Risk Northwest Coastal Areas of Portugal," *Proceedings Littoral'94,* European Coastal Zone Association for Science and Technology (EUROCOAST-Portugal), pp. 981-996.

Gomes, N., Peixeiro, L. C., Vieira, M. C., and Valle, A. S., 1985. "Planification Integrée et Conception des Nouveaux Ports de Pêche Portugais," *Proceedings 26th International Navigation Congress,* Brussels, Section II, Subject 2, pp. 121-142.

Ligteringen, J., Ramos, F. S., Meer, J. W., and Rita, M., 1994. "West Breakwater Sines Definitive Rehabilitation, General Concept," Reconstruction of the West Breakwater at Port Sines, Portugal, A.S.C.E., ISBN 0-7844-0044-X/94, pp. 168-186.

Loureiro, A., 1904/1910. "Os Portos Marítimos de Portugal e Ilhas Adjacentes." Lisboa. Imprensa Nacional.

Magoon, O. T., 1994. "Overview of the Sines West Breakwater Reconstruction," Reconstruction of the West Breakwater at Port Sines, Portugal, A.S.C.E., ISBN 0-7844-0044-X/94, pp.1-2

Mettam, J., 1976. "Design of Main Breakwater at Sines Harbour," *Proceedings 15th Coastal Engineering Conference*, Honolulu, pp. 2499-2518.

Oliveira, F. A. S., and Jong, H., 1994. "West Breakwater Rehabilitation and Construction Supervision," Reconstruction of the West Breakwater at Port Sines, Portugal, A.S.C.E., ISBN 0-7844-0044-X/94, pp. 327-407.

Oliveira, I. B., Valle, A. J. and Miranda, F. C., 1982. "Littoral Problems in the Portuguese West Coast," *Proceedings 18th Coastal Engineering Conference*, Cape Town, pp. 1950--1969.

Oliveira, I. B., 1992. "Port of Lisbon. Improvement of the Access Conditions Through the Tagus Estuary Entrance," *Proceedings 23rd Coastal Engineering Conference*, Venice, pp. 2745-2757.

Peixeiro, L. C., and Queirós, A. M., 1990. "Design and Construction of Berthing Facilities of a Multipurpose General Cargo Terminal," *Proceedings 27th International Navigation Congress,* Osaka, Section II, Subject 3, pp. 113-118.

Silva, M. A. G., 1985. "Coal Port Facilities in Portugal," *Proceedings 26th International Navigation Congress,* Brussels, Section II, Subject 1, pp. 231-236.

Vera-Cruz, D., 1972, "Artificial Nourishment of Copacabana Beach," *Proceedings 13th Coastal Engineering Conference*, Vancouver, pp. 1451-1463.

THE HISTORY OF COASTAL ENGINEERING IN SOUTH AFRICA

edited by D H SWART, P O Box 395, Pretoria, South Africa.

and compiled by

D Bilse, Portnet, Durban; **A E F Heydorn**, private consultant, Stellenbosch; **G Mocke**, CSIR, Stellenbosch; **G K Prestedge** and **M Coetzee**, Watermeyer Prestedge Retief, Cape Town; **W A M Botes**, WAM Technology, Stellenbosch; **F Kapp**, Entech, Stellenbosch; **D Phelp**, **K S Russell**, **J S Schoonees** and **M L Gründlingh**, all CSIR, Stellenbosch.

Paper reviewed for content by :
G de F Retief, WPR, Cape Town; **A R Wijnberg**, A R Wijnberg, Inc., Mossel Bay; **J A Zwamborn**, private consultant, Stellenbosch, with editorial inputs by **J H Spencer-Jones**, Cape Town.

SYNOPSIS

The history and development of South Africa has been strongly linked to the sea over the past 500 years, since the first visits by the early Portuguese navigators in the late 15th and early 16th centuries. The earliest coastal structures were associated with the development of ports and harbours, and the country now has six commercial ports equipped with modern facilities, as well as a number of smaller ports used mainly by the fishing fleets. Some of these have presented particular challenges, because of the littoral drift and its impact on adjacent beaches. Since the 1950s, a series of excellent facilities and amenities, including tidal pools and small craft marinas, have been developed for recreation, while the management of effluent has been managed through the construction of ocean outfalls. On the South African west coast, and extending into Namibia, rich coastal and offshore marine diamond deposits occur. Innovative approaches have been developed to allow the mining of these deposits. In all of these developments, up-to-date, accurate environmental information is a key input to engineering design, and appropriate measuring networks and data management techniques and facilities have been developed to allow the use of this information. Over the last about 50 years coastal engineers and marine and environmental scientists in South Africa have developed methodologies that allows a holistic approach to the technological support of the judicious development of the coastal and marine resources of South Africa

The paper starts with a historical perspective, then deals with harbours, major coastal development schemes, recreational developments, coastal engineering research and education, coastal zone management and it ends with a salute to the ASCE.

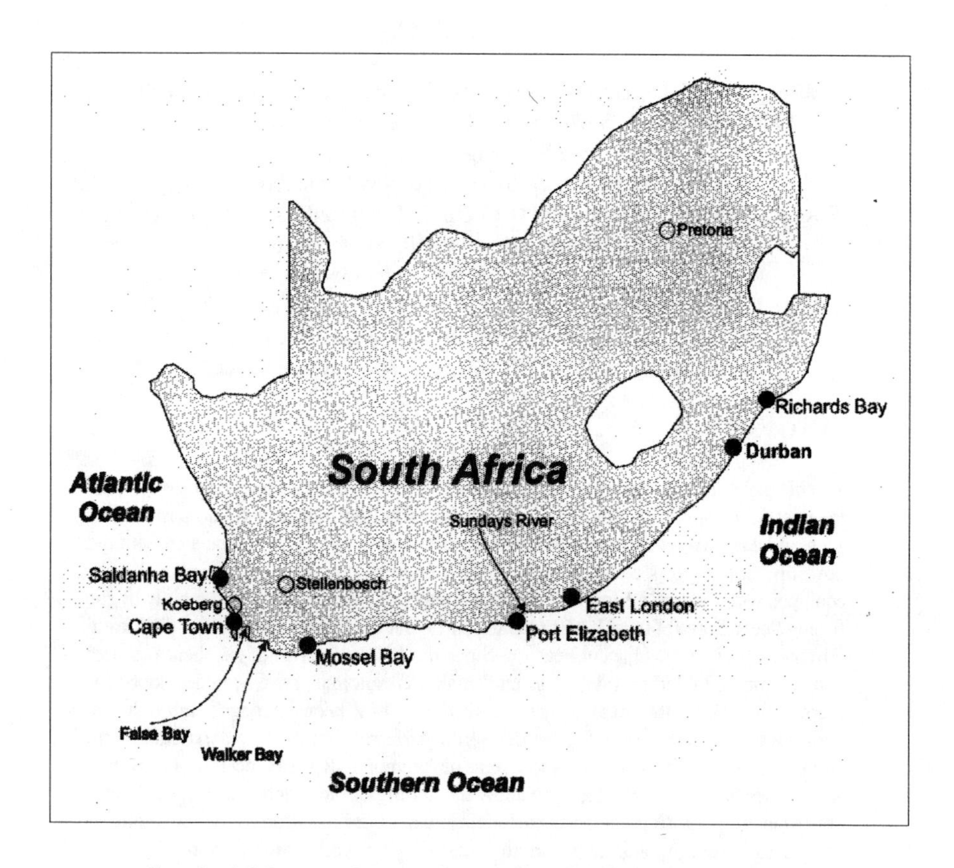

South Africa at the southern tip of Africa is bordered by oceanic watermasses on three sides

SOUTH AFRICA - ON THE EASTERN SEA ROUTES

South Africa's development has been, and still is, strongly linked to the sea. Along its coast runs one of the world's major seaways, between Europe and the east, and since it was first colonised, it has served as a transit point for ships trading on this route. Coastal engineering may, then, be regarded as one of the oldest engineering disciplines in South Africa, encompassing applications from the almost continuous - and still on-going - growth of ports, to the present day impacts resulting from factors such as population growth, tourism and environmental change.

The coastal currents and winds around Africa have played a major role in its maritime exploration. The earliest reference to a circumnavigation of the continent, although unconfirmed, was that by Herodotus, who in about 600 BC wrote that the Pharaoh Necho, then at war with the Syrians and wishing to combine his Mediterranean and Red Sea fleets, caused a fleet of ships manned by Phoenicians to sail from the Erythraean (Red) Sea and return through the Pillars of Hercules (Straits of Gibraltar). The journey is reputed to have taken three years, and winds and currents make such a voyage in square-rigged vessels a possibility.

Other records exist of voyages along the coast of Africa, but the first reliable accounts in South African maritime history are those of the Portuguese navigators of the 15th century. The first of these was Bartholomew Diaz, who in 1487-88 landed at various places along the coast, including Mossel Bay. He had rounded the Cape out of sight (only sighting it on his return trip), but because of his rough passage, named it *The Cape of Storms*. On his return to Portugal, it was renamed *The Cape of Good Hope* , as it promised a sea route to the East Indies, so long a goal of the earlier explorers.

The voyages of Diaz were followed by those of da Gama in 1497 and de Saldanha in 1503, while the coast of southern Africa was first surveyed by Perestrello in 1576. On 18 July 1580, Sir Francis Drake rounded the Cape homeward bound on his circumnavigation of the world, and the opening of this sea route led to extensive trade with the east coast of Africa and the far east. However, it was not until after the formation of the Dutch East India Company in the early 17th century that a serious attempt was made to establish a permanent port of call at the Cape for the replenishment and repair of passing ships.

South Africa has a coastline some 3 000 km long, extending from the desert coast of Namibia to tropical Mozambique. The character of this coast is determined by a number of factors, including the geomorphology of coastal regions; the influence of three major marine water bodies off the southern African continent - the Indian, Atlantic and Southern Oceans; the air-sea interaction generated by these water bodies, which has a decisive influence upon the climate of southern Africa; the land-sea interaction, which shapes beaches, dunes and estuaries; the impact of a multitude of human activities which have transformed large areas of the coast through development and exploitation of resources (Branch and Branch, 1981; Heydorn et al., 1992, Zwamborn and Swart, 1988). Long stretches of sandy beaches are interspersed by rocky sectors. The typical littoral drift directions are "up" the coast from Cape Town along both seaboards, as driven by the strong south westerly swells arriving from 1000s of kilometres away in the Southern Ocean. Net drift rates vary with local conditions and coastal alignment but could be well over 1 million cubic metres per year.

HARBOURS

South Africa's 3 000 km long coastline is rugged and inhospitable, with few natural bays or estuaries. Those locations offering the best protection, together with the opportunity to replenish water and food, gradually emerged as potential harbours.

There are seven commercial harbours along the coast, and each has developed its own character, based on its natural hinterland. These are: Durban (the second largest port in Africa in terms of cargo handled); Richards Bay (190 km north of Durban, built for the export of coal); East London (the only river port); Port Elizabeth (a major mineral ore export harbour); Table Bay, Cape Town (the oldest port); Saldanha Bay (built for the export of iron ore); and lastly Mossel Bay (on the Cape south coast). These ports, which are administered by Portnet, a division of Transnet, currently handle more than 135 million tons of cargo and more than 13 400 vessels with a gross registered tonnage of more than 520 million tons annually, and each operates as an independent business.

Portnet is responsible for the provision and maintenance of the basic infrastructure, marine services and navigational safety of the ports. A dredging service was established in 1988 as an independent service centre within Portnet, to contract to all the ports.

Each port is introduced briefly below, but aspects of the newer, deeper port of Richards Bay are given in more detail to highlight some of the technological innovations which have transformed South African ports.

Numerous other small harbours, most of which were located in river mouths, also developed along the coastline, including Port Alfred, Port St Johns, Port Shepstone, Knysna and Port Nolloth. These were important in the coastal trade and were active up to the 1930s, when improvements in road and rail connections and the continuing increase in the size of freighters caused their decline.

The South African fishing fleet, consisting of vessels of various sizes and operating mainly on the pelagic fishing grounds off the west coast of South Africa, utilises the 13 shallow draft harbours which are controlled by the relevant provincial authorities. More recently, a number of pleasure craft marinas linked to residential developments have also been developed.

Table Bay - 'Tavern of the Seas'

The first harbour to be established in South Africa was in Table Bay, and dates from April 1652, when Jan van Riebeeck landed on its shores to set up a transit station for ships of the Dutch East India Company. The voyage - one way - from Holland to the Indies then averaged about 6 months, with appalling loss of men and ships.

Table Bay is bounded by Table Mountain in the south and a northerly running shoreline to the north, but is otherwise exposed to the prevailing winter north-westerly gales, which penetrate the bay. It therefore offers only limited protection for anchorage, and indeed many ships were wrecked before a permanent harbour was built.

A number of wooden jetties, dating from 1656, were built, but an uncompleted rubble mound breakwater, which was started in 1743 but soon abandoned, is regarded as the first significant coastal structure in South Africa. This was financed by 'taxing' farmers delivering produce to the town, through the use of their wagons to haul stone from the quarry to the breakwater.

Nevertheless, one of the original jetties was to last for a considerable period, though much altered and repaired. In 1798, General J. H. Craig, in command of the occupying British forces, wrote of it as a '... rickety, worm-eaten jetty ... to provide for the landing and embarkation of both passengers and cargo' (Veitch, 1994). By 1809, almost all the tonnage duty imposed on callers using the Table Bay anchorage was being spent on keeping it in a reasonable state of repair.

Subsequently numerous plans were put forward for better harbour facilities, but it was only in 1860 that a 546m long breakwater and excavation of a dock was commenced. The work was supervised by Sir John Coode, with the breakwater constructed on the 'pierre perdue' principle by convicts, and the dock, known as the Alfred Basin, was opened in 1870. In 1905, this was extended with the addition of the Victoria Basin, catering for ships up to 13 500 tons.

After the First World War, trade to South Africa picked up and between 1929 and 1945 the breakwater was extended further and the Duncan Dock developed. This provided an additional 1 830 m of 12 m deep quay wall, which was built mainly of mass blockwork.

Initially a problem of wave resonance arose in the rectangular basin, which was overcome by modifying the basin shape and entrance configuration following pioneering work in this field by Basil Wilson, harbour engineer for Cape Town in the 1950's, but the changing wave pattern deflected off the new harbour walls had the effect of scouring the southern part of the coastline opposite and causing progressive erosion of the beaches there (Veitch, 1994).

Construction of the Ben Schoeman Basin, seaward of the Duncan Dock, for container handling was commenced in 1969, and it became fully operational in 1978.

Cape Town is strategically positioned on the world's sea routes and also serves as a hub port for West African countries. Exports include maize, deciduous fruits and frozen products. The port also contains the largest 'graving dock' in South Africa, the Sturrock dock, with a maximum allowable beam of 41.15 m, determined originally by the dimensions of the battleships during World War 2.

Durban

Several attempts were made to establish a port at Durban, following the sighting by Vasco da Gama of the coast which he named Natal on Christmas Day, 1497, but the single biggest obstacle was a sand bar across the entrance to the bay.

Such bars occur across the entrances of every river on the south-east coast of Africa, and result from northerly littoral drift fed by the silt-laden rivers along that seaboard. In the case of Durban, a further (but lesser) factor is the prevailing onshore wind, which blows directly into the harbour entrance (Pearson, 1995).

The port eventually began to develop in 1824, after the brig *Salisbury* had run over the bar in heavy weather and its captain, James King, charted the bay. The first survey of the entrance was undertaken by Haure in 1831.

In 1850, the first Harbour Engineer, John Milne, initiated construction of two breakwaters designed to force the flow of water down the channel and scour the entrance. The rock for these was quarried on the Bluff and transported to a loading jetty by rail trucks pulled by a team of oxen along timber rails. This was the first railway in South Africa, and a fresh team of oxen was 'encouraged' to swim the channel to the Bluff each week (Bender, 1988). In the event, only one of the breakwaters was partially completed and while a survey showed a depth of 6.4 m in 1854, this decreased progressively thereafter.

In 1877 Sir John Coode published a detailed description of the stability of the tidal entrance. He undertook an 'estimate' of the tidal prism (quoting the tidal volume to be 18 794 737 yd^3) and by comparison with the entrance areas at Dunkirk, Calais, le Havre, Madras and Algiers, suggested that the entrance width should be reduced from 800 feet to 600 feet, with a flow area between 1 300 to 1 400 yd^2, to produce a stable entrance.

Many alternative schemes were presented, but in 1882 a new attempt at the construction of piers was initiated, based on a joint Milne/Coode scheme under Edward Innes, a former pupil of Coode's, and by 1886 these were completed, providing a 150 m channel of 3.2 m average water depth. Dredging vessels were purchased in 1889, and by 1900 the depth had been increased to 5.9 m. Thus far dredging had been confined to the removal of obstructions after gales, but in 1902 systematic dredging of the entrance and of a sand trap at Cave Rock updrift of the breakwater was commenced.

This virtually solved the problem of the bar and by 1906 there was a depth of 10 m in the entrance. Since then, this practice has continued, with updated equipment but essentially following the same principles, with an average of 700 000 m^3/yr having been dredged in the entrance and sand trap.Over the years, the breakwaters have been considerably strengthened and improved, and by 1944-45, the breakwater length had reached 850 m (as currently). Between 1982-91, the block armouring was reinforced with 18 t dolosse.

Aerial view of Cape Town Harbour ca 1992

Aerial view of Richards Bay Harbour entrance ca 1982

Durban is the busiest port in South Africa, based on its proximity to the country's financial and industrial hub in Gauteng, and is a full services general cargo port. The container terminal presently handles in excess of 900 000 TEUs annually, with extensive expansion anticipated.

Port Elizabeth

Algoa Bay, on which Port Elizabeth was founded, was visited by Diaz in 1488, but it provided scant shelter or trade for ships, and was little used until the British settlers landed there in 1820.

The first jetty was commenced in 1837, using the remains of a wreck as the support for its deck. This jetty, measuring 147 m of timber jetty and 63 m of masonry approach, was completed in 1841, but was totally destroyed during a storm in 1843.

In the 1860s, a 400 m long solid breakwater was built at the mouth of the Bakens River, but proved to be a failure due to the build-up of sand to the south of the structure and eventual shoaling in the lee.

The main breakwater was commenced in 1922 and completed in 1933. This was constructed from the top of an existing piled Dom Pedro jetty, which was filled with quarry material only after completion of the breakwater. A further expansion of the port commenced in 1975, including a 335 m curved extension of the breakwater and a 3 km approach channel, 900 m in width and of 14.5 m dredged depth. Windblown sand was reduced by the planting of grass and trees.

Port Elizabeth is close to large farming areas and also has a major fruit terminal. Other commodities handled include motor vehicles and manganese.

East London

East London is the only South African river port and is located on the Buffalo River.

The natural harbour, formed in the mouth of the river, was founded in 1689 by a survey party sent out from Cape Town by Simon van der Stel, but like South Africa's other east coast ports, was hindered by a sand bar blocking the entrance.

The opening of the port dates back to 1893, when dredging had sufficiently deepened the entrance to allow ships of up to 8 000 tons to enter the port.

In 1927, construction commenced of the present turning basin, which was completed by 1937, and subsequently enlarged between 1959-61. Regular dredging is needed to maintain channel depths.

East London is mainly a grain exporter. The port is also renowned as the birth place of the dolos armour unit, discussed later in this paper.

Richards Bay

Richards Bay, which is named after Rear-Admiral Sir Frederick Richards, who landed troops there in 1879 during the Zulu War, was developed as a harbour between 1972-76, principally for the export of coal.

During the early 1900s, the British undertook hydrographic surveys along the coast to locate a possible bunker station for the Royal Navy. In 1902, the bay was surveyed by Cathcart Methven, the Harbour Engineer of Durban, who rated it as having more development potential than Durban, and he proceeded to design a harbour, though this scheme was never realised.

Richards Bay lies in the original salt water lagoon at the mouth of the Mhlathuze River and the harbour development followed one of the first integrated environmental engineering studies at this scale performed in South Africa. In order to preserve the abundant bird and aquatic life of the area, the river and southern half of the lagoon were separated from the port by a berm wall. This wall also provides road and rail access to the seaward side of the port. A new opening was dredged to the sea to maintain the tidal cycle in the lagoon, which was declared a nature reserve. The berm wall thereby reduces the dredging necessary in the port and prevents pollution from spreading to the reserve, see Zwamborn and Cawood (1974).

The entrance channel, dredged on the alignment of an old submerged gorge, has a length of 3.5 km, width 400 m and depth 24 m outside the breakwaters, and length of 6.1 km, width 30 m and depths 24 m to 19 m inside the breakwaters. The breakwaters are rubble mound structures protected by dolos blocks ranging from 5 to 30 t, with the main south breakwater extending 1.3 km offshore and the north breakwater 0.6 km offshore. Dredging of the entrance channel and harbour area involved a total volume of 160 million m^3 of spoil material and a further 35 million m^3 of dredged material used for reclamation.

As at Durban, the northbound littoral drift (up to 800 000 m^3/yr) is intercepted by a sand trap located adjacent to the south breakwater, of which approximately 400 000 m^3 is bypassed by a dredge by being pumped onto the north beach, and the remainder is dumped 5.5 km out to sea. Originally two hopper dredges (with hopper capacity of 4 000 tons each) fitted with bow pipes were needed to maintain the sand trap. By optimisation of the shape of the sand trap, extending the dredger working hours and making use of sidepipe trailing, the trap is now maintained with only one dredge.

The original channel depths were derived for ships with drafts of 17 m. Subsequently, on the basis of field monitoring, physical model testing and simulation of wave-induced ship motions, allowance criteria have been derived for ships with drafts in excess of 17 m, in the range 150 000 dwt to 300 000 dwt.

A fully automated 'ship operational system' has been implemented at the port, by linking an environmental data recording system, including a wave direction buoy recorder, weather station, tide recorder and current meter to PC-based software, to allow automatic data processing, analysis and interpretation.

Richards Bay handles approximately 50% of South Africa's total cargo tonnage, and its coal terminal, which is the largest in the world, has a capacity of 54 million tons. Specialised handling equipment include a woodchip loader and a pneumatic ship unloader while bulk products are stored in silos and sheds whilst open areas can also hold bulk products before being loaded onto the vessels. A new combi terminal allows for the grouping of different cargos destined for one vessel, eliminating the need to move the vessel during loading.

Another recent innovation and a first in South Africa towards making the operation of the port more efficient has been the use of a helicopter for the transfer of pilots to and for visiting ships. A twin engine Bell 212 helicopter is used, which is capable of hovering on one engine and is fitted with emergency floatation gear. It is also fitted with a personnel winch in the doorway, which is essential, as more than half of the ships have decks which are too obstructed for a helicopter to land.

Saldanha Bay

Saldanha Bay, which is named after Antonio de Saldanha (although he never visited it), was visited by the Dutch explorer van Spilbergen in 1601. Despite providing the best natural anchorage along the entire coastline of South Africa, it was not developed as it lacked a source of fresh water.

Nevertheless, it was the scene of major conflict between the Dutch, the French and the British, exploiting the seals and later guano, and one report records that during August 1844, some three hundred ships were concentrated inside the

Bay. Subsequently, it became the centre of the fishing and whaling industries along South Africa's west coast.

The present harbour was developed for the export of iron ore from the Sishen mine (some 860 km distant by rail) and constructed between 1973-76. Construction included an iron ore jetty, protected from wave penetration from the open sea behind a sand breakwater linking Marcus Island with the headland at the entrance to the Bay. The jetty consists of 25 caissons founded at a depth of 28 m. The 2 km long breakwater is of somewhat unique construction, being built as a sand spending beach (of about 20 million m^3 of sand), and with the natural shape of the adjacent North Bay, forms a bight of semi-elliptic shape for the absorption and retention of wave energy. A 3.2 km causeway was constructed to the base of the jetty to gain access to deep water, which carries access roads, services and the ore conveyor.

The port also has a conventional quay of 250 m length and 14 m depth, but plans are being made to extend its length to handle exports from a steel plant, which is currently under construction.

The port has a capacity for ships up to 350 000 dwt.

Mossel Bay

Mossel Bay's history as a port dates back to 1488 when Diaz landed here. Since then, Mossel Bay has served as a fishing port, with only limited commercial cargo activity. Towards the end of the 19th century it became the export port for wool and ostrich feathers. In more recent times the oil industry and two offshore mooring points operate within the port limits. Further development of the oil and gas fields, and the possible export of by-products will have a big influence on the growth of the area as a whole and the port, in particular.

Sand bypassing at harbours

Considerable progress has been made on the study of the sedimentation of ports and the effects of dredging since the design of the port of Richards Bay in the early 1970s. Especially the East Coast ports of Richards Bay, Durban and East London have to contend with sedimentation problems (WPR, 1996).

An evaluation of modern dredging and sand bypassing techniques applicable to South Africa was carried out by Coppoolse and Schoonees (1992), which led to the redesign the sand traps at Richards Bay, Durban and East London.This was followed by studies on the sedimentation of harbour entrance channels and their trapping efficiencies and the characteristics of mud processes.

At Richards Bay the sediment transport regime was investigated and the best position for discharging the bypassed sand onto the north beach was determined (Swart, 1981; CSIR, 1985, 1994a). Subsequently, the effects of pumping dredge spoil on the beach were modelled (CSIR,1993 and Coppoolse *et al*, 1994), while also a number of studies were carried out which provided insight into the sediment transport regime and addressed the feasibility of using a deep-water dump site for storing sand (Laubscher *et al* 1991).

The sedimentary regime at East London is quite different from the other ports, as it is located in a river. The location and layout of a number of proposed new sand traps were determined to intercept the main sources of sediment deposition in the harbour and entrance channel, and the optimum dimensions of these were derived in terms of theoretical sand trapping efficiency and practical dredging efficiency (CSIR, 1994b). Based mainly on the current patterns and the carrying capacity of the offshore currents, a new (closer) dredge-material dump site was proposed. The location and layout was determined conceptually of a spur on the main breakwater with the main purpose of trapping stones which damage the armour units. Sedimentation on the inside of the southern (main) breakwater was also investigated in detail (CSIR,1995a). Most recently, the practical and economical feasibility of reducing the amount of sediment transported towards the Port of East London by means of a structure attached to the main breakwater was investigated. This would significantly reduce the annual amount of maintenance dredging required with substantial direct costs savings.

Durban beach maintenance : The progressive erosion of the beaches of Durban, one of the most popular South African beach resorts, can be traced back to 1850 when harbour entrance channel works interrupted the natural longshore drift of sediment.

Various attempts, including the construction of groynes, were made to control this, but none were successful until 1938 when, following recommendations by a Belgian engineer G. Nijhoff, a scheme incorporating a series of long, low groynes to prevent further erosion in conjunction with sand bypassing (with sand dredged from Cave Rock bight and pumped to the beach) for beach restoration was implemented (Nijhoff, 1935). These were the first of this kind of beach protection measure to be introduced in South Africa.

One of the all-time classical situation analyses re the Durban beach and harbour problem is that by Kinmont (1955), which covers coastal engineering theory, field measurements, practical solutions and also finances.

A comprehensive range of field, laboratory and theoretical investigations (CSIR, 1963, 1967, 1976) provided quantitative insight into the prevailing sea conditions (Barnett, 1982, Campbell *et al*,1985). Some of the principal findings of these studies were that:

- an underwater mound, of crest elevation around -8 m below low-water level at spring tide (LWOST) and located about 1 200 m offshore, offered little protection against beach erosion, although it was effective as storm protection;
- the set of impermeable groynes located in the central beach area induce local erosion and should be replaced by low-level permeable structures; and,
- following an initial beach fill of 600 000 m^3, an annual maintenance pumping volume of at least 100 000 m^3 should be bypassed onto the beaches from harbour dredging.

During the early 1980s the old groynes were demolished and new low level, permeable groynes constructed which, with an annual pumping volume of the order of 275 000 m^3, have been found to maintain the beaches in a stable configuration (CSIR, 1995b). This maintenance volume is less than 50% of the estimated annual littoral drift quantity arriving updrift of the harbour.

Recently, studies have been carried out to examine the detailed feasibility of using embedded jet pumps to create a permanent trap updrift of the harbour that can operate as required even during storms. The discharge would be pumped via an under-channel tunnel to the downdrift beaches to restore the continuity of transport interrupted during most of this century.

MAJOR COASTAL DEVELOPMENT SCHEMES

South Africa's coastline is subjected to a diversity of nearshore processes and physical conditions as a result of the prevailing winds and currents, and the most sensitive ecological areas are the estuaries and shallow inshore areas, which are also the breeding habitats of the greater majority of marine organisms. This has presented particular challenges in coastal developments, which have ranged from beach maintenance to offshore diamond mining support, effluent outfalls, and the development of the Koeberg nuclear power station. Large projects such as this have advanced coastal engineering in South Africa significantly.

Roman Rock lighthouse

Of South Africa's 48 lighthouses, only one, at the isolated Roman Rock off Simon's Town in False Bay, is sea-based (Williams, 1993).

The lighthouse, which was made of cast iron segments bolted together to form a conical tower 14.5 m high, was built between 1857-61, which included almost 1 000 hours of work on the rock itself because of the difficulty of access in all but the best of weathers. As the sea washes over the rock, even at low water in fine weather, the foundation was formed by cutting a deep circular trench in the rock and filling the lower three metres of the structure with concrete. However, even before the foundation had been completely filled, cracks developed in the ground tier of the plates, and soon thereafter, the concrete in the base was excavated and replaced with rubble masonry of granite in Portland cement and a 1.2 m thick granite wall was built around the outside of the base to a height of 2 m.

Despite doubts on the safety and stability of the lighthouse when it was first completed, it still stands today.

Port Elizabeth beachfront development

Like in Durban, the development of the harbour at Port Elizabeth cut off the alongshore transport, which resulted in serious downdrift erosion of the main beaches (situated south of the harbour). A total of 130 m of beach has been lost over the past 30 years, while also enormous accretion updrift of the harbour occurred, resulting in the creation of King's Beach. This area is accreted to its maximum capacity and studies are also underway to examine the bypassing of the alongshore transport to downdrift beaches to reinstate the continuity of transport.

Despite a net alongshore rate of 150 000 m^3/yr, other beaches updrift of the harbour continued to be plagued by periodic exposure of bedrock. Beaches on this rocky section of coast are small pocket beaches with inadequate depth of sediment to accommodate beach profile variations without frequent exposure of bedrock. The use of a submerged panel groyne concept was developed to mimic the natural existing headland conditions as closely as possible (Prestedge, 1992). The structure effectively creates a submerged reef, but is relocated further seaward to create a deeper pocket of sand on the beach. The system permits the overtopping of sand once the beach has accreted to the required levels without deflection of sediment into deeper water offshore. A pier superstructure overhangs the submerged groyne, providing a recreational facility, and the columns provide the support structure for the infill panels, which were designed to be capable of being raised incrementally over time to mitigate any downdrift impacts during the initial period of beach filling. As part of the project a theory was developed to determine the required length of a groyne in a wide breaker zone. The superstructure was constructed onshore and jacked over the piers using the incremental launching technique as used on bridge construction, while access for construction of the piers was obtained from the cantilevered end of the advancing deck construction. The structure, the Shark Rock pier, has become the focal point of development of other tourist facilities in the area.

Offshore diamond mining support

Coastal engineering decision support for diamond mining operations along and offshore of the west coast of southern Africa has been provided for the past three decades. Such advice has primarily related to coastline and sand seawall stability considerations in support of back beach mining excavations (CSIR, 1979, Möller and Swart, 1988).

Through the judicious use of sand overburden emanating from mining operations, reclamation initiatives near Oranjemund, Namibia, have resulted in a seaward coastline advance of up to 300 m, despite the highly dynamic wave environment prevailing.

Such operations have been supported by mathematical modelling of coastline and beach profile evolution (Smith *et al*, 1994), and this is also being applied to the prediction of likely shoreline progradation rates in the event of groyne construction. The mathematical modelling of shoreline progradation and turbid plume behaviour has also served to quantify the environmental implications of discharging diamond mining overburden material in the marine environment. A more recent application relates to the proposed use of a dredge to mine the overburden, where the coastline response to material discharge onto the beach is important for determining seepage rates into the dredge area.

A further application of coastal engineering principles relates to the prediction of deposition areas for diamands themselves in the nearshore and offshore areas, the diamonds having been brought to the coast by major rivers during glacial periods with sea-levels very different from those at present and transported alongshore by the then prevailing wave conditions. Hit rates on the location of diamond deposition areas are better than conventional prospecting methods by about 50 per cent

Cooling water intake basin and outfall
Koeberg Nuclear Power station

Shark Rock Pier, Port Elizabeth

Effluent outfalls

Since 1965, 13 deep sea outfalls have been constructed in South Africa, which discharge industrial and sewage effluents in excess of 600 000 m^3/day (approximately 330 000 and 270 000 m^3/day respectively). A recent example of one of these is a 1.7 km long HDPE pipeline with diffuser, which was concreted into a rock trench, at Green Point, Cape Town, and which replaced the 280 m remnant of a damaged 1700 m HDPE surface outfall installed on the same site (KPR, 1990).

As a result of experience gained in the design, operation and monitoring of outfalls along such a diverse coastline, a multi-disciplinary management strategy has evolved, which includes:

- ❑ technological and scientific expertise i.e. engineering design, monitoring of physical processes, environmental impact assessments and monitoring programmes;
- ❑ water quality guidelines for all coastal beneficial uses;
- ❑ pollution control procedures with appropriate legislation; and
- ❑ integrated environmental management procedures to establish and maintain the perspective between environmental, health, social and economic issues.

The development of the South African water quality management strategy is an ongoing process, taking into account international technological and environmental developments. Its components include:

- **Statutory requirements**: Design criteria with respect to marine water quality, as laid down in the guidelines (DWAF 1990, and currently under revision) are determined for each site. For the determination of the environmental objectives, ambient water quality conditions, the ecology and sediment chemistry are taken into account and extensive field surveys are conducted to determine these. These also form the basis for all future impact assessments.

- **Engineering design, construction and operation**: Baseline surveys are conducted to determine the nearshore processes that may influence the structural design as well as the mixing and transport of the effluents, and include current, wave, wind, stratification and dispersion measurements. Normally the data are acquired over a period of at least 12 months. Detailed geophysical surveys include bathymetry, side-scan sonar, sub-bottom profiling, probing and drilling. The optimisation of the hydraulic system and the determination of the achievable dilutions and transport are calculated using models based on South African conditions.

- **Environmental monitoring and impact assessment**: Extensive monitoring, including water quality and ecological monitoring and using tracers such as rhodamine B, are conducted at each outfall site. The data now available, collected over the past three decades, provides a valuable source of information, not only to assess short, medium and long term impacts, but also as a basis for the improvement of sampling and analysis techniques. It has also improved understanding of the physical mixing processes and enabled development of appropriate modelling techniques.

Outfalls remain a viable option for effluent disposal in South Africa without appearing to cause environmental degradation at current volumes, but this may change as volumes increase. Their use will also be negated when water reclamation becomes an economically viable option.

Koeberg nuclear power station

The Koeberg nuclear power station was constructed in a shallow embayment on an open stretch of coastline some 35 km north of Cape Town, on a site exposed to an area of almost the highest wave energy on the country's entire coastline. Seawater at the rate of 80 m^3/sec is used for cooling the two 1 000 MW reactors.

During the initial design studies in the early 1970s, mathematical models were developed to predict dispersion of the warm water plume due to cooling water returning to the sea, and also models to simulate the coastal sediment dynamics and rate of sedimentation of the intake basin (Fleming, 1977, Prestedge and Hutchinson, 1976). The harbour-type intake basin was designed as a stilling basin in which suspended sediment could settle, enabling sediment-free water to be abstracted for cooling.

Another first associated with this project was the use of sand bitumen in the core of the rubble mound breakwater to prevent penetration of warm water between the outfall south of the basin and the intakes within the basin, radar images were also used to record wave directions and an outfall channel structure was developed to dissipate wave energy with minimal head loss to the outfall flow (Witthaus *et al* , 1982).

RECREATIONAL DEVELOPMENTS

Background

With South Africa's roots growing from the sea and with a climate conducive to outdoor activities, it could be expected that the population should also direct its recreational activities to the sea, including swimming, boating and angling. Additionally, the ecological and physical diversity of the coastline gives it great visual appeal, making it attractive for residential development and tourism.

The coast has only limited protected areas for recreation, but demands for additional facilities have only become apparent with the rapidly increasing population in the latter part of this century.

Tidal pools and bathing facilities

The first developments along the South African coast directed at providing recreational facilities were tidal pools, which were constructed mainly on rocky shores where swimming from the beach was difficult. Some 80 tidal pools exist along the coast, of which about 50 are in the Western and Eastern Cape provinces and 30 in KwaZulu-Natal. These pools vary in size from several tens of square metres to over 30 000 m^2.

These started initially as merely closed openings between rocks, but proceeded to the engineered design of pools to promote water circulation and limit sand ingress.

In 1980 the CSIR launched a data collection project on tidal swimming pools along the coast. This study indicated a number of factors which should be considered in the design of a recreational tidal pool, including needs which must be provided; siting, to obtain proper founding and limit sand entry; shape, to ensure water replenishment and quality; safety; and maintenance (Scholtz and Bosman, 1981)

Notwithstanding this study, some pools designed subsequently were found to be lacking in terms of water replenishment, as well as safety and maintenance, and recent engineering has concentrated on the upgrading of existing pools.

Closely related to swimming facilities is the safety of bathers, and South Africa has played a leading role in developing shark deterrent measures. The Natal Sharks Board has undertaken extensive research in this field, and besides maintaining shark nets along the KwaZulu-Natal coast, has recently developed an electronic shark deterrent, which as yet has to find routine application.

Small craft harbours and marinas

Pleasure boating in South Africa was initially limited to beach launching, followed by the construction of boat ramps for the launching of power boats and smaller sailing craft. These developed fairly rapidly and ski-boat clubs now far outnumber those of other forms of boating (with a large number based on inland dams) even though formalised harbour developments for pleasure craft lagged world developments, taking off only recently.

The history of small craft harbours in South Africa dates back to the establishment of the Royal Cape Yacht Club in Cape Town in 1905, and more recently the False Bay Yacht Club in Simon's Town in 1958 and the Point Yacht Club in Durban in 1978. These clubs are located in established harbours. Subsequently the Algoa Yacht Club (Port Elizabeth), East London Yacht Club, Hout Bay Yacht Club and Zululand Yacht Club (Richards Bay) were established, following the same pattern. Initially, moorings were generally trot or swing moorings, depending on the area available and the number of boats to be moored. In the late 1970s the Royal Cape Yacht Club started developing walk-on moorings, which were anchored on blocks and chains, and some of the other clubs, pressed for water area, followed soon afterwards. In 1989, the Royal Cape Yacht Club developed a separate basin in which all moorings were anchored in piles, and subsequent developments have followed this pattern.

The first residential marina in South Africa was developed in the early 1960s at St Francis Bay. This was followed by a private marina development, Port Owen, near the Berg River mouth at Velddrift, which forms part of a large residential development incorporating both waterfront living units as well as a public area with walk-on moorings and facilities. This marina had sea access provided through an artificial river mouth constructed previously for commercial fishing activities.

Since the late 1980s, the demand for residential development, coupled with seafront living and access to boating facilities, has increased, with developments at Club Mykonos in Saldanha Bay, Marina Martinique in Aston Bay and Royal Alfred Marina in Port Alfred. Marina Martinique was the first marina with a raised water level and a lock for sea-going craft. Recent developments include a small craft harbour development at Granger Bay, Cape Town, with moorings for some 90 boats, entailing the upgrading of an existing small craft harbour, and a marina project at Port St Francis near Port Elizabeth, using Core-Loc units as armour, and which includes mooring for 200 fishing and recreational boats as well as a 1 ha reclaimed island.

CEASAR Field Exercise, Walker Bay
SOUTH AFRICA

Coastal development project at Milnerton near Cape Town

Waterfront developments

Another development of the 1990s in South Africa is the revitalisation of old harbours to provide waterfront recreation and living.

The first of these is the Victoria and Alfred Waterfront in Cape Town, which is being developed in phases around the Victoria and Alfred Basins, with facilities including shops, hotels, restaurants and an aquarium. A marina is under construction and residential units will follow.The basins themselves remain a working harbour, while also the historic character of the original harbour is being retained. This, with currently over 16 million visitors annually, and planned redevelopments at other harbours, for example at Durban, also serve as an important base for tourism development.

The development of pleasure facilities along the South African coast is now at an exciting point. The increasing affluence and growing tourist industry will put increased pressure on coastal developments, and due to the more suitable sites having already been developed, more ingenuity in design and construction methods will result.

COASTAL ENGINEERING RESEARCH AND EDUCATION

Much of the coastline, however, is exposed to the highly energetic south-westerly swells originating in the Southern Ocean, and this wave action controls the orientation and configuration of prominent coastal features such as the crenulate-shaped coastal embayments. Another effect is the interaction of these swells with the Agulhas current, which gives rise to the 'freak' waves that have sunk or structurally damaged many ships.

The earliest measurements associated with the ocean around South Africa were the tidal measurements made by the astronomers Thomas Maclear and John Herschel in False Bay and Table Bay beginning in 1834, but oceanography as a science has only developed significantly since the late 1950s.

Coastal dynamics

Field measurement exercises aimed at developing a better understanding of coastal dynamics have been carried out at a number of sites around the coast, including a comprehensive series of measurements at two coastal embayments on the South African southern coast.

A common theme in such studies has been the quantification of processes in the highly dynamic nearshore environment, particularly with regard to the complex surf zone region, and they have constituted a fundamental element in the design of harbour and beach development schemes along the coast (Zwamborn, 1970, Zwamborn *et al*, 1972).

Some of the earliest detailed field and laboratory experiments carried out in South Africa were directed at elucidating features of the complex circulation and mixing regime in the surf zone. In a pioneering set of observations, fluorescent dye was used to identify dominant longshore current and rip-current features at coastal sites near Durban (Harris, 1961). In this, and a later study directed at quantifying turbulent mixing processes in the surf zone (Harris *et al*, 1964), the primordial importance of wave breaker magnitudes, directions and breaker induced turbulence was apparent.

Nearshore circulation patterns and corresponding morphological response were investigated through a series of 2D and 3D physical model studies (Jordaan, 1961). These experiments reproduced primary and secondary circulation features, including rip-currents, as well as sand bar formation processes. Further field observations and laboratory experiments quantified characteristic nearshore circulation patterns and dimensions, with special reference to rip-current spacings, pulsation periods and flow volumes (Harris, 1969).

The Sundays River experiment carried out in 1983 was the first of a number of major field exercises directed at quantifying coastal dynamical processes (Swart, 1984). The study site, a south-facing crenulate-shaped beach on the south eastern coast near Port Elizabeth, is directly exposed to dominant southerly wave directions. Measurements included offshore and surf zone waves (Waverider and gauges), currents (dye and drogues), sediment characteristics and bed levels. Sediment suspension concentrations and transport rates were estimated using sampling bottles and bamboo poles. As the first in a series, the exercise proved a useful trial for instrumentation and experimental procedures for later exercises.

Subsequently the CAESAR series of field experiments were carried out at Walker Bay, an extensive sandy bay on the southern coast near Cape Town and likewise exposed to dominant southerly waves, and at Strand, a mildly sloping beach located in False Bay (Swart, 1988; Schoonees, 1991; Coppoolse and Schoonees, 1992).

Amongst others, in connection with the Walker Bay studies, the magnitudes of breaking wave heights as a relative function of local water depth (H/h) as they propagate across the surf zone were investigated (Nelson and Gonsalves,1992). They found that a linear relationship for H/h is not evident even for the relatively mildly sloping study beach, with H/h progressively reducing below the straight line relationship with increasing distance inshore of the breaker line.

The underlying mechanism appears to be primarily attributable to variable time dependent effective water depths as a function of infra-gravity motions.

Theoretical analyses were used to deepen the understanding of the nature of trends in suspended sediment concentrations highlighting the contribution to the above of wave breaker turbulence (Mocke and Smith, 1992; Smith and Mocke, 1993).

An inverse modelling investigation on prototype scale Delta Flume '93 measurements, underlined the importance of wave breaking related transition zone effects (Mocke *et al*, 1994). Further comparisons with these measurements demonstrated the utility of a turbulence based model for predicting cross-shore sediment transport fluxes and thus beach profile response to wave attack (Smith *et al*, 1995).

In an exercise directed at obtaining improved quantitative information on surf zone hydrodynamic processes, including wave roller geometry and aeration scales, a non-intrusive video arrangement has been used to image waves breaking on a laboratory beach (Govender *et al*, 1996). Using sub-pixel resolution techniques, detailed internal properties such as velocity vector and vorticity flow fields are determined using particle image velocimetry (PIV). Water levels and roller properties are imaged using a process that analyses video greyscale intensities.

Recognising the importance of properly quantifying wave kinematics, particularly for the computation of sediment transport in the nearshore zone, a higher order wave theory has been derived (Swart and Loubser, 1978, Swart, 1978). The variable-order cosinusodal or Vocoidal wave theory was shown to be comparable with Dean's stream function theory in representing measured orbital velocity distributions (Swart and Loubser,1979). Subsequently, this work was extended to account for the effects of a sloping bed (Swart and Crowley, 1988), and has proved to be versatile in predicting wave orbital velocities and moments (Nairn, 1991, Soulsby *et al*, 1993).

Pioneering work on the prediction of on-offshore sediment movement has formed the basis over the years for many innovations and improvements to the original model (Swart,1974), and was regularly used in coastal design.In a comprehensive study undertaken regarding longshore transport, field data were collected from the literature and for typical South African conditions (Schoonees and Theron, 1993). In addition a new longshore transport predictor was derived in terms of the applied wave power and extensively calibrated (Schoonees, 1996). This formulation also proved to be promising for the prediction of the longshore transport of coarse material. Of particular importance in this regard was an investigation into the incipient motion criterion for coarse material. This followed on earlier work (Lenhoff, 1982) in which a dimensional analysis of existing data was performed and an empirical formula for the onset of movement for sand-sized material was derived.

A characteristic feature of much of the South African littoral active zone is coastal dunes, and a predictive technique for aeolian sediment transport based on local wind statistics has been formulated (Swart, 1986a, 1986b). The technique essentially entails using 18 available aeolian transport predictors available in the scientific literature and on the basis of statistical analysis selecting the most likely wind-driven sand transport rate. This was later adapted to incorporate aerodynamic interactions with plant communities in the dune field (Barwell and Burns, 1989).

The scientific basis for coastal zone management

It is clear of course that legislation, even if formulated in an optimal manner, cannot be effective if it is not based on sound scientific information and if such information is not translated into guidelines which are understandable to the layman. In this field South Africa has perhaps been more successful in the creation of cohesive legislation. The reason lies in the existence of a number of statutory research institutions and the effective co-ordination of their work with that of universities and non-statutory research bodies. Space does not allow detailed mention of this network of scientific activity and a broad brush sketch will have to suffice.

Marine and coastal research in South Africa is celebrating its 100-year centenary in 1996. In essence this started with government sponsored fisheries research of which Dr J.D. Gilchrist is considered to be the founder. Initial work was focused on the study of individual species of organisms of commercial importance. This approach was gradually broadened by the governmental Sea Fisheries Research Institute into studies of the physical environment and interactions between various populations of exploitable organisms within the overall food-network of the sea.

In the 1970's and 1980's inputs from Universities, especially the University of Cape Town, brought substantial sophistication to this work, including energy-flow studies, sea-atmosphere and land-sea interactions. The results of this collaborative and cohesive research initiative over a time span of decades provided a thorough understanding of the natural processes governing the various sectors of the South African coast. The CSIR and the Foundation for Research Development (FRD) played a key role in the co-ordination and funding of much of this work. Important components included:

- studies of rocky shores of the South African coastline (Universities of Cape Town, Transkei and the Oceanographic Research Institute, Durban);
- studies of sandy shores of the South African coastline and sediment dynamics thereof (CSIR, University of Cape Town and University of Port Elizabeth);
- project-orientated coastal engineering studies,(e.g. for the development of Richards Bay harbour in collaboration with biological, chemical and physical science (CSIR);
- marine pollution and outfall studies (Sea Fisheries Research Institute, CSIR and Universities);
- compilation of synopses of available information on South African estuaries and recommendations for their management (Oceanographic Research Institute, Durban and CSIR);
- development of Integrated Environmental Management (IEM) and Environmental Impact Assessment (EIA) procedures (University of Cape Town and CSIR);
- aquacultural research (University of Rhodes and CSIR);
- taxonomy of marine estuarine and freshwater fish (JLB Smith Institute of Ichthyology, Museums, Oceanographic Research Institute, Durban and Universities);
- research into the recreational fisheries along the South African coast (Oceanographic Research Institute, Durban, Sea Fisheries Research Institute, Universities and Museums);
- research into the commercial fisheries along the South African coast (Sea Fisheries Research Institute in collaboration with Universities);
- physical oceanographic and climatological studies (Sea Fisheries Research Institute, University of Cape Town and CSIR);
- marine geological studies, *inter alia* relevant to exploration of marine gas and oil fields (Geological Survey, CSIR, Universities);
- planktological and microbiological studies (Sea Fisheries Research Institute, Universities, CSIR);
- marine law and sociological studies in the coastal zone (Universities of Cape Town and Natal).

This list is by no means complete but is gives an indication of the scope and collaboration in the marine and coastal research fields which has been conducted in South Africa.

Energy generation

South Africa has amongst the highest wave energy coastlines in the world, but of the various methods of extracting energy from the ocean, only current energy and to a lesser extent wave energy appear to have potential, and investigations have been made of these (Retief *et al*, 1984). However, at present at least, such schemes are not cost-effective.

The dolos

The dolos armour unit is perhaps the single most important contribution to coastal engineering that has been made in South Africa.

The dolos (an Afrikaans term referring to the knucklebone of the sheep or goat) was designed by Eric Merrifield, Harbour Engineer of East London, and his approach was to find a unit with a high void-to-solid ratio, which interlocked well and which was easy to manufacture. The resultant design takes the form of 'H' with one arm twisted through 90^0.

Merrifield never patented the design, believing it to belong to all mankind, and subsequently further research has been undertaken on these units, which are to be found in use worldwide (Bunt, 1977).

Education

Coastal engineering is only cursorily included in undergraduate engineering courses at the Universities of Cape Town and Stellenbosch, but both these Universities provide post-graduate courses in Coastal Engineering. Coastal engineering teaching in South Africa was initiated by prof Andre Coetsee at the University of Stellenbosch. Marine Sciences is taught at the Universities of Cape Town, Port Elizabeth and Durban and also at Rhodes University in Grahamstown. The University of Cape Town also has strong departments in Marine Law and Environmental Management and Assessment.

Although plagued by financial strictures, the South African Network for Coastal and Ocean Research (SANCOR), a joint venture of the South African Department of Environmental Affairs and the Foundation for Research Development (FRD), provides a discussion forum for some 60 South African organisations involved in coastal development, and provides coordination of the marine science and engineering research efforts in South Africa.

The funding situation

Some mention must be made of funding procedures for coastal engineering and science research in South Africa. Up to the mid-1970's the bulk of the funding came from governmental sources, mainly in the form of parliamentary grants to Universities, the CSIR, the FRD or to other research institutes via the CSIR and FRD. As social and political pressure increased, the priorities of the South African Government changed progressively towards social upliftment, housing and health projects. This forced most research institutions to look towards the private sector for funding. This was and is being achieved in two ways. Both the CSIR and other institutions were forced to re-orientate themselves along market-driven lines, and, the funding role of non-governmental fund-raising organisations such as the World-Wide Fund for Nature (WWF-SA) increased enormously. The changes in funding mechanisms had both positive and negative consequences. Positive in that coastal engineers and scientists were forced to concentrate their efforts on problems of direct relevance to effective coastal zone management. Negative in that, inevitably, fundamental aspects of engineering and science receive less attention although recent indications are that collaborative ventures with key international institutes are reversing this trend.

MARINE ENVIRONMENTAL DATA MANAGEMENT

Considering the large amount of data involved with the activities described above, it is not surprising that the management of this data has required a concerted effort.

The parameter of greatest importance in the South African coastal engineering context is waves. Historically, waves were observed visually and recorded manually by officers on merchant vessels plying the coastal sea routes. This information, along with reports on weather and sea temperature, extends back to the mid 1800s, and is stored by the Southern African Data Centre for Oceanography (SADCO). Every attempt is made to have this data set as up to date as possible, and the most recent records are normally not more than a few months behind time. Although the quality of the data is insufficient for accurately estimating maritime design conditions, its temporal and spatial coverage make it useful in areas where alternative coverage is limited, or if long-term statistics are required.

High quality wave information (wave height and period) is presently collected from locally manufactured accelerometer-type wave buoys. A number of these are deployed at harbours around the coast and are directly linked via dial-up modems to the base station in Stellenbosch. The quality of this data is rigorously controlled, before being entered on the database. This data is suitable for design specifications for all maritime structures. The most authoritative reference on wave climate and design parameters for the South African coast is Rossouw (1989).

Another parameter of importance is wind. The largest collector of wind data in the region is the South African Weather Bureau, where measurements collected by manual or automatic weather stations are stored. In isolated cases, and for site-specific investigations, automatic weather stations are also deployed in coastal regions by organisations including the CSIR, the Sea Fisheries Research Institute, Portnet, Mossgas, Eskom and others.

Data on currents has also been collected in connection with investigations on the Agulhas and Benguela currents and the prediction of wave conditions around the coast. Measurements of current velocities have been made largely with self-recording, moored current meters, and data is stored by the individual collecting organisations, including the CSIR, the Sea Fisheries Research Institute and the Institute for Maritime Technology.

Bathymetrical data (sea bottom topography, sediments) has not been automated and digitised to the same extent as waves, wind or currents. The result is that this information has not been subjected to the same degree of computer management as have the other parameters. Nevertheless, extensive measurements have been collected by the SA Hydrographer, mineral exploration organisations, the CSIR and universities.

A large amount of marine environmental data has been, and is being, collected by environmental satellites, including sea surface temperatures, colours and wave and wind parameters. For the South African region, this data is collected by the Satellite Application Centre of the CSIR, and its management requires sophisticated storage techniques, such as high-density optical tapes.

The subject of data management is becoming increasingly important in South Africa, as elsewhere, with the higher, and in some cases continuous, rates of data collection and the increasing demands for its use. With limited financial and infrastructural resources at their disposal, this is challenging data managers to find new ways of storing data, creating rapid-access archiving facilities, and streamlining the data flow.

COASTAL ZONE MANAGEMENT

The ecological diversity of this coastline gives it great visual appeal and makes the coastal regions attractive for residential development and tourism. Over and above this the relative abundance of water along most of the coastline is inducive to economic activities such as agriculture and forestry. The harbour cities, Richards Bay, Durban, East London, Port Elizabeth, Cape Town and Saldanha, are attractive to industrial development.It is not surprising therefore, that in keeping with other countries in the world, an estimated 60% of South Africa's human population has concentrated along the coast.

South Africa's population has rapidly grown from 20 million in the mid-1900's to its present 40 million and by the year 2020 it is expected to reach the 80 million level.It is understandable therefore that coastal zone management is becoming increasingly more important.

The legal situation

Historically,the status of the coastal zone has been governed by the all embracing Sea Shore Act 21 of 1935. However, this was found to lack the legislation capable of controlling the increasing development in the coastal zone. Although there was strong motivation for promulgation of a separate Coastal Zone Management Act to supersede the Sea-Shore Act, this did not materialise, and control of the coast was incorporated in the Environment Conservation Act 100 of 1982 and in the revised Environment Act 73, promulgated in 1989 (Rabie, 1992). However, this remains inadequate in terms of legislation which will ensure cohesive control of the wide spectrum of human activities in the coastal zone, and the need for a separate Coastal Zone Management Act will become ever more acute with further human and developmental pressures.

Nevertheless, one positive outcome was that the Council for the Environment, a statutory body established in 1983 to advise the Minister of Environmental Affairs, produced two significant documents containing guidelines for coastal planning and for the protection of coastal land-forms (Council for the Environment, 1989 and 1991), though the commitment to their adherence by all authorities is lacking.

Implementation of coastal zone management in South Africa

It is necessary to comment on the implementation of coastal zone management in South Africa.

From the above text it will be seen that an excellent information base has been built up and that lack of scientifically based information is not a constraint. The difficulties lie at two planes :

- the lack of cohesive legislation specifically formulated to meet the needs of coastal zone management in the "new" South Africa, its burgeoning human population and concomitant demands for residential- work- and recreational facilities on the coast;

- the translation of scientific results into understandable guidelines to those who are actually responsible for implementing coastal zone management, i.e. the local authorities right down to bulldozer drivers.

Against this background, and with a sound scientific base of understanding of marine and coastal science built up over the past century (since the first government-sponsored fisheries research was undertaken in 1896), a major coastal zone management policy initiative is being launched with funding from the British Overseas Development Agency.

The objectives of this initiative, which was developed through a negotiated process by all key national coastal interest groups, are to promote integrated management and collaboration amongst all users of the coastal zone, and in the first two year phase, interested and affected parties will formulate a practically implementable policy, while in the second three year phase, attention will be focused on effective implementation of the policy. In these respects it offers a new, and ground breaking, approach to policy development in South Africa.

The initiative promises to provide a robust framework for addressing the many complex issues that have thwarted coastal zone management efforts in the past. However, while protection of the coastal zone is essential for social, economic and environmental reasons, a balance will have to be struck between meeting basic needs through development on the one hand and conservation on the other.

CONCLUDING REMARK

South African coastal engineers have for more than thirty years been regular attendants at the International Conferences on Coastal Engineering, where they have had the opportunity of testing their ideas and findings. Equally, the exchange forum created by these ICCE's has been one of the most important formative forces towards creating a small, but healthy community of coastal engineers in South Africa.

In November 1982 the ICCE came to Cape Town and the influx of some 400 coastal engineering specialists afforded South African coastal engineers a first-hand opportunity to learn from leaders in the field. Networks established at that time have in most cases been maintained to this day.

We salute the American Society for Civil Engineers for its vision in maintaining the ICCE as an institution and look forward to a long and mutually rewarding association with fellow coastal engineers world wide.

REFERENCES

Barnett, K A (1982). Durban Beaches Reclamation: Practical aspects. Proc. 18th Coastal Engineering Conference, South Africa.

Barwell, L and Burns, M E R (1989). Sediment budgets and plant/wind interactions. Coastal Zone '89, South Carolina.

Bender, C (1988). *Who Saved Natal?* The story of the Victorian Harbour Engineers of Colonial Port Natal. Published by the Author.

Branch, G and Branch, M. (1981). *The living shores of southern Africa.* Cape Town. C. Struik. 272 pp.

Bunt, E A (1977). Some highlights of engineering research in South Africa. In *A History of Scientific Endeavour in South Africa*, Ed. A C Brown, Royal Society of South Africa, p. 427.

Campbell, N P, MacLeod, D C and Swart, DH (1985). Bypassing and Beach nourishment scheme at Durban. 26th Intl. Navigation Congress, Brussels.

Coppoolse, RC and Schoonees, J S (1992). Evaluation of modern dredging/sand bypassing techniques applicable for South African conditions. CEDA-PIANC Conference Amsterdam, Paper D3.

Coppoolse, R C, Schoonees, J S and Botes, W A M (1994). Physical impacts of the disposal of dredger spoil at Richards Bay, South Africa. 28th Congress, PIANC, Section II, Subject 1, pp73-85, Seville, Spain.

Council for the Environment (1989). *A policy for coastal zone management in the Republic of South Africa.* Part 1. *Principles and objectives.* Pretoria, South Africa. 11 pp

Council for the Environment (1991). *A policy for coastal zone management in the Republic of South Africa. Part* 2. *Guidelines for coastal land-use.* Pretoria. South Africa. 94 pp.

CSIR (1963). Durban harbour siltation and beach erosion investigation, Parts 1 and 2, CSIR Contract Report C MEG 558, Pretoria, South Africa.

CSIR (1967). Durban Beach Protection: Interim Report No. 5, CSIR Report MEG 528, Pretoria, South Africa.

CSIR (1976). Management of the beaches in the Durban Bight, Volumes 1 and 2, CSIR Report C/SEA 7622, Stellenbosch, South Africa.

CSIR (1979). Oranjemund Beach Study. CSIR Report C/SEA 7935, Stellenbosch, South Africa.

CSIR (1985). Richards Bay harbour development field studies. Progress report June 1983 to May 1985. CSIR Report C/SEA 8626, Stellenbosch.

CSIR (1993). Physical impacts of the disposal of dredger spoil at Richards Bay CSIR Report EMAS-C 93028. Stellenbosch.

CSIR (1994a). Erosion and sedimentation problems at Richards Bay. CSIR Report EMAS-C94003. Stellenbosch.

CSIR (1994b). Oos-Londenhawe: Sedementasie, sandvangput en storting. CSIR Report EMAS-C 94020, Stellenbosch.

CSIR (1995a). Port of East London: Sedimentation on the inside of the southern breakwater. CSIR Report EMAS-C 95034, Stellenbosch.

CSIR (1995b). Durban beach monitoring progress report: July 1993 to June 1994 Volume 1: Main report. CSIR Report EMAS-C 95039/1.

DWAF (1990). Water Quality Guidelines for the South African Coastal Zone; Department of Water Affairs and Forestry

Fleming, C A (1977). The development and application of a mathematical sediment transport model. PhD Thesis. University of Reading.

Govender, K, Mocke, G P, Alport, M, Smith, F, Hough, G, Pelletier, L (1996). Measurement and modelling of water levels and flow fields in the surf zone. Proc. 6th Intl. Symp. on flow modelling and turbulence measurements, Talahasee, to be published.

Harris, T F W (1961). The near shore circulation of water. CSIR Sympos No. 52, CSIR Report P-4104.

Harris, T F W (1969). Near shore circulations: Field observations and experimental investigations of an underlying cause in wave tanks. CSIR Report S.37.

Harris, T F W, Jordaan, J M, McMurray, W R, Verwey, C J and F P Anderson (1964). Mixing in the surf zone. Int. Conf. On Water Pollution res., London. CSIR Report P-2817.

Heydorn, A.E.F. Glasewski, J.I. and Glavovic, B.C. (1992). The coastal zone. In: *Environmental management in South Africa*. R.F. Fuggle and M.A. Rabie (eds). Cape Town Juta & Co., Ltd. Pp 669 - 689.

Jordaan, J M (1961). Basic model studies of near shore wave action. CSIR Symp No. 52. CSIR Report P-4104.

Kinmont, A (1955). Beach erosion and protection. In Journal of the Institute of Municipal Engineers, Vol. 1, No 8.

KPR (1990). Green Point Sewage Disposal; Review and Comparative Appraisal of Land and Marine Options; Kapp, Prestedge and Retief

Laubscher, W I, Coppoolse, R C, Schoonees, J S and Swart, DH (1991). A calibrated longshore transport model for Richards Bay. Coastal Sediments '91, ASCE, Volume 1: 197-211, Seattle, Washington.

Lenhoff, L (1982). "Incipient motion of particles under oscillatory flow". Proc 18th Coast Eng Conf, 2: 1555-1568.

Mocke, G P, Reniers, A, Smith, G (1994). A surf zone parameter sensitivity analysis on LIPIID suspended sediment and return flow measurements. Proc. Coastal Dynamics '94, Barcelona.

Mocke, G P and Smith, G G (1992). Wave breaker turbulence as a mechanism for sediment suspension. Proc 23rd ICCE, ASCE, pp 2279-2292.

Möller, J P and Swart, D H (1988). Extreme erosion event on an artificial beach. Proc of the 21 st Coastal Eng. Conf., ASCE, pp 1882-1896.

Nairn, R B (1991). Description of the advanced near shore profile model. PhD thesis, Imperial College, London

Nelson, R C and Gonsalves J (1992). Surf zone transformation of wave height to water depth ratios. Coastal Engeering., 17, 49-70.

Nijhoff, G P (1935). Shore Protection and Beach Restoration at Durban.

Pearson, Tony (1995). *African Keyport, Accucut Books.*

Prestedge, G K (1992). Shark Rock Pier and Submerged Groyne. Shore and Beach, Journal of the American Shore and Beach Preservation Association, Vol. 60, No. 3.

Prestedge, G K and Hutchinson I P G (1976). A Mathematical Sediment Model for a Sea Water Intake Basin. Interdisciplinary Conference on Fresh and Marine Water Research in South Africa, Port Elizabeth.

Rabie, MA (1992) Environmental conservation act. In. *Environmental management in South Africa.* Cape Town Juta & Co., Ltd. Pp 99 - 119

Retief. G, Muller F P J, Prestedge G K, Geustyn L C and Swart D H (1984). Detailed Design of a Wave Energy Conversation Plant. 19th Int. Conf. on Coastal Engineering. Houston, Texas.

Rossouw J (1989) Design waves for the South African coastline; PhD dissertation, University of Stellenbosch

Scholtz DJP and Bosman DE (1981). A study of tidal swimming pools along the South African coast: Summary and interpolation report, CSIR Technical report T/SEA 8136

Schoonees, JS (1991). 'Field measurements of suspended sediment concentrations in the surf zone', Euromech 262 - 'Sand transport in rivers, estuaries and the sea', Soulsby, R L, Bettess, R (Eds), Balkema, Rotterdam.

Schoonees, J S (1996). Longshore transport in terms of the applied wave power approach. PhD Thesis (to be published), University of Stellenbosch

Schoonees, J S, Theron, A K (1993), "Review of the field-data base for longshore sediment transport', Coastal Engineering, 19, pp 1-25.

Smith, G G and Mocke, G P (1993). "Sediment suspension by turbulence in the surf-zone". Proc. Euromech 310, Le Havre.

Smith, G G, Mocke G P and D H Swart (1994). Modelling and analysis techniques to aid Mining operations on the Namibian coastline. Proceedings of the 24th Coastal Eng. Conf., ASCE.

Soulsby, R L, Hamm, L, Klopman, G, Myrhaug, D, Simons, R R and Thomas, G P (1993). Wave-current interaction within and outside the bottom boundary layer. Coastal Engineering, 21, pp 41-69.

Swart, D H (1974). "Offshore sediment transport and equilibrium beach profiles", D Sc dissertation, Delft University of Technology, The Netherlands.

Swart, D H (1978). Vocoidal water wave theory; Volume 1; Derivation CSIR Research Report 357, 137 pp, Stellenbosch.

Swart, DH (1981). Effect of Richards Bay harbour development on the adjacent coastline. 25th Congress, PIANC, Section II, Vol. 5: 899-917, Edinburgh

Swart, D H (1984). 'Sediment dynamics field experiment: Sunday's River', Proc. 19th Int. Conf. On Coastal Eng., Houston, September 1984, Vol. II, p 1371-1385, ASCE.

Swart, D H (1986a). 'Physical environmental interactions in the Sunday's River/Schelmhoek area',CSIR Report 568.

Swart, D H (1986b). Prediction of wind-driven transport rates, Proceedings 20th Coastal Eng. Conf., Taipei, Taiwan.

Swart, D H (1988). CAESAR coastal area and sediment applied research: Surf zone experiments in South Africa, in De Graauw and Hamm (Eds)

Swart, D H and Loubser, C.C. (1978). Vocoidal Theory for all non-breaking waves. Proc. 16th International Conference on Coastal Engineering, Hamburg.

Swart, D H and Loubser, C.C. (1979). Vocoidal water wave theory; Volume 2; Verification; CSIR Research Report 360, 129 pp, Stellenbosch.

Swart, D H and Crowley, J B (1988). Generalized wave theory for a sloping bottom. In: Proc 21st Inst Conf on Coastal Eng, Malaga. ASCE, pp. 181-203.

Veitch, N (1994). *Waterfront and Harbour: Cape Town's Link with the Sea*, Human & Rousseau.

Williams, H (1993). *Southern Lights: Lighthouses of Southern Africa*. William Waterman Publications.

Witthaus, K G, Retief, G de F, Prestedge G K and Huskins, L R (1982). Dissipation of Wave Energy in a Seawater Outfall Channel. 18th Intl. Conference on Coastal Engineering, Cape Town.

WPR-Watermeyer Prestedge Retief (1996). Feasibility Studies of the use of Fixed Sand Bypassing at the Ports of Port Elizabeth, Durban and Richards Bay, in progress.

Zwamborn, J A (1970).Coastal Engineering Research CSIR Report R/SEA 7003.

Zwamborn, J A, Russell, KS and Nicholson J (1972). Coastal Engineering measurements. CSIR re. ME 1148.

Zwamborn JA and Cawood CH (1974). Major port developments at Richards Bay with due regard to preserving the natural environment. Trans. South African Institute of Civil Engineers, Vol 16, No 2.

Zwamborn JA and Swart DH (1988) South African artificial structures and shorelines. Edited by HJ Walker, Kluver Academic Publications

HISTORY OF COASTAL ENGINEERING IN SPAIN

M.A. Losada, Member, ASCE, R. Medina, C. Vidal and I.J. Losada, Associate Member, ASCE[1]

ABSTRACT. This chapter is devoted to the history of coastal engineering in Spain, a country strongly related to the sea. Spain has over 8000 km of coastline and four different marine climates. The chapter includes a summary of the main events occurred in Spain in the field of coastal engineering in the last 3000 years. In the last twenty five years the coastal engineering field has been very active, due to the economical expansion and the tourism boom. Among the most important contributions are the approval of two laws one related to harbor organization, the other to coastal protection. Moreover, a Regulatory Guide Programme of Maritime Works and the Spanish network of measurement and storage of wave data have been initiated. The investment in coastal protection during the last five years overpassed the investment of countries like USA in the last 40 years. Most of the coastal and harbor engineering work developed during the last years follows the methodology proposed by Prof. Iribarren. In this chapter some of his work and circumstances under which he had to work are presented. Nowadays, his legate is well represented in several University Research Groups and Research Centers. It seems that coastal engineering in Spain is a growing field, and that the future of this civil engineering branch is very promising.

INTRODUCTION

The history of Spain is strongly related to the sea. The Iberian Peninsula has played an important role in history because of its location between the Mediterranean Sea and the Atlantic Ocean and between Africa and Europe. Consequently, coastal engineering as a part of ocean engineering, has played an important role in the Spanish civil engineering.

This chapter is dedicated to the history of coastal engineering in Spain with special emphasis on the past twenty-five years. However, and in order to give the reader an overall justification of the Spanish heritage, some details of the history of Spain are included.

The chapter is organized as follows. First, a general description of the Spanish coastline and its climate is presented. Next, a summary of the history of Spain and its relationship to the sea, such as the development of harbors in the New World, is given.

[1] Ocean & Coastal Research Group. Universidad de Cantabria.
E.T.S. Ingenieros de Caminos, C. y P. Avd. de los Castros, s/n. 39005 Santander (Spain)

The next section is dedicated to Spanish coastal engineering in the second half of the 20th Century. Special attention is given to harbor and coastal development. The former is strongly associated with the development of the Spanish economy. The latter was induced and determined by the tourism boom which occurred in the 1960s. As a consequence of this large exploitation of the coastline during the past ten years, coastal engineering has become one of the most important branches of civil engineering. During the second half of this century, the main contributions to coastal engineering can be synthesized in the following aspects:

1. A new Law of Coastal Protection.
2. A new Law of Harbor Organization.
3. A Network for Maritime Climate Recording.
4. A Regulatory Guide Program for Maritime Work.
5. Investment Plans for Harbor and Coastal Development.

All of this development has been possible because of the Spanish maritime work and construction and of the methodological contribution by engineers such as Iribarren. Consequently, some of Iribarren's contributions are detailed. Moreover, Iribarren is considered the father of a new generation of coastal engineers and associated applied research centers in Spain.

Finally, a section with some details of the maritime work and coastal protection built in the last few years is given. The reader will learn that in Spain there are more than 300 major harbors, on average 3.5 harbors every 100 km of coast and over 180 km of breakwaters. In order to give an idea of the magnitude of the effort necessary to achieve these works, two examples of historical harbor development are given.

MORPHOLOGY AND CLIMATE OF THE SPANISH COASTLINE

In this section, a brief description of the morphology of the Spanish coastline, including the Balearic and Canary Islands, and some features of the wind and wave climate will be given.

Morphology

The Iberian Peninsula (Fig. 1), is located on the southwestern tip of Europe between Tarifa and the Estaca de Bares headland and the Creus headland and Roca headland. The Atlantic Ocean washes the northern, western and southwestern coasts, and the Mediterranean Sea washes the southern and eastern coasts.

Portugal and Spain make up the Iberian Peninsula. Portugal occupies part of the western and southwestern Atlantic coasts. Also included in Spanish territory are the Balearic Islands in the Mediterranean Sea, the Canary Islands in the Atlantic Ocean, and the North African cities of Ceuta and Melilla.

Spain, thanks to its unique peninsular and insular geography, has the longest coastline of any country in the European Union. Of the 26,000 km of coastline shared by the countries of the European Union, 8,000 km correspond to peninsular Spain and its islands. Sixty percent of the coastline is located along the Mediterranean sea and 40% along the Atlantic Ocean. Almost 3,000 km of coastline correspond to the Canary and Balearic Islands.

Cliffs stretch for more than 4,500 km of coastline, about 60% of the whole seaboard. They can be found all along the Spanish coast, particularly in the north, northeast, and northwest, and in the south, and southeast, and on the islands. Almost 25% of the Spanish coast corresponds to natural beaches, mostly sandy beaches, while swamps and wetlands occupy 15% of the coast.

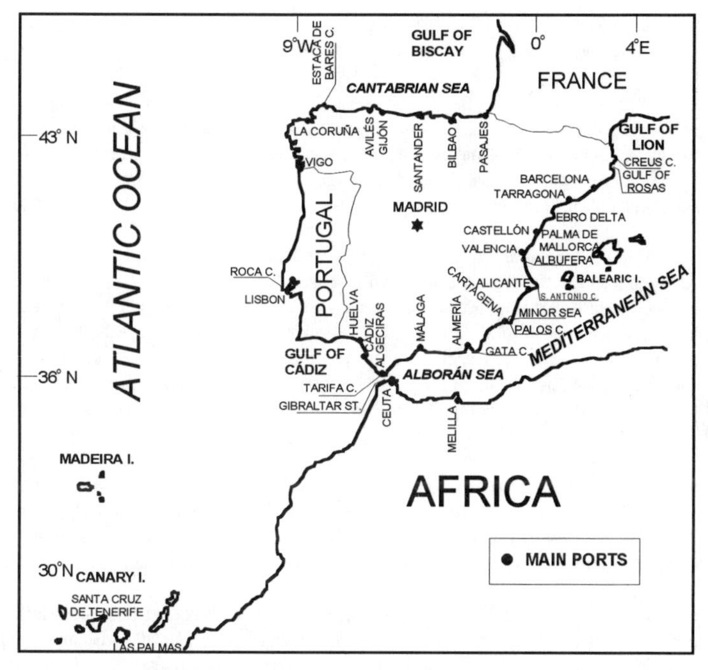

Fig. 1 Location of the Iberian Peninsula and Spain

From a morphological point of view, the northern or Cantabrian coast is comprised mostly of cliffs, interspersed pocket beaches of fine sand. The spring tidal range is 4.5 m, and the North Atlantic storms and swells periodically rebuild the beaches. At points where short rivers reach the coast, small and medium-sized estuaries appear, which are generally partially closed by long sand spits. Inside these estuaries, natural wetlands, ports and beaches share the space even though industrial and urban activities have suffocated some of them (Pasajes, Bilbao, Aviles).

The Northwestern or Galician Coast is mostly rocky cliffs, but a tectonic sinking of the land submerged the coastal river valleys, creating long, deep estuaries, called *rías*.

Along the coast and inside the *rías*, hundreds of small pocket beaches of fine and medium-sized sand dot the coast. Inside the *rías*, human activity is intense: fishing and commercial ports, cities and villages are found along the coast.

The Southwestern Coast (Gulf of Cadiz) is a sandy coast with a low cliff formed during the Miocene Era and currently under a regressive process. The long, straight beaches stretch tens of kilometers between the mouths of the rivers. At these points, long spits tend to close the navigation channels. Inland, dunes which are partially stabilized by vegetation are dominant. Some of the areas, such as the coast of the Doñana Natural Park, are still untouched by humans. South of Cadiz, the coast rises and stretches of cliffs appear, increasing towards the Strait of Gibraltar.

The southern or Alboran Coast (Costa del Sol) starts in the Gibraltar area, being low and sandy with long fine sand beaches. To the east of Malaga, the proximity of the Penibetica Mountains to the coast generates cliffs and pocket beaches of generally coarse sand or shingles. In some places, littoral plains, the emerged part of river deltas created by torrential rivers, are the main features of the coastline. In these areas, long beaches border the coast. The tidal range is of the order of tens of centimetres and the wave energy is moderately low.

Generally speaking the morphology of the southeastern Coast is similar to that of the southern Coast. Sometimes, in the areas where the mountains are far from the coast, a coastal plain permits the formation of long, straight beaches of medium or fine sand. In Murcia, sandy littoral barriers as found in the Minor Sea, enclose swamp areas. Around the large headlands of Gata, San Antonio and Palos, where the mountains sink into the sea, the coast consists of mostly high cliffs with pocket beaches.

North of the San Antonio headland, the mountain range retreats leaving a wide, low littoral plain built up by river sediment, which finishes in a low, straight, sandy coast. The longest beaches in Spain, which have very active littoral drift and which are fed by river sediments, can be found between Denia and Barcelona. In some places, littoral barriers have closed lagoons such as the Albufera of Valencia. The regulation of rivers by dams has decreased the volume of sediment that reaches the coast, and many beach stabilization and nourishment programs have been carried out. In the north, the Ebro Delta region forms a very special area of changing coastline with plenty of beaches, swamps, wetlands, and spits.

To the North of Barcelona, the coast becomes progressively rocky and, again, pocket beaches of medium sand share the coast with the cliffs. The area called "Costa Brava" stretches north to France except for the Gulf of Rosas which is the only interruption of the highlands. Near the French border, the Pyrenees sink into the sea, forming the high intricate coast of the Creus headland area.

The Balearic Coast is, in general, a rocky but not too high coast with pocket beaches. Mallorca, the biggest of the islands, has two large areas located north and south of the islands, with ample bays and low, sandy coasts. The northwestern coast of Mallorca, with a mountain range rising from the sea up to more than 1,400 m, is one of the most spectacular coastal landscapes in Spain.

Thanks to its volcanic origin, the Canary Islands have a very special morphology. Because of the youthfulness of the coast, the continental shelf is practically nonexistent, and the coastal slope drops directly to abyssal depths. Because of this characteristic, sandy beaches are scarce, and most of them are composed of volcanic rock shingle. In some areas where the continental shelf widens, sandy beaches of fine aeolic sand from the nearby Sahara Desert can be found. On the biggest islands, long stretches of the coast, especially in the north and northwest, attacked by North Atlantic swells, sink directly into the sea forming high cliffs of startling beauty.

The Marine Climate of Spain

The climatological characteristics of the Iberian Peninsula are a function of its latitude in the Northern Hemisphere, superficial extension and mountain ranges, and location on the western edge of Europe, between the Mediterranean Sea and the Atlantic Ocean.

The Iberian Peninsula is affected by the polar front that separates subtropical and polar air masses. This polar front oscillates north and south during the year. In winter during the transitional periods from fall to winter and from winter to spring, the polar mass occupies low latitudes, and the polar front is located over the peninsula, producing unstable weather. The extratropical Atlantic storms approach the coasts from the northwest, west, and southwest, carrying strong winds and waves to these coasts. Some of these storms cross through the Gulf of Leon to eastern coasts. Finally, and less frequently, some of the storms cross the Strait of Gibraltar and move along the northern coast of Africa, carrying bad weather to the southern and southeastern coasts. In the summer, low activity of the polar front and its position in higher latitudes makes the weather, generally speaking, more stable.

All year long, the trajectory of North Atlantic storms generate swells that reach the northern and northwestern coasts. In winter, when height persistent winds blow over fetches that stretch thousands of kilometers over the Atlantic, very high sea states may be generated, which reach the coasts with significant wave heights of up to 10 m. In the Mediterranean, short fetches and the fugacity of storms make swells very rare, but sea waves can provoke great damage to coastal beaches and are always dangerous for shipping activities.

The Balearic Islands have the same maritime weather as the Eastern Mediterranean coast. The case is different for the Canary Islands. Because of to their low latitude, they are scarcely directly affected by North Atlantic storms, but their swells are more frequent and are responsible for some of the higher waves that attack the northern part of the islands. Most of the year, the islands receive light or moderate trade winds from the northeast. On some occasions, always during the winter, once every 6-8 years, the polar front moves southwards to form a low near the island, blowing strong winds from the southwest, west, or northwest, and causing severe damage to beaches and ports.

Tides in Spain are semidiurnal. Spring tidal ranges are around 4.5 m in the north, 4 m in the northwest, 3.5 m in the southwest, and around 3 m in the Canary Islands, which are all located in the Atlantic Ocean. In the Mediterranean, tidal range is moderate, less than 0.5 m during spring tides, and many times the changes of sea level are obscured by the meteorological tides.

Information about wave climate is coordinated by the Programa de Clima Marítimo (PCM), (Maritime Weather Program), created in 1983 which coordinates the data taken with a network of wave buoys, radar, and tidal gauges located all around the coasts of Spain (Fig. 2).

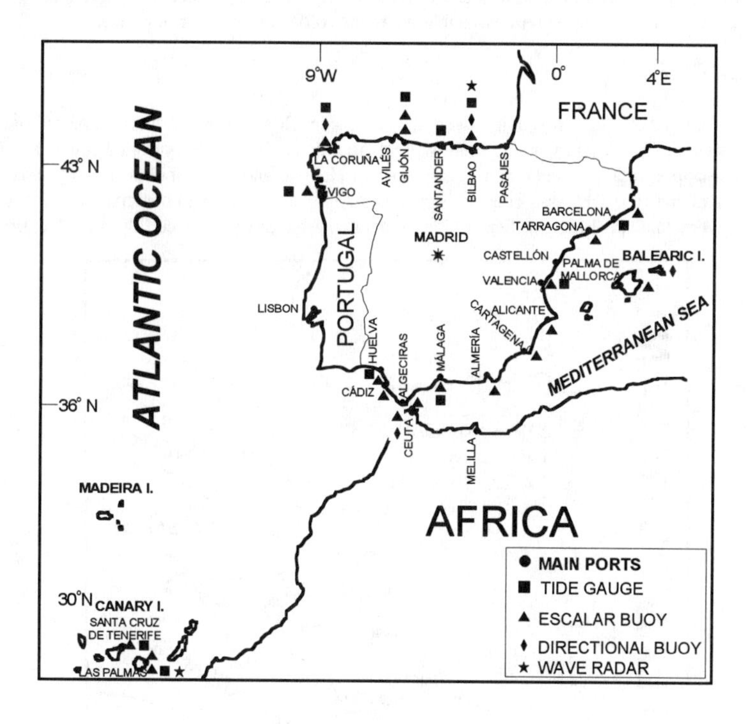

Fig. 2 Location of REMRO wave buoys

The network of wave buoys was started in 1972 by the former Dirección General de Puertos y Costas, under the REMRO Program (Spanish Network of Measurement and Storage of Wave Data), (Suárez Bores, 1974). Today, the network is the responsability of the public enterprise Puertos del Estado, (State Ports of Spain) and is maintained by the CEDEX-CEPYC (National Coastal Engineering Laboratory). The analysis and publication of data is organized by the PCM. The network has 21 scalar WaveRider buoys and 4 directional WaveScan buoys. In 1994, the PCM installed a radar for directional wave measurement on the northern coast. In the future, the PCM will complete scalar data from the buoys with the directional information given by a network of radars on the coast. Specific data from the buoys or general statistics are available to the public in magnetic or paper format.

Another source of wave data coordinated by the PCM is the visual data taken from ships. Date, hour, atmospheric pressure, wind speed and direction, wave height and period, and sea and swell direction are available in raw form or in tables.

The publications Regulatory Guides for Maritime Works, ROM 0.3-91, and ROM 0.4-95 summarize the basic wave and wind statistics for the design of coastal structures taken from instrumental and visual data. As will be discussed below, ROM stands for Regulatory Guides for Maritime Works.

THE HISTORY OF SPAIN AND ITS COASTS

The history of Spain is strongly related to the sea. The location of the Iberian Peninsula, in the western coast of the Mediterranean Sea, as well as between Africa and Europe, and the richness of its inland made its eastern and southern coasts valuable areas for colonization and commerce.

The Ancient Age

In the 8th century B.C., the Phoenicians were the first to develop harbors on Iberian coasts. The main colony was Gadir (now Cadiz), on the southwestern Atlantic coast, which controlled commerce, based on the rich copper, mercury, gold, and silver mines in Huelva, and at the mouth of the Guadalquivir River, door of the rich Tartessos country (nowadays Andalucia). Other Phoenician colonies, of less influence, were Malaka (Malaga), Sexi (Almunecar) and Abdera (Adra) (see Fig. 3 for location).

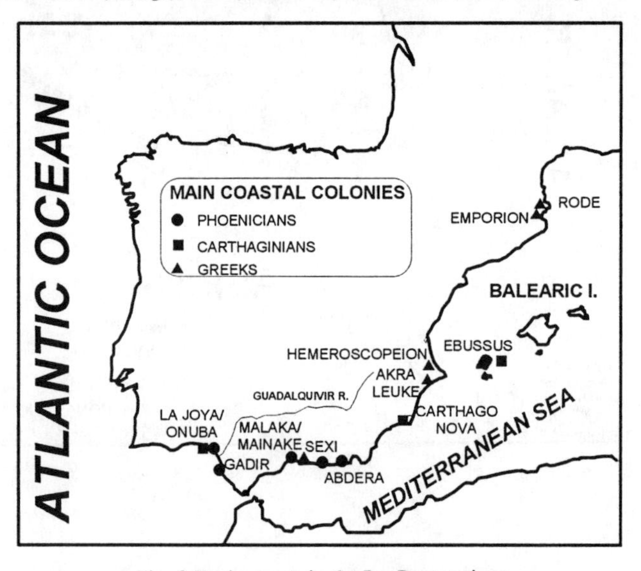

Fig. 3 Iberian ports in the Pre-Roman times

After the 6th century B.C., the Carthaginians took control of the southeastern Mediterranean Sea. The foundation of Qarthadasat (after the Romans, Cartago Nova and now Cartagena) are from that age. Other Carthaginian ports were Ebussus (Ibiza) on the Balearic Island of the same name and Onuba (Huelva), at the mouth of the Tinto and Odiel Rivers.

Simultaneously, the Greeks started their expansion through the Western Mediterranean. The most important colony in Iberia was Emporion (Ampurias) in the Gulf of Roses. Originally, the colony was on one island, the ancient Village of Paliapolis (see Fig. 4), but because of the increasing population, a new city, Neapolis, on the Southern side of the Ancient Port, was created. Today, it is located inland due to littoral drift sedimentation.

Emporion was the first city conquered by the Romans in Iberia, which occurred in 218 B.C. That marked the initiation of the decline of the Carthaginian Empire and the rise of the Roman Empire. Emperor Caesar Augustus in the 1st century A.D. occupied the whole of Iberia. The Roman city of Emporion was built inland of the Greek cities. The port was protected by a vertical breakwater made of rectangular stone blocks and filled with hydraulic conglomerate. The vertical part of the breakwater was 6 m wide and 7 m high, and was built over a rubble foundation. The remaining 85 m of this breakwater are some of the best preserved Roman harbor works, (Fig. 4).

The Romans organized Hispania into provinces in their search for the most effective systematic exploitation of its resources. They exported through their ports cinnabar (ore of mercury) from Almaden, weavings from Ampurias, salt, salted fish and garum (marinade) from all of the southern ports of the peninsula; esparto, silver and lead from Cartagena; dried figs from Ibiza; gold, silver and olive oil from the Baetica province; wine and wheat from the Tarraconensis province; and gold from the northwest. To give us an idea of the power of that commerce, mostly related to Rome, Mount Testaccio, in Rome, was built with 40 million amphoras of oil, wine and garum mostly from Hispania. In the 2nd century A.D., half of the wine and oil consumed in Rome was from Hispania. The boats sailed around the coasts of Hispania to Alicante, the Balearic Islands, the Bonifacio Strait and finally Ostia, the port of Rome.

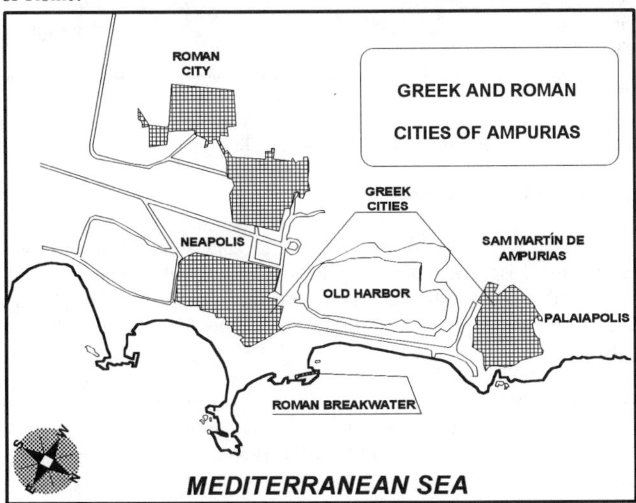

Fig. 4 Greek and Roman colonies of Ampurias

Roman ports were widespread throughout Hispania (Fig. 5). In the north, Portuondo, Flaviobriga (Castro Urdiales), Portus Victoriae (Santander), Portus Veseiasueca (San Vicente de la Barquera), and Noega (Gijón) were small military strongholds, related to military control and commerce with

Britannia. On the northwestern coast, the Flavium Brigantium (La Coruña) stood out, whose lighthouse, The Tower of Hercules, 35 m high, built of stone and hydraulic conglomerate, still guides the boats which navigate the rough seas of the northwest. On the western coast, the Lusitania province (now mostly Portugal), the ports of Calem (Oporto), Olisipo (Lisbon) and Caetobriga (Setubal) stood out. On the southwest, Onuba, Gades (Cadiz) and the fluvial port of Hispalis (Sevilla) can be pointed out. On the Mediterranean coast, Malaca (Malaga), Carthago Nova (Cartagena), Tarraco (Tarragona) and Emporiae (Ampurias) together with Gades witnessed most of the intense commerce in the Mediterranean area.

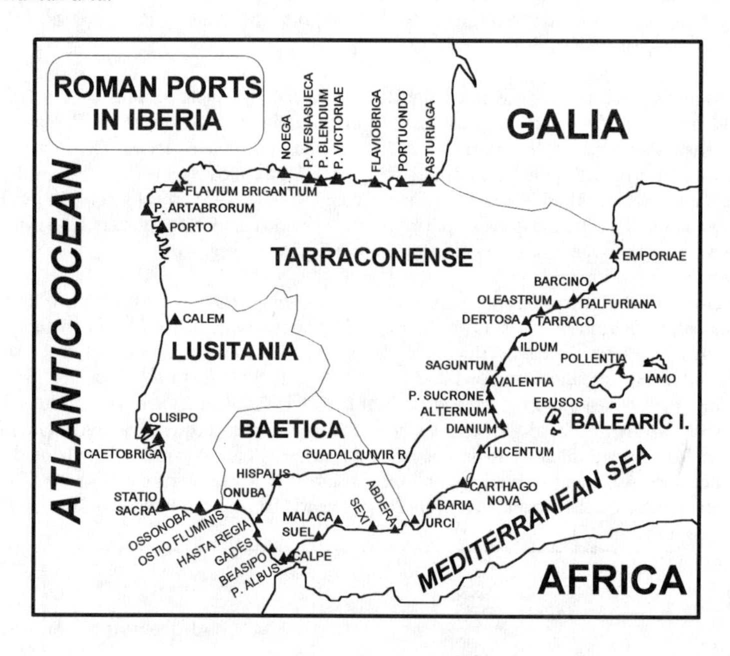

Fig. 5 The Iberian Peninsula in Roman times

From the Romans to the Middle Ages

The Roman Empire declined after the 4th century A.D., and Celtic peoples (Suevos, Vandals and Alanos) spread through Hispania. After them and with a treaty with Rome, the Germanic Visigoths conquered most of Hispania (the Suevos held Galicia). The slow decline of the Roman Empire which ended in A.D. 476 with the last Roman emperor, converted the Visigoths into the hegemonic people of western Europe.

The Visigothic times were not good times for maritime commerce nor for maritime engineering. On the Mediterranean coast, commercial transactions were in the hands of oriental merchants who had operational warehouses in the principal southern and eastern ports. North Africa, Italy, and the Orient fed the Visigothic kingdom with priceless articles from the Orient which were exchanged for wheat, oil, and wine.

In the 8th century, the Muslim people of the Damascan Empire invaded Hispania, and the Visigoths, considerably weakened by internal conflicts, were swept from most of Hispania, resisting only in the North. For 700 years, two civilizations in continuous conflict divided Hispania, separated by a no man's land. In the East and South, the Muslim government improved civil organization and, as a result, commerce and civilization flourished.

The Hispano-Muslin engineers improved the fluvial ports of Tortosa (Ebro River) and Seville (Guadalquivir River) and the coastal ports of Denia, Alicante, Malaga or Almeria. The ships of their fleet, which were the mainstay of their sea supremacy, were built in special covered shipyards called *atarazanas*.

The Middle Ages

In the 10th century, Iberia was divided as seen in Figure 6. The north and northwest were dominated by the Christian kingdom of Leon, and the counties of Castille, Navarra, Aragon and Catalonia (this last controlled by the Franks). The only ports were those in the north, northwest and northeast (Barcelona) and maritime commerce was very scarce. The ports of the Mediterranean and Andalusia progressively fell under their control and increased wealth improved commerce. By the middle of the 13th century, only the Muslim kingdom of Granada stood against Christian progress, and that situation lasted for more than 100 years after King Fernando and Queen Isabel conquered Granada in 1492.

Throughout the 12th and 13th centuries, the ports of the northern coast, mainly San Vicente de la Barquera, Santander, Laredo, and Castro Urdiales (known as "the Four Cities") were the contact points for the kingdom of Castille and Leon with the ports of western Europe. Bordeaun, La Rochelle, Bristol, Southampton, Bruges, Hamburg, and Lübek were the main shipping destinations. By that time, maritime traffic in the English Channel and the Bay of Biscay was controlled by the "Four Cities" ports. In the 14th century, other ports in the Basque Provinces mainly Bilbao, besides the Nervion river, joined the competition. Cargos of wool, wheat and wine from Castille salted fish and iron ore, steel and weapons from neighbouring mines, foundries and forges were the main exports while weavings and general cargo were the main imports. Also, the shipyards for the fleet were of capital importance.

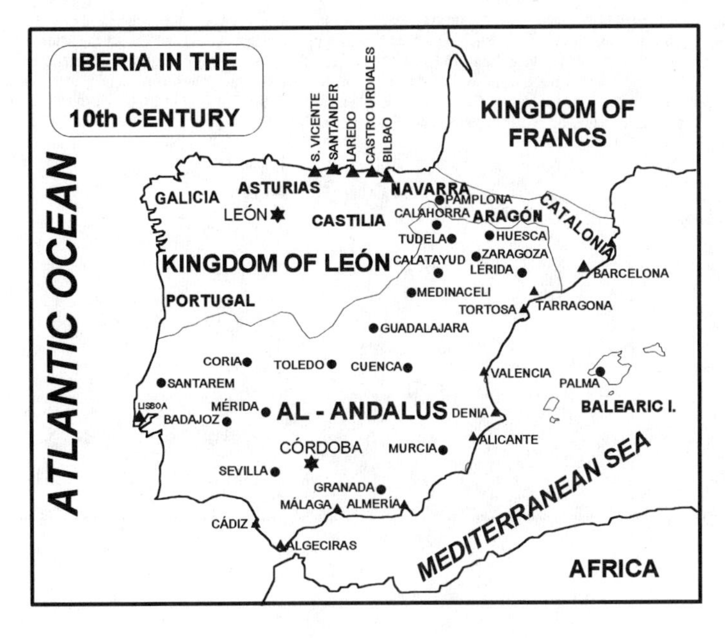

Fig. 6 The Iberian Peninsula in the 10th Century

The Spanish Empire and the New World

After the discovery of America, the foundation in Seville of the "Casa de Contratación", a kind of chamber of commerce for business with the Americas, moved the economic control of Spain to the Guadalquivir area: Seville, Sanlucar de Barrameda and, finally, Cadiz which in 1717 became the new *Casa de Contratación.*

During the Spanish Empire in the 16th century, the Netherlands and Belgium were part of the Empire territories and their ports were controlled by the Spanish government. In America and the Philippines, colonization produced a large development of maritime engineering. New ports were built throughout Central and South America on the Atlantic and Pacific coasts such as: Veracruz (Mexico), Cartagena (Colombia), Maracaibo (Venezuela), San Juan (Puerto Rico), Havana (Cuba), Montevideo (Uruguay), Buenos Aires (Argentina), Valparaiso (Chile), Callao (Peru), and Manila (The Philippines), among others, (CEHOPU, 1994). Moreover, new techniques on coastal and harbor engineering were introduced: new methods to avoid wood deterioration due to biofouling, dredging, piling, and breakwater construction. For example, the first known coastal protection in Spain using a field of groins was built in the port of Callao (Peru) in 1728 to avoid coastal retreat (which endangered the defense walls) and to stabilize the harbor entrance, Fig. 7.

The introduction of steam power and steel construction in the 18th century made the construction of powerful engines possible, like the first bucket-ladder dredge, designed by the engineer Agustin de Bethancourt in 1791, Fig. 8, or the "Titan" crane, Fig. 9, installing 50 T-shaped concrete blocks for the Santa Cruz de Tenerife breakwater in 1888.

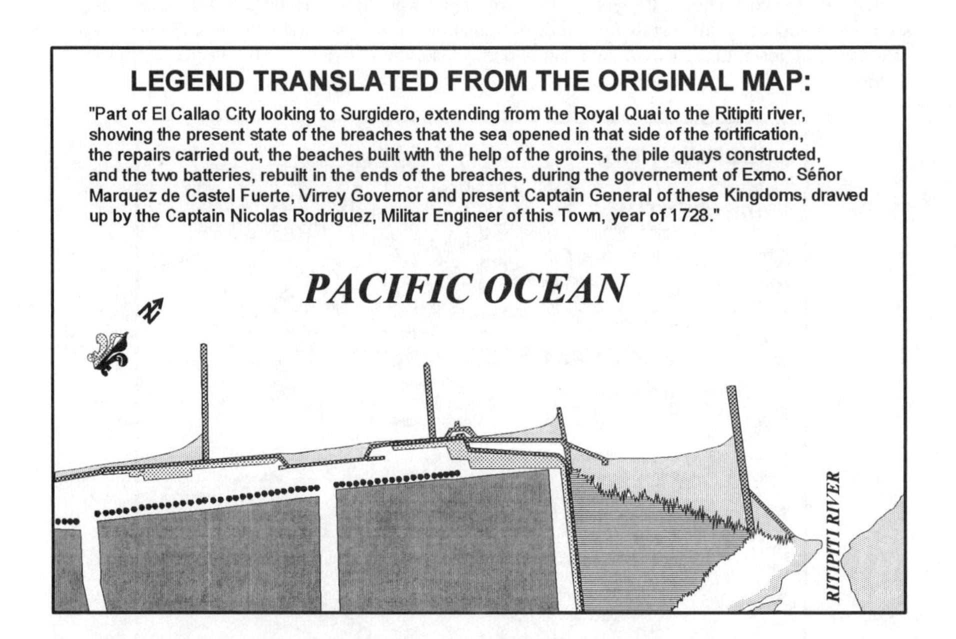

Fig. 7 Coastal stabilization works in the port of Callao, 1728

Fig. 8 First steam bucket-ladder dredge 1791

Fig. 9 "Titan" crane installing 50 T. concrete blocks. 1888.

COASTAL ENGINEERING DURING THE SECOND PART OF THE 20th CENTURY

During the 20th century, Spanish ports and coastal engineering continued growing, especially in the field of maritime works against wind waves for the development of harbors and of coastal protection. Only after the tourism boom, which started in the 1960s and its unwanted consequences to the natural resources, did beach engineering assume an important role in Spanish coastal engineering.

Harbors in Spain

There are about 300 harbors in Spain. Half of these harbors are located along the Mediterranean coast and half along the Atlantic coast. Galicia, in the North-West, is the region with the most harbors, about 20% of the total, mainly fishing harbors. On average there are 3.5 harbors every 100 km along the Spanish coast. In the Mediterranean Sea this number increases up to 4.3 harbors every 100 km Catalonia in the north-east, with 7 harbors every 100 km, has the highest density of harbors per unit length of coast, (MOPTMA, 1988).

In order to give an idea of the evolution of the investment in the ports system, the State Ports of Spain has doubled investment in the last 10 years (about $ 500 million U.S.). Thirty-eight percent of this investment is being used in breakwaters and mooring lines.

The former organization for the administration of harbors became obsolete and a new law of harbors was approved by the Spanish Congress. A public enterprise "Puertos del Estado" (State Ports of Spain) was created. Some details are given below.

Spain holds a privileged place in regards to the distribution of the world's tourism. It is the third most-visited country in the world. In 1994, for instance, 61.4 million visitors came to Spain, and the income from tourism reached ($26 billion U.S.). Tourism is, consequently, a primordial sector for Spanish economy. In fact, it forms a major source of income in foreign currency, it represents 8% of the Gross Domestic Product (GDP), and 1.4 million jobs depend on it (11% of the entired working population), (MOPTMA, 1993).

The Spanish coast and beaches, are basic sources of Spanish economy since 80% of the tourists are concentrated on the seashore, while only 20% visit the interior of the country.

The fact that 35% of the Spanish population lives on a coastal strip 5 km wide, with a population density four times higher than the national average during the period without tourism, gives us an idea of the pressure that the littoral zone has to bear. During the tourist season, the population density increases up to 12 times the national average, (MOPTMA, 1993).

Furthermore, the tendency of the population to concentrate at the sea side, especially since the tourist boom that began in the 1960s and 1970s, implies the necessity to supply this development with infraestructure such as power, sewage, and road systems, and more specifically with marinas and waterfronts. From 1960 to 1985, most of the coastal engineering works were projects against waves with very few projects on beach restoration.

Moreover, the degradation of the coastline was severe, and the natural cycle of many beaches was broken by uncontrolled large-scale urban development on a very narrow strip of coast near the shore.

Because former legislation proved to be insufficient in preventing the degradation of the littoral zone, new legislation on coasts was passed in 1988 in order to prevent the improper use of this valuable, but delicate strip of territory as well as to recover lost coast, (MOPTMA, 1993). Moreover, a *Plan de Costas* (Coastal Plan) was defined and developed during the last nine years. Some of the work executed within the *Plan de Costas* will be detailed in a further section.

THE LAW OF COASTS AND HARBORS

In order to understand the coastal and harbor work undertaken during the past decades in Spain, it is necessary to explain the new Spanish legislation of coasts and harbors. This legislation delineates the competencies and responsibilities on maritime works that can be carried out by central and local authorities along the Spanish coastline.

Legislation on Coasts

Spanish cultural and legal traditions have always considered coastal areas to belong to everyone and that their ultimate situation should naturally be one of free public access, use and enjoyment for common purposes. Hence, from ancient Roman and medieval laws to the present Spanish law has protected the seashore by making it a part of public domain.

Alfonso X (also known as Alfonso the Wise) who commissioned in the 13th century, the ("Código de las Siete Partidas", 1955) (The Code of the Seven Partitions) stated that: "the things that communally belong to all creatures living in this world are: the air, the rainwater, the sea, and its shore. No building can be undertaken on the seashore that could restrain the communal use of the people" (Partition 5, 9th Title, 7th Law). It is noticeable that some of the United States, such as Texas, are still using as law of the coasts some of the sections of the "Código de las Siete Partidas" as they were written in the 13th century with adenda being made to in the 16th century.

Following this medieval law, the present Legislation on Coasts, was passed in 1988 to end the degradation of the littoral. In a special way, it protects the most fragile area of the coast: the maritime-land public domain. This law assures its physical integrity as well as free and public access for everyone.

The Public domain is a coastal strip that stretches as far as the waves can reach during the worst storms in the area, and it includes beaches, dunes, cliffs, swamps, and wetlands. No type of commercial activities can be allowed on this property. Its natural utility is common use in harmony with nature, such as walking, swimming, or simply relaxing.

The Legislation on Coasts extends its influence to private lands adjacent to public property. In order to prevent any activity being carried out on these lands which could be harmful to its space of high environmental value, private property is restricted in three areas. The first area, "the right of way" is 6 m wide and can be extended to 20 m. It must always be open to public pedestrian use. The following area, *the right of protection*, is 100 m wide and can be expanded another 100 m. In this area, only the necessary facilities for public use and enjoyment on the coast are allowed. The third area, named *of influence*, is at least 500 m wide, and constructions can be made on it provided that limitations on building density are respected. The above-mentioned concepts are shown in Fig. 10.

The Legislation on Coasts also distributes the competencies concerning the seashore between the civil service, self-governing communities, and municipal governments.

A) Competencies of the Civil Service

With regard to the Civil Service, the Ministry of Public Works, Transportation, and Environment has the responsibility to manage and protect the maritime-land public domain and the right-of-way, as well as the right to shore protection. Within the Ministry, the State Coast Office exercises its own competencies in the matter of coasts. It grants concessions for permanent work in the public domain and the authorizations in the right-of-way area. In addition, the State Coast Office is carrying out the determination of the boundaries of the public domain, and, is in this way, incorporating lands into public property to widen it.

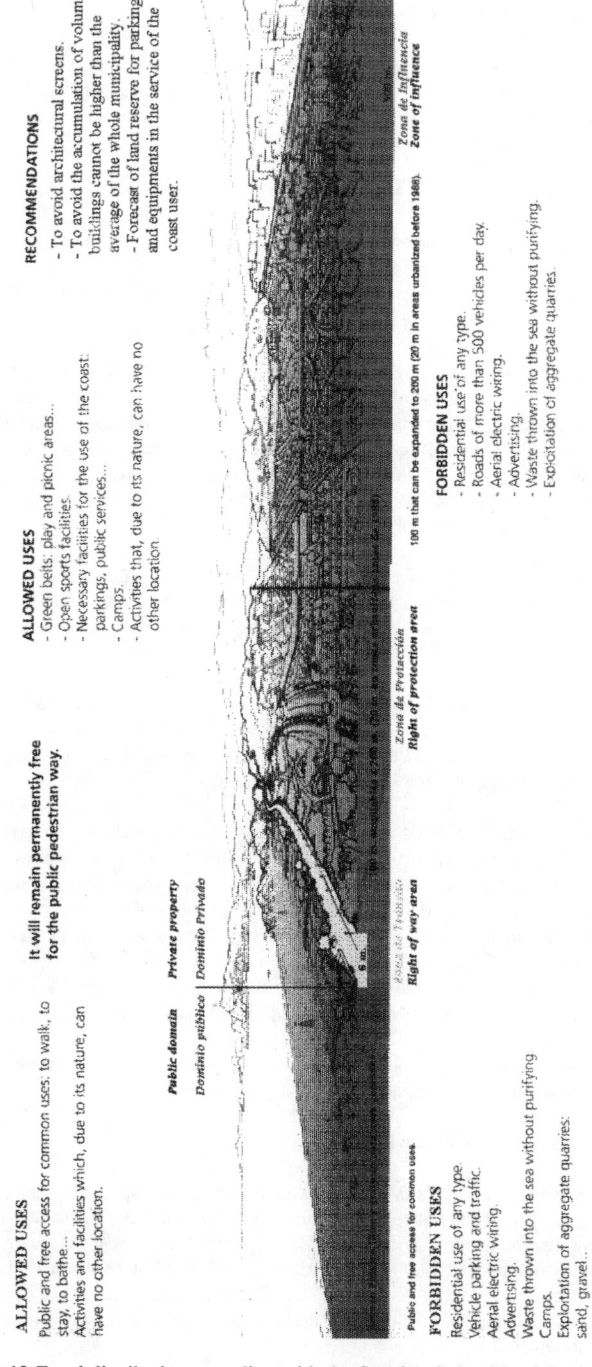

Fig. 10 Zonal distribution according with the Spanish Coastal Legislation

The Civil Service is also responsible for conducting the necessary work concerning the shore protection, preservation, and use of public domain. This includes creative beach recovery work, as well as seashore access work.

B) Competencies of the Self-Governing Communities

The competencies of the self-governing communities with regard to the regional development of the coast, are the following:

They authorize the construction of yachting harbors, provided that they have no negative influence on other stretches of their coast. It is their responsibility to authorize proper waste disposal and to delimit areas for sea farming. They also approve facilities in protected areas provided that they respect guidelines established in the Legislation on Coasts (services of general interest). At the same time, they also have to report on the limitations carried out by the Civil Service.

C) Competencies of the Municipal Governments

The municipal government takes care of the operation of seasonal services on the coast. It must assure the surveillance of the beaches to safeguard the lives of people coming to the seashore. In the same way, it is responsible for the cleanliness and hygiene of the beaches. It also issues reports on authorizations and grants to occupy the maritime-land public domain.

Legislation on Harbors

The Spanish Constitution provides that: "The State has exclusive responsibility over coastal lighting, maritime signalling and parts of general interest..." This responsibility is assumed by the Ministry of Public Works, Transportation, and Environment through its "State Ports of Spain" organization (Puertos del Estado) and the Port Authorities, Fig. 11.

State Ports are defined as being ports of *general interest*, meaning that they meet the following criteria: (1) they serve as a base for international commercial maritime activity, (2) their zone of commercial influence extends to several regions, (3) they serve industries which are of strategic importance to the national economy, (4) their annual commercial maritime traffic is of sufficiently high volume, or the type of traffic handled plays an essential role in the general economic activity of the state, and (5) their particular technical or geographical conditions are vital to the security of maritime traffic.

These ports must also have the necessary physical and technical qualities, as well as the required degree of specialization, to efficiently handle all types of traffic and ships in large volumes, thereby allowing them to compete with the major ports of Europe.

The principal aim of State Ports of Spain is to ensure that the ports contribute efficiently to the economic development of the country. Supporting initiatives, ensuring the smooth functioning of procedures, drawing up plans and designing strategies of combined action and optimizing management efficiency are just a few of the functions carried out by the State Ports of Spain. This implies revising the port policy of the government and serving as intermediary between the ports and the central administration.

Besides State Ports, there are a great number of local communities and private ports. Although only 30% of all the Spanish ports are state ports, this percentage changes if we look at the length of the quays (50%) or at the tonnage of goods imported into Spain (80%). State Ports also concentrate most of the investments made in Spanish harbors.

Since 1994, State Ports of Spain no longer depends on state budgets as a source of financing, not even for services such as maritime aids to navigation. Nowadays, State Ports service cover not only running costs and maintenance but also new construction, including major infrastructure projects such as breakwaters and dredging. These works will be analyzed below.

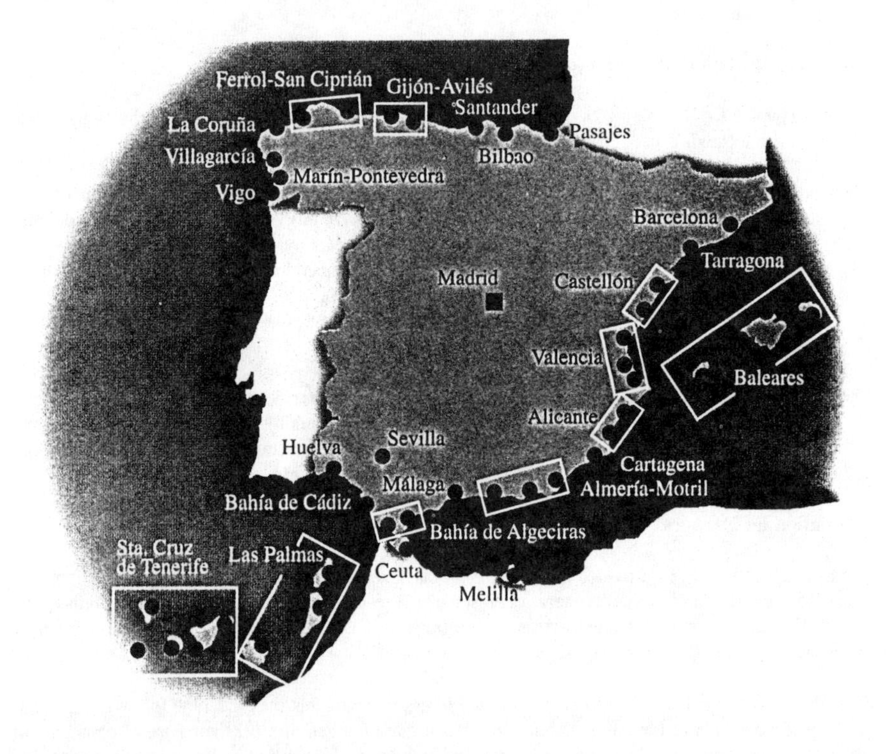

Fig. 11 Port Authorities in Spain

THE SPANISH REGULATORY GUIDE PROGRAM

The State Coast Office of the Ministry of Public Works and Transportation, (MOPTMA), and the public enterprise *Puertos del Estado* (State Ports of Spain) is developing the Regulatory Guide Program for Maritime Works. The main objective of the Program is to establish methodological guides for each of the maritime work areas, defining a minimum level of requirements and helping in the application of the European Normative in Spain.

The Regulatory Guides are written as a set of technical criteria, which are not mandatory, but which help the designers and constructors achieve the defined quality level. The Technology and Regulation Department, depending on State Ports, is responsible for the edition of each Guide, which is written by specific technical commissions. It is expected that in the near future and after several years of operating with the Regulatory Guides, they will become, properly corrected and augmented the Spanish Instruction for Maritime and Related Works.

In 1990, the first Regulatory Guide of the Program ROM 0.2-90: *Acciones en el Proyecto de Obras Marítimas y Portuarias* (Actions in the Design of Maritime and Harbour Works) was published. In addition to the ROM 0.2-90, today the following guides are available:

ROM 0.3-91: Acciones medio ambientales I: oleaje
Anejo I: Clima marítimo en el litoral español
(Waves, Annexe I: Wave climate on the Spanish coast).

ROM 0.5-94: Recomendaciones geotécnicas para obras marítimas,
(Geotechnical recommendations for maritime works),

ROM 4.1-94: Recomendaciones para el proyecto y construcción de pavimientos portuarios.
(Design and construction of harbor pavements)

ROM 0.4-95: Acciones climáticas II: viento
Anejo I: Atlas de viento en el litoral español
(Climatic effects II: wind, Appendix I: Wind Atlas of the Spanish Littoral zone).

The following guidelines are presently being written:

ROM -97: Recomendaciones para el proyecto y construcción de obras de abrigo.
(Recommendations for the design and construction of maritime works against wind waves),

ROM -96: Recomendaciones para la utilización de escolleras en las obras marítimas.
(Recommendations for the use of rubble-mound in maritimes works.)

These guides are also published in English.

In ROM 0.2-90, the available information and the criteria used to define effects on the Maritime and Harbor Works, as in local Spanish conditions, are described.

In ROM 0.3-91, the available information and the criteria for characterizing wind waves in deep water along the Spanish coastline for operating conditions as well as for extreme conditions is summarized. It helps to quickly and precisely define design wave climate.

In ROM 4.1-94, the available information and criteria for the construction of pavements in harbor areas is summarized.

In ROM 0.5-94, a geotechnical reference frame and criteria for the analysis of the most common geotechnical problems which occur in the construction of maritime works are defined. The variety of construction conditions occurring in Maritime Works, are analyzed under the same procedures yielding homogeneous input and output.

ROM 0.4-95 is an expansion of ROM 0.2-90 and is entirely devoted to the wind as one of the main actions in Maritime Works. It includes an Atlas of the Wind Regime and of Extreme Wind Regime which allows a quick evaluation of the wind conditions along the Spanish coastline.

ROM -97 presently being written, will include guides to be applied under Design and Construction processes such as:

(1) Field data recording and experimental analysis.
(2) Definition of design actions.
(3) Methods for analyzing anality and stability.
(4) Construction procedures and maintenance.

It is expected to be published by the end of 1997.

Since 1990, the application of the ROM is being extended to all the Groups undertaking maritime works, public or private, not only in Spain, but in several countries around the world.

SPANISH ENGINEERING IN THE SECOND HALF OF THE 20th CENTURY

This chapter is dedicated to Prof. Ramón Iribarren, who developed a methodology for the design of harbors. He is considered the father of the present generation of Spanish coastal engineers. Moreover, a short description of research and applied coastal engineering center is given.

Prof. Ramón Iribarren (1900-1967)

The professional career of Prof. Iribarren lasted forty years, from 1927 to 1967, first as a Harbor Engineer in Guipuzcoa, his home region, and lastly as a professor at the Civil Engineering School in Madrid. For fifteen years, he was the Director of the Laboratory of Ports, which is now known as CEPYC-CEDEX.

In order to understand his contribution to the field of coastal engineering, it is necessary to point out that he worked during a very difficult period of the 20th century. In the period from 1936-39, Spain was involved in the Civil War. These events influenced his knowledge, observations, and reasoning as he was without "outside" input. The beauty of his work lies in a combination of intuition, empiricism, and theory.

The contribution of Prof. Iribarren may be split into two parts. One is the overall methodology for the design of harbors and maritime works against wind waves, for which he established in a step-by-step procedure. This methodology is still used in Spain, and it can be said, without any doubt, that all the construction done in Spain over the last forty years follows this methodology. The second part is the specific solution he gave to some of the questions which arose during the application of previous methodology. Obviously, some of his solutions have been clearly surpassed by others, but some of them can still be applied with confidence.

The overall methodology is summarized in his book, (*Oleaje y Diques* (Waves and Breakwaters), 1964), which he wrote with his colleague, Casto Nogales (1908-1985).

Following this methodology, an engineer can: (1) define the wave climate off the Spanish coast and at any location on the shore; (2) analyze wave breaking-reflection conditions; (3) design and check the stability of a mound breakwater with or without a crown, submerged or emerged, of a vertical wall or a coastal protection; (4) study harbor tranquility problems associated with short and long waves; and (5) understand the main construction procedures and related problems.

He provided specific answers to each of the topics required in his days, giving engineers adequate tools. Some of them, are listed below.

Wave propagation diagram: "The wave plan method"

Iribarren started to develop a graphic method for wave propagation in 1929 which he applied for the first time in the expansion of the Motrico harbor, (in the Guipuzcoa region), a small fishing village in the North of Spain. After using it in several places and checking its accuracy, he published his method in the *Revista de Obras Públicas* (Journal of Public Works) (ROP) in January, 1941 and reviewed it in 1946. This paper was published in the *Dock and Harbor Authority* in 1942, in Tecnica (Portugal) in 1945 and in *Annales des Ponts et Chausses* in 1946. A complete description of his method was presented at the (17th International Navigation Congress held in Lisbon, Iribarren, 1949).

The most noticeable aspect of this method is that he considered the propagation of a "wave unit" instead of a single point as the ray method does. He defined this unit as a parallelepiped, which in planform is a square of side L/4 with a depth equal to the average depth of the corners and the central point of the square, where L is the local length wave, Fig. 12. This method includes several refinements such a diffraction at an isolated breakwater or a gap, wave reflection, and wave breaking. All the wave parameters were calculated under the frame work of the trocoidal wave theory.

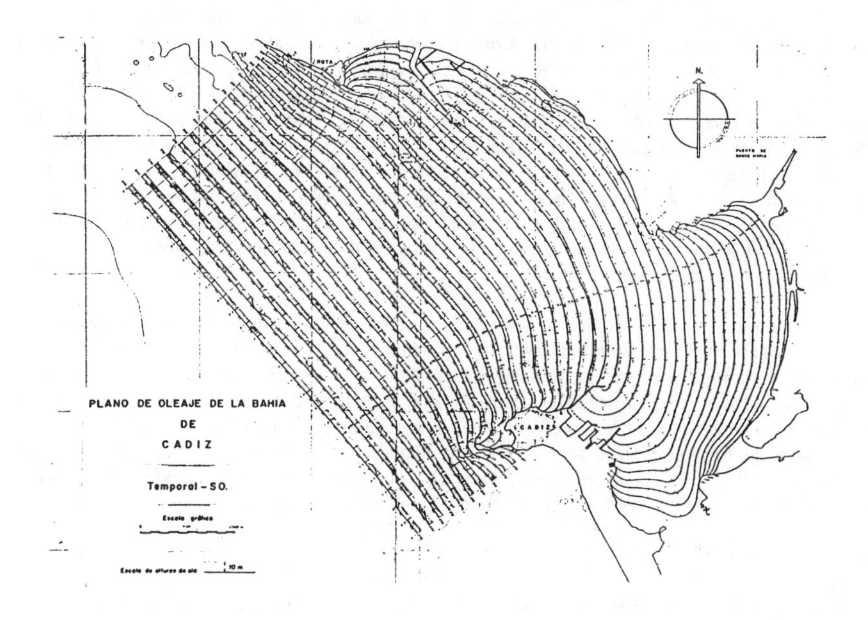

Fig. 12 Wave propagation using Iribarren's method

Wave breaking versus wave reflection

One of his best-known contributions is the evaluation of the critical average slope for which an incident clearly breaks or is clearly reflected. He published his theoretical development as well as his experimental study in the ROP, (Iribarren and Nogales, 1950). In this paper, he introduced the parameter,

$$Ir = \tan \alpha / \sqrt{(H/L)}$$

where α is the slope angle and H and L the wave height and wave length. Respectively. He proposed the value 2.3 to separate wave reflection from wave breaking. (Battjes, 1974), proposed to call this parameter "Iribarren's number" or the surf similarity parameter. Later, this wave behaviour was called "collapsing breaker", and it has been shown that this parameter defines the critical flow conditions for the stability of armor units on a slope.

Breakwater stability

Following an old Spanish tradition, Iribarren carried out research on the stability of the armor units of mound breakwaters. Before him, there was a long list of people who faced this problem and gave different formulae. Perhaps (E. Castro, 1934), opened the main patterns of research by developing the first theoretical approach to the problem. In 1938, Iribarren proposed a preliminary formula which related the weight of the armor unit to the cube of the incoming wave height. Moreover, the mound slope and the armor's relative density were taken into account. By observing the behaviour of the units on constructed breakwaters, he determined a stability parameter that depends on the type of unit.

In the early 1950s, after the construction of the first flume at the *Laboratorio de Puertos* in Madrid, he performed a large number of experiments in order to study armor stability, (Iribarren, 1951). He reviewed the theoretical analysis of the formula introducing a new parameter of the evaluation of friction and interlocking forces and their transmission through the contact points between units. Next, he determined the number of surrounding pieces involved in the stability of each armor unit. Through an ingenious experimental device, he arrived at the number five on average. Moreover, he established the size of the damaged area. Later on, (Losada et al., 1979) defined this level of damage as Iribarren's Damage, state between Initiation of Damage and Destruction. If this engineering approach had been employed in the past, some of the "new" structurally weak armor units would never have been used, and damage like that which occurred in Sines, San Ciprian, and elsewhere, would not have happened. Once the friction coefficient was established, he evaluated the parameters depending on the type of armor unit and the level of damage.

However, if research on the stability of armor units was important to coastal engineering in Spain, the criteria he established for the preliminary design of rubble mound breakwaters and vertical walls were even more important. Many of them are still used today by changing monochromatic design wave height to significant design wave height. For example, using these criteria, after the evaluation of the weight of the armor unit, the width of the main layer and of all the secondary layers, the crest level of the core, the berm height and width, the crown dimensions, and the toe berm size can be defined. In summary, he proposed engineering guidelines which have been used for over thirty years in Spain, a period of time in which more than one hundred breakwaters was built.

Harbor resonance

Most of the fishing harbors on the Cantabrian Coast (The Bay of Biscay) used to suffer problems of harbor resonance. Generally speaking, they are located in shallow water (5 - 10 m depth) and have a rectangular plane shape with vertical walls. Iribarren faced the problem of harbor resonance in several harbors in Guipuzcoa, and particularly in Motrico in 1932. He measured the harbor oscillations and proposed to dredge the basin to change the harbor oscillation period. Some of his work on this topic can be found in (Iribarren et al., 1948) and (Iribarren et al., 1958).

Nearshore circulation

In 1940, Iribarren analyzed the tidal inlet of the River (Bidasoa), which is located on the border between France and Spain. He designed a jetty, which was built on the Spanish side, allowing the development of a beach, which is now Hondarribia Beach. During the design period, Iribarren tried to understand the nearshore circulation along perforated tubes. He postulated that the circulation inside the surf zone was driven by the gradient of the set-up. Although this picture was not complete he understood the main idea, because on the spit, the long waves arrive at the beach almost parallel to the coastline, and the beach is located in the shallow area of Cape Higuer.

Research and Applied Centers

Following Iribarren's legacy with the encouragement and funding of the Administration, particularly the *Ministerio de Obras Públicas, Transporte y Medio Ambiente (MOPTMA)* (Ministry of Public Works, Transportation and Environment) and the *Ministerio de Educación y Ciencia* (Ministry of Education and Science) several centers and university groups are working in coastal engineering. Ascribed to the MOPTMA is the *Laboratorio de Puertos y Costas*, CEDEX in Madrid which has large facilities including a directional wave basin. There are several research groups associated with different Spanish Universities: University of Cantabria ((Losada, 1990), (Medina et al., 1994), (Vidal et al., 1995), (Losada et al., 1995)), Catalonia University of Technology ((Arcilla, A.S. et al., 1990), (Rivero, F.J. et al., 1995), (Jiménez, J. et al., 1993), (Rodríguez, A. et al., 1995)), and Polytechnical University of Valencia ((Medina, J.R. et al., 1985), (Medina, J.R. et al., 1990), (Medina, J.R. et al., 1994)). These Universities have their own lab facilities. The Polytechnical University of Madrid has the oldest chair in the field of coastal engineering, initially occupied by Prof. Iribarren. Today the chair is occupied by Prof. Suárez Bores who introduced short-term and long-term analysis of irregular waves in Spain. He defined the Wave Regime (1967) as the annual distribution of significant wave height and developed a method to evaluate it from wind regimes for short fetches.

Furthermore, he defined the extreme wave regime or storm regime as the distribution of the most severe storms in N years considering one storm per year. Based on the storm regime, he developed a risk analysis (1969) for the maritime works against wind waves.

SOME EXAMPLES OF COASTAL AND HARBOR WORKS

In this section a brief summary of the main coastal and harbor works undertaken in Spain during the past few decades is presented.

Coastal Works

The Spanish coast is the basic source of Spanish economy.

In this economic framework, Spain is conducting a 5-year program (1993 to 1997) to restore existing beaches and build new ones. To give us an idea of the magnitude of this investment it should be pointed out that this amound is greater than the money spent in the U.S. on beach-restoration during the past 40 years. This plan, conducted by the *Dirección General de Costas* (*DGC*) (State Coast Office) of the MOPTMA develops the necessary steps to restrain the destruction of the littoral and restore its natural values.

The works carried out by the *DGC* along the 8,000 km of Spanish littoral have stopped the damage of coastal areas and have increased the dry area of many beaches. To carry out this work, very often improperly located buildings invading the beach have to be torn down. In addition to the investment in coastal protection and beach restoration pedestrian and recreational zones were designated and constructed for purposes such as strolling, swimming and relaxation. The creation of new beaches and the restoration of other coastal zones, such as swamps, dune systems, cliffs or salt marshes were also carried out. On rocky stretches of coast, work has been undertaken to facilitate access to the sea and to develop recreational areas.

The projects provided for in the Coastal Plan aim at improving environmental quality of the coast and, therefore, quality of life. Furthermore, they contribute to the economic revitalization of the area.

Some examples undertaken during the Coastal Plan

1. Demolition of the thatched huts located in the Elche and Guardamar del Segura Municipalities (Alicante).

Legal procedures were initiated in 1982 against the thatched huts in the Pinet, Rebollo and Pesqueros beaches (Municipalities of Elche and Guardamar del Segura) and a ruling was issued by the Valencia Territorial Courts on February 10, 1989. The demolition operations started on May 9 of the same year and included the cleaning and sifting of the sand, (Fig. 13a and Fig. 13b). Budget: Pesetas 26,7 million.

Fig. 13a Pinet, Rebollo and Pesqueros beaches before the coastal action

Fig. 13b Pinet, Rebollo and Pesqueros beaches after the coastal action

2. Regeneration of Maresme beaches: Premia de Marmongat (Barcelona).

The Maresme coast, from the outlet of the Tordera river up to Mongat was sandy in its natural state. The older inhabitants of the area remember how they used to play on the beach as children.

The situation changed drastically with the construction of the port of Arenys de Mar, blocking the principal source of sediments for the beaches in the area, proceeding from the Tordera river. Therefore, while one could say that the beaches to the North of the port are generally quite healthy, to the South they clearly suffered severe damage. The stretch restored with this action comprised the municipalities of Mongat, El Masnou, Premiá de Mar, and to a lesser extent, Badalona, and meant the returning to this coast (of approximately 7 km) on the natural appearance it had 20 or 30 years ago, before the degeneration process commenced. The work consisted of the construction of a breakwater to prevent the accretion of the port of Premiá de Mar, and in the artificial supply of 2,045,000 m^3 of sand, Fig. 14. Budget: 509.8 million pesetas.

Fig. 14 Maresme beaches before and after the coastal action

Breakwaters in Spain

The following information about Spanish breakwaters has been extracted from the publication (*Catálogo de Obras de Abrigo de España* (Catalog of Spanish Protective Works), 1988).

Breakwater length

The total length of Spanish breakwaters is approximately 180 km. About 51% of them are on the Atlantic coast while the remaining 49% are located on the Mediterranean coast, Fig. 15. The average length of all the breakwaters is 625 m with small differences between the Atlantic and Mediterranean coast. However, there are some breakwaters that by far exceed this length. For instance, the King Juan Carlos I breakwater in Huelva in the Gulf of Cadiz has a length of 15 km.

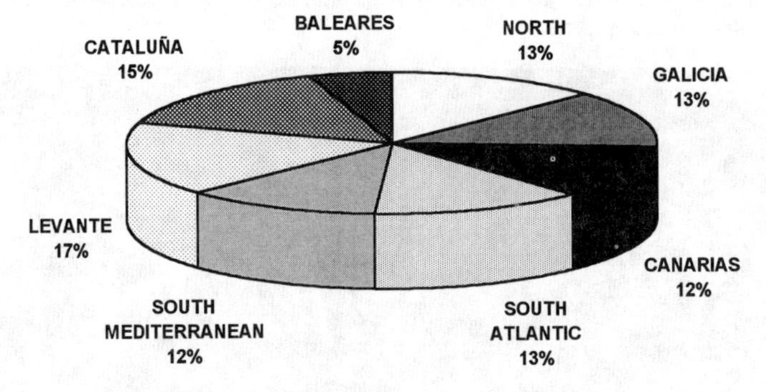

Fig. 15 Breakwater distribution along the Spanish coast

The mean breakwater length changes depending on its location in the harbor layout. Breakwaters parallel to the coast, about 66% of the total number of breakwaters, have a mean length of 700 m. Perpendicular breakwaters (groins), about 20%, have a mean length of 430 m. Detached breakwaters have a mean length of 375 m. There are also significant differences in the length according to the function of the harbor. Sports and fishing harbor breakwaters have a mean length of 500 m, while commercial harbor breakwaters have a mean length of 1,300 m, Fig. 16.

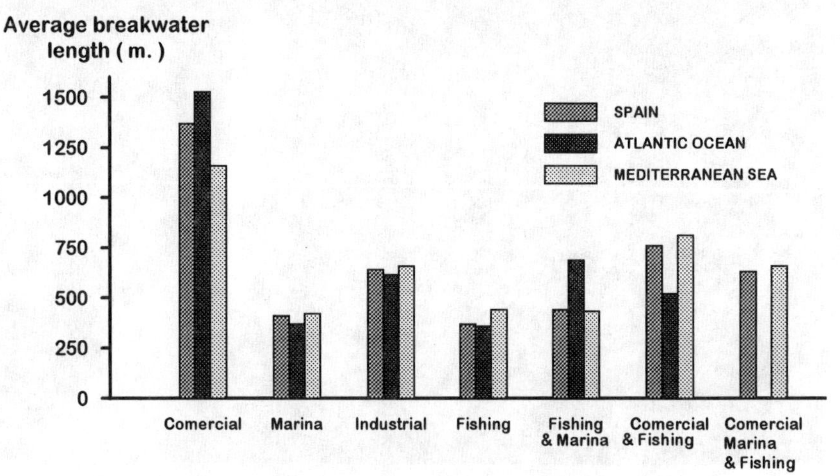

Fig. 16 Average breakwater length versus harbor function

Breakwater water depth

About 42% of the breakwaters are located in water depths of more than 10 m, Fig. 17. The percentage is larger on the Atlantic coast, 61% of the total, and smaller on the Mediterranean coast, 33%. The Canary Islands have the greatest number of breakwaters constructed in water depths greater than 25 m and these breakwaters are also longer, 23 km in those depths. Four breakwaters are located at a 50 m depth, and one is located at a 78 m water depth (Santa Cruz de Tenerife harbor).

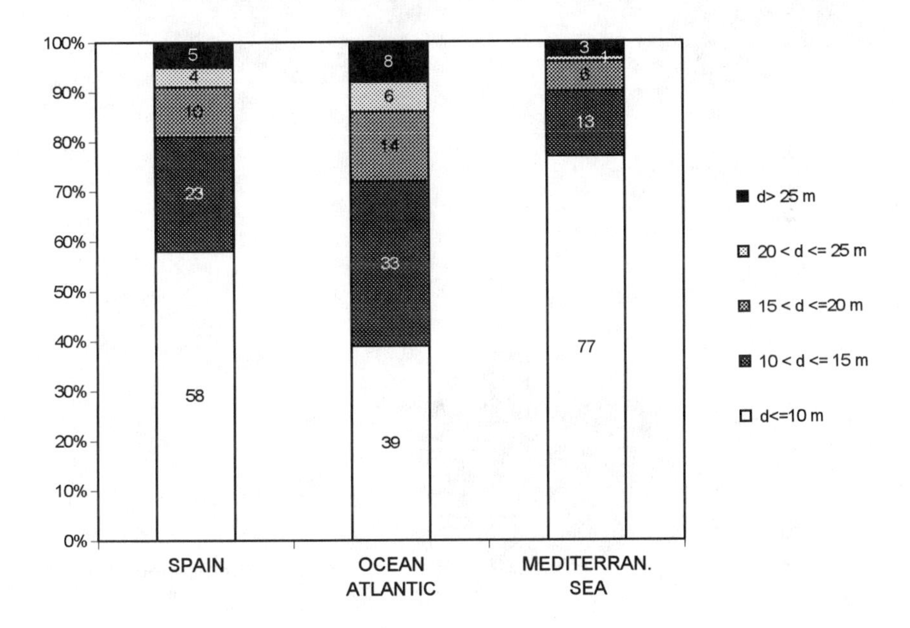

Fig. 17 Breakwater water depth vs. location along the Spanish coast

Number, length, water depth of breakwater sections

The mean number of different breakwater cross-sections within a breakwater is 1.67. This number decreases on the Mediterranean coast up to 1.5 and increases on the Atlantic coast up to 2.0. The mean length of a cross-section is 310 m while this figure is 289 m on the Mediterranean Coast and 336 m on the Atlantic coast. According to the water depth 49% of the sections are located in areas of less than 10 m water depth, 22% between 10 - 15 m 12% between 15 - 20 m, 7% between 20 - 25 m and 10% in areas of more than 25 m of water depth.

Typology

Most of the Spanish breakwaters (83%) are mound breakwaters, Fig. 18. The percentage is larger on the Mediterranean coast, with 88%, and smaller on the Atlantic coast, 79%. Vertical breakwaters represent 8% of the total breakwaters. Composite breakwaters can be found in Galicia (in the Northwest) where they represent 12% of the total, but on the Mediterranean coast, this type is almost absent (0.5%). On the other hand, low-crested breakwaters can be found only on the Mediterranean coast where 4% of the breakwaters are of that kind. Furthermore, low-crested breakwaters are always in areas where the water depth is less than 10 m.

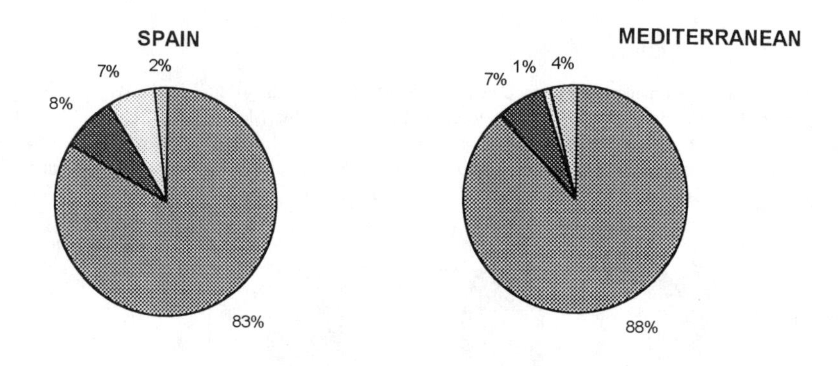

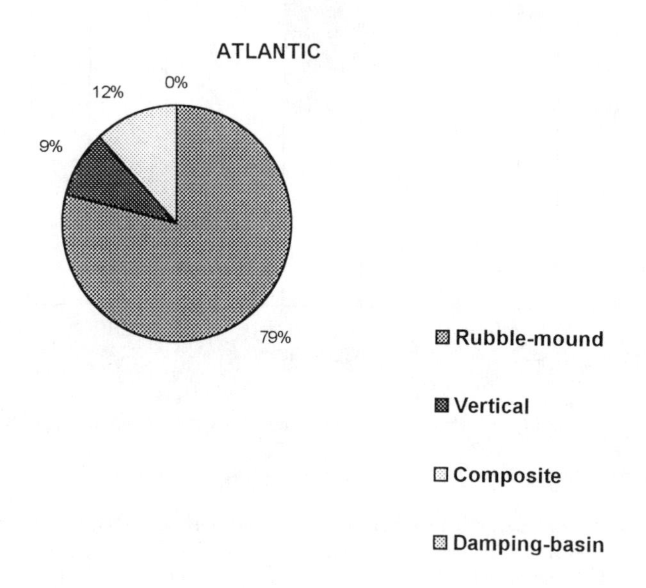

Fig. 18 Breakwater typology distribution along the Spanish coast

Damage

Including minor and major damage, 38% of the sections have suffered some damage (44% on the Atlantic coast and 32% on the Mediterranean coast). This parameter reaches a minimum in the Canary Islands with 7% and a maximum in Galicia with 57%, Fig. 19. About half of the above-mentioned percentages refer to major average length of an area of major damage is about 340 m larger than the mean length of a typical cross-section 310 m. Damage percentages have a maximum at sections located in a water depth between 10 - 15 m.

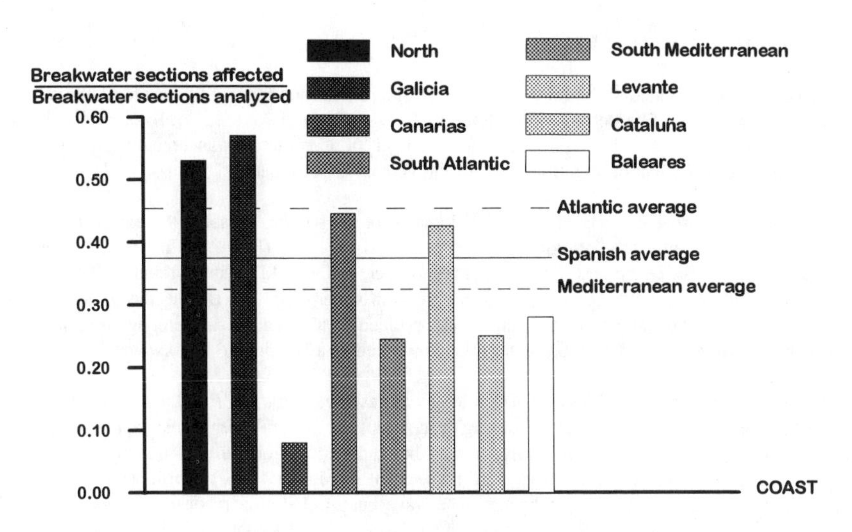

Fig. 19 Breakwater damage distribution along the Spanish coast

Although all types of breakwaters have suffered damage, this has been especially significant for composite and low-crested breakwaters respectively 60% and 51% of the length of these types of breakwaters have resulted in major damage. Fourteen percent of the mound breakwaters (118 km) have also been affected by major damage. Because the mound breakwater is the most popular cross-section, 125 km of mound breakwater damage (minor and major) represent 75% of the total damage.

The Twenty-first Coastal Engineering Conference

In June 20-25, 1988 the twenty-first coastal engineering conference was held in Torremolinos, Malaga, Spain a world known tourist village in the Costa del Sol, under the auspicies of the coastal engineering research council of the American society of civil engineers. It was organized by the *Colegio de Ingenieros de Caminos, Canales y Puertos* of Spain and cosponsored by the Directorate General for Ports and Coasts of the Spanish Ministry of Public Works, and the Polytechnical Universities of Madrid, Barcelona, Santander and Valencia.

During the conference, 225 papers were presented and over 1500 people attended the sessions. The proceedings of the conference were published in three volumes edited by Billy L. Edge.

The conference was a great technical success, but also a great social event, thanks to the dense and well organized programme of activities developed after the daily sessions. Felipe Martínez, at that time Director of the *Laboratorio de Puertos y Costas* was the ultimate responsable of the organization.

SOME EXAMPLES OF HARBOR DEVELOPMENT

This section is dedicated to the development of two Spanish harbors: the Commercial Harbor of Bilbao, and the Fishing Harbor of Luarca. The main idea of this section is to describe how two harbors, different in size and activity (commercial, fishing) have envolved during their history. Consequences of the new requirements have been a transformation in their configuration and coastal defense structures.

The Commercial Harbor of Bilbao

The city of Bilbao, located at the northeast coast of Spain, was founded in 1300 as an Administrative Center for the control of harbor activities along the Nervion River and the Bay of Bilbao. In 1511, the Consulate of Bilbao, an old version of the Chamber of Commerce was created. In 1872, the Administration of the harbor was transfered to the Federal Government.

At that time, the entrance bar limited the development of the harbor. To solve the entrance problem the construction of the jetty of Portugalete was started in 1877, (Fig. 20). In 1901, the Harbor Authority finished the construction of the east breakwater, and, in 1902, King Alfonso XII placed the first stone of the breakwater *Dique de Santurce* in 20 m water (Fig. 21). During the construction, a storm destroyed part of the breakwater, and it was decided to start the construction again leewards of the destroyed structure, under its protection which worked as a submerged breakwater.

In the early 1970s, a new breakwater 2,500 m long was designed, *Dique de Pta. Lucero*, in 33 m water depth. Similar to the *Dique de Santurce*, during the construction, several storms delayed the completion of the works for several years. Actually, it may be said that the quantity of quarry used for the construction of the core was enough to build it twice. In December 1976, a storm with $Hs > 8.5$ m damaged several sections of the breakwater. The wave buoy failed after recording a wave height of 16 m. The breakwater was rebuilt with a new main layer of 150 Tn concrete blocks (Fig. 22).

Fig. 20 Portugalete pier, 1887

Fig. 21 Santurce breakwater under construction, 1902

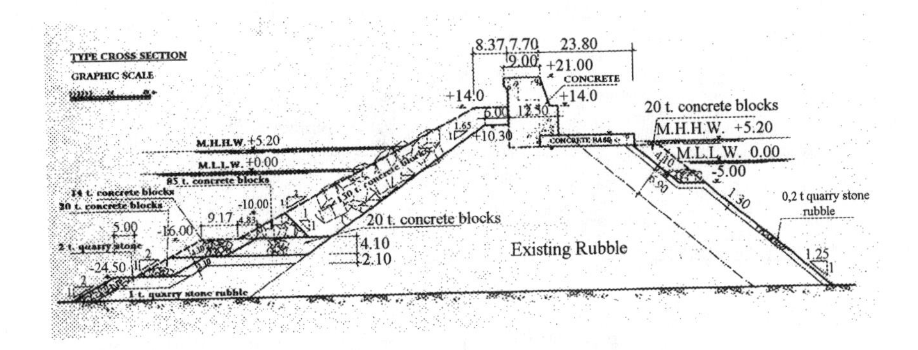

Fig. 22 Punta Lucero breakwater under construction, 1902

Nowadays, a new 3,150 m long breakwater is being built in the leeside of the *Dique de Pta. Lucero* (Fig. 23). The cross section of the breakwater is the traditional section used in Spain, following Iribarren's methodology: a main layer with a screen wall. The armor units are concrete 100-Tn. blocks (Fig. 24). The head of the breakwater is built with a caisson of approximately 29 m length. Again, during the construction, it has suffered some damages.

Bilbao is a very good example of the difficulties coastal engineers are facing to provide adequate protection against the wind waves generated in the Bay of Biscay.

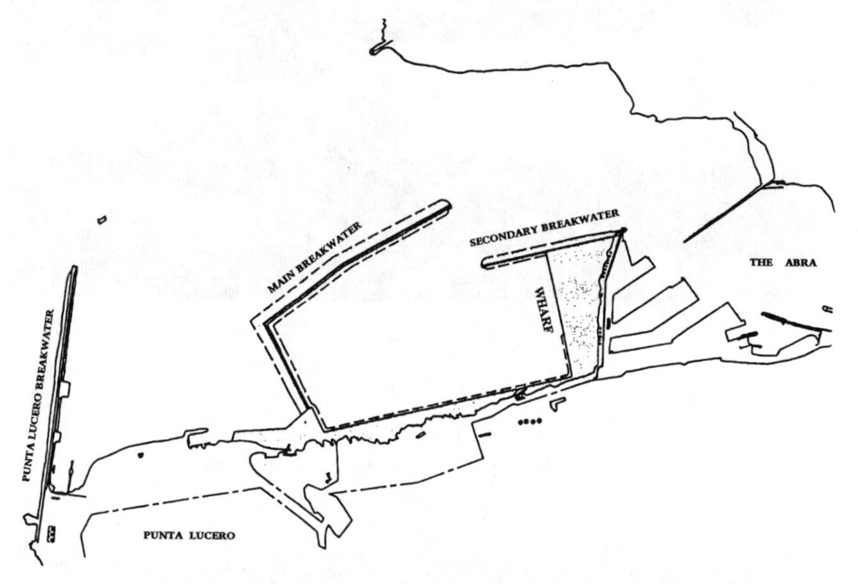

Fig. 23 Development of the Port of Bilbao in the outer Abra. Main breakwater cross-section

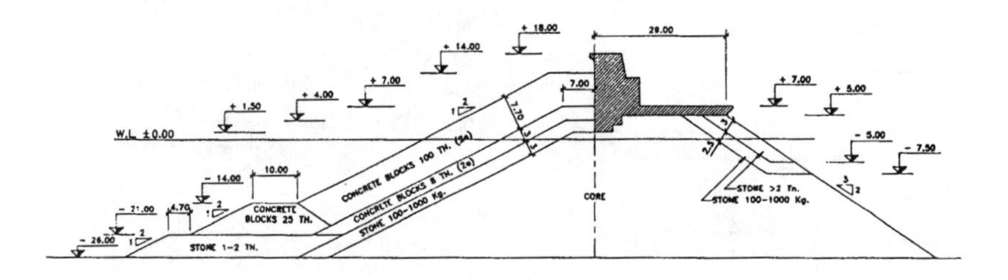

Fig. 24 Development of the Port of Bilbao in the outher Abra. Main breakwater cross-section

The Fishing Harbor of Luarca

Luarca is a fishing village on the Cantabrian Sea whose economy has been strongly dependent on fishing for over a thousand years. In this respect, the use of small rowing boats for whale fishing is well documented and now, of course, the activity extends over other species and into the North Atlantic Ocean.

The harbor is located inside a medium size bay protected by a vertical breakwater built in the 1930s. In the 1960s, the breakwater was reinforced with 20-Tn blocks. In 1988, a new mooring line was built, and the mooring area was enlarged. Main reasons for the new configuration of the mooring area were resonance problems and the requirement of a higher mooring surface. To protect the new harbour the following sequential construction was undertaken: (1) A vertical porous screen was placed at the basin entrance; (2) a sloping porous protection was added to the walkway inside the harbor, and (3) the old breakwater was rebuilt.

The main problems of the old breakwater can be summarized as follows: (1) Functionality Minimum harbour protection, because of insufficient height (2.5 m over high tide level), length and width; (2) Stability - Although the vertical wall was stable (even if it was completely overtopped by big waves) it had grooves and cavities along it, making maintenance necessary every year; (3) Slope stability - Protection blocks, built in the 1960s had almost disappeared, leaving a few units which are rounded or cracked. Furthermore, certain restrictions were imposed on the design and construction of the new breakwater. To avoid the transportation of materials and equipments through the village, a new access had to be built on a cliff. And as much as possible, the beautiful landscape resulting from the combination of the old harbour, the graveyard, the village, the cliffs and the sea had to be taken into account.

In order to fullfil the previous requirements, it was decided to design a breakwater which avoided overtopping mainly by width rather than by height. Furthermore, once the work was finished the new access was transformed into a pedestrian way. Consequently, the retaining wall and neighbourhood were built with high-quality materials, imitating the local rocky cliff. The old vertical wall was maintained in such a way that it is still seen from the village. The transition of that wall and the new crown has been clearly indicated by using steel clamps. Finally, a high tower with suitable shape was built on the breakwater head.

The final breakwater section is shown in Fig. 25, where the most characteristic figures are also given. The breakwater was enlarged 25 m. Under design storm conditions, which correspond to a maximum wave height limited by bottom, H = 9 m there is an overall reduction coefficient of K = 0.1 at the entrance to the basin. In order to avoid reflection of waves from the wall which protects the new access area, it was reinforced with 20-Tn blocks. Figures 26a and b show Luarca harbor before and after construction.

Luarca is an example of the extensive programmes that the Spanish self-governing Communities at the coast have been carrying out in the past decade in order to improve the fishing harbor operation conditions.

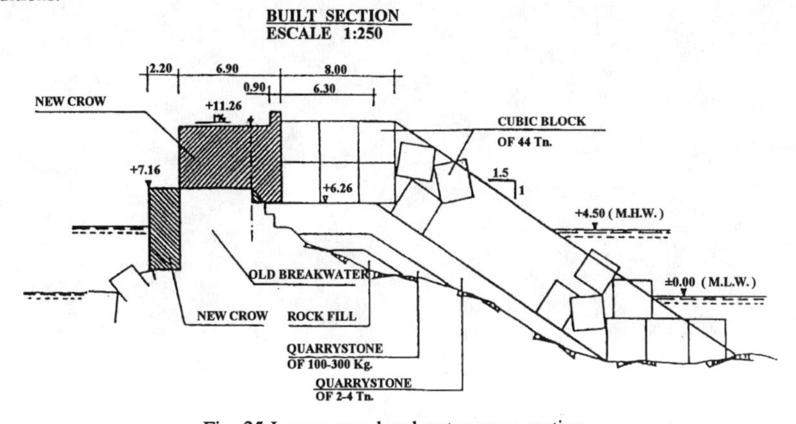

Fig. 25 Luarca new breakwater cross-section

Fig. 26a Luarca old breakwater under storm conditions.

Fig. 26b Luarca harbor today.

CONCLUSIONS

This chapter is devoted to the history of coastal engineering in Spain, a country strongly related to the sea. Eight thousand kilometers of coastline under at least four different climatologies are a clear indication of the difficulties the Spanish coastal engineers are facing.

In the last twenty five years because of the economical expansion and the tourism boom, coastal engineering in Spain was one of the most important branches of civil engineering.

In order to respond properly to this demand, the coastal engineers developed several aspects of their professional activity, which may be summarized as follows:

1. A new Law of Coastal Protection.
2. A new Law of Harbor Organization.
3. A Regulatory Guide Program.
4. A Network for Maritime Climate Recording.
5. Investment Plans for Harbor and Coastal Development.
6. Creation or improving of several Applied Research Centers and University Coastal Engineering Programmes.

To have an idea of the effort carried out by the Spanish coastal engineers it is pointed out that the investment of the *Plan de Costas* during the period 1993-1997 is greater that the amound spent by the USA on beach restoration programme in the last 40 years.

Most of the maritime work developed in Spain in the last twenty five years has been design and built following the methodology established by Prof. Iribarren. Moreover, Iribarren can be considered the father of a new generation of coastal engineers and associated applied research centers. After more then 300 major harbors, on average 3.5 harbors every 100 km of coast and over 180 km of breakwaters being built, it can be said that Spanish coastal engineers have a very large experience in the design and construction of maritime works against waves and in coastal protection.

ACKNOWLEDGEMENTS

The authors would like to thank State Ports of Spain, Maritime Climate Program, State Coast Office and Harbor Authority of Bilbao for providing valuable information, figures, and photographs.

REFERENCES

Arcilla, A.S. (1990). "Surf-zone hydrodynamics". CIMNE Pineridge Press, Barcelona, ISBN 84-404-6502-5, 310 pp.

Battjes, J. (1974). "Surf similarity". *Proc. 14th Coastal Engineering Conference*. ASCE, Copenhagen. pp. 466 -480.

Castro, E. (1934). "Diques de escollera". *Revista de Obras Públicas*. Feb., pp. 72 - 78 (In Spanish).

CEHOPU (1994). "Puertos Españoles en la historia". MOPTMA. Madrid. 389 pp. (In Spanish)

(1555). "Código de las siete partidas de Alfonso X el Sabio". Edición en facsimil. Salamanca. (In Spanish)

Iribarren, R. (1938). "Una fórmula para el cálculo de diques de escollera". *Ed. Bermejilla y Hª*. Pasajes. 21 pp. (In Spanish).

Iribarren, R. (1941). "Obras de abrigo de los puertos". *Revista de Obras Públicas*. nº 1709, pp. 13 - 25. (In Spanish)

Iribarren, R. (1946). "Planos de oleaje. Influencia de la pendiente del fondo en la altura de ola". *Revista de Obras Públicas*. Madrid, 14 pp. (In Spanish).

Irribarren, R. y Nogales, C. (1948). "Corrientes y oscilaciones de resacas en el interior de los puertos". *Revista de Obras Públicas.* Abril, pp. 150 - 157. (In Spanish).

Irribarren, R. (1949). "Protecion des Ports". *Proc. XVII International Navigation Congress.* Section II. Chapter 4. Lissabon. (In French).

Iribarren, R. y Nogales, C. (1950). "Talud límite entre la rotura y la reflexión de las olas". *Revista de Obras Públicas,* nº 2818, pp. 65-72. (In Spanish)

Iribarren, R. y Nogales, C. (1950). "Generalización de la fórmula para el cálculo de los diques de escollera y comprobación de sus coeficientes". *Revista de Obras Públicas,* nº 2818, pp. 227 -239. (In Spanish)

Irribarren, R. and Nogales, C. (1951). "Generalization of a formula for rubble-mound breakwater calculation". *The Buletin of Beach Erosion Board.* Vol. 5, No. 1.

Irribarren, R., Nogales, C. y Fernández, P. (1958). "Onda de resaca en los puertos. Ensayos de resonancia en modelos reducidos". *Revista de Obras Públicas.* Feb., pp. 69-81. (In Spanish)

Iribarren, R. and Nogales, C. (1964). *Obras marítimas. Oleaje y diques.* Editorial Dossat, S.A. Madrid. 376 pp. (In Spanish)

Jiménez, J. and Arcilla, A.S. (1993). "Medium-term coastal response at the Ebro delta, Spain". *Marine Geology,* 114, pp. 105 - 118.

Losada, M.A. and Giménez-Curto, L.A. (1979). "The joint effect of the wave height and period on the stability of rubble mound breakwaters using Iribarren's number". *Coastal Engineering.* Vol. 3, No. 2. pp. 77 - 97.

Losada, M.A. (1990). "Recent developments in the design of mound breakwaters". In *Handbook of Coastal and Ocean Engineering.* Vol I. Edit.: J. B. Herbich. Gulf Publishing Company, Houston. pp. 940 - 1045.

Losada, M.A., Ortín, F., Vidal, C. and del Busto, J.A. (1992). "Damage and repair work performed on some mound breakwaters in Spain in coastal structures and breakwaters". *Institution of Civil Engineers.* pp. 321-334.

Losada, M,A., Martín, F.L., Medina, R. (1995). "Wave kinematics and dynamics in front of reflective structures". *In Wave Forces on Inclined and Vertical Wall Structures.* ASCE, New York, pp. 282-310.

Medina, J.R., Aguilar, J., and Diez, J.J. (1985). "Distortions associated with random sea simulators". *Journal of Waterway, Port, Coastal and Ocean Engineering,* ASCE, 111(4), pp. 603-627.

Medina, J.R., and Hudspeth, R.T. (1990). "A review of analyses of ocean wave groups". *Coastal Engineering,* 14 (1990), pp. 515-542.

Medina, J.R., Hudspeth, R.T., and Fassardi, C. (1994). "Breakwater armor damage due to wave groups". *Journal of Waterway, Port, Coastal and Ocean Engineering,* ASCE, 120 (2), pp. 179-198.

Medina, R., Losada, M.A., Losada, I.J. and Vidal, C. (1994). "Temporal and spatial relationship between sediment grain size and beach profile". *Marine Geology.* Vol. 118, pp. 195 -" 206.

Ministerio de Obras Públicas y Transportes (1988). *Catálogo de obras de abrigo en España.* Serie Monografías. Centro de Publicaciones, Madrid, Spain. (In Spanish).

Ministerio de Obras Públicas y Transportes (1993). *Recovering the coast.* Serie Monografías. Centro de Publicaciones, Madrid, Spain.

Puertos del Estado (1990). ROM 0.2-90. *Actions in the design of maritime and harbour works.* MOPTMA. Madrid. 264 pp.

Puertos del Estado (1991). ROM 0.3-91. *Waves. Annexe I: Wave climate on the Spanish coast.* MOPTMA. Madrid. 76 pp.

Puertos del Estado (1992 - 1996). *Recomendaciones para obras marítimas.* Six volumes. Ministerio de Obras Públicas y Transportes. Madrid. (In Spanish)

Puertos del Estado (1994). ROM 4.1-94. *Design and construction of harbour pavements.* MOPTMA. Madrid. 164 pp.

Puertos del Estado (1995). ROM 0.5-94. *Geotechnical recommendations for maritime works.* MOPTMA. Madrid. 430 pp.

Rivero, F.J. and Arcilla, A.S. (1995). "On the vertical distribution of <uw>". *Coastal Engineering*, 25, pp. 137-152.

Rodríguez, A., Arcilla, A.S., Redonds, J.M., Bahía, E. and Sierra, J.P. (1995). "Pollutant dispersion in the nearshore region: modelling and measurements". *Water Science Technology*. Vol. 32, No. 9-10, pp. 169 - 178.

Suárez Bores, P. (1969). *Apuntes del oleaje*. E.T.S. I. de Caminos, Canales y Puertos de Madrid. (In Spanish).

Suárez Bores, P. (1967). *Estructura del oleaje*. Ministerio de Obras Públicas, Vo. 4-B. Madrid. (In Spanish).

Suárez Bores, P. (1974). "Sea observation in coastal areas. The Spanish offshore network". *Waves 74*. New Orleans.

Suárez Bores, P. (1979). "Multivariate risk and optimum economic analysis in maritime structures". *Revista de Obras Públicas*. Vol. pp. 315.

Vidal, C., Losada, M.A., Mansard, E.P.D. (1995). "Stability of low-crested roubble-mound breakwater heads". *J. of Waterway, Ports Coastal and Ocean Engineering*, ASCE. Vol. 121, No. 2, pp. 114-122.

HISTORY OF COASTAL ENGINEERING IN TAIWAN

Ching-Ton, Kuo[1]

ABSTRACT: Coastal engineering in Taiwan started from historical needs. Coastal environments of Taiwan is first introduced to illustrated the natural characteristics of coastline around the island. The development of this interdisciplinary field is then revisited in the sequences of time. Progresses of coastal engineering work are reported for land reclamation, coastal defense, harbor constructions and coastal zone management. Before the closing remark, coastal research program and activities of integrated efforts are described.

COASTAL ENVIROMENT OF TAIWAN

Taiwan is an island surrounded by the Pacific Ocean and Taiwan Strait, with Pen-Hu islands (Pescadores) located between China mainland and Taiwan island. It has a coastline of 1,600Km long and area of 36,000Km^2. The interior geography is primarily mountainous area and a narrow band of plain along the western coast where 21 million of population live. The geographical characteristics of Taiwan coast can be broadly classified into four different coastal regions. The southern end of island has a coal reef coast, the northern and north-eastern coast is rocky. The eastern coast facing the Pacific Ocean is a sandy-pebble beach with a steep slope of 1/7~1/10. The western coast facing Taiwan Strait is sandy . More Specifically, the south-western coast that bordered by sand barriers with shallow lagoons on the inside extending for more than 150Km. The middle-western coast has a very gently slopping foreshore of 1/500~ 1/2,000 with a large tidal range. The shoreline between high and low tides may extend 4Km long, which is so called the tidal land.
The tidal waves move from the Pacific Ocean that swing from the southern and

---

1) Professor, Department of Hydraulics and Ocean Engineering, National Cheng-Kung University, Tainan, TAIWAN.

northern ends of island, then move to middle western coast along Taiwan Strait. The mean tidal range at east coast is 0.95m, 0.64m at northern end, 0.52m at south end and increases to 3.7m at middle of western coast.

Predominant waves are generated by monsoon in winter from NNE~NE direction during October to March. The average significant wave height in winter is 1.8m with wave period of 7.5sec along eastern and northern coasts. But when the wind blows continuously over two days, the waves could grow to a significant wave height of 6m with wave period of 9sec. Waves in Taiwan Strait are always smaller than those of northern coast by 20% due to the limited fetch of Taiwan Strait. The average significant wave height in winter is 1.3m with period of 5.7sec in Taiwan Strait. Moreover, the wave height toward the south-western coast generally further decreases due to sheltering of the island itself.. Therefore, it is always calm in winter with the average significant wave height of only 0.9m and period of 5.3sec. In summer, swells approach from the Pacific Ocean to the eastern coast. The dominant wave direction is EW and the average significant wave height is 1.2m with period of 8.0sec. In southern coast, swells arrive from the South China Sea with SW direction. There, the water is generally calm in summer, with the average significant wave height of 0.75m and period of 5.4sec. But south-west monsoon winds usually induce large swells of 2m wave height, which are the primary causes of coastal erosion of southern coast. Typhoon- generated strong seas and swells with storm surges impinges on coast margin of this island and always cause severest damage. The significant wave height of 13.9m and period of 14.1sec was recorded in 1990 by Typhoon Ofelia on eastern coast. The predicted extermal waves by typhoons of different return periods are listed in Table 1.

Table 1. Extermal waves by typhoons with different return periods

Return period(yrs)	10		25		50	
Location	Wave height (m)	wave period (sec)	wave height (m)	wave period (sec)	wave height (m)	wave period (sec)
northern coast	9.4	12.3	11.2	13.4	12.6	14.2
north-western coast	6.3	10.0	7.5	11.0	11.8	9.7
mid-western coast	5.2	9.1	6.2	9.9	6.9	11.3
south-western coast	6.6	10.8	7.7	11.6	8.5	12.2
eastern coast	9.5	12.6	11.3	13.8	12.6	14.6

A HISTORY RIVEW

Depleted by the World War II, Taiwan‘s economy remained devastated with national per capita income at around US$150 till 1960s. Agriculture was the main economic activity. Sugar, banana, and rice are the main exporting products. Then, there were only three commercial harbors, with Keelung on the north, Kaohsiung on the south and Hwalien on the mid-east. There were less than ten small fishing harbors scattered around the island and Pen-hu islands.

Wuchi (or Niikata) harbor was one of those small fishing harbor which lies in mid-western coast of Taiwan at that time. It could only host 20-30 crafts, and could only be accessed through tidal channels during a high tide. During the period of Japanese occupation of Taiwan, the Japanese started their field investigation and planing work in 1936. They intended to make it as a base for their southward-expansion activities and as a commercial harbor to import steel and iron from Southeast Asia. They establisthed a 2.3 kilometer mound foundation with ten caisson-breakwaters on the foreshore and dredged a 7 meter wind mavigation channel. Howerer, the whole project was suspended because of financial difficulty during the war.

Construction of harbor along sandy coast especially on a tidal flat, is a difficult task. During the 1950s, Wuchi harbor, which was later renamed as Taichung Harbor, was evaluated for possible expansion to a commercial harbor under the General Dajia River Development Project. Professor Lin-wu Tan, who was then the Director of Engineering Office of Taichung Harbor under Keelung Harbor Authority, went to Japan for inquiry and invited Professor Takeshi Ijima , who was then served in the Port and Harbor Research Institute (PHRI), Japan, to Taiwan for technical consultation on field investigation of sand drifts and hydraulic model tests. The new and advanced knowledge of coastal Engineering was then introduced in to Taiwan. It was under his advice that a 80m×1m×1.2m wave flume and a 40m×16m×0.8m wave basin was first constructed in 1959, with Professor In-Ben-Dai in charge of the sediment transport model tests of Taichung Harbor.

As Taiwan's industrial technology was not yet well developed then, an 10 horsepower hinge type wave generator was installed. It was equipped with a variable speed gear box and an adjustable chain belt to adjust wave period. Waves were measured by resistance type wave sensors and recorded by two elements pen-write oscillograph. In order to simulate topographical impacts of Taichung Harbor, a 1/80 beach slope was adopt as to measure the change of beach profile under wave actions. The change of tidal level was also considered as a variable in the model tests.

In 1963, wave generating system was greatly improved as a new Nyerpic wave generator, contributed by the United Nations, was installed . It was under this improved wave flume that the run-up and armor cover-layer stability tests of Shintsu reclamation dike were carried out. This wave flume was later also used in other wave studies, such as wave deformation in shallow waters, especially on breaking wave and wave transformation in surf zone. Shortly, a wind generator was added for the study on generation of wind waves in shallow waters. A capacitance wave sensor was upgraded with silicon-covered wire to improve non-linearity. All these research results were published on ICCES and Conferences of Coastal Engineering in Japan during the period of 1963-1972.

On the other hand, the wave basin built in 1959 was equipped with hinge type wave generators with a overflow weir to control tidal levels. The initial layout of the Taichung Harbor model was set up to determine the most effective arrangement of breakwaters to stop along shore sand drift. The sediment transport rate was estimated around $4x10^6M^3/yr$.

The information of coastal engineering was scarce and difficult to acquire during 1960s. There were only a few photocopies of research papers published in the US, copies of translation in Japanese of the first and second ICCE Proceedings, and some

Proceedings of Conference on Coastal Engineering in Japan. Nevertheless, in the following years, the Proceeding of ICCE was bought successively. We were also later awarded with some very valuable references and publications by the US Army Corps of Engineers.

The courses on coastal engineering was first offered in the Department of Civil Engineering of Chengkung University in 1958, which is several years behind that in the US. It was essential to Taiwan, that is surrounded by sea and needs to solve her coastal engineering problems badly. Fortunately, the developments of research in coastal engineering were soon immensely improved and a number of engineers and students engage in this field.

In 1968, emeritus Professor Masashi Homma of Tokyo University was invited to teach courses of advanced fluid mechanics and coastal engineering for one semester after his retirement. In 1975, Professor C. L. Bretschneider and Dr. T. T. Lee from University of Hawaii were also invited as visiting professors to teach courses of wave forecasting and advanced coastal engineering. These internationally famous scholars indeed contributed tremendously to the initial development of research programs of Coastal Engineering in Taiwan.

Tidal-land reclamation is highly valuable for the over-populated Taiwan. As early as the 15th Century, Taiwanese ancestors had started reclamation of the wetland surrounding the Tai-chan inlet for agrculuature and salt pan usage. The Japanese also tried to use the reclaimed lands to sugarcane plantations near mid-western coastline such as Shinkung and Hampao. The scale of area was small and mostly concentrated on some very shallow water areas. In 1958, under the leadership of Professor Cheng Kin and the assistance from Mr. Huang Tsai-fan, tidal-lands of 1,112 hectares were reclaimed for growing rice by building 1 ： 6 slope of gravel dike along +1.0m elevation line. The model test was performed in the wave flume at National Cheng-Kung University.

In 1964, in order to built a second entrance for Kaoshiung Harbor, a model test facility was set up on site to test the stability of cylindrical caisson-breakwater and littoral drift problem. Dr. Yoshiuki Ito of PHRI, Japan, was invited for consultation on is investigation and research work.

In 1970, another hydraulic laboratory was set up to handle model tests for the construction of the Taichung Harobr. Dr. Shochi Sato and Dr. Nobuo Tanaka of PHRI, Japan, was also invited to assist in laying out the harbor and the model tests. The first stage of the construction of Taichung Harbor started in October, 1973 and began operation in October, 1976 with breakwaters of 6,530 meters long.

The eastern coast of Taiwan hosts most of the fishing resources of the island. However, the construction of harbor is challenged by the massive destructive waves and sever sand drift caused by typhoons. The first fishing harbor built in 1957 near Dawu and was damaged in the following year, because of the failures of breakwaters and siltation by typhoon. It was the reason why Cheng-Kung University was entrusted with the model testing project in 1962 for resolutions. These kinds of critical challenges have stimulated further development of Coastal Engineering in Taiwan.

Storm surges and waves caused by typhoons often bring along with devastating flood and coastal erosion. Between 1967 and 1969, the western coast was flooded by Typhoons Gorilla, Elaine, Viola, Else and Flossy. In 1971, there were also four typhoons, including Typhoon Lucia, which hit Taiwan with a wind speed of 35m/s and resulted in sever damages. Nearly 100 kilometers of seawalls were damaged, causing flooding of coastal farms and villages, and immense loss of property. About 300 kilometers of badly-constructed old sandy or brick masonry seawalls were the main reason for this disastrous damage.

In 1973, Taiwan Bureau of Hydraulic Engineering Division proposed a reconstruction project to fortify the seawalls around the island, and in fact more than 525 kilometers of seawalls were built since then, under a very scarce resources condition. There were very few coastal engineers then. On top of that, there were no observation data available on storm surges and waves. Engineers could only design seawalls by empirical prediction and experiences.

As far as research equipment is concerned, wave gauges were difficult to obtain and maintain before 1970s, due to financial shortage. There were only two wave gauges, one at Kaoshiung and Taichung Harbor, separately. One is pressure type and the other is ultra-sonic type. In 1980, Central Weather Bureau installed four ultra-sonic type wave gauges at Pitochao, Tonchi Island, Shaoliuchiou and Chengkung, for long-term wave observation. Wave-rider type wave meter was set up in Suow and Hwalien Harbors. However, only short-term data were available due to the damage caused by typhoons. Based on these data many improvements have been planned and constructed.

COASTAL ENGINEERING WORKS IN TAIWAN

The coastal engineering works in Taiwan has four different types, such as land reclamation, coastal defense, harbor construction and coastal zone management.

a.)Land reclamation

The modern coastal engineering works in Taiwan first started in 1942. The Japanese engineer reclaimed the Lunwei tidal land at now Yunlin County of Taiwan. The reclamation was done by constructing dikes at back shore areas. Between 1960s to 1970s, several land reclamations were planned and constructed at river mouths such as Shingthu, Auku, Tzenwen, Wankung, Yupu and Taishi reclamations. The modern knowledge of coastal hydrodynamics was applied to planning and designing of the coastal structures for land reclamation. Wave refraction, shoaling and hindcast calculation were used to analyze sea conditions. Usually, a mild slope(1:6~1:7) sea dike was constructed by using sand core with gravels of 30cm diameter for the cover layer. The dike position was usually located at the mean high water level. Tidal lands of 18,683 hectares were reclaimed for agricultural and aquacultural uses. This helped to solve the shortage of food brought by the increase in population. In order to develop these tidal lands, ten tidal stations were built, but no wave measurements were done. Empirical predictions and several model tests were conducted to investigate wave run-up, overtopping, cover layer stability and scouring of seawalls, etc.

From 1970s to 1980s, the land reclamation continued successfully. Due to the improvement of the coastal engineering experience and land resources restrain, the dike is mostly located at mean water level. In most cases, the fill program was not introduced to the reclamation project. Because the reclamation was located at tidal areas, the tidal basin closure's problem was the very most concerned within this period.

From 1980s to 1990s, the Changhua Coastal Industrial Park and Yunlin Offshore Industrial Estate projects were initiated. These two projects were huge land reclamation project for industrial uses. The fill program was first introduced to the reclamation projects. The areas of Changhua Project and Yunlin Project are 3,000 hectares and around 10,000 hectares, respectively. The seawall of Yunlin Project is located at water of -10 meter depth, which is the deepest reclamation project in Taiwan. In the 1990s, the tread of land reclamation is still in highly demand in Taiwan. Several reclamation projects are proposed for basic industry, such as Pinnan Industrial Park, Kuangyin Offshore Industrial Park, Tainan industrial technology Park and Shingshan Project , etc.

Recently, the reclamation project has also concentrated on landfill problems. The Nanshing Project at Kaohsiung City is a good example. The Project plans to reclaim about 300 hectares of land by dumping the ashes and construction disposals. A similar project is also under planning and designing, such as Bali Project near Taipei County.

The coastal engineering work has expanded in sizes and numbers in Taiwan. Recent projects under planning include the Master Plan of Kaoshiung Harbor for the 21th Century, 2,500 hectares of Pinnam Land Reclamation Project, 1,000 hectares of Shiangshan Land Reclamation, Landfill Project of Kaohsiung City and Dapengwan Recreation Development Project, etc.

b)Coastal Defense

The sand beach and dune type coast exist on the western coast of Taiwan due to sedimentation process of rivers. Most coastal areas are flat and protected by the dikes. At present, most of sandy coasts are eroded due to shortage of sand budget blocked by the soil conservation and dam construction on the upstream of rivers and mining of river sand as housing construction material. Typhoons and storm surges occur every year at various places in Taiwan. In most places of Taiwan, coastal defense system is strongly needed to protect the villages.

The coastal defense system was built in last 20 years and under continuous development. Seawall system is the most used type for coastal defense system. Usually seawalls are built with sandy core with rubble protection or concrete revetment. Concrete blocks are used for wave energy dissipation and toe protection. The slope of seawalls is usually around 1:1.5 to 1:2. A milder slope of 1:6 is also used in designed. The total length of seawall is more than 500 kilometers.

Due to erosions of the coasts, the groin type system were built at Changhua Coast, Kaoshiung Coast and Tansui Coast. Usually, groins are constructed with rubble mound type around 100m in length. Recently, land subsidence occurred at central and southern regions of the western coastal areas of Taiwan. The sea water level raises relatively to the land subsidence. Hence, erosion of coasts is more serious in these areas. In order to protect the seawalls in these regions, detached breakwaters had been used for coastal protection work at Wenfeng, Chaetin, Suileeheng and Hawmaili, etc. After two years of construction, tombolos or salients were observed behind the detached breakwaters.

The dike was also built for flood defense along the rivers of Taiwan. Due to the land subsidence, the alluvial rivers of Taiwan have become an estuary-type rivers. In response to land subsidence, the dike and storm surge defense system are under evaluation at these areas. Storm surge barrier and , tidal gate system are investigated for flood defense system for this new condition.

c).Harbor Construction

Harbors exist almost in every seashore county in Taiwan. There are four different types of harbors in Taiwan, namely, international port, national port, fishing harbor and marina, according to its function. The Kaohsiung, Keelung, Taichung, Suao and Hwalien are five international ports in Taiwan. The Kaohsiung Port with two entrances is the biggest harbor in Taiwan. The breakwaters of the second entrance of Kaohsiung harbor was built by cylindrical type caisson-breakwater.

The Taichung Harbor was built on a sand beach with breakwaters. Now the breakwaters were extended to -27 m of water depth. A caisson as large as 25m long was used for breakwaters. The significant wave height is 6.5 m for design condition. Suow Harbor was built in 1978. The breakwaters were built to -20 m of water depth with design wave height of 13 m. Dollas of 50 tons with steel bar were used for its own protection. In order to permit remarkable growth of modern-day demands, a new Keelung Harbor is being planned. The breakwaters may be located along the -40m of water depth contour.

There are 132 fishing harbors around Taiwan's islamds. Usually, fishing harbors can be classified in three type, small fishing harbor for nearshore fishing, the second class harbors for offshore fishing and the first rank harbor for deep-sea fishery. In most cases, the small fishing harbors are located within surf zone. Badouzi and Hsinda fishing harbors are two typical of the second rank fishing harbors. Anpin and Hsinda deep-sea fishing harbor have a navigation channel of 8 m in depth and consist of a wharf some 4,500m long to host those large fishing ships of around 5,000 tons. The longshore sediment transport is a problem for fishing harbor layouts planning . Numerical methods and physical models are used to evaluate functions of the harbors in the process of planning and designs. Dissipating-type quay walls are frequently used in small fishing harbor in order to reduce the fluctuation of water level.

In 1990s, marina becomes popular in Taiwan. A lot of small marina were built in Penhu islands which are abounded with recreation resources. Lungdon at north-eastern corner of Taiwan is the newly built marina. The design and construction of marina in Taiwan are still in an experimental basis. The breakwaters of London marina were damaged by a typhoon. The estimated wave height could have been over 14 meters. The sands were deposited in the basin due to stronger currents in that event.

d). Coastal management

A vast of low energy shoreline has been reclaimed in the last two decades and still in progress. The lose of wetlands has led not only to serious hydraulic modification of coastal areas but also to the coastal ecosystem. Sand beaches and barriers are more liable to move. While coastal erosion becomes more severe, views of natural sea are lost. When a natural coastal system is disturbed, mangroves are reduced and seabirds tend to be fewer then before. There are many parties, including coastal scientists, are concerned about these changes and attack government policies reverent to development on wetlands, tidal lands and barrier islands. The need of coastal zone management, to provide sustainable utilization of coastal resources, is strongly desired. Planning guidelines for uses of coastal zone development and maps of coastal zone management master plans have recently been published. The licensing system of coastal development has been approved by the government and EIA is also required for a development project. The draft of Coast Act has been proposed to manage the coastal zone. Coastal management project of storm hazards and flooding hazards are also under investigation.

The Kuandu Natural Park project is a typical case for estuary preservation program . The Taipei City has decided to acquire this land for naturally preserved park with 400 million US dollars. The Chiku wetland and Aoku plodde were preserved for wild birds. The Shichau, next to the newly developed Tainan Technology Park, estuary was also planned for natural preservation.

COASTAL RESEARCH PROGRAM

Since 1957, the knowledge of coastal engineering has been introduced in Taiwan. The number of engineers and students engage in this field has increased and renders a great contribution to the economic development of this country. The takeoff of Taiwan's economy during the 1970s encouraged more students to go abroad, including Japan, United States, Holland and Germany, etc. to study coastal engineering and acquire their Ph.D. and master degrees. Up to now, there are more than 60 researchers with Ph.D. degrees in Taiwan, most of them acquired their degrees from internationally famous universities. They joined engineering firms or universities after their returns to Taiwan. Courses on coastal engineering are offered widely in main universities. Coastal laboratories were set up. Numbers of graduate students majoring in coastal engineering increase rapidly. Research activities in coastal engineering enter a new burgeoning era.

In 1970s, the courses of coastal engineering has consequently been offered in Department of civil engineering in Chung-Hsing. University and Taiwan Ocean University, and successively to Chiao-Tung University, Taiwan University, etc. in 1980s. Most of these universities have equipped wave flumes and/or wave basins, as shown in Table-2 for basic studies and applied researches as well as to serve hydraulic models' tests. The largest wave flume has a dimension of 100m x 3m x 2m and the largest wave basin has a size of 70m x 40m x1.2m.

Institute of Harbor and Marine Technology was established in 1981, under the supervision of Taiwan Provincial Transportation Department, and was charged with main research activities in areas of harbor structures, management and coastal engineering. It is one of the best well-equipped hydraulic laboratory. There have a 100 m x 1.5 m x 2.0 m wind flume, two wave basins of 55 m x 40 m and 62 m x 60 m, respectively.

Tainan Hydraulic Laboratory, NCKU, also has well-equipped facilities. There major research works have undertaken field investigations and model tests of land reclamation sponsored by the Bureau of Industry, Ministry of Economics. A large scale coastal hydraulic laboratory is under construction at NCKU.

Most of research activities in universities sponsored by National Science Council. Recent coastal engineering research projects have two emphases. One group concentrates on the utilization of coastal space, which included five teams; wind-wave-current interaction, ocean pollution, marine farm engineering, offshore structures, promenade coastal structures and coastal topographic changes. Another group locks into coastal erosion protection under the Project of Natural Disaster Prevention. Results of research efforts are published in conference proceedings or technical journals.

The first conference on Coastal Engineering was summoned in 1976 in Taiwan. Since then, conferences have been held annually in October. The number of papers published in the conference proceedings has increased remarkably from 18 in the first conference to the 104 papers in last year as shown in Figure-2. Most of these papers are written in Chinese. Also some papers concerning coastal engineering are published in biannual Hydraulic Engineering Conference in Taiwan, as show in Fig-3, and also in Transaction of Institution of Civil and Hydraulic Engineering etc.

There also some text books published in Taiwan concern on coastal engineering such as :

1. Chow, C. L ： Ocean Engineering(in Chinese) , 1981, Dep. River and Ocean Engg. NTOU.
2. Chow, C. L. and Han, W. E. ： Fishing Harbor Engineering(in Chinese), 1992, Dep. River and Ocean Engg. NTOU.
3. Department. of Transportation ： Standard of Harbor Structures Design(in Chinese), 1981.
4. How, H. S. ： Planning and Design of Coastal and Harbor Engineering(in Chinese), 1983, HMTRI.
5. Kuo, C. T. ： Coastal Engineering, 2ed(in Chinese), 1989, SICCH.
6. Tang, L. W. ： Planning and Design of coastal Engineering(in Chinese), 1972, JCCR.
7. Tang, L. W. ： Harbor and Coastal Engineering(in Chinese), 1989, SICCH.

CLOSING REMARKS

The field of coastal engineering in Taiwan has been initiated by needs. It started from few individuals with the backgrounds of river hydraulics. As the society of Taiwan grows, the field of coastal engineering expands not only in its activities of research and engineering but also in the number of participants. The education program has speeded up the technical level while citizen awareness promotes environmental conservation. The tilted-view of economical development in coastal zone has been moved into a more healthy direction by taking coastal ecosystem into accounts.

In the field of coastal engineering, my colleagues and myself in Taiwan has learned a great deal from the past and from overseas. We are now facing even more challenges in this interdisciplinary field of coastal engineering. Non-technical measures such economical analysis of insurance and techniques in other fields such as remote sensing and automatic communication system are being integrated into team efforts. Of course, the traditional areas associated with hydrodynamics remain as a strong driving machine for coastal engineering in Taiwan. Personally, I feel that a necessary move needed for the future is to promote publication of research results and engineering project progress in English.

The importance of coastal zones management for the rapidly increasing population and economic development is evident in Taiwan. It is undoubtedly noted that a great deal of researches and studies is still necessary to tackle the challenges from the sea. We do appreciate any chances of corporation of colleagues around the world in this exciting field.

ACKNOWLEDGMENTS

The author wishes to express his appreciation to Dr. C. Z. Jung, Deputy Director of IHMT, Dr. C. S. Kung, Manager of SEC and Prof. T. K. Tsay for their helps in preparation of materials and review of this paper.

REFERENCES

'93 Years-Book of Fishing Port.,1993. Fishery Mag.

Bureou of Taiwan Hyd. Engg.,1973."Report of seawall reconstraction plane in Taiwan.".

Dai, E.B.,1959."Report on hydraulic model test of Taichung Harbor." Rep. No.4,Tainan Hyd. Lab.

Mintry of Internal Affair,1995."Rport of Integrated Planning of Taiwan's coastal zone."

Tan,L.W. & Kuo, C.T.,1961."Report on the model test of Shintsu tidal-land's dike." Tainan Hyd. Lab.

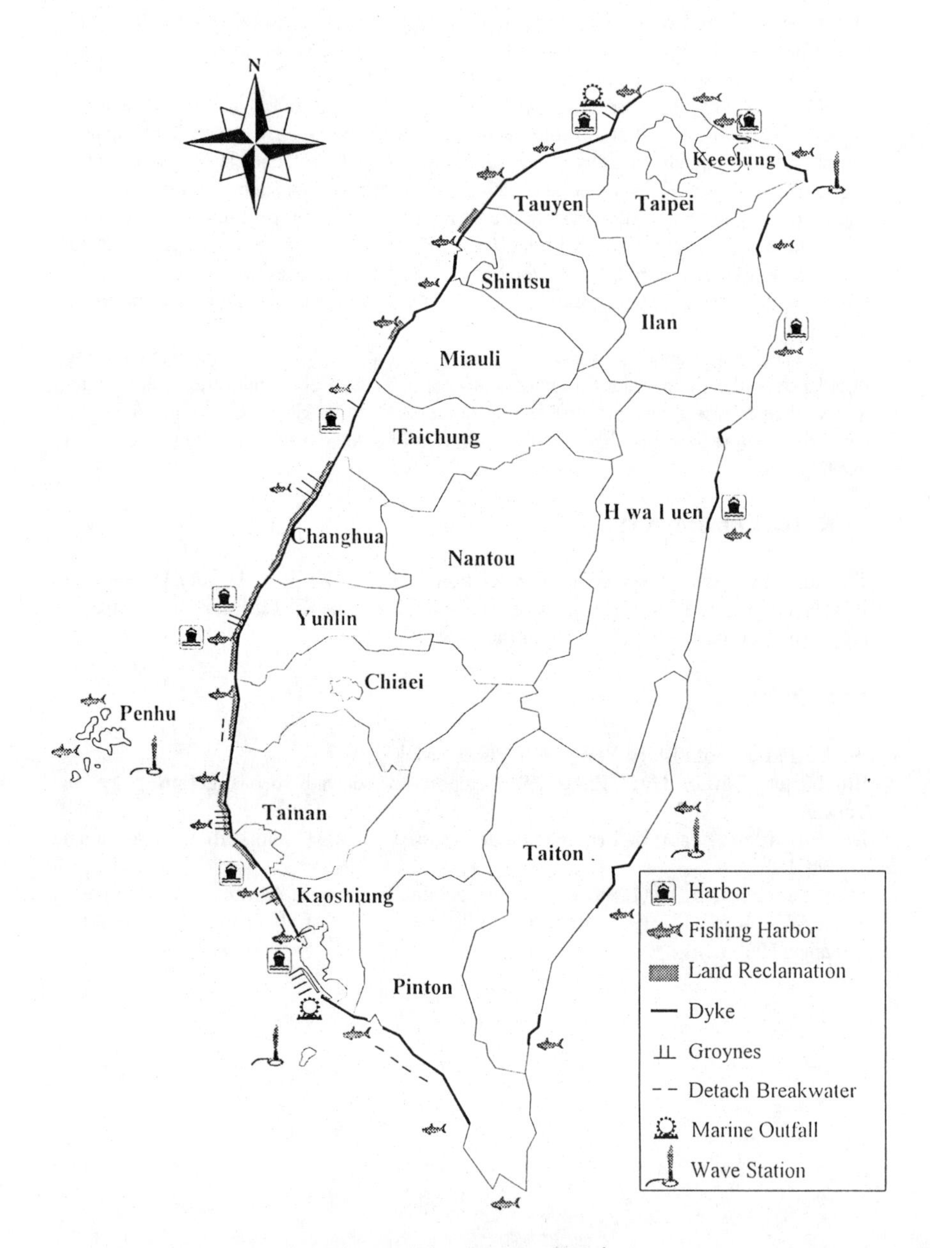

Fig. 1 The Map of Taiwan

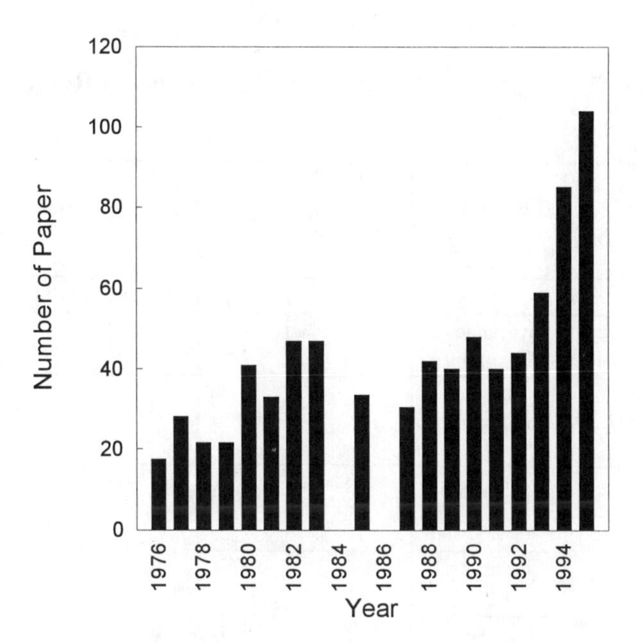

Fig-2 Histogram of papers presented in Taiwanese Conference on Coastal Engineering

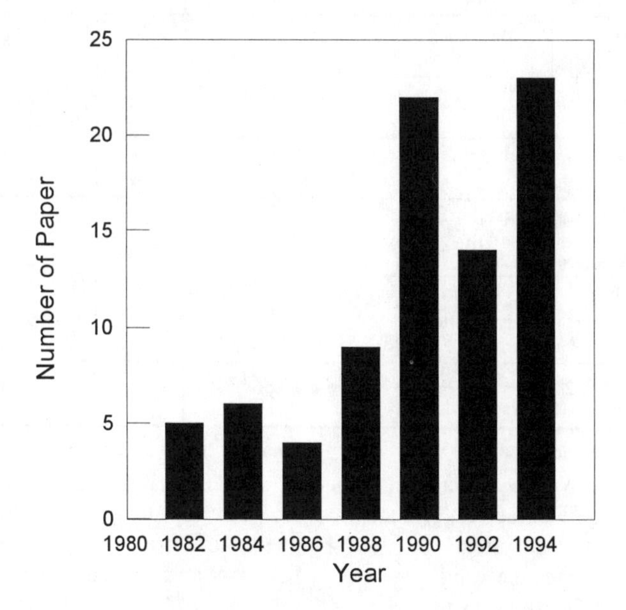

Fig-3 Histogram of papers on coastal engineering presented in Taiwanese biannual Conference on hydraulic engineering

Table 2. List of Institutes Involved in Coastal Research

Institute or University	Department	No. Faculty in Coastal Eng.	course offered Coastal Eng.	No. Graduated		Main Facilities
				Master	Doctor	
National Cheng-Kung University	Hydraulic & OceanEng.	16	1957	305	15	a,b,c,d,e,f ,g,k
National Sun-Yat-Sen University	Marine Environment	10	1989	7	--	,b,c
National Taiwan University	Civil Eng.	3	1988	7	--	c
	Ocean Inst.	1	1975	10	--	b,e,g,k
	Naval Architecture & Ocean	3	1984	12	--	a,b,c,e,k
National Taiwan Ocean University	Harbor & River Eng.	10	1978	100	2	a,b,c,e,g, k
National Chung-Hsing University	Civil Eng.	2	1973	20	--	a,c,d,g
National Chiao-Tung University	Civil Eng.	3	1977	29	4	a,b,c
Feng-Chia University	Hydraulic Eng.	2	1980	6	--	a,e
Institutc of Harbor & Marine Technology	Coastal Eng. Div.	13	1981	--	--	a,b,c,d,e, g,h,I,j,k
National Cheng-Kung University	Tainan Hydraulics Lab.	15	1957	--	--	a,b,c,d,f, g,j,k
Industrial Tech. Res. Institute	Energy & ResoucesLab.	17	1988	--	--	

Remark：a- wave basin
b-Wind-wave-current flume
c-Fluid mechanics flume
d-LVD system
e-Irregular wave generator
f-Density flow test basin
g-Computer center
h-Automatic weather station
i-Automatic tidal station
j-Offshore observation pile platform
k-Waverider, current meter, etc.

HISTORY OF COASTAL ENGINEERING IN THE USA

Robert L. Wiegel[1], Hon. M. ASCE, and Thorndike Saville, Jr.[2], F. ASCE

ABSTRACT

The term coastal engineering was first used in print in 1950, but components have been practiced in the USA since its founding. Early engineering and coastal works were concerned with navigation. Use of beaches for recreation began after the Civil War, particularly in New Jersey, and expanded rapidly after WW I. Books on waves and coastal structures, shore processes and tides were published in 1904, in 1919 and in 1926, but there were few technical papers about what is now considered coastal engineering. The late 1920's and early 1930's saw two organizations founded that had major impact on coastal engineering development: the American Shore and Beach Preservation Association and the Beach Erosion Board. The combination of engineering and science, and the development of modern analysis began in WW II for amphibious operation needs. The state of the art and practice, plus case histories, were presented in October 1950 at the First Conference on Coastal Engineering. Rapid expansion of data collection, analysis techniques and projects occurred in the 1950's and 1960's, with large numbers of works built, conferences and publications. Academic and research programs were developed at several universities, hydraulic model studies were advanced, and the use of numerical models started. The late 1960's and 1970's brought concerns with environmental impact, and recognition of the importance of coastal management. Becoming integrated into coastal engineering, these led to expansion of coastal regulations, much of which has required input of coastal data, science and engineering. Rehabilitation of projects and regions has become more important and extensive, sometimes requiring regional planning. Recent history has been left for others to write later to permit a historical perspective. Owing to space limitations, the interrelationships between advances in this and other countries have not been included.

---

[1]Professor Emeritus, University of California, Berkeley, CA 94720, USA

[2]Formerly Technical Director, USACE CERC, Bethesda, MD 20816, USA

INTRODUCTION

The term coastal engineering was probably first used in print in October 1950, at the Institute on Coastal Engineering held in Long Beach, California, planned by Morrough P. O'Brien and J.W. Johnson, organized by University of California Extension (Johnson, 1994). The purpose was to help engineers by summarizing the state of the art and science related to the planning and design of coastal works; each of the 35 lectures was by invitation. The written versions were published as the *Proceedings of First Conference on Coastal Engineering*, Council on Wave Research [established shortly after the institute was held] of The Engineering Foundation, edited by J.W. Johnson. The proceedings were dedicated to the memory of Professor Boris A. Bakhmeteff, "who inspired the formation of the Council on Wave Research." Most of the subjects in the present history were included in it.

In the proceeding's Preface, Morrough P. O'Brien wrote:
"A word about the term 'Coastal Engineering' is perhaps in order here. It is not a new or separate branch of engineering and there is no implication intended that a new breed of engineer, and a new society is in the making. Coastal Engineering is primarily a branch of Civil Engineering which leans heavily on the sciences of oceanography, meteorology, fluid mechanics, electronics, structural mechanics, and others {such as geology - RLW & TS}. However, it is also true that the design of coastal works does involve many criteria which are foreign to other phases of civil engineering and the novices in this field should proceed with caution."
This is still good advice.

Prior to that time there were coastal works, and books and papers on many aspects of coastal engineering, or activities with a substantial coastal engineering component: beach engineering, coast and hydrographic surveying, coast protection engineering, dredging, harbor engineering, lighthouse engineering, maritime engineering, port engineering, shore protection and improvement works, tidal hydraulics, etc. The term coastal engineering includes these and other types of works such as: coastal wetland protection and rehabilitation, estuaries, ocean outfalls {including mixing processes and water quality} and saltwater intakes, piers, pollution control, sand dune protection and rehabilitation. Coastal engineering is a broad and complex subject, interacting with economics and planning; owing to this, we have included coastal regulatory activities which require engineering/ scientific input.

In this paper, a chronological review is given, followed by a historical survey of a few special topics, then the history of a few important organizations, then a list of awards and the individuals or projects receiving them. Anecdotes are given to show what instigated important areas of research and development. A few natural disasters are mentioned, as learning from studies of them has been important to the improvement of coastal engineering analysis, practice, construction codes, and zoning regulations. Most of the concepts of coastal works are old. What is new is our growing capability for analysis and the understanding of forcings and processes, and the improvement and increase in collecting and presenting data. This is not a history of just technical studies, but includes events, projects and organizations, and we tried to give credit to those persons who were responsible for the work. The chronological review is not strictly in order, although mostly so, as it was found the inclusion of a few later studies or works helped to round out coverage of some subjects. These chronological digressions are usually within square brackets [...].

Engineering and science relies on work done throughout the world by many people, and often there are interactions. A scholarly history would trace these roots and interactions, but there is not space for this, nor are we professional historians. Also, it would be nearly impossible; the number of papers published in the 24 ICCE proceedings to date is about 2300, and this is just one source. Probably because of the combination of the age of the authors, and the time it would take to "mine" the great number of technical papers published in the most recent several decades, we have written more about the earlier work, especially through the two decades after WW II. We were a small number of engineers and scientists working in the coastal field. We knew each other, were colleagues who often collaborated, and often attended the same technical meetings. We crossed back and forth between engineering and science, between theory and practice, and between laboratory and field work. Owing to page and time limitations, this history is not complete, and probably largely reflects our own experience during these decades, which limits the breadth of coverage. We have surely overlooked some important works, and apologize for this. We decided the history since the early 1980's should be left for others to write in the future to permit a historical perspective, but a few things are included as we believe them necessary to the overall picture. Some are of disastrous storms which were well documented by measurements and observations. Also, some recent papers document earlier work, and put it in context with current practice. While writing this history, the authors became more aware than ever that most of the growth along the coast which was largely responsible for the need for coastal engineering, occurred during their lifetime.

Histories have been written about aspects of coastal engineering. J.W. Johnson (1974) gave one at the 14th ICCE, mostly on U.S. work. He wrote about the following subjects: (a) forecasting and hindcasting of sea and swell; (b) transformation of waves in shoaling water; (c) littoral processes; (d) wave forces on structures; and (e) tidal estuaries. He cited 122 references. Per Bruun (1972) wrote a history for the 13th ICCE, which was international in scope. The history of the USACE Beach Erosion Board was written by Mary-Louise Quinn (1977), and the history of CERC, 1963-1983, written by Moore and Moore (1991). John Lockett (1972) wrote a history of the USACE Committee on Tidal Hydraulics from 1949 to 1971. Wiegel wrote a short history of wave measurement and analysis in 1974. The USACE Los Angeles District (1986) prepared an oral history of coastal engineering in southern California from 1930 through 1981, with the resources being four coastal engineers with extensive experience in that region: James W. Dunham, William J. Herron, Jr., Omar Lillevang, Kenneth A. Peel. There are also histories of coastal works in a geographical region, such as Clarence Wicker's (1951) history of the New Jersey shore and Craig Leidersdorf and his colleagues' (1994) history of human intervention with beaches of Santa Monica Bay, CA. Much information on coastal projects is given in Parkman's (1978) history of the Army Engineers in New England, 1775-1975. Most advances are made by, or are due to individuals, or sometimes committees. In the coastal engineering field, three engineers had a great impact - Morrough P. O'Brien, Joe W. Johnson, and Joseph M. Caldwell. There are brief biographical sketches of O'Brien and Johnson (Wiegel, 1987; 1988), and a short sketch on O'Brien's contributions to coastal engineering (Johnson (1987).

To obtain a better understanding of the technical history of subsets of coastal engineering, survey papers of the subjects can be referred to. These are usually international in content, as they should be, but they cover much coastal engineering in the U.S.

The major reporting on USA coastal engineering has been in the ICCE proceedings, ASCE journals and specialty conference proceedings, the ASBPA journal, and publications of the BEB and CERC (for annotated bibliographies of the BEB/CERC see Allen and Spooner, 1968; Szuwalski and Wagner, 1984).

Some abbreviations used in this paper are: American Shore and Beach Preservation Association (ASBPA); American Society of Civil Engineers (ASCE); ASCE Hydraulics Division (HY); ASCE Waterway, Port, Coastal and Ocean Division (WW); Beach Erosion Board (BEB); California Institute of Technology (CIT); Coastal Engineering Research Center (CERC); Coastal Zone Conferences (CZs); Code for Federal Regulations (CFR); Committee on Tidal Hydraulics (CTH); Federal Emergency Management Agency (FEMA); Field Research Facility (FRF); Field Wave Gaging Program (FWGP); International Conference on Coastal Engineering (ICCE); Marine Board (MB); Massachusetts Institute of Technology (MIT); National Academy of Engineering (NAE); National Academy of Sciences (NAS); National Advisory Committee on Oceans and Atmosphere (NACOA); National Data Buoy Center (NDBC); National Flood Insurance Program (NFIP); National Ocean Survey (NOS); National Oceanic and Atmospheric Administration (NOAA); National Research Council (NRC); National Science Foundation (NSF); Office of Naval Research (ONR); Scripps Institution of Oceanography, University of California (SIO); Texas A & M University (TAMU); University of California at Berkeley (UCB); U.S. Army Corps of Engineers (USACE), U.S. Coast and Geodetic Survey (USC&GS); U.S. Geological Survey (USGS), U.S. Navy (USN); Waterways Experiment Station (WES); Woods Hole Oceanographic Institute (WHOI); World War II (WW II).

CHRONOLOGICAL REVIEW

Early Activities

Early coastal engineering activities in the USA were mostly for ports and navigation. Federal interest and participation in this aspect was because of an implied power derived from its direct power from what is commonly referred to as the "interstate commerce clause" of the Constitution. Federal work started in May 1824 after Congress voted appropriations for 3 navigation projects, one of which was the repair of Long Beach at Plymouth, Massachusetts (Parkman, 1978). This project was to improve the long narrow sand spit which protected Plymouth Harbor. Congress did not specify who was to do the work, but the President assigned it to the USACE. In the following decade or so surveys were made and navigation projects undertaken at the direction of Congress, by the USACE or the Topographic Bureau {later Topographic Corps}. From the early 1850's until the end of the Civil War almost no such work was done. Beginning with the River and Harbor Act of June 1866 many examinations and surveys were made by direction of Congress. Many reports were unfavorable. [In 1913 Congress "legislated that requests for reviews of reports, to determine if any modifications should be made, might be submitted directly to the Chief of Engineers merely by resolutions of the appropriate committees" (Parkman, 1978).] In 1852 Congress created a Lighthouse Board, under the Secretary of the Treasury, composed of army engineers, naval officers and civilians. Valuable technical information is available in annual reports of the USACE Chief of Engineers, the U.S. Coast and Geodetic Survey {formerly U.S. Coast Survey established in 1816, now the National Ocean Survey in NOAA}, and the U.S. Geological Survey. Navigation charts, topographic maps, tidal information {measurements and predictions of tidal elevations and currents; tidal datums (e.g. Harris, 1894 and later)}, and scientific reports were made available. This type of information is still being obtained and published. The long term measurements have been very important in the study of changes in relative mean sea level (see, for example, Hicks, 1981). [For an example of a recent use of historic survey maps, see Thompson (1994).] [Tidal measurements have been continued, providing information to coastal engineers and scientists. Survey data are updated on occasions.] Beach erosion control was started, using various shore protection works {groins, revetments, seawalls} constructed by local agencies or private interests. Many were built along the New Jersey shore in the 1880's and 1890's. Federal interest in beaches came later, in the 1930's, owing largely to the growing knowledge that coastal navigation projects such as jetties and dredging at harbor entrances and breakwaters usually had substantial effects on contiguous shores, accretion updrift and erosion downdrift (Brooke, 1934).

The first oil obtained from under coastal waters was produced in 1898 from a well about 200 ft from shore, drilled from a pier at Summerland, near Santa Barbara, CA (Anon., 1995).

Some of the first quantitative studies of waves and coastal structures were made by members of the U.S. Army Corps of Engineers in the 1880's, 1890's and early 1900's, as reported in Corps reports, and in what was probably the first book in the USA on waves and their effects on coastal structures: *Wave Action in Relation to Engineering Structures*, by D.D. Gaillard, originally printed in 1904. Gaillard was interested in measuring wave-induced forces and the characteristics of shallow water waves, because as he stated: "As a rule, an engineering structure subject to wave action is exposed only to the attack of shallow water waves; yet strange as it may seem, the number of recorded measurements of waves of this class is insignificant when compared with that of deep-water waves." Many of the wave force measurements he reported on were made using a marine dynamometer of the type developed and used about a half century earlier by Thomas Stevenson, the Scottish lighthouse engineer {whose son was the famous author Robert Louis Stevenson}. These measurements were made at various times and places by three Corps of Engineers officers: Lt. Col. Henry M. Robert, Capt. D.D. Gaillard, and Lt. C.H. McKinstry. Gaillard presented a review of the wave theories and wave observations that had been made previously by British, French and German engineers and scientists, and then gave detailed accounts of his own observations of the generation of waves by winds, wave transformation in shoaling water {including breakers}, wave transformation by opposing currents, and wave reflection. The observations on beach changes made by Gaillard was part of the construction program by Capt. Black and Lt. Gaillard of groins both north and south of St. Augustine Inlet, FL in the 1880's (Brooke, 1934). These observations and Gaillard's description of them should be "must reading" for coastal engineers today.

Development of beaches for recreational purposes started in the early to mid 19th Century {and expanded rapidly in the post Civil War period} along the New Jersey shore, because of its excellent beaches and summertime warm ocean water and climate, and because the beaches were near the metropolitan areas of New York and Philadelphia and accessible by boat or wagon, then by rail (Quinn, 1977). The first developments were at Cape May, then Long Branch and Atlantic City (Wicker, 1951). Piers and boardwalks were built [at a later date, the famous "Boardwalk at Atlantic City"]. Many of the beaches are on barrier islands, with numerous tidal inlets. The hydraulics and sediment movement of tidal inlets have been and are of major importance {navigation, beaches, wetlands} and interest to U.S. coastal engineers owing in part to the great number of inlets in the USA. It is a very complicated and interesting phenomenon, not yet adequately understood.

At the turn of the century, on 8 September 1900, a great hurricane struck Galveston, Texas, killing about 6,000 people and destroying about 3,600 buildings. A major hurricane and storm surge protection project was designed, and constructed in 1902-1905: the Galveston Grade-raising, Seawall and Revetment, which is a landmark in coastal engineering (Anon., 1902; Davis, 1952; Wiegel, 1991). The Board of Engineers was chaired by retired Brigadier General Henry M. Robert, former Chief of Engineers {for just one day}. He became famous for his *Pocket Manual of Rules of Order*, originally published in 1876, commonly known as "Robert's Rules of Order." [Incidently, these Rules of Order are still useful in public hearings for projects, as well as in parliamentary meetings.] Owing to the increase in population along the Atlantic and Gulf coasts where hurricanes are common, the characteristics of this type of storm, storm surge and wave setup, runup and overtopping, and how to decrease loss of life and property damage resulting from them is of prime importance to coastal engineers, scientists and planners, and to governments. This has been, and is, an active subject for field measurements and research. [Many recent projects, and coastal zone and construction regulations in these regions have hurricane and storm surge protection features. The National Flood Insurance Program, administered by FEMA, has a major concern and influence in this regard.]

The first detached breakwater was built in 1905 at Venice, California as a navigation project to provide shelter for boats operating from the Venice Pier. This type of structure has been found to be a useful device for beach protection owing largely to the tombolo effect, and will be discussed in more detail subsequently.

An interesting anecdote is related to the calculations made in 1908 of the maximum wave height at the entrance to the harbor at Manitowoc, Wisconsin {using Stevenson's empirical fetch- wave height relationship}, and the reduction in height within the harbor. They were made by lst Lieutenant Douglas A. MacArthur, Corps of Engineers, later to become General of the Army (Anon., 1985).

The 1920's; ASBPA Formation; Coastal Development; Coastal Subsidence

In the 1920's there were pioneering activities in regard to beaches and coastal processes such as: the work done within the New Jersey Board of Commerce and Trade by its Engineering Advisory Board on Coast Erosion {under the leadership of J. Spencer Smith and Victor Gelineau}; the formation and work of the National Research Council's Committee on Shoreline Studies; the creation of the American Shore and Beach Preservation Association; the establishment and work of the USACE Board of Sand Movement and Beach Erosion {abolished at the establishment of the Beach Erosion Board in July 1930}. Some engineers and planners in these organizations recognized that beaches form the greatest "national park" system in the country, accessible to all, and that it was in the public interest to protect them and to utilize effectively this national resource. However, most beach users were visitors on holiday, who had little knowledge of what occurred during hurricanes or winter storms {such as "northeasters"}, and little interest in funding studies and works. [The problem of funding is still with us.] The first beach nourishment project was designed and constructed at Coney Island, New York in 1922, together with its famous boardwalk (Farley, 1923; Dornhelm, 1995). The esplanade and seawall at Ocean Beach, San Francisco, California, was designed and built in segments from 1915 through most of the 1920's (O'Shaughnessy, 1924; O'Doherty, 1994); this was one of the early projects resulting from the growth in automobile use. Automobile travel in New York led to Moses' completion of Jones Beach, NY in 1929, and Orchard Beach in the Bronx, NY in 1936; both included major beach nourishment (Moses, 1943; Renshaw, 1969; Perlez, 1986). They are urban beaches, readily and cheaply accessible.

Lake Worth Entrance, FL (also called Port of Palm Beach Entrance) was cut in 1918 to access the waters of Lake Worth. Bakers Haulover Inlet was cut by local interests in 1925 through the Miami Beach barrier island, Florida, so that tidal flow would provide fresh ocean water to Biscayne Bay to improve the poor water quality that then existed in the bay (von Oesen, 1973). In 1927 South Lake Worth Inlet, Florida was dredged through the barrier island to permit tidal flow through it to relieve the stagnant water conditions in South Lake Worth (Caldwell, 1950).

Few papers were published prior to the 1930's about engineering aspects of beaches. Only five papers dealing with coastal erosion and mitigation appeared in *Transactions of the ASCE* in the sixty years from 1867 through 1936 {sic} (Saville, 1942). A comment by Henry S. Sharp (1927) describes the state of the art:
"Conditions vary so widely from place to place that rule-of-thumb methods are sure to give a large percentage of failures, and a structure successful at one place may be a dismal failure at another. On the other hand, the engineer who wishes to attack his problem scientifically finds that science has done very little to help him. He is almost entirely without trustworthy facts, and must work up his data from hasty studies of his own."
[Note: The situation started to improve in the 1930's, but it was the work done during World War II for amphibious operations that brought about the use of the combination of engineering and science. This work, and the measurement and analysis of data has continued to expand. Experience has shown, however, that nearly every site is different in the combination {"mix"} of processes, forcings, and boundary conditions, so that common solutions are rare.]

An important book had been published in 1919, *Shore Processes and Shoreline Development*, by Professor Douglas W. Johnson of Columbia University, and a popular book by H.A. Marmer of the USC&GS, *The Tide* published in 1926. [Johnson amongst other things was one of the first to study mean sea level changes (1929); he died in Florida on 24 Feb. 1944, at the age of 65.] Johnson was a geologist/ geographer, and was the first chair of the National Research Council Committee on Shoreline Studies. The later activities of this committee were based on three main considerations (NRC, 1928, as given in Quinn, 1977):
"(a) The great social and economic importance of the beaches of our {the U.S.} Atlantic and Gulf coasts; (b) the constant attrition of valuable lands which in many localities is resulting in large economic losses, and in some cases threatened the continued existence of communities; and (c) the present empirical basis of engineering practice in shore protection, which has been found inadequate to meet the situation."

One reason there were few papers was that the coastal population was small then, and there was little interest. It was the development of automobiles and highway systems that made beaches accessible to the general population who resided inland. The Malibu coast of southern California was not accessible until the county road was built in 1921, which was superseded by the Pacific Coast Highway in 1929 (Pfeifer, 1985). The development of Malibu began in 1928, with exclusive beach front 30-ft wide lots leased to movie stars - 5 year leases at $1 per front foot per month. [Details of coastal engineering developments in southern California are available in Herron (1983), USACE Los Angeles District (1986), and Wiegel (1994).] Miami Beach, FL had almost no development prior to 1910. The Collins Bridge was opened in 1913, and the

County Causeway in 1920. The 1920 census counted 644 residents living in Miami Beach and the 1930 census counted 6,494. Major growth occurred in the 1930's, and later. With development, problems occurred owing to the encroachment on beaches with buildings and infrastructure. As an example, consider the comment by the then City Engineer of Miami Beach, Morris Lipp (1955): "After the war, many hotel owners who had insufficient vacant land for pool facilities requested the City Council to permit them to construct new bulkheads 75 feet seaward of the existing one ... In most instances, the new bulkhead location was seaward of the existing Mean High Water line." At Santa Barbara, CA in about 1890 the sand dunes were scraped off, and a boulevard built on top with a plank retaining wall installed to hold the sand in place, with a strip of beach left between the wall and the ocean; in 1912 the boulevard was widened 10 feet on the seaward side and a new more substantial wall constructed (Lawton, 1936). At about the turn of the century, and a little later, the "Great Highway" at Ocean Beach, San Francisco, CA, was built by paving over the surface of the sand dunes (Olmstead and Olmstead, 1979). Another example is the construction in the 1920's and 1930's of buildings and a street on the ocean side of the base of the seacliff at Oceanside, CA. [A little later, post-WW II, the Federal interstate highway program greatly increased accessibility to the coast of residents of inland regions.]

Coastal subsidence occurred in the Long Beach/ Terminal Island, CA region, 1926-1968, associated with the withdrawal of oil and gas from the Wilmington Oil Field, with the greatest amount nearly 30 ft (Allen and Mayuga, 1969). [Subsidence due to withdrawal of oil/ gas/ water has been important elsewhere, of at least several feet (Habel, 1978; NRC Committee on Engrg. Implications of Changes in Relative Mean Sea Level, 1987; Wiegel, 1994; USACE, Los Angeles District, 1995).]

The 1930's. BEB Formation; Shore and Beach; Coastal Studies and Works; Littoral Cells; Tidal Inlets; Estuaries; Hydraulic Laboratory Studies

The Board on Sand Movement and Beach Erosion (BSMBE) was established by the USACE in January 1929, with Col. William J. Barden as senior member, and Douglas W. Johnson and Thorndike Saville as civilian advisers. A plan for field studies was prepared by Professor Johnson (Quinn, 1977, p. 20). The first work was done, May 1929 - Sept. 1930, on the beach adjacent to a jetty built at Far Rockaway, Long Island, NY and along parts of the New Jersey shore, by Morrough P. O'Brien and lst Lt. Leland H. Hewitt and staff. This included collecting sand samples and analyzing them for size. The Beach Erosion Board, USACE, was established on 18 Sept. 1930, and the BSMBE discontinued. In 1930 the BEB contracted with O'Brien - "to make a reconnaissance of beaches, inlets and harbors of the Pacific coast from the Strait of Juan de Fuca at the Washington-Canadian border to the Tijuana Slough at the California-Mexican border" (Quinn, 1977). A 7-volume report was prepared. In 1932, field studies of water particle motions in breakers and waves nearshore {using a small current meter} and sand movement were made along the New Jersey coast, based at Long Branch. Suspended sand samples were collected, and underwater observations made by a diver. Some measurements were made of alongshore currents. Also, field measurements were made at Pensacola, FL during the summer of 1932. Many of their data and findings were published in *Interim Report of the Beach Erosion Board* (1933). This report is interesting to read today. Funding of the BEB was meager, but they performed a number of field investigations and studies of coastal regions in cooperation with requesting State agencies, on a cost sharing basis. A small wave tank {24 ft long by 12 ft wide by 3 ft deep} was completed in October 1932; a larger one {85 ft long by 14 ft wide by 4 ft deep} was built in 1937, both at Fort Humphrey, VA [later, renamed Fort Belvoir].

The ASBPA started publishing its journal *Shore and Beach* in 1933.

During 1931-1934, a series of 5 detached rubblemound breakwaters, each 300 ft long, with 136-ft gaps, were built at Winthrop, Massachusetts for shore protection (Hale, 1936). In 1934, a 2,000-ft long shore-parallel detached rubblemound breakwater was constructed at Santa Monica, CA to provide a nearshore harbor. The "tombolo effect" caused an accumulation of sand on the updrift side and in its lee, and erosion downdrift of it (Eaton, 1959; Johnson, 1959). [Experience gained at this structure was used in the design of the detached breakwater/ harbor entrance protection/ sand bypassing system at Channel Islands Harbor, CA (Herron and Harris, 1966).] The first sand bypassing systems at navigation projects were developed and put in operation in the mid-1930's, one mobile and one fixed: Santa Barbara, CA (Penfield, 1960) and South Lake Worth Inlet, FL (Caldwell, 1951). Possibly the first use of wave refraction drawings for quantitative evaluation of wave height and direction along a section of shore and the resulting sand transport were made in 1936 for the Santa Barbara project (O'Brien, 1950). In 1939 a beach nourishment project was constructed on Waikiki Beach, Hawaii. During the middle of the 1930's (in the economic depression), Federal relief funds were available to local governments for public works, but there were few applications for coastal works, primarily because there were no plans available (Saville, 1942). Some local agencies started to develop shoreline plans, such as Los Angeles for the Santa Monica Bay region (A.G. Johnson, 1950).

In the 1930's a number of technical studies of beach processes were made, and the results published. A few small wave tanks and hydraulic model basins were built and used, at the BEB, and at UCB, and a small tank at MIT. One study made was of the Columbia River tidal estuary (O'Brien, 1935; 1972). Experiments on beach deformation by waves {including the effect of seawalls} were made by the BEB in the mid-1930's (BEB, 1936). Possibly the first project model was for St. Andrews Bay, FL in 1933, for which waves were generated by a laborer using a hand crank to wind up and release on a timed basis a hinged gate (USACE, WES, 1933). One of the first 3 movable bed models of an actual project, Santa Barbara, CA breakwater and sand trap, was made by Lapsley (1937) in a small basin at UCB. Another, with tides, was done at WES by Joseph Caldwell (1937) of Ballona Creek outlet jetties, Venice, CA. Both studies showed coastal structures trapped the littoral drift of sand. Three theses were done at MIT on seawalls and beaches in the late 1930's (e.g. Reynolds, 1937). The first measurements of mass transport by water waves were made by Mitchim

(1940) in a UCB tank. An experiment on the profiles of a sandy beach in front of a vertical seawall was made by Dorland (1940); he measured and reported the median diameter, specific gravity and settling velocity of the sand {also used by the BEB}. [In O'Brien's last paper "Beach at a Seawall: Completion of a 1940 Experiment" (1987), he reported on this study, showing the regimes of "storm profile" and "normal profile" could be defined by the deep water wave steepness and the ratio of sand settling velocity to the product of gravity and the wave period, the parameter developed by Robert G. Dean in his work on equilibrium beach profiles.]

Measurements of beach profile changes were made by LaFond (1939) along, but 10 ft outboard of {to avoid scour holes} the SIO pier at La Jolla, CA, together with tides and waves. These were made once a week for 2 years, twice a week for 1 year, and daily from Oct. 1937 to Oct. 1938.

The concept of littoral cells, quantitative data on alongshore sand transport rates, and the decrease of sand delivered to the coast owing to the construction of dams were mentioned in studies of the Santa Barbara and nearby regions (O'Brien, 1936), and used in detail by J.W. Handin (1951) in his studies of southern California beaches. The term "river of sand" appears in the 1933 *Interim Report of the BEB* (USACE BEB, 1933, #5/10; Quinn, 1977). The speed of movement downdrift of a sudden input of sand from a periodic bypass operation or a river flood, or an "erosion sand wave" following the building of a breakwater {or jetty} or the cessation of sand bypassing, was found to be important. [These concepts were developed extensively by Professor Douglas L. Inman and his colleagues at the Scripps Institution of Oceanography, University of California (Inman and Bagnold, 1963; Inman and Frautschy, 1965, Komar and Inman, 1970; Inman, 1987), together with the concept of quantitative sand budgets for a cell {sand sources and sinks}, which proved to be very useful, and have become common practice.]

[Identification of the natural mineral content, grain shape and size of beach sands, together with the same data for nearshore sands, sand dunes, continental shelf sands, river sands, and seacliff and bedrock mineralogy is useful in determining sources and sinks. This was commented on by Gaillard (1904, p. 204) in regard to siliceous beach sand on Florida's northern east coast, and its possible origin. Some studies were: Martens (1935) on southern Atlantic Coast beaches; Trask (1952) in the vicinity of Santa Barbara, CA; Trask (1959) and Wilde and his colleagues along the coast of central and northern California (Wilde, 1965; Wilde et al., 1969; Lee et al., 1971; Minard, Jr., 1971; Wilde et al., 1973; Wilde and Case, 1977); Taney (1961) and Williams and Meisburger (1987) along the Atlantic coast of Long Island, NY; Clemens and Komar (1988) along the Oregon coast. Also, studies were made of the statistical properties of beach sediments by Krumbein (1954) leading to the means of specifying sand properties for beach fill (Krumbein, 1957; Krumbein and James, 1965; James, 1974). By the 1960's, there were a number of federal programs dealing with sedimentology of the continental shelf (Taney, 1966). One interesting conclusion reached by investigators of northern California and Oregon beaches was that relict sands in present day pocket beaches between headlands were from various sources during times of lower mean sea levels, when the present headlands did not exist, and that the sand moved onshore after the last ice age (Minard, Jr., 1971; Clemens and Komar, 1988).]

In addition to studying beach sand mineralogy, Martens (1934) described the characteristics of beaches/ sand/ water surface tension in pores that resulted in a beach with good trafficability properties. [Later, considerable work was done by others on beach trafficability (e.g., Horonjeff et al., 1953).]

As mentioned previously, tidal inlets {natural and constructed} and estuaries are very important in the USA, owing in part to the large number of them. They are complex nonlinear coupled systems. The processes are difficult and expensive to study, and sufficient measured data relatively sparse. Analytical studies of inlet hydraulics were made by Col. E.I. Brown (1928), followed by the important empirical relationship between the area of the tidal entrance and the tidal prism of the inland bay or estuary shown by O'Brien (1931). [The analytical work was extended by Francis Escoffier (1940). Pillsbury (1940) summarized the knowledge of tidal hydraulics at that time. Many studies were made in the 1950's and later; some are described in a subsequent section.]

As a part of the investigations of a proposed salt water barrier in San Francisco Bay, CA, to prevent sea water from moving into the San Joaquin and Sacramento Rivers in the Delta, model tests using fresh/salt water were made at UCB. A model law for densimetric flows was developed by O'Brien and Cherno (1934). In a discussion of the paper, Garbis H. Keulegan (1934) presented a more rigorous derivation of the model law {densimetric Froude Number} that should be used. [At a later date, Don {D.W.} Pritchard (1952, 1955), and Arthur Ippen and his colleagues did much work along these lines (Ippen, ed. 1966).]

Hurricane of 21 September 1938

On 21 September 1938, a great hurricane struck portions of the northern New Jersey coast, Long Island, NY, {crossing near Fire Island Inlet}, and southern New England (Anon., 1939; BEB, 1939; Brooks,, 1939; Brown, 1939; Hale, 1939; Smithsonian, 1939). [Recently a 1-hour documentary on this event was presented on TV by the Public Broadcasting System.] There was not then the warning system in effect that has since been developed with modern equipment {satellites, radar, hurricane aircraft}, theory, computer based high speed analysis and prediction, and communication and disaster preparedness systems. More than 600 lives were lost and over $400 million {in 1939 $} in damages (Smithsonian, 1939). Thousands of homes, commercial buildings, and boats were destroyed or damaged, and infrastructure was severely damaged {railroads, roads, utilities, docks, piers, seawalls, breakwaters, Coast Guard stations}. The combined storm surge, wave setup, runup and overtopping was very high, and there were extensive dune erosion and breakthroughs {10 new inlets formed between Fire Island and East Hampton, some of which were closed by artificial means shortly after the hurricane}. A comprehensive pictorial record of the effects of this hurricane is available (Federal Writers' Project, 1938). This event and Hurricane Carol of 31 August 1954 were largely responsible for the decision to design, construct, and operate the New England Hurricane and Coastal Storm Surge Barriers, which are discussed subsequently.

The 1940's. BEB Move; WW II/ Amphibious Operations; Wave Forecasting; Beaches and Surf; Wave Measurements and Analysis; Surf Beat; Sea Sled; Sand Transport/ Scale Effects

The BEB was moved into its new facility on the Dalecarlia Reservation, Washington D.C. in March 1940 (Quinn, 1977). The BEB published *Summary of the Theory of Oscillatory Waves* by O'Brien (1942), a useful compilation of theory.

Great advances were made in the combined engineering and science of coastal processes, activities and works during World War II. Nearly everything developed for military applications has since been beneficial in coastal engineers' work for civilian needs. {Much of the work on waves was given in a National Bureau of Standards Symposium in 1951, and published in the proceedings: *Gravity Waves* (1952).} Studies of waves {including their generation by winds}, beaches, and landing craft performance in the surf were made by personnel of the University of California's Department of Engineering at Berkeley and Scripps Institution of Oceanography, members of the Woods Hole Oceanographic Institute (Johnson, 1994), and at the Beach Erosion Board in support of U.S. amphibious operations. Much of this was done for the USN Bu. Ships, with Roger R. Revelle, then a naval officer, as liaison. {For some details see the oral history of Morrough P. O'Brien (1988).} These operations required amongst other things {such as tides, currents, and wave climate data}: forecasting waves at sea and their transformation over a beach or reef into surf {including littoral currents}; knowledge of beach characteristics; the interaction of surf and beaches {beach erosion and accretion, formation of bars - cross-shore sand transport}; knowledge of the behavior of landing craft, landing ships, and amphibious vehicles in the surf and at the beach face, and beach trafficability, during both the assault and supply phases (Seiwell, 1947; Bates; 1949; Thompson, 1949). A course was given at SIO on sea and swell forecasting for Naval aerologists, one of whom was Warren C. Thompson {then, a Lt., USNR}, who later participated in a number of beach/surf reconnaissance surveys and forecasting operations in the Pacific. Lt. D.W. Pritchard (AUS) made forecasts for the U.S. D-day landings on 6 June 1944 in Normandy, France, and he and Lt. R.O. Reid (AUS) made wave forecasts for U.S. Army commands for the logistical operations of unloading supplies at Omaha Beach through the summer and fall of 1944 (Bates, 1949). Pritchard and Reid were Army Air Corps Meteorologists who received their training at UCLA with wave and surf forecasting lectures being given by Prof. H.U. Sverdrup of SIO (Pritchard, 1995; Reid, 1995). Pritchard, Reid, and Thompson subsequently had life-long professional careers in oceanography and coastal science.

During WW II the BEB facilities were used for wartime applications, such as studies on landing craft capability, water depth determination {Keulegan's study on bar depths (1945) was of particular importance}, movable breakwaters {including sunken Liberty ships and the artificial harbors used in the Normandy landings}, and a proposed floating aerodrome to be used in mid-Atlantic for aircraft refueling. The staff prepared studies of beach and nearshore characteristics worldwide for war-connected operations. WES facilities were likewise used for similar war applications, including importantly a study of the stability of the Normandy artificial harbor breakwaters (Tiffany, 1968; Quinn, 1977).

The principles developed at that time for forecasting wind waves and swell, and breakers and surf are given in two USN Hydrographic Office publications (1943; 1944), and a revision in 1951. The basic physical relationships and techniques for forecasting waves generated by wind were published by Sverdrup and Munk in 1946, and updated with additional data by Bretschneider in 1951 - the SMB (Sverdrup, Munk, Bretschneider) wave forecasting curves. Bretschneider (1965) gives a detailed history of these developments, and many others, which improved forecasting techniques during the following two decades. O'Brien's intuitive understanding of fluid mechanics and his thorough knowledge of similitude and model laws led to his suggesting to Sverdrup that ocean wave forecasting procedures should be based on the empirically determined relationships between the dimensionless parameters gH/U^2 and gL/U^2 and the independent variables gF/U^2 and gt/U, where g is the acceleration of gravity, H is the wave height {generally taken to be the "significant wave height"}, U is the wind speed, L is a measure of the wave length {related to the average period of the highest one-third of the waves}, F is the fetch, and t is the duration of the wind flow over the fetch. This is the classic SMB relationship. The original formulation is in a memo from O'Brien to H.U. Sverdrup and W.H. Munk, on 31 Dec. 1943, with an expanded version dated 5 Sept. 1944; the first use of this nondimensional presentation appears in *SIO Wave Report No. 30*, 7 Dec. 1944. {See also, the oral history of O'Brien (1988).}

The concepts of wave spectra and wave height distribution functions were known by those developing forecasting procedures for operational use. This approach is described in *Proposed Uniform Procedure for Observing Waves and Interpreting Instrument Records* (SIO, Dec. 1944), which includes O'Brien's statement in his letter to H.U. Sverdrup (Feb.,1944):
"The variability is not... the result of superposition of waves from different storms but is the consequence of the spectrum from a single fetch or the interference of waves from different portions of the same storm. If a single generating area sends out a spectrum, then the variability is as much a characteristic of the wave train as the height and period of the predominant wave."
The wave research group at SIO developed an approximate curve of the percentage of occurrence of relative wave height for sea and for swell (*SIO Wave Report No. 68*, 1947; reproduced in Bates, 1949). The relationship between what they termed the "significant wave height" {the average of the highest one-third of the waves, which is

still in general use} and the mean wave height was shown. The term came into being as follows (from *SIO Wave Project Rept. No. 26*, Dec. 1944, as given in Snodgrass and Putz, 1958):
"A comparison of the visual and instrumental values indicated '...the natural tendency for the observer...to record not the average wave height but a wave height based on some kind of average of the highest waves. The general experience is that an observer will give the value for the wave height which represents the average of the highest 20 to 40 percent of the waves'...The average height of the highest one-third of the waves, $H_{1/3}$, was therefore suggested as the characteristic {or significant} wave height. 'Characteristic wave period' was given a corresponding definition as the average period of the highest one-third waves."

This work continued throughout the 1940's and afterwards. It included the development of scientifically based wave forecasting and hindcasting procedures, wave measurement and analysis programs {requiring instrument development}, a rational foundation for the calculation of wave induced forces on structures, both fixed and floating, and the functional design of temporary harbors.

[A statistical description based on specifications of a random sea was developed later by Pierson (1952) and Neumann (1953). Practical methodology was given by Pierson, Neumann, and James (1955), resulting in a spectrum description as well as significant wave. Theoretical developments by Miles (1957) {discussed later}, and by Phillips (1957) in the U.K. {Phillips has been at John Hopkins Univ. since 1957), allowed much more advanced systems to be developed. Kinsman (1965) gives a good description of wave generation/ propagation, integrating the various theories. During this time, too, applications of the SMB method were being successfully made to ever more complicated situations such as hurricanes (Bretschneider, 1957). Advances have been made in understanding the physics of wave generation, including wave-wave interaction, and large scale computer programs have been developed {at CERC, the Weather Service, the U.S. Navy, and various universities and private companies} and are used particularly for forecasting and for obtaining climatic data. Details of the third generation of wave modelling is given in the final report of the wave modelling group {WAM}, which includes nonlinear wave-wave interactions and dissipation (see the review of their book by Miles, 1995). {The SMB method remains useful in much local design in bays, lakes and reservoirs because it can be handled easily to yield useful estimates of significant height and period.}]

The development of wave recorders by Professor R.G. Folsom and his colleagues is described in his 1949 paper "Measurement of Ocean Waves," and by Frank E. Snodgrass (1950). A step resistance wave gage was developed by the BEB for field use (Caldwell, 1948), based on circuitry developed at WES in about 1940 for small scale hydraulic model measurements. A wave gage and spectral analyzer system was developed at WHOI by Klebba (1949). {Additional treatment on wave measurement and analysis is given in subsequent sections.}

Techniques were developed for the analysis of wave refraction {both wave front - which had been used in the mid-1930's by O'Brien (1950) - and orthogonal methods}, including the effects of currents, trapping of waves, and refraction of wave groups (Johnson, 1947; Johnson, O'Brien and Isaacs, 1948; Isaacs et al., 1951; Pierson, 1951; Williams and Isaacs, 1952), diffraction by breakwaters (Blue and Johnson, 1949; Carr and Stelzriede, 1952; Wiegel, 1962), diffraction by islands (Arthur, 1946; 1951), shoaling and breaking on beaches. The wave diffraction studies were largely an outgrowth of the work done in the U.K. for the development and operation of the Mulberry "B" mobile breakwater for the June 1944 landings in Normandy, France (Institution of Civil Engineers, 1948). [Several decades were to pass before numerical methods were developed for computer calculation of refraction and plotting (Harrison and Wilson, 1964; Wilson, 1966), and some limitations discussed (Coudert and Raichlen, 1970). Later, the theory of linear refraction of directional spectra was developed, which overcame some of the difficulties inherent in the refraction of one-direction uniform periodic waves (Collins, et al., 1981).]

[Problems associated with refraction in a convergence {of orthogonals} zone were studied theoretically and in hydraulic laboratories, which showed the need for combined refraction/ diffraction theory (e.g., Pierson, 1951; Whalin, 1971). Whalin observed and measured nonlinear wave instabilities {the longer the wave relative to the water depth, the greater the transfer of energy to higher harmonics}, of the type described in subsequent work by Harms (1975) and also to a limited extent by earlier work of Horikawa and Wiegel (1959). Rudolph Savage of the BEB had observed this earlier. Harms also described the formation of cross-waves {standing waves along the generator, with half the frequency of the wave generator}, under certain conditions; these had been observed by a number of investigators. These studies indicate one of the fundamental difficulties in the use of hydraulic models.]

Wave energy dissipation in deep water is relatively small (e.g. Snodgrass et al., 1966), but dissipation by bottom friction and percolation in shallow water is important. Theory and experimental work was done for a sandy bottom by Putnam and Johnson (1949), Putnam (1949), and Bretschneider and Reid (1954). Reid and Kajiura (1957) solved the coupled problem of wave damping over a permeable seabed. Dissipation of wave energy moving over a mud bottom is more difficult to analyze, although fishermen in India had long taken advantage of it (e.g. Foda, 1989).

Breaker-induced littoral currents were important in amphibious operations. They are 3-dimensional, including seaward flowing rip currents and "undertow", and alongshore currents (e.g. Shepard and Inman, 1951). They caused landing craft to broach. Theory for alongshore currents {we use this term rather than the commonly used "longshore"} was developed by Putnam, Munk and Traylor (1949), and field and laboratory tests were performed. Two theories were derived, one based on energy considerations, and the other on momentum flux. [Galvin (1967) reviewed both theory and data. There is doubt about some of the laboratory data, so numerical results should be used with caution. Later, after an extensive investigation, Galvin (1991) recommended the laboratory experiment data not be used, as it seems the published breaker height and direction were not experimental, but derived.] Few measurements were made of littoral currents in the ocean, by timing the movement of dye patches, as measurement was difficult. A BEB field party used the technique in the 1940's and 1950's, and it was used later in the LEO program {discussed in a subsequent section}. Field studies at Seagrove, FL were made by Sonu (1972) of the 3-dimensional characteristics of the surfzone, including setdown, setup and rip currents, and a theoretical advance on these phenomena was made by Noda (1974) which accounted for some of the observations. As examples of more recent work, see Liu and Dalrymple (1978) and the Nearshore Sediment Transport Study (Seymour, ed. 1989).]

That waves breaking at an angle to a beach, and on the beach face produce {swash} transport sand along the shore was known to D.D. Gaillard, D.W. Johnson, Morrough P. O'Brien and others, but few quantitative data were available. These data were obtained by surveys of sand trapped on the updrift side of jetties or long groins (Gaillard, 1904; USACE BEB, 1933; O'Brien, 1939; J.W. Johnson, 1959). [It is interesting to note that what is probably the most detailed set of measurements of sand accumulation at a "total trap" was made at Santa Barbara about 50 years later in the Nearshore Sediment Transport Study (Dean, 1989 a,b)]. Relationships were developed between the alongshore component of wave energy and the littoral transport rate for sand (Krumbein, 1944; Saville, Jr., 1950; Eaton, 1951; Johnson, 1952; Watts, 1953; Caldwell, 1956; Savage, 1959). Semi-empirical formulas and theories were developed, which are in use today (Eaton, 1951; Watts, 1954; Caldwell, 1956; Inman and Bagnold, 1963; Komar and Inman, 1970; Galvin, 1972; Dean, 1973; Komar, 1976; USACE CERC, 1984 and earlier editions of both SPM and TR4). Eaton (1951) and Caldwell (1956) write about the "littoral drift factor" formula which was developed in the USACE Los Angeles District, and is the basis for the SPM formula. This was probably based upon the section on "average work factor" in the SIO (1947) report to the USACE Los Angeles District. They state that the "...sand transport by wave action it appears that wave work rather than wave height is the significant parameter." In the SIO report, an equation for the time rate of work {wave power} is developed in terms of significant wave height, associated significant wave period, and the frequency distribution of these in terms of the summation of $H_s^2 T_s$ per unit length of beach. [Bailard (1984) modified this equation to include the ratio of wave orbital velocity magnitude to sand settling velocity, and investigated the effect of bottom slope. There is uncertainty in regard to the correlation coefficient used in practice with the equations, and research is still being done. For a recent update, see Schoonees and Theron (1995). They compare many field measurements with the "SPM formula", its predecessors and subsequent equations. It is interesting to note the relative good fit of the equation of Caldwell with field data.]

Field studies were made of Pacific Coast beaches by John Isaacs and Willard {"Bill"} Bascom and colleagues in the mid 1940's (O'Brien and Johnson, 1947; Johnson and Isaacs, 1948), and in the late 1940's by Robert Wiegel with Marine Corps and Naval personnel for the U.S. Marine Corps and ONR as part of a study of amphibious vehicles and craft performance in the surf (Wiegel, 1951, 1952) - surveying beach profiles through the surf, measuring and observing wave characteristics, and collecting and measuring sand samples. One finding was an approximate relationship among sand size, exposure to waves, and beach foreshore slope (Bascom, 1951; Wiegel, 1964). The BEB had proposed field research of this type in the 1930's, but was unfunded at that time. Field groups were established by the BEB in Florida in 1947, New Jersey in 1948, and California in 1948 (Quinn, 1977).

Characteristics of breaking waves on a beach {spilling, plunging, surging} were studied in a hydraulic laboratory and in the field (Iversen, 1952; 1953; Patrick and Wiegel, 1955; later, Galvin, 1968; 1969; Weggel, 1972; Van Dorn, 1978). Munk (1949) presented his ideas on how solitary wave theory might be applied to breaker kinematics. Wiegel (1960) thought that shallow water cnoidal wave theory {Korteweg-deVries equation} would be more useful, certainly just seaward of the breaker zone, and developed practical means for calculating their shape and kinematics. [Recall Gaillard's comment about the lack of measurement of shallow water waves. Measurement of shallow water waves, and analysis of records, is more difficult than in deep water owing in part that the concept of linear superposition is not appropriate. Theory is much more difficult than for deep water waves.] Parameters {wave steepness and beach slope} useful to predict when waves will reflect, when they will break, and wave runup on a structure and on a beach are discussed in a subsequent section. [The association of breaker type with both wave steepness and beach slope was observed, but the terms dissipative system and reflective system were not used until later (e.g. Guza and Inman, 1975). It was recognized that crucial to any theory of wave breaking and runup is a reliable estimate of the rate and spatial distribution of energy dissipation. Theoretical, laboratory and field studies have been made of this process. Sobey (1993) presents a recent comparison of four well known estimates of the dissipation coefficient.]

The relatively long period oscillations {2 to 3-minute periods} that occur in the surf zone were named "surf beats" by Munk (1949) in his presentation of theory and measurements. He also mentioned what is now called wave setup. Munk states: "To anyone wading in shallow water it is apparent that a rapid succession of high breakers temporarily raises the water level. John Isaacs {personal communication} has observed such changes in sea level .... at Twin Rocks Beach, Oregon." It is of interest that a nearly simultaneous but independent study was undertaken by Tucker (1950) in England.

The sea sled was developed in 1945 by John Isaacs to improve the accuracy of surveying beach profiles through the surf zone (Isaacs, 1945; Warner, 1946; O'Brien and Johnson, 1947; Wiegel, 1951). It was equipped with a "flapper" and "dog" so it could be moved by wave action seaward through the surf zone, towing a cable which was used to pull it back to shore, after the "dog" was released. It could also be installed seaward of the breaker line and moved by wave action to shore. More accurate profiles could be obtained with it than with the fathometer mounted on a

DUKW and lead line used in the surf for most WW II surveys. A non-self-propelled sled {Sears Sea Sled} was developed in the Baltimore District in 1947 and successfully used for surveys of Ocean City, MD. The sled was towed seaward by a line attached to a truck winch on shore and a survey boat offshore. The system was used for subsequent Corps surveys (Kolessar and Reynolds, 1966). [This type of device is still useful for beach surveys with the sled taken to sea in a boat and then pulled to shore by cable (Langley, 1992; Grosskopf and Kraus, 1994). Later, the Coastal Research Amphibious Buggy (CRAB), a 3-legged vehicle which can traverse the surf zone was developed by CERC for nearshore surveys (Birkemeier, et al., 1981). Beach profile surveys are important in coastal engineering, and studies of their accuracy were made by Saville and Caldwell (1953), and many years later by Magoon and Sarlin (1970), by Dean (1989 a,b), and by Grosskopf and Kraus (1994). The importance of beach profile surveys in monitoring beach nourishment projects has been discussed recently by Weggel (1995) and the NRC Committee on Beach Nourishment and Protection (1995).]

The fact that sand moves cross-shore {offshore-onshore} as well as alongshore was known, but few measurements were available. Profiles were measured as part of the amphibious operations studies, and are available in UCB and SIO technical reports, but only a few were published (Wiegel, Patrick and Kimberley, 1954; Komar, 1978). Profiles were routinely taken, and historic contour comparisons made where possible for the cooperative beach erosion studies made by the USACE through its Districts and the BEB, and are published in House Documents reporting these studies. Where areas have been restudied over the years, these profiles become a valuable resource. Shepard (1950) and Inman and Rusnak (1956) also published their measurements. Hydraulic laboratory studies were made of cross-shore sand movement at the BEB (BEB, 1936; Rector, 1954) including the effects of tide (Watts and Dearduff, 1954) and varying wave period (Watts, 1954), and at UCB (Scott, 1954). [It was recognized that there were scale effects (see Wiegel, 1964), probably related to the ratio of the magnitude of water particle velocity to the settling velocity of the sand particles (e.g. Dean, 1983; O'Brien, 1987). Large scale experiments at the BEB (Saville, Jr., 1957) showed onshore movement as well as offshore under prototype waves, and by comparison with 1:10 sand models proved the suspected scale effect existed, and that the steepness ratio of about 0.025 as the dividing point of on/ offshore movement was not valid for prototype, and strongly affected by some parameter probably related to wave height. Subsequent BEB tests with coal showed sediment settling velocity to be one of the scaling parameters (Saville, Jr. and Watts, 1969); Saville, Jr., 1980). Other relatively early scale effect tests were done by Noda (1972).]

[Studies were made of sand movement in the nearshore using tracers and SCUBA divers (Inman and Chamberlain, 1959; Duane and Judge, 1969; Komar and Inman, 1970). Among other things, the investigators were interested in what depth of water sand moved (see, also, later work by Hallermeier, 1981; Hands and Allison, 1991). Many years later more extensive field work was done, some by Inman and his colleagues at SIO, the Nearshore Sediment Transport Study (Seymour, ed., 1989), and many studies made at the CERC Field Research Facility at Duck, NC. The relationship between forcings and sediment transport in the surf zone is very complex, especially during storms. Also, cross-shore and alongshore transport occur simultaneously with interactions, and nonlinear interactions with the forcings, a coupled system.]

Observations were made of the trafficability of beaches, and one conclusion was (as reported in Horonjeff et al., 1953) "... the saturated sand near the water's edge was caused to liquify locally due to the vibrations produced by vehicular traffic. As a result the vehicle would sink and become stuck." A new professor at UCB, H. Bolton Seed, continued to work on this phenomenon, and developed the basic theory of liquefaction caused by earthquake vibrations, and how to design for the problem. Liquefaction is important in some fills for coastal works, for example during the 1994 Northridge Earthquake, the center of the northern mole at king Harbor, Redondo Beach, CA, suffered subsidence of 2-3 feet, with damage to structures (Dames & Moore, 1994).

Analysis techniques were developed for both functional and structural design of coastal works. Major beach nourishment projects were constructed in southern California and in New Jersey. For a listing of many projects, see Hall (1952) and recent papers by Wiegel (1994) and Sudar et al. (1995), and a subsequent section.

During the latter half of the 1940's academic graduate courses on aspects of coastal engineering were given at a few universities. Research was done at several universities funded by the USN and the USACE BEB {research contracts as we now know them were started by the BEB in about 1949 and by the USN at an earlier date}, and at the BEB. Many of these are referred to elsewhere, but as an example, studies were done at CIT on the causes and effects on moored ships of harbor oscillations by Professor Robert Knapp (1951) and his colleagues. Other work was done on mobile breakwaters by John Carr (1952). Use of hydraulic laboratory facilities to study waves and their effects on beaches and structures expanded. Use of linear wave theory was simplified by calculation of the functions of d/L vs. d/L_0, the solution of the dispersion equation, and the preparation of tables by Wiegel (1948 a,b, 1954) at the suggestion of J.D. Isaacs, under contract with USN Bu. Ships. [These tables have been largely superseded by algorithms for computer calculations, but with less accuracy (Eckart, 1952; Hunt, 1979; Fenton and McKee, 1990).] Wiegel (1950) developed a parallel wire resistance wave gage/ oscillograph system for hydraulic laboratory studies in 1947-1948, under the guidance of Prof. H.A. Einstein, based on a CIT device. A similar device was developed at the BEB at about the same time.

Earthquake/Tsunami, 1 April 1946; Subsequent Developments

On 1 April 1946, an earthquake occurred in the Aleutian Trench, Alaska, which generated a large tsunami {probably caused by a massive underwater landslide triggered by the earthquake} which travelled across the Pacific Ocean, and caused loss of life and much coastal damage in Hawaii, and damage along sections of the California coast (Bascom, 1946; Isaacs, 1946; Shepard et al., 1950). Coastal engineers and scientists made field investigations of damages and developed hypotheses and theories for their generation, travel and runup; e.g. Kaplan (1955). Development of tsunami generation theory was done by Keller (1961), and extended by Noda (1971). [Later, when high speed digital computers became available, William Garcia (1972) applied ABMAC {Arbitrary Boundary Marker and Cell} to solve the problem for the 2-dimensional case, and made hydraulic laboratory tests which verified it. For survey papers of theories and field data, see Wiegel, 1970; 1976.] The Seismic Sea Wave Warning System was established in Hawaii in August 1948 by the USC&GS to provide real time warnings to the Hawaiian Islands of probable tsunamis generated from distant earthquake sources. [Owing to the danger and importance of tsunamis, it was later expanded to include systems of other countries in and bordering the Pacific Ocean (Cox, 1968). It is currently named the International Tsunami Warning System, an international cooperative operation. The International Tsunami Information Center is located in Hawaii, sponsored by NOAA and UNESCO's Intergovernmental Oceanographic Commission.]

Wave Measurement, Analysis, and Climate: The 1940's, 1950's and Later

The concepts of ocean wave spectra and distribution functions were known during WW II, by the British {who developed a technique for spectral analysis of wave records} and by U.S. scientists {see Klebba (1949); note that probably the first spectral record of hurricane waves, 15 Sept. 1946, is reproduced in this paper}. The first quantitative data obtained in the U.S. on relationships between maximum wave height, significant wave height, average wave height, etc., useful later in developing mathematically expressed wave height distribution functions, were by Seiwell (1948) and Wiegel (1949; later, Wiegel and Kukk, 1957). The theoretical work done in the U.K. was by Michael Longuet-Higgins and his colleagues, and in the U.S. by Robert Putz (1952; 1954) and Frank Snodgrass at UCB, by Neumann and Pierson at New York University (1953; 1957), and by Bretschneider (1959). Putz (1954) showed the Rayleigh Distribution to be one of a family of distribution functions useful in describing water waves. The relevant parameter to identify the specific distribution was the ratio of the number of wave maxima and minima, N_0, to the number of zero crossings {that is, the number of times the wave time history passed through the line representing the mean value of the time history, N_1}. When this value was nearly unity, the Rayleigh Distribution was a good approximation. This ratio, together with the mean ordinate and the number of zero crossings gives the data necessary to specify the waves. Later theoretical work on distribution functions by Putz (1954; also 1957, summarized in Snodgrass and Putz, 1958) made use of the papers on random noise by S.O. Rice (1944 and 1945). Studies were made of the distribution of wave periods and wave lengths, and the joint probabilities of heights and periods (Bretschneider, 1959). [For an update on the status of knowledge of this, see Sobey (1992), who also shows the importance of spectral shape to joint probabilities.]

Theory for an electronic analyzer was developed by Pierson (1954) under contract to the BEB, and a magnetic tape recorder and analyzer designed by Chang

(1955). Snodgrass and Putz (1958) developed theory and hardware for a wave height and frequency meter that reported the output of combined electronic and hydraulically damped meters which "analyzed" the signal from a wave gage installed seaward of the surf zone, on a continuous basis. The outputs were the "zero-crossing period," and a measure of the mean absolute ordinates, which they showed theoretically to be related to the square root of the total power in the spectrum, and thus to the significant wave height. The system was tested in the ocean for nearly a month and found to give results that were satisfactory compared with the then standard method of wave record analysis. The first major study of low frequency ocean wave spectra was by Snodgrass and Tucker (1959). Frank {F.E.} Snodgrass was a major innovator of wave, tsunami, and tide gages; some historical information on his work has been given in his obituary by Walter Munk (1985). [The state of the art for one-dimensional wave spectra was presented at the Conference on Ocean Wave Spectra, Easton, Maryland, 1-4 May 1961, sponsored by the USN Oceanographic Office {formerly Hydrographic Office}, and the NAS/NRC (1961).]

Measuring waves and analyzing data to obtain directional spectra with sufficient resolution is difficult in the ocean and in the hydraulic laboratory (Mobarek, 1965; Borgman, 1969, 1976; Panicker and Borgman, 1970) and expensive. [Lack of data with adequate resolution remains a major reason for our difficulties in the study of coastal processes and the design of coastal works (NRC Committee on Coastal Engineering Measurement Systems, 1989).] An early experiment was done using a 3-gage array off southern California, with the output digitally recorded for computer analysis (Munk et al., 1963). This work of Snodgrass was the pioneering one in the digital recording of ocean waves (Munk, 1985). [Development of models of directional wave spectra was done by Borgman (e.g. 1969, 1976) and others. A conference on directional spectra applications was held in Berkeley, CA in 1981, by the NRC Marine Board (Wiegel, 1982). There are now many directional wave buoys in operation.]

Visual observations of wave direction had been made, but attempts at measuring the direction(s) were not successful. One of the earliest experiments was using a Rayleigh Disk (Hall, 1950).

Wave climate data are important in most aspects of coastal engineering. [Wave climate and the probability of occurrence of wave height and period and their relationship to lightering operations from ship to shore were vital for the WW II Normandy logistical operations (Seiwell, 1947; Bates, 1949).] The need for long term wave measurements was recognized by coastal engineers and scientists. A few wave recorders were installed at several locations on a relatively short term basis (e.g. Wiegel, 1949; Wiegel and Kukk, 1957). The USACE BEB installed two step resistance wave gages at Atlantic City, NJ in April 1948 (Williams, 1969), and expanded their program with this and other types of gages (Peacock, 1974). A major problem with measurement programs then was the slowness in analyzing wave records, which had to be done manually. Equipment and techniques were improved so analyses could be made by computer. Funding was not available for a measurement program such as existed for tides {for navigational use}.

Wave forecasting/ hindcasting techniques developed during WW II were used by a group at SIO (1947) for USACE Los Angeles District, to develop a wave climate for the 3-year interval 1936-1938, using Northern Hemisphere daily synoptic historical weather maps. {One of the meteorologists was Warren Thompson.} {A comparison of hindcast waves with actual recorded waves at 2 California locations (Isaacs and Saville, 1949) gave credence to the reliability of making statistical wave hindcasts from weather maps.} A rudimentary part of a wave climate for a region was searching old newspapers for accounts of major coastal wave events, and then hindcasting wave data {height, period, direction, duration} for these storms. The data were used in the design of coastal projects in southern California (Marine Advisers, 1960, for USACE, Los Angeles District). [For an update, see USACE Los Angeles District report (by Noble Consultants, 1995).] The BEB, as part of its efforts to determine wave climate for the entire US coast, made 3-year hindcasts for the Great Lakes (Saville, 1953), North Atlantic (Saville, 1954; Neuman and James, 1955), and Gulf of Mexico (Bretschneider and Gaul, 1956). Comparison of these hindcast statistics with observations at a nearshore gage at Long Branch, NJ showed good agreement at lower wave heights, with divergence at the higher waves, the divergence depending on assumptions made about refraction. In the Great Lakes studies, a 12-year hindcast was also made for just the severe storms, those expected to generate waves exceeding 8 feet in {significant} height. The statistics for the Gulf of Mexico used relationships developed for the Gulf's broad shallow shelf (Bretschneider, 1956), taking into account loss of energy by friction and percolation (Bretschneider and Reid, 1954). The USACE contracted with National Marine Consultants (1960): "To compile deep-water wave statistics based on meteorological records and charts for the years 1956, 1957, and 1958. The statistics to be compiled are wave height, wave direction and wave period and are to be presented as monthly and annual averages." As in the earlier studies, the then state of the art wave hindcasting techniques were used for 7 carefully selected deep water stations along the California coast. The sea and swell tables and annual sea and swell roses were used for many years in the design of coastal projects. [Later, California's Department of Navigation and Ocean Development {DNOD} contracted with Meteorology International Inc. {MII} to develop computer programs and compile statistics for 6 deep water stations along the California coast, using the USN Fleet Numerical Weather Central {FNWC}, Monterey synoptic wave hindcasts/ forecasts which had been made since 1946 for the northern hemisphere (Habel, 1977). The specifications were developed by Professor Warren C. Thompson. The data were based on hindcasts for 1951-1970 and operational forecasts for 1971-1974. The output was in 5 volumes (e.g. Met. Inter., Inc., 1977).]

In the early 1950's, the USACE BEB started to obtain surf data using visual observations. The U.S. Coast Guard agreed to work with the BEB with Coast Guard personnel making observations, the Cooperative Wave Observation Program. It was started in 1954 at 27 Coast Guard stations (Darling and Dumm, 1967). The number of stations was gradually reduced to 16, as many of the original stations were no longer manned. In 1967, the Littoral Environmental Observations {LEO} program was started (Berg, 1968; Schneider, 1981). The original program consisted of a cooperative effort of the California State Department of Beaches and Parks and the USACE CERC, with observations made by State employees at 50 sites along the California coast. It was expanded to other states, on a volunteer basis.

[The BEB initiated a wave measurement program in April 1948, with gages in New Jersey. The first gage installed on the Pacific Coast, at Huntington Beach, CA operated {more or less} continuously for over 30 years. {The original strip charts and digitized data are no longer available (USACE, Los Angeles District, 1993, p. 4-I-8); however, some significant wave height and period data for 19 wave gage sites are summarized in Thompson (1977).} The program was ambitiously budgeted for nearly 100 gages, with several in deep water, but never funded, and the most operating at any one time were about 10 to 15. The earlier gage program is described by Darling and Dumm (1967). In the late 1960's, the data started to be directly recorded at CERC, transmitted from the gages by leased telephone lines (Williams, 1969). In the mid-1970's a wave measurement program was started jointly by the USACE and the CA Dept. of Boating and Waterways (Seymour and Sessions, 1976; Seymour, 1987). This system used computer analysis of digitized data. Data reports were published monthly (SIO, 1995), and annual compilations also published. {Note: these monthly data reports were published until 1995; starting January 1995 the monthly *Compendium of U.S. Wave Data Summary Statistics* have been published jointly by USACE WES and NOAA.} The USACE has other wave gages {part of its Field Wave Gaging Program}, and NOAA's NDBC has an extensive network, including many directional wave gages (McGehee and Hemsley, 1994); this national network is described briefly in a subsequent section.]

[The USACE has a program in which improved wave hindcasts were made for the 20-year interval 1956 to 1975 for locations along U.S. coasts, the "Wave Information Study" {WIS};(e.g. Corson et al., 1982). Regional wave climates, including direction data, were calculated in this program. Experience in using the data and a comparison of hindcasts for a later year (1988) with data from NOAA NDBC buoys and nearshore USACE wave gages showed modifications to the hindcast model were needed (Anon., 1992). Also, separate hindcasts were needed for hurricanes (Tracy and Hubertz, 1990), and in southern California local wind generated waves and the effects of nearby islands were required (Jensen et al., 1992), and the hindcasting of southern swell - which has proved to be difficult (Hubertz et al., 1994).]

{As an incidental note, the largest waves reliably reported until now were from the USS Ramapo in the North Pacific Ocean in 1933 when trained Naval officers used nautical instruments {e.g., sextant} and known distances and elevations of the ship to measure a series of waves over 100 feet in height, the highest being 112 feet (Whitemarsh, 1934; Gower and Jones, 1994).}

The 1950's. Conferences; Marinas; Facilities; Academic Programs

In the 1950's and 1960's there was a rapid expansion of coastal engineering activities. As an example, the first conference on coastal engineering was held in Long Beach, CA in 1950. [The 25th ICCE is scheduled for 1996, for which this history has been prepared]. The first conference on coastal engineering instruments and data analysis was organized by Wiegel, and held in Berkeley, CA in 1955. [The second on wave measurement and analysis was held in New Orleans, LA, in 1974 (Edge and Magoon, 1975), and the third in New Orleans, LA, in 1993 (Magoon and Hemsley, 1993).] The first ASCE specialty conference was held in Princeton, NJ in October 1958, on the subjects of sand bypassing and groins, organized by the Committee on Coastal Engineering and chaired by ADM. W. Mack Angus; but there were no published proceedings. The first ASCE conference with published proceedings {separate from the ICCEs} was held in Santa Barbara, CA, in 1965, under the Committee on Coastal Engineering, with T. Saville, Jr. as chair (Saville, Jr., compiler, 1965). This type of activity continues to expand.

There was a rapid growth in the demand for marinas for recreational boating, which has continued unabated.

Many hydraulic laboratory facilities were built at this time, and these are described in a subsequent section.

In the 1950's, academic graduate programs on coastal/offshore engineering were developed at about a half dozen universities. These were closely related to research projects sponsored by USN ONR, NSF, and USACE BEB. Several of these had ties with Naval Architecture departments. A few "short courses" were given through university extension services to provide an opportunity for engineers to become familiar with the new developments in coastal/offshore engineering. Later, this activity was also performed by professional societies, and others.

Students and visiting scholars from other countries came to several universities to learn about advances made in coastal engineering since the start of WW II, and to take advantage of the research facilities. Some came as Fulbright scholars and others through a variety of sources. This was an excellent mechanism for the exchange of ideas, and was mutually beneficial; we learned from each other.

Storm Surge; Beach Nourishment; Shore Protection Planning and Design

Storm surge is important along the Atlantic and Gulf coasts, as shown dramatically by the disastrous hurricanes in Galveston, 1900, Lake Okeechobee, 1928, and Long Island/ New England, 1938. With the enlargement of the Lake Okeechobee levees {as a part of the Central and Southern Florida Water Plan} and the development of offshore oil drilling in the Gulf of Mexico, both the USACE and private industry increased their efforts at surge prediction (e.g., Haurwitz, 1951; Saville, Jr., 1952; Reid and Bretschneider, 1953). Some of these used what at the time was the most extensive group of wind and water level data during a hurricane ever taken, at Lake

Okeechobee for the hurricanes of August 1949 and October 1950 (USACE, Jacksonville District, 1955). Studies were made in wind/ water flumes (e.g., Keulegan, 1951) and small water bodies (e.g. van Dorn, 1953). The Great Lakes suffer from wind tide {setup} also, particularly Lake Erie where seiching afterwards can maintain a problem long after the winds cease, transferring it from end to end of the lake (e.g. Hunt, 1959). Early theoretical work by Keulegan (1951) was followed by the work by Platzman (e.g., 1963). With the long string of major hurricanes in the mid-1950's to early 1960's (Harris, 1963), particularly Carol, 1954, Hazel, 1954, Donna, 1960, and Carla, 1961, a major effort was started by both the USACE and the Weather Bureau for both warning and protective measures. Work by Harris, Jelesnianski, Myers and others at the Weather Bureau resulted in standard storm parameters, the Standard Project Hurricane (U.S. National Weather Service, 1979). Several surge prediction models were developed, with differences resulting partly from different needs of the modelers {USACE for protection, Weather Service for warning, FEMA for insurance, etc.}. See, for example, Reid and Bodine (1968), Jelesnianski (1972; 1974), Wanstrath (1977), Reid et al. (1977), and Tetra Tech (1981). A comparative study of many of these was made by the USACE Committee on Tidal Hydraulics (1980). All models have been updated in the past decade. Improved prediction schemes have been accompanied by vastly improved observations {routine ship reports by 1928, aircraft reconnaissance and radar after WW II, and satellite reconnaissance about 1953} and communication systems. These coupled with both local and federal interest have resulted in present day warning systems, and evacuation programs, which have largely prevented loss of life despite increasingly higher density population of coastal areas (e.g., Barnes, 1995). These are not discussed in detail, as being outside normal coastal engineering; but they are important adjuncts.

Based on their experience with beach nourishment (Hall, Jr., 1952), the USACE stated in a report to Congress (U.S. Congress, 1953): "Where conditions permit, probably the best means of protecting a beach or shoreline against erosion of any type is to introduce a sandfill between the shoreline to be protected and the ocean and then to maintain that protective fill against long-term erosion." This was an important recommendation; many beach nourishment projects have been constructed, and others planned (Sudar et al., 1995; NRC Committee on Beach Nourishment and Protection, 1995).

The book *Shore Protection Planning and Design* was published in June 1954 by the USACE BEB, revised in Feb. 1957 and in May 1961. The manual was the coordinated work of many, primarily Kenneth P. Peel and Kenneth Kaplan, with sections also written by R.H. Allen, C.T. Fray, R.L. Harris, W.J. Herron, T. Saville, Jr., W.H. Vesper and L.L Watkins. Substantial contributions were made by R.O. Eaton and J.V. Hall. Revisions to the first "draft" were compiled by R.A. Jachowski and G.M. Watts, and the report was edited by A.C. Rayner and R.L. Rector. It had worldwide influence on the design of coastal projects. [It was superseded by the *Shore Protection Manual* {SPM} in 1973. Both had several later revised editions, the latest SPM being 1984.]

A Bit on Breakwaters, Harbors and Moorings

Most breakwaters and jetties in the U.S. are the rubble mound type. Robert Y. Hudson (1953) improved the design formula originally developed by the Spanish engineer R. Iribarren, making it dimensionaly homogeneous. The relationship Hudson developed, with input from extensive hydraulic laboratory tests and some prototype observations, has been widely used for many years. Information on the model laws are given in the early reports, and in the USACE CERC report *Coastal Hydraulic Models* (Hudson, et al., 1979). Use of the Hudson formula and hydraulic model data were extended later to include cast concrete armor units. Detailed procedures and design data are in the latest *Shore Protection Manual* (USACE CERC, 1984), and updated in reports and papers in the past decade.

Generally, adequate stone was economically available from quarries. However, as the size of armor stone required is proportional to the cube of the wave height to which they are subjected, there are sites where adequate size armor stone is not available. This led to the use of cast concrete armor units, and owing to their interlocking and hydraulic characteristics, units of smaller weight than the stone could be used, which reduced the amount of material required. The design wave often was the wave that would break in the depth of water in which a breakwater or jetty section was founded. There were problems at some sites where it was necessary to use very large units owing to the size of the waves, such as the seaward ends of the jetties at Humboldt Bay, CA and segments of the Crescent City, CA breakwater. There seemed to be a scale effect between model and prototype. Orville T. Magoon studied these and other failures, (Magoon et al., 1974; Edge and Magoon, 1979), and in 1980 established and chaired the Rubble Mound Structures Committee of the ASCE WW Division, which is still active. Details of problems encountered, lessons learned, and evolving practices are given in a subsequent section.

Owing to the prevalence of rubble mound structures in the U.S., few studies have been made of waves breaking against vertical walls. However, as mentioned earlier, some of the first {in the late 1800's} coastal engineering studies were of these forces. After WW II, investigations were started again, but in hydraulic laboratory facilities rather than in the ocean or Great Lakes. The first was by John Carr and his colleagues at CIT (1954) and Ross (1953) at the BEB, followed by Leendertse at the USN Civil Engineering Laboratory (1961), Weggel and Maxwell (1970), and Kamel (1970). [Little work has been done in the U.S. since, although it would be useful in the design of seawalls and buildings subject to this type of force during hurricane and "northeaster" storm surge, wave setup and runup conditions, and for estimates needed by FEMA to establish coastal high hazard areas.]

A major problem in harbors is the motion of moored ships and boats. Coastal engineers are concerned with this as well as with breakwaters, entrances, channels, waves, tides, currents, and winds (ASCE Task Committee on Small Craft Harbors, 1969). In some ports, e.g. Long Beach, CA, a significant problem is harbor surging, {long period oscillations} (Hudson, 1947; Vanoni and Carr 1950), possibly related to surf beat (Munk, 1949). Time-lapse {1 frame per second} 16-mm movies were taken of the harbor at Monterey, CA, and other locations by W.N. Bascom and by J.W. Johnson which show this clearly. In other harbors ordinary swell penetrating from the ocean, or wind waves generated in the harbor, or ship waves is a problem. The required reduction of wave height for a harbor to be functionally acceptable depends on wave period as well as height, and the response of moored vessels. [Standards are still a matter of debate.] Mooring systems are strongly nonlinear, and difficult to analyze and to design. Data on seiches in harbors {their cause and characteristics} are needed. The work of Knapp has been referred to. A theory describing the longitudinal motion of a moored ship forced by standing waves was developed by Basil Wilson (1951), and extended by Abramson and Wilson (1955). Techniques were worked out for use in hydraulic model studies of this nonlinear process by Wiegel and his colleagues (Wiegel, Beebe and Dilley, 1957; Wiegel, Clough, Dilley and Williams, 1959). These, and other model studies showed that coupled motions were induced by nonlinear actions (Wiegel, et al., 1959), probably chaotic for some conditions. Full scale studies were done in parallel with these by J.T. O'Brien and his colleagues at the USN Civil Engineering Laboratory, Port Hueneme, CA (J.T. O'Brien, 1955; J.T. O'Brien and Kuchenreuther, 1958). Additional theoretical development, hydraulic laboratory model tests and field measurements were made by Fredric Raichlen (1968). In addition to continued model studies at WES on a number of ports/ harbors, work on wave-induced harbor oscillations was done by McNown (1952), Kravtchenko and McNown (1955), LeMehaute (1962); Raichlen and Ippen (1965), Miles (1970), and Lee (1971).

Wave Setup and Setdown

At a meeting of the USACE Hurricane Coordinating Committee several members {J.M. Caldwell, J.B. McAleer, H.B. Simmons, T. Saville, Jr., C.F. Wicker} suggested the possibility that a 3-foot discrepancy in water level elevations measured at Narragansett Pier and Narragansett Bay, RI made during the 1939 hurricane might have resulted from a wave-induced water level rise at the pier which would not have occurred in the more open waters of the bay entrance. Tests run at the BEB (Fairchild, 1958) confirmed this and more extensive tests (Saville, Jr., 1961) indicated a value of 10-20% of the wave height for normal beach slopes, with a small decrease in water level immediately seaward of the breaker. The terms wave setup and wave setdown were coined for this phenomenon. Interestingly, the ratio also fits Munk's {and Isaacs'} earlier observations (Munk, 1949) and studies of surf beat, although the phenomena are probably not the same {but, it is difficult to separate the phenomena in the irregular motions of the surf on a ocean beach (e.g., Herbers et al., 1995)}. The same phenomenon had been recognized too in tests by Savage (1957), and Saville (1957) showed that with wide (the horizontal distance normal to the shore) berms the runup is associated with the depth of water setup on the berm rather than the wave height itself; that after a certain width of berm, the runup remains essentially constant regardless of how much wider the berm is made. Dorrestein (1962) and Bowen, Inman and Simmons(1968) have provided further data and Hwang and Divoky (1970) have further developed the theory {based on the "radiation stress" theory of Longuet-Higgins and Stewart}. [Sonu (1972) measured wave setup and littoral currents {such as rip currents)} in 3 dimensions at a Gulf of Mexico beach, Seagrove, FL. Guza and Thornton (1981) measured setdown, setup and runup at Torrey Pines Beach, near San Diego, CA obtaining a wave setup value of about 17% of the wave height.]

Wave Runup and Overtopping

In designing rubble mound breakwaters, revetments or levees, it is necessary to calculate wave runup on the seaward slope {and overtopping, if this occurs}. In the design of revetments and levees, it is almost always necessary to prevent overtopping. In 1952 the BEB initiated a long-term program on wave runup and overtopping. Laboratory tests were made at UCB (Granthem, 1953), WES (Saville, Jr. and Caldwell, 1953), and at BEB (Saville, Jr., 1955, 1956; Savage, 1958, 1959). A few experiments were made at UCB on the effect of wind (Sibul and Tickner, 1955). Data on runup on rubble structures were also obtained at WES by Hudson (1959). Design curves derived from these data as shown by Saville, Jr. and Savage were used in *BEB TR 4* and *SPM* for many years to relate relative runup to wave steepness and structure slope as a function of structure depth {a measure of wave breaking location, and probably type}. Hunt (1959) also used these data for his runup formula incorporating what is now known as the "surf similarity parameter", or the Iribarren Number.{Hunt presented results of the relationship between wave steepness, the structure slope, and breaker characteristics, including a description of conditions for which waves would reflect simply, when surging type waves would occur, when waves would break on a structure, and when they would break seaward of the structure; whether the system was reflective or dissipative. He referred to the pioneering works of Miche and of Iribarren and Nogales.} [The same concept was later used for beaches by Guza and Inman (1975).] Large scale tests were undertaken by the BEB in its Large Wave Tank to investigate possible scale effects, partly in conjunction with hurricane levee design for Lake Okeechobee, FL (Saville, Jr., 1958, 1987); a derived scale-effect curve was published in *BEB TR 4* (1961). [Stoa (1978) also discussed these data in relation to scale effects, and presents a revised and extended correction curve. A systematic study of various theoretical approaches was made by LeMehaute et al. (1968) for periodic and solitary waves. This study included the work of many earlier investigators, and original work by the authors, and is a good base for understanding recent work. Stoa (1978) and Ahrens (1977) suggest roughness and porosity correction factors. Seelig (1980) gives curves for runup on rubble structures by a number of different authors using the method of Ahrens and McCartney (1975). Saville, Jr. (1958) presented a method for determining runup on composite slopes. Saville, Jr. (1962) has shown the runup distribution for waves in a generating area to be roughly the same as that for wave heights, and a Rayleigh distribution has been proposed by Ahrens (1977) and others. Guza and Thornton (1981) measured runup as well as setdown and setup on a beach, Torrey Pines Beach {near San Diego}, CA. Carlson (1984) obtained field data for a dissipative wave/beach system and analyzed them, of wind waves uncontaminated by swell at 2 locations in San Francisco Bay, CA. As recognized by all investigators, the rate at which wave energy is dissipated in the surfzone is crucial to the problem (see, for example, Sobey, 1993). The 1980's have brought a new surge of effort, particularly looking at theory (e.g. Kobayashi, 1986) and the applicability to irregular waves, including maxima (e.g., Resio, 1987) - which we shall leave to the next history update.]

Overtopping calculations were made initially by hand calculator using the model data of Saville, Jr (1955) and a gross simplification of the wave distribution given by Bretschneider (1959), calculating estimated overtopping values for a series of wave components in a wave train {using different scales to obtain more values}, and adding these proportionately according to Bretschneider's distribution. Cross and Sollitt (1971) developed a semi-empirical model, and Weggel (1976) derived a formula based expression. [Madsen and White (1976) developed a model for predicting transmission coefficients. Seelig (1980) ran a further set of experiments, and further developed the Madsen-White model. As with runup, much theoretical work was done in the 1980's {e.g., Kobayashi and others at the University of Delaware}, left to a future history update.]

The 1950's and 1960's. Some "Firsts"; SCUBA; Tension-leg; Sand Movement; Geotextiles; Digital Computer Use; Wave Generation; Storm Surge Theory; Breakwaters; Impulsively Generated Waves; "Mach-reflection"; "Waves Across the Pacific"; Risk Criteria

The first use in the U.S. of SCUBA (Self-contained Underwater Breathing Apparatus) for coastal engineering and science studies began in the early 1950's. It is a valuable tool for many purposes (Dill, 1955). It was used extensively at SIO for measuring sand changes in the nearshore (e.g., Inman and Rusnak, 1956).

Willard Bascom and his colleagues at SIO developed a deep water instrument station, which had an anchor and a cable which was kept taut by a large underwater buoy to keep the cable in high tension (Bascom, 1955). It is likely the "grandfather" of the tension leg platforms now used by the offshore oil industry.

Fundamental theoretical and laboratory studies of sediment motion under oscillating {simulated wave} conditions were started by Professor H.A. Einstein and his students at UCB, applying the concepts he developed for sand transport in rivers; this included the conditions necessary for the threshold motion of sand, and the formation and development of ripples (Einstein, 1948, 1972; Manohar, 1955; Kalkanis, 1964). Professor Peter S. Eagleson and his students at MIT started basic studies of the mechanics of discrete spherical particles by shoaling waves (Eagleson et al., 1958). Much of this work was under contract to the BEB. [Komar and Miller (1974) advanced this work about two decades later.]

Some of the first work on edge waves {original theory by G.G. Stokes in the mid-19th century} was done by Walter Munk, Frank Snodgrass and George Carrier in 1956. [Later, theoretical and laboratory studies were made of edge waves and cusps by Guza and Davis (1974); Guza and Inman (1975); and progressive edge waves by Yeh (1985).]

During a field trip of the 6th Conference on Coastal Engineering in Florida (1957), some delegates {including Per Bruun, Joe Caldwell, J.W. Johnson, M.P. O'Brien, Thorndike Saville, and Thorndike Saville, Jr.} met Mel Greiser, the owner of a beach home just south of Palm Beach. He was President of Carthage Mills, Inc., and they had recently developed a plastic filter cloth, "Filter X" {what is now called a geotextile}. It was being used on an experimental basis in front of his house - a porous plastic filter cloth laid on top of a prepared sand slope on the frontal dune, with stone or other units placed on top to form a revetment. Its purpose is to relieve water pressure within the sand dune and to prevent loss of sand through it. This material has been used subsequently for vertical seawalls, beneath jetties and breakwaters, and other coastal works. [Details for coastal engineering design were given at the 10th ICCE by Robert Barrett (1966). Use of this type of material is now common practice.]

John Miles developed his shear flow theory of the generation of surface waves by wind - an instability theory, with rapid growth of the waves (1957; 1959; and others). This was a major breakthrough in our understanding of an important wave generating mechanism by wind blowing over the water surface. A digital computer was required to calculate the functions.

Following the work of Kaplan (1953), Basil W. Wilson (1961) developed an improved procedure for forecasting/ hindcasting waves generated by a moving fetch with variable winds - a major advance, which also involved an early use of a digital computer.

Owing to the development and availability of high speed digital computers, numerical models became practical {see examples above}. They were rapidly adopted in the design of offshore structures. Computers were used to calculate tables of functions of several nonlinear wave theories for use in the design of offshore and coastal structures: Stokes 3rd order (Skjelbreia, 1959); Stokes 5th order (Skjelbreia and Henderson, 1960); Stream Function (Borgman and Chappalear, 1957); Cnoidal (Wiegel, 1960; Laitone, 1960, 1963; Masch and Wiegel, 1961) [For later work on the Korteweg-deVries Eq., which is necessary for shallow water, see Segur (1973) and Hammack and Segur (1974, 1978). For a recent presentation of how to make a rational selection of steady wave theory, see Sobey et al.(1987). Later, major computer codes were developed for wave forecasting and hindcasting by the USN and by CERC/ WES {the WIS program (Resio, 1981; 1982; - and Tracy, 1983); tsunamis (Hwang, et al., 1972; Houston and Garcia, 1978 for a FEMA contract to WES); hurricane storm surge (Reid et al., 1968, 1977; Bodine, 1971; Jelesnianski, 1972, 1974 {SPLASH - Special Program to List Amplitudes of Surges from Hurricanes}; Wanstrath, 1977 for USACE; Tetra Tech, 1981 for FEMA; NRC evaluation, 1983); refraction (Street, et al., 1969;

O'Reilly, 1993). The original big push for hurricane storm surge prediction was by the U.S. Weather Service for use in hurricane warnings. The development of many of the more recent models was driven by the needs of FEMA for the National Flood Insurance Program. Application to coastal sediment transport came at a later date and will be mentioned subsequently. Extensive development of 3-dimensional models of estuaries and coastal seas was done by Jan J. Leendertse, Shiao-Kung Liu and their colleagues at Rand Corporation (e.g. 1973, and subsequent reports).]

Hydraulic laboratory studies were made, and simple theories developed for submerged breakwaters, spaced-pile breakwaters, curtain breakwaters, flapper breakwaters, floating breakwaters, hovering breakwaters, and pneumatic and hydraulic breakwaters (e.g. Johnson et al., 1951; Patrick, 1951; Wiegel, 1961; Wiegel et al., 1962). Most were for possible use in amphibious operations, but some concepts were investigated later for other purposes. [Several rubber tire breakwaters were built in the 1970's for marinas, but there were structural problems during storms. This led to the development of the pipe-tire floating breakwater (Harms et al., 1980; Harms, 1980; Nelson et al., 1983). Two types of floating breakwaters, a pipe-tire unit and a concrete unit were tested in Puget Sound, near Seattle, WA. This was an interesting test, as it was made at a site where it was expected that during the 2-year test, the units would be subjected to waves as severe as the design conditions at the project site. A modified version of the concrete unit was installed at Friday Harbor, WA (Nelson et al., 1983; USACE Seattle District, 1991). A review of many types or floating breakwaters suggested or tried was made by Hales (1981). A recent summary of types and installations has been made by a PIANC committee, chaired by John Oliver (1994).]

Impulsively generated waves were investigated by Egbert {J.E.} Prins while a Fulbright Scholar at UCB (Prins, 1958, a,b). The purpose was to learn more about tsunami generation and characteristics for a range of ratios of source length to water depth and source amplitude to water depth. For small values of source length to water depth, the theories of Cauchy, of Poisson, of Penney, of Unoki and Nakano, and of Kranzer and Keller (1955) were useful. However, for a source length several times the water depth, a "solitary wave" was generated; for longer sources what are now called "solitons" formed, and for even longer sources bores formed. Galvin (1972) investigated waves breaking in shallow water which for some conditions had characteristics similar to the waves observed by Prins for long sources relative to water depth; Galvin referred to these as solitons. [A rather common phenomenon, but little studied. The formation of multiple crests results in shorter wave periods within the surf zone, or over a reef (Wiegel, 1964; 1990; Galvin, 1990).]

John Isaacs mentioned to Wiegel he had observed waves in shallow water reflecting from a breakwater that seemed to have a characteristic analogous to the Mach-stem reflection of a shock wave in air moving over the ground {sonic boom}. Wiegel and several graduate students performed hydraulic laboratory studies, first with solitary waves, and later with shallow water periodic waves, and found this to be the case. The height of the stem increased as it moved along a vertical wall breakwater

(Perroud, 1957; Nielsen, 1962; Wiegel, 1964). [Theory came later than observations, first for solitary waves (Miles, 1977; Melville, 1980), then for periodic shallow water waves (Yue and Mei, 1980; Yoon and Liu, 1989). The term "stem waves" is also used now.] This phenomenon can be seen in a few photographs of the 1 April 1946 tsunami at Hilo, Hawaii. They have been observed at the concrete storm drain at Waikiki, HI, along the rock revetment on the railroad embankment opposite the entrance to Humboldt Bay, CA, and along the south jetty at Anaheim Bay, CA, where they are important in the alongshore transport of sand. Also, Gaillard seems to have observed them along the west breakwater {about a mile in length} at Oswego, NY. He states (1904, p. 202): "During westerly gales the waves coming from that direction struck this breakwater at a small angle, and were in part reflected and in part crowded against the breakwater, accumulating size as they travelled, until reaching the east breakwater, they passed in solid volume over the tops of the snubbing posts 15 feet above the water level."

In the mid 1960's an important oceanographic study was made of waves generated in the Southern Ocean and the South Indian Ocean, in the south 40's and 50's latitudes, with the swell travelling northward past New Zealand, across the equator, past the Hawaiian Islands, all the way to the Aleutian Islands in Alaska (Snodgrass et al., 1966). This was an influential scientific study of wave dispersion, scattering, and energy dissipation. {A documentary movie *Waves Across the Pacific* was made of this experiment.} It was also helpful to coastal engineering in developing an understanding of the "southern swell" component of the coastal wave climate and the resulting alongshore sand transport in southern California (see Wiegel, 1949; Hubertz et al., 1994).

Risk criteria and probabilistic design are an important aspect of engineering. A pioneering work on risk criteria in regard to design waves and total damage was done by Leon Borgman in 1963.

Wave-Induced Forces on Pile-Supported Coastal Structures

Directional drilling for hydrocarbons from piers in southern California and offshore platforms in shallow coastal waters in Louisiana led to the need for better quantitative understanding of wave induced forces on structures. Work on this subject, originally sponsored by the USN Bu. Yards and Docks for their needs in the design of pile supported port structures, was expanded by several oil companies. Walter Munk (1948) gave an equation for drag forces exerted by waves on structure components. This approach was extended at UCB by Morison, O'Brien, Johnson and Schaaf (1950), to include both drag and inertial forces and their equation is the one that has been used in the design of nearly every space-frame type offshore structure in the world. It is generally known as the "Morison Equation," but should be referred to as the "MOJS Equation." MacCamy and Fuchs (1954) developed a linear theory for calculating wave-induced inertial forces in an inviscid fluid, for right circular cylinders that have diameters ranging from small compared with the wave length to diameters that exceed the wave length. Their theory is the standard with which numerical models are compared. Wiegel, Beebe and Moon (1957) made measurements in 1953/1954 at the end of a pier at Davenport, CA which was fully exposed to Pacific Ocean waves, for near breaking and breaking waves. The coefficients of drag and mass {inertia} were the first prototype data available in the published literature, and were used in the design of many offshore structures. Work on wave-induced forces was also done at that time in the Gulf of Mexico by Robert Reid (1956) and his colleagues at TAMU. The equations and coefficients mentioned above are also used in designing pile support systems of beach houses which are subject to possible flooding and wave attack during hurricanes and other severe coastal storms.

The studies of Keulegan and Carpenter (1958) and McNown and Keulegan (1959) were very important in identifying a parameter which is as important in unsteady flow as Reynolds Number is for steady flow; it is now known as the Keulegan-Carpenter Number. During full scale studies in the ocean, Wiegel et al. (1957) had observed the importance of eddy formation and shedding on the forces and motions of the test pile. [This phenomenon was studied in detail at a later date, with the term "lift forces" {really, "transverse forces"} given to these periodic forces (Bidde, 1971; Wiegel and Delmonte, 1972).]

In the mid 1960's, Borgman (1967) developed a theory for spectral analysis of wave-induced forces on a pile and gave a method for calculating forces for a given wave spectrum. Later, he developed a theory for directional spectra wave loading (Borgman and Yfantis, 1979). Techniques were developed to calculate forces and motions of both rigid and flexible space-frame offshore structures (Penzien, et al., 1972; Berge, 1973; Berge and Penzien, 1974). Work on wave and current induced forces expanded rapidly with the increase in the number of offshore platforms for exploration and production of hydrocarbons. Many papers on this subject appear in the proceedings of the annual {beginning in 1969} Offshore Technology Conferences held in Houston, TX. {It will not be considered further in this history, as it is now generally classified "offshore engineering/ technology". A few references are: Heidemann et al. (1979); Chakrabarti (1980); Garrison (1980, 1990); Sarpkaya and Cakal (1983); Nath (1984); Amer. Petroleum Inst. {API} (1993). There are several books on the subject.}

People enjoy piers that extend through the surf zone, for fishing, walking along and sitting on benches to observe the activities and the surf. The piles are subject to forces exerted by breaking waves, and the decks if not high enough are subject to wave-induced uplift forces. [The Huntington Beach, CA pier was completely rebuilt in 1992; major repair work had been done on four previous occasions after severe damage by storms. It was nominated for the ASCE Outstanding Civil Engineering Achievement Award in 1993 (Anon., 1993).] Early experiments with piles in both the ocean and in a wave tank identified two force components, the slowly varying "normal" forces of the type considered in the paragraphs above, and a quasi-impact load which is a function of the breaker type and the precise location between the pile and the breaker at time of breaking (Snodgrass et al., 1951; Hall, 1957). [It was not until many years later that a theory became available to explain the observations (Wiegel, 1982). It was based in part on a concept by Kaplan and Silbert (1976) for wave impact loads on horizontal structural members in the "splash zone" of offshore platforms. In designing a pier, details are needed of the shape and kinematics of breaking waves, and bottom profile changes during storms.] Theory of the vertical loading on the pier deck by waves (the "dock problem") was investigated by Stoker (1957). Hydraulic laboratory studies were made by Elghamry (1971, based on his 1963 Ph.D. thesis), and laboratory and theoretical studies by French (1979). A numerical model was developed by Lee and Lai (1986). [It is interesting to note that due in part to the uncertainty in estimating uplift forces on a pier deck, the USACE CERC pier at Duck, NC, was constructed with grates in the deck to relieve the pressure. During the "northeaster" storm of 24-25 Oct. 1982, 37 of the grates were displaced (USACE CERC FRF, Oct. 1982); they performed as expected.

In designing piers, the variation in bottom depths with changing wave conditions must be considered (LaFond, 1939; Shepard, 1950; Berkemeier et al., 1981). Also, a pier may cause changes in a sandy bottom along and near the pier, as has been measured at the USACE pier (Berkemeier et al., 1981; Miller et al., 1983).

CERC and CERB Established, 1963; BEB Abolished

In November 1963, the BEB was abolished and replaced by the Coastal Engineering Research Board (CERB) and the Coastal Engineering Research Center (CERC). [Ten years later, in 1973, CERC was moved from the Delecarlia Reservation in Washington D.C. to Fort Belvoir, VA; twenty years later, in 1983, CERC was moved from Fort Belvoir, VA to WES at Vicksburg, MS.]

Pollution Control: A Start

Wastes {residential, industrial, heat from power plants} and storm water runoffs were usually discharged into the ocean, bays and estuaries, sometimes with serious effects to human health and environment (NACOA, 1981; Brooks and Krier, 1981). Pollution control in coastal waters requires extensive environmental monitoring before and after project construction, and detailed design analysis (NACOA, 1981). Intakes and outfalls are often major coastal works, sometimes affecting navigation projects (Lillevang, 1966) and nearby beaches. In one case, the Hyperion Sewage Treatment facility and ocean outfall at Santa Monica Bay, CA there was a beneficial byproduct, the nourishment of adjacent beaches with sand excavated for the project (Pardee, 1960). The design and construction of a major outfall requires much coastal engineering input (Murphy et al., 1978; Grace, 1978). Predictions of effluent paths, and near-source, intermediate, and far field mixing require data from pre-project monitoring of environmental conditions. Mixing of effluent with seawater using diffusers to obtain dilution was the first method of mitigating effects (Rawn and Bowerman, 1950; Pearson, 1958; Rawn, Bowerman and Brooks, 1960). Additional work {more treatment or prevention, especially of toxic wastes} has become increasingly important (NACOA, 1981). Studies were made of water circulation and mixing processes in tidal estuaries by Pritchard and his colleagues (1952, 1955, 1959; 1969). The First International Conference on Waste Disposal in the Marine Environment was held in Berkeley, CA in 1959, and the proceedings published (Pearson, ed., 1960). Environmental biological, chemical and physical studies have been made, with perhaps the most detailed and long term being the Southern California Coastal Water Research Project, started in 1969 and still in operation (e.g., Bascom, 1981). Work on mixing of buoyant flows was done at CIT by Norman Brooks and his colleagues (e.g. Brooks and Koh, 1965; List and Imberger, 1973), at MIT by Arthur Ippen, Don Harleman and their colleagues (e.g. Ippen, ed., 1966; Stolzenbach and Harleman, 1971), and at UCB by Wiegel and his colleagues (e.g. Abraham, 1960; Jen et al., 1966; Liseth, 1972; Buhler, 1977; Safaie, 1979; Gosink, 1982; Wiegel and Doyle, 1987). An influential book on mixing in coastal and inland waters, including estuaries, was written by Fischer et al. (1979). Submarine pipeline integrity is important from a functional standpoint and for pollution control, such as preventing structural failure of an oil pipeline from an offshore terminal to shore; analysis of wave induced forces, and possible breakout of partially or completely buried pipes are critical (Bowie and Wiegel, 1977; Grace, 1978; Foda et al., 1991).

Books Published, 1950's and 1960's

In the late 1950's and the 1960's several books were published which had major impacts on teaching, research and practice: *Water Waves* by J.J. Stoker (1957), *Surface Waves* by John V. Wehausen and E.V. Laitone (1960), *Design and Construction of Ports and Marine Structures* by Alonzo DeF. Quinn (1961), *Oceanographical Engineering* by Robert L. Wiegel (1964), *Waves and Beaches* by Willard Bascom (1964), *Wind Waves: Their Generation and Propagation on the Ocean Surface* by Blair Kinsman (1965), *Estuary and Coastal Hydrodynamics* by Arthur Ippen (and eight coauthors, 1966), and *Principles of Physical Oceanography* by Gerhard Neumann and Willard J. Pierson, Jr. (1966).

Disastrous Coastal Events and Follow-ups, 1960's

The "Ash Wednesday Storm of 1962" battered the Atlantic Coast from Montauk Point on Long Island, NY to Florida (Podufaly, 1962). It has been classified an "extreme northeaster" by Dolan and Davis (1992) owing to the combination of its duration {through 5 consecutive high tides}, wave size, storm surge height, and high astronomical tides. It caused extensive erosion of beaches and dunes, overwashes and breakthroughs of the barrier islands, and damaged or destroyed many buildings and much infrastructure. Several USACE Districts, either with their own staff or by contractors, made surveys of beaches and dunes, and identified the damages and estimated the emergency construction required. Colonel Podufaly (1962) presents the findings, and describes the rehabilitation work done to dune and beach to provide protection from a "10-year storm." This used the "Caldwell Section," designed by J.M.

Caldwell, USACE BEB, based on essentially "around the clock" emergency tests in the BEB Large Wave Tank. A sketch of the section, with dimensions is in Podufaly's paper. It is probably the basis for the cross-section designs used in many subsequent beach nourishment/ storm surge protection projects. {The value of sand dunes has long been recognized, and research has been done by the USACE BEB on dune creation and stabilization by fences and vegetation, including field experiments along part of the North Carolina coast (Savage and Woodhouse, Jr., 1968).} The aftermath of this storm brought about a cooperative BEB-Weather Bureau study to attempt prediction of beach/ shore damage that might occur for a particular storm, and the establishment of vulnerability charts relating predicted damage at a specific beach to predicted storm character which then could be used in a warning system (Darling, 1965). FEMA would later take over this type of effort.

A great earthquake (Richter Magnitude 8.4 to 8.6; Moment Magnitude 9.2) occurred on 29 March 1964 in Alaska, with its epicenter in the Prince William Sound region, slightly northeast of Whittier. It generated a very large tsunami of oceanic scale, and local tsunamis created by high speed underwater landslides (NAS, 1972). Numerical models were developed to calculate its generation, travel and runup along the coasts of the Pacific Ocean. Uplift of large areas, both land and underwater, occurred in the Prince William Sound region, and there was considerable subsidence in the Kodiak Island/ Cook Inlet/ Anchorage region, and in the Whittier region of Prince William Sound. Several tens of thousands of square miles underwent tectonic uplift or subsidence. Changes in the location of the shoreline occurred along many miles. There was mass movement of deltas. Coastal engineers should be able to learn from this and similar events, to estimate possible effects of a major change in mean sea level due to global warming or other cause.

Theoretical and hydraulic laboratory studies of tsunamis were made, and data collected of actual events in the interval between the April 1946 Aleutian Trench and the 1964 Alaskan earthquakes and tsunamis. Work was accelerated after the 1964 Alaskan event, as often is the case after a natural disaster occurs which affects adversely lives and property of many people. Much of this was summarized by Wiegel in chapters on tsunamis in two earthquake engineering books (1970; 1976). Two catalogs of tsunami occurrence in the Pacific Ocean were prepared, one at the Univ. of Hawaii and one at the International Tsunami Information Center (Iida et al. 1967; Pararas-Carayannis et al., 1982). The International Tsunami Information Center, in Honolulu, Hawaii has published its *Newsletter* {now named the *Tsunami Newsletter*} since 1967, which has a wealth of information on tsunami warnings, occurrences, conferences and publications. FEMA had numerical predictions made of tsunami runup probabilities along much of the Pacific Coast done under contract to USACE WES (Garcia and Houston, 1975; Houston and Garcia, 1977; 1978). Theoretical and laboratory studies were made which clarified some tsunami generation and transmission characteristics (Hammack and Segur, 1974; 1978). Much work done for the generation of waves by explosions also applies to tsunamis (e.g., Van Dorn, Le Mehaute and Hwang, 1968; Smith, 1967. Other work was done on the forces exerted on structures by impulsively generated waves running over land (e.g. Cross, 1967). For a summary of "tsunami engineering", see Camfield (1980).

[It is important to collect information on natural disasters shortly after their occurrence, to document events and effects. The NRC Committee on Natural Disasters was established in 1971. It was derived from the NAE Committee on Earthquake Engineering, established in 1966, which had been expanded to include wind and other natural disasters. On 1 May 1992, the NRC Committee on Natural Disasters and another NRC Committee on Earthquake Engineering were combined to form the NRC Board on Natural Disasters (Clarke, 1995). Reports useful to coastal engineering have been prepared of events such as Hurricane Iwa, Hawaii in 1982 (Chiu, et al., 1983) and the coastal erosion and storm damage in California during the "El Nino Winter" of 1982-1983 (Dean et al., 1984).]

Hurricanes and severe coastal storms cause coastal flooding in many sections of the Atlantic and Gulf coasts, with loss of life and substantial damages to property. Engineering and economic studies made in the late 1950's by the USACE New England Division led to the planning, construction and operation of 5 hurricane/coastal storm surge barriers on the southern coast of New England {New Bedford-Fairhaven, MA; Fox Point, RI; New London, CT; Pawcatuck, CT; and Stamford, CT} (McAleer and Townsend, 1958). They were mostly built in the mid-1960's. The projects are combinations of dikes, walls, navigation gates, conduit gates, other gates, pumping stations, and other appurtenant structures and equipment (Perdikis, 1967). The gates have been operated {closed and re-opened} many times since project completions (Wiegel, 1993). In addition to saving lives, it was estimated the total damages by flooding prevented since project completions through 1992 was about $35 million. The projects are an excellent example of cooperation and operational coordination of personnel of the Corps, other Federal agencies, and local interests to prevent or substantially reduce harmful effects of natural disasters.

Growth of Coastal Environmental Awareness

Gradually an awareness arose of possible effects of coastal projects on the biologic as well as the physical environment. Some early serious studies in the U.S. were those of USACE CERC in the late 1960's on the effect of beach fills and use of offshore sources on the surf clam, and {under CERC contract} effects of waste disposal in New York Bight. An annotated bibliography of early reports is given by Pullen et al. (1978). Increasing concern led to further studies on the whole gamut of coastal projects, particularly on the effect of dredged material disposal. [The Corps has continued its interest in the environment. In 1992, Lt. Gen. Arthur E. Williams, Chief of Engineers, stated at the 57th meeting of CERB (reported in McGehee and Hemsley, 1994, p. 208): "Most environmental problems involve a chemical and/or biological component and a physical processes component. A holistic approach must consider both components and the way they interact. To assess environmental risk, we must consider potential threat or exposure and resulting impacts ... I also want you to recommend other areas where partnerships between environmental and coastal communities can demonstrate our clear commitment to protecting quality of our environment and move the Corps into the forefront of environmental responsibility."]

Late 1960's and Early 1970's. Analysis; Offshore Technology Conferences; Littoral Currents; Sea Grant; Shore Protection Manual; CERC Move

We give here an example of the development of an analysis process. A thermal-electric power plant cooling water system was being designed in the 1960's for a rocky {no sand} coastal site, using an excavated open channel intake with a pair of jetties. Ocean wave action on the intake pumps had to be minimized. Dick {R.O.} Eaton told the designers he had observed that for a harbor with a dredged channel extending considerably seaward of the jetties there was a significant decrease in wave heights in the harbor. A hydraulic model study confirmed this. Later, theoretical and hydraulic laboratory studies were made of this refraction/ diffraction phenomenon, and concepts developed for use in future analyses (Hilaly, 1969).

The first Offshore Technology Conference (OTC) was held in Houston, Texas in May 1969, and has been held annually since. Proceedings/Preprints are available for each conference. There is no well defined boundary between coastal and offshore engineering. Most of the theory, data and practice in regard to wave climate, wave characteristics and their effects on structures are important for both.

Field and theoretical studies were made in the late 1960's and early 1970's of littoral currents (e.g. Komar and Inman, 1970; Thornton, 1970; Komar, 1976). Possible mechanisms for the formation of rip currents and beach cusps were investigated (Bowen, 1969; Bowen and Inman, 1969, 1971; Sonu, 1972; Noda, 1974).

In 1966 Congress established the National Sea Grant Program, as a part of NOAA. The first awards were made two years later; funding for fiscal years 1968-1970 was more than $20 million (Abel, 1970; Assoc. of Sea Grant Institutions, 1971). Some grants have been for projects of interest to coastal engineers.

[In 1976, Congress modified the Sea Grant Act to include national projects. The first was the Nearshore Sediment Transport Study {NSTS} of coastal sediments, during 1977-1982. It was originally funded by NOAA's Office of Sea Grant, with latter portions partially funded by USACE, ONR, USGS, and the National Park Service. The total cost was about $4 million. The Steering Committee members were: Richard J. Seymour {Chair and Project Manager}, Robert G. Dean, Douglas L. Inman and Edward B. Thornton. The objective of the NSTS was "... the development and the experimental verification of hydrodynamic laws governing the transport of marine sediments in the flow fields occurring in coastal waters." (Duane et al., 1989). Owing to budget and analyses restraints, "..only two major field experiments {Torrey Pines and Santa Barbara in California} and a limited scope validation experiment {Rudee Inlet, Virginia} were completed." Studies were made by investigators from 7 universities, institutions and government agencies. Four workshops were held during the investigations. Many results were published in papers in the book *Nearshore Sediment Transport* (Seymour, ed., 1989).]

On 19-20 May 1973 the USACE CERC was moved to Fort Belvoir, VA. The *Shore Protection Manual* {successor to *BEB Tech. Rept. No. 4*, and based on it} was published in 1973 in 3 volumes under the general editorship of R.J. Jachowski and technical editorship of J. Richard Weggel, with significant technical contributions from many of the CERC staff. It was revised and published in 1975, in 1977, and again in 1984 {in 2 vols.}. The 1984 revision is still in use, and is almost a "standard" for practicing coastal engineers. [In addition to this book {a replacement is currently being written}, CERC has issued since 1979, as needed, short notes, usually 1 to 6 pages, *Coastal Engineering Technical Notes {CETN}* on the following subjects: waves and coastal flooding; beach behavior and restoration; coastal structures; inlet and estuary channels; coastal ecology; misc.]

Multiple Use of the Coast; Coastal Zone Management; Surfing; Rehabilitation/ Beach Nourishment

Pressures caused by multiple uses of the coast, usually with conflicting requirements, and the need for environmentally sound practices, led to the start of coastal zone management. The Federal Coastal Zone Management Act of 1972 was a milestone. The broad systems approach, which encompasses coastal engineering, was the basis for the conference "Coastal Zone '78", organized by Orville T. Magoon, and his co-chairs {Jon T. Moore, Billy L. Edge, John R. Clark and Robert W. Knecht}, in San Francisco, CA in 1978 (Magoon et al., ed., 1978)); the 8th was in Tampa, FL in 1995. At the other end of the spectrum, there have been conferences and workshops on specialized parts of coastal engineering, many sponsored by the ASCE Waterway, Port, Coastal and Ocean Division, such as Waves'74 {co-chaired by Billy L. Edge and Orville T. Magoon}, Modeling'75, Ports'77 {chaired by William Herron, Jr.}, Coastal Sediments'77 {co-chaired by George Watts and Billy L. Edge}, Coastal Structures'83 {chaired by Robert M. Sorensen}, World Marina '91 {chaired by Victor Adorian}. Subsequent specialty conferences on coastal sediments were organized by Nicholas C. Kraus and his colleagues {New Orleans, LA, 12-14 May 1987; Seattle, WA, 25-27 June 1991}. ASCE published the proceedings.

Surfing {board, canoe, body} is an important recreational activity. Hawaiians enjoyed the sport for hundreds of years, and at about the turn of the century, visitors to Hawaii started to use surfboards at Waikiki. The sport spread to southern California in 1907, then to northern California, and later to other regions. The boards were long and heavy, but with development of lighter and more maneuverable boards, and the invention of the wetsuit {by Hugh Bradner of SIO}, the sport became very popular (Johnstone, 1964). A few coastal engineers who were also surfers, such as Homer Johnstone and James R. {Kimo} Walker, began research to find what combination of waves, bathymetry, bottom conditions, shore configuration, winds, currents, tides, made some surf sites superior to others. This included evaluation of access routes, board recovery areas, and alongside channels in rider return areas. Responding to requests of the surfing community, and others, the Hawaii State Legislature requested a report on Hawaiian shoreline projects and their effects on natural surf conditions. In 1971, the Legislature reaffirmed the need "... to provide improved protection and stability to Hawaiian beaches and the preservation and enhancement of surfing areas.." (*Senate Standing Report No. 436*, as in Kelly, 1973). A study of surfing parameters was made by the Dept. of Ocean Engrg., Univ. Hawaii (Walker et al., 1972; Kelly, 1973; Walker, 1974). Technical considerations for preservation and enhancement of surf sites are given by Walker et al. (1972). Preliminary designs of artificial reefs or natural site improvement have been made (Walker et al., 1972), but none constructed in the ocean, at least in the U.S. (Walker, 1995).

Interest and demand for the rehabilitation of degraded coastal areas continues to grow. One result of the recognition of the need for environmentally sound practice has been the increased use of beach nourishment (Sudar et al., 1995). Beach nourishment was first used in 1922, with many projects since (Hall, Jr., 1952; Davison et al., 1992). A good example is the project at Miami Beach, FL and contiguous beaches to the north (Wiegel, 1992). It consisted of widening the beach, building sand dunes and planting them with native vegetation, providing adequate public access and pedestrian walkovers, a timber promenade along the Miami Beach section and a curving concrete walkway along the Bal Harbour section. In recent Atlantic and Gulf Coast project designs, use has been made of the concept of equilibrium beach profiles, pioneered by D.W. Johnson (1919) and Per Bruun (1954; 1973) and extended substantially by Robert Dean and his colleagues at the University of Florida (e.g. 1991; Dean and Yoo, 1993). [Recently, other equilibrium profile shapes have been proposed by Komar and McDougal (1993), and by Inman and colleagues (1993). In 1902, Prof. N.M. Fenneman proposed an equilibrium subaqueous profile: "A compound curve which is concave near the shore, passing through a line of little or no curvature, to a convex front. Where this front rests upon the bottom below the reach of currents, the descent merges into the more level bottom by another concave curve, due to deposition from suspension."] Data on "closure depths" are also required (Hallemeier, 1981, 1983; Birkemeier, 1985). Experience with beach nourishment has shown that "erosion hotspots" occur; it is important to understand why, and how to remedy the problem. More coverage on beach nourishment is in a later section. [Owing to extensive use of this type of project, questions about them, and media attention, a study was made by the NRC Committee on Beach Nourishment and Protection; their report is *Beach Nourishment and Protection* (1995).]

Mid 1970's to Early 1980's. CERC Pier; Books; Breakwaters; Wave Measurements

The USACE decided a pier was needed to facilitate study of beach, surf, and nearshore processes. The pier, 1841 ft long {deck 15 ft wide, 25.4 ft above mean sea level and 3-ft dia. steel piles, with removable gratings in the deck} was built on the

Outer Banks {barrier island} at Duck, NC, a little north of Kitty Hawk. It was completed in August 1976, and the laboratory and office building completed in March 1980; it is the CERC Field Research Facility {FRF} (Birkemeier et al., 1981). A basic environmental measurement program {beach and nearshore profiles, waves, tides, currents, meteorological measurements} has been in operation since the facility opened. Monthly data reports have been issued since January 1981 (e.g. Baron et al., 1995). Several large scale multi-agency/ university studies have been made: ARSLOE (Baer and Vincent, 1983), DUCK 82 (Mason et al., 1984), DUCK 85 (Mason et al., 1987), SUPERDUCK in 1986 (Crowson et al., 1988), and DUCK 94.

In the 1970's and early 1980's several new books on coastal engineering, dredging, or waves/tides were published: Huston (1970); LeMehaute (1976); Wood, 1976; Sorensen, (1978); Mei, 1983; Dean and Dalrymple (1984).

Worldwide, breakwaters were being built in deeper water than before, and at sites exposed to severe wave action. On 26 Feb. 1978, waves caused extensive damage to the breakwater at Port Sines, Portugal. This will be discussed in a subsequent section, as the event led to studies which caused a "watershed change" in the design and construction of rubble mound breakwaters. This was triggered by the ASCE publication of the report of the NSF funded {mostly} Port Sines Investigating Panel of coastal engineers (Edge - Chair, Baird, Caldwell, Fairweather, Magoon and Treadwell, 1982).

In the mid-1970's a new start was made on a long-term wave recording and analysis program (Seymour and Sessions, 1976; Seymour, 1987), enlarging significantly the old network of gages, placing them on a more permanent basis, with improved real-time analysis and compilation. {At about the same time the USACE began a major new program of hindcasting waves using spectral methodology, improved wind estimates, and computers, the Wave Information Study {WIS} (e.g. Corson et al., 1982)}. It would eventually cover the entire U.S. coast, including the Great Lakes. A major problem was measuring wave direction with adequate resolution, but headway was made (Seymour and Higgins, 1978). Coastal data networks were established in Florida {CDN} and in Virginia {VCIP}. The USACE CERC started its Network for Engineering Monitoring of the Ocean {NEMO}, and NOAA started the NDBC. [There are many directional wave measuring buoys installed. A brief outline of the USACE and NOAA programs are given by McGehee and Hemsley (1994). The USACE "..is principally interested in the wave conditions and water levels at coastal projects and most of its stations are in shallow {10 -20 m} water." "NDBC has been measuring waves under its moored buoy program since the mid 1970's in support of its basic mission to provide through NWS, warning and forecast services to the public." The two agencies coordinate their programs as there are of necessity areas of overlap. Some details of operational systems, measurement systems and plans for future wave measurement projects are described by Steele and Mettlack (1994).]

Coastal Regulations - FEMA/ Florida; The Shoreline; Coastal Law

In the 1970's and 1980's coastal regulations were established by some states and by the Federal government. The Federal government through the FEMA administered National Flood Insurance Program {NFIP} developed regulations in regard to storm surge and tsunami flooding {mentioned also in other sections; for an interesting study of coastal flooding see Wood (1976)}. From 1975 to 1980 FEMA published the first Flood Insurance Rate Maps {FIRMs}, delineating Special Flood Hazard and Coastal High Hazard {V zone} boundaries (Pajak, 1995). {Definitions (FEMA, 1990; CFR {Code for Federal Regulations}, 1993) - Area of Special Flood Hazard ".. is the land in the flood plain within a community subject to a one percent or greater chance of flooding in any given year. The area may be designated as A zone..." (*Index 44 CFR*, Section 59.1); Coastal High Hazard Area "... means an area of special flood hazard extending from offshore to the inland limit of a primary frontal dune along an open coast and any other area subject to high velocity wave action from storms or seismic sources." (*Index 44 CFR*, Section 59.1)}. The criteria for "high velocity wave action" is a 3-ft or higher wave (USACE Galveston District, 1975; NAS, 1977; NRC, 1983; FEMA, 1986; Pajak, 1995) and procedures are recommended for mapping the inland limit of the 3-foot wave. Both USACE and NRC reports to FEMA stress that the criteria for combined storm surge and wave action were for the Atlantic, Gulf and Great Lake Coasts, as the storm surge and waves are of common origin; they recommend it not be used along the Pacific Coast, Alaska coasts or Hawaii "...where flood levels and waves are not necessarily directly related" (NAS, 1977).

The intent of the NFIP is to "...reduce future damage and provide owners with protection from financial losses through an insurance mechanism that allows a premium to be paid by those most in need of this protection. This program is based on the agreement that if a community will practice sound floodplain management, the Federal Government will make flood insurance available." (Pajak, 1995). FEMA has continued to work on coastal flooding, wave action, and storm-induced erosion through contracts with NAS, CERC, and others. This resulted in several numerical models, mentioned elsewhere. Guidelines and specifications for wave elevation determination and V Zone mapping have been modified based in part on these studies, with the most up to date version scheduled for publication in 1996 (Pajak, 1995). Dune erosion along the Atlantic and Gulf Coasts has also been studied (FEMA, 1989). Birkemeier et al. (1987) conducted an evaluation for FEMA of the various dune erosion models available in 1985-1986.

Substantial base data and coastal engineering analyses are needed to provide reliable quantitative information to meet the requirements, such as water elevations {including wave runup and overtopping}, coastal erosion and scour, and wave-induced forces on structures. Several numerical models have been described previously {a few were developed under contract with FEMA}. Others are: EDUNE {Equilibrium Dune Erosion} (Kriebel and Dean, 1985; Kriebel, 1986); SBEACH {Storm-induced BEAch CHange} (Larson and Kraus, 1989, 1990; Kraus and Wise (1993); GENESIS (Hanson and Kraus, 1989). In the future this must be done from the standpoint of risk analysis; for example, Monte Carlo modeling (Le Mehaute et al. 1981).

In some states, setback lines are used as a regulatory tool, the locations based on physical evidence and reasoning. Florida is used as an example. It has 3 different legal lines along the coast: erosion control line {ECL}, coastal construction setback line, and the coastal construction control line {CCCL}. According to Peter H.F. Graber in the Florida article (1981) of his series "The Law of the Coast in a Clamshell", the ECL becomes a permanently fixed boundary line, and after the line has been recorded:
"...title to all lands seaward of the ...line shall be deemed to be vested in the state...[and] the common law shall no longer operate to increase or decrease the proportions of any upland property lying landward of such line, either by accretion or erosion or by any other natural or artificial processes." {Quotation by Graber from *Beach and Shore Preservation Act*, *State of Florida Statutes*, Section 161.191 (1), (2) 1980 Supp.}

Graber presents information on the CCCLs in his Footnote 81, stating:
"These construction control lines are to be established by the Department of Natural Resources on a county-by county-basis 'along the sand beaches ...fronting on the Atlantic Ocean and the Gulf of Mexico.' Fla. Stat. Sec. 161.053 (1) (1980 Supp.)., These engineered 'lines shall be established so as to define that portion of the beach-dune system which is subject to severe fluctuations based on a 100-year storm surge or other predictable weather conditions, and so as to define the area within which special structural design consideration is required to insure protection of the beach-dune system, any proposed structure, and adjacent properties, rather than to define a seaward limit for upland structures.'"

The general goals of Florida are given in the *1987 Florida Comprehensive Beach Management Act* (Tait, 1991). There were earlier acts: *Florida Coastal Management Act of 1978* and *Florida Coastal Mapping Act of 1974* (Graber, 1981).

In establishing boundaries along beaches and other tidelands {the shore and sea boundaries problem} well defined tidal datums and knowledge of laws and judicial decisions are necessary (Shalowitz, 1962; Harris, 1981). In addition to establishment of tidal datums {which has been done in great detail}, data on the location of the beach surface is also needed, and this fluctuates. Anyone who has observed an ocean beach for a year or so has noticed how much the profiles and shoreline {MSL or some other datum} location change with varying wave conditions. The location of the mean high tide {or mean sea level} line moves back and forth over a zone, and often there is also long-term erosion or accretion. Much work has been done and many court decisions written on tidal datums, but not on beach fluctuations, and this is very important to coastal engineers (Shepard, 1950; Johnson, 1971; O'Brien, 1982; Bokuniewicz, 1981; Bokuniewicz and Schubel, 1987). The problem of obtaining reliable quantitative data on long-term erosion rates of beaches from a small number of aerial photographs and ground surveys taken only occasionally, has been studied (e.g. Dolan et al., 1980; Leatherman, 1983; Anders and Byrnes, 1991).

[Charts, maps, and tidal datums are important in coastal engineering. Care must be exercised in using older ones, as effective 1 January 1989 NOAA/NOS changed their chart datum and tidal datum along the U.S. east coast from Mean Low Water {MLW} to Mean Lower Low Water {MLLW}, the same as used along the Pacific Coast. Also, NGVD is being replaced by the North American Vertical Datum of 1988 {NAVD 1988} (U.S. Dept. Commerce, NOAA, NOS, 1992; Zilkowski et al., 1992). FEMA (1992) is converting the National Flood Insurance Program to the NAVD 1988. NGVD and NAVD may have nearly the same value as Mean Sea Level {MSL} in some locations, but may differ by a half foot or so elsewhere. MSL has a physical significance in coastal engineering, while NGVD and NAVD 1988 are for surveying convenience.]

Coastal engineers work closely with government agencies and planners, and with private interests. Much of what they do is associated with laws, regulations, permitting and codes. Each coastal state has different laws; detailed information has been given by Peter H.F. Graber in *Shore & Beach*, in his "Law of the Coast in a Clamshell" series, the first in the October 1980 issue and the 26th in the April 1989 issue.

The 1980's. CERC Moved to WES; Global Warming/ Relative Mean Sea Level Change; O'Brien's Death; Misc.

Little is written here of the 1980's, as it is best left to a future update for perspective; this includes preservation and rehabilitation of coastal wetlands.

CERC was moved from Fort Belvoir to WES at Vicksburg, MS in 1983.

Worldwide interest developed in the 1980's on the subject of possible global warming with accompanying mean sea level rise. Estimates have been made of possible rise, which continue to change (e.g. Titus and Narayanan, 1995) and scenarios given of what might be done to mitigate harmful effects along coasts. Change in relative mean sea level {sea level rise and/or subsidence} had been observed and studied for many decades and some studies of the effects on beaches (D.W. Johnson, 1929; Fairbridge and Newman, 1968; Allen and Mayrega, 1970; Habel, 1978; Hicks, 1981; Hicks et al., 1983; Barnett, 1983; Wiegel, 1994). {This shows the importance of long term coastal surveys and tide measurements.} An NRC Marine Board committee was established; results of their deliberations are given in *Responding to Changes in Sea Level: Engineering Implications* (1987). [For an update on the present status of global warming, and estimates of possible future sea level rise due to it, see Titus and Narayanan (1995).] {Much can be learned by studies of coastal regions where subsidence has occurred owing to the removal of underground oil/ gas/ water (e.g. Habel, 1978; USACE Los Angeles District, 1995), and by tectonic movement.}

The *Journal of Coastal Research* began publication in 1985.

Morrough P. O'Brien died on 28 July 1988, in Cuernavaca, Mexico, at the age of 85. He started to work on coastal engineering in 1929, during the summer under contract to the USACE BSMBE, and continued his interest in the subject throughout the rest of his life. In addition to his own work, he motivated many others to make coastal engineering studies. A symposium was held in his honor in March 1987, and the papers published in the combined July-October 1987 *Shore & Beach*. The ASBPA Morrough P. O'Brien Award for Outstanding Achievement in Shore and Beach Preservation was established in his honor.

In the 1980's and later there was development of numerical models of coastal sediment transport, beach and nearshore changes. These, together with the numerical models of wave generation, storm surges, tsunamis, refraction and diffraction were needed for many purposes, including input to FEMA's regulations and the needs of coastal state governments, as mentioned previously.

In the late 1980's, the NRC Committee on Coastal Erosion Zone Management was established in response to a request from the FEMA/FIA. It "...was asked to provide advice on appropriate erosion management strategies, supporting data needs, and applicable methodologies to administer these strategies through the National Flood Insurance Program" (NRC, 1990). Their findings and recommendations on erosion hazards are given in *Managing Coastal Erosion*. This is an important part of coastal zone management, requiring substantial coastal engineering input.

Well Documented Coastal Storms

During 17-18 January 1988, an intense storm struck southern California, causing substantial damage along the coast. Based on a statistical analysis of quantitative wave data, Seymour (1989, p. 12) estimated the storm had "...a recurrence interval of not less than 100-200 years..." A workshop was held at SIO in La Jolla, CA on 18 Jan. 1989. Presentations were made on the storm, waves, sea level, beach response, damages to buildings, breakwaters, boats and infrastructure, effects on the Tijuana Estuary, and the performance of a moored small ocean semisubmersible platform. The papers were published in a dedicated issue of *Shore & Beach* with Richard Seymour, guest editor (1989).

Hurricane Hugo (a Class IV hurricane) caused great damage in the Virgin Islands and Puerto Rico, then made landfall near Sullivan's Island, SC on 21 Sept. 1989 at 2300 EST. "The hurricane caused 13 directly related deaths, 22 indirectly related deaths, and injured several hundred people in South Carolina. Damage within the Palmetto State from Hugo has been estimated to exceed $7 billion" (Davidson, Dean and Edge, 1990). The storm characteristics and coastal effects {breakthroughs of barrier islands, buildings, boats and infrastructure destroyed or damaged} were well documented. A workshop was held in Folly Beach, SC on 21-22 May 1990, attended by a small group of coastal engineers, scientists, and planners. The proceedings are in the October 1990 issue of *Shore & Beach*, with guest co-editors Margaret Davidson,

Robert Dean and Billy Edge.

The 2 January 1992 storm that struck Ocean City, Maryland, and the adjacent coast is important for several reasons. It is probably the event for which we have the most quantitative data on waves {including directional information}, sea levels, beach profile changes {both shortly before and just after, with surveys made by USACE and the State of Maryland}. Furthermore, it had a recent beach nourishment project, and there was much interest in its performance. A series of papers are in a dedicated issue of *Shore & Beach*, including one on a numerical simulation of the erosion (Nicholas C. Kraus, guest editor, 1993).

How well do coastal projects perform, functionally, structurally, and economically? Owing to the costs of monitoring, and the seemingly lack of interest of politicians to have follow-up studies made of government funded projects, little monitoring is done. Ocean City, MD is an exception. What should be monitored, and how it should be done, has been discussed in the NRC (1995) report, and by Weggel (1995). {See also Stronge (1994) and Houston (1995).}

Importance of Recreation/Tourism to the Economy

Recreation is important to the economy - job creation, positive flow of tourist dollars, benefits to the nation's social fabric. Tourism is now the largest industry worldwide, and coastal related tourism is a major factor in the U.S. economy (Stronge, 1994; Houston, 1995). An address to the ASBPA many years ago by L.H. Weir (1933) "How Modern Economic Progress Has Multiplied the Demand for Outdoor Recreation" was printed in the first issue of *Shore and Beach*. He referred to the "Age of Leisure" he expected to occur owing to the great increase in productivity in manufacturing goods and in constructing infrastructure. This age has not come about as rapidly as he thought, owing to the economic depression, many wars which drained the economies of the world, and other factors - but it is here now for a great many people. In addition to vacationers, many retirees live on or near the coast, with a major impact on the types of buildings, infrastructure and job market in those localities (Stronge, 1994). This has led to synergic activities of new companies founded locally by some retirees, and the movement of young relatives and others to these regions. The trickle of tourists, retirees, and others became a strong flow. Pressure on the coast is great. In addition to big increases in the use of shores and beaches, there have been changes in user requirements, resulting in changes in public policies and infrastructure.

Coastal System Planning

Substantial advances have been made in our understanding of the physical, chemical and biological coastal environment, and in data collection. In spite of all the work done during the past seven decades, there is still a need for a more complete understanding of the processes and their mix at a particular site, together with long-term reliable measurements, analyses and presentation. There are uncertainties in our understanding of processes, forcings, boundary conditions, trends and their "mix". More planning must be made of coastal systems rather than just components. The value of regional planning and coastal programs had long been recognized in a number of USACE studies, and recently has been undertaken anew in two regions by others (Noble Consultants, 1989; SANDAG, 1995). Risk analyses are needed of projects and their interactions, such as beach nourishment and tidal entrances, and the findings clearly explained to people who are directly concerned. More information on post-construction {or non-construction, if this was the case} benefits and costs is needed, and analyses made of indirect or secondary effects/ interactions, possibly unintentional, of the projects with other interests of coastal communities.

A FEW SPECIAL TOPICS

Hydraulic Laboratory Facilities/ Wave Tanks/ Scale Effects

A few small wave tanks and basins were built at several universities, BEB, and WES, starting in the 1930's. An important early WES facility was a flume 119 ft long, 5 ft wide, 4 ft deep used for breakwater stability tests beginning about 1940. Details of the ones at UCB are given by Snyder et al. (1958), and at the BEB by Rayner and Simmons (1964). The Coast Model Test Basin at the BEB {150 ft by 300 ft, 3 ft deep} with tidal capability was used from the early 1950's for movable bed beach research. A number of tanks and model basins existed at WES with a wide variety of tidal and wave capabilities, primarily for physical model testing, particularly after WW II {see later paragraph}. The first "prototype" wave test facility was the Large Wave Tank {635 ft long, 20 ft deep, 15 ft wide} of the BEB in Washington D.C., dedicated in 1955; some uses of it are described in the following paragraph. A large, but smaller tank was built at UCB in the mid-1950's, with a bulkhead type mechanical wave generator {200 ft long, by 8 ft wide by 6 ft deep}. The first programmable irregular wave generator {probably the first wave spectra generator, and funded by NSF} was designed, constructed and used in this tank, renamed the Ship Model Towing Tank {designed by Jerry Cuthbert, William C. Webster and O. Sibul between late 1959 and 196l} (Cuthbert, 1959; Webster, 1994); it was also used for coastal engineering studies. Another wave tank with a wave generator, and also with a removable cover, and with a large blower to generate wind waves was built at UCB. Also, a basin {150 ft long, by 64 ft wide by 2.5 ft deep} was built at UCB, which had five large blowers installed in a 60 ft long by 12 ft wide section to study the generation of waves by wind with directional spectra (Mobarek, 1965). It should be noted that many investigators of wave motions in long tanks observed instabilities in the waves, and in wide basins observed cross waves that formed under some conditions. These are discussed in detail by Harms (1975), for work he did in this basin. A covered channel 130 ft long by 6 ft wide by 4 ft deep channel with a wave generator and blower at one end, and a 26 ft by 19 ft diffraction basin at the other end was built at the Univ. of Florida in 1957 (Bruun et al., 1958). The model basin at UF was built in the late 1950's and early 1960's, with the first 50 ft of the "snake" type wave generator installed in 1961-62, and an additional 21 ft installed in 1972 (Dean, 1995). This device was used to generate waves with a range of angles. A large wave tank was built at Oregon State University in 1972-73 {342 ft long, by 12 ft wide by 15 ft deep}; the tank is now in a building. [Much later, in 1985, the Directional Spectra Wave Generator was installed at WES/ CERC in a 96 ft by 114 ft {water depth up to 2 ft} section of a larger basin.

The world's first "prototype" wave test facility, mentioned above, could generate waves 6 feet high with breakers sometimes reaching 7-1/2 feet (Rayner and Simmons, 1964). The earliest tests involved equilibrium beach profiles and cross-shore sand movement (Saville, Jr., 1957; Kraus and Larson, 1988), including data on suspended sediment in the surf zone (Fairchild, 1959). The second major project was on wave run-up and overtopping, primarily for design of Lake Okeechobee levees, but with broader application (Saville, Jr., 1958; Saville, Jr., McClendon and Cochran, 1962), as used in *BEB Technical Report No. 4*. Interspersed were tests of forces on a pile (Ross, 1959) funded by Humble Oil Co., with most data held proprietary by them. Other major early tests involved the stability of rubblemound structures, including those by Bretschneider in 1960-62 which were the first indication that wave period might have an influence. All testing examined the effect of model scale; indeed, a major purpose of most studies was to replicate at a much larger scale {usually 10:1} a few selected tests from a much broader suite done at small scale to check on possible scale effects. All tests involved only monochromatic waves. When CERC was moved to Fort Belvoir in 1973, the facility was duplicated, still with basically only monochromatic waves although the period and height could be varied during a test by varying the period which in turn varied the height in an inverse direction; but no spectral wave generation was possible. With the move of CERC to WES in Vicksburg, MS, in 1983 this facility was "mothballed".

Other large scale, or prototype facilities built specifically for experimental purposes {rather than facilities built for a specific purpose and then used on an opportunistic basis} include the Prototype Experimental Groin at Pt. Mugu, CA, operated by the BEB for about 10 years in the 1970's, and the CERC Field Research at Duck, NC, described in another section.

Coastal Hydraulic Models

A few uses of hydraulic models have been mentioned elsewhere in this history in regard to breakwaters, moored ships, tidal entrances, wind-generated waves with directional spectra. Some other uses are described below.

WES has been in the business of modelling specific projects to obtain specific answers to design questions since its inception in 1929. It is the largest hydraulic modelling laboratory in the country. Its work has covered the gamut of coastal problems involving shores, harbors, and inlets/ entrances; and use, navigation, and protection thereof (e.g. Whalin et al., 1977). Models are both 2 and 3-dimensional and involve waves, tides, currents, and occasionally wind. The early models of WES,

BEB, UCB and elsewhere have already been referenced. Model techniques have advanced over the years from wave generation by a person turning a crank to raise and lower a plunger to highly sophisticated electronically-controlled computer-driven 3-dimensional spectral generators. Tidal models have gone from simple fresh water flows to extremely complicated fresh and salt water inflows from various sources with calibrated roughnesses to model accurately complex tidal phenomena over and through complex hydrography under distorted scales. Studies involve both fixed and movable beds, and can include hurricane surge. Different laboratories use somewhat different techniques. The CERC report on modelling (Hudson et al., 1979) summarizes the usage at WES and CERC, including some discussion of model laws. Because of the expense and time required for physical models, there has been a major development of numerical computer modelling, and many model projects today are a hybrid mix of the two {See, for example, Butler and Durham (1976).} Each has its advantages and disadvantages, but together they often produce a better product.

Similitude relations have been developed and discussed by many, and a summary is given in Hudson et al. (1979). Relations for movable bed studies are still somewhat indefinite, though successful models can be, and have been, made. Work at Tetra Tech, Inc. underlies many of our procedures in movable bed modelling (Fan and LeMehaute, 1969; Le Mehaute, 1970, 1976 - and O'Brien's discussion, 1977; Noda, 1972).

In tidal models, Simmons and Hudson of WES determined experimentally that distorted scale models required salt water to preserve Froudian scales (WES, 1942; 1949). This was confirmed theoretically by Keulegan who made a series of 13 notable reports for WES on modelling (Keulegan, 1946-1958). Simmons and Bobb (WES, 1953, 1956) developed roughness strips to reproduce accurately both vertical and lateral current distributions. These and other modelling concerns are discussed by Herrmann, Jr. and Letter, Jr. (1990), from which portions of the above and following are abstracted.

The Chesapeake Bay Model in Maryland, operated by the USACE from its 1975 construction until its 1983 deactivation, was the largest tidal model ever built. It was about 1100 ft long with a maximum width of 700 ft, and covered an 8-acre paved area under a 14.5-acre shelter. Model results were used to estimate effects of channels and other improvements on salinity and general circulation, dispersion from spills, effects of dredged material disposal, results of fresh water inflow, and even assistance in recovery of accident victims. But high costs, partly associated with the long time required to reach stability prior to a test, resulted in its replacement by a 3-dimensional numerical hydrodynamic model, which has the advantage of being able to include wind effects. Another very large model is that for the San Francisco Bay-Delta, built in 1956-57 in Sausalito, CA. In size, 320 by 400 feet, it covers the area from 17 miles seaward, the entire San Francisco Bay, San Pablo Bay, Suisun Bay, and much of the Sacramento-San Joaquin Delta. At a horizontal scale of 1:1000 and a distortion of 10, it has been used to study hydraulic changes from barriers, reclamation, and dredging;

water quality and pollution, sediment disposal from dredging; currents and salinity; and other factors of bay operations (USACE San Francisco District, undated).

The most extensive model study of a coastal intake and discharge cooling water system for a thermal-electric generating plant {4 pumps, 1,000 cfs of cooling water for each pump, a total of 4,000 cfs} was made for the Pacific Gas & Electric Co. Diablo Canyon Plant, by UCB (Wiegel and Doyle, Jr., 1987). The intake was within a harbor constructed on the open coast by building two breakwaters, and the warm water was discharged into the ocean at the shore through an open concrete channel. The physical model had an undistorted geometric scale, with the flows and temperatures based on densimetric Froude Number scaling. Long term environmental measurements were made at the site by PG&E, beginning in 1966. The model was constructed in May 1974, with the first series of tests performed during 1975-1976, then 1984-1987, and specialized studies made in between these dates. Extensive model and field measurements were made of intake and discharge temperatures and discharge velocities for a variety of tide, flow and temperature conditions. The comparisons of model and field data were reasonably good, and showed that the following had to be modelled very carefully: mass flux and buoyancy flux of the warm water discharge, geometry of the site, details of the bathymetry, bottom roughness, waves, currents, and tides.

Tidal Entrances and Estuaries

After WW II work began again on tidal entrances and estuaries. The width, depth, and length of entrances, and the currents within them are important to navigation and sediment transport. Sometimes entrances close, sometimes they open, and some entrances remain open. Coastal engineers and planners want to be able to predict what will happen to a particular entrance under varying conditions, including wave conditions at the entrance. Field data are needed. The work of O'Brien (1931) on the relationship between cross sectional area of an entrance and the tidal prism was extended by Tom Jarrett (1976), and modified by Escoffier (1977) to include littoral transport. Keulegan (1967) developed the concept of repletion coefficient to determine entrance and exit losses. C.L. Vincent and his colleagues examined many charts and aerial photographs to obtain data on U.S. tidal inlet geometry and stability (Vincent et al., 1980; 1991). Information on tidal currents in entrances was collected by Joe Caldwell in 1955, who found that the maximum mean velocity during the tidal cycle varied from inlet to inlet (see also, Escoffier, 1940; O'Brien, 1969; Jarrett, 1976; Bruun et al., 1978). Costa and Isaacs (1977) gave an enlightening discussion on the highly nonlinear nature of sand transport in an entrance.

The analytical work done in the 1930's was extended by Keulegan and Hall in 1950, Keulegan in 1951, O'Brien and Dean in 1972, Sorensen in 1977, and van de Kreeke in 1992, as well as others. Einstein and Fuchs (1955) presented a thorough review of the basic nonlinear differential equations, boundary conditions and approximations made by others in the U.S., of the numerical computation of tides and tidal currents in entrances and estuaries. Leendertse (1967) pioneered work in numerical modelling of tidal hydraulics, which has led to many later versions. His application to Jamaica Bay, NY used an alternating direction implicit solution scheme. Lumped-parameter models were developed, by Huval and Wintergast in 1977 and by Mayor-Mora in 1974. A linearized analysis, useful in early steps of project planning, was given by Walton and Escoffier in 1981. A detailed overview is given by Mehta and Joshi (1988), based on their participation in the ASCE Task Committee on Tidal Hydraulics.

Keulegan's work (1946-1958), though done for density current modelling, defined theoretically many characteristics of the salinity wedge, the mixing and stresses associated with it, and the pertinent velocities. It was recognized by the early 1950's that the salinity wedge played a major role in sedimentation rate and location in estuaries, with Simmons (1955) developing a flow predominance determination of the null zone subject to rapid shoaling. Harleman and Ippen (1967) proposed a velocity predominance method.

Waves and wind transport sand towards and into inlets, and tidal currents transport sand out of an inlet to the ocean, or into the bay. Bruun and Gerritson (1960) studied this problem, and developed the Bruun-Gerritsen Stability Parameter for estimating whether or not an entrance would remain open, or close. Johnson (1973) presented data on the behavior of some Pacific Coast inlets, and a detailed study was made by Rice (1974) of the mouth of the Russian River, CA. This type of work was extended by Goodwin and Williams in 1991. Sand must be bypassed at many entrances, for the maintenance of the entrance navigation width and depth, and the preservation of the downdrift beach.

Even a small amount of fresh water flow into an estuary can be important in regard to the open/close regime, shown by Escoffier and Walton (1979).

There are still major difficulties in applying the theories and techniques described in the technical literature. According to van der Kreeke (1992): "Determination of the exact closure for a real inlet requires a full-fledged two dimensional model for the hydrodynamics of the inlet and the bay. This is beyond the budget of most inlet studies. Instead recourse is taken to lumped-parameter models. However, the assumption of a uniformly fluctuating bay level, that is the basis for these models, is usually not satisfied. Consequently the accuracy of the resulting closure curves is marginal."

Beach Nourishment/Sand Bypassing

Beach nourishment is the placement of sand, gravel or cobbles on a section of coast to increase its width and/or elevation. This includes the beach, dunes, and contiguous nearshore. It provides additional recreational area and protection of nearby buildings and infrastructure. At many locations where there are `navigation structures {jetties, dredged inlets, breakwaters}, it is necessary to bypass sand, using either mobile or fixed systems. {The negative effects of many entrances on downdrift beaches has been investigated by several coastal engineers (e.g. Brooke, 1934; Dean and Walton, Jr., 1973; Dean, 1986).} The first beach nourishment project was at Coney Island, NY in 1922. This, and a number of other projects have been mentioned earlier in this paper. A study made in the early 1950's gives information of the earlier projects (Hall, Jr., 1952), and a later one by Davison, Nicholls and Leatherman (1992) updates Hall's. The projects in southern California, many using "sand of opportunity", have been described by Flick (1993) and by Wiegel (1994). Nearly all the material, usually sand, used in nourishment is from a nearby source {offshore, entrances, bays, river beds, dunes}, but in a few cases is manufactured by crushing rock.

In the 1930's on two occasions, sand was deposited by a dredge in a long low mound seaward of the breakers, in the hope that the sand would be moved to the beach, but it did not. One was at Santa Barbara, CA, and the other at Long Branch, NJ (Hall, 1950). Many decades later a satisfactory method was developed to calculate whether or not the sand in such a mound would move onshore, or whether it would remain where placed; this advance in coastal engineering was made by E.B. Hands and Allison (1991). Recently, this method has been used, successfully, at Silver Strand State Park {near San Diego}, CA (Andrassy, 1991), and off the mouth of the Santa Ana River, CA (Walker and Brodeur, 1993).

Engineering and scientific considerations needed for the design of projects are distributed throughout this history.

Beach nourishment has been considered in great detail by a committee of the NRC Marine Board, the Committee on Beach Nourishment and Protection. They wrote a report on the reasons for the need of beach nourishment {including hybrid projects}, design procedures, construction methods, and performance: *Beach Nourishment and Protection* (1995).

Some details of 56 projects in which the Federal government participated have been summarized in the paper by Sudar et al. (1995): "Shore Protection Projects of the U.S. Army Corps of Engineers", which gives data on locations, quantities of initial fill, and amount or renourishment (if it occurred).

Sand Dunes and Vegetation

Sand dunes and ridges have long been recognized as a means of shore protection and as beneficial sand reservoirs for the coast. Although sporadic attempts have been made to build or stabilize dunes along U.S. coasts, the first major program was in the mid-1930's on the barrier island beaches of North Carolina where the CCC {Civilian Conservation Corps} and the National Park Service used both fences and vegetation to construct and stabilize much of the island shore of the Outer Banks. A similar program was carried out in Oregon in the 1930's, particularly on Clatsop Spit by the Soil Conservation Service (e.g. Meyers and Chester, 1977) {capitalizing on remains of 1910 and 1926-1928 work by the Forest Service and Bureau of Plant Industry}. Davis (1957) summarizes much of this earlier work, including costs. The NC dune areas were deteriorated after hurricanes in the early mid-1950's, and the State and the BEB cooperated on studies of fence effectiveness and use of vegetation, resulting in vastly increased knowledge of how and what to use (Savage, 1963; Woodhouse and Hanes, 1967; Woodhouse, Seneca and Broome, 1976). At the same time, similar work was going on in Rhode Island (Jagschitz and Bell, 1966a, 1966b). Woodhouse (1978) is the first comprehensive report on dune building and stabilization of dunes throughout the U.S. coastal areas. Studies were likewise made on the use of marsh grasses to stabilize shore and nearshore areas (e.g. Woodhouse, Seneca and Broome, 1974; Garbisch, Woller, and McCallum, 1975). These studies were extended to dredged material disposal stabilization (e.g., Knutson, 1975; USACE San Francisco District, 1976). Much of what has been done is summarized in two CERC reports and guidelines (Woodhouse, 1979; Knutson and Woodhouse, 1983). The use of marsh plants behind small detached breakwaters has now become common in areas of {generally} low waves, such as Chesapeake Bay. Interestingly, use of the marsh grass as a stabilizer makes a project "non-structural" despite the conjunctive use of a {usually rock} breakwater.

Rubble Mound Breakwaters

On 26 Feb. 1978, nearly every cast concrete armor unit on the Port Sines, Portugal breakwater broke during a major storm. A panel of engineers {Edge - Chair, Baird, Caldwell, Magoon, and Treadwell, partially funded by the NSF}, visited the site on two occasions at the invitation of the Portuguese authorities. They worked together to document the design, construction and performance of the breakwater, and the oceanographic conditions at the time of failure. The panel report, published by the ASCE, *Failure of the Breakwater at Port Sines, Portugal* (Port Sines Investigating Panel, 1982) made available to the coastal engineering community a large amount of data, obtained while memories were still fresh and documents available. The most important scenario proposed by the panel was: "The dolos units rocked, broke and moved in the armor layer under the wave conditions that existed during the storm of February 26, 1978" (Baird, et al., 1980). The failure of the armor units was considered shortly after by Professor William G. Godden (1981). He presented his concepts of the scale effect that exists when modeling both the material properties of the units and dimensions of water waves. He lectured on this at a "short course" on recent developments in ocean engineering at UCB in January 1981. All model tests that had been made of rubble mound breakwaters with cast concrete armor units neglected the fact that the model scaling of the material properties of the units is different than the Froude Number scaling for the waves. In hydraulic model tests "failure" had been measured by the number of units displaced, not by breakage. If water is used in the hydraulic model, then the model armor unit:".. must have the same density as the prototype, its strength and E {modulus of elasticity} values must be reduced in the same proportion as the linear scale and its surface friction must have the same value as the prototype blocks." This requires a fragile model of the units, which was not done prior to then. Consideration must be given to the fact that there is a difference in density of fresh water used in model tests and the density of sea water, for breakwaters in the ocean (see also, Hudson, 1953).

A "watershed change" occurred, with numerical stress analyses, laboratory and prototype studies made. A number of instrumented 42-ton dollose were installed on the Crescent City, CA, breakwater and monitored for a number of years (Kendall and Melby, 1992), which showed the gradual build up of static stresses in the dollose. One result of the studies made by the USACE at CERC of both prototype and model units {including hydraulic model and numerical studies} was the development of a new type concrete armor unit (Turk and Melby, 1995).

A great amount of work has been done in recent years which we will leave to the future history update. However, several books published by the ASCE resulting from workshops or conferences on rubble mound structures should be noted: *Proceedings of Structures '83*, ed. by Weggel (1983); *Stresses in Concrete Armor Units*, ed. by Davidson and Magoon (1989); *Durability of Stone for Rubblemound Breakwaters*, ed. by Magoon and Baird (1991).

Detached Breakwaters

The first few detached breakwaters built in the U.S. have been mentioned at appropriate locations in the chronological review. Some details on these, and more recent ones are given in the paper by Chasten et al. (1994). Others have been constructed by private interests, particularly in protected waters {e.g. Chesapeake Bay} and often in conjunction with beach or marsh grass plantings. Functional design methods are available in a USACE CERC technical report by Julie Dean Rosati (1990). Structural design methods are distributed throughout a great many publications, some of which have cited previously in this history.

ORGANIZATIONS

ASBPA (American Shore and Beach Preservation Association)

In the early 1920's the Committee on Shoreline Studies was formed in the National Research Council with D.W. Johnson of Columbia University as its first chair. Cdr. {later Capt., then RADM} R.S. Patton of the USG&GS {later, its Director} became chair, with other members being D.W. Johnson, Isaiah Bowman {President of the American Geographical Society; later Chairman of the National Research Council}, and Nevin M. Fenneman of Univ. of Cincinnati. One of their recommendations was the need of an organized effort to deal with coastal problems. Officials of the state of New Jersey under the leadership of Governor Moore took action. A meeting of 85 delegates, representing 16 coastal states, met first at Asbury Park, NJ on 14-15 Oct. 1926 with subsequent meetings at Norfolk,VA, and Washington D.C. The ASBPA was founded as a result of this. The constitution and bylaws were adopted at a meeting in the NRC Building in Washington D.C. on 8 Dec. 1926. The ASBPA was eventually incorporated, in New Jersey on 26 July 1933.

The objectives of the ASBPA are:
"This Association is formed in recognition of the fact that the shores of our oceans, lakes and rivers constitute important assets for promoting the health and physical well-being of the people of this nation, and that their contiguity to our great centers of population affords an opportunity for wholesome and necessary rest and recreation not equally available in any other form. The purpose of the Association is to bring together for cooperation and mutual helpfulness the many agencies, interests and individuals concerned with the protection and proper utilization of these lands, and in all legitimate ways to foster that sound, far-sighted and economical development and preservation of the lands which will aid in placing their benefits within the reach of the largest possible number of people in accordance with the ideals of a democratic nation."

Within a few years, the ASBPA was working with appropriate people in Washington D.C., through their local State representatives, which led to legislation that resulted in the formation of the USACE Beach Erosion Board.

Wave Research Council/ Coastal Engineering Research Council

The first conference on coastal engineering was held in Long Beach, CA in Oct. 1950 under auspices of the University of California's Dept. of Engineering at Berkeley and the Extension Division. It was planned by Morrough P. O'Brien and J.W. Johnson. Each paper was by invitation, chosen so that the program represented the state of the art. Shortly after, at the urging of Boris A. Bakhmeteff, the Council on Wave Research was established by The Engineering Foundation, with O'Brien as Chairman, Johnson as Secretary, and Wiegel as executive engineer {later, vice-chair}. Approval of its formation was helped by Thorndike Saville, who at that time was on the Engineering Foundation Committee on Research. Funds were made available to publish the proceedings, and the Council was responsible for organizing additional conferences and publishing the proceedings. The 2nd {1951} conference was held in Houston, TX, with case studies of the Gulf of Mexico coast included, followed by the 3rd (1952) in Cambridge, MA, emphasizing the U.S. Atlantic Coast, and the 4th (1953) in Chicago, IL, with many papers on problems along the Great Lakes. Two papers were given in Chicago by engineers from The Netherlands on the great and terrible North Sea storm of 1 Feb. 1953, and of the plans being made for reconstruction and public safety works. Pierre Danel of France suggested another conference was needed to emphasis European works and practice, and the 5th (1954) was held in Grenoble, France, making the conferences international, which they have been since. In 1964, the Council was transferred to the ASCE, and broadened in scope, becoming the Coastal Engineering Research Council. At the 16th ICCE {1978} in Hamburg, Germany, O'Brien and Johnson retired, and Wiegel became Chair, Billy L. Edge secretary and Orville T. Magoon vice-chair. Wiegel retired at the end of the 23rd ICCE in Venice, Italy, Oct. 1992, and Robert G. Dean became Chair.

The locations and dates of the ICCEs are: 1,1950, Long Beach, CA; 2, 1951, Houston, TX; 3, 1952, Cambridge, MA; 4, 1953, Chicago, IL; 5, 1954, Grenoble, France; 6, 1957, Gainesville, Palm Beach and Miami, FL; 7, 1960, The Hague, The Netherlands; 8, 1962, Mexico City, Mexico; 9, 1964, Lisbon, Portugal; 10, 1966, Tokyo, Japan; 11, 1968, London, England; 12, 1970, Washington, D.C.; 13, 1972, Vancouver, B.C., Canada; 14, 1974, Copenhagen, Denmark; 15, 1976, Honolulu, HI; 16, 1978, Hamburg, Federal Republic of Germany; 17, 1980, Sydney, Australia; 18, 1982, Capetown, Republic of South Africa; 19, 1984, Houston, TX; 20, 1986, Taipei, Taiwan, ROC; 21, 1988, Malaga, Costa del Sol, Spain; 22, 1990, Delft, The Netherlands; 23, 1992, Venice, Italy; 24, 1994, Kobe, Japan; 25, 1996, Orlando, FL.

BEB/ CERC-CERB

As a result of public interest and its own concern with shore problems {particularly associated with navigation}, and expressed concern by the ASBPA, the USACE established on 23 January 1929 the Board on Sand Movement and Beach Erosion. This Board consisted of four officers {Col. Barden, Senior Member, Col. Pillsbury, LTC Dent, and Major Somervell, Recorder} and had appointed two civilians as consultants {Douglas Johnson and Thorndike Saville}. Investigations were made of shoreline problems related to federal coastal navigation works or on military reservations; to overcome the lack of basic data, Douglas Johnson prepared a plan for field studies involving some 30 different experiments. Field sites were established along the New Jersey coast under the direction of Lt. Hewitt, with Morrough P. O'Brien as a consultant. The *Interim Report*, referred to earlier (BEB, 1933) resulted from these, and is well worth reading still for its insight into shore processes. Colony's (1930) report on sand origins along the NJ and Long Island shores has findings still of value today. A bibliography (Hafercorn, 1929) was a comprehensive listing of appropriate technical literature to that date. With further public interest, again stimulated by the ASBPA and the state of New Jersey, Congress on 3 July 1930 established the Beach Erosion Board with seven members, 4 military and 3 civilian, to make studies "devising effective means of preventing erosion ... by waves and currents". Studies were to be cooperative, with the state involved funding one-half the study cost. Moneys for construction were not included. The first Board consisted of Col Barden, Senior Member, Col. E.I. Brown, LTC E.J. Dent, and Major G.R. Young, as military, and Richard K. Hale, MA, Victor J. Gelineau, NJ, and Thorndike Saville, NC, as civilians. The staff organization became the U.S. Coastal Engineering Research Center {CERC} in 1963, and the Coastal Engineering Research Board {CERB} established with 4 Corps officers and 3 civilian members. It has been responsible for much of the research in this country on coastal engineering, particularly in the early years. See Quinn (1977) and Moore and Moore (1991) for more detailed history.

American Society of Civil Engineers (ASCE)

Hydraulics Division {Authorized April 19, 1938; merged with other divisions into Water Resources Division, October 1, 1994}. The Committee on Tidal Hydraulics is largely concerned with coastal engineering and science.

Technical Council on Ocean Engineering {Authorized Dec. 2, 1966; absorbed into WW Division, Oct. 1976}. While activities were primarily related to the relatively deep ocean, some touched on coastal work. The Council organized several "Civil Engineering in the Ocean" conferences.

Waterway, Port, Coastal and Ocean Division {Authorized as Waterways Division June 16, 1942; Harbors added February 13, 1956; Coastal Engineering added April 6-7, 1970; Ocean added, complete title modified July 10, 1976. Most of the technical committees {and one administrative committee - the Coastal Engineering Research Council} are concerned with coastal engineering, particularly the Committee on Coastal Engineering.

Committee on Tidal Hydraulics, USACE

The Committee on Tidal Hydraulics {CTH} was established within the USACE in 1948 to advise on, and supervise research in tidal hydraulics and tidal modelling, with the added duty to advise Corps offices on project planning and design in the tidal area. It has produced a number of reports, notable among them two on evaluation of knowledge (CTH, 1950; Wicker, 1965), and extended bibliography (CTH, 1954, with 10 supplements through 1987). Its history is given by Lockett (1972).

NAE Committee on Ocean Engineering/ NRC Marine Board

"First organized under the NAE in 1965 as the Committee on Ocean Engineering of the National Academy of Engineering, it was placed under the NRC, expanded and renamed the Marine Board in 1970. On July 1, 1982, the board merged with the Maritime Transportation Research Board (MTRB) to be operated under the Commission on Engineering and Technical Systems as one organizational and fiscal unit." (Facsimile, 28 April 1995 from Charles Bookman, Executive Director, MB).

The Board's technical and policy analyses and other activities on behalf of the federal government and the public have three general goals (NRC MB, 1995):
".. to improve and expand the base of technology and knowledge in ocean and coastal development and uses;
...to ensure that planning, design, and other engineering activities result in marine structures and systems that are safe, environmentally sound, and economically competitive; and
...to enhance the value and usefulness of the marine engineering and technology base as a decision-making tool in policy development and program planning and management."

NRC Committee on Natural Disasters/ Board on Natural Disasters

This was established in 1971, derived from the NAE Committee on Earthquake Engineering {established in 1966} which had been expanded to include wind and other natural disasters. On 1 May 1992, the NRC Committee on Natural Disasters and another NRC Committee on Earthquake Engineering were combined to form the NRC Board on Natural Disasters.

NRC Ocean Studies Board

"The board {organized in 1983} contributes to the advancement of the scientific understanding of the ocean by maintaining oversight of the health of the ocean sciences and the stimulation of their progress; fostering the application of scientific knowledge to the wise use of the ocean and its resources; providing leadership for the formulation of national and international marine policy and clarifying scientific issues that affect ocean policy; and addressing marine science issues involved in cooperative international oceanographic research an in improvement of technical assistance." (NAS/NAE/IOM/NRC, 1995). The OSB replaced the Ocean Affairs Board and a series of NAS Committees on Oceanography (NASCO). The most important effect for coastal engineering was increased federal funding resulting from inclusion in NASCO reports and efforts in the 1960's.

AWARDS

International Coastal Engineering Award, ASCE

The award is made annually to an individual who has made a significant contribution to the advancement of coastal engineering in the manner of engineering design, teaching, professional leadership, construction, research, planning, or a combination thereof. Awards made to: Bernard LeMehaute, 1979; William J. Herron, Jr., 1980; Kiyoshi Horikawa, 1981; Morrough P. O'Brien, 1982; Robert G. Dean, 1983; Michael S. Longuet-Higgins, 1984; Robert L. Wiegel, 1985; Eco Wiebe Bijker, 1986; Joe W. Johnson, 1987; Douglas L. Inman, 1988; Yoshimi Goda, 1989; Jurjen A. Battjes, 1990; Thorndike Saville, Jr., 1991; Fredrick L.W. Tang, 1992; John B. Herbich, 1993; Leon E. Borgman, 1994; Chiang C. Mei, 1995.

John G, Moffatt - Frank E. Nichol Harbor and Coastal Engineering Award, ASCE

The award is made annually to a member of the ASCE who has made a definite contribution in the fields of harbor and coastal engineering. This contribution may have been made either in the form of written presentation or notable performance. Awards made to: Robert L. Wiegel, 1978; Thorndike Saville, Jr., 1979; Omar J. Lillevang, 1980; Joe W. Johnson, 1981; James W. Dunham, 1982; Basil W. Wilson, 1983; Charles C. Calhoun, Jr., 1984; John H. Nath, 1985; Eugene H. Harlow, 1986; Robert G. Dean, 1987; Charles L. Bretschneider, 1988; Robert Hoffmaster, 1989; Orville T.

Magoon, 1990; Ray B. Krone, 1991; Chiang C. Mei, 1992; J. Richard Weggel, 1993; Fredric Raichlen, 1994; Jay B. Weidler, 1995.

Morrough P. O'Brien Award for Outstanding Achievement in Shore and Beach Preservation, ASBPA

The award is made not more frequently than annually to a person {not a project} for an outstanding record in achieving the objectives of the ASBPA and/or in direct contributions to the ASBPA. Awards made to: J.W. Johnson, 1988; Peter H.F. Graber, 1989; Henry M. von Oesen and Paul S. Denison, 1990; Allen G. Ten Broek, 1991; Gerald J. Giefer, 1992; Billy L. Edge, 1993; Orville T. Magoon, 1994; Robert L. Wiegel, 1995.

Outstanding Coastal Project Award, ASBPA

The award is made in recognition of a completed, successful coastal project, with a minimum of five years elapsed since completion. Awards made to: Miami Beach, Florida, Beach Replenishment Project, 1985; Lakeview Park, Ohio Beach Erosion Project {in Lorrain}, 1988; Galveston, Texas Grade-raising, Seawall and Embankment, 1990; O'Shaughnessy Seawall and Esplanade, San Francisco, California, 1992; Coney Island, New York, Public Beach and Boardwalk Improvement of 1922-23, 1994; Santa Barbara, CA sand bypass system 1996.

Joe W. Johnson Outstanding Beach Preservation Award, California Shore and Beach Preservation Association

Awards made to: Robert L. Wiegel, 1993; Douglas L. Inman, 1995.

Acknowledgements

The authors wish to thank very much their colleagues Robert G. Dean, Joe W. Johnson and Nicholas C. Kraus for their valuable comments on drafts of this history. Appreciation is expressed for help received from the USACE CERC, particularly Frederick E. Camfield, to William H. McAnally, Jr., WES Hydraulics Laboratory, and to D. Lee Harris. As noted at the start, there are some areas that are skimped or even omitted {dredging is an obvious one}. For these lacks, we are sorry, but everything could not be included - and we drew from our own experience, which {obviously} is not all inclusive. There have been great advances in this field in the past 50 years, and we have been fortunate to have been a part of it.

REFERENCES

Abel, Robert B., "Opening Comment" in *Sea Grant 70s*, 1(1), Sept. 1970, 1.

Abraham, G., "Jet Diffusion in Liquid of Greater Density," *Jour. Hyd. Div., Proc. ASCE*, 86(HY6), June 1960, 1-13.

Abramson, H. Norman and Basil W. Wilson, "A Further Analysis of the Longitudinal Response of Moored

Vessels to Sea Oscillations," *Proc. Midwestern Conf. of Fluid and Solid Mechanics, Purdue Univ., Sept. 1955*, 236-25l.

Ahrens, J. and B.L. McCartney, "Wave Period Effect on the Stability of Riprap," *Proc. Civil Engrg. in the Oceans III*, ASCE, 1975.

Ahrens, J., "Prediction of Irregular Wave Run-up", USACE CERC *Coastal Engrg. Tech. Aid* 77-2, 1977.

Allen, R.H. and E.L. Spooner, *Annotated Bibliography of BEB and CERC Publications*, USACE, CERC, Misc. Pub. No. 1-68, 1968.

Allen, D.R. and M.N. Mayuga, "The Mechanics of Compaction and Rebound of Wilmington Oil Field, Long Beach, California," *Land Subsidence: Proc. of the Tokyo Symposium, Sept. 1969*, UNESCO, II, 1970, 410-429.

American Petroleum Institute, *Recommended Practice for Planning, Designing and Constructing Fixed Offshore Platforms - Working Stress Design*, API Recommended Practice 2A-WSD (RP 2A-WSD), 20th Ed., July 1993.

Anders, Fred J. and Mark R. Byrnes, "Accuracy of Shoreline Change Rates as Determined from Maps and Aerial Photographs," *Shore & Beach*, 59(1), Jan. 1991, 17-26.

Andrassy, Christopher J., "Monitoring of a Nearshore Disposal Mound at Silver Strand State Park," *Coastal Sediments '91, Proc. of a Specialty Conf., Seattle, WA, 25-27 June 1991*, ed. by Nicholas C. Kraus et al, ASCE, II, 1970-1984.

Anon., "Plans for the Protection of Galveston from Floods," *Engineering News*, XLVII(17), April 24, 1902, 343-344.

Anon., "Reports on Hurricane Damage for the U.S. Engineers Districts of New York, Providence and Boston," *Shore and Beach*, 7(1), Jan. 1939, 29-37.

Anon., "A Bit of History," *Shore & Beach*, 53(2), April 1985, 32.

Anon., "Revised Wave Information Study (WIS) Results for the U.S. Atlantic Coast 1956-1975," *Coastal Engrg. Tech. Note*, CERC CETN I-51, June 1992, 7pp.

Anon., "Thirteen Projects, Large and Small, in Running for ASCE's Top Award," *ASCE News*, 18(3), March 1993, 1-2.

Anon., Hydrocarbons, Seances, and the Move Offshore," *Bechtel Briefs*, 50(2), July 1995, 1.

Arthur, Robert S., "Refraction of Waves by Islands and Shoals with Circular Bottom Contours," *Trans. Amer. Geophys. Union*, 27(2), April 1946, 168-177.

Arthur, R.S., "Variability in Direction of Wave Travel," *Ocean Surface Waves, Annals of the New York Acad. Sci.*, 51(3), May 1949, 511-522.

Arthur, Robert S. "The Effect of Islands on Surface Waves," *Bulletin of the Scripps Inst. of Ocean.*, Univ. of Calif., 6(1) 1951, 1-16.

ASCE Task Committee on Small Craft Harbors, *Report on Small Craft Harbors*, ASCE Manuals and Reports on Engrg. Practice No. 50, 1969, 139 pp.

Assoc. of Sea Grant Program Institutions, *Newsletter*, pub. by the Grad. School of Oceanography, Univ. Rhode Island, I(1), starting Jan. 1971.

Baer, L. and C.L. Vincent, "Atlantic Remote Sensing Land Ocean Experiment (ARSLOE): Overview," *IEEE Jour. Oceanic Engrg.*, OE-8(4), Oct. 1983, 201-205.

Bailard, James A., "A Simplified Model for Longshore Sediment Transport," *Nineteenth Coastal Engrg. Conf. Proc. of the Inter. Conf., 3-7 Sept. 1984, Houston, TX*, ASCE, II, 1985, 1454-1470.

Baird, William F., Joseph M. Caldwell, Billy L. Edge, Orville T. Magoon and Donald K. Treadwell, "Report on the Damage to the Sines Breakwater, Portugal," *Proc. 17th Coastal Engrg. Conf., March 23-28, 1980, Sydney, Australia*, ASCE, II, 1981, 3036-3077.

Barnes, Jay, *North Carolina's Hurricane History*, Univ. North Carolina Press, 1995, 206 pp.

Barnett, T.P., "Recent Changes in Sea Level and Their Possible Causes," *Climate Change*, 5, 1983, 15-38.

Baron, Clifford et al., *Preliminary Data Summary for Jan. 1995, CERC Field Research Facility*, USACE CERC, 1995, 27 pp.

Barrett, Robert J., "Use of Plastic Filters in Coastal Structures," *Proc. of 10th Conf. on Coastal Engrg., Tokyo, Japan, Sept. 1966*, ASCE, ed. by J.W. Johnson, II, 1048-1067.

Bascom, W., *Effect of Seismic Sea Wave on California Coast*, UCB, IER, Tech. Rept. 3-204, 16 April 1946.

Bascom, Willard J., "The Relationship Between Sand Size and Beach-face Slope," *Trans. Amer. Geophys.*

Union, 32(6), 1951, 866-874.

Bascom, Willard J., "A Deep Water Instrument Station ", *Proc. 1st Conf. Coastal Engrg. Instruments*," ed. by Robert L. Wiegel, Council on Wave Res., The Engrg. Found., 1956, 297-301.

Bascom, Willard, *Waves and Beaches*, 1964, Anchor Books, Garden City, NY, 267 pp.

Bascom, Willard (ed.), *Coastal Water Research Project Biennial Report 1979-1980*, Southern California Water Research Project, Long Beach, CA, 1981, 363 pp.

Bates, Charles C., "Utilization of Wave Forecasting in the Invasion of Normandy, Burma, and Japan," *Annals of the New York Acad. of Sci.*, 51(3), May 1949, 545-569., with Discussion by Warren C. Thompson, 569-572.

BEB, "Inspection of Beaches in Path of the Hurricane of Sept. 21, 1938," *Shore and Beach*, 7(1), Jan. 1939, 43-48.

Berg, Dennis W., "Systematic Collection of Beach Data," *Proc. 11th Conf. Coastal Engrg., London, England, Sept. 1968*, ASCE, 1969, 273-297.

Berge, Bent, *Three Dimensional Stochastic Response of Offshore Towers to Wave Action*, Ph.D. thesis, UCB, Dept. Civil Engrg., Nov. 1973, 148 pp.

Berge, Bent and J. Penzien, "Three Dimensional Stochastic Response of Offshore Towers to Wave Forces," *Preprints: Offshore Tech. Conf., Houston, TX, 6-8 May 1974*, II (OTC 2050), 173-190.

Bidde, D.D., "Laboratory Study of Lift Forces on Circular Piles," *Jour. Waterways, Harbors, and Coastal Div., Proc. ASCE*, 97(WW4), Nov. 1971, 595-614.

Birkemeier, W.A., A.E. DeWall, C.S. Gorbics and H.C. Miller, *A Users Guide to CERC's Field Research Facility*, USACE CERC Misc. Rept. No. 81-7, Oct. 1981, 118 pp.

Birkemeier, W.A., "Field Data on Seaward Limit of Profile Change," *Jour. Waterway, Port, Coastal and Ocean Engrg.*, ASCE, 3(3), 1985, 598-602.

Birkemeier, W.A., N.C. Kraus, N.W. Scheffner, S.C. Knowles, *Feasibility Study of Quantitative Erosion Models for Use by the Federal Emergency Management Agency in the Prediction of Coastal Flooding*, USACE, CERC, 1987, 82pp and Appendix.

Blue, F.L., Jr., and J.W. Johnson, "Diffraction of Water Waves Passing Through a Breakwater Gap," *Trans. Amer. Geophys. Union*, 30(5), Oct. 1949, 705-718.

Bodine, B.R., *Storm Surge on the Open Coast: Fundamentals and Simplified Prediction*, USACE CERC, TM-35, May 1971.

Bokuniewicz, H.J., "The Seasonal Beach at East Hampton, New York," *Shore & Beach*, 43(3), July 1981, 28-33.

Bokuniewicz, H.J. and J.R. Schubel, "The Vicissitudes of Long Island Beaches, New York," *Shore & Beach*, 55(3/4), July/Oct. 1987, 71-75.

Borgman, L.E. and J.E. Chappalear, "The Use of the Stokes-Struik Approximation for Waves of Finite Height," *Proc. 6th Conf. on Coastal Engrg., Florida, Dec. 1957*, Council on Wave Res., The Engrg. Found., 1958, 252-280.

Borgman, Leon E., "Risk Criteria," *Jour. Waterways and Harbors Div., Proc. ASCE*, 81(WW3), Aug. 1963, 1-36.

Borgman, Leon E., "Spectral Analysis of Ocean Wave Forces on Piling," *Jour. Waterways and Harbors Div., Proc. ASCE*, 9(WW2), May 1967, 129-156.

Borgman, Leon E., "Directional Spectra Models for Design Use for Surface Waves," *Preprints, Offshore Tech. Conf., Houston, TX, 18-21 May 1969*, I(1069), 721-746.

Borgman. Leon E. *The Statistical Anatomy of Ocean Wave Spectra*, USACE CERC Tech. Paper No. 76-10, July 1976, 102 pp.

Borgman, Leon E. and E. Yfantis, "Three-Dimensional Character of Waves and Wave Forces," *Civil Engrg. in the Oceans IV*, ASCE, II, 1979, 791-804.

Bowen, A.J., D.L. Inman and V.P. Simmons, "Wave 'Set-Down' and Set-Up,' *Jour. Geophys. Res.*, 73(8), 5 April 1968, 2569-2577.

Bowen, A.J. "Rip Currents. 1. Theoretical Investigations," *Jour. Geophys. Res.*, 74, 1969, 5467-5478.

Bowen, A.J. and D.L. Inman, "Rip Currents. 2. Laboratory and Field Observations," *Jour. Geophys. Res.*, 74, 1969, 1313-1320.

Bowen, A.J. and D.L. Inman, "Edge Waves and Crescentic Bars," *Jour. Geophys. Res.*, 76 (36), 1971, 8662-8671.

Bowie, George L. and R.L. Wiegel, *Marine Pipelines: An Annotated Bibliography*, USACE, CERC, Misc. Rept. 77-2, 1977, 58 pp.
Bretschneider, Charles L.,"Revised Wave Forecasting Relationships," *Proc. 2nd Conf. on Coastal Engrg., Houston, TX, Nov. 1951*, Council on Wave Res., The Engrg. Found., 1952, 1-5.
Bretschneider, C.L. and R.O. Reid, *Modification of Wave Height Due to Bottom Friction, Percolation, and Refraction*, USACE BEB Tech. Memo. No. 45, Oct. 1954, 36 pp.
Bretschneider, C.L., *Wave Forecasting Relationships for the Gulf of Mexico*, USACE, BEB TM 84, 1956.
Bretschneider, C.L. and R.D. Gaul, *Wave Statistics for the Gulf of Mexico off Brownsville, Texas*, USACE, BEB, TM 85, 1956 (also TM 86, *Caplen, TX*, TM 87, *Burwood, LA*, TM 88, *Apalachicola, FL* and TM 89, *Tampa Bay, FL*, all 1956).
Bretschneider, C.L., "Hurricane Design Wave Prediction," *J. WW Div., Proc. ASCE*, Paper 1238, May 1957; also in *Trans. ASCE, 124, 1959*.
Bretschneider, Charles L., *Wave Variability and Wave Spectra for Wind-generated Waves*, USACE BEB Tech. Memo. No. 118, Aug. 1959, 192 pp.
Bretschneider, Charles L., *Generation of Waves by Wind: State of the Art*, National Engrg. Science Co., Wash. D.C., Contract Nonr-4177, 15 Jan. 1965, 96 pp.
Brooke, Col. Mark M., "Shore Preservation in Florida", with discussion, *Shore and Beach*, 2(4), Oct. 1934, 151-154.
Brooks, Charles F., "Meteorology of the September Hurricane," *Shore and Beach*, 7(1), Jan. 1939, 27-29.
Brooks, N.H. and R.C.Y. Koh, "Discharge of Sewage Effluent from a Line Source into a Stratified Ocean," *IAHR, 11th Congress, Leningrad, USSR, 1965*, II (2.19), 9 pp.
Brooks, Norman H. and James E. Krier, "Alternative Strategies for Ocean Disposal of Municipal Wastewater and Sludge," background paper for *Symposium on Engineering Aspects of Using the Assimilative Capacity of the Oceans, 23-24 June 1981, Lewes, DE*, 38 pp.
Brown, E.I., "Inlets on Sandy Coasts," *Proc. ASCE*, 54, 1928, 505-553.
Brown, E.I., "Report on Rhode Island Beaches," *Shore and Beach*, 7(1), Jan. 1939, 40-42.
Bruun, Per, *Coastal Erosion and Development of Beach Profiles*, USACE BEB, Tech. Memo. No. 44, June 1954, 79 pp.
Bruun, Per, J.J. Leendertse and L.W. Cover, "Florida's Coastal Engineering Wave Tank," *Engrg. Progress at Univ. Florida*, 12(7), Leaflet No. 99, July 1958, 15 pp.
Bruun, Per and F. Gerritsen, "Stability of Coastal Inlets," *Proc. 7th Conf. on Coastal Engrg., The Hague, The Netherlands, Aug. 1960*, ASCE, Vol. I, 386-417.
Bruun, Per, "The History and Philosophy of Coastal Protection," *Proc. 13th Coastal Engrg. Conf., July 10-14, 1972, Vancouver, B.C., Canada*, ASCE, 1973, 33-74.
Bruun, Per, A.J. Mehta, and I.G. Jonsson, *Stability of Tidal Inlets - Theory and Engineering*, Elsevier Scientific Publishing Co., 1978, 510 pp. Note: sections of this book were written by A.J. Mehta, E. Ozsoy, I.G. Jonsson, and J.A. Purpura.
Bruun, Per, "The Bruun Rule of Erosion by Sea-Level Rise: A Discussion on Large-Scale Two-and Three-Dimensional Uses," *Jour. of Coastal Res.*, 4(4), Fall 1988, 627-648.
Buhler, Johannes, "On Buoyant Surface Layers Generated by Wastewater Discharged from Submerged Diffusers," *IAHR, Proc. 17th Congress, 15-19 Aug. 1977, Baden-Baden, Fed. Rep. Germany*, 1(A42), 1977, 325-332.
Butler, H.L. and D.L. Durham, "Applications of Numerical Modelling to Coastal Engineering Problems", *Proc. 1976 Army Numerical Analysis and Computers Conf.*, 1976.
Caldwell, Joseph M. and E.P. Fortson, *A Model Study of Maintenance Works at Ballona Creek Outlet, Venice, California*, USACE, WES, Paper No. 18, June 1937, 44 pp.
Caldwell, Joseph M., *An Ocean Wave Measuring Instrument*, USACE, BEB, TM 6, 1948.
Caldwell, Joseph M., "By-passing Sand at South Lake Worth Inlet, Florida," *Proc. 1st Conf. Coastal Engrg., Long Beach, CA, Oct. 1950*, ed. by J.W. Johnson, Council on Wave Res., The Engrg. Found., 1951, 320-325.
Caldwell, Joseph M., "Tidal Currents at Inlets in the United States," *Proc. Hyd. Div., ASCE*, 81(Sep. No. 716), 1955, 12 pp.
Caldwell, Joseph H., *Wave Action and Sand Movement Near Anaheim Bay, California*, USACE BEB Tech. Memo. No. 68, Feb. 1956, 21 pp.

Camfield, F.E., *Tsunami Engineering*, Special Rept. No. 6, USACE, CERC, 1980.
Carlson, Christopher T., "Field Studies of Run-up on Dissipative Beaches," *19th Coastal Engrg. Conf., Proc. of the Inter. Conf., 3-7 Sept. 1984, Houston, TX*, ASCE, I, 1985, 399-414.
Carr, John H., "Mobile Breakwaters," *Proc. 2nd Conf. on Coastal Engrg., Houston, Texas, 1951*, Council on Wave Res., The Engrg. Foundation, 1952, 281-295.
Carr, J.H. and M.E. Stelzriede, "Diffraction of Water Waves by Breakwaters," *Proc. U.S. National Bureau of Standards Symposium on Gravity Waves*, U.S. NBS Circular 521, Nov. 1952, 109-125.
Carr, John H., *Breaking Wave Forces on Plane Barriers*, CIT Hydro. Lab., Contract NOy-12561, Rept. No. E-11.3, Nov. 1954.
Chakrabarti, Subrata K., "Inline Forces on Fixed Vertical Cylinder in Waves," *Jour. Waterways, Port, Coastal and Ocean Engrg. Div., Proc. ASCE*, 106(WW2) May 1980, 145-155.
Chang, S.S., *A Magnetic Tape Wave Recorder and Energy Spectrum Analyzer for the Analysis of Ocean Wave Records*, USACE, BEB, TM 58, 1955
Chasten, Monica A., John W. McCormick and Julie D. Rosati, "Detached Breakwaters for Shoreline and Wetland Stabilization," *Shore & Beach*, 62(2), April 1994, 17-22.
Chiu, Arthur N.L., Luis E. Escalante, J. Kenneth Mitchell, Dale C. Perry, Thomas A. Schroeder and Todd Walton, *Hurricane Iwa, Hawaii, November 23, 1982*, NRC Committee on Natural Disasters, National Acad. Press, Wash. D.C., 1983, 129 pp.
Clarke, Caroline, NRC Board on Natural Disasters, *Personal Communication*, Facsimile to R.L. Wiegel, 2 May 1994, 1 p., and tel. call on 3 May 1995.
Clemens, Karen E. and Paul D. Komar, "Oregon Beach-sand Composition Produced by the Mixing of Sediments Under A Transgressing Sea," *Jour. Sedimentary Petrology*, 58(3), May 1988, 519-529.
Code of Federal Regulations (CFR), *Title 44 - Emergency Management and Assistance*, Office of the Federal Register, Revised Oct. 1, 1993, Sect. 59.1, 221-222.
Collins, J. Ian, Wen-Li Chiang and Frank Wu, "Refraction of Directional Spectra," *Directional Wave Spectra Applications: Proc. of Conf., 14-16 Sept. 1981, Berkeley, CA*, ASCE, 1982, 251-268.
Colony, R.J., *Report to the Board on Sand Movement and Beach Erosion on the Source of Sand on Long Island and New Jersey Beaches*, USACE, BSMBE, Dec. 1930.
Committee on Tidal Hydraulics, *Evaluation of Present State of Knowledge of Factors Affecting Tidal Hydraulics and Related Phenomena*, USACE, WES, CTH Rept. No. 1, 1950.
Committee on Tidal Hydraulics, *Bibliography on Tidal Hydraulics*, CTH Rept. No. 2, USACE WES, 1954, {with 10 supplements through 1987}.
Corson, W.D., D.T. Resio, R.M. Brooks, B.A. Ebersole, R.E. Jensen, D.S. Ragsdale, and B.A. Tracy, *Atlantic Coast Hindcast, Phase II Wave Information*, USACE WES, WIS Rept. 6, March 1982, 1186 pp.
Costa, G.L. and J.D. Isaacs, "The Modification of Sand Transport in Tidal Inlets, " *Coastal Sediments '77, Charleston, SC, Nov. 2-4, 1977*, ASCE, 946-965.
Coudert, J.F. and F. Raichlen, "Wave Refraction Near San Pedro Bay, California," *Jour. WW Div., Proc. ASCE*, (WW3), 1970
Cox, Doak C., *Performance of the Seismic Sea Wave Warning System, 1948-1967*, Hawaii Inst. of Geophysics, Univ. of Hawaii, HIG-68-2, March 1968, 69 pp and appendices.
Crowson, Ronald A, William A. Birkemeier, Harriet M. Klein, Herman C. Miller, *Superduck Nearshore Processes Experiment: Summary of Studies, CECR Field Research Facility*, USACE CERC Tech. Rept. CERC-88-12, 1988, 116 pp.
Cross, R.H., "Tsunami Surge Forces on Coastal Structures," *Jour. Waterways and Harbors Div., Proc. ASCE*, 93(WW4), 1967, 201-231.
Cross, R.H. and C. Sollitt, *Wave Transmission by Overtopping*, Tech. Note 15, MIT, Ralph M. Parsons Lab.,197l.
Cuthbert, Jerry W., *Considerations in the Design of an Irregular Wave Generator*, UCB Dept. Naval Architecture, May 1959, 32 pp and app.
Dalrymple, R.A. and P.L.-F. Liu, "Waves on Soft Muds: A Two Layer Fluid Model," *Jour. Geophys. Res.*, 8, 1978, 1121-1131.
Dames & Moore, *The Northridge Earthquake, January 17, 1994*, special report, 1994, 29 pp.
Darling, John M., "Study of Pilot Beaches in New England for the Improvement of Coastal Storm

Warning," CERC *Bulletin*, Vol. II, 1965-66.
Darling, J. and D.G. Dumm, *The Wave Record Program of CERC*, USACE, CERC, Misc. Paper 1-67, 1967
Darling, John M., "Surf Observations Along the United States' Coasts," *Jour. Waterways and Harbors Div., Proc. ASCE*, 94(WW1), Feb. 1968, 11-21.
Davidson, D.D. and Orville T. Magoon (ed.), *Stresses in Concrete Armor Units*, derived from a Seminar at USACE WES/CERC, Vicksburg, MS, 7-8 Nov. 1989, ASCE, 422 pp.
Davidson, Margaret A., Robert G. Dean and Billy L. Edge (guest ed.), *Shore & Beach*, 58(4), Oct. 1990, 83 pp.
Davis, Albert B., Jr., "History of the Galveston Seawall," *Proc. 2nd Conf. on Coastal Engrg., Houston, Texas, Nov. 1951*, ed. by J.W. Johnson, Council on Wave Res., The Engrg. Found., 1952, 268-280.
Davis, John H., *Dune Formation and Stabilization by Vegetation and Plantings*, USACE, BEB, TM No. 101, 1957.
Davison, A. Todd, Robert J. Nicholls and Stephen P. Leatherman, "Beach Nourishment as a Coastal Management Tool: An Annotated Bibliography on Developments Associated with the Artificial Nourishment of Beaches," *Jour. Coastal Res.*, 8(4), Summer 1992, 141-193.
Dean, Robert G., "Stream Function Representation of Nonlinear Ocean Waves," *Jour. Geophys. Res.*, 70(18), 1965, 4561-4572.
Dean, Robert G., "Heuristic Models of Sand Transport in the Surf Zone," *Conf. on Engrg. Dynamics in the Surf Zone, Sydney, Aust., 1973*, The Inst. of Engineers of Australia, 1973, 208-214.
Dean, Robert G. and Todd L. Walton, Jr., "Sediment Transport Processes in the Vicinity of Inlets with Special Reference to Sand Trapping", *Second Inter. Estuarine Res. Conf., Myrtle Beach, SC, Oct. 1973.*
Dean, R.G., *Evaluation and Development of Water Wave Theories for Engineering Application. Vol. II, Tabulation of Dimensionless Stream Function Theory Variables*, USACE CERC Special Rept. No. 1, Nov. 1974, 534 pp.
Dean, R.G., "Principles of Beach Nourishment," *CRC Handbook of Coastal Processes and Erosion*, CRC Press, Boca Raton, FL, 1983, 217 pp.
Dean, Robert G. and Robert A. Dalrymple, *Water Wave Mechanics for Engineers and Scientists*, Prentice-Hall, Inc., 1984, 353 pp.
Dean, Robert G., George A. Armstrong and Nicholas C. Sitar, *California Coastal Erosion and Storm Damage During the Winter of 1982-83*, NRC Committee on Natural Disasters, National Academy Press, Wash. D.C., 1984, 74 pp.
Dean, Robert G., "Eroding Shorelines Impose Costly Choices", *Geotimes*, 33(5), May 1986, 9-11.
Dean, Robert G., "Additional Sediment Input to the Nearshore Region," *Shore & Beach*, 55(3-4), July-Oct. 1987, 76-81.
Dean, Robert G., "Measuring Longshore Transport with Traps," *Nearshore Sediment Transport*, Plenum Press, NY, 1989(a), 313-336.
Dean, Robert G., "Measuring the Nearshore Morphology. B. Offshore surveys," *Nearshore Sediment Transport*, Plenum Press, NY, 1989(b), 43-50.
Dean, Robert G., "Equilibrium Beach Profiles: Characteristics and Applications," *Jour. Coastal Res.*, 7(1), 1991, 53-84.
Dean, Robert G. and C-H. Yoo, "Predictability of Beach Nourishment Performance," in the volume on *Beach Nourishment Engrg. and Management Considerations, Coastal Zone '93, New Orleans, LA*, D.K. Stauble and N.C. Kraus (ed.), ASCE, 1993, 86-102.
Dean, Robert G., *Personal Communication by Facsimile*, to R.L. Wiegel, 31 July 1995.
Dill, Robert F., "The Use of Free Diving Equipment as a Tool in Coastal Engineering and Scientific Studies," *Proc. First Conf. on Coastal Engrg. Instruments*, ed. by Robert L. Wiegel, Council on Wave Res., The Engrg. Found., Berkeley, CA, Oct. 31-Nov. 2, 1955, 272-296.
Dolan, Robert, Bruce P. Hayden, Paul May, Suzette May, "The Reliability of Shoreline Change Measurements from Aerial Photographs," *Shore & Beach*, 48(4), Oct. 1980, 22-29.
Dolan, Robert and Robert E. Davis, "Rating Northeasters," *Mariners Weather Log*, NOAA, 36(1), Winter 1992, 4-11.
Dorland, G.M., *Equilibrium Sand Slopes in Front of Seawall*, M.S. Thesis, Civil Engrg., UCB, 1940.
Dornhelm, Richard B., "The Coney Island Public Beach and Boardwalk Improvement of 1923," *Shore &*

Beach, 63(1), Jan. 1995, 7-11.

Dorrestein, R., "Wave Set-up on a Beach", *Proc. 2nd Tech. Conf. on Hurricanes*, National Hurricane Res. Rept. No. 50, 1962, 230-241.

Duane, D.B. and C.W. Judge, *Radioactive Sand Tracer Study, Point Conception, California*, USACE, BEB, Misc. Paper 2-69, 1969.

Duane, D.B., R.J. Seymour, and A.G. Alexiou, "Introduction", *Nearshore Sediment Transport*, Richard J. Seymour, ed., Plenum Press, NY and London, 1989, 1-6.

Eagleson, P.S., R.G. Dean, and L.A. Peralta, *The Mechanics of the Motion of Discrete Spherical Bottom Sediment Particles Due to Shoaling Waves*, USACE BEB, Tech. Memo. No. 104, Feb. 1958, 41 pp and app.

Eaton, R.O., "Littoral Processes on Sandy Coasts," *Proc. 1st Conf. on Coastal Engrg., Long Beach, Calif., Oct. 1950*, Council on Wave Res., The Engrg. Found.,1951, pp 140-154.

Eaton, R.O., "Some Examples of Large Scale Shore Protection Projects," *Shore and Beach*, 27(1), June 1959, 8-13.

Eckart, Carl, "The Propagation of Gravity Waves from Deep to Shallow Water," *Gravity Waves. Proc. NBS Symposium on Gravity Waves, 18-20 June 1951*, NBS Circular 521, Nov. 1952, 165-174.

Edge, Billy L. and Orville T. Magoon, (ed.), *Proc. International Symposium on Ocean Wave Measurement and Analysis, Sept. 9-11, 1974, New Orleans, LA, U.S.A.*, ASCE, 1975, 2-vols., 913 and 302 pp.

Edge, Billy L. and Orville T. Magoon, "A review of Recent Damages to Coastal Structures," *Coastal Structures '79. A Specialty Conf. of the Design, Construction, Maintenance and Performance of Port and Coastal Structures, March 14-16, 1979, Alexandria, VA*, ASCE, I, 1979, 333-349.

Edge, Billy L. {Chair}, W.F. Baird, J.M. Caldwell, V. Fairweather, O.T. Magoon, D. Treadwell, *Failure of the Breakwater at Port Sines, Portugal*, ASCE, 1982, 278 pp.

Edge, Billy L., Editor, *ICCE Proceedings*, ASCE, 1980-present.

Einstein, Hans Albert, "Movement of Beach Sands by Water Waves," *Trans. Amer. Geophys. Union*, 29(5), Oct. 1948, 653-655.

Einstein, H.A. and R.A. Fuchs, "Computation of Tides and Tidal Currents - United States Practice," *Proc. ASCE*, 81(Sep. 715), June 1955, 715-1 to 715-17.

Einstein, H.A., "A Basic Description of Sediment Transport on Beaches," *Waves on Beaches and Resulting Sediment Transport*, Academic Press, Ny, NY, 1972, 53-93.

Elghamry, Osman A., "Uplift Forces on Platform Decks," *Preprints: 1971 Offshore Technology Conf., Houston, TX*, I(OTC 1381), 1971, I-537 to I-548; based on his 1963 Ph.D. thesis at UCB.

Escoffier, F.F., "The Stability of Tidal Inlets," *Shore and Beach*, 8(4), Oct. 1940, 114-115.

Escoffier, Francis F., *Hydraulics and Stability of Tidal Inlets*, USACE, CERC, GITI Rept. 13, 1977.

Escoffier, Francis F. and Todd L. Walton, Jr., "Inlet Stability Solutions for Tributary Inflow," *Jour. Waterway, Port, Coastal and Ocean Div., Proc. ASCE*, 95(WW 4), Nov. 1979, 341-355.

Fairbridge, R.W. and W.S. Newman, "Postglacial Coastal Subsidence of the New York Area," *Zeitschr. f. Geomorph., N.F.*, 12(3), 1968, 296-317.

Fairchild, J.C., "Model Study of Wave Set-up Induced by Hurricane Waves at Narragansett Pier, Rhode Island," *Bull. BEB*, 12, 1958.

Fairchild, J.C., *Suspended Sampling in Laboratory Wave Action*, USACE, BEB, TM 115, 1959

Fan, L-N. and B. LeMehaute, *Coastal Movable Bed Scale Model Technology*, Rept. TC-131, Tetra Tech, Inc., Pasadena, CA, 1969.

Farley, P.P., "Coney Island Public Beach and Boardwalk Improvement," Paper 136, *The Municipal Engineers Journal*, 9(4), 1923, 136.1-136.32, with discussions.

Federal Writers' Project, *New England Hurricane*, Federal Works Administration, Hale, Cushman and Flint, 1938

FEMA, *Assessment of Current Procedures Used for the Identification of Coastal High Hazard Areas (V Zones)*, Office of Risk Assessment (ORA), Federal Insurance Administration (FIA), Sept. 1986, 99 pp and appendix.

FEMA, *Basis of Assessment Procedures for Dune Erosion in Coastal Flood Insurance Studies*, prepared by Dewberry and Davis, Jan. 1989, 57 pp and appendices.

FEMA, *National Flood Insurance Program (Regulations for Floodplain Management and Flood Hazard Identification), Index 44 CFR, Revised, 1 Oct. 1990*, 245-381.

FEMA, *Converting the National Flood Insurance Program to the North American Vertical Datum of 1988. Guidelines for Community Officials, Engineers and Surveyors*, FIA-20, June 1992, 21 pp and appendix.

Fenneman, M.M., "Development of the Profile of Equilibrium of the Subaqueous Shore Terrace," *Jour. of Geology*, 10(1), Jan.-Feb. 1902, 1-31.

Fenton, J.D. and McKee, W.D., "On Calculating the Lengths of Water Waves," *Coastal Engrg.*, 14, 1990, 499-513.

Fischer, Hugo B., E. John List, Robert C.Y. Koh, Jorg Imberger, and Norman H. Brooks, *Mixing in Inland and Coastal Waters*, Academic Press, Inc., 1979, 481 pp.

Flick, Reinhard E., "The Myth and Reality of Southern California Beaches," *Shore & Beach*, 61(3), July 1993, 3-13.

Foda, Mostafa A., "Sideband Damping of Water Waves Over A Soft Bed", *Jour. Fluid Mech.*, 201, 1989, 189-210.

Foda, Mostafa A., Jo Y.-H. Chang and Adrian W.-K. Law, "Wave-induced Breakout of Half-buried Marine Pipes," *Jour. of Waterway, Port, Coastal and Ocean Engrg.*, 116(2), March/April 1990, 267-286.

Foda, Mostafa A., James R. Hunt and Hsien-Ter Chou, "A Nonlinear Model for the Fluidization of Marine Mud by Waves," *Jour. Geophys. Res.*, 98(4), 15 April 1993, 7039-7047.

Folsom, R.G., "Measurements of Ocean Waves," *Trans. Amer. Geophys. Union*, 30(5), Oct. 1949, 691-699.

French, J.A., "Wave Uplift Pressures on Horizontal Platforms," *Civil Eng. in the Oceans IV*, San Francisco, CA, Sept. 10-12, 1979, ASCE, I, 187-202.

Gaillard, D.D., *Wave Action in Relation to Engineering Structures*, The Engineer School, USACE, 1904, reprinted in 1935 and in 1945, 218 pp.

Galvin, C.J., Jr., "Longshore Current Velocity: A Review of Theory and Data," *Reviews of Geophysics*, 5(3), 1967

Galvin, C.J., Jr., "Breaker Type Classification on Three Laboratory Beaches," *Jour. Geophys.* Res., 73(12), 15 June 1968, 3651-3659.

Galvin, C.J., Jr. "Breaker Travel and Choice of Design Wave Height," *Jour. Waterways and Harbors Div., Proc. ASCE*, 95(WW 2), May 1969, 175-200.

Galvin, Cyril J., Jr.,"A Gross Longshore Transport Rate Formula," *Proc. 13th Coastal Engrg. Conf., July 10-14, 1972, Vancouver, B.C., Canada*, ASCE, 953-970.

Galvin, Cyril J., Jr., "Wave Breaking in Shallow Water," *Waves on Beaches and Resulting Sediment Transport*, Academic Press, Inc., 1972, 413-456.

Galvin, Cyril, Jr., "Shore & Beach Observations. Transformation of Swell Over a Reef: Solitons," *Shore & Beach*, 58(3), July 1990, 31.

Galvin, Cyril, Jr. "Longshore Currents in Two Laboratory Studies: Relevance to Theory," *Jour. Waterway, Port, Coastal and Ocean Engrg.*, 117(1), Jan./Feb. 1991, 44-59.

Garbisch, E.W., Jr., P.B. Woller and R.J. McCallum, *Salt Marsh Establishment and Development*, USACE, CERC, TM 52, 1975.

Garcia, Andrew W. and James R. Houston, *Type 16 Flood Insurance Study: Tsunami Predictions for Monterey and San Francisco Bays and Puget Sound*, prepared for Federal Insurance Adm., Wash. D.C., USACE WES, Tech. Rept. H-75-17, Nov. 1975, 263 pp.

Garcia, W.J., Jr., *A Study of Water Waves Generated by Tectonic Displacements*, Ph.D. thesis, UCB, Dept. of Civil Engrg., 1972, 114 pp.

Garrison, C.J., "A Review of Drag and Inertia Forces on Circular Cylinders," *Proc. 1980 Offshore Tech. Conf., May 5-8, 1980, Houston, TX*, II (OTC 3760) 205-218.

Garrison, C.J., "Drag and Inertia Forces on Circular Cylinders in Harmonic Flow," *Jour. Waterway, Port, Coastal and Ocean Engrg.*, ASCE, 116(2), March/April 1990, 169-190.

Godden, W.G., "Breakwater Modeling Problems," *Recent Developments in Ocean Engrg., 24-27 Feb. 1981*, a short course, UCB, Engrg. Extension, preprint, 16pp.

Goodwin, Peter and Philip B. Williams, "Short-term Characteristics of Coastal Lagoon Entrances in California," *Coastal Sediments '91, Seattle, WA, June 25-27, 1991*, ed. by Nicholas C. Kraus et al., ASCE, I, 1192-1206.

Gosink, J.P., "Thermal Front Formation from Buoyant Jets," *Jour. Hyd. Div., Proc. ASCE*, 108(HY2), Feb.

1982, 252-257.

Gower, Jim and David Jones, "Canadian West Coast Giant Waves", *Mariners Weather Log*, 38(2), Spring 1994, 4-8.

Graber, Peter H.F., "The Law of the Sea in a Clamshell", in 26 parts, *Shore & Beach*, from October 1980 and the last in April 1989.

Graber, Peter H.F., "The Law of the Sea in a Clamshell. Part IV: The Florida Approach," *Shore & Beach*, 49(3), July 1981, 13-20.

Grace, R.A., *Marine Outfall Systems: Planning, Design and Construction*, Prentice-Hall, Inc., Englewood Cliffs, NJ, 1978.

Granthem, K.N., "Wave Run-up on Sloping Structures," *Trans. AGU*, 34(5), 1953.

Grosskopf, William G. and Nicholas C. Kraus, "Guidelines for Surveying Beach Nourishment Projects," *Shore & Beach*, 62(2), April 1994, 9-16.

Guza, Robert T. and Russ E. Davis, "Excitation of Edge Waves by Waves Incidence on a Beach," *Jour. Geophys. Res.*, 79(9), March 1974.

Guza, Robert T. and D.L. Inman, "Edge Waves and Beach Cusps," *Jour. Geophys. Res.*, 80(21), 20 July 1975, 2997-3012.

Guza, R.T., and E.B. Thornton, "Wave Set-up on a Natural Beach," *Jour. Geophy. Res.*, 86(C5), 20 May 1981, 4133-4137.

Habel, John S., "Ocean Wave Statistics for the California Coast," *Shore & Beach*, 45(3), July 1977, 3-9.

Habel, John S., *Shoreline Subsidence and Sand Loss*, Internal Rept. , CA State DNOD, 1978, 5 pp and attach.

Hafercorn, H.E., *Sand Movement, Beaches and Kindred Subjects, A Bibliography*, USACE, The Engineer School, Fort Humphrey, VA, 1929

Hale, Richard K., "Shore Protective Work at Winthrop, Massachusetts," *Shore and Beach*, 6(3), July 1938, 92-95.

Hale, Richard K., "Effect of the Hurricane on Massachusetts Shores," *Shore and Beach*, 7(1), Jan. 1939, 37-40.

Hales, Lyndell Z., *Floating Breakwaters: State-of-the-Art Literature Review*, USACE, CERC, TR No. 81-1, 1981.

Hall, J.V., Jr., *Test of Nourishment of the Shore by Offshore Deposition of Sand*, USACE, BEB, TM 17, 1950.

Hall, J.V., Jr., *The Rayleigh Disk as a Wave Direction Indicator*, USACE, BEB, TM 18, 1950.

Hall, J.V., Jr., "Artificially Nourished and Constructed Beaches," *Proc. 3rd Conf. on Coastal Engrg., Cambridge, Mass., Oct. 1952*, Council on Wave Res., The Engrg. Found., 1953, 119-136. Also, USACE BEB Tech. Memo. No. 29, Dec. 1952.

Hall, M.A., *Laboratory Study of Breaking Wave Force*, Tech. Rept.: Series 5, Issue 2, Inst. Eng. Res., UCB, Jan. 1957; also BEB Tech. Memo. No. 106, USACE, 1958.

Hallermeier, R.J., *Seaward Limit of Significant Sand Transport by Waves*, USACE CERC CETA 81-2, 1981.

Hallermeier, R.J., "Sand Transport Limits in Coastal Structure Design," *Coastal Structures '83, 9-11 March 1983, Arlington, VA*, ASCE, 703-716.

Hammack, Joseph L. and Harvey Segur, "The Korteweg-deVries Equation and Water Waves. Part 2., Comparison with Experiments," *Jour. Fluid Mech.*, 65 (2), 1974, 289-314.

Hammack, Joseph L. and Harvey Segur, "The Korteweg-deVries Equation and Water Waves. Part 3. Oscillatory Waves, " *Jour. Fluid Mech.*, 84(2), 1978, 337-358.

Handin, J.W., *The Source, Transportation and Deposition of Beach Sand in Southern California*, USACE, BEB, Tech. Memo. 22, March 1951, 125 pp.

Hands, E.B., "Unprecedented Migration of a Submerged Mound off the Alabama Coast," *Proc. 12th Annual Conf. of the Western Dredging Assoc. and the 24th Annual Texas A & M Dredging Seminar, 15-17 May 1991, Las Vegas, NV*, 20 pp.

Hands, E.B., and M.C. Allison, "Mound Migration in Deeper Water and Methods of Categorizing Active and Stable Berms," *Proc. Coastal Sediments '91, Seattle, WA, 25-27 June 1991*, ASCE, 1991, 1985-1999.

Hanson, H. and N.C. Kraus, *GENESIS: Generalized Model for Simulating Shoreline Change, Report 1,*

Tech. Ref., USACE CERC, Tech. Rept. CERC-89-19, Dec. 1989.

Harleman, Donald R. and A.T. Ippen, *Two-dimensional Aspects of Salinity Intrusion in Estuaries: Analysis of Salinity and Velocity Distributions*, USACE, CTH, Tech. Bull. No. 13, 1967.

Harms, Volker W., *Diffraction of Water Waves by Cylindrical Structures of Arbitrary Shape*, Ph.D. thesis, UCB Dept. Civil Engrg., Dec. 1975, 339 pp.

Harms, V.W. and J.J. Westerink, *Wave Transmission and Mooring Force Characteristics of Pipe-Tire Floating Breakwaters*, Lawrence Berkeley Lab., UCB, Rept. No. LBL-11778, 1980.

Harms, Volker W., "Floating Breakwater Performance Comparison," *Proc. 17th Coastal Engrg. Conf., March 23-28, 1980, Sydney, Australia*, ASCE, III, 1981, 2137-2158.

Harris, D.L., *Characteristics of the Hurricane Surge*, US Weather Bureau Tech. Paper No. 48, 1963.

Harris, D.L., *Tides and Tidal Datums in the United States*, Special Rept. No. 7, USACE, CERC, 1981.

Harris, R.A., *Manual of Tides*, USC&GS, Report(s) 1894, 1897, 1900, 1904 and 1907, Wash. D.C.

Harrison, W. and W.S. Wilson, *Development of a Method for Numerical Computation of Wave Refraction*, USACE, CERC, TM 6, 1964

Haurwitz, B., *The Slope of Lake Surfaces Under Variable Wind Stresses*, USACE BEB TM 25, 1951.

Heidemann, John C., Odd A. Olsen and Per I. Johansson, "Local Wave Force Coefficients," *Civil Engrg. in the Oceans IV*, ASCE, II, 1979, 684-699.

Herbers, T.H.C., Steve Elgar, R.T. Guza and W.C. O'Reilly, "Infragravity-Frequency (0.005-0.05 Hz) Motions on the Shelf. Part II: Free Waves", *Jour. Physical Oceanography*, 25 (6,I), June 1995, 1063-1079.

Herrmann, Frank A., Jr. and Joseph V. Letter, Jr., "Advancement in Tidal Hydraulics", in *50th Anniversary of the Hydraulics Division, 1938-1988*, A.M. Alsaffer {ed.}, ASCE, 1990, 36-80.

Herron, William J. and Robert L. Harris, "Littoral Bypassing and Beach Restoration in the Vicinity of Port Hueneme, California," *Proc. 10th Conf. on Coastal Engrg., Tokyo, Japan, Sept. 1966*, ASCE, I, 651-675.

Herron, William J., "The Influence of Man Upon the Shoreline of Southern California," *Shore & Beach*, 51(3), July 1983, 17-27.

Hicks, S.D., "Long Period Sea Level Variations for the United States Through 1978," *Shore & Beach*, 49(2), April 1981, 26-29.

Hicks, S.D., H.A. Debaugh, Jr., and L.E. Hickman, Jr., *Sea Level Variations for the United States, 1955-1980*, NOAA, 1983, 170 pp.

Hilaly, Nabil, "Water Waves Over a Rectangular Channel Through a Reef," *Jour. Waterways and Harbor Div., Proc. ASCE*, 95(WW1), Feb. 1969, 77-94.

Horikawa, Kiyoshi, and R.L. Wiegel, *Secondary Wave Crest Formation*, UCB, College Engrg., Tech. Rept. Ser. 89(4), Feb. 1959, 62 pp.

Horonjeff, R., H.B. Seed, C.J. Van Til, R.L. Wiegel and P,D. Trask, *A Review and Evaluation of Research Related to Trafficability of Beaches*, UCB IER Rept. 59-1, prepared for the U.S. Naval Research and Evaluation Laboratory, Bu. Yards and Docks, USN July 1953, 79 pp.

Houston, James R. and Andrew W. Garcia, *Tsunami Elevation Predictions from Numerical Models*, ASCE Fall Convention and Exhibition, San Francisco, CA, 17-21 Oct. 1977, Preprint No. 2996, 25 pp.

Houston, James R. and Andrew W. Garcia, *Type 16 Flood Insurance Study: Tsunami Predictions for the West Coast of the Continental United States*, USACE WES, Tech. Rept. H-78-26, Dec. 1978, 69 pp.

Houston, James R., "Coastal Forum II. Beach Nourishment," *Shore & Beach*, 63(1), Jan. 1995, pp 21-24.

Howell, Gary L., "Measurements of Forces on Dolos Armor Units at Prototype Scale," *21st Coastal Engrg. Conf. Proc. Inter. Conf., 20-25 June 1988, Costa del Sol - Malaga, Spain*, ASCE, 3, 1989, 2355-2369.

Hubertz, J.M., J.B. Payne and P.D. Farrar, *Hindcasting Swell from the Southern Ocean Along the U.S. Pacific Coast*, USACE CERC, WIS Rept. DRAFT, 1994, 41 pp and appendix.

Hudson, R.Y., *Model Study of Wave and Surge Action, Naval Operating Base, Terminal Island, San Pedro Harbor, California*, USACE, WES, TM 2-237, 1947.

Hudson, R.Y., "Wave Forces on Breakwaters," *Trans. ASCE*, 118, 1953, 653-674.

Hudson, R.Y., "Laboratory Investigation of Rubble Mound Structures", *Jour. Waterways & Harbors Div., Proc. ASCE*, 85(WW3), 1959.

Hudson, Robert Y., Frank A. Herrmann, Jr., Richard A. Sager, Robert W. Whalin, Garbis H. Keulegan,

Claude E. Chatham, Jr., and Lyndell Z. Hales, *Coastal Hydraulic Models*, USACE CERC, Spec. Rept. No. 5, May 1979, 531 pp.
Hunt, Ira A. Jr., "Design of Seawalls and Breakwaters," *Jour. Waterways and Harbors Div., Proc. ASCE*, 85(WW3), Sept. 1959, 123-151.
Hunt, Ira A., Jr., *Winds, Wind Set-up, and Seiches on Lake Erie*, US Lake Survey, USACE, 1959.
Hunt, John N., "Direct Solution of Wave Dispersion Equation," *Jour. Waterway, Port, Coastal and Ocean Engrg. Div., Proc. ASCE*, 105(WW4), Nov. 1979, 457-459.
Huston, John , *Hydraulic Dredging*, Cornell Maritime Press, Inc., 1970, 318 pp.
Huval, C.J. and G.L. Wintergerst, *Simplified Numerical (Lumped Parameter) Simulation. Appendix 4, Comparison of Numerical and Physical Hydraulic Models, Masonboro Inlet, North Carolina*, USACE, CERC GITI Rept. 6, June 1977, 115 pp.
Hwang, Li-San and David Divoky, "Breaking Wave Setup and Decay on Gentle Slopes," *Proc. 12th Coastal Engrg. Conf., Sept. 13-18, 1970, Wash. D.C.*, ASCE,I, 377-389.
Hwang, L.-S., H.L. Butler, and D. Divoky, "Tsunami Model: Generation and Open-sea Characteristics," *Bull. Seismological Soc. Amer.*, 62(6), 1972, 1579-1596.
Iida, Kumizi, Doak C. Cox, and George Pararas-Carayannis, *Preliminary Catalog of Tsunamis Occurring in the Pacific Ocean*, Univ. Hawaii, Hawaii Inst. of Geophysics, HIG-67-10, Aug. 1967, 2-vols.
Inman, D.L. and G.S. Rusnak, *Changes in Sand Level on the Beach and Shelf at La Jolla, California*, USACE BEB, Tech. Memo. No. 82, July 1956, 30 pp and appendices.
Inman, D.L. and T.K. Chamberlain, "Tracing Sand Movement with Irradiated Quartz," *Jour. Geophys. Res.*, 64(1), Jan. 1959, 41-47.
Inman, Douglas L. and R.A. Bagnold, "Littoral Processes," *The Sea: Ideas and Observations*, Vol.3, Interscience Publishers, NY, 1963, 529-553.
Inman, Douglas L. and Jeffrey D. Frautschy, "Littoral Processes and the Development of Shorelines," *Coastal Engrg.: Santa Barbara Specialty Conf., Oct. 1965*, ASCE, 591-593.
Inman, Douglas L., "Accretion and Erosion Waves on Beaches," *Shore & Beach*, 55(3-4), July-Oct. 1987, 61-66.
Inman, D.L., M.H.S. Elwany and S.A. Jenkins, "Shorerise and Bar-berm Beach Profiles on Ocean Beaches," *Jour. Geophys. Res.*, 98(C10), 1993, 18181-18199.
Institution of Civil Engineers, *The Civil Engineer in War. A Symposium of Papers on War-Time Engineering Problems, Vol. 2. Docks and Harbours*, London, U.K., 1948, 450 pp.
Ippen, Arthur T.(and eight coauthors), *Estuary and Coastal Hydrodynamics*, McGraw-Hill, Inc., 1966, 744 pp.
Isaacs, John D., *Report on Beach Survey with Sea Sled and Mast at Pismo Beach,California*, UCB, College of Engrg., Contract NObs-16290, Tech. Rept. HE 116-135, 12 July 1945, 2 pp.
Isaacs, J.D., *Field Report of the Tsunami of April 1, 1946*, UCB, Dept. Engrg., May 3, 1946.
Isaacs, John D. and Thorndike Saville, Jr., "A Comparison Between Recorded and Forecast Waves on the Pacific Coast", *Ocean Surface Waves: Annals of the New York Acad. Sci.*, 51(3), May 1949, 502-510.
Isaacs, J.D., E.A. Williams and C. Eckart, "Reflection of Surface Waves by Deep Water," *Trans. Amer. Geophys. Union*, 32(1), Feb. 1951, 37-40.
Iversen, H.W., "Laboratory Study of Breakers," *Proc. U.S. National Bureau of Standards Symposium on Gravity Waves*, NBS Circular 521, Nov. 1952, 9-32.
Iversen, H.W., "Waves and Beaches in Shoaling Water," *Proc. 3rd Conf. on Coastal Engrg., Cambridge, MA, Oct. 1952*, Council on Wave Res., The Engrg. Found., 1953, 1-12.
Jagschitz, J.A. and R.S. Bell, *Restoration and Retention of Coastal Dunes with Fences and Vegetation*, Bull. 382, RI Agricultural Expt. Station, Kingston, RI, 1966a.
Jagschitz, J.A. and R.S. Bell, *American Beach Grass - Establishment, Fertilization and Seeding*, Bull. 383, RI Agricultural Expt. Station, Kingston, RI, 1966b.
James, W.R., "Beach Fill Stability and Borrow Material Texture," *Proc. 14th ICCE, Copenhagen, Denmark, 1974*, ASCE, 1975
Jarrett, J.T., *Tidal Prism - Inlet Area Relationships*, USACE, CERC GITI Rept. No. 3, 1976, 32 pp.
Jelesnianski, Chester P., *SPLASH (Special Program to List Amplitudes of Surges from Hurricanes), Part I - Landfall Storms*, NOAA Tech. Memo. NWS TDL-46, 1972.
Jelesnianski, Chester P., *SPLASH (Special Program to List Amplitudes of Surges from Hurricanes), Part*

II - General Track and Variant Storm Conditions, NOAA Tech. Memo. NWS TDL-52, March 1974.
Jen, Y., R.L. Wiegel, and I. Mobarek, "Surface Discharge of Horizontal Warm Water Jet", *Jour. Power Div., Proc. ASCE*, 92(PO2), April 1966, 1-30.
Jensen, R.E., J.M. Hubertz, E.F. Thompson, R.D. Reinhard, B.J. Borup, W.A. Brandon, J.B. Payne, R.M. Brooks, D.S. McAney, *Southern California Hindcast Wave Information*, USACE CERC WIS Rept. 20, Dec. 1992, 249 pp
Johnson, A.G., "Santa Monica Bay Shoreline Development Plans," *Proc. of 1st Conf. on Coastal Engrg., Long Beach, Calif., Oct. 1950*, ed. by J.W. Johnson, Council on Wave Res., The Engrg. Found., 1951, 271-276.
Johnson, Douglas W., *Shore Processes and Shoreline Development*, Wiley and Sons, New York, NY, 1919 {reprinted 1938}.
Johnson, Douglas W., *Studies of Mean Sea Level*, Bull. No. 70, National Acad. Press, Wash. D.C., 1929.
Johnson, J.W., "The Refraction of Surface Waves by Currents," *Trans. Amer. Geophys. Union*, 28(6), Dec. 1947, 867-874.
Johnson, J.W., M.P. O'Brien, and J.D. Isaacs, *Graphical Construction of Wave Refraction Diagrams*, USN Hydrographic Office, H.O. Pub. No. 605, Jan. 1948, 45 pp.
Johnson, J.W. and J.D. Isaacs, "Action and Effect of Waves," *Western Construction News*, April 1948.
Johnson, J.W. (ed.), *Proceedings of First Conference on Coastal Engineering, Long Beach, California, October 1950*, Council on Wave Research, The Engrg. Foundation, 1951, 334 pp.
Johnson, J.W., Editor, *Coastal Engrg. Proc. (incl. ICCEs)*, 1950-1978.
Johnson, J.W., R.A. Fuchs and J.R. Morison, "The Damping Action of Submerged Breakwaters," *Trans. Amer. Geophys. Union*, 32(5), Oct. 1951, 704-718.
Johnson, J.W., "Sand Transport by Littoral Currents," *Proc. 5th Hyd. Conf., State Univ. of Iowa*, Studies in Engrg. Bull. 34, June 1952.
Johnson, J.W., "The Littoral Drift Problem at Shoreline Harbors," *Trans. ASCE*, 124, 1959, 525-546.
Johnson, J.W., "The Supply and Loss of Sand to the Coast," *Jour. Waterways and Harbors Div., Proc. ASCE*, 85(WW3), Sept. 1959, 227-251.
Johnson, J.W., "The Significance of Seasonal Beach Changes in Tidal Boundaries," *Shore and Beach*, 39(1), April 1971, 26-31.
Johnson, Joe W., "Characteristics and Behavior of Pacific Coast Tidal Inlets," *Jour. Waterways, Harbors, and Coastal Engrg. Div., Proc. ASCE*, 99(WW3), Aug. 1973, 325-339.
Johnson, J.W., "History of Some Aspects of Modern Coastal Engineering," *Proc. of 14th Coastal Engrg. Conf., June 24-28, 1974, Denmark*, ASCE, I, 1975, 21-44.
Johnson, J.W., "A Bit of History on Some Important Contributions to Coastal Engrg. by Morrough P. O'Brien," *Shore & Beach*, 55(3-4), July/Oct. 1987, 15-18.
Johnson, J.W., "The Beginning of the Coastal Engineering Program at Berkeley", *Shore & Beach*, 62(3), July 1994, 9-10.
Johnstone, Homer, "California Surfing," *Shore and Beach*, 32(1), April 1964, 30-36.
Kalkanis, George, *Transport of Bed Material Due to Wave Action*, USACE CERC, Tech. Memo. No. 2, Feb. 1964, 38 pp and app.
Kamel, Adel M., "Shock Pressure on Coastal Structures," *Jour. Waterway, Harbor, and Coastal Eng. Div., Proc. ASCE*, 96(WW3), Aug. 1970, 689-699; "Closure," 98(WW2), May 1972, 266-271.
Kaplan, K., *Analysis of Moving Fetches for Wave Forecasting*, USACE, BEB, TM No. 35, 1953.
Kaplan, K., *Generalized Laboratory Study of Tsunami Run-up*, USACE, BEB, TM 60, 1955
Kaplan, Paul and Mark N. Silbert, "Impact Forces on Platform Horizontal Members in the Splash Zone," *Proc. 1976 Offshore Tech. Conf., May 3-6, Houston, TX*, l(2498), 749-758.
Keller, J.B., "Tsunamis - Water Waves Produced by Earthquakes," *Proc. Tsunami Meetings, 10th Pacific Science Congress, Aug.-Sept. 1961, Honolulu, HI*, IUGG Monograph No. 24, Paris, France, 1961, 154-166.
Kelly, John, *Surf Parameters, Final Report. Part II. Social and Historical Dimensions*, Univ. Hawaii, Look Lab., Tech. Rept. 73-33, 251 pp, Nov. 1973.
Kendall, Thomas R. and Jeffrey A. Melby, "Movement and Static Stress in Dolosse: Six Years of Field Monitoring at Crescent City," *Proc. of the 23rd Inter. Conf. on Coastal Engrg., Oct. 4-9 1992, Venice, Italy*, ASCE, 1285-1298.

Keulegan, G.H., "Discussion of 'Model Law for Motion of Salt Water Through Fresh,'" by Morrough P. O'Brien and John Cherno, *Trans. ASCE*, 99, 1934, 602-606.

Keulegan, G.H., *Depths of Offshore Bars*, USACE, BEB, TM 8, 1945.

Keulegan, G.H. and W.C. Krumbein, "Stable Configuration of Bottom Slope in Shallow Water and Its Bearing on Geological Process," *Trans. Amer. Geophys. Union*, 30(6), 1949.

Keulegan, Garbis H. and J.V. Hall, Jr., "A Formula for the Calculation of the Tidal Discharge through an Inlet," *Bull. of the BEB*, USACE, 4(1), Jan. 1, 1950, 15-29.

Keulegan, Garbis, H., *Tidal Flow in Estuaries. Water-level Fluctuations of Basins in Communication With Seas*, U.S. NBS Rept. 1146, 10 Sept. 1951, 32 pp.

Keulegan, Garbis H., "Hydrodynamic Evaluation of Storms on Lake Erie," *Proc. 2nd Conf. on Coastal Engrg., Houston, Texas, Nov. 1951*, Council on Wave Res., The Engrg. Found., 1951.

Keulegan, Garbis H., "Wind Tides in Small Closed Channels," *Jour. Research, National Bu. Standards*, RP2207 46(5), May 1951, 358-381.

Keulegan, Garbis H., *First through 13th Progress Reports on Model Laws for Density Currents*, U.S. National Bu. Standards, Hyd. Lab., Wash. D.C., April 1946 through 1 April 1958.

Keulegan, Garbis H. and Lloyd H. Carpenter, "Forces on Cylinders and Plates in an Oscillating Fluid," *Jour. of Res.*, NBS, 60(5), May 1958, 423-440.

Keulegan, Garbis H., *Tidal Flow in Entrances: Water-Level Fluctuations of Basins in Communication with Seas*, Tech. Bull. No. 14, USACE, CTH, 1967.

Kinsman, Blair, *Wind Waves: Their Generation and Propagation on the Ocean Surface*, Prentice-Hall, Inc., 1965, 676 pp.

Klebba, A.A., "Details of Shore-based Wave Recorder and Ocean Wave Analyzer," *Ocean Surface Waves, Annals of the New York Acad. Sci.*, 51(3), 13 May 1949, 533-544.

Knapp, R.T., *Determination of Wave, Surge, and Ship Motion, U.S. Naval Station, Long Beach, Calif.*, CIT, Pasadena, CA, Final Rept. of Contract Noy-13116, Bu. Yards and Docks, USN, May 1951, 96 pp.

Knutson, P.L., "The Use of Dredged Material for the Development of Inter-tidal Marshlands", *Proc. MTS/EEE Ocean '75 Conf., San Diego, CA, 1975.*

Knutson, P.L. and W.W. Woodhouse, *Shore Stabilization with Salt Marsh Development*, USACE, CERC, Special Rept. No. 9, 1983.

Kobayashi, K., "Prediction of Wave Run-up and Rip-rap Stability," *Proc. 20th Coastal Engrg. Conf., Nov. 1986, Taipei, Taiwan, ROC*, ASCE, 1987, 1958-1971.

Kolessar, Michael A. and J.L. Reynolds, "The Sears Sea Sled for Surveying in the Surf Zone", *Bulletin and Summary Report of Research Progress, 1965-66*, Vol. II, USACE CERC, 1966.

Komar, Paul D. and Douglas L. Inman, "Longshore Sand Transport on Beaches," *Jour. of Geophys. Res.*, 75(30), 20 Oct. 1970, 5914-5927.

Komar, Paul D. and M.C. Miller, "Sediment Threshold Under Oscillatory Waves," *Proc. 14th Coastal Engrg. Conf., 24-28 June 1974, Copenhagen, Denmark*, ASCE, 2, 1975, 756-775.

Komar, Paul D., *Beach Processes and Sedimentation*, Prentice-Hall, Inc., 1976, 429 pp.

Komar, Paul D., "Beach Profiles on the Oregon and Washington Coasts Obtained with an Amphibious DUKW," *Shore & Beach*, 46(3), July 1978, 27-33.

Komar, P.D. and W.G. McDougal, "The Analysis of Exponential Beach Profiles," *Jour. Coastal Res.*, 10(1), 1993, 59-69.

Kranzer, H.C. and J.B. Keller, *Water Waves Produced by Explosions*, New York Univ., Inst. Math. Sci., IMM-NYU-222, Sept. 1955.

Kraus, Nicholas C. (ed.), *Coastal Sediments '87, New Orleans, LA, 12-14 May 1987*, ASCE, 2 v., 2177 pp.

Kraus, Nicholas C. and M. Larson, Beach Profile Change Measured in the Tank for Large Waves, 1956-57 and 1962, Tech. Rept. CERC 88-6, USACE, 1988.

Kraus, Nicholas C. (ed., *Coastal Sediments '91, Seattle, WA, 25-27 June 1991*, ASCE, 2 v.

Kraus, N.C., M. Larson and D.L. Kriebel, "Evaluation of Beach Erosion and Accretion Predictors," *Proc. Coastal Sediments '91, Seattle, WA, 25-27 June 1991*, ASCE, 572-587.

Kraus, Nicholas C. (guest ed.), *Shore & Beach*, 61(1), Jan. 1993, 44 pp.

Kraus, Nicholas C. and Randall A. Wise, "Simulation of January 4, 1992 Storm Erosion at Atlantic City, Maryland," *Shore & Beach*, 61(1), Jan. 1993, 34-41.

Kravtchenko, Julien and John S. McNown, "Seiche in Rectangular Ports," *Quarterly of Applied Math.*, 13 (1), April 1955, 19-26.

Kriebel, D.L. and Robert G. Dean, "Numerical Simulation of Time-dependent Beach and Dune Erosion," *Coastal Engrg.*, 9, 1985, 221-245.

Kriebel, D.L., "Verification Study of a Dune Erosion Model," *Shore & Beach*, 54(3), July 1986, 13-21.

Krumbein, W.C., *Shore Currents and Sand Movement on a Model Beach*, USACE, BEB, TM 7, 1944.

Krumbein, W.C., *Statistical Significance of Beach Sampling Methods*, USACE, BEB, TM 50, 1954

Krumbein, W.C., *A Method for Specification of Sand for Beach Fills*, USACE, BEB, TM 102, 1957

Krumbein, W.C. and W.R. James, *A Lognormal Size Distribution Model for Estimating Stability of Beach Fill Material*, USACE, CERC TM 161, 1965

LaFond, Eugene, "Sand Movements Near the Beach in Relation to Tides and Waves," *Proc. 6th Pacific Sci. Congress*, 1939, 795-799.

Laitone, E.V., "The Second Approximation to Cnoidal and Solitary Waves," *Jour. Fluid Mech.*, 9, 1960, 430-444.

Laitone, E.V., *Higher Approximations to Non-linear Water Waves and the Limiting Heights of Cnoidal, Solitary and Stokes Waves*, UASCE BEB, Tech. Memo. No. 133, Feb. 1963, 106pp.

Langley, T.B., "Sea Sled Surveying Through the Surf Zone," *Shore & Beach*, 60(2), April 1992, 16-19.

Lapsley, William W., *Sand Movement and Beach Erosion*, M.S. thesis in Civil Engrg., UCB, 1937.

Larson, M. and N.C. Kraus, *SBEACH: Numerical Model for Simulating Storm-Induced Beach Change. Report 1: Empirical Foundation and Model Development*, USACE, CERC, Tech. Rept. No. CERC-89-9, 1989.

Larson, M., N.C. Kraus, and Mark R. Byrnes, *SBEACH: Numerical Model for Simulating Storm-Induced Beach Change. Report 2, Numerical Foundation and Model Tests*, USACE, CERC, Tech. Rept. No. CERC-89-9, 1990.

Lawton, C.H., "Santa Barbara Beach," *Shore and Beach*, 4(2), April 1936, 63.

Leatherman, S.P., "Shoreline Mapping: A Comparison of Techniques," *Shore & Beach*, 51(3), July 1983, 28-33.

Lee, J., T. Yancy, M. Glogoczowski, and P. Wilde, *Recent Sediments of the Central California Continental Shelf, Pigeon Point to Sand Hills Bluffs. Part B. Mineralological Data*, UCB Hyd. Lab., Tech. Rept. HEL 2-31, July 1971, 55 pp.

Lee, Jiin-Jen, "Wave Induced Oscillations in Harbors of Arbitrary Shape," *Jour. Fluid Mech.*, 45, 1971, 375-393.

Lee, Jiin-Jen and C.P. Lai, "Wave Uplift on Platforms or Docks in Variable Depth," *20th Coastal Engrg. Conf., Proc. of the Inter. Conf., Nov. 9-14, 1986, Taipei, Taiwan*, ASCE, III, 1987, 2023-2034.

Leendertse, J.J., *Forces Induced by Breaking Water Waves on a Vertical Wall*, USN Civil Engrg. Lab., Port Hueneme, CA, Tech. Rept. 092, 9 March 1961, 80 pp.

Leendertse, J.J., *Aspects of a Computational Model for Well-mixed Estuaries and Coastal Seas*, RM 5294-PM, Rand Corp., Santa Monica, CA, 1967.

Leendertse, Jan J., Richard C. Alexander, and Shiao-Kung Liu, *A Three-Dimensional Model for Estuaries and Coastal Seas: Vol. I, Principles of Computation*, Rand Corp., R-1417-OWRR, Dec. 1973, 57 pp.

Leidersdorf, Craig B., Ricky C. Hollar, and Gregory Woodell, "Human Intervention with the Beaches of Santa Monica Bay, California," *Shore & Beach*, 62(3), July 1994, 29-38.

LeMehaute, B., "Theory of Wave Agitation in a Harbor," *Trans. ASCE*, 127, 1962, 364-383.

LeMehaute, Bernard, Robert Y.C. Koh and Li-San Hwang, "A Synthesis on Wave Run-up," *Jour. Waterways and Harbors, Proc. ASCE*, 94(WW1), Feb. 1968, 77-92.

LeMehaute, Bernard, "A Comparison of Fluvial and Coastal Similitude", *Proc. 12th Conf. of Coastal Engrg., Washington, D.C., 1970*, ASCE.

LeMehaute, Bernard, *An Introduction to Hydrodynamics and Water Waves*, Springer-Verlag, New York, 1976, 322 pp.

LeMehaute, Bernard, "Similitude in Coastal Engineering", *Jour. Waterways, Harbors and Coastal Engrg. Div., Proc. ASCE*, 102(WW3), Aug. 1976.

LeMehaute, B., J.D.Wang and C.C. Lu,"Monte Carlo Simulation of Wave Climatology for Shoreline Processes," *Proc. Conf. on Directional Wave Spectra Application, Berkeley, CA*, ASCE, 1981.

LeMehaute, B. and J.D. Wang, "Wave Spectrum Changes on a Sloped Beach," *Jour. Waterways,, Port,*

Coastal and Ocean Div., Proc. ASCE, 108(WW1), 1982, 33-47.

Lillevang, Omar J., "Coastal Power Generation," *Shore and Beach*, 34(2), Oct. 1966, 2-8.

Lipp, Morris N., "Legal Problems of Miami Beach Shore Preservation", *Shore and Beach*, 23(1), April 1955, 13-15.

Liseth, P., "Mixing of Merging Buoyant Jets from a Manifold in Stagnant Receiving Water of Uniform Density," *Proc. 6th Inter. Conf. on Advances in Water Pollution Res., Jerusalem, Israel, 18-23 June 1972*, IAWPR, Pergamon Press, 1973, 921-936.

List, E.J. and Jorg Imberger, "Turbulent Entrainment in Buoyant Jets and Plumes," *Jour. Hyd. Div., Proc. ASCE*, 99(HY9), Sept. 1973, 1461-1474.

Liu, P. L.-F. and R.A. Dalrymple, "Bottom Frictional Stresses and Longshore Currents Due to Waves with Large Angles of Incidence," *Jour. Mar. Res.*, 36(2), 1978, 357-375.

Lockett, John B., *History of the Corps of Engineers Committee on Tidal Hydraulics (January 1949 to 1971)*, USACE Committee on Tidal Hydraulics Tech. Bull. No. 18, June 1972, 9 pp and 8 appendices.

MacCamy, R.C. and R.A. Fuchs, *Wave Forces on Piles*, Tech. Memo. No. 69, BEB, USACE, Dec. 1954, 17pp.

Madsen, O.S. and S.M. White, *Reflection and Transmission Characteristics of Porous Rubble-mound Breakwaters*, USACE, CERC, Misc. Rept. No. 76-5, 1976.

Magoon, Orville T. and W.O. Sarlin, "Effect of Long-period Waves on Hydrographic Surveys," *Proc. 12th ICCE, Washington D.C., 13-18 Sept. 1970*, ASCE, II, 2251-2266.

Magoon, Orville T., Robert L. Sloan and Gary L. Foote, "Damages to Coastal Structures," *Proc. 14th Coastal Engrg. Conf., June 24-28, 1974, Copenhagen, Denmark*, ASCE, III, 1975, 1655-1676.

Magoon, Orville T. et al. (ed.), *Coastal Zone'78, San Francisco, CA, March 14-16, 1978*. Proc. of a Specialty Conf., ASCE, 4-v., 3091 pp.

Magoon, Orville T. and W.F. Baird, *Durability of Stone for Rubblemound Breakwaters*, Derived from a Workshop, Cleveland, OH, 22-23 May 1991, ASCE, 1992, 277 pp.

Magoon, Orville T. and J. Michael Hemsley, (ed.), *Ocean Wave Measurement and Analysis: Proc. of the 2nd International Symposium, New Orleans, LA, July 25-28, 1993*, ASCE, 1054 pp.

Manohar, Madhav, *Mechanics of Bottom Sediment Movement Due to Wave Action*, USACE BEB, TM 75, June 1955, 121 pp, based on Ph.D. thesis.

Marmer, H.A., *The Tide*, D. Appleton and Co., New York, and London, 1926, 282 pp.

Marine Advisers, *Design Waves for Proposed Small Craft Harbor at Dana Point, California*, prepared for USACE Los Angeles Dist., March 1960, 23 pp and appendix. Also appendix to USACE Los Angeles Dist., Sept. 1961.

Martens, James H.C., "Characteristics of Beaches Suitable for Motoring," *Shore and Beach*, 2(3), July 1934, 122-124.

Martens, James H.C., "Beach Sands Between Charleston, South Carolina, and Miami, Florida," *Bull. Geological Soc. Amer.*, 46, 31 Oct. 1935, 1563-1596 and 7 figs.

Masch, Frank D. and Robert L. Wiegel, *Cnoidal Waves: Tables of Functions*, Council on Wave Res., The Engrg. Found., Berkeley, CA, March 1961, 129 pp.

Mason, C. A.H. Sallenger, R.A. Holiman, and W.A. Birkemeier, "DUCK 82 - A Coastal Storm Process Experiment," *Nineteenth Coastal Engrg. Conf. Proc. Inter. Conf., 3-7 Sept. 1984, Houston, TX*, ASCE, II, 1985, 1913-1928.

Mason, C., W.A. Birkemeier and P.A. Howd, "Overview of DUCK 85 Nearshore Process Experiment," *Proc. Coastal Sediments'87*, ASCE, 1987, 818-833.

Mayor-Mora, Ramiro, "Hydraulics of Tidal Inlets on Sandy Coasts," *Proc. of the 14th Coastal Engrg. Conf., July 24-28, 1974, Copenhagen, Denmark*, ASCE, II, 1975, 1524-1545.

McAleer, John B. and George E. Townsend, "Hurricane Protection Planning in New England," *Jour. of the Hyd. Div., Proc. ASCE*, 84(HY4), Aug. 1958, 1-36.

McGehee, David D. and J. Michael Hemsley, "Implementing a National Wave Monitoring Network - Some Lessons and Plans," *Ocean Wave Measurement and Analysis. Proc. 2nd Inter. Symposium, Honoring Prof. Robert L. Wiegel, New Orleans, LA, 25-28 July 1993*, ASCE, 1994, 208-222.

McNown, J.S., "Waves and Seiche in Idealized Ports," *Gravity Wave Symposium, U.S. National Bureau of Standards*, NBS Circular No. 521, 1952, 153-164.

McNown, J.S. and G.H. Keulegan, "Vortex Formation and Resistance in Periodic Motion," *Jour. Engrg.*

Mech., Div., Proc. ASCE, 85(EM1), Jan. 1959, 1-6.

Mehta, Ashish J. and Prakash B. Joshi, "Tidal Inlet Hydraulics," *Jour. of Hyd. Engrg.*, ASCE, 114(11), Nov. 1988, 1321-1338.

Mei, Chiang C., *Applied Dynamics of Ocean Surface Waves*, John Wiley & Sons, 1983, 740 pp.

Melville, W.K., "On the Mach Reflexion of a Solitary Wave," *Jour. of Fluid Mech.*, 98, 1980, 285-297.

Meteorology International Inc., *Deep-water Wave Statistics for the California Coast, Station 51, Lat. 33.5 N - Long 120.4 W,* State of California, The Resources Agency, Dept. Navigation and Ocean Development (DNOD), Feb. 1977, 41 pp and numerous tables.

Meyers, Andrea and Ann L. Chester, "The Stabilization of Clatsop Plains, Oregon", *Shore & Beach*, 45(4), Oct. 1977, 34-41.

Miles, John W., "On the Generation of Surface Waves by Shear Flows," *Jour. Fluid Mech.*, 3(3), Nov. 1957, 185-204.

Miles, John W., "On the Generation of Surface Waves by Shear Flows, Part 2," *Jour. Fluid Mech.*, 6(4), 1959, 568-582.

Miles, John W., "Resonant Response of Harbors (The Harbor Paradox Revisited)," *Eighth Symposium, Naval Hydrodynamics. Hydrodynamics in the Ocean Environment, Aug. 24-28, 1970, Pasadena, CA*, USN ONR, ACR-179, 1970, 95-115.

Miles, J.W., "Diffraction of Solitary waves," *Jour. Appl. Math. Phy.*, 28, 1977, 889-902.

Miles, J.W., "Review of Dynamics and Modelling of Ocean Waves by G.J. Komen et al." *Science*, 270(5234), 13 Oct. 1995, 320.

Miller, H.C., W.A. Birkemeier, and A.E. DeWall, "Effects of CERC Research Pier on Nearshore Processes," *Coastal Structures '83, 9-11 March 1983, Arlington, VA*, ASCE, 769-784.

Minard, Claude R., Jr., *Quaternary Beaches and Coasts Between the Russian River and Drakes Bay, California*, UCB Hyd. Lab. Tech. Rept. HEL 2-35, Aug. 1971, 205 pp.

Mitchim, C.F., "Oscillatory Waves in Deep Water," *The Military Engineer*, 32(182), March-April 1940, 107-109.

Mobarek, Ismail El-Sayed, "Directional Spectra of Laboratory Wind Waves," *Jour. Waterways and Harbors Div., Proc. ASCE*, 91(WW3), Aug. 1965, 91-116.

Modeling '75. Symposium on Modeling Techniques, 3-5 Sept. 1975, San Francisco, CA, ASCE, 2-vols., 1975, 1686 pp.

Moore, Jamie W. and Dorthy P. Moore, *History of the Coastal Engineering Research Center, 1963-1983*, USACE, WES, Vol. 1: WES Laboratory History Series, 1991, 112 pp.

Morison, J.R., M.P. O'Brien, J.W. Johnson and S.A. Schaaf, "The Forces Exerted by Surface Waves on Piles," *Jour. of Petroleum Tech., Pet. Branch, Amer. Inst. of Mining and Metal. Engineers*, 2(5), May 1950, 149-154.

Moses, Robert, "Post-War Beach Problems", *Shore and Beach*, 11(2), Oct. 1943, 35-41.

Munk, W.H., "Wave Action on Structures," Amer. Inst. of Mining and Metal. Engineers, Tech. Pub. No. 2322, Class G., *Pet. Trans.*, March 1948.

Munk, Walter H., "The Solitary Wave Theory and Its Application to Surf Problems," *Ocean Surface Waves, Annals of the New York Acad. of Sci.* 51(3), May 1949, 376-424.

Munk, W.H., "Surf Beats," *Trans. Amer. Geophys. Union*, 30(6), Dec. 1949, 849-854.

Munk, W.H., F.E. Snodgrass, and George Carrier, "Edge Waves on the Continental Shelf," *Science*, 23(3187), Jan. 27, 1956, 137-132.

Munk, W.H., G.R. Miller, F.E. Snodgrass and N.F. Barber, "Directional Recording of Swell from Distant Storms," *Phil. Trans., Royal Soc. of London*, Ser. A, 255(1062), 18 April 1963, 505-584.

Munk, W.H., "Frank Snodgrass, 1920-1985," *EOS*, 66(45), 5 Nov. 1985, 753.

Murphy, G.J., J.A. Belvedere, B.J. Van Wide and P.H. Gilbert, "San Francisco Southwest Ocean Outfall - Design Considerations," *Preprints, Third Australian Tunneling Conf., Sydney, Australia, 12-15 Sept,. 1978*, 11-16.

Nath, John H., "Marine Roughened Cylinder Wave Force Coefficients," *19th Coastal Engrg. Conf., Proc. of the Inter. Conf., 3-7 Sept. 1984, Houston, TX*, ASCE, III, 1985, 2710-2724.

NAS, Committee on the Alaska Earthquake, *The Great Alaska Earthquake, 1964: Oceanographic and Coastal Engineering*, National Acad. Press, 1972, 556 pp.

NAS, NRC, *Ocean Wave Spectra: Proceedings of a Conference, Easton, Maryland, May 1-4, 1961*,

Prentice-Hall, Inc., 1963, 357 pp.

NAS, Panel on Wave Action Effects Associated with Storm Surges, *Methodology for Calculating Wave Action Effects Associated with Storm Surges*, Wash. D.C., 1977, 29 pp.

NAS/NAE/IOM/NRC, *Organization and Members, 1995, NAS/NAE/IOM/NRC*, Washington, D.C., p. 203.

National Advisory Committee on Oceans and Atmosphere (NACOA), *The Role of the Ocean in a Waste Management Strategy. A Special Report to The President and Congress*, Wash. D.C., Jan. 1981, 103 pp and appendices.

National Bureau of Standards, *Gravity Waves: Proc. of NBS Semicentennial Symposium on Gravity Waves, June 18-20, 1951*, NBS Circular 521, 28 Nov. 1952, 287 pp.

National Marine Consultants, *Wave Statistics for Seven Deep Water Stations Along the California Coast*, prepared for USACE Los Angeles Dist. and San Francisco Dist., Dec. 1960, 20 pp and extensive tables and figs.

NRC, Committee on Beach Nourishment and Protection, *Beach Nourishment and Protection*, Marine Board, National Acad. Press, Dec. 1995, 336 pp.

NRC, Committee on Coastal Engineering Measurement Systems, *Measuring and Understanding Coastal Processes for Engineering Purposes*, National Acad. Press, Wash. D.C., 1989, 119 pp.

NRC, Committee on Coastal Erosion Zone Management, *Managing Coastal Erosion*, Water Science and Technology Board and Marine Board, National Acad. Press, 1990, 182 pp.

NRC, Committee on Engrg. Implications of Changes in Relative Mean Sea Level, *Responding to Changes in Sea Level: Engrg. Implications*, MB, National Acad. Press, 1987, 148 pp.

NRC, Committee on Flooding from Hurricanes, *Evaluation of FEMA Model for Estimating Potential Coastal Flooding from Hurricanes and Its Application to Lee County, Florida*, National Acad. Press, Wash., D.C., 1983, 154 pp.

NRC, Division of Geology and Geography, *Annual Report for the Year 1927-1928*, App. C, p. 6.

NRC, Marine Board, *Marine Board Annual Report: 1994*, National Acad. Press, Wash. D.C., 1995, 59 pp.

NRC, *Measuring Ocean Waves. Proceedings of a Symposium and Workshop on Wave Measurement Technology*, National Acad. Press, Wash. D.C., 1982.

Nelson, Eric, Donald Christensen and A. David Schuldt, "Floating Breakwater Prototype Test Program," *Coastal Structures '83, 9-11 March 1983, Arlington, VA*, ASCE, 433-446.

Neumann, Gerhard, *On Ocean Wave Spectra and a New Method of Forecasting Wind-generated Sea*, USACE, BEB, Tech. Memo. No. 43, Dec. 1953, 43 pp.

Neumann, Gerhard and R.W. James, *North Atlantic Coast Wave Statistics Hindcast by the Wave Spectrum Method*, USACE, BEB, TM 57, 1955.

Neumann, Gerhard and Willard J. Pierson, "A Detailed Comparison of Wave Spectra and Wave Forecasting Methods," *Deutsche Hydrograph. Z.*, Vol. 10, Nos. 3,4, 1957, 73-92, 134-146.

Neumann, Gerhard and Willard J. Pierson, Jr., *Principles of Physical Oceanography*, Prentice-Hall, Inc., 1966, 545 pp.

Nielsen, A.H., *Diffraction of Periodic Waves Along a Vertical Breakwater for Small Angles of Incidence*, UCB IER Tech. Rept. HEL 1-2, 1962.

Noble Consultants, *Coastal Sand Management Plan, Santa Barbara/ Ventura County Coastline, CA, Main Report*, prepared for BEACON (Beach Erosion Authority for Control Operations and Nourishment), 14 July 1989, 186 pp.

Noble Consultants, Inc., *Nearshore Hydrodynamic Factors and Wave Study of the Orange County Coast. Draft*. Prepared for USACE Los Angeles District, March 1995, various pagination.

Noda, E.K., "Fourier Analysis of Transient Wave System," *Jour. WW Div., Proc. ASCE*, 97(WW4), Nov. 1971, 663-670.

Noda, E.K., "Equilibrium Beach Profile-Model Relationship," *Jour. WW Div., Proc. ASCE*, 98(WW4), 1972

Noda, E.K., "Wave-induced Nearshore Circulation," *Jour. Geophys. Res.*, 79(27), 1974, 4097-4106.

O'Brien, John T., "Forces on Moored Ships Due to Wave Action in Basins," *Conf. on Ships and Waves, Hoboken, NJ, Oct. 1954*, Council on Wave Res., The Engrg. Found. and the Soc. Naval Architects and Marine Engineers, 1955, 455-473.

O'Brien, J.T. and D.I. Kuchenreuther, *Forces Induced by Waves on the Moored U.S.S. Norton Sound*

(AUM-1), USN Civil Eng. Lab., Port Hueneme, CA, Tech. Memo. No. M-129, April 1958.

O'Brien, Morrough P., "Estuary Tidal Prisms Related to Entrance Area," *Civil Engrg.*, 1(8), May 1931, 738-739.

O'Brien, Morrough P. and John Cherno, "Model Law for Motion of Salt Water Through Fresh" *Trans. ASCE*, 99, 1934, 576-594, with discussions and closure, 595-609.

O'Brien, Morrough P., "Models of Estuaries," *Trans. Amer. Geophys. Union*, Pt. 2, 1935, 485-492.

O'Brien, Morrough P., "The Coast of California as a Beach Erosion Laboratory," *Shore and Beach*, 4(3), July 1936, 74-79.

O'Brien, Morrough P., "Beach Restoration at Santa Barbara. II. Engineering Aspects and Measures," *Shore and Beach*, 7(3), July 1939, 92-97.

O'Brien, Morrough P., *A Summary of the Theory of Oscillatory Waves*, USACE BEB, Tech. Rept. No. 2, 1942, 43 pp.

O'Brien, Morrough P., *Letter to H.U. Sverdrup, 31 December 1943*; also is an appendix in *Memorandum Concerning Wind, Waves, and Swells. A Basic Method of Forecasting* - by H.U. Sverdrup and W.H. Munk, College of Engrg., UCB, Memo. No. HE 116-7, 5 Sept. 1944, 7 pp.

O'Brien, Morrough P., *Letter to H.U. Sverdrup, 28 Feb. 1944*; also in SIO *Proposed Uniform Procedure for Observing Waves and Interpreting Instrument Records*, Wave Project Rept. No. 26, 12 Dec. 1944, 22 pp.

O'Brien, M.P. and J.W. Johnson, "Wartime Research on Waves and Surf," *The Military Engineer*, June 1947, pp 1-6.

O'Brien, Morrough P., "Wave Refraction at Long Beach and Santa Barbara, California", *Bull. BEB*, USACE, BEB, 4(1), Jan. 1950, 1-12.

O'Brien, Morrough P., "Preface," *Proc. of 1st Conf. on Coastal Engineering, Long Beach, Calif., Oct. 1950*, Council on Wave Res., The Engrg. Found., 1951, v.

O'Brien, Morrough P., "Equilibrium Flow Areas of Tidal Inlets on Sandy Coasts," *Jour. Waterways and Harbors Div., Proc. ASCE*, 95(WW1), Feb. 1969, 43-52.

O'Brien, Morrough P., "Field and Laboratory Studies, Navigation Channels of the Columbia River Estuary," *Proc. 13th Coastal Engrg. Conf., July 10-14, 1972, Vancouver, B.C., Canada*, ASCE, 2475-2498.

O'Brien, Morrough P. and Robert G. Dean, "Hydraulic and Sedimentary Stability of Coastal Inlets," *Proc. 13th Coastal Engrg. Conf., July 10-14, 1972, Vancouver, B.C., Canada*, ASCE, II, 761-780.

O'Brien, Morrough P., "Discussion of LeMehaute 'Similitude in Coastal Engrg.'", *Jour. Waterways, Port, Coastal and Ocean Engrg. Div., Proc. ASCE*, 103(WW3), Aug. 1977.

O'Brien, Morrough P., "Editorial. Our Wandering High-tide Lines." *Shore & Beach*, 50(4), Oct. 1982, 2-3.

O'Brien, Morrough P., "The Beach at a Seawall: Completion of a 1940 Experiment," *Shore & Beach*, 55(3-4), July-Oct. 1987, 54-55.

O'Brien, Morrough P., *Morrough P. O'Brien: Dean of the College of Engineering and Consultant to General Electric*, an oral history conducted 1986-88 by Marilyn Ziebarth, Regional Oral History Office, The Bancroft Library, University of California at Berkeley, 1988, 312 pp.

O'Doherty, John, "Michael Maurice O'Shaughnessy: San Francisco City Engineer, 1912-1932," *Shore & Beach*, 62(4), Oct. 1994, 17-23.

Offshore Technology Conference, Houston, TX, *Preprints/Proceedings*, annually since 1969.

Oliver, John G., et al., "Floating Breakwaters: A Practical Guide for Design and Construction", PIANC Working Group 13, PTC II, *Supplement to Bulletin 85, PIANC*, 1995.

Olmstead, Roger and Nancy Olmstead, *Ocean Beach Study: A Survey of Historic Maps and Photographs*, report for the City of San Francisco Wastewater Program, 23 Feb. 1979, 48 pp and plates.

O'Reilly, William C. and R.T. Guza, "A Comparison of Two Spectral Wave Models in the Southern California Bight," *Coastal Engrg.* 19, 1993, 263-282.

O'Reilly, William C., "The Southern California Wave Climate: Effects of Islands and Bathymetry," *Shore & Beach*, 61(3), July 1993, 14-19.

O'Shaughnessy, M.M., "Ocean Beach Esplanade, San Francisco, California," Paper No. 1539, *Trans. ASCE*, LXXVII, 1924, 492-505, with discussions and closure, pp 506-534.

Pajak, Mary Jean, "National Flood Insurance Program Identification of Coastal High Hazard Areas and Information on Minimum Buildings Standards for Structures in Coastal High Hazard Areas," *Personal*

Communication: Letter to Robert Wiegel of 24 April 1995, 9pp.

Panicker, N.N. and L.E. Borgman, "Directional Spectra from Wave Gage Arrays," *Proc. 12th Coastal Engrg. Conf., Sept. 13-18, 1970, Wash. D.C.*, ASCE, I, 117-136.

Pararas-Carayannis, B. Dong, R. Farmer, *Annotated Tsunami Bibliography, 1962-1976*, International Tsunami Information Center, NOAA, Aug. 1982, no page numbers.

Pardee, Lyall A., "Beach Development and Pollution Control by City of Los Angeles in Hyperion - Venice Area," *Shore and Beach*, 28(2), Oct. 1960, 16-19.

Parkman, Aubrey, *Army Engineers in New England, 1775-1975*, USACE, New England Div., U.S. Gov't. Printing Office, 1978, 319 pp.

Patrick, D.A., *Model Study of Amphibious Breakwaters*, UCB Inst. Engrg. Res., Tech. Rept. HEL 3-331, 1951, 10 pp and appendices.

Patrick, D.A. and R.L.Wiegel, "Amphibian Tractors in the Surf," *Proc. First Conf. on Ships and Waves*, The Engrg. Found. Council on Wave Res. and Amer. Soc. Naval Architects and Marine Engineers, 1955, pp 397-422.

Peacock, Harold G., "CERC Field Wave Gaging Program," *Proc. Inter. Symposium on Ocean Wave Measurement and Analysis, New Orleans, LA, Sept. 9-11, 1974*, ASCE, 170-185.

Pearson, Ermin A., "Submarine Waste Disposal Installations," *Proc. 6th Conf. on Coastal Engrg., Florida, Dec. 1957*, Council on Wave Res., The Engrg. Found., 1958, 586-606.

Pearson, E.A. (editor), *Waste Disposal in the Marine Environment: Proc. of the First International Conf., Berkeley, CA. July 22-25, 1959*, Pergamon Press, 1960, 569 pp.

Penfield, Wallace C., "The Oldest Periodic Beach Nourishment Project," *Shore and Beach*, 28(1), April 1960, 9-15.

Penzien, Joseph, Maharaj K. Kaul and Bent Berge, "Stochastic Response of Offshore Towers to Random Sea Waves and Strong Motion Earthquakes,"*Computers & Structures*, 2, 1972, 733-756.

Perdikis, Harry S., "Hurricane Flood Protection in the United States," *Jour. Waterways and Harbors Div., Proc. ASCE*, 93(WW1), Feb. 1967, 1-24.

Perlez, Jane, "50 Years: A Place in the Sun in the Bronx," *The New York Times*, 17 Aug. 1986.

Perroud, Paul H., *The Solitary Wave Reflection Along a Straight Vertical Wall at Oblique Incidence*, Ph.D. thesis, UCB: also, IER Tech. Rept. 99-3, 1957.

Pfeifer, Luanne, *The Malibu Story*, The Malibu Lagoon Museum, 1985, 39 pp.

Phillips, O.M., "On the Generation of Waves by Turbulent Winds," *J. Fluid Mech.*, 2(5), July 1957, 417-445.

Pierson, Willard J., Jr. *The Interpretation of Crossed Orthogonals in Wave Refraction Phenomena*, USACE, BEB, Tech. Memo. No.21, Jan. 1951, 83 pp.

Pierson, Willard J., Jr., *A Unified Mathematical Theory for the Analysis, Propagation and Refraction of Storm Generated Ocean Surface Waves, Parts I, II and III*, New York Univ. College of Engrg., Res. Div., 1952.

Pierson, Willard J., Jr., *An Electronic Wave Spectrum Analyser and Its Use in Engineering Problems*, USACE, BEB, TM 56, Oct. 1954, 91 pp.

Pierson, Willard J., Jr., Gerhard Neumann, and Richard W. James, *Practical Methods for Observing and Forecasting Ocean Waves by Means of Wave Spectra and Statistics*, USN Hydro. Office, H.O. Pub. No. 603, 1955, 284 pp.

Pillsbury, G.B., *Tidal Hydraulics*, USACE Professional Paper No. 34, U.S. Gov't. Printing Office, 1940.

Platzman, G.W., "The Dynamic Prediction of Wind Tides on Lake Erie, *Meteorological Monographs*, 4(26), 1963.

Podufaly, E.T., "Operation Five-High," *Shore and Beach*, 30(2), Oct. 1962, 9-18.

Port Sines Investigating Panel (Edge - Chair, Baird, Caldwell, Fairweather, Magoon, Treadwell), *Failure of the Breakwater at Port Sines, Portugal*, ASCE, 1982, 278 pp.

Prins, J.E., "Characteristics of Waves Generated by a Local Disturbance," *Trans. Amer. Geophys. Union*, 39(5), Oct. 1958a, 865-874.

Prins, J.E., "Water Waves Due to a Local Disturbance," *Proc. 6th Conf. on Coastal Engrg., Florida, Dec. 1957*, Council on Wave Res., The Engrg. Found., 1958b, 147-162.

Pritchard, D.W., "Estuarine Hydrography," *Advances in Geophysics,* Academic Press Inc., NY, NY, 1952, 243-280.

Pritchard, D.W., "Estuarine Circulation Patterns," *Proc. ASCE, Hyd. Div.*, 81 (Sep. 717), June 1955, 1-11.

Pritchard, Donald W., "The Movement and Mixing of Contaminants in Tidal Estuaries," *Proc. 1st Inter. Conf. of Waste Disposal in the Marine Environment, July 22-25, 1959, Berkeley, CA*, Pergamon Press, 1960, 512-526.

Pritchard, Donald W., "Dispersion and Flushing of Pollutants in Estuaries, *Jour. Hyd. Div., Proc. ASCE*, 95 (HY1), Jan. 1969, 115-124.

Pritchard, D.W., *Personal Communication*, by telephone, 6 June 1995.

Pullen, E.J., R.M. Yancey, P.L. Knutson and A.K. Hurme, *An Annotated Bibliography of CERC Coastal Ecology Research*, MR No. 78-2, USACE, CERC, 1978.

Putnam, J.A. and J.W. Johnson, "The Dissipation of Wave Energy by Bottom Friction," *Trans. Amer. Geophys. Union*, 30(1), Feb. 1949, 67-74.

Putnam, J.A., "Loss of Wave Energy Due to Percolation in a Permeable Sea Bottom," *Trans. Amer. Geophys. Union*, 30(3), June 1949, 349-357.

Putnam, J.A., W.H. Munk and M.A. Traylor, "The Prediction of Longshore Currents," *Trans. Amer. Geophys. Union*, 30(3), June 1949, 337-345.

Putz, R.R., "Statistical Distribution for Ocean Waves," *Trans. Amer. Geophys. Union*, 33(5), Oct. 1952, 685-692.

Putz, R.R., "Statistical Analysis of Wave Records," *Proc. 4th Conf. on Coastal Engrg., Chicago, Ill., Oct. 1953*, ed. by J.W. Johnson, Council on Wave Res., The Engrg. Foundation, 1954, 13-24.

Putz, R.R., *A Method for the Measurement of the Correlation Function and Ordinate Distribution for Two Time-history Functions*, Tech. Rept. Series 61, Issue 11, Inst. Engrg. Res., UCB ,1957 (unpub.).

Quinn, Alonzo deF., *Design and Construction of Ports and Marine Structures*, McGraw Hill Book Company, Inc., New York, 1961, 529 pp.

Quinn, Mary-Louise, *The History of the Beach Erosion Board, U.S. Army Corps of Engineers*, Misc. Rept. No. 77-9, USACE CERC, Aug. 1977, 181 pp.

Raichlen, Fredric and Arthur T. Ippen, "Wave Induced Oscillations in Harbors," *Jour. Hyd. Div., Proc. ASCE*, 91(HY2), March 1965, 1-26.

Raichlen, Fredric, "The Motion of Small Boats in Standing Waves," *Proc. 11th Conf. on Coastal Engrg., London, England, Sept. 1968*, ASCE, 1969, 1531-1554.

Rawn, A.M. and F.R. Bowerman, "Factors Influencing and Affecting the Location of Sewer Ocean Outfalls," *Proc. of the 1st Conf. on Coastal Engrg., Long Beach, CA, Oct. 1959*, Council on Wave Res., The Engrg. Foundation, 1951, 186-191.

Rawn, A.M., F.R. Bowerman and Norman H. Brooks, "Diffusers for Disposal of Sewage in Sea Water," *Jour. Sanitary Engrg. Div., Proc. ASCE*, 86(SA2), March 1960, 65-105.

Rayner, A.C. and G.W. Simmons, *Summary of Capabilities*, USACE, CERC, Misc. Paper 3-64, 1964.

Rector, Ralph L., *Laboratory Study of Equilibrium Profiles of Beaches*, USACE BEB, Tech. Memo. 41, Aug. 1954, 38 pp.

Reid, R.O. and C.L. Bretschneider, *The Design Wave in Deep or Shallow Water, Storm Tide, and Forces on Vertical Piling and Large Submerged Objects*, Texas A & M Univ., Dept. of Ocean., A Tech. Rept. to USN Bu Yards and Docks and ONR, and the USACE BEB, Oct. 1953, 36 pp, graphs and appendices.

Reid, Robert O., *Analysis of Wave Force Experiments at Caplen, Texas*, Texas A & M Univ., Dept. of Ocean., Tech. Rept. No. 38-4, Jan. 1956.

Reid, R.O. and K. Kaijura, "On the Damping of Gravity Waves Over a Permeable Seabed," *Trans. Amer. Geophys. Union*, 38, 1957.

Reid, R.O. and B.R. Bodine, "Numerical Model for Storm Surges in Galveston Bay," *Jour. Waterways and Harbors Div., Proc. ASCE*, 94(WW1), 1968, 33-57.

Reid, R.O., Andrew C. Vastano and Thomas J. Reid, *Development of Surge II Program With Application to the Sabine-Calcasieu Area for Hurricane Carla and Design Hurricanes*, USACE CERC, Tech. Paper No. 77-13, Nov. 1977, 218 pp.

Reid, Robert O., *Personal Communication*, by telephone, 5 June 1995.

Renshaw, Clarence, "The Beaches of Long Island," *Shore and Beach*, (37)2, Oct. 1969, 50-59.

Resio, D.T., "The Estimation of Wind Wave Generation in a Discrete Model", *Jour. Physical Oceanography*, II, 1981, 510-525.

Resio, D.T., *The Estimation of Wind-Wave Generation in a Discrete Spectral Model*, USACE, CERC,

Rept. WIS 5, March 1982.

Resio, D.T., *A Numerical Model for Wind-Wave Prediction in Deep Water*, USACE, CERC, Rept. WIS 12, Jan. 1983.

Resio, D.T., *Extreme Run-up Statistics on Natural Beaches*, USACE, CERC Misc. Paper 87-11, 1987.

Reynolds, Kenneth Cass, *An Experimental Investigation of the Reliability of Models for Determining Wave Action on Sea Walls*, D.Sc. thesis, MIT, 1937.

Rice, Merritt, "Closure Conditions: Mouth of the Russian River," *Shore and Beach*, 42 (1), April 1974, 15-20.

Rice, S.O., "Mathematical Analysis of Random Noise, Parts I-IV," *Bell System Tech. Jour.* 23 and 24, 1944-1945, 282-332 and 46-156.

Rosati, J.D., *Functional Design of Breakwaters for Shore Protection: Empirical Methods*, Tech. Rept. CERC-90-15, USACE, CERC, 1990, 43 pp.

Ross, C.W., "Shock Pressure of Breaking Waves," *Proc. 4th Conf. on Coastal Engrg., Chicago, IL, Oct. 1953*, Council on Wave Res., The Engrg. Found., 1954, 323-332.

Ross, C.W., *Large Scale Tests of Wave Forces on Piling (Preliminary Report)*, USACE, BEB, TM 111, 1959

Safaie, Bijan, "Mixing of Buoyant Surface Jet Over Sloping Bottom," *Jour. Waterway, Port, Coastal and Ocean Div., Proc. ASCE*, 105(WW4), Nov. 1979, 357-372.

SANDAG {San Diego Association of Governments}, "Shoreline Preservation Strategies for the San Diego Region," *Shore & Beach*, 63(2), April 1995, 17-30.

Sarpkaya, Turgut and Ibrahim Cakal, "A Comprehensive Sensitivity Analysis of the OTS Data," *Proc. 1983 Offshore Tech. Conf., 2-5 May 1983, Houston, TX*, 3(OTC 4616) 317-325.

Savage, Rudolph P., "Model Tests of Wave Run-up for Hurricane Protection Project," USACE, BEB, *Bull. BEB*, 11, 1957.

Savage, Rudolph P., "Wave Run-up on Roughened and Permeable Slopes", *Jour. Waterways & Harbors Div., Proc. ASCE*, 84(WW3), 1958.

Savage, Rudolph P., *Laboratory Data on Wave Run-up on Roughened and Permeable Slopes*, USACE, BEB TM 109, 1959.

Savage, Ruldoph P., *Laboratory Study of the Effect of Groins on the Rate of Littoral Transport*, USACE BEB, Tech. Memo. No. 114, June 1959, 56 pp.

Savage, R.P., "Experimental Study of Dune Building with Sand Fences," *Proc. 8th Conf. Coastal Engrg., Mexico City, Mexico*, The Engrg. Foundation, 1963.

Savage, Rudolph P. and W.W.Woodhouse, Jr., "Creation and Stabilization of Coastal Barrier Dunes," *Proc. 11th Conf. Coastal Engrg., London, England, Sept. 1968*, ASCE, 1, 1969, 671-700.

Saville, Thorndike, "Future Federal Policy in Coastal Erosion?," reprinted in *Shore and Beach*, 10(2), Oct. 1942, 35-40.

Saville, Thorndike, Jr., "Model Study of Sand Transport Along an Infinitely Long, Straight Beach," *Trans. Amer. Geophys. Union*, 31(4), Aug. 1950.

Saville, Thorndike, Jr., *Wind Set-up and Waves in Shallow Water*, USACE, BEB TM 27, 1952.

Saville, Thorndike, Jr. and Joseph M. Caldwell, "Accuracy of Hydrographic Surveying in and Near the Surf Zone," *Proc. of 3rd Conf. on Coastal Engrg., Cambridge, MA, 1952*, Council of Wave Res., The Engrg. Found., 1953, 31-47.

Saville, Thorndike, Jr. and J.M. Caldwell, "Experimental Study of Wave Overtopping on Shore Structures", *Proc. Minnesota Inter. Hyd. Conf.*, IAHR, 1953.

Saville, Thorndike, Jr., "Hindcast Wave Statistics for the Great Lakes," *Proc. 4th Conf. on Coastal Engrg., Chicago, Illinois, 1953*, Council on Wave Res., The Engrg. Found., 1954.

Saville, Thorndike, Jr., *North Atlantic Coast Wave Statistics Hindcast by Bretschneider-Revised Sverdrup-Munk Method*, USACE, BEB, TM 55, 1954.

Saville, Thorndike, Jr., *Laboratory Data on Wave Run-up and Overtopping on Shore Structures*, USACE, BEB, TM 64, 1955.

Saville, Thorndike, Jr., "Wave Run-up on Shore Structures", *Jour. Waterways and Harbors Div., Proc. ASCE*, 82(WW2), 1956. Also, *Trans. ASCE*, 123, 1958.

Saville, Thorndike, Jr., "Scale Effects in Two Dimensional Beach Studies," *Trans. 7th General Meeting, Inter. Assoc. Hyd. Res.*, (IAHR), 1957, A3-1 to A3-10.

Saville, Thorndike, Jr., "Wave Run-up on Composite Slopes," *Proc. 6th Conf. on Coastal Engrg., Gainesville, FL, Dec. 1957*, Council on Wave Res., The Engrg. Found., 1958, 691-699.
Saville, Thorndike, Jr., *Large Scale Model Tests of Wave Run-up and Overtopping, Lake Okeechobee Section*, unpublished m.s., USACE, BEB, 1958.
Saville, Thorndike, Jr., "Experimental Determination of Wave Set-up," *Proc. 2nd Tech. Conf. on Hurricanes, June 27-30, 1961, Miami Beach, FLA.*, U.S. Dept. Commerce, Weather Bu., National Hurricane Res. Project, Rept. No. 50, I, March 1962, 242-252.
Saville, Thorndike, Jr., "An Approximation of the Wave Run-up Frequency Distribution", *Proc. 8th Conf. on Coastal Engrg., Mexico City, Mexico, Nov. 1962*, Council on Wave Res., The Engrg. Found., 1963.
Saville, Thorndike, Jr., E.W. McClendon and A.L. Cochran, "Freeboard Allowance for Inland Reservoirs," *Jour. WW Div., Proc. ASCE*, May 1962.
Saville, Thorndike, Jr., compiler, *Coastal Engineering, Santa Barbara Specialty Conference, October 1965*, ASCE, 1005 pp.
Saville, Thorndike, Jr. and G.M. Watts, "Coastal Regime," U.S. paper for SII, S2, *22nd International Congress, Paris, France*, PIANC, 1969.
Saville, Thorndike, Jr., "Comparison of Scaled Beach Deformation Tests Using Sand and Coal", *Abstracts-in-Depth, 17th Inter. Conf. on Coastal Engrg., Sydney, Australia, 23-28 March 1980*, The Inst. of Civil Engrs., Australia, 1980, p. 439.
Saville, Thorndike, Jr., "Early Large-scale Experiments on Wave Runup," *Shore & Beach*, 55(3-4), July-Oct. 1987, 101-108.
Schneider, Christine, *The Littoral Environmental Observation (LEO) Data Collection Program*, USACE CERC, CETA No. 81-5, March 1981, 23 pp.
Schoonees, J.S. and A.K. Theron, "Accuracy and Applicability of the SPM Longshore Transport Formula," *Coastal Engrg. 1994: Proc. 24th Inter. Conf., 23-28 Oct. 1995, Kobe, Japan*, ASCE, 3, 1995, 2595-2609.
Scott, Theodore, *Sand Movement by Waves*, USACE BEB Tech. Memo. No. 48, Aug. 1954, 37pp.
Scripps Inst. Oceanography, *Non-dimensional Presentation of Generation and Decay of Waves*, SIO Wave Rept. No. 30, 7 Dec. 1944, 6 pp and plates.
Scripps Inst. Oceanography, *Proposed Uniform Procedure for Observing Waves and Interpreting Instrument Records*, SIO Wave Project Rept. No. 26, 12 Dec. 1944, 19 pp and app.
Scripps Inst. Oceanography, Univ. of Calif., *A Statistical Study of Wave Conditions at Five Open Sea Localities Along the California Coast*, prepared for USACE, Los Angeles Office, Contract W-04-353-Eng-1951, SIO Wave Rept. No. 68, 1 July 1947, 34 pp and app.
Scripps Inst. Oceanography, Univ. of Calif. at San Diego, *Coastal Data Information Program Monthly Report, January 1995.* A cooperative program of USACE and CA Dept. of Boating and Waterways. Monthly Summary Rept. No. 227, SIO Ref. 95-08. 15 Feb. 1995.
Seelig, W.N., *Two-Dimensional Tests of Wave Transmission and Reflection Characteristics of Laboratory Breakwaters*, USACE CERC TR 80-1, 1980.
Segur, Harvey, "The Korteweg-deVries Equation and Water Waves. Part 1. Solutions of the Equation," *Jour. Fluid Mech.*, 59, 1973, 721-736.
Seiwell, H.R., "Military Oceanography in World War II," *Military Engineer*, 39(259), May 1947, 202-210.
Seiwell, H.R., "Results of Research on Surface Waves of the Western North Atlantic," *Paper Phys. Oceanog. Met.*, 10(4), 1948.
Seymour, R.J. and M.H. Sessions, "Regional Network for Coastal Engineering Data," *Proc. 15th Coastal Engrg. Conf., 11-17 July 1976, Honolulu, HI*, ASCE, I, 1977, 60-71.
Seymour, R.J. and A.L. Higgins, "Continuous Estimation of Longshore Sand Transport," *Coastal Zone '78*, ASCE, III, 2308-2318.
Seymour, Richard J., R.R. Strange, R.D. Cayan, and R.A. Ratham, "Influence of El Ninos in California Wave Climate," *Proc. 19th Coastal Engrg. Conf., 3-7 Sept. 1984, Houston, TX*, ASCE, I, 577-592.
Seymour, Richard J., "Collecting Long-term Wave Records," *Shore & Beach*, 55(3-4), July-Oct. 1987, 109-112.
Seymour, Richard J., (guest ed.), *Shore & Beach*, 57(4), Oct. 1989, 48 pp.
Seymour, Richard J. (ed.), *Nearshore Sediment Transport*, Plenum Press, New York, NY, 1989, 418 pp.
Seymour, Richard J., "Wave Observations in the Storm of 17-18 January, 1988," *Shore & Beach*, 57(4),

Oct. 1989, 10-13.

Shalowitz, Aaron L., *Shoreline and Sea Boundaries*, 2-vols., U.S. Dept. Commerce, USC&GS Pub. No. 10-1, 1962, 420 and 749 pp.

Sharp, Henry S., "Artificial Beach Construction in the Vicinity of New York," *The Scientific Monthly*, 25, July 1927, 34-39.

Shepard, F.P., G.A. MacDonald, and D.C. Cox, "The Tsunami of April 1, 1946," *Bull. Scripps Inst. Oceanography*, 5(6), 1950, 391-528.

Shepard, F.P., *Longshore-bars and Longshore-troughs*, USACE BEB, Tech. Memo. No. 15, Jan. 1950, 32 pp.

Shepard, F.P., *Beach Cycles in Southern California*, USACE BEB Tech. Memo. No. 20, 1950, 26 pp.

Shepard, F.P. and Douglas L. Inman, "Nearshore Circulation," *Proc. 1st Conf. on Coastal Engrg., Long Beach, CA, Oct. 1950*, Council on Wave Res., The Engrg. Foundation, 1951, 50-59.

Sibul, O.J. and E.G. Tickner, *A Model Study of the Run-up of Wind-Generated Waves*, USACE, BEB TM 67, 1955.

Simmons, Henry B., "Some Effects of Upland Discharge to Estuarine Hydraulics", *Proc. ASCE*, 81(Sep. 792), 1955.

Simmons, Henry B., "Application and Limitations of Estuary Models in Pollution Analysis," *Proc. 1st International Conf. on Waste Disposal in the Marine Environment, July 22-25, 1959, Berkeley, CA*, Pergamon Press, 1960, 540-546.

Skjelbreia, Lars, *Gravity Waves, Stokes Third Order Approximation: Tables of Functions*, Council on Wave Res., The Engrg. Found., Berkeley, CA, June 1959, 337 pp.

Skjelbreia, L. and J.A. Henderson, "Fifth Order Gravity Wave Theory," *Proc. 7th Conf. on Coastal Engrg., The Hague, The Netherlands, Aug. 1960*, Council on Wave Res., The Engrg. Found., 1961, 184-196.

Smith, R.A., *Annotated Bibliography on Water Waves Caused by Explosions - 1946-66*, DASIAC Special Rept. 58, DASA Info. and Analysis Center, April 1967.

Smithsonian Institution, *Smithsonian Report for 1939*, Pub. No. 3563, 1939, 241-251.

Snodgrass, F.E., "Wave Recorders, " *Proc. 1st Conf. on Coastal Eng., Long Beach, CA, Oct. 1950*, Council on Wave Res., The Engrg. Found., 1951, 69-81.

Snodgrass, F.E. and E.K. Rice, and Michael Hall, *Wave Forces on Piling (Monterey Field Test)*, Tech. Rept. No. 35-4, Inst. Engrg. Res., UCB, June 1951, 9 pp and 18 figs.

Snodgrass, Frank E. and Robert R. Putz, "A Wave Height and Frequency Meter," *Proc. 6th Conf. Coastal Engrg., Florida, Dec. 1957*, Council on Wave Res., The Engrg. Found., 1958, 209-230.

Snodgrass, F.E. and M.J. Tucker, "Spectra of Low Frequency Waves," *Bull. Scripps Inst. Oceanogr.*, 7(4), 1959, 283-367.

Snodgrass, F.W., G.W. Groves, K.F. Hasselmann, G.R. Miller, W.H. Munk, and W.H. Powers, "Propagation of Ocean Swell Across the Pacific Ocean," *Phil. Trans., Royal Soc. (London)*, Ser. A, 259, 1966, 431-497.

Snyder, C.M., "Model Study of Hydraulic Breakwater Over a Reef," *Jour. Waterways and Harbors Div., Proc. ASCE*, 85(WW1), March 1959, 41-68.

Snyder, C.M., R.L. Wiegel and K.J. Bermel, "Laboratory Facilities for Studying Water Gravity Wave Phenomena," *Proc. 6th Conf. Coastal Engrg., Florida, Dec. 1957*, Council on Wave Res., The Engrg. Found., 1958, 231-252.

Sobey, Rodney J., Peter Goodwin, Robert J. Thieke, and Robert J. Westbery, Jr., "Application of Stokes, Cnoidal, and Fourier Wave Theory," *Jour. of Waterway, Port, Coastal and Ocean Engrg.*, ASCE, 113(6), Nov. 1987, 565-587.

Sobey, Rodney J., "The Distribution of Zero-crossing Wave Heights and Periods in a Stationary Sea State," *Ocean Engrg.*, 19(2), 1992, 101-118.

Sobey, Rodney J., "Quantifying Coastal and Ocean Processes," *Proc. 11th Aust. Conf. of Coastal and Ocean Engrg., Brisbane, Australia*, Inst. of Engineers Australia, Aug. 1993, 1-10.

Sonu, C.J., "Field Observation of Nearshore Circulation and Meandering Currents," *Jour. Geophys. Res.*, 77(18), 28 June 1972, 3232-3247.

Sorensen, Robert M., *Procedures for Preliminary Analysis of Tidal Inlet Hydraulics and Stability*, USACE, CERC Coastal Engrg. Tech. Aid No. 77-8, Dec. 1977, 20 pp.

Sorensen, Robert M., *Basic Coastal Engineering*, John Wiley & Sons, NY, 1978, 227 pp.

Steele, K.E. and Theodore Mettlach, "NDBC Wave Data - Current and Planned," *Ocean Wave Measurement and Analysis. Proc. 2nd. International Symposium, Honoring Professor Robert L. Wiegel, New Orleans, LA, 25-28 July 1993*, ASCE, 1994, 198-207.

Stoa, P.N., *Reanalysis of Wave Run-up on Structures and Beaches*, USACE, CERC, Tech. Aid 77-2, 1978.

Stoker, J.J., *Water Waves*, Interscience Pub., Inc., New York, 1957.

Stolzenbach, K., and D.R. Harleman, *An Analytical and Experimental Investigation of Surface Discharges of Heated Water*, Rept. No. 135, Ralph M. Parsons Lab., MIT, 1971.

Street, R.L., T. Mogel and B. Perry, *Computation of the Littoral Regime of the Shore of San Francisco County, California by Automatic Data Processing Methods*, prepared for USACE San Francisco District, 1969.

Stronge, W.B., "Beaches, Tourism and Economic Development," *Shore & Beach*, 62(2), April 1994, 6-8.

Studds, Robert F.A., Rear Admiral, "Coast and Geodetic Survey Data - An Aid to the Coastal Engineer," *Proc. of 1st Conf. on Coastal Engrg., Long Beach, Calif., Oct. 1950*, ed. by J.W. Johnson, Council on Wave Res., The Engrg. Found., 1951, 102-125.

Sudar, R. Anne, Joan Pope, Ted Hillyer, and John Crumm, "Shore Protection Projects of the U.S. Army Corps of Engineers, " *Shore & Beach*, 63(2), April 1995, 3-16.

Sverdrup, H.U. and Walter H. Munk, "Empirical and Theoretical Relations between Wind, Sea and Swell," *Trans. Amer. Geophys. Union*, 27, 1946, 823-827.

Sverdrup, H.U. and W.H. Munk, *Wind, Sea, and Swell: Theory of Relations for Forecasting*, USN Hydrographic Office, H.O. Pub. No. 601, March 1948, 44 pp.

Szuwalski, Andre and S. Wagner, *Bibliography of Publications Prior to July 1983 of the Coastal Engineering Research Center and the Beach Erosion Board*, USACE, CERC, March 1984.

Tait, Stan, "Florida's Comprehensive Beach Management Law," *Shore & Beach*, 59(4), Oct. 1991, 24-27.

Taney, N.E., *Littoral Materials of the South Shore of Long Island, NY*, USACE, BEB, TM 129, 1961.

Taney, N.E., *Interagency Conference on Continental Shelf Research*, USACE, BEB, Misc. Paper No. 1-66, 1966.

Tetra Tech, Inc., *Coastal Flooding Storm Surge Model. Part 1, Methodology; Part 2, User's Guide; Part 3, Codes*. For Federal Emergency Management Agency (FEMA), Wash. D.C., 1981.

Thieke, Robert J. and Rodney J. Sobey, "Cross-shore Wave Transformation and Mean Flow Circulation," *Coastal Engrg.*, 14, 1990, 387-415.

Thompson, Edward F., *Wave Climate at Selected Localities Along U.S. Coasts*, USACE, CERC, TR 77-1, Jan. 1977, 364 pp.

Thompson, Warren C., "Discussion of: Utilization of Wave Forecasting in the Invasions of Normandy, Burma and Japan", by Charles C. Bates, *Annals of the New York Acad. of Sci.*, 51(3), May 1949, 569-572.

Thompson, Warren C., "Shoreline Geomorphology of the Oxnard Plain from Early U.S. Coast Survey Maps", *Shore & Beach*, 62(3), July 1994, 39-50.

Thornton, E.B., "Variation of Longshore Current Across the Surf Zone," *Proc. 12th Conf. Coastal Engrg., Sept. 13-18, 1970, Wash. D.C.*, ASCE, I, 1971, 292-308.

Tiffany, Joseph B., *History of the Waterways Experiment Station*, USACE, WES, June 1968.

Titus, James G. and Vijay K. Narayanan, *The Probability of Sea Level Rise*, U.S. Environmental Protection Agency, Oct. 1995, 186 pp.

Tracy, Barbara A. and Jon M. Hubertz, *Hindcast Hurricane Swell for the Coast of Southern California*, USACE CERC, WIS Rept. 21, Nov. 1990, 40 pp.

Trask, Parker D., *Source of Beach Sand at Santa Barbara, California as Indicated by Mineral Grain Studies*, USACE BEB, TM 28, Oct. 1952, 24 pp.

Trask, Parker D., *Beaches Near San Francisco, California, 1956-1957*, USACE, BEB, TM 110, April 1959, 89 pp.

Tucker, M.J., "Surf Beats: Seawaves of 1 to 5 Min. Period," *Proc. Royal Soc. London*, Ser. A, 202(1071), 22 Aug. 1950, 565-573.

Turk, George F. and Jeffrey A. Melby, "Core-Loc™: A Major Development in Concrete Armor," *The REMR Bull.*, 12(1), Jan. 1995, 1-5.

USACE, BEB, *Interim Report of Beach Erosion Board*, Office of the Chief of Engineers, Washington,

D.C., April 15, 1933.

USACE, BEB, *Wave Tank Experiments on Sand Movement*, 2 vol., Dec. 1936.

USACE, BEB, *Shore Protection Planning and Design*, Tech. Rept. No. 4, 1954, revised 1957, 1961 and 1966.

USACE, CERC, *Coastal Engineering Technical Notes*, (CETN), issued as needed, starting in 1979.

USACE, CERC, FRF, *Basic Environmental Data Summary, Oct. 1982, CERC FRF, Duck, NC*, 33 pp.

USACE, CERC, *Shore Protection Manual*, U.S. Government Printing Office, Washington, D.C., in 3 vol., 1973, revised in 1975 and 1977, and 1984 (in 2 vol.).

USACE, Committee on Tidal Hydraulics, *Evaluation of Numerical Storm Surge Models*, OCE, CTH, Tech. Bull. 21, 1980,

USACE, Galveston District, *Guidelines for Identifying Coastal High Hazard Zones*, June 1975, 26 pp and exhibits and appendices.

USACE, Jacksonville District, *Waves and Wind Tides in Shallow Lakes and Reservoirs, Summary Report*, Project CW 167, 1955.

USACE, Los Angeles District, *Survey Report for Navigation. Dana Point Harbor, Dana Point, California*, 15 Sept. 1961, 24 pp and plates and appendices.

USACE, Los Angeles District, *Oral History of Coastal Engineering Activities in Southern California, 1930-1981* (James W. Dunham, William J. Herron, Jr., Omar Lillevang, Kenneth A. Peel), Jan. 1986, 254 pp.

USACE, Los Angeles District, *Existing State of Orange County Coast*, CCSTWS, Rept. 93-1, Final Rept., April 1993.

USACE, Los Angeles District, *Nearshore Hydrodynamic Factors and Wave Study of the Orange County Coast. Draft*. Prepared by Noble Consultants, Inc., March 1995.

USACE, Los Angeles District, *Huntington Cliffs Reconnaissance Report, Huntington Beach, Orange County, California: Tech. Appen.*, March 1995.

USACE, San Francisco District, *San Francisco Bay and Estuary Dredge Disposal Study,* Appendix K "Marsh Development", April 1976.

USACE, San Francisco District, *San Francisco Bay-Delta Tidal Hydraulic Model*, a brochure, 14 pp, undated.

USACE, Seattle District, "Friday Harbor, Washington," *Information Paper*, 1 Jan. 1991, 1 p and 2 attachments.

USACE, WES, *Report on Progress to Date, Expt. on Ship CHannel Relocation, St. Andrews Bay, Fla.*, TM 13-2, 6 March 1933.

USACE, WES, *Model Study of Salt Water Intrusion, Lower Mississippi River: Preliminary Report*, 8 May 1942.

USACE, WES, *Plans for Improvement of Navigation Conditions and Elimination of Shoaling in Savannah Harbor, Georgia*, WES TM 2-268, 1949.

USACE, WES, *Roughness Standards for Hydraulic Models*, WES, TM 2-364, Rept. No. 1, 1953.

USACE, WES, *Delaware River Model: Hydraulic and Salinity Verification*, WES, TM 2-237, Rept. No. 1, 1956.

U.S. Congress, *Appendix II, Coast of California, Point Mugu to San Pedro Breakwater, Beach Erosion Control Study*, Letter from the Secretary of the Army transmitting a Letter from the Chief of Engineers...83rd Congress, 2nd Session, House Doc. 277, November 3, 1953, 178 pp plus plates.

U.S. Department of Commerce, NOAA/NOS, *North American Vertical Datum of 1988 (NAVD)*, 1992, a one-page note.

U.S. National Weather Service, Hydrometeorological Section, *Interim Report: Characteristics of the Probable Maximum Hurricane, Atlantic and Gulf Coasts of the U.S.*, HUR 7-97, Wash. D.C., May 1968.

U.S. National Weather Service, *Meteorological Criteria for Standard Project Hurricane and Probable Maximum Hurricane Windfields, Gulf and East Coasts of the US*, NOAA Tech. Rept. NWS 23, 1979.

USN Hydrographic Office, *Wind Waves and Swell: Principles in Forecasting*, H.O. Misc. No. 11,275, March 1943, 61 pp.

USN Hydrographic Office, *Breakers and Surf: Principles in Forecasting*, H.O. 234, Nov. 1944, 52 pp and plates.

USN Hydrographic Office, *Techniques for Forecasting Wind Waves and Swell*, H.O. Pub. No. 604, 1951, 37 pp and plates.

van de Kreeke, J., "Stability of Tidal Inlets: Escoffier's Analysis," *Shore & Beach*, 60(1), Jan. 1992, 9-12.

Van Dorn, W.G., "Wind Stress on an Artificial Pond," *Jour. Marine Res.*, 12, 1953.

Van Dorn, W.G., B. LeMehaute, L-S. Hwang, *Handbook of Explosion-Generated Water Waves, Vol. I - State of the Art*, Rept. No. TC-130, Tetra Tech Inc., Pasadena, CA, 1968.

Van Dorn, W.G., "Breaking Invariants in Shoaling Waves," *Jour. Geophys. Res.*, 83 (C6), 20 June 1978, 2981-2988.

Vanoni, Vito A. and John C. Carr, "Harbor Surging," *Proc. 1st Conf. on Coastal Engrg., Long Beach, CA, Oct. 1950*, Council on Wave Res., The Engrg. Found., 1951, 60-68.

Vincent, Charles L. and William D. Corson, *The Geometry of Selected U.S. Tidal Inlets*, USACE, CERC GITI Report 20, May 1980, 163 pp.

Vincent, C. Linwood, William D. Corson, Kathryn J. Gingerich, *Stability of Selected United States Tidal Inlets*, USACE, CERC, GITI Report 21, Sept. 1991, 167 pp.

von Oesen, Henry M.,"A Beach Restoration Study: Bal Harbour Village, Florida," *Shore & Beach*, 41(2), Oct. 1973, 3-4.

Walker, James R., Robert Q. Palmer and Joseph K. Kuhea, "Recreational Surfing on Hawaiian Reefs," *Proc. 13th Coastal Engrg. Conf., 10-14 July 1972, Vancouver, B.C., Canada*, ASCE, III, 1973, 2609-2628.

Walker, James R., *Recreational Surf Parameters*, Univ. of Hawaii, Look Lab., Tech. Rept. No. 30, Feb. 1974, 311 pp and appendix.

Walker, James R., R.A. Nathan, R.J. Seymour and R.R. Strange III, "Coastal Design Criteria in Southern California," *Nineteenth Coastal Engrg. Conf., Proc. of the International Conf., 3-7 Sept. 1984, Houston, TX*, ASCE, III, 2827-2841.

Walker, James R. and Susan M. Brodeur, "The California Beach Nourishment Success Story," *Proc. 1993 National Conf. of Beach Preservation Technology, 10-12 Feb. 1993, St. Petersburg, FL*, Florida Shore & Beach Preservation Assoc., May 1993, 239-258.

Walker, James R., *Personal Communication*, 23 May 1995.

Walton, T.L. and Francis F. Escoffier, "Linearized Solution to Inlet Equation with Inertia," *Jour. of the Waterway, Port, Coastal and Ocean Div., Proc. ASCE*,107(WW3), Aug. 1981, 191-195, [with discussion by David B. Kink, Jr.108(WW3), Aug. 1982, 441-442, and Closure by Walton and Escoffier, 109(WW1), Feb. 1983, 142].

Wanstrath, J.J., *Nearshore Numerical Storm Surge and Tidal Simulation*, Tech. Rept. H-77-17, USACE, WES, 1977.

Warner, S.E., *Report on Sea Sled Investigations*, UCB, College of Engrg., Contract NObs- 2490, Tech. Rept. 116-136, 118 March 1946, 8 pp.

Watts, George M., *A Study of Sand Movement at South Lake Worth Inlet, Florida*, USACE, BEB, TM 42, 1953.

Watts, George M. and R.R. Dearduff, *Laboratory Study of Effect of Tidal Action on Wave-formed Beach Profiles*, USACE, BEB, TM 52, 1954.

Watts, George M., *Laboratory Study of Effect of Varying Wave Periods on Beach Profiles*, USACE, BEB, TM 53, 1954.

Webster, William C., *Personal Communication*, 1994.

Weggel, J. Richard and W.H.C. Maxwell, "Experimental Study of Breaking Wave Pressures," *Preprints: 1970 Offshore Tech. Conf., Houston, TX, 22-24 April 1970*, II (OTC 1244),II-175 to II-188.

Weggel, J. Richard, "Maximum Breaker Height," *Jour. Waterways, Harbors and Coastal Eng. Div., Proc. ASCE*, 98(WW4), Nov. 1972, 529-548.

Weggel, J. Richard "Wave Overtopping Equation," *Proc. 15th Conf. on Coastal Engrg., Honolulu, Hawaii*, ASCE, 1976.

Weggel, J. Richard (ed.), *Proc. of Coastal Structures '83, March 9-11, 1983, Arlington, VA*, ASCE, 1983, 1012 pp.

Weggel, J. Richard, "A Primer on Monitoring Beach Nourishment Projects," *Shore & Beach*, 73(3), July 1995, 20-24.

Wehausen, John V. and E.V. Laitone, "Surface Waves," *Handbuch der Phisik*, Springer Verlag, 9(3),

1960, 446-778.

Weir, L.H., "How Modern Economic Progress has Multiplied the Demand for Outdoor Recreation," *Shore and Beach*, 1(1), April 1933, 13-15.

Whalin, R.W., *The Limit of Applicability of Linear Wave Refraction Theory in a Convergence Zone*, USACE WES, Res. Rept. H-71-3, Dec. 1971, 156 pp.

Whalin, R.W., R.A. Sagar, C.E. Chatham, D.D. Davidson, C.L. Vincent, and G.M. Fisackerly, "Coastal Engineering at the Waterways Experiment Station", *Shore & Beach*, 45(4), Oct. 1977, 18-27.

Whitemarsh, R.P., "Great Sea Waves", *Proc. Naval Institute*, 60, 1934, p. 1100.

Wicker, C.F., "History of New Jersey Coastline," *Proc. 1st Conf. on Coastal Engrg., Long Beach, Calif., Oct. 1950*, Council on Wave Res., The Engrg. Found., 1951, 299-319.

Wicker, C.F. (ed.), *Evaluation of Present State of Knowledge Affecting Tidal Hydraulics and Related Phenomena*, Rept. No. 3, Committee on Tidal Hyd., USACE, WES, 1965.

Wiegel, Robert L., *Gravity Waves: Tables of Functions of d/L and d/L_o*, Dept. Engrg., UCB, Tech. Rept. HE-116-265, Jan. 1948. Also, USACE, Bull. BEB, Special Issue No. 1, July 1947. Reprinted by Council on Wave Research, The Engrg. Found., 1954, 30 pp.

Wiegel, Robert L., "An Analysis of Data from Wave Recorders on the Pacific Coast of the United States," *Trans. Amer. Geophys. Union*, 30(5), 1949, 700-704.

Wiegel, Robert L., "Experimental Study of Surface Waves in Shoaling Water," *Trans. Amer. Geophys. Union*, 31(3), June 1950, 377-385.

Wiegel, Robert L. (Project Engineer and Ed.), *Summary Report of Amphibious Studies for the Period 1 Jan. 1949 to 31 Dec. 1950*, UCB, College of Engrg., ONR Contract N7onr 29519, Tech. Rept. HE-155-45, Feb. 1951, 232 pp.

Wiegel, Robert L. (Project Engineer and Ed.), *Manual of Amphibious Oceanography*, Pentagon Printing Office, 1952, 1750 pp.

Wiegel, Robert L., D.A. Patrick and H.L. Kimberley, "Wave, Longshore Current, and Beach Profile Records for Santa Margarita River Beach, Oceanside, California, 1949," *Trans. Amer. Geophys. Union*, 35(6), Dec. 1954, 887-896.

Wiegel, Robert L., (ed.), *Proc. of the First Conf. on Coastal Engrg. Instruments, Berkeley, Calif., Oct. 31 -Nov. 2, 1955*, Council on Wave Res., The Engrg. Found., 1956, 302 pp.

Wiegel, Robert L. and J. Kukk, "Wave Measurements Along the California Coast," *Trans. Amer. Geophys. Union*, 38(5), Oct. 1957, 667-674.

Wiegel, Robert L., K.E. Beebe and James Moon, "Ocean Wave Forces on Circular Cylindrical Piles," *Jour. Hyd. Div., Proc. ASCE*, 83(HY2), Paper 1199, April 1957, 36pp.

Wiegel, Robert L., K.E. Beebe and R.A. Dilley, "Model Study of the Dynamics of an LSM Moored in Waves," *Proc. 6th Conf. on Coastal Engrg., FL, Dec. 1957*, Council on Wave Res., The Engrg. Found., 1957, 844-877.

Wiegel, Robert L., R.W. Clough, R.A. Dilley and J.B. Williams, "Model Study of Floating Drydock Mooring Forces," *Intern. Shipbuilding Progress*, 6(56), April 1959, 147-159.

Wiegel, Robert L., "Transmission of Waves Past a Rigid Vertical Thin Barrier," *Jour. Waterways and Harbors Div., Proc. ASCE*, 86(WWl), Paper 2413, March 1960.

Wiegel, Robert L., "A Presentation of Cnoidal Wave Theory for Practical Applications," *Jour. Fluid Mech.*, 7(2), 1960, 273-286.

Wiegel, Robert L., "Closely Spaced Piles as a Breakwater," *Dock and Harbour Authority*, 42(491), Sept. 1961, p. 150.

Wiegel, Robert L., "Diffraction of Waves by Semi-infinite Breakwaters", *Jour. Hyd. Div., Proc. ASCE*, 88(HY1), Jan. 1962, 27-44.

Wiegel, Robert L., H.W. Shen and J.D. Cumming, "Hovering Breakwater," *Jour. Waterways and Harbors Div., Proc. ASCE*, 88(WW2), May 1962, 23-50.

Wiegel, Robert L., *Oceanographical Engineering*, Prentice-Hall, Inc. 1964, 532 pp.

Wiegel, Robert L., "Water Wave Equivalent of Mach-Reflection," *Proc. of 9th Conf. on Coastal Engrg., Lisbon, Portugal, June 1964*, ASCE, 1965, 82-102.

Wiegel, Robert L., "Tsunamis," *Earthquake Engineering*, ed. by Robert L. Wiegel, Prentice-Hall, Inc., 1970, pp 253-306.

Wiegel, Robert L. and R.C. Delmonte, "Wave-induced Eddies and 'Lift' Forces on Circular Cylinders,"

Proc. 9th Symposium on Naval Hydrodynamics, Paris, France, Aug. 20-25, 1972, Vol. 1: Unconventional Ships and Ocean Engrg., ed. by R. Brad and A. Castera, ACR-203, USN ONR, 1974, U.S. Govt. Printing Office, 761-791.

Wiegel, Robert L., "Engineers' Concern with Waves, and Their Measurement," *Proc. of the International Symposium on Ocean Wave Measurement and Analysis, Sept. 9-11, 1974, New Orleans, LA*, , ed. by Billy L. Edge and Orville T. Magoon, ASCE, 1975, II, 1-22.

Wiegel, Robert L., "Tsunamis," *Seismic Risk and Engineering Decision*, ed. by C. Lomnitz and E. Rosenbleuth, Elsevier Scientific Publishing Co., 1976, 225-286.

Wiegel, Robert L. (editor), *Directional Wave Spectra Applications '81: Proceedings, April 1981, Berkeley, CA*, ASCE, 1982, 550 pp.

Wiegel, Robert L., "Forces Induced by Breakers on Piles," *Proc. 18th Coastal Engrg. Conf., 14-19 Nov. 1982, Cape Town Republic of South Africa*, ASCE, II, 1983, 1699-1715.

Wiegel, Robert L., "Biographical Sketch of Morrough P. O'Brien," *Shore & Beach*, 55(3-4), July/Oct. 1987, 6-14.

Wiegel, Robert L. and M.J. Doyle, Jr., "Cooling by Ocean Water; Model/ Field Comparison," *Shore & Beach*, 55(3-4), July-Oct. 1987, 38-53.

Wiegel, Robert L., "Biographical Sketch of Joe W. Johnson," *Shore & Beach*, 56(4), Oct. 1988, 7-12.

Wiegel, Robert L., "Shore & Beach Observations. Transformation of Swell Over a Reef," *Shore & Beach*, 58(2), April 1990, 31 and cover.

Wiegel, Robert L., "The Coast-Line - III. Protection of Galveston, Texas, from Overflows by Gulf Storms: Grade-Raising, Seawall and Embankment. ASBPA Coastal Project Award for 1990", *Shore & Beach*, 59(1), Jan. 1991, 4-10.

Wiegel, Robert L., "Dade County, Florida, Beach Nourishment and Hurricane Surge Protection," *Shore & Beach*, 60(4), Oct. 1992, 2-28.

Wiegel, Robert L., "Hurricane and Coastal Storm Surge Barriers in New England," *Shore & Beach*, 61(2), April 1993, 30-49.

Wiegel, Robert L., "Ocean Beach Nourishment on the USA Pacific Coast," *Shore & Beach*, 62(1), Jan. 1994, 11-36.

Wilde, Pat, *Recent Sediments of the Monterey Deep-Sea Fan*, UCB Hyd. Engrg. Lab., Tech Rept. HEL 2-13, May 1965, 155 pp.

Wilde, P., C. Isselhardt, L. Osuch and T. Yancy, *Recent Sediments of Bolinas Bay, California. Part C. Interpretation and Summary of Results*, UCB Hyd. Engrg. Lab., Tech. Rept. HEL 2-23, Dec. 1969, 86pp.

Wilde, P., J. Lee, T. Yancy and M. Glogozczowski, *Recent Sediments of the Central California Continental Shelf, Pillar Point to Pigeon Point. Part C. Interpretation and Summary of Results*, UCB Hyd. Eng. Lab., Tech. Rept HEL 2-38, Oct. 1973, 83pp.

Wilde, Pat and Charles W. Case, "Techniques for Predicting Sediment Transport in the Marine Environment Using Natural Heavy Mineral Tracers," *Shore & Beach*, 45(2), April 1977, 25-30.

Williams, E. Allan and John D. Isaacs, "The Refraction of Groups and the Waves Which they Generate in Shallow Water," *Trans. Amer. Geophys. Union*, 33(4), Aug. 1952, 523-530.

Williams, Leo C., *CERC Wave Gages*, USACE CERC, TM 30, Dec. 1969, 117 pp.

Williams, S.J. and E.P. Meisburger, "Sand Sources for the Transgressive Barrier Coast of Long Island, New York: Evidence for Landward Transport of Shelf Sediments," *Coastal Sediments '87, New Orleans, LA, 12-14 May 1987*, ASCE, 1517-1532.

Wilson, Basil W., "Ship Response to Range Action in Harbor Basins," *Trans. ASCE*, 116, 1951, 1129-1157.

Wilson, Basil W., "Deep Water Wave Generation by Moving Wind Systems," *Jour. Waterways and Harbors Div., Proc. ASCE*, 87(WW2), May 1961, 113-141.

Wilson, W. Stanley, *A Method for Calculating and Plotting Surface Wave Rays*, USACE, CERC, Tech. Memo. No. 17, 1966.

Winton, T.C. and A.J. Mehta, "Dynamic Model for Closure of Small Inlets Due to Storm-induced Littoral Drift," *IAHR Congress, New Delhi, India, 1-7 Feb. 1981. Proc., Subject B*, III, Inter. Assoc. Hyd. Res., (IAHR), 153-159.

Wood, Fergus J., *The Strategic Role of Perigean Spring Tides In Nautical History and North American*

Coastal Flooding, 1635-1976, U.S. Dept. Commerce, NOAA, Superintendent of Documents, 1976, 538 pp.

Woodhouse, W.W., Jr., and R.E. Hanes, *Dune Stablization with Vegetation on the Outer Banks of North Carolina*, USACE, CERC, TM No.22, 1967.

Woodhouse, W.W., Jr., E.D. Seneca and S.W. Broome, *Propagation of Spartina Alterniflora for Substrate Stabiliization and Salt Marsh Development*, USACE, CERC, TM No. 46, 1974.

Woodhouse, W.W., Jr., E.D. Seneca and S.W. Broome, *Ten Years of Development of Man-initiated Coastal Barrier Dunes in North Carolina*, Bull. No. 453, NC Agricultural Expt. Station, Raleigh, NC, 1976.

Woodhouse, W.W., Jr., *Dune Building and Stabilization with Vegetation*, USACE, CERC, Special Rept. No. 3, 1978.

Woodhouse, W.W., Jr., *Building Salt Marshes Along the Coasts of the Continental United States*, Special Rept. No. 4, USACE, CERC, 1979.

World Marina '91. Proc. of the 1st International Conf., 4-8 Sept. 1991, Long Beach, CA, ASCE, 1991, 764 pp.

Yancy, T.E., *Recent Sediments of Monterey Bay, California*, UCB Hyd. Engrg. Lab. Tech. Rept., HEL 2-18, July 1968, 145 pp.

Yeh, Harry H., "Nonlinear Progressive Edge Waves: Their Instability and Evolution," *Jour. Fluid Mech.*, 152, 1985, 479-499.

Yoon, Sung B. and Philip L.-F. Liu, "Stem Waves Along Breakwater," *Jour. Waterway, Port , Coastal and Ocean Engrg.*, 115 (5), Sept. 1989, 635-648.

Yue, D.K.P., and C.C. Mei, "Forward Diffraction of Stokes Waves by Thin Wedge," *Jour. Fluid. Mech.*, 99, 1980, 33-52.

Zilkowski, David B., John H. Richards, and Gary M. Young, "Special Report. Results of the General Adjustment of the North American Vertical Datum of 1988," *Surveying and Land Information Systems*, 52(3), 1992, 133-149.

SUBJECT INDEX

Page number refers to first page of paper

AUTHOR INDEX

Page number refers to first page of paper